CLAY SCIENCE FOR ENGINEERING

PROCEEDINGS OF THE INTERNATIONAL SYMPOSIUM ON SUCTION, SWELLING,
PERMEABILITY AND STRUCTURE OF CLAYS – IS-SHIZUOKA 2001
SHIZUOKA/JAPAN/11 – 13 JANUARY 2001

Clay Science for Engineering

Edited by

K. Adachi
Shibaura Institute of Technology, Tokyo, Japan

M. Fukue
Tokai University, Shimizu, Japan

A.A. BALKEMA/ROTTERDAM/BROOKFIELD/2001

Photo cover: Microscopic view of clays

Transferred to Digital Printing 2010

The texts of the various papers in this volume were set individually by typists under the supervision of each of the authors concerned.

Published by
A.A. Balkema, P.O. Box 1675, 3000 BR Rotterdam, Netherlands
Fax: +31.10.413.5947; E-mail: balkema@balkema.nl; Internet site: www.balkema.nl
A.A. Balkema Publishers, 2252 Ridge Road, Brookfield, VT 05036-9704, USA
Fax: 802.276.3837; E-mail: info@ashgate.com

ISBN 90 5809 175 9

Publisher's Note
The publisher has gone to great lengths to ensure the quality of this reprint
but points out that some imperfections in the original may be apparent.

Clay Science for Engineering, Adachi & Fukue (eds) © 2001 Balkema, Rotterdam, ISBN 90 5809 175 9

Table of contents

2 *Permeability and contaminant transport*

3 Adsorption and desorption in clays

4 Structure of clays

Clay Science for Engineering, Adachi & Fukue (eds) © 2001 Balkema, Rotterdam, ISBN 90 5809 175 9

Preface

As a result of anthropogenic activities, generated waste materials and waste products find their way into the environment. These pose direct threats to human health and the environment. In order to reduce and/or eliminate these threats, proper waste management and disposal techniques are required. The use of clay soils as liners and barrier systems to contain waste streams and waste materials has been proposed and/or adopted in many different parts of the world. It is apparent that in order for these clay materials to be used in an effective and proper manner in a waste containment barrier system, it is necessary to fully comprehend the contributing role of the fundamental properties and characteristics of the clay materials in the containment systems.

This book contains peer reviewed papers presented at the 'International Symposium on Suction, Swelling, Permeability and Structure of Clays' held at the Granship in Shizuoka, Japan, in January 2001. The symposium was organized to promote the exchange of the latest developments in research and practice in the disciplines of geotechnical engineering, clay science and geochemistry concerned with the study and determination of the fundamental properties of clay liner and buffer materials. The papers in this book have been grouped together into five major topics: Suction and swelling; Permeability and contaminant transport; Structure of clays; Soil interactions; and Clay liner and buffer materials.

The symposium was organized by the Japanese Geotechnical Society in collaboration with the Geo-Institute of ASCE, the Clay Science Society of Japan and the Japanese Society of Irrigation, Drainage and Reclamation Engineering. We would like to express our appreciation to Dr K. Kita of the secretariat at Tokai University and Professors T. Miyazaki and H. Kazama for undertaking the role of vice-chairmen of the symposium. Finally, we wish to extend our sincere thanks to the members of the organizing, executive and advisory committees, especially Professor R. N. Yong for his kind cooperation in preparing and organizing the symposium.

K. Adachi
M. Fukue

Clay Science for Engineering, Adachi & Fukue (eds) © 2001 Balkema, Rotterdam, ISBN 90 5809 175 9

Organization

ORGANIZING COMMITTEE

K.Adachi, *Chairman*
T.Miyazaki, *Vice-Chairman*
H.Kazama, *Vice-Chairman*
M.Fukue, *Secretary General*
K.Kita, *Secretary*

Members:

H.Abe	H.Ishikawa	N.Toride
T.Akae	Y.Kohgo	I.Towhata
A.Asaoka	H.Komine	H.Tsuchiya
K.Fujii	T.Sato	T.Uno
S.Imamura	T.Sugii	S.Wada
T.Ishii	I.Sugiyama	

EXECUTIVE COMMITTEE

Members:

S.Anma	Y.Jitsukata	N.Morishima
Ö.Aydan	H.Kato	T.Nakamura
M.Hayashi	T.Kawakami	T.Ohba
M.Ishizuka	Y.Kubota	M.Yamanishi
N.Jitsukata		

INTERNATIONAL ADVISORY COMMITTEE

Members:

A.S.Balasubramanium	H.Kawakami	R.Pusch
G.E.Blight	C.Mulligan	F.C.Townsend
D.G.Fredlund	M.Nakano	W.Wolski
S.Jefferis	H.W.Olsen	R.N.Yong
D.Karube		

Acknowledgements

The editors would like to express their sincere gratitude to the following persons, who helped to review the manuscripts and to improve the technical standard of the papers in this proceedings.

H.Abe	R.Kitamura	M.Otsubo
Y.Adachi	A.Kobayashi	R.Pusch
T.Akae	Y.Kohgo	T.Sato
Ö.Aydan	H.Komine	M.Shimizu
G.E.Blight	J.Konishi	T.Shogaki
D.G.Fredlund	T.Koumoto	T.Sugii
K.Fujii	T.Miyazaki	I.Sugiyama
S.Horiuchi	M.Mizoguchi	H.Suzuki
T.Igarashi	Y.Mori	Y.Takeshita
S.Imamura	T.Morii	N.Toride
T.Ishida	T.Moriwaki	I.Towhata
M.Ishiguro	C.N.Mulligan	H.Tsuchiya
T.Ishii	M.Nakano	S.Wada
H.Ishikawa	H.Narioka	Y.Watabe
K.Iwasaki	K.Nishida	W.Wolski
S.Jefferis	M.Nishigaki	K.Yasukawa
H.Kawakami	T.Nishimura	R.Yatabe
S.Kato	T.Nishimura	R.N.Yong
T.Katsumi	H.W.Olsen	
H.Kazama	K.Onitsuka	

Special lectures

Clay Science for Engineering, Adachi & Fukue (eds) © 2001 Balkema, Rotterdam, ISBN 90 5809 175 9

Water infiltration and suction in clay in relation to swelling and/or expansion

M. Nakano
Department of Agriculture, Kobe University, Japan

ABSTRACT: This paper presents that in expansive clays containing montmorillonites the generalized water flow equation can be described regarding a solid volume as a reference. The water retention curves are derived by calculating both volumetric water content and the chemical potential of water in a stacking model of the 2:1 layer. It is pointed out that molecular dynamics simulation is useful to analyze the characteristics of water retention curves, and that a homogenization analysis also is useful to solve the flow problems in an inhomogeneous porous body.

1 INTRODUCTION

Water infiltration into expansive clays, for example, bentonite, occurs a change of bulk density profile in clays during infiltration and as a consequence of wetting. In clay confined with a rigid wall, the bulk density rapidly decreases near a small water inlet end and the local swelling in the wetted parts produces a compensating compression in the remaining non-wetted portion (Nofziger & Swartzendruber 1976). In the unconfined clay, the bulk density in the whole part in where water flow reaches decreases due to free local swelling and consequently, volume expansion occurs toward the unconstrained surfaces (Nakano et al. 1986). So, the analysis of water infiltration in expansive clays requires to be performed in full consideration of a change in the local bulk density and in volume of materials. A local change of bulk density during infiltration leads to the distinguished concepts in describing mass conservation's law for water and in applying the Darcy' law to water flow in clays.

Previous studies indicated that the Darcy's law was a law on the water flow along a surface of clay particles at rest, and that a flux observed in a spatial coordinate system which was fixed to the body of expansive clay should be distinguished from a flux decided by the Darcy' equation (Biot 1955, Zaslavsky 1964, Raats & Klute 1968a, 1968b, and Yong 1973).

A generalized mass conservation for water was described in principal by using the material coordinate of the solid phase, by which implied a modified scale of length as an alternative coordinate to length in a Cartesian coordinate system of fixed scale (Hartley & Crank 1949, Fatt & Goldstick 1965, Philip 1968 and Smiles & Rosenthal 1968). As a result, a generalized governing equation for water flow included a velocity of solid particles. In addition, the analysis and/or prediction of the water content profiles required to add some relations, for example, the water retention curves and/or the experimental relation between void ratio and water content, to the governing equation. Especially, the water retention curves were needed.

In expansive clay, a major component of water content is considered to be water adsorbed on the surface of clay minerals. Various kinds of forces act in water. It is reasonable to represent the water retention curves as a relation between water content and the chemical potential. Water content could be geometrically calculated by assuming a stacking model of a few plates of the 2:1 layers as clay particles (Grim 1953). The chemical potential decrement of water could be obtained by considering three components: that is (1) the decrement produced by solute, (2) the decrement produced by the electric field in which the ions produced the electric density and (3) the decrement produced by the van der Waals attractive force (Bolt, 1955, Bolt and Miller, 1958, Taylor, 1959, Levine and Bell, 1960, 1964, Babcock, 1963, Iwata, 1974, Ravina and Gur, 1978, Gur et al., 1978, Frahm and Diekson, 1979, Iwata 1974a, 1974b, Israelachvili 1991, Fujii and Nakano 1984).

Recently, some new trials were reported in relation to predicting the chemical potential of water (Kawamura 1992) and to solving the flow problems (Ichikawa and Kawamura et al 1999).

In the following it will be attempted to report the outline of significant advances in the field of soil science for the past fifty years.

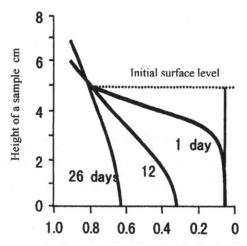

Fig. 1 Change of volumetric water content with time. Bentonite. Bulk density 0.846 g/cm³

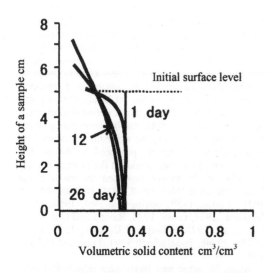

Fig. 2 Change of volumetric solid content with time for same sample as in Fig. 1.

2 FLOW EQUATIONS

2.1 *Mass conservation law for water*

Regarding a solid volume as a reference, the generalized mass balance for water in a fixed volume with respect to the Cartesian coordinate can be written as:

$$\frac{\partial \vartheta}{\partial t} = -\frac{1}{\sigma} \, divq \qquad (1)$$

where ϑ is the volume of water per the unit volume of solid particles, σ is the volume of solid particles per the fixed unit volume, q is the water flux. In such system there exists some relations between some obtainable variables in experiment, for example, ϑ, σ, the volumetric water content θ and the porosity n:

$$\vartheta = \frac{\theta}{\sigma} = \frac{\theta}{1-n} \qquad (2)$$

The equation (1) is equivalent to an equation proposed by Philip & Rosenthal (1968) if the material coordinate $dm_i = \sigma \, dx_i$ is introduced.

Substituting a relation $\vartheta = \theta / \sigma$ into a left hand side of equation (1) and transforming yield

$$\frac{\partial \theta}{\partial t} - \frac{\theta}{\sigma} \frac{\partial \sigma}{\partial t} = -divq \qquad (3)$$

Yong (1973) showed a similar equation to equation (3) in a mathematical form.

Regarding the derivative in equation (3) as the generalized material derivative, equation (3) is equivalent to Raats's equation (Raats & Klute 1968a)

2.2 *Governing equations for water flow*

An average water flux q_w in the fixed unit area with respect to a fixed volume can be described by using an average liquid water flux defined by Darcy's law q_l, an average flux of solid particles q_s and the vapor flux q_v.

$$q_w = q_l + \frac{\theta}{\sigma} q_s + q_v \qquad (4)$$

However, the measurement of Darcy's flux q_l and vapor flux q_v are in a major difficulty. The only measurement of a flux q_w is possible in a spatial coordinate system at rest.

The Darcy's law may be written by using the only water potential term because the magnitude of gravitational force term is extremely small in expansive clays.

$$q = -kgrad\phi \qquad (5)$$

where k is the hydraulic conductivity, ϕ is the water potential.

According to Yong (1973), the solid particle flux q_s may be proposed as:

$$q_s = -k_s grad\phi_s \qquad (6)$$

The mass balance for solid particles was written as:

$$\frac{\partial\sigma}{\partial t} = -divq_s \qquad (7)$$

where k_s is the clay particle conductivity, ϕ_s is the internal pressure responsible for movement of clay particles.

Let us consider q in equation (3) as $q=q_l + q_v$. So substituting equation (4) into equation (3) and transforming yield

$$\frac{\partial\theta}{\partial t} = -divq_w + q_s grad(\frac{\theta}{\sigma}) \qquad (8)$$

If q_s is considered to be small, equation (8) can be reduced to

$$\frac{\partial\theta}{\partial t} = -divq_w \qquad (9)$$

Writing the observed flux q_w as

$$q_w = -k_w grad\phi_w \qquad (10)$$

and introducing a relation such as $grad\,\phi_w = grad\,\phi$, equations (9) and (10) yields

$$\frac{\partial\theta}{\partial t} = divk_w grad\phi \qquad (11)$$

Supposing $grad\,\phi_s = -grad\,\phi$ on ϕ_s, substituting equation (6) into equations (7) yields

$$\frac{\partial\sigma}{\partial t} = -divk_s grad\phi \qquad (12)$$

Applying a water retention curve which is a relation between water potential and volumetric water content, equations (11) and (12) can be transformed to the simple diffusion-type equations respect to θ and σ, which include variables $D_w=k_w d\,\phi/d\,\theta$ and $D_s=k_s d\,\phi/d\,\sigma$, respectively (Chijimatsu et al 2000).

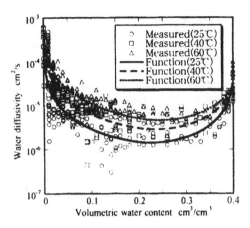

Fig.3 Water diffusivity. Bentonite, Bulk density 1.6 g/cm³ (Chijimatsu et al 2000)

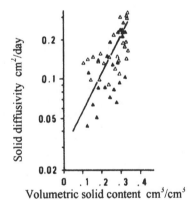

Fig. 4 Solid diffusivity. Bentonite. Bulk density 0.846 g/cm³

3 SUCTION AND CHEMICAL POTENTIAL

3.1 Suction

The water potential ϕ_w can be considered as the decrement of chemical potential to a standard value. In the microscopic sense, the chemical potential becomes same values at the whole location around solid phase. However, the forces acting to water molecules at a location in water are different depending on distances from the surface of solid phase.

The forces are the surface tension, the osmotic pressure, the electric field, the van der Waals force and the internal pressure. The internal pressure contributes to equalize the chemical potential at the whole location in water. Thus, the chemical potential at a location in water is represented by the sum of each decrements produced by each forces acting there. In a case where there is the air-water interface in expansive clays, the chemical potential of water is represented by the effect of surface tension and those are called as the matrix potential.

Suction is equivalent to the value of the matrix potential per unit volume of water, which has a dimension of pressure (Bolt and Miller 1958, Iwata et al. 1995).

Water is considered to be the adsorbed water on the external surface of clay particles and/or almost the interlayer water in compacted clay. Water molecules interact to ions and to montmorillonite minerals.

The measurements of their high suction are considered to be difficult in the tensiometer reading (Groenevelt and Bolt 1972). But it is possible in principal. There exists an epoch-making experiment and an ambitious attempt (Ridley and Burland 1995).

The measurement of chemical potential is easy using a condition that the liquid phase and the vapor phase of water are in equilibrium. There exist the pressure plate and /or membrane apparatuses.

In addition, in a microscopic sense, the local swelling pressure is considered to be equivalent to the difference in the chemical potential between inside and outside of a clay particle (Iwata 1995) and thus, it can be obtained by using the water retention curves (Warkentin et al 1957, Low 1980).

The prediction of chemical potential also is theoretically possible over the whole range of water content from the saturation to the air-dried condition as follows.

3.2 The volumetric water content

Let us assume a stacking model of the 2:1 layers of montmorillonite as structure of expansive clays. In addition, let us consider that the adsorbed water also consists of the stack of thin layers of water with the thickness 0.276 nm which is equal to a diameter of water molecule, and that water molecules and ions were regularly arranged in each thin layers of water.

Volumetric water content can be defined as the sum of an amount of water adsorbed on the external surface of the stacks and an amount of the interlayer water. As an amount of water adsorbed at the sides of the stacks and an amount of water held by the other minerals such as quarts are considered to be small, volumetric water content is written as:

$$\theta = (\frac{810}{2L})\{2n+(L-1)m\}\delta\rho_b R \qquad (13)$$

where θ is volumetric water content, L is the number of the stacked layer, n and m are the number of molecular layers in water adsorbed on the external surface and in the interlayer water, respectively, δ is the diameter of a water molecule, ρ_b is the bulk density of clay, and R is the montmorillonite content in clay by weight.

3.3 The decrement produced by solute

The decrement of chemical potential μ_o produced by solutes in water can be estimated as

$$\mu_o = -kTCv \qquad (14)$$

where k is the Boltzmann constant, T is temperature, C is the concentration of cations in water and v is the partial volume of water.

The concentration C can be obtained as a function of distance from the surface of the 2:1 layers by solving the following Poisson equation with the Boltzmann's law.

$$grad\psi = -\frac{4\pi}{\varepsilon}\rho \qquad (15)$$

$$\rho = \sum z_i en_i \qquad (16)$$

$$n_i = n_{i0}\exp\left(-\frac{z_ie\psi}{kT}\right) \qquad (17)$$

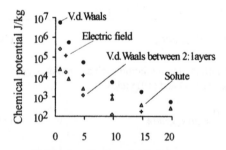

The number of water molecular layers

Fig. 5 Comparison of chemical potentials

where Ψ is the electrical potential, ρ is the density of volume charge, ε is the relative dielectric constant, n_i is the number of the ith ion per unit water volume, n_{i0} is the number of the ith ion at $\Psi=0$, z_i is the valence and e is the electronic charge.

3.4 The decrement produced by the electric field

When there exist water molecules in the electric field caused by the charge of cations, the polarization of water molecules is induced. Thus, motion of water molecules is restricted and the chemical potential decreases (Israelachvili 1991).

The decrement of chemical potential μ_e at a position in each thin water layer could be obtained as

$$\mu_e = -\frac{v}{\varepsilon_0}\int_0^D P\,dD \qquad (18)$$

Introducing the following relations

$$P = D - \varepsilon_0 E, \qquad D = \varepsilon\varepsilon_0 E \qquad (19)$$

Equation (18) yields

$$\mu_e = -v\varepsilon_0\left\{\frac{1}{2}(\varepsilon E)^2 - \int_0^{\varepsilon E} E\,d(\varepsilon E)\right\} \qquad (20)$$

where ε_o is the dielectric constant of vacuum, v is the specific volume of water, P is the magnitude of polarization, D is the electric displacement and E is the electric density.

The electric density E can be obtained by using the solution of the Poisson equation mentioned above and by using a relation $E = -grad\psi$.

3.5 The decrement produced by van der Waals force

The motion of each water molecules is also restricted due to the van der Waals attractive force and consequently the chemical potential of water decreases.

The van der Waals attractive energy E_w between two points is described as

$$E_w = -\frac{\Lambda_{A,B}}{r^6} \qquad (21)$$

The energy between a point and a plate can be obtained by integrating the force between two points over the whole volume of a plate extending indefinitely (Israelachvili 1991).

Thus, the decrement of chemical potential of water can be obtained by multiplying the van der Waals interaction energy between a water molecule and a sheet of the 2:1 layer of clay minerals by the number

of water molecules contained in the unit mass of water.

The Hamaker constant $\Lambda_{A,B}$ between materials A and B can be estimated by the following equations

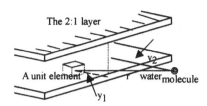

The 2:1 layer

A unit element   water molecule

Fig. 6 . Water molecule at edge of montmorillonite

$$\Lambda_{A,B} = \frac{3}{2}\alpha_A\alpha_B\frac{I_A I_B}{I_A + I_B} \qquad (22)$$

$$\alpha = \frac{3}{4\pi n}\frac{R_e^2 - 1}{R_e^2 + 2} \qquad (23)$$

where α is the polarizability, I is the ionization potential, R_e is the refraction index, and n is the number of molecules contained in the unit volume of clay minerals. The ionization potentials can be estimated by using the component ratio of atoms in water and in clay, respectively.

Using the polarizability of water $1.444\times10^{-30}m^3$, the ionization potential $6.536\times10^{-18}J$ for water and $6.166\times10^{-18}J$ for clay and the refraction index of clay 1.519 for montmorillonite,

The chemical potential of water μ_f at the center of surface of a 2:1 layer can be described as

$$\mu_f = -8.696\times10^{-24}\left\{\frac{1}{d^3} - \frac{1}{(d+\delta)^3}\right\} \qquad (24)$$

where d is the distance between a water molecule and a 2:1 layer and δ is the thickness of a 2:1 layer.

3.6 The increase produced by van der Waals force

The van der Waals attractive force between adjacent 2:1 layers applies pressure to the interlayer water and consequently causes the increase of the chemical potential in interlayer water. The increase can be obtained by multiplying the van der Waals attractive force between adjacent 2:1 layers by the specific volume of water. The van der Waals attractive energy in this case is given integrating those between a water molecule and a 2:1 layer from the distance d between adjacent 2:1 layers to the distance $d + \delta$.

Thus, the increase of chemical potential in the interlayer water μ_p can be obtained by multiplying by the specific volume of water.

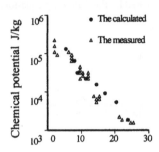

Fig. 7 The calculated water retention curve. L=4. Bulk density 1.0 Mg/m3

$$\eta_p = 4.238 \times 10^{-24} \left\{ -\frac{2}{(d+\delta)^3} + \frac{1}{d^3} + \frac{1}{(d+2\delta)^3} \right\}$$
(25)

3.7 Water adsorbed at the edge of clay particles

Assuming that water surrounds the edge of clay particles under the action of the van der Waals attractive force and consequently chemical potential of the surrounding water also decreases. The decrement can be obtained integrating the van der Waala attractive force between a water molecule located outside the sideline of clay particles and the 2:1 layer over the whole volume of two 2:1 layers.

Considering the decrement of chemical potential μ_s at a point on a midline of the interlayer water, it can be described as.

$$\mu_s = -3.933 \times 10^{-49} \int_{\infty}^{r} \left\{ \sum \frac{y_1 + y_2}{r^8} \right\} dr$$
(26)

where y_1 is the distance between the sideline and an unit element of the 2:1 layer, y_2 is the distance between a water molecule and the sideline taken on a midline of the interlayer water and r is the distance between a water molecule and an unit element of the 2:1 layers.

3.8 The procedures predicting water retention curves

Estimating the water retention curves from the above ideas is conducted as follows: (1) assuming values of L, n and m, volumetric water content is calculated at first. (2) Assuming values of s, which is a number of the thin layers of water adsorbed at the edge, the chemical potential of water on the external surface, in the interlayer and at the edge are individually calculated. (3) Then, the largest value is selected from among them and it is regarded as the calculated chemical potential at the calculated volumetric water content. (4) Changing values of L, n and m, this procedure is repeated. (5) Then, the pairs of chemical potential versus volumetric water content are decided.

3.9 Analysis by the molecular dynamics

Recently, the internal energy of water also was estimated by using molecular dynamics simulation (MD) in an (NPT)-ensemble because all of atoms interact each other in the mixture of water and clay minerals. The introduced inter-atomic potential for all atom-atom pairs was composed of the Coulomb potential, the short-range repulsion potential, the van der Waals potential and the Morse potential applied only to the interaction between O-atoms and H-atoms.

Kawamura (1992) and Kumagai et al (1994) proposed that in a case of the problem of two bodies the potential energy of system was written as

$$U_{ij}(r_{ij}) = \frac{z_i z_j e^2}{4\pi\varepsilon_0 r_{ij}} + f_0 (b_i + b_j) \exp\left\{ \frac{a_i + a_j - r_{ij}}{b_i + b_j} \right\} - \frac{c_i c_j}{r_{ij}^6}$$
$$+ D_{ij} \left[\exp\left\{ -2\beta_{ij} \left(r_{ij} - r_{ij}^* \right) \right\} - 2\exp\left\{ -\beta_{ij} \left(r_{ij} - r_{ij}^* \right) \right\} \right]$$
(27)

For water molecules, the following potential was added as the problem of three bodies regarding an angle made by two O-H bonds

$$U_{ijk}\left(\theta_{ijk}, r_{ji}, r_{jk} \right) = f_k \left[1 - \cos\left\{ 2(\theta_{ijk} - \theta_0) \right\} \right] (k_i k_j)^{\frac{1}{2}}$$
(28)

Table 1 Number of thin water layers

	Volumetric water content cm^3/cm^3							
Location	0.053	0.066	0.094	0.133	0.159	0.186	0.212	0.239
External n	2	1	2	2	3	4	5	6
Interlayer m	0	1	1	2	2	2	2	2
Edge s	0	1	2	2	3	4	6	6

8

$$k_i = \frac{1}{\exp\{g_r(r_{ij} - r_m)\} + 1} \qquad (29)$$

where U is the potential energy, r is the inter-atomic distance, θ is the angle of H-O-H structure, k is the effective range of the three body potential, and (f_0, f_k, a, b, c, r^*), and (D, β, g_r, r_m) are the parameters decided so as to reproduced structure.

4 UNSATURATED HYDRAULIC CONDUCTIVITY AND PERMEABILITY

4.1 *The measurements*

In the infiltration experiments the changes of both volumetric water content profiles, θ -profiles, and bulk density profiles, σ -profiles, are obtained with time. From the data the fluxes q_w and q_s can be estimated at a position in the experimental column. The gradient of both θ and σ also are obtained, respectively. Using equation (10) and equation (6), $D_w = k_w \, d\phi/d\theta$ and $D_s = k_s \, d\phi/d\sigma$ are obtained, respectively.

The obtained D_w and D_s are useful for estimation of the changes of both θ -profiles and σ -profiles with time in other materials.

However, it should be noted that these variables D_w and D_s change not only with the bulk density but also with both temperature and the concentration of solutes, because the water retention curves change with them (Chijimatsu et al 2000).

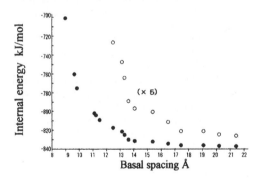

Fig. 8 Internal energy of interlayer water (by Kawamura)

4.2 *The recent alternative flow analysis*

In the above treatment, it assumes that expansive clays are to be homogeneous. However, the actual materials are of inhomogeneous with a periodic structure. Water flow appears in the pores of materials.

Ichikawa et all (1999) proposed that it was possible to solve the problem by applying the method of homogenization analysis (HA). Their treatment was as the followings.

The Navier-Stokes equation with respect to the global coordinate system x_i was written as

$$-\frac{\partial P^\varepsilon}{\partial x_i} + \eta \frac{\partial^2 V_i^\varepsilon}{\partial x_k \partial x_k} + F_i = 0 \qquad (30)$$

$$\frac{\partial V_i^\varepsilon}{\partial x_i} = 0 \qquad (31)$$

where V_i^ε is the velocity vector with the shearing viscosity η, P is the pressure, F_i is the body force vector.

The local coordinate system y was introduce by $y = x/\varepsilon$ and the asymptotic expansion was performed as

$$V_i^\varepsilon(x) = \varepsilon^2 V_i^0(x,y) + \varepsilon^3 V_i^1(x,y) + \cdots \cdots$$
$$P^\varepsilon(x) = P^0(x,y) + \varepsilon P^1(x,y) + \cdots \cdots \qquad (32)$$

$$\frac{\partial}{\partial x_i} \Rightarrow \frac{\partial}{\partial x_i} + \frac{1}{\varepsilon} \frac{\partial}{\partial y_i} \qquad (33)$$

where there existed the following periodic relations

$$V_i^\alpha(x,y) = V_i^\alpha(x, y+Y)$$
$$P^\alpha(x,y) = P^\alpha(x, y+Y) \qquad (\alpha = 0, 1, 2, \cdots) \qquad (34)$$

Substituting equations (32) and (33) into equation (30), taking $\varepsilon \to 0$, and using the following variables for the ε^0-term,

$$V_i^0 = (F_k(x) - \frac{\partial P^0(x)}{\partial x_k}) v_i^k(y)$$
$$P^1 = (F_k(x) - \frac{\partial P^0(x)}{\partial x_k}) p^k(y) \qquad (35)$$

The micro–scale partial differential equation of only y could be written as

$$-\frac{\partial p^k}{\partial y_i} + \eta \frac{\partial^2 v_i^k}{\partial y_j \partial y_j} + \delta_{ik} = 0 \qquad (36)$$

$$\frac{\partial v_i^k}{\partial y_i} = 0 \qquad (37)$$

9

where $v_{ik}(y)$ and $p_k(y)$ were the the characteristic velocity and the characteristic pressure, respectively, δ_{ik} is the Kronecker's delta.

An averaging operation for equation (35) yielded

$$\tilde{V}_i^0 = K_{ji}(F_j - \frac{\partial P^0}{\partial x_j})$$

$$K_{ji} = \tilde{v}_i^j = \frac{1}{|Y|}\int_Y v_i^j dy \qquad (38)$$

where $\tilde{V}_i^0$ is the averaged mass velocity in the unit cell, $|Y|$ is the volume of the unit cell representing a periodic structure.

Calculating the true pressure P and velocity V_i in the first order approximation in equation (32), a following relation was obtained as

$$\tilde{V}_i^\varepsilon \cong \varepsilon^2 \tilde{V}_i^0 \qquad (39)$$

Considering that $V_{i\varepsilon}$ was equivalent to the average velocity indicated by the empirical Darcy's law, the Darcy's permeability K_{ij}' was obtained as

$$K_{ij}' = \varepsilon^2 \rho g K_{ij} \qquad (40)$$

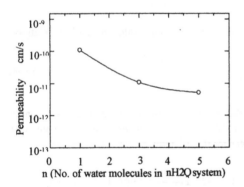

Fig. 9 Calculated permeability
(Saturated density $1.8 g/cm^3$. The nH_2O-system means $Na_{1/3}Al_2(Si_{11/3}Al_{1/3})O_{10}(OH)_2 nH_2O$. When n increases, the void of inter-particles becomes small (Ichikawa et al 1999))

In this system, it is necessary to assume a model of pores in a unit cell Y in order to calculate K_{ij}. The stacking model of plates was useful for predicting the permeability of expansive clays containing montmorillonite.

In this calculating, the viscosity of interlayer water was needed. Ichikawa and Kawamra et al (1999)

applied a result obtained by molecular dynamics simulation using equations (27) to (29) to calculating permeability. The viscosity was expressed as a function of distance from the flat montmorillonite surface. In the 3-4 nm layer on the surface, the viscosity rapidly changed from about 28.0 at surface to 2.5 Pa ·s with distances from the surface.

5 CONCLUSIONS

1. The generalized mass balance for water in a fixed volume with respect to the Cartesian coordinate can be written regarding a solid volume as a reference.
2. It is possible to predict theoretically the chemical potential of the adsorbed and the inter-layer water in expansive clays containing montmorillonite by considering the van der Waals forces.
3. The molecular dynamics simulation (MD) appears to give a more strict solution to the analysis of water retention curves in near future.
4. Flow analysis taking an inhomogeneous structure in expansive clays in consideration appears to be possible by applying a homogenization analysis (HA).

REFERENCES

Babcock, K.L. 1963. Theory of chemical properties of soil colloidal systems at equilibrium. *Hilgardia.* 34:417-542.
Biot, M. A. 1955. Theory of elasticity and consolidation for a porous anisotropic solid. *J. Appl. Phys.* 26: 182-185.
Bolt, G.H. 1955. Analysis of the validity of the Gouy-Chapman theory of the electric double layer. *J. Colloid Sci.* 10:206-218.
Bolt, G.H. and R.D. Miller 1958. Calculation of total and component potentials of water. *Trans. Am. Geophys. Union.* 39:917-928.
Booth, F. 1961. The electric constant of water and saturation effect. *J. Chem. Phys.* 19:391-394.
Chijimatsu, M., T. Fujita, A. Kobayashi and M. Nakano 2000. Experiment and validation of numerical simulation of coupled thermal, hydraulic and mechanical behavior. *Int. J. for Numer. And Anal. Methods I Geomechanics.* 24: 403-424.
Fatt, I. and T. K. Goldstick 1965. Dynamics of water transport in swelling membranes. *J. Colloid Sci.* 20: 962-989.
Frahm, J. and S. Diekmann 1979. Nunerical calculation of diffuse double layer properties for spherical colloidal particles by means of a modified nonlinearlize Poisson-Boltzmann equation. *J. Colloid Interface Sci.,* 70:440-447.

Fujii, K and M. Nakano 1984. Chemical potential of water adsorbed to Bentonite. *Trans. Jpn. Soc. Irrig. Drain. and Reclam. Eng.* 112:43-53 (in Japanese).

Groenevelt, P. H. and G. H. Bolt 1972. Water retention in soil. *Soil Sci.* 113: 238-245.

Grim, R. E., 1953. Clay mineralogy. *McGraw-Hill, New york* 43-77, 161-189.

Guy, Y., Ravina, I. and Babchin, A.J., 1978. On the electrical double layer theory 2: The Poisson-Boltzmann equation including hydration forces. *J. Colloid Interface Sci..* 64:333-341.

Hartley, G. S. and J. Crank 1949. Some fundamental definitions and concepts in diffusion processes. *Trans. Faraday Soc.* 45: 801-818.

Ichikawa, Y., K. Kawamura, M. Nakano, K. Kitayama and H. Kawamura 1999. Unified molecular dynamics and homogenization analysis for bentonite behavior: current results and future possibilities. *Eng. Geology* 54: 21-31.

Israelachvili, J. N., 1991. Intermolecular and surface forces. Second ed., *Academic Press.*

Iwata, S., 1974a. Thermodynamics of soil water 3: *Soil Sci.*, 117: 87-93.

Iwata, S., 1974b. Thermodynamics of soil water 4: *Soil Sci.*, 117:135-139.

Iwata, S., Tabuchi, T. and Warkentin, B. P., 1995. Soil-water interactions. Second ed., *Marcel Dekker, New York*, 13-39, 96-104.

Kawamura, K. 1992. Inter-atomic potential models for molecular dynamics simulations of multicomponent oxides. In Molecular Dynamics simulations. *Springer-Verlag, Berlin.*

Kawamura, K., Y. Ichikawa, M. Nakano, K. Kitayama and H. Kawamura 1999. Swelling properties of smectite up to 90 ℃ in situX-ray diffraction experiments and molecular dynamic simulations. *Eng. Geology* 54: 75-79.

Kumagai, N., K. Kawamura and T. Yokokawa 1994. An inter-atomic potential model for H_2O: Application to water and ice polymorphs. *Molecular Simulations.* 12: 3-6, 177-186.

Levine, S. and Bell, G.M., 1960, Theory of a modified Poisson-Boltzmann: the volume of hydrated ions. *J. Phys. Chem.*, 64:1188-1195.

Low, P. F. 1976. Viscosity of interlayer water in montmorillonite. Soil Sci. Soc. A. J. 40: 500-504.

Low,P. F. 1980. The swelling of clay: II Montmorillonites. *Soil Sci. Soc. Am. J.* 44:667-676.

Marshall, C.E., 1949. The colloid chemistry of the silicate minerals. *Academic Press, New York.*

Nakano, M., Y. Amemiya and K. Fujii 1986. Saturated and unsaturated hydraulic conductivity of swelling clays. *Soil Sci.* 141(1): 1-6.

Nofziger, D. L. and D. Swartzendruber 1976. Water content and bulk density during wetting of a bentonite-silt column. *Soil Sci. Soc. Am. J.* 40: 345-348.

Philip, J. R. 1968. Kinetics of sorption and volume change in clay-colloid pastes. *Aust. J. Soil Res.* 6: 249-267.

Ravina, I. and Gur, Y., 1978. Application of the electrical double layer theory to predict ion adsorption in mixed ionic systems. *Soil Sci.*, 125:204-209.

Raats, P. A. C. and A. Klute 1968a. Transport in soils: The balance of mass. *Soil Sci. Soc. Am.* 32: 161-166.

Raats, P. A. C. and A. Klute 1968b. Transport in soils: The balance of momentum. *Soil Sci. Soc. Am.* 32: 452-456.

Raats, P. A. C. 1968. Forces acting upon the solid phase of a porous medium. *J. Appl. Math. and Phys.* 19: 606-613.

Ridley, A. M. and J. B. Burland 1995. A pore pressure probe for the in situ measurement of soil suction. *Proc. of Conf. of Advances in Site Investigation Practice. I. C. E. London.*

Smiles, D. E. and M. J. Rosenthal 1968. The movement of water in swelling materials. *Aust. J. Soil Res.* 6: 237-248.

Tayler, A. W., 1959. Concentration of ions at the surface of clays. *J. Am.Ceram.Soc.*, 42:182-184.

Warkentin, B. P., G. H. Bolt 1957. Swelling pressure of montmorillonite. *Soil Sci. Soc. Am. Proc.* 21: 495-497.

Yong, R. N. 1973. On the physics of unsaturated flow in expansive soils. *Proc. of 3rd Int. Conf. on Expansive soils, Haifa, Israel.* 2: 1-9.

Yong, R. N. and B. P.Warkentin 1974. Soil properties and behaviour. *Elsevier publishing Co. Ltd., Amsterdam.*

Zaslavsky, D 1964. Saturated and unsaturated flow equation in an unstable porous medium. *Soil Sci.* 98: 317-321.

Clay Science for Engineering, Adachi & Fukue (eds) © 2001 Balkema, Rotterdam, ISBN 90 5809 175 9

Interactions in clays in relation to geoenvironmental engineering

R. N. Yong
Geoenvironmental Research Centre, University of Wales, Cardiff, UK

ABSTRACT: The interactions in clay soils are examined relative to their influence on some typical Geoenvironmental Engineering situations. Particle-to-particle together with particle-porewater interactions, and the various interactions involving the functional groups of soils and contaminants are seen to be the major elements which control many of the physico-chemical and transmission properties and performance of the soils in relation to many Geoenvironmental Engineering problems.

1 INTRODUCTION

In general, interactions in clay soils occur between: (a) soil particles, (b) soil particles and porewater, (c) solutes and molecules in the porewater. The level of scrutiny of the factors involved and results of these interactions will depend on the nature of the problem at hand, and on the desired purpose of the scrutiny.

The interactions of interest in a typical clay soil in respect to Geoenvironmental engineering are those which impact directly on the physical-mechanical, physico-chemical and transmission properties of the soil. In respect to the physical-mechanical properties of a soil, it is often considered that a basic element in the development of these properties is its structure. By that, it is meant that the microstructure and macrostructure of the soil are direct reflections of interactions between soil fractions in the presence of an aqueous phase. For the purpose of this discussion, the various soil fractions that interact and combine to form a general soil structure are shown in Fig. 1. The definition of soil structure and soil fabric will be addressed in a subsequent section.

Included amongst the contributions of soil structure to the physical-mechanical properties and characteristics of the soil are soil suction, swelling performance and pressures, permeability, and features which control short and long mechanical stability of the soil.

Whilst soil structure is also an important consideration in regard to the physico-chemical and transmission properties a clay soil, the nature of the reactive surfaces of the soil fractions contribute significantly to the development of these properties. These feature prominently in the transport and fate of contaminants – as shown in the sketch in Fig. 2 which portrays a simple view of the interactions.

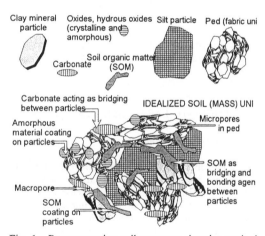

Fig. 1 - Representative soil structure showing typical soil fractions and their distribution (from Yong, 2000).

The principal issues that need to be considered in this discussion are:
(1) What is meant by "interactions";
(2) The various types and manner of interactions;
(3) The basic elements in a soil-water system which participate in the "interaction" processes, and the various mechanisms of interaction;
(4) Interactions resulting in the development of soil structure and physico-chemical properties;

(5) Interactions of particular interest in contaminant transport processes and leachate management.

The intent of this presentation is to provide an overall view of interparticle action – in terms of physical and physico-chemical interactions. As such, we will consider particle-to-particle and particle-group interactions as physical and also physico-chemical "interparticle" actions. We will note that the terminology of "groups" is used to denote soil structural units consisting of peds, clusters, domains, aggregate groups etc. To illustrate the interaction phenomena, selected geoenvironmental engineering problems will be examined.

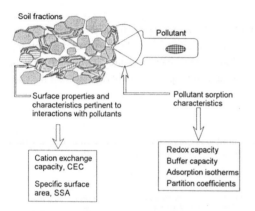

Fig. 2 - Elements contributing to basic interaction between contaminants, soil and water (from Yong, 2000).

2 PARTICLE-TO-PARTICLE INTERACTIONS

2.1 Basic considerations

The various interactions of concern in a clay soil can be viewed in simple terms as those occurring between: (A) soil fractions (i.e. particle-to-particle interactions), (B) soil fractions and porewater, and (C) solutes (atoms and molecules) in the porewater. On a macro scale, one could consider particle-to-particle interactions to be physical in nature. However, on a micro level, and particularly in clay soils, these particle-to-particle interactions involve both the physical and chemical features of the soil fractions. Accordingly, it is more convenient to consider interactions in terms of macro- and micro-interactions.

The various basic elements in macro- and micro-interaction processes in a soil-water system (clay soil) considered in this discussion are shown in Figs. 1 and 2. These elements consist of:

1] The clay soil fractions which include clay minerals, clay-sized primary minerals, amorphous materials (the various oxides and oxyhydroxides), soil organic matter, carbonates and sulphates;
2] The aqueous phase represented by the porewater. In the natural clay soil, the various dissolved solutes (ions, molecules, etc) which define the chemistry of the porewater are present as natural constituents.
3] Contaminants (pollutants and non-pollutants) introduced into the clay soil (soil-water system). These include both inorganic and organic chemicals.

Whilst not directly shown in the interactions between all of these elements, the processes associated with the metabolism of microorganism will also contribute to the status of the clay soil at any one time. These interactions are well illustrated in the nitrogen and sulphur cycles in the biosphere which have been extensively studied and reviewed. Of interest are those portions of the various cycles attributable to the activities of microorganisms in the soil-water system. Both aerobic and anaerobic microorganisms contribute to the biochemical reactions and the fate of the pollutants. While *eukaryotes* (cells with true nucleus; generally $< 2 \mu m$) and *prokaryotes* (cells without true nucleus; generally $> 2 \mu m$) utilize terminal electron acceptors, prokaryotes constitute the larger proportion of microorganisms that utilize a greater range of terminal electron acceptors found in the soil – e.g. NO_3^-, SO_4^{2-}, Fe^{3+}.

From Fig. 1, two groups of interactions are evident: (a) those contributing to the development of the "resident" soil structure, and (b) those involving the surface features of the soil solids. The former is a direct outcome of particle-to-particle interactions, and impacts directly on the integrity of the soil-water system and the short and long term stability of the soil mass. The latter derives from micro-interactions between particles and porewater and solutes within the porewater, and concerns itself with the various processes which will determine the stability of the soil mass and the transport and fate of contaminants.

It can be reasonably argued that the development of soil structure is determined by micro-interactions of the surfaces of the soil solids, and is dependent on the nature of the soil fractions and the chemistry of the porewater. The term *resident soil structure* has been used to mean the soil structure at any one time – i.e. soil structure determined at any one particular time through soil investigative techniques. This acknowledges that the structure of a soil is not fixed in time and space, and that regional controls have the ability to cause changes in soil structure. By considering (a) and (b) separately, one has the capability to evaluate: (i) interactions that contribute directly to soil structure and the integrity of a clay soil – particularly in respect to the physical-mechanical aspects of the soil, and (ii) interactions involving the surfaces of the soil particles with solutes in the porewater (contaminants etc.) – in respect to the physico-chemical performance of the soil and the processes associated with contaminant transport and fate.

It is useful to note also that surface interactions established between soil fractions result in conditions which will define or change soil structure through distribution of the various soil fractions. In essence, these interactions influence and/or control the manner in which the various soil fractions are distributed in the soil. Of particular concern are the physico-chemical and surface properties of the resultant soil structure obtained therefrom. The other kind of surface interactions, i.e. interactions between the surfaces of soil fractions and solutes in the porewater, are central to the control of the fate of contaminants and other solutes in the porewater. The various properties and characteristics pertinent to these interactions are shown in Fig. 2. These will be addressed in a later Section.

2.2 Particle interactions and suspended solids in slime ponds

Interactions between particles, i.e. particle-to-particle interactions, in relation to problems and/or situations in Geoenvironmental engineering can be found for example in discharge waste tailings and slime ponds associated with the mineral extraction and processing industry, and in liner-barrier systems. In the case of tailing ponds associated with the mining and mineral processing industry, the interaction of surface-active soil particles produces a slime or sludge which is essentially a soil-fluid suspension (suspended solids) commonly known as a "solids suspension".

Solids suspensions obtained as discharge from processes associated with waste tailings have many "names". Some of these include:
slime ponds – phosphate and tin mining,
sludge ponds – processing of tar sands,
red mud ponds – bauxite mining and processing.

The common feature to all these ponds is the inability of the soil solids in the pond to sediment (settle) – as shown in the solids concentration profile in the sketch in Fig. 3.The concentration of suspended solids in the "stagnant" zone shown in the Figure varies from about 12% (by weight) to about 42% depending on the type of pond (Yong. 1984).

Calculations of the volume of water associated with a gram of soil particle in equilibrium in an aqueous phase, based on type of soil fraction and double-layer interactions can be made using the DDL models. These can be compared with measurements of equilibrium solids concentrations obtained in soil-suspension experiments. The results obtained by Yong and Sethi (1978) are shown in Table 1 for some typical soil fractions. These results are expressed in terms of the equilibrium volume of

water per unit weight of soil. The terminology "equilibrium suspension volume" is used for this purpose and the units are given in terms of cc/g of soil. The void ratios shown in the third column of Table 1 have been calculated from the measured equilibrium volumes.

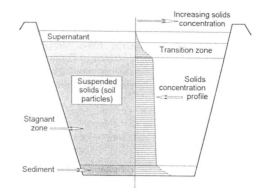

Fig. 3 - Schematic diagram of slime pond showing solids (particle) concentration profile and the various characteristic zones. (From Yong, 1984).

Table 1 - Equilibrium suspension volumes from soil suspension experiments.

Soil fraction	Eq. Vol cc/g	Void ratio
Kaolinite	1.3	3.4
Illite	3.1	8.2
Montmorillonite	21.5	57
Amorphous Fe_2O_3	20.5	82
Mica	3.0	7.9

Yong (1984) has shown that one can obtain good correlation with calculated and measured equilibrium solids concentration for the stagnant region of a slime pond – with a knowledge of the composition of the soil solids. In the actual cases examined, predicted solids concentrations were compared with actual solids concentrations obtained from samples in the stagnant zone (Fig. 3) for phosphatic slimes, aggregate slimes, tin mining slimes, beneficiation slurry (slimes), tar sand sludges, etc. Except for the aggregate slimes obtained from aggregate recovery of aggregate loams, a comparison of the predicted solids concentration with actual measured values showed good accord. The ratio of predicted to measured solids concentration (predicted:measured) varied from 0.96 to 1.05.

2.3 Dispersion stability and suspended solids

The calculations used to predict the total amount of water associated the soil solids in the stagnant zone shown in Fig. 3 are generally based on DDL concepts of particle interaction. It is useful to identify the governing mechanisms responsible for the recalcitrant settling performance of the suspended solids which constituted the stagnant zone. In doing so, one can develop counter-active procedures which would "release" the solids from the pseudo suspended state, promoting settling of the solids and clarification of the waste tailings pond. The use of flocculants is a good example of such counter-active measures.

Instead of using DDL concepts, the stability of the suspended solids in the stagnant zone can be viewed as the result of interactions of the soil solids with the molecules in the fluid medium (Brownian motion) – as in the method of study and analysis used by Yong and Wagh (1985) to explain the diametrically opposite character of the settling velocity curves between kaolinite soil suspensions and a bauxite red mud samples for solids concentrations ranging from 5 to 25 weight/weight (Fig. 4).

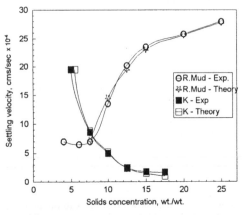

Fig. 4 - Calculated and measured settling velocities for kaolinite soil suspension and bauxite red mud.(R.Mud = red mud, K = kaolinite slurry, Exp. = measured values, and Theory = calculated values.)

Consideration of the interactions between particles and molecules in the fluid medium in terms of interparticle collision require the tracking of the evolution of particle distributions over time and space. Thus, if a number of particles at some position r with velocity v at some time t can be identified by some specific distribution function $f(r,v,t)$, the average velocity can be given as:

$$v^*(r,t) = \frac{1}{n}\int v\,f(r,v,t)\,dv \qquad (1)$$

where v^* is the average velocity and n is the concentration of particles (i.e. number density of particles). If m denotes the mass of a particle, the particle energy $e(v,t)$ will be given as:

$$e(v,t) = \frac{1}{n}\int \frac{1}{2}mv^2 f(r,v,t)\,dv \qquad (2)$$

If we consider two-particle collision phenomena as a means of analysis, we can invoke the Boltzmann microscopic evolution relationship (Chapman and Cowling, 1960). This permits us to describe the time-evolution of the distribution function due to external forces and interparticle collisions. To account for the effects of high concentrations of particles where higher order collisions can occur, simultaneous multiple collisions need to be considered in the calculations. The relationship obtained is seen as:

$$\frac{\partial f}{\partial t} + v\cdot\frac{\partial f}{\partial r} + \frac{F}{m}\cdot\frac{\partial f}{\partial v} = C(f) \qquad (3)$$

where f represents the distribution function, F = net external force, and $C(f)$ is a function which represents interparticle collision. This is commonly referred to as the *interparticle collision term* By using the relaxation time approximation procedure, and using τ and f^{eq} to represent the relaxation time of interparticle collision and local equilibrium function respectively, the interparticle collision term will be obtained as:

$$-\frac{f(r,v,t) - f^{eq}(r,v,t)}{\tau} \qquad (4)$$

The local equilibrium function f^{eq} is the equilibrium Maxwell-Boltzmann distribution function and is given in the form:

$$f^{eq}(r,v,t) = n\left(\frac{m}{2\pi\kappa T}\right)^{\frac{3}{2}}\exp\left[-\frac{mv^2}{2\kappa T}\right] \qquad (5)$$

where κ is the Boltzmann constant, and T is the absolute temperature.

In the study reported by Yong and Wagh (1985), they show that:

$$f(v) = f^{eq}(v) + g\left(1 - \frac{v_z}{V}\right)\frac{m\tau}{\kappa T}v_z f^{eq} \qquad (6)$$

where g represents the gravitational term, v_z = velocity in the z (downward) direction, and V is the Stokesian velocity. With the Boltzmann relaxation time approximation, the relationship for v_z will be obtained as:

$$v_z = \frac{1}{8}\left(\frac{m}{2\pi \kappa T}\right)^{\frac{1}{2}} \cdot V^2\left[1 - \frac{2\tau n\kappa T}{\eta}\frac{\tau gmV}{\kappa T} + \frac{3mng}{\eta}V\tau^2\right] \quad (7)$$

The question of whether the preceding considerations of particle interaction is useful can be answered by examining the results shown in Fig. 4. If one uses simple physical models which consider only Stokesian relationships, it should be clear that the relaxation time should vary as the reciprocal of particle concentration. However, as seen from the results in Fig. 4, this is not true for the case of the red mud sample obtained from a bauxite waste discharge site in Jamaica. The greater presence of fine particles and their surface activities render the last term in equation (7) very dominant – in contrast to the kaolinite slurry where this term is relatively insignificant.

Simple and measurable indices can be used to characterize the interparticle forces (and their changes), e.g. ESP (exchangeable sodium percentage), SAR (sodium adsorption ratio), pH, dielectric dispersion and zeta potential ζ. Whilst these tests are generally conducted with soil suspensions, the information obtained therefrom can be used in a mechanistic sense to provide interpretations and evaluations on the processes contributing to the "suspended" state of the solids in the stagnant zone. Fig. 5 shows the relationship between dispersion stability of clay soils (such as those that can be found in the stagnant zones) and their average zeta potential ζ.

The total energy of interaction between the suspended particles in relation to particle spacing, and expressed in terms of the zeta potential is the result of the following: (a) repulsion energy arising from Brownian motion, and (b) attractive energy due to the London-van der Waals forces. We recognize that one of the principal factors contributing to the ability of clay soil fractions to remain "suspended" in a soil solution is due to the surface active nature of the soil particles. In swelling soils where the double layer is well developed, interlayer swelling plays a dominant role. In the case of kaolinite soils however, the dispersibility of the clay soil in the absence of interlayer phenomena is to a very large extent dependent on interactions between the soil particles and the chemical constituents in the porewater.

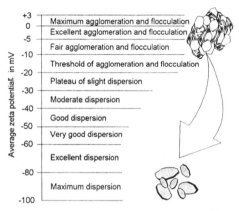

Fig. 5 - Ranges of average zeta potential ζ and dispersion stability for clay soils.

Instead of interlayer interactions, interparticle (particle-to-particle) actions are the more dominant mechanisms of interactions for low swelling or non swelling clays. Desorption or removal of adsorbed ions from the clay particles' surfaces will increase the double-layer thickness associated with the soil particles, producing thereby a particle-particle "swelling" phenomenon. Thus for example, the desorption of Cl^- and Na^+ in a low salt content range from kaolinite particles will increase the double layer and hence promote repulsion between particles. The actions of potential determining ions such as bicarbonates, carbonates, hydroxides etc. will also contribute significantly to the net repulsive activity between the particles.

2.4 Soil structure and physical integrity

The physical integrity of clay soils in engineered barriers can be compromised because of detrimental interactions between soil solids. These (interactions) result from changes in the structure of the soil. For this discussion, we define soil structure to include both the geometric arrangement of particles (fabric) and the result of interactions established between particles (intrinsic energy of the soil-water system). By this definition, changes in soil structure can occur as a result of: (a) changes in the arrangement of particles, and/or (b) changes in the various sets of interactions which establish the internal energy of the soil-water system. It is the latter that is of particular concern in geoenvironmental engineering.

Changes in the various internal energy components can arise due to natural sets of forces attributable to regional controls and anthropogenic activities. It is clear that any change in the soil structure will result in changes in the mechanical and physico-chemical properties of a soil. Additionally, the surface properties of the soil will also be affected by the resultant soil structure changes. A good demonstration of this can be found in the creep performance of a plastic lacustrine shown in Fig. 6.

The creep results shown in the diagram indicate a low creep rate which is increased dramatically after the introduction of a sodium silicate leachant after 9 days (12,615 minutes) of creep testing. Changes in the internal energy due to introduction of the leachant reflect on the structural integrity of the sample. Microstructure and double-layer interactions between particles and peds (microstructural units) are affected with the introduction of the leachant. Since double-layer interactions are influenced by soil composition and aqueous phase chemistry, calculations can be made to determine the energies of interaction using simplified models.

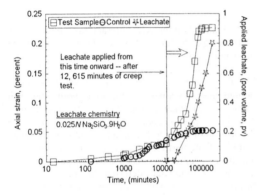

Fig. 6 – Effect of leachate chemistry on the creep performance of a plastic lacustrine clay (from Yong *et al.*, 1985).

To determine the resistance of the soil to the applied creep load as a result of particle-to-particle interaction, at least two types of models are available. These are the DDL and the DLVO models. They provide one with the tools to calculate the resultant energies of interaction between particles These models have been used to predict the behaviour of ideal systems. When any significant in simplification of physical reality for computational expediency is made, the calculated values can best be used for qualitative comparisons and predictions. A review of these models can be found in the many textbooks dealing with soil properties and behaviour.

3 INTERACTIONS AND SOIL INTEGRITY

3.1 *Initial considerations*

The representative soil structural unit given in Fig. 1 shows most of the basic soil fraction units and the results of interactions between these units – i.e. resultant soil structural unit.The "SOM" shown in Fig. 1 which represents "soil organic matter", together with the amorphous materials have the capability of coating the mineral and silt-sized particles – thus changing not only the surface charge characteristics of the soil structural unit, but its interaction with other soil structural units. Additionally, SOM, carbonates and amorphous materials can also form bridging and cementation bonds between soil particles. Amongst the various interaction conditions favouring any or all of these capabilities of the soil fractions SOM, carbonates and amorphous materials to function as coatings and bridges, the pH and Eh of the micro-environment are considered to be the most significant.

3.2 *Particle coatings and discrete units*

Both specific surface area (SSA) and cation exchange capacity (CEC), which are directly linked to soil structure are sensitive to the manner in which the SOM and amorphous materials are distributed in the soil. This can be directly inferred from the results of experiments conducted to determine the influence of amorphous iron oxide (ferrihydrite) on the particle surface interactions of kaolinite and illite minerals – as reported by Yong and Ohtsubo (1987) and Ohtsubo *et al.*, (1991). The results of zeta potential determinations are presented in Fig. 7 for the case of interactions between: (a) kaolinite and iron oxides (denoted as K+Oxide) , and (b) illite and iron oxides (denoted as I+Oxide) in relation to pH – for a 5% (w/w) ferrihydrite presence in the mineral-oxide mixture.

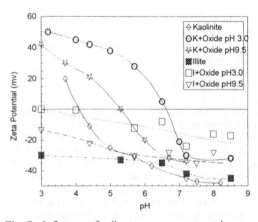

Fig. 7 - Influence of soil structure on measured zeta potential for kaolinite and illite mixed with iron oxide and prepared at pH 3.0 or pH 9.5.

The nature (sign) of the net surface charge of the ferrihydrite and mineral particle plays a significant role in the determination of whether the ferrihydrite coats the mineral particles. At pH 3.0, the surface charge for the ferrihydrite is predominantly positive – as are the edges of the kaolinite mineral particles. Positive edge-charge characterization for illites can, at best, be considered to be minimal. Thus, mixtures of ferrihydrite with kaolinite formed at pH 3.0 will show the ferrihydrite coating the particles' surfaces whilst generally leaving the edges of the particles uncoated. The same cannot be said for the edges of the illite particles. Some coating of the edges of the illite particles, in addition to surface coating, is also expected.

At pH 9.5, the surface charges of ferrihydrite are predominantly negative. The isoelectric point (iep) for the ferrihydrite is estimated to be between 7 and 9, whereas the iep for the kaolinite is between 4.0 to 4.5, and "undetermined – but considerably less than 2" for the illite. Kaolinite-ferrihydrite mixtures formed at pH 9.5 will not produce the same structure as those formed at pH 3.0 because the interactions between the surface charges of both kaolinite and ferrihydrite are between positive charges. The pattern is similar for illite and ferrihydrite mixtures formed at pH 9.5.

The determination of the zeta potential ς at various pH values of the ferrihydrite-mineral particle mixtures (Fig. 7) provides information concerning the interactions between the soil fractions. mineral particles and ferrihydrite. Since the zeta potential ς is a direct reflection of the interaction developed between the diffuse ion-layer and the outer Helmholtz plane (ohp) during electrophoretic flow, its variation in relation to a changing pH environment informs one about the nature of the interactions in the clay-water system. Partial or total coating of mineral particles, together with the existence of discrete entities of the ferrihydrite can occur dependent on whether the soil structure was formed at low or high pH environments. Considering the zeta potential ς to be a measure of the viscous resistance at the ohp interface during electrophoretic flow, it is useful to view the equivalent performance obtained as viscous shear resistance (Bingham yield stress) – (shown in Fig. 8).

The basic lesson learnt in respect to the results shown in Figs. 7 and 8 concerns the manner in which the amorphous materials are distributed in the combined (mixture) mineral-ferrihydrite soil. By extrapolation, one might argue that SOM which could coat the particles or act as discrete units would also produce similar soil structure characteristics. In the final analysis, the manner of distribution or utilization of these materials can significantly alter the net surface charge charges and also the specific surface area of the soil. These are evident in the Bingham yield stress relationships shown in Fig. 8.

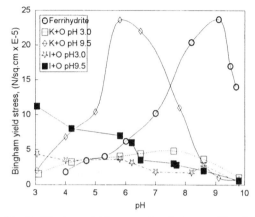

Fig. 8 - Bingham yield stresses in relation to pH, and conditions for formation of the iron oxide complexes.

The results of consolidation tests on a marine clay shown in Fig. 9 compares the consolidation performance of the natural clay against companion samples where Fe_2O_3 and Al_2O_3 (denoted as "sans Al and Fe) were removed by leaching, and also against other samples where SiO_2 (denoted as "sans Si") was removed. Leaching of companion samples by a buffer solution at pH 4 for removal of the Fe_2O_3 and Al_2O_3 and at pH 10.5 for removal of the SiO_2 took from 140 to 240 days.

Since the weight percentage in the natural marine clay of the SiO_2 (3.6%) and the combined Al_2O_3 (0.9%) and Fe_2O_3 (3.4%) weight percentages were not too far apart, it is useful to know how removal of these amorphous materials affected the properties and behaviour of the soil. For ease in comparison, the ordinate in Fig. 9 shows the ratio of the void ratio e in relation to the initial void ratio e_o, (e/e_o). For the particular test results shown in Fig. 9, the compressibility of the "sans Si" increased by 2.2 times (from 0.44 to 0.98) whereas it decreased from 0.44 to 0.30 for the "sans Al+Fe" sample. The preconsolidation pressure decreased for both the "sans" cases, from 450 kpa for the natural clay soil to 300 kpa for "sans Al+Fe" to 220 kpa for "sans Si".

Fall cone tests conducted on the samples show a decrease of shear strength from 131 kpa for the natural clay to 39.2 kpa for the "sans Al+Fe" and 67.7 kpa for the "sans Si" The more useful information is in the demonstrated sensitivities which increased from a medium-low sensitivity (8.5) for the natural clay to about 14 for the "sans Al+Fe" to "infinite" for the "sans Si".

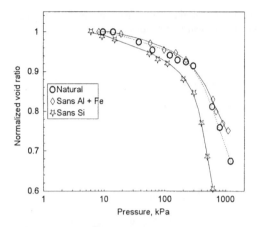

Fig. 9 - Controlled gradient consolidation test results for marine clay.

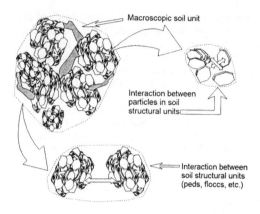

Fig. 10 - Macroscopic soil unit (REV) consisting of mesoscopic units (peds, floccs, etc.) and microscopic units (particles).

From a soil structure point of view, it is useful to note that the interactions between the amorphous materials and the crystalline particles and layer lattice minerals control both the soil properties and behaviour of the soil. The effect of removal of the amorphous materials is also seen to be dependent on the type of amorphous materials – as shown by Yong et al., (1980) and Ohtsubo et al., (1991). Geoenvironmental engineering implications arising from the above include the effects of leaching on soil-engineered barriers, stability of side slopes and substrate bearing materials, transmission properties, and overall integrity of the impacted soil materials.

3.3 Macrostructure and microstructure

To evaluate macroscopic interactions involving soil particles and their aggregate units (peds, domains, clusters, etc.), it is useful to obtain an appreciation of the contributions made to the overall macrostructure by the microstructural units (peds, clusters, etc.) that form the macrostructure. The basic macroscopic soil structural unit shown in Fig. 10 consists of interacting mesoscopic units (peds, domains, floccs, etc.) and microscopic units (particles). The overall stability and integrity of the macroscopic unit is the result of the various interactions occurring mesoscopically and microscopically.

To be in a position to deal with physico-chemical forces and relationships between meso- and micro structural units, at least two general approaches can be used. The first approach deals mainly with the mechanics of the system and the physical interactions between particles and peds, whereas the second approach utilizes concepts from thermodynamics for expression of the internal energy of the micro and macro structural units.

In the first approach, the stress-strain relations for the micro soil structural units are given as:

$$\epsilon_{ij} = a_{ijkl}\,\xi_{kl}\;;\qquad \xi_{ij} = b_{ijkl}\,\epsilon_{kl} \qquad (8)$$

and the corresponding stress-strain relations for the macro unit are given as:

$$e_{ij} = A_{ijkl}\,\sigma_{kl}\;;\qquad \sigma_{ij} = B_{ijkl}\,e_{kl} \qquad (9)$$

where ε and e are the micro and macro strains, and ξ and σ are the micro and macro stresses, and the "compliances" for the micro and macro units are represented by a, b, A, and B respectively. Taking into account the varied number and nature of the micro soil structural units that combine to form the macro unit, a complete statistical description of the physical constants and the micro stresses and strains the entire macro domain of interest is generally required. The use of fluctuating tensor components accounts for the second moments of the distributions. Thus the difference in strain energy due to the fluctuating tensors will be given as:

$$\Delta u = \frac{1}{2} <a_{ijkl}> <\xi_{ij}^{*}\xi_{kl}^{*}> + \frac{1}{2} <a_{ijkl}^{*}\xi_{ij}^{*}\xi_{kl}^{*}>$$
$$= \frac{1}{2} <a_{ijkl}> R_{ijkl} + \Delta u^{*} \qquad (10)$$

where the asterisks * represent average values, and R_{ijkl} is a correlation tensor which represents the random tensor field which is a second order characteristic of a random order variable. If x represents a vector position of a micro unit in the macro volume, and r represents the separation distance between two micro soil structural units, then:

Whilst the micro mechanics approach for determination of the integrity of the macro soil structural unit is given as:

$$R_{ijkl} = < \xi_{ij}^*(x) \cdot \xi_{kl}(x+r) > \qquad (11)$$

It can be shown that since the fluctuating quantities represent random stress and strain fields, and are in fact the second moments of random variables, the degree of inhomogeneity of the unit can be related to the correlation tensor R_{ijkl} and a mathematical characteristic φ_{ijkl} given as a fourth order tensor. Thus:

$$\Delta u = \frac{1}{2} [< a_{ijkl} > + \varphi_{ijkl}] R_{ijkl} \qquad (12)$$

We note that φ_{ijkl} is a measure of the heterogeneity of the material. A flow function can be written as:

$$F [\sigma_{ij}, T, e_{ij}^P, \beta_1, \beta_2, ..., \beta_n, K_1, ... K_n] \geq 0 \qquad (13)$$

where $K_1, ... K_n$ are physical parameters associated with the micro structure. $\beta_1, ... \beta_n$ constitute a set of parameters associated with micro deformations, and are obtained during deformation or unloading processes as non holonomous differential expressions with quantities depending on the loading process, in the form of some functional.

3.4 *Structure and swelling clays*

The difficulties in ascribing actual physical values and properties to the various parameters described previously become compounded when one deals with clay soils that contain swelling clay minerals. The sketch shown in Fig. 11 provides an overall view of a clay ped consisting of numerous 2:1 layer-lattice mineral particles. For convenience, the microstructural units which could include peds, clusters, aggregate groups, etc. will be identified as "peds".

The development of soil properties in relation to micro- and macrostructures depends on the interactions established between the particles in the peds and the interactions of the peds. For non-swelling 1:1 layer-lattice clay minerals, interactions will be between particles, and sorbed water will be the hydration water surrounding individual particles or peds. For non-swelling 2:1 layer-lattice clay

minerals, the sorbed water in the interlayers will be the hydrated layers. Interlayer expansion will be limited to the hydration layers. For swelling soils, greater interlayer expansion (beyond hydration) will be obtained via forces which are well described by the diffuse double-layer (DDL) models.

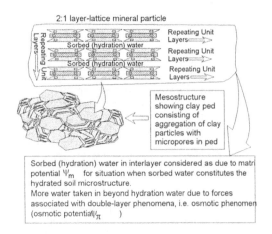

Fig. 11 - Interactions between particles within a clay ped – contributing to ped structure, properties and characteristics.

The corresponding thermodynamic potentials or "force-equivalents" resulting from these interactions are the matric and osmotic potentials (or matric and osmotic suctions). These impact directly on various geoenvironmental engineering situations. This is shown in the schematic diagram of Fig. 12 for the example of water movement controlled by internal forces – and the resultant impact on properties and response behaviour of geoenvironmental engineering interests.

The results of interactions between the various micro- and macrostructural units, expressed in terms of the soil-water potential ψ, have also been viewed in terms of physical-mechanical equivalences – e.g. soil suction. Perhaps the most "popular" application of the soil-water potential ψ and soil-suction concepts in Geoenvironmental engineering is in evaluation and prediction of transport processes in soils. Prominent amongst these are: (a) transport of contaminants in problems associated with leachate management and buried wastes, (b) moisture movement in partly saturated soils, and (c) coupled heat and moisture transport in backfill-buffers and barrier systems.

Denoting t as "time", D as the diffusion coefficient, and x as the spatial coordinate, the simple one-dimensional moisture transport relationship given as:

$$\frac{\partial \theta}{\partial t} = \frac{\partial}{\partial x}\left(D(\theta) \frac{\partial \theta}{\partial x} \right) \qquad (14)$$

implicitly assumes that ψ is a single-valued function of the volumetric water content θ of the soil, and that $D(\theta) = k(\theta)[\partial \psi/\partial \theta]$. Whilst this has application to homogeneous soils with little or no change in pore geometry upon wetting and moisture transport, complications arise when this relationship is applied to swelling soils. Soil volume expansion occurs with water uptake if the soil is not constrained. Volume expansion is a direct function of the water uptake and soil structure. However, if volume expansion is not allowed, the swelling pressures developed will be consistent with DDL model predictions – as shown in results reported by Yong (1999) in Fig. 13.

The experimental results shown in Fig. 13 for a Na-montmorillonite saturated with a 10^{-3} MNaCl, are compared with calculations from DLVO theory, Gouy Chapman model, and results reported by Bolt (1956) and van Olphen (1977). In the last case (van Olphen) the results have been interpreted from pressures required to remove the first four water layers next to a typical montmorillonite unit. We can conclude from the results that swelling performance is a function of the osmotic potential ψ_{π}. More detailed comparisons of the results can be found in Yong (1999).

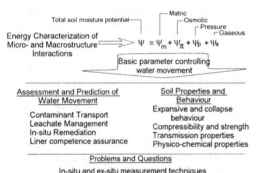

Fig. 12 - Interactions in micro- macrostructures of clay soil and impact on properties and performance relevant to geoenvironmental engineering.

For swelling soils at "dry" initial conditions, water uptake in the interlayer spaces due to hydration forces will provide for a different form of water structure, and volume expansion due to this phenomenon is defined as crystalline swelling due to ψ_m. van Olphen (1977) maintains that at interlayer spacings of up to about 1 nm, the dominant processes

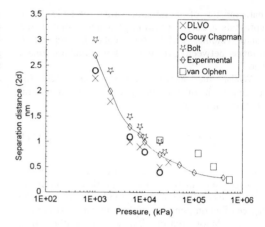

Fig. 13 - Comparison of calculated results with measured swelling pressures for swelling clay (from Yong, 1999).

responsible for hydration and resultant interlayer separation (2d distance shown in Fig. 13) derive from the adsorption energy of water at clay surfaces."Sorption" of water (hydration) is due to to the presence of charge sites and exchangeable ions in the interlayers, and the attractions between water molecules and the polar surface groups. Interlayer expansion from water uptake between about 1.0nm and 2.2nm will be *crystalline swelling*. Yong (1999) has noted that sorption forces associated with the matric potential ψ_m, are responsible for crystalline swelling, and that interactions described by the osmotic potential ψ_{π} are the primary mechanisms for interlayer or interlamellar swelling beyond crystalline swelling (i.e. double-layer swelling).

The influence of the interactions between micro and macro soil structure, particularly in swelling soils, in relation to the various transport processes can be seen in the collection of studies reported by Pusch et al., (1999). The studies on soil microstructure evolution (Pusch, 1999) and the various forms of soil micro structure (Fukue et al., 1999) as deduced from electrical conductivity measurements, show the importance of recognizing micro and macro structure when evaluating the performance of the soil. These impact on performance of the macro soil material in regard to hydraulic conductivity (Pusch and Schomburg, 1999), and transport modelling (Thomas and Cleall, 1999).

4 SOIL-CONTAMINANT INTERACTIONS

4.1 *Interactions and partitioning of contaminants*

The fate of contaminants in soils is a function of the properties of the contaminants and soil fractions, and the various interaction mechanisms established thereby. Much obviously also depends on the nature and distribution of the dissolved solutes in the porewater. The *fate descriptors* that are generally considered include: Persistence, Accumulation, Transport, and Disappearance of the contaminants under consideration. The various reactions and controls involved in consideration of the fate descriptors include the reactions in the porewater, and sorption mechanisms and transport processes contributing to the partitioning of the contaminants.

4.2 *Interactions and pE-pH*

The porewater in soils is in reality a complex aqueous regime consisting of naturally occurring salts and inorganic and organic contaminants. "Equilibrium" in the soil-contaminant-water system involves reactions between the surface reactive groups of the soil fractions and the chemical constituents in the porewater. as by the chemistry of the porewater – together with the reactions provoked by biologically mediated chemical processes. The chemical interactions between contaminants and soil fractions include acid-base reactions (including hydrolysis), speciation, complexation, precipitation and fixation, are some of the many manifestations of the interactions. Accordingly, the redox environment is a major element in the control of the fate of the contaminants in the soil (Fig. 14).

The *pE*-pH diagram in Fig. 14 shows that at a *pE* 4, Fe exists as Fe^{3+} at the lower pH values and that with increasing pH, precipitation of the Fe occurs resulting in the formation of Fe(III) hydroxides ($Fe(OH)_3$). Lowered values of *pE* at the higher pH values will result in precipitates of Fe(II). Using "A" and "B" to represent "reactant" and "product", the influence of pH, *Eh* or *pE* of the porewater and the various contaminants in the porewater can be assessed using the Nernst equation as follows:

$$pE = 16.92\,Eh$$
$$= E^o + \left(\frac{RT}{nF}\right) \ln \frac{[A]^a[H_2O]^w}{[B]^b[H^+]^h} \qquad (15)$$

The superscripts *a, b, w, h,* in the above equation refer to number of moles of reactant, product, water, and hydrogen ions respectively.

The redox reactions pertaining to the conditions shown by the uppermost and lowermost sloping boundaries are given as follows:

$$2H_2O \leftrightarrow O_2(g) + 4H^+ + 4e^-$$
$$2H_2O + 2e^- \leftrightarrow H_2(g) + 2OH^- \qquad (16)$$

where the first equation refers to oxidation and the second refers to reduction. The upper and lower sloping boundaries in the diagram define the limit of water stability.

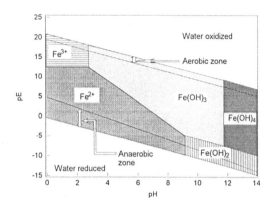

Fig. 14 - *pE*-pH chart for Fe with maximum soluble Fe concentration of 10^{-6} moles (adapted from Fetter, 1993).

4.3 *Interactions and partitioning*

Changes in both pH and *pE* values due either to reactions or external circumstances affect interactions between contaminants and soil fractions. A good example of this is the precipitation of metals as pH increases beyond the precipitation pH of the metals. Partitioning of heavy metals results from interactions that include: (a) sorption of the metals by soil fractions, (b) speciation and complexing with inorganic and organic ligands, and (c) precipitation of the metals.

A not too well appreciated result in the study of multi-component heavy metal contamination of soils is the effect of the presence of other metal species on the precipitation pH of individual metals. Fig. 15, using information from MacDonald (1994) shows the soluble-precipitate transition performance of 3 heavy metals acting singly and as equal mixtures of the three. The most significant change in the precipitation performance can be seen in Zn, where onset of precipitation of the single species solution at around pH 4.5 has been increased to about pH 6.4 for a mixture (Zn-mix) of equal parts of Zn, Cu and Pb.

The presence of multi-component species of metals and various organic/inorganic ligands plus organic chemical contaminants in contaminated soils sets up a whole host of interactions which can be difficult to fully delineate. Thus, experimental procedures which seek to establish physically measured partitioning of contaminants have been used as a means to assess total interaction effects between contaminants and soil fractions. Chief

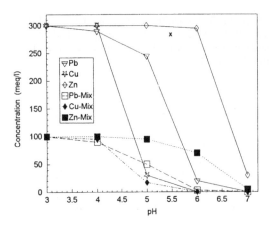

Fig. 15 - Precipitation of heavy metals as single solutions and as mixtures of the three metals in equal proportions (Data from MacDonald, 1994).

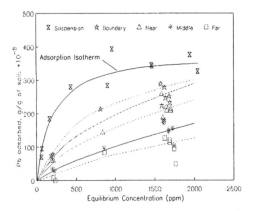

Fig. 16 - Adsorption isotherm and adsorption characteristic curves for Pb with kaolinite soil (data from Cabral, 1992 and Weber, 1991).

amongst these is the production of adsorption isotherms – generally obtained through experiments conducted with soil suspensions and target contaminants.

As can be seen in Fig. 16, the uppermost curve which has been identified as the adsorption isotherm is derived from soil-suspension tests on a kaolinite soil. The other curves are sorption results from leaching tests on compact samples. These are adsorption characteristic curves which represent adsorption of the target contaminant over a specified time period. We expect that the adsorption characteristic curve for the "boundary" of the compact sample (i.e. interface boundary between leachant and sample) represents the maximum Pb sorption capacity of the compact kaolinite, and hence would further expect that all the other

adsorption characteristic curves would (with time) match the "boundary" curve.

The differences between the adsorption characteristic curves and the adsorption isotherm reflect the differences in the results of interaction between the contaminants in the Pb leachant and the soil fractions. Since the soil type is common to both types of tests, the key element which differentiates between the two types of curves is the amount of exposed soil particle surfaces. This simple example demonstrates that assessment of partitioning of contaminants in soils must consider: (a) single as opposed to multi-component contaminant leachants, (b) proportion of soil particle surfaces exposed to interaction with the contaminants, and (c) all other factors associated with the kinetics of reactions such as temperature, time, pH etc. However, since all other factors are generally considered to remain constant (i.e. item (c) remaining constant), items (a) and (b) are the factors that need serious attention and consideration. In particular, transport analyses will be significantly impacted with an improper choice of partition coefficients.

4.4 *Organic chemicals and partitioning*

The interactions between organic chemical contaminants and soil fractions require considerations of both abiotic and biotic processes. The reactions include hydrolysis, photolysis and biodegradation. At the intermolecular level, the interactions include: London-van der Waals forces, hydrophobic reactions, hydrogen bonding and charge transfer, ligand and ion exchanges and valence bondings associated with chemisorption. Probably one of the more significant interaction and bonding mechanism is the hydrophobic bonding. This is particularly true if the soil contains soil organic matter – since the functional groups of the soil organic matter are closely similar to most of the organic chemical contaminants.

Many organic molecules are positively charged by protonation and hence the surface acidity of soil particles are significant factors in the sorption of the ionizable organic molecules. Water solubility plays an important role in partitioning. Fig. 17 shows the adsorption isotherms for naphthalene, 2-methyl naphthalene and 2-naphthol with the same kaolinite soil used for tests in Fig. 16.

As might be deduced, from the adsorption isotherms shown in the Figure, the water solubilities of naphthalene ($C_{10}H_8$) and 2-methyl naphthalene ($C_{11}H_{10}$) are closely similar (around 30 mg/L) whilst the water solubility of the 2-naphthol ($C_{10}H_8O$)is about 25 to 30 times larger (around 850 mg/L). The greater the water solubility of the organic chemical, the greater is the capability for the chemical to remain soluble in water. This reduces the sorption potential of the chemical vis-a-vis soil-sorption interactions.

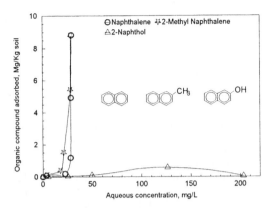

Fig. 17 - Adsorption isotherms for some PAHs with kaolinite. (Data from Hibbeln, 1996).

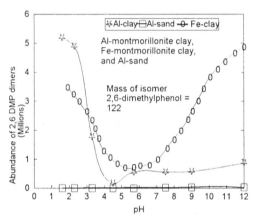

Fig. 18 - Oxidation of 2,6-dimethylphenol (mass 242) by Al- and Fe-montmorillonite and Al-sand. (Data from Yong, et al., 1997).

The k_{ow} octanol-water partition coefficient has been found to be relatively well correlated with water solubility and also with soil sorption coefficients. The relationship for the n-octanol water partition coefficient k_{ow} has been expressed in terms of the solubility S by Chiou et al., (1982) as:

$$\log k_{ow} = 4.5 - 0.75 \log S \quad (ppm) \qquad (17)$$

Organic chemicals with k_{ow} less than about 10 are generally considered to be relatively hydrophillic, i.e. they have large water solubilities and low soil sorption coefficients. On the other hand, compounds with k_{ow} values greater than 10^4 are considered to be very hydrophobic, and will demonstrate low water solubilities.

4.5 Interactions and electron transfer

Probably, one of the most important and interesting sets of interactions are those which lead to transformation of the contaminants and soil fractions. The major transformations are generally associated with organic chemical compounds, and involve abiotic and biotic transformation processes. The abiotic transformation processes include such chemical reactions as oxidation-reduction and hydrolysis. The transformed products obtained thereby are generally other types of organic chemical compounds.

It is interesting to note that abiotic transformation processes can occur with or without benefit of electron transfer. A common example of this is non-reductive chemical reactions which involve attacks by nucleophiles on electrophiles.

A neutral hydrolysis reaction can be viewed as a nucleophilic attack on an electrophile, i.e. a water molecule replaces groups of atoms in the organic chemical. This reaction also allows for an OH^- ion to replace another atom in the organic chemical. Where catalytic activity is obtained by the H^+ and OH^- ions, one obtains acid-catalyzed and base-catalyzed hydrolysis. The distinction between these and neutral hydrolysis reactions is important inasmuch as the kinetics of hydrolysis are directly related to the process involved.

In respect to organic chemical compounds, abiotic transformation processes are relatively minor in comparison to biotic transformation processes. These biotic processes are biologically mediated transformation reactions which may or may not include associated chemical reactions, and are the only types of transformation processes which can lead to mineralization of the target organic chemical compound.

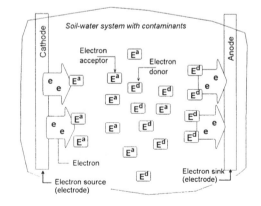

Fig. 19 - Insertion of active electrodes into soil with resultant charge transfer between electrode and electron acceptor and electron donor and electrode.

Distinguishing between redox reactions that occur abiotically and under biotic conditions can quite often be difficult. It is often argued that such a distinction is not always useful since the number of functional groups of the target organic chemicals that can be oxidized or reduced abiotically is not considered to be significant in comparison to those that can be oxidized or reduced under biotic conditions. Since some clay soils demonstrate good electron acceptor capabilities, one can expect transformations of susceptible organic chemical compounds when interaction between the two occurs. The recent results reported by Yong et al, (1997) shown in Fig. 18 demonstrate the effectiveness of the oxidizing capability of the Fe(III) of the montmorillonite clay.

A significant contribution to the interactions involving electron transfer involve microorganisms. Of particular interest is the phenomena of nitrogen and sulphur cycles in the biosphere since these have an impact on the interactions between contaminants and soil fractions. While nitrogen N can exist in various valence forms (from +5 to -3), the reduced form is the one which contributes most to the interactions in soil. Similarly, sulphur S which can also exist in valence form ranging from +6 to -2, the reduced form is also of greater significance in the interactions. Thus for example, eukaryotes and prokaryotes utilize terminal electron acceptors such as NO_3^-, SO_4^{2-} and Fe^{3+}. In addition, oxidized nitrogen compounds obtained as a result of dissimilatory nitrate reduction are also terminal electron acceptors for microorganisms under anoxic conditions.

Probably one of the more dramatic results of the effects of S-oxidizing bacteria is the production of acid mine drainage – a problem which is common in metal mining industry in many parts of the world. Acid production and transport in the subsurface constitutes a major problem.

5 INTERACTIONS AND POLLUTION MITIGATION

5.1 *Basic considerations*

The principal focus in remediation of contaminated ground is pollution mitigation. This takes the form of either removal of the contaminants, or neutralizing them so that they do not pose any further threat to human health and the environment. Effective removal of the contaminants requires detachment of the contaminants from the soil fractions. This is a process which can be called "de-bonding", i.e. neutralization of the interactions between the contaminants and the surfaces of the soil particles.

To illustrate the role of interactions in mitigation of pollution, we cite two examples of recent remediation techniques: (a) the use of electrokinetics, and (b) utilization of permeable reactive walls.

5.2 *Interactions and electrodics*

Application of electrokinetics in remediation of contaminated sites capitalizes on the development of electro-osmotic flow in an electric field manipulated by the operator. Significant chemical transformations can occur when active electrodes are inserted into a clay soil. To obtain a better view of the various electrochemical reactions, one should examine the transfer mechanisms associated with electrontation and de-electronation. The basic elements of such transfers are shown in Fig. 19.

Electrons emitted from the cathode to the electron acceptors in the soil respond to the interaction process defined as electronation, and the resultant effect in respect to ions as electron acceptors is reduction of the positive charges of the ions. At the anode, the interaction process of de-electronation occurs resulting in oxidation of the electron donors. One observes from the preceding that both oxidation and reduction of the target contaminants occur during the application of electrokinetic remediation as a treatment process. Additionally, dissociation reactions occurring in the porewater of the soil will increase the concentrations of H^+ ions at the anode and OH^- ions at the cathode.

Application of electrokinetic phenomena in soils for removal of contaminants benefits from the interactions modelled generally by DDL theory. Efficiencies and capabilities for contaminant removal via electro-osmotic processes can be deduced directly from double-layer theory. The basic elements of the process and interactions are shown in Fig. 20 and are consistent with general DDL models – with the diffuse-ion layer bulk (electro-osmotic) flow represented as a planar sheet with thickness given as the inverse of the Debye-Hückel parameter χ.

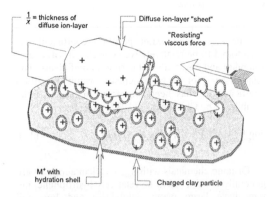

Fig. 20 - Electro-osmotic flow due to imposition of an electric field.

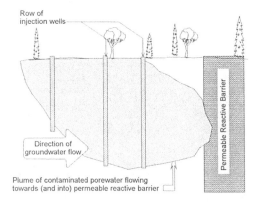

Fig. 21 - Permeable reactive barrier used as a treatment wall to intercept contaminated porewater.

5.3 *Interactions and treatment walls*

The recent use of "treatment" walls as intercepting permeable reactive barriers, for treatment of contaminated porewater is probably the one of the best demonstrations of the use knowledge of interactions between contaminants and soil fractions. Several types of interactions are generally involved – depending not only on the treatment material in the wall, but also on the treatment of the contaminated ground.

The sketch shown in Fig. 21 illustrates the different options available. Chemical treatment of the contaminated site via injection wells for example, to "release" the contaminants into the water (porewater) is a "popular" option. Other options could include bioremediation techniques. Removal of the contaminants "deposited" in the porewater relies on the material contained in wall. Reactions sought (in the wall) as a means for contaminant removal include sorption, precipitation, transformation, complexation, oxidation, reduction, abiotic transformation and biotic transformation, and biodegradation. Obviously, choice of treatment material for inclusion in the wall, together with control on permeability such that interactions can occur to their fullest extent are strict design requirements.

By channeling the flow of water to the permeable reactive barrier using channel barriers, one is ensured of capturing all the contaminated water. Treatment materials included in the reactive barrier system for interactions that would provide the various treatments for the inorganic and organic chemical contaminants include:

- Chelating agents, catalysts, microorganisms, zero valent metals, zeolite, ferrous hydroxides, carbonates, activated carbon, alumina, nutrients, phosphates, etc.

6 CONCLUDING REMARKS

Of necessity, the remarks made in this presentation have been brief and in many instances, lacking in the kinds of desired detail. What has been addressed has been the many aspects of interactions between particles and groups of particles which constitute a soil mass. These interactions are physical, physico-chemical, chemical, and biochemical.

We can view the interactions solely as physical for many situations which fulfil "engineering" purposes. However, in the case of many of the problems in geoenvironmental engineering where contaminants are involved, the basic sets of interactions revolve around electron transfer between the many constituents that populate a soil mass. In the final analysis, for such problems (i.e. geoenvironmental problems), the mandate for geotechnical engineers is the design, construction and implementation of facilities and processes that will benefit the public – i.e. minimize and/or eliminate the potential threats to human health and the environment. Much has yet to be learnt about the various kinds of interactions in clay soils. However, in the interim, it behooves us to pay attention to the interactions described herein (and in other published literature) inasmuch as these have been shown to contribute significantly to the physical, physico-chemical and chemical behaviour of the soils.

REFERENCES

Bolt, G.H., 1956, Physico-chemical analysis of compressibility of pure clays, *Geotechnique* 6:86-93.

Cabral, A.R., 1992, A study of compatibility to heavy metal transport in permeability testing, *Ph.D. Thesis*, McGill University.

Chapman, S., and Cowling, T.G., 1960, *The mathematical theory of non-uniform gases.* Cambridge University Press.

Chiou, G.T., Schmedding, D.W., and Manes, M., 1982, Partition of organic compounds on octanol-water system, *Environ. Sci. Technol.*, 16:4-10.

Fetter, C.W., 1993, *Contaminant hydrogeology*, Macmillan Publishing co., New York, 458p.

Fukue, M., Minato, R., Horibe, H., and Taya, N., 1999, The micro-structures of clay given by resistivity measurements, *Jour. Engr. Geology*, 54:43-54.

MacDonald, E.M., 1994, Aspects of competitive adsorption and precipitation of heavy metals by a clay soil, *M.Eng. Thesis*, McGill University.

Ohtsubo, M., Yoshimura, A., Wada, S, and Yong, R.N., 1991, Particle interaction and rheology of illite-iron oxide complexes, *Clays and Clay Minerals*, 39:347-354.

Pusch, R., Yong, R.N., and Grindrod, P. (Eds.), 1999, *Microstructural modrelling with special emphasis on the use of clays for waste isolation*, Special Issue, Jour. Engr. Geology, Vol. 54, Nos. 1 and 2.

Pusch, R., and Schomburg, J., 1999, Impact of microstructure on the hydraulic conductivity of undisturbed and artificially prepared smectite clay, *Jour. Engr. Geology*, 54:167-172.

Thomas, H.R., and Cleall, P., 1999, Inclusion of expansive clay behaviour in coupled thermo hydraulic mechanical models, *Jour. Engr., Geology*, 54:93-108.

Weber, L., 1991, The permeability and adsorption capability of kaolinite and bentonite clays under heavy metal leaching, *M.Eng. Thesis*, McGill University.

van Olphen, H., 1977, *An introduction to clay colloid chemistry*, 2nd. edition, John Wiley and Sons, N.Y., 318p.

Yong, R.N., and Sethi, A.J., 1978, Mineral particle interaction control of tar sand sludge stability, *Journal Canadian Petroleum Technology*, 17(4):1-8.

Yong, R.N., 1984, Particle interaction and stability of suspended solids, *Proc. ASCE Symp. On Sedimentation Consolidation Models*, R.N.Yong and F.C.Townsend (eds.)

Yong, R.N., and Wagh, A.S., 1985, Dispersion stability of suspended solids in an aqueous medium. *In Flocculation, sedimentation and consolidation*, B.M.Moudgil and P. Somasundaran (eds.), National Science Foundation, pp.307-326.

Yong, R.N., Elmonayeri, D.A., and Chong, T.S., 1985, The effect of leaching on the integrity of a natural clay. *Jour. Engr. Geology*, 21:279-299.

Yong, R.N., and Ohtsubo, M., 1987, Interparticle actions and rheology of kaolinite-amorphous iron hydroxide complex, *Applied Clay Science*, 2:63-81.

Yong, R.N., Desjardins, S., Farant, J.P., and Simon, P., 1997, Influence of pH and exchangeable cation on oxidation of methylphenols by a montmorillonite clay, *App. Clay Sci.*, 12:93-110.

Yong, R.N., 1999, Soil suction and soil-water potentials in swelling clays in engineered clay barriers, *Jour. Engr. Geology*, 54:3-13.

Yong, R.N., 2000, *Contaminated soils: Pollutant fate and mitigation*. Lewis Publishers, Boca Raton.

Clay Science for Engineering, Adachi & Fukue (eds) © 2001 Balkema, Rotterdam, ISBN 90 5809 175 9

Clay liners for waste landfill

Masashi Kamon
Disaster Prevention Research Institute, Kyoto University, Uji, Japan

Takeshi Katsumi
Department of Civil Engineering, Ritsumeikan University, Kusatsu, Japan

Abstract: Waste containment is one of the most important issues in waste management practice. General and legal review necessitates the usage of the liner systems and barrier materials, and emphasizes the applicability of clay liners (either natural, compacted, or geosynthetic clay liners). Difference between Japan and western countries on the regulation of bottom liner system is presented based on the performance of contaminant transport. Several basic properties of clays are examined in view of their applicability to liner material. Critical issues in conducting the laboratory and in-situ hydraulic conductivity testing are summarized, and the specification and construction quality control to obtain the compacted clay liners of low hydraulic conductivity are presented. Since the annual rainfall in Japan is relatively large and the ground water level is also high, the reduction of shear strength could be possible, The above aspect is investigated to discuss the stability of canyon-type waste landfills that are common in Japan.

1 INTRODUCTION

The quantity of waste generation is increasing day by day which leads to an increase in the frequency of illegal dumping of waste. The quality of waste is often lowered due to the contamination by toxic substances. Disposal sites are being filled faster than originally predicted, and isolation and safe-containment systems should be established in order to avoid the contamination of surrounding sites and groundwater.

The major objectives in the safe disposal and containment of any type of waste include as follows (Mitchell 1997):

(1) Construction of liners, floors, walls, and covers that adequately limit the spread of pollutants and the infiltration of surface waters,
(2) Containment, collection and removal of leakage from landfills,
(3) Isolation, containment of spills,
(4) Control, collection, and removal or utilization of landfill gases,
(5) Maintenance of landfill stability,
(6) Monitoring to assure that the necessary long-term performance is being achieved.

The typical configuration of a landfill is a combination of below-ground and above-ground disposal, utilizing a bottom liner/drainage system and a final cover system as shown in Fig. 1 (Daniel 1998).

One of the most important factors of many bottom liner and cover systems for landfills is to achieve a low hydraulic conductivity for compacted clay liner (CCL). Geomembrane (GM) liner used to be often applied to contain wastes in Japan. Recently, the regulation on the facility of waste landfill has been amended, and thus, the applications of CCLs have just been started in practice to decrease the leakage of leachate.

Unlike Japan, most western countries have regulated their landfill liners with emphasizing the importance of clays as well as GMs. Thus, numerous researches have been conducted on clay liners. Summary of them can be found in the text books by Daniel (1993), Rowe et al. (1995), and Daniel and Koerner (1995), and in Manassero et al. (1997), Rowe (1998), Daniel (1998), and Benson (2000) as lecture papers.

Fig.1   Components of waste landfill

The purpose of this paper is to present the applicability and excellent performance of clays as waste landfill liners.

2 GENERAL AND LEGAL ASPECTS

2.1 *Liner Systems and Materials*

Various types of bottom liner systems have been proposed, regulated, and utilized, and could be selected based on the risk levels of sites conditions as shown in Fig. 2. Sophisticated bottom liner systems usually consist of barrier (liners) and collection layers. Collection layers are meant to collect and remove the leachate generated from wastes, and to reduce the hydraulic gradient across the liner due to the decrease of water head above the liner such that the risks will be significantly decreased. Coarse soils (e.g., gravels) have been used for the collection layer, but recently geosynthetics (geotextiles, geopipes, geocomposites) are often used (Rowe 1998). Barrier (liner) layer overlain by a collection layer is a key component to contain the pollutant. Single GM liner with a thin GM sheet above the base soil is the simplest liner system. Single clay liner, that is the liner constructed with clay soils, is also used. The most common type of liner system for municipal solid waste (MSW) in North America and Europe is the composite liner, which consists of a GM underlain by a clay layer. Double liner systems are those which consist of two single (either GM or clay) or composite liners. Among them, double composite liners (GM-clay and GM-clay) are used for more hazardous waste landfills due to the highest redundancy against the risk of leaching.

GMs are thin polymer sheets. Most common polymers are high-density polyethylene (HDPE), low-density polyethylene (LDPE), and polyvinyl chloride (PVC). Although GMs are practically impervious, there are several critical issues pointed out: (1) holes and punctures are unavoidable during the construction and landfill operations, (2) improper seaming to connect two GM sheets, (3) diffusion transport might occur through the intact GM if the pollutant is organics, and (4) possible material deterioration (Rowe 1998, Benson 2000).

Unlike GMs, clays are unlikely to deteriorate for long duration. In addition, since the clay liner thickness should be more (50-100 cm for CCLs) than GM liners, redundancy can be more expected, and thus, clay is a key material for liners. If the natural clay layer continuously maintains a low hydraulic conductivity, it could be utilized as a clay liner material. However, CCLs are becoming more common. The advantageous of clay liners are: (1) low hydraulic conductivity, (2) high adsorption capacity against chemicals and retardation of chemical transport, and (3) less expensive material if the clay can be obtained nearby. The disadvantage of clay liners are: (1) to need laboratory tests to screen unsuitable materials and to obtain the design specification, and (2) difficulty in construction. However, Benson et al. (1999) proved that CCLs can be constructed with a broad variety of clayey soils using proper construction specification and quality control.

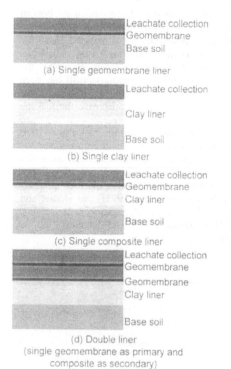

(a) Single geomembrane liner

(b) Single clay liner

(c) Single composite liner

(d) Double liner
(single geomembrane as primary and composite as secondary)

Fig. 2   Cross sections of typical bottom liner systems

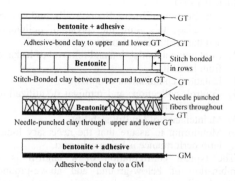

Fig. 3   Cross sections of currently available geosynthetic clay liners (GCLs)

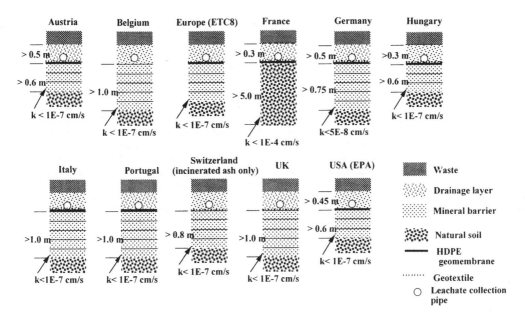

Fig. 4   Regulations of bottom liner systems for MSW landfill in several countries

Geosynthetic clay liners (GCLs) are the recently developed materials, which are thin prefabricated clay liners consisting of sodium bentonite encased between two geotextiles or glued to a GM. Examples of currently available GCLs are shown in Fig. 3. The bentonite layer provides the extremely low hydraulic conductivity to water (~10^{-9} cm/s), due to their high swelling and water retention capacity. Since GCLs are factory-manufactured, construction quality control is not necessary unlike CCLs. In addition, GCLs can save more air space because they are thin (5-10 mm). Thus, GCLs have been recently used in bottom liner systems either as a substitute to CCL or together with CCL. The critical issues of GCLs are: (1) chemical imcompatibility, (2) equivalency of the performance to CCLs, (3) redundancy, and (4) low shear strengths.

2.2 Performance and Regulation

Regulations or guidelines on bottom liner systems for MSW landfill in North America and Europe are summarized in Fig. 4. The main points of these regulations are whether a GM shall be placed or not, and how thick and how low in hydraulic conductivity the clay liner must be, although there are other several factors influencing the pollutant transport through the liners. In particular, diffusion and adsorption, as well as advection that is governed by the hydraulic conductivity, are the important factors to control the chemical transport in clay liners. The purpose of having hydraulic conductivity as the regulated values (e.g., 10^{-7} cm/s for most countries)

is probably that it is low enough that the diffusion transport could be dominant relative to advection. For the more hazardous waste landfill, each country has more strict codes. For example, the United States (EPA) requires a double composite liner (single GM as the primary liner, and composite liner as the secondary in which the clay liner is thicker than 90 cm with hydraulic conductivity lower than 10^{-7} cm/s) for the hazardous waste landfill.

Table 1   Leakage rate per unit area through various types of liners (Giroud et al. 1994)

Types of liners	Rate of leachate in litres per hectare per day (L/h/d)
Mineral liner	100,000 L/h/d for a soil liner having $k \sim 10^{-5}$ cm/s
	10,000 L/h/d for a soil liner having $k \sim 10^{-6}$ cm/s
	1,000 L/h/d for a soil liner having $k \sim 10^{-7}$ cm/s
GM (GM) liner	100 L/h/d for a GM installed with strict construction quality assurance and containing a typical amount of defects
Simple composite liner	10 L/h/d for a composite liner with soil having $k \sim 10^{-6}$ to 10^{-5} cm/s
	0.1 L/h/d for a composite liner with soil having $k \sim 10^{-7}$ cm/s
Double composite liner	~0.1 L/h/d for a double composite liner where each of the two liners is constructed with soil having $k \sim 10^{-7}$ cm/s

In Japan, the waste landfills are categorized into (1) least-controlled landfill, (2) controlled landfill, and (3) strictly-controlled landfill, in accordance with the type of wastes. Least-controlled landfills are for wastes that are neither hazardous nor degradable (e.g., soils, glasses). Strictly-controlled landfills, where the hazardous wastes are disposed of, require cement concrete layer to contain the waste. Numbers of this type of landfills are however very limited. Controlled landfills are for the most common non-hazardous but degradable wastes (including MSW and MSW incinerator ash). Until the regulation on controlled landfill was improved in 1998, no specified structures or values had been regulated. Thus, the old waste landfills do not have any sophisticated bottom liner, or have only single GM liner with little engineering concern. The new regulation permits that a natural clay ≥ 5 m in thickness with hydraulic conductivity ≤ 10^{-5} cm/s could be utilized as a liner. Otherwise, it requires one of three following liner systems: (1) two GMs which sandwich a non-woven fabric or other cushion material, (2) a GM underlain by an asphalt-concrete layer ≥ 5 cm in thickness with hydraulic conductivity ≤ 10^{-7} cm/s, or (3) a GM underlain by a clay liner ≥ 50 cm in thickness and having hydraulic conductivity ≤ 10^{-6} cm/s. It is clear that these codes are not enough from the engineering point of view (Katsumi et al. 1999a, Kamon 1999 and 2000). For example, the hydraulic conductivity of CCL is one order of

Table 2   Summary of clay mineral characteristics

Type	Sub-Group and Schematic Structure	Mineral	Basal Spacing	Size	Cation Exchange Cap. (meq / 100g)	Specific Gravity	Specific Surface (m² / g)
Amorphous	Allophane	Allophanes		0.05-1			
1 : 1	Kaolinite	Kaolinite	0.72 nm	0.1-4 μm × 0.05-2 μm	3 - 15	2.60 - 2.68	10 - 20
		Halloysite (Hydrated)	1.01 nm	0.07 μm dia 1 μm long	5 - 40	2.0 - 2.2	35 - 70
1 : 2	Smectite	Montmorillonite	0.96 nm	> 0.001 μm × up to 10 μm	80 - 150	2.35 - 2.7	700 - 840
	Hydrous Mica (Illite)	Illites	1 nm	0.003 - 0.1 μm · × up to 10 μm	10 - 40	2.6 - 3.0	65 - 100
	Vermiculite	Vermiculite	1.05 - 1.4 nm	< 2 μm	100 - 150		870
2 : 1 : 1	Chlorite	Chlorite	1.4 nm	1 μm	10 - 40	2.6 - 2.96	
Chain Structure	Attapulgite	Double Silica Chains		Max, 4 - 5 μm × 0.005 - 0.01 Width = 2t	20 - 30	2.1	

magnitude higher than the values regulated in North America and Europe. Chemical leakage through different liner systems is significantly affected by the hydraulic conductivity of CCLs rather than other contaminant transport parameters, such as retardation factor or dispersion coefficient. In addition, the methods to measure the hydraulic conductivity are not mentioned in the regulation. No authorized test methods to measure such low hydraulic conductivity exist in Japan. Also, little discussion on the applicability of marine clay as a liner has been provided, although coastal landfill is becoming popular in the metropolitan areas of Japan. Kamon et al. (2001) attempt to consider the chemical and bio-chemical effects of marine clay on chemical transport. These issues are crucial to be discussed.

To compare the effectiveness of different liner systems or alternatives to a regulatory prescribed liner, a performance-based analysis is necessary. Several performance-based methods have been proposed (e.g., Giroud and Bonaparte 1989, Giroud et al. 1994, Foose et al. 1996, Giroud et al. 1998, Rowe 1998, Katsumi et al. 2000). Table 1 shows the rate of leakage through several different liner systems computed by Giroud et al. (1994). In addition to the rate of leakage, solute flux or concentration of an individual solute species can be used as performance criteria. Katsumi et al. (2000) presented a simplified performance-based method by using typical spreadsheet applications.

3 PROPERTIES OF CLAY SOILS

3.1 Characteristics of Clay Minerals

The controlling factors of clay soils are clay mineralogy, which determine the sizes, shapes, and surface characteristics of the particles in the soil (Mitchell 1993). The clay mineral characteristics are summarized in Table 2. The representative clay minerals like smectites, illites, and kaolinites are alumino-silicates. Their mechanical properties are caused by their mineralogical structure, small particle size, and inter-surface activities. Basic properties of these clay materials could be specified as follows:

(1) high consistency
(2) large specific surface
(3) large cation exchange capacity
(4) low unit weight in compaction
(5) low shearing strength
(6) low hydraulic conductivity

Cation exchange is the most important factors of clay soil behavior. Under given environmental conditions (temperature, pressure, pH, chemical and biological composition of the pore water), clay adsorbs cations of specific type and the adsorbed

amount will balance the charge deficiency of clay particles. Exchange reactions can occur in response to changes in the environmental conditions which induce significant changes in the mechanical properties of soil. The consistency of each clay mineral is directly affected by this cation exchange. Table 3 shows examples of the change of consistency of clay soils under the different cation conditions. Values of hydraulic conductivities at the liquid limit for several soils are given by Nagaraj et al. (1991) as shown in Table 4. The approximately equal hydraulic conductivities at the liquid limit can be explained by the concept of microstructure of the soils, in particular, that the average adsorbed water layer thickness is about the same for all particle surfaces. This means that the greater the specific surface, the greater the total amount of water required to satisfy the conditions at the liquid limit. After the consolidation or compaction, the void of each soil is decreased depending on its surface environment, and then its engineering properties are changed drastically.

Table 3 Change of consistency of clay minerals by cation exchange

Clay mineral	Exchangeable cation	Liquid limit (%)	Plastic limit (%)	Plasticity index
Smectite	Na	710	54	656
	K	660	98	562
	Ca	510	81	429
	Mg	410	60	350
	Fe	290	75	215
Illite	Na	120	53	67
	K	120	60	60
	Ca	100	45	55
	Mg	95	46	49
	Fe	110	49	61
Kaolinite	Na	53	32	21
	K	49	29	20
	Ca	38	27	11
	Mg	54	31	23
	Fe	59	37	22

Table 4 Hydraulic conductivity at liquid limit for several clays

Soil type	Liquid limit, w_L (%)	Void ratio at liquid limit, e_L	Hydraulic conductivity (10^{-7} cm/s)
Bentonite	330	9.240	1.28
Bentonite + sand	215	5.910	2.65
Natural marine soil	106	2.798	2.56
Air dried marine soil	84	2.234	2.42
Oven dried marine soil	60	1.644	2.63
Brown soil	62	1.674	2.83

Generally, the containment function of clay barriers is controlled by three parameters: advection, diffusion, and adsorption. Advection is characterized by hydraulic conductivity (k) and hydraulic gradient (i). Diffusion is based on the concentration gradient that includes chemicals in the leachates. Adsorption occurs at the clay surface, which is negatively charged.

3.2 Waste-Soil Interactions, Chemical Compatibility, and Chemical Resistant Clay

Adverse effect of chemicals on the properties of clays that are used for landfill barrier is the most important subject. Many researchers reported that, even though the hydraulic conductivity is low when the clays are permeated with pure water, permeation with chemical solutions will result in the increase (sometimes significant) in hydraulic conductivity. These effects can be explained by the change of the soil fabric, and categorized into (1) dissolution of clay particle and chemical compounds, (2) development of diffuse double layer, and (3) osmotic swelling for smectite clay. Mitchell and Madsen (1987), Shackelford (1994), and Shackelford et al. (2000) summarized the clay-chemical interactions and chemical compatibility of clay.

Dissolutions of soil components result from strong acids and bases. Strong acids promote dissolving of carbonates, iron oxides, and alumina octahedral layers of clay minerals. Bases promote dissolving of the silica tetrahedral layers. These effects can cause the increase in hydraulic conductivity, although re-precipitation of dissolved compounds might clog the pore and decrease the hydraulic conductivity (Mitchell and Madsen 1987).

Mesri and Olson (1971) showed that the hydraulic conductivity is largest for non-polar fluids, smaller for polar fluids of low dielectric constant (ethyl alcohol and methyl alcohol), and lowest for water, which is polar and has a high dielectric constant (~80). Similar results explaining the effects of dielectric constant were reported by other researchers (e.g., Bowders and Daniel 1987, Mitchell and Madsen 1987, Fernandez and Quigley 1988, Acar and Olivieri 1989, Shackelford 1994). However, they showed that the significant increase in hydraulic conductivity did not occur when the concentration of organic chemical is lower than 50%. This is because dilution with water leads to an increase in the dielectric constant.

Direct permeation with organic chemicals is not likely in practice, where clay liners are applied. Fernandez and Quigley (1985) conducted two different hydraulic conductivity tests on clayey soils; one was permeated first with water and then benzene, and the other was permeated first with water, then with ethanol, and finally with benzene. The former test result showed that the concentration of benzene of effluent reached 50% after only 0.28 pore volume of flow of benzene, and reached 100% at two pore volumes of flow. Only 8% of the water that existed initially in the soil was replaced by polar benzene. Organic liquids flow dominantly through the large pores, and most of water in the micro pores remains in the soil. The increase in hydraulic conductivity is negligible. The other case (permeated with water, ethanol, and benzene) exhibited the quite different results. A large portion of water was replaced, and hydraulic conductivity significantly increased. This is attributed to the solubility of ethanol to water. Benzene can replace easily the dilute ethanol in soil pores, although it cannot replace the water.

Osmotic swelling is an important phenomenon for smectite clay. When dry smectite is hydrated, water molecules are strongly attracted to internal and external clay surfaces during hydration phase. After this hydration process, if the exchange cations of smectite clay are monovalent, the region of interlayer may retain numerous layers of water molecules during osmotic phase (Norrish and Quirk 1954). This condition can be observed as significant amount of swelling, and is typically observed when Na-bentonites are hydrated with deionized water. When the smectite having monovalent at the exchange site is permeated with the solution containing low concentration of monovalent cations, large fraction of water is attracted to clays, less mobile water is available for water flow, and the hydraulic conductivity is low. When polyvalent cations exists at exchange sites, osmotic phase does not occur, and less swelling is observed. Divalent and trivalent cations in permeants result in a significant increase of the hydraulic conductivity of GCLs, much greater than monovalent cations (Katsumi et al. 1999b, Shackelford et al. 2000).

Chemically-resistant clays have been recently developed to improve the compatibility of clays, in particular of bentonite (Lo et al. 1994 and 1997, Onikata et al. 1996 and 1999). For example, Onikata et al. (1999) developed a chemically resistant bentonite, called multi-swellable bentonite (MSB), which is made from a natural bentonite mixed with propylene carbonate as swelling activation agent. Propylene carbonate can enter the interlayer region of smectite, attracts numerous numbers of water molecules, and consequently results in the strong swelling power even if the permeant contains polyvalent cations or high concentration of monovalents. A series of hydraulic conductivity tests showed that MSB yields one or two orders of magnitude lower hydraulic conductivity than natural bentonite when the permeant contains calcium cations (Lin et al. 2000).

4 EVALUATION OF LOW HYDRAULIC CONDUCTIVITY

4.1 *Laboratory Hydraulic Conductivity*

Laboratory hydraulic conductivity tests are conducted to assess the material suitability of clays as a liner and to obtain the design and construction specification for the clay soil to be used. Sometimes, laboratory tests are performed for the construction quality control of CCL or material quality control for GCL. Compared to the relatively permeable material (such as sandy soil), different attentions need when the tests are conducted on low hydraulic conductivity materials, such as clay liner. For example, American Society for Testing and Materials (ASTM) established a standard test method for low hydraulic conductivity soil (ASTM D 5084), while no standard test method is established and only little discussion has been conducted in Japan. Important factors affecting the hydraulic conductivity values are the type of permeameter, effective stress, hydraulic gradient, size of the specimen, type and chemistry of the permeant, and termination criteria. These effects on the hydraulic conductivity were recently summarized by Benson et al. (1994), Daniel (1994), and Shackelford (1994) for compacted clays, and by Daniel et al. (1997), Petrov et al. (1997b), and Shackelford et al. (2000) for GCLs. Selected factors affecting the hydraulic conductivity are described as follows. Discussion herein is focused on the saturated hydraulic conductivity, and a comprehensive review on the unsaturated hydraulic conductivity is provided by Benson and Gribb (1997).

4.1.1 Type of permeameter

One critical issue for the hydraulic conductivity tests is to select the types of permeameters. The permeameters typically used are (1) rigid-wall permeameter, (2) flexible-wall permeameter, and (3) consolidation-cell. When the clay soil of low hydraulic conductivity is used for the test, high hydraulic gradient has to be induced to the specimen to reduce the test duration. However, such high hydraulic gradient results in the large seepage force, which might induce the sidewall leakage (leakage between the specimen and the ring) if the rigid-wall permeameter is used. Further, if the specimen is likely to shrink (due to the effect of permeant chemicals), the sidewall leakage is unavoidable. Unlike rigid-wall permeameters, flexible-wall permeameters are effective to minimize the sidewall leakage, even for test specimens with rough sidewalls. A typical schematic diagram of a flexible-

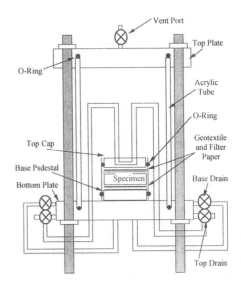

Fig. 5   Schematic of flexible-wall permeameter

wall permeameter is shown in Fig. 5. The test specimen is confined with end caps having porous disks on the top and bottom and by a latex membrane on the sides. The cell is filled with water and pressurized to press the membrane, and therefore sidewall leakage is eliminated. There are other several advantages of flexible-permeameters: backpressure usage to saturate the specimen, control of principal stresses, and reduction of the testing duration (Daniel 1994). Disadvantages of flexible-wall permeameter are relatively high cost and more complicated operation. Usually, flexible-wall permeameter requires three pressure positions (for cell, influent, and effluent pressures), which costs more. If only the gravity force is used for these pressures, the permeameter system can be more simplified with less cost, but the longer testing duration is needed or the specimen must be thinner to obtain high hydraulic gradient under the limited cell and influent pressures.

Consolidation cell is used to obtain the hydraulic conductivity, either by calculating the consolidation rate or by inducing the seepage directly. The errors on hydraulic conductivity values calculated from the consolidation rate are likely to exist. Olson (1986) indicated that the hydraulic conductivity calculated from the rate of consolidation is almost always smaller than the direct measurement of hydraulic conductivity. In addition, advantages and limitations of direct measurement of hydraulic conductivity from consolidation-cell are similar to those of rigid-wall permeameter.

When tests are conducted to measure the hydraulic conductivity of clays permeated with chemical liquids (chemical solutes, pure organics, real/simulated waste leachate), rigid-wall permeameter and consolidometer will tend to overestimate the hydraulic conductivity because clay is likely to shrink or crack. Flexible-wall permeameter will be the best system to evaluate the hydraulic conductivity with chemicals.

To calculate the hydraulic conductivity, thickness of the specimen must be determined, in particular when the soil specimen is compressible (such as sludge-state material) or under swelling (such as GCL). Cathetometer is used to measure the thickness from outside of the cell. The use of triaxial compression cell of flexible-wall permeameter, instead of the flexible-wall permeameter, provides the direct measurement of thickness (height), but results in the extremely high cost.

In conclusion, flexible-wall permeameter is considered most appropriate. In Japan, many studies were however conducted by using rigid-wall type permeameters, because bentonite mixtures were used in most cases, which might swell and does not induce the sidewall leakage even when the high hydraulic gradient was applied (e.g., Imamura et al. 1996).

4.1.2 Effective stress and hydraulic gradient

If the hydraulic conductivity test is performed by using a flexible-wall permeameter, principal effective stress (average across the specimen), except the extremely low confining pressure, can be controlled by the cell pressure and hydraulic gradient. Hydraulic conductivity of compressible soils or soils containing the fractures and macro-pores are very sensitive to the changes in effective stress. Higher effective stress reduces the hydraulic conductivity. Thus, the selection of appropriate stress level is important. Most appropriate way is to conduct the test under the stress condition that represents the field situation.

High hydraulic gradient is often used to reduce the test duration. Negative effects of high hydraulic gradient are considered that (1) the large seepage force results in consolidation of the specimen and decrease the hydraulic conductivity, and (2) piping of fine particles due to the large seepage force increases the hydraulic conductivity. ASTM D 5084 recommends a maximum hydraulic gradient of 30 for media with low hydraulic conductivity ($k \leq 10^{-7}$ cm/s).

There are several results reported on the effect of hydraulic gradient on hydraulic conductivity of compacted clays and GCLs. Fox (1996) derived the equations to obtain the pore water pressure and effective stress distributions across the specimen during hydraulic conductivity testing, which indicated that excessive hydraulic gradients can cause reductions in the measured hydraulic

conductivity, in particular for normally consolidated soils with high compressibility, such as soft clays and soil-bentonite slurries. Imamura et al. (1996) conducted a series of long-term rigid-wall permeameter tests on the compacted sand/bentonite specimens, and the test results indicated that a two order increase in the hydraulic conductivity (10^{-9} cm/s to 10^{-7} cm/s) occurred after more than 25 pore volumes of flow of $Ca(OH)_2$ solution (600 ppm as Ca^{++} concentration) under a hydraulic gradient of 800. Further, a decrease in the hydraulic gradient from 800 to 30 has resulted in the recovering decrease of hydraulic conductivity (10^{-7} cm/s to 10^{-9} cm/s) again. Thus, Imamura et al. (1996) concluded that this increase in hydraulic conductivity under the high hydraulic gradient may be attributed to the piping of fine particles.

Although ASTM D 5084 recommends a hydraulic gradient lower than 30, hydraulic gradients used for measuring the hydraulic conductivity of GCLs typically ranged from 50 to 550. Rad et al. (1994) reported that the hydraulic conductivity of a GCL was not affected by a hydraulic gradient of 2800. Hydraulic gradient has relatively minor effect on the hydraulic conductivity if the mean effective stress is constant (Petrov et al. 1997a, Shackelford et al. 2000). The recommendation of a hydraulic gradient lower than 30 is probably effective for the ordinary compacted clay specimen, having the specimen length or thickness of 116 mm (the height of Proctor mold), and Shackelford et al. (2000) indicated that the higher hydraulic gradient can be applied for the thinner specimens. Figure 6 indicates the increase in effective stress at the effluent end of the specimen due to the hydraulic gradient for various specimen thicknesses. For the compacted clay specimen having the length of 116 mm, a hydraulic gradient of 30 causes 17 kPa increase in effective stress at the effluent end. For the same increase in effective stress, a hydraulic gradient of 342 can be applied to a 10

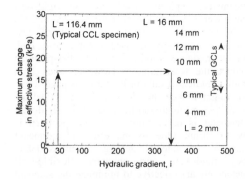

Fig. 6 Change in effective stress due to applied hydraulic gradient for different specimen thickness (Shackelford et al. 2000)

mm-thick GCL. However, it does not necessarily assure the use of high hydraulic gradient for extremely thin-cut specimens of compacted clay which is usually used as a thick layer in practice. Since the higher hydraulic gradient across GCLs installed in the field might occur because it is extremely thinner than CCL, the use of such high hydraulic gradient for the laboratory test is considered reasonable for thin liner materials to some extent.

4.1.3 Termination criteria

Long duration test is usually needed to conduct the hydraulic conductivity test on low-permeable materials. Provided a compacted clay specimen having a thickness of 116 mm and hydraulic conductivity of 10^{-7} cm/s is tested under a hydraulic gradient of 30, test duration longer than 40 days is needed to achieve only one pore volume of flow through the specimen. Also, among the soil properties, none varies over so wide range as the hydraulic conductivity, which makes it difficult to decide when the test could be terminated. Assuming the same hydraulic gradients and the same specimen lengths for two different specimens having different hydraulic conductivities, the test duration required to obtain the same volume of flow is simply proportional to the hydraulic conductivity. Thus, the termination criterion is another critical issue in conducting the hydraulic conductivity test. Typical termination criteria are [1] equality of the inflow and outflow rates ($\leq$ 25%), and (2) measurement of a steady hydraulic conductivity (four or more consecutive measurements within 25% to 50% of the mean) (Daniel 1994). For the chemical compatibility test, where the chemical solutions and waste leachates are used as a permeant, permeation of a minimum of two pore volumes of flow has to be achieved, and similarity between the chemical composition of the effluent and the influent has to be ensured (Daniel 1994, Shackelford 1994, Shackelford et al. 1999). To establish the similarity in the chemical composition of effluent and influent sometimes requires an extremely long duration (Bowders 1988, Imamura et al. 1996, Shackelford et al. 2000). For example, Imamura et al. (1996) reported that the hydraulic conductivity of a compacted sand-bentonite specimen permeated with $Ca(OH)_2$ solution increases by more than two orders of magnitude after more than 25 pore volumes of flow corresponding to three years of permeation.

For practical indicators of chemical equilibrium, an experimental and theoretical study by Shackelford et al. (1999) suggests that electric conductivity (EC) measurements can be used as an index of the chemical composition of electrolyte solutions. Measuring the electric conductivity is simple,

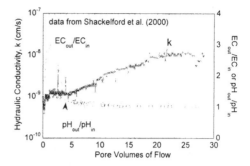

Fig. 7 Long-term hydraulic conductivity test results for a needle-punched GCL permeated with 0.012 M $CaCl_2$ solution (Katsumi et al. 1999b)

inexpensive, and rapid. The results of a long-term hydraulic conductivity test on a needle punched GCL with a dilute (0.0125 M) $CaCl_2$ solution is shown in Fig. 7 (Katsumi et al. 1999b, Shackelford et al. 2000). This data illustrate how misleading can occur. The hydraulic conductivity was low and steady for the first 5 pore volumes of flow, which satisfy some of the termination criteria except for chemical equilibrium. However, the hydraulic conductivity increased 25 times larger after approximately 30 pore volumes of flow. The similarity of EC between effluent and influent was not yet achieved at 5 pore volumes of flow. The hydraulic conductivity started to increase after the similarity of EC was achieved, which means that measuring EC of outflow is effective to draw an appropriate termination criterion.

In measuring a low hydraulic conductivity, to permeate soils with deaired water is the best way. Although water contains approximately 8 mg/L of dissolved oxygen at atmospheric pressure, properly deaired water will contain less than 1 to 2 mg/L (Daniel et al. 1997).

4.2 In-situ Hydraulic Conductivity

Hydraulic conductivity of the specimens sampled from the field does not necessarily represent the in-situ hydraulic conductivity, because macropores control the flow of water at field scale while the small specimen does not always contain these pores. Daniel (1984) addressed that the field-scale hydraulic conductivity of clay liners could be orders of magnitude higher than the hydraulic conductivity measured on small specimens in the laboratory. Thus, measuring the hydraulic conductivity in the field with an appropriate scale is important. In the United States, several methods to measure the field hydraulic conductivity have been proposed, and the

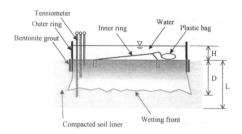

Fig. 8 Schematic of sealed double-ring infiltrometer (SDRI) test (Wang and Benson 1995)

most common of them are: (1) sealed double-ring infiltrometer (SDRI) test and (2) two-stage borehole (TSB) test.

A schematic diagram of SDRI test is shown in Fig. 8. Inner ring is to permeate the compacted clay. Hydraulic conductivity (k) can be computed from the infiltration rate (I) and hydraulic gradient (i). There are three methods to obtain the hydraulic gradient: apparent gradient method, wetting front method, and suction head method (Daniel and Koerner 1995). Apparent gradient method assumes that the clay liner is fully saturated with water over the entire depth (L), although it is not realistic and provides too conservative results ($i = (H+L)/L$, H is the depth of water above the surface of liner). Wetting front method considers the depth of saturated zone (D) and to assume that the water pressure at wetting front be zero ($i = (H+D)/D$). The depth of wetting front is measured by using tensiometers over time. Although the suction at wetting front is not taken into account, the wetting front method provides conservative results, but the errors should be so small that this method could be used practically enough. Suction head method is used to measure the suction at wetting front (S) to be added to the static head of water to calculate the hydraulic gradient ($i = (H+D+S)/D$). Method to obtain the suction head is presented by Wang and Benson (1995).

TSB test was developed since SDRI costs and needs a long testing duration (a month or more) (Benson 2000). A schematic of TSB test is shown in Fig. 9. The base of the borehole is flush to result in the vertical flow, in Stage 1. After the equilibrium is achieved, the borehole is extended by using a auger to a depth about 1.5 times of the inside diameter of the casing. Then, another falling head infiltration test is conducted. From the two apparent hydraulic conductivities (K_1 and K_2) obtained from Stages 1 and 2, vertical (K_v) and horizontal (K_h) hydraulic conductivity can be computed.

Both SDRI and TSB provide essentially zero confinement pressure, which is unlikely in the field after the landfill is ongoing or completed. Thus,

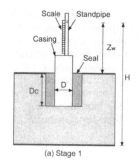

(a) Stage 1

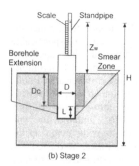

(b) Stage 2

Fig. 9 Schematic of two-stage borehole (TSB) test (Benson 2000)

laboratory hydraulic conductivity test on the specimens colleted from the field during the construction of clay liners is conducted to provide the proper effective stress. The laboratory specimens for hydraulic conductivity testing must be large enough to represent the liners existing in the field. Results of the SDRI tests using inner rings having different sizes (0.6 m x 0.6 m, 0.9 m x 0.9 m, and 1.2 m x 1.2 m) and laboratory hydraulic conductivity tests on the specimens with different sizes (460, 305, 152, and 76 mm in diameters) trimmed from the SDRI test spot showed that block specimens for the hydraulic conductivity tests needs to have 300 mm or more in diameter (Benson et al. 1994). Benson et al. (1997) reported the results of four different hydraulic conductivity test methods (SDRI, TSB, and laboratory tests on large block and on small specimens) performed, and concluded that large-scale test either from field tests (SDRI and TSB) or laboratory test yield similar values of hydraulic conductivity.

4.3 Construction of CCL

4.3.1 Specification

To construct a clay liner of low hydraulic conductivity, key issues such as selection of a suitable soil material and provision of an appropriate construction specification under the given soil material are needed. Daniel (1993) presented that the minimum requirement recommended to achieve a hydraulic conductivity lower than 10^{-7} cm/s for most soil liner materials are: percentage fines ≥ 20-30%, plasticity index ≥ 7-10%, percentage gravel ≤ 30%, and maximum particle size = 25-50 mm.

Water content of the soil, method of compaction, and compaction effort have major influence on the hydraulic conductivity of compacted clay liners (CCLs). Early experiences showed that low hydraulic conductivity can be easily achieved when the soil is compacted wet of optimum water content with a high level of kneading-type of compaction energy (Mitchell et al. 1965). This is because the influence of clods is eliminated when the compaction is performed at wet of optimum due to the re-arrangement of clay particles (Mitchell et al. 1965, Benson and Daniel 1990). Following these conclusions, Daniel and Benson (1990) established the specification of water content-density criteria for CCL to achieve a low hydraulic conductivity.

Typical construction specification for compact soils (e.g., for embankment construction) requires water content fall within a specified range and that percent compaction is equal to or exceeds a specified minimum value. However, the water content-density zone providing low hydraulic conductivity is different from this typical specification as shown in Fig. 10 (Daniel and Benson 1990). Figure 10(b) provides the hydraulic conductivity values for the specimens whose water content-density data are plotted in Fig. 10(a). It is clear that the maximum dry density does not result in the lowest hydraulic conductivity, and minimum hydraulic conductivity is achieved when the water content-density plot on or above the line of optimums, as shown in Fig. 10(c). In addition, Fig. 10(d) considers the minimum value of dry density to ensure the adequate shear strength of CCL.

4.3.2 Field performance and quality control

The most important question is whether the CCLs could be really constructed in the field by reasonable construction methods with the soil materials that can be obtained. An extensive study analyzing the hydraulic conductivity data collected from 85 different sites in North America reported by Daniel (1998) and Benson et al. (1999) answers this question.

The data reported in Benson et al. (1999) are: index properties, laboratory and field hydraulic conductivities, and construction specifications. Seventy four percent of the test sites meet the hydraulic conductivity criterion ($\leq 10^{-7}$ cm/s). Their findings are: (1) a broad variety of clayey soils can be used to construct CCLs having hydraulic conductivity lower than 10^{-7} cm/s, (2) compaction

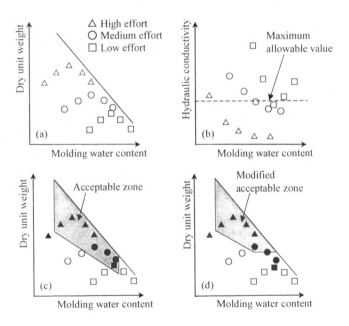

Fig. 10   Procedure for determining acceptable compaction zone to achieve low hydraulic conductivity (Daniel and Benson 1990)

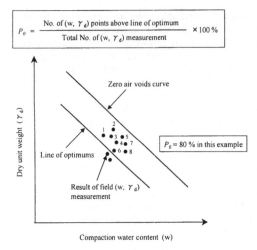

$$P_0 = \frac{\text{No. of } (w, \gamma_d) \text{ points above line of optimum}}{\text{Total No. of } (w, \gamma_d) \text{ measurement}} \times 100\%$$

Zero air voids curve

Dry unit weight (γ_d)

Line of optimums

$P_0 = 80\%$ in this example

Result of field (w, γ_d) measurement

Compaction water content (w)

Fig. 11   Definition of P_0 (percentage of field compaction data wet of the line of optimums) for construction quality control (Benson et al. 1999)

should be ensured wet of the line of optimums, (3) percent compaction, laboratory hydraulic conductivity values, and index properties are not necessarily the important factors to achieve the specified hydraulic conductivity, and (4) thicker liners or liners having many compaction lifts tend to have lower in-situ hydraulic conductivity.

Figure 11 shows the method for construction quality control presented by Benson et al. (1999). Several water content-density data are collected from a site, and plotted as shown in Fig. 11. Percentage of field compaction data wet of the line of optimums (P_0) can be calculated. For the case shown in Fig. 11, 8 of 10 plots are wet of the line of optimums, therefore P_0 is 80%. Test results showed that, if P_0 is greater than 70-80% for a site, the in-situ hydraulic conductivity can be assured lower than 10^{-7} cm/s.

In conclusion, practical lessons to construct CCLs are: (1) to screen the unsuitable soils for CCL by index property tests, and conduct the laboratory hydraulic conductivity test to obtain the construction specification, (2) to avoid using the conventional compaction specification and to ensure wet of the line of optimums, and (3) to ensure 70-80% of the field values of water content-density points above the line of optimums (Benson et al. 1999).

5 SHEAR STRENGTH AND LANDFILL STABILITY

Clay liners are used to prevent leachate from polluting the nearby environment. Generally, due to the large amount of precipitation of rainfall in Japan, groundwater level in canyons is high. Thus, the shear

strengths of clay liners are expected relatively low. There are several slope failures in Municipal waste landfills in USA (Mitchell et al. 1990, Seed et al. 1990, Eid et al. 2000, Stark et al. 2000). Therefore, the stability of a landfill along clay liners should be taken into account prior to its construction.

Water migration and pF tests were conducted on clay liner specimens to evaluate the water migration characteristics between liners and underlying ground. Interface and/or internal direct shear tests were conducted on clay liner specimens with various water contents obtained from the above tests. Stability of typical canyon solid waste landfills along clay liners was analyzed with the shear strength parameters obtained from the shear strength tests (Kamon et al. 2000).

5.1 Experimental

5.1.1 Water migration test

Water migration tests were conducted to evaluate water migration characteristics between liners and base soil. Two types of bentonite mixtures (sand-bentonite and clay- bentonite) were used as liner material, and Toyoura sand was used for the base soil. Sand-bentonite is the mixture of 15% Na-bentonite and 85% Toyoura sand in dry weight, and clay-bentonite is one of 15% bentonite and 85% Fukakusa clay. These bentonite mixtures were compacted at the optimum water content (w_{opt}), that are 17% for sand-bentonite and 30% for clay-bentonite.

A polyvinyl chloride (PVC) mold with 50 mm inside diameter and 100 mm height was used for the test. Toyoura sand specimens compacted at 10, 15, 20 or 25% water contents were placed into the lower half part of the mold, and the compacted bentonite mixtures were placed into the upper part of the mold contacted with the sand. Top of the mold was sealed with a wrap to prevent the water evaporation. Static force of 3.1 kPa was applied to simulate the weight of liners. After 3, 7, and 21 days, each specimen was cut into 10 equally-thick slices from top to bottom, and the water content for each slices were measured.

Figure 12 shows the results of clay-bentonite and base soil after 3 day testing. The water content of all clay-bentonite specimens increased, the most at lower and the least at the middle part of specimens. The water contents at the upper part of specimens increased higher than the middle part of specimens that was possibly due to the affection of wrap at the top. On the contrary, the water contents of all slices from base soil decreased, the most at the upper and the least at lower part of specimens. Results of other conducted cases have similar tendencies, which can be summarized as below: (1) water moved from the base soil to liner due to capillarity, and (2) most of the water moved from base soil to liner within the first 3 days. After 21 days, the water contents of

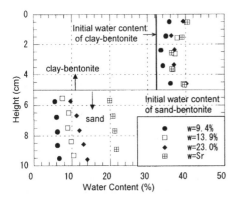

Fig.12 Water migration test results

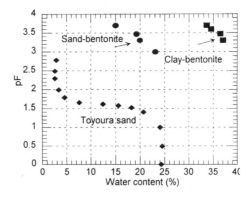

Fig. 13 pF test results

clay-bentonite placed on the base soil having water contents of 10, 15, 20 and 25% reached 36, 36, 38 and 40%, and the water contents of sand-bentonite increased to 22, 24, 24, and 30%, respectively.

5.1.2 pF test

Triaxial compressive test equipment was used for the pF test. Specimens of 10 mm height and 50 mm diameter were prepared using PVC molds. After the specimen was saturated with 100 kPa back pressure in the triaxial cell, a saturated ceramics plate (500 kPa of air entry value) was placed at the bottom of the specimen. Air pressures of 200, 300, 400, and 500 kPa were applied respectively.

Eight specimens were tested and their results are shown in Fig. 13. Compared with that of Toyoura sand reported by CPAUG (1997), pF values of bentonite mixtures are higher. The pF values of clay-bentonite and sand- bentonite compacted at w_{opt} are about 3.7 and 3.5 respectively, while that of compacted sand at water contents greater than 5% is less than 1.8. This is why water can migrate from sand to liner material during water migration tests.

Due to the great amount of precipitation, the water content of underlying ground at canyons in Japan is considered high. Therefore, water can move from the underlying base to liners, which results in high water content for liner material.

5.1.3 Direct shear strength test

Shear tests were performed using a standard direct shear test apparatus. Internal shear test on bentonite mixtures and interface shear test between bentonite mixtures and porous stone were conducted. Porous stone was used to simulate the base soil. Specimens of bentonite mixtures of 10 mm height and 60 mm diameter were compacted at w_{opt}. Tests were conducted under 3 different initial conditions (Table 5), including (1) bentonite mixture of w_{opt} and dry porous stones, (2) bentonite mixture having a water content (w_m) determined from the water migration test results, and saturated (but not submerged) porous stone, (3) bentonite mixture having a water content of w_m and submerged porous stone. Shear strain rate of 0.05 mm/min under the normal stresses of 100, 200, 400, and 500 kPa was used.

Table 5 Conditions and results of direct shear tests

Material	Case	Initial conditions		Results	
		Specimen	Porous stone	Cohesion (kPa)	Friction angle (degree)
Clay-bentonite internal	(1)	w_{opt} =30.0%	Dry	141	21.4
	(2)	w_m =38-40%	Unsubmerged	40	23.4
	(3)	w_m =38-40%	Submerged	31	24.1
Clay-bentonite/ porous stone interface	(1)	w_{opt} =30.0%	Dry	77	24.7
	(2)	w_m =38-40%	Unsubmerged	13	26.3
	(3)	w_m =38-40%	Submerged	16	27.0
Sand-bentonite internal	(1)	w_{opt} =16.4%	Dry	111	36.6
	(2)	w_m =24-26%	Unsubmerged	67	38.8
	(3)	w_m =30%	Submerged	66	38.2
Sand-bentonite/ porous stone interface	(1)	w_{opt} =16.4%	Dry	15	34.4
	(2)	w_m =24-26%	Unsubmerged	11	35.5
	(3)	w_m =30%	Submerged	9	34.6

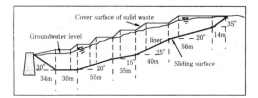

Fig.14 Cross section of canyon solid waste landfill

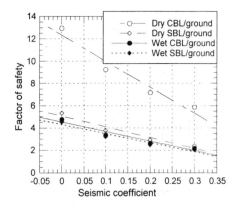

Fig. 15 Variation of FOS with horizontal seismic coefficient (k_h)

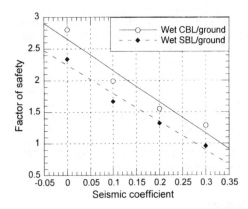

Fig. 16 Variation of FOS for wet liner interfaces with horizontal seismic coefficient (kh)

Table 5 summarizes the test results. Friction angles of clay-bentonite specimens under case (1), (2), and (3) were 21.4, 23.4, and 24.1 degrees, and cohesions were 141, 40, and 31 kPa respectively. Friction angles of clay-bentonite obtained under the 3 cases were similar, but cohesions of clay-bentonite under case (1) was much greater than that obtained for cases of (2) and (3). Similar tendencies were obtained for sand-bentonite internal shear strength and sand- bentonite/porous stone and clay-bentonite/porous stone interface shear strengths.

5.2 *Stability Analysis of Waste Landfill*

The assumed cross section of a canyon waste landfill used for stability analysis is shown in Fig. 14. Wet and dry unit weights of 15.5 kN/m³ and 12.0 kN/m³, and shear strength parameters of c = 0 kPa, ϕ = 43 degree for solid waste were assumed. Groundwater level was presumed under two conditions: high enough to near cover surface and low enough to be below liners at any point. All sliding surfaces are considered to occur along liners consisting of bentonite mixtures except one sliding surface at the slope toe near the retaining dam occurs in solid waste. The slope was divided into 13 slices according to the variation of inclination of underlying ground and cover surface.

Based on the direct shear test results shown in Table 5 and the presumed conditions above, the shear strength parameters of clay-bentonite/porous stone interface and sand- bentonite/porous stone interface under the case (1) and (3) conditions were selected for calculation. The calculation including dry and wet liner systems was based on Janbu's Generalized Procedure of Slice.

The factors of safety (FOS) for both wet and dry liner interfaces were calculated first when groundwater was presumed to be below liners. Figure 15 shows the results of the stability. The factors of safety for wet liner interfaces are less than those for dry liner interfaces with various horizontal seismic coefficients. Thus, the landfill faces to lose the stability when dry liners become wet.

The factors of safety for wet liner interfaces were calculated secondary when groundwater level was presumed to near cover surface as shown in Fig.14. Figure 16 shows the results of this case. The Factors of safety for wet clay-bentonite liner (CBL) interface are greater than those for wet sand-bentonite liner (SBL) interface with various horizontal seismic coefficients. Therefore, the landfill is considered to be more stable using CBL than SBL. When horizontal seismic coefficient k_h = 0, the factors of safety for both CBL and SBL interfaces are greater than 2.3. The landfill is stable when earthquake does not occur. When horizontal seismic coefficient k_h >0.28, the factor of safety for wet SBL interface is less than 1.0, and when k_h >0.33, the factor of safety for wet CBL interface is less than 1.0. The landfill is unstable either with wet CLB when earthquake magnitude is great enough to produce k_h >0.33 or with wet SLB when earthquake magnitude is great enough to produce k_h >0.28.

6 CONCLUSIONS

This paper discussed some geotechnical issues on the application of clay liners for waste landfill. The applicability and excellent performance of clay liners were presented. Results obtained are summarized as follows:

(1) General and legal review of the liner systems and barrier materials was conducted and the applicability of clay liners is clearly shown based on the several basic properties of clays.

(2) There are still remaining issues on the difference between Japan and western countries regarding the regulation of bottom liner system, and thus the need to be amended the Japanese criteria of landfill structure is pointed out.

(3) The compatibility of the various hydraulic conductivity tests is examined. The advantageous of flexible-wall permeameter test is listed up with suitable effective stress and hydraulic gradient applied.

(4) Critical issues in conducting in-situ hydraulic conductivity testing for CCLs are summarized, and the specification of SDRI and TSB tests and construction quality control to obtain the compacted clay liners having low hydraulic conductivity are presented.

(5) Possible reduction of shear strength of CCLs is investigated for heavy rainfall and increase of the groundwater level conditions. Water movement to CCLs due to capillarity results in large increase in water CCLs. Solid waste landfills at gentle slope (not greater than 20 degree) do not fail in the both cases of CBL and SBL even with horizontal seismic coefficient k_h < 0.28. The stability of canyon-type waste landfills could be confirmed more safety using CBL than SBL.

Consequently, waste containment system having clay liners could be effectively used for waste landfill.

ACKNOWLEDGEMENTS

Helpful comments and discussions have been provided to the authors by Professor Craig H. Benson (University of Wisconsin-Madison) and Professor Charles D. Shackelford (Colorado State University). Their contributions were greatly appreciated. Some of the drawings in this paper were prepared by Mr. T. Inui and Mr. K. Endo.

REFERENCES

Acar, Y.B. and Olivieri, I. (1989): Pore fluid effects on the fabric and hydraulic conductivity of laboratory-compacted clay, *Transportation Research Record 1219*, TRB, pp.144-159.

Benson, C.H. (2000): Liners and covers for waste containment, *Creation of New Geo-Environment, Fourth Kansai International Geotechnical Forum*, JGS Kansai Branch, pp.1-40.

Benson, C.H. and Daniel, D.E. (1990): Influence of clods on the hydraulic conductivity of compacted clay, *Journal of Geotechnical Engineering*, ASCE, Vol.116, No.8, pp.1231-1248.

Benson, C.H. and Gribb, M.M. (1997): Measuring unsaturated hydraulic conductivity in the laboratory and field, *Unsaturated Soil Engineering Practice*, S.L. Houston and F.G. Fredlund (eds.), ASCE, pp.113-168.

Benson, C.H., Hardinato, F.S., and Motan, E.S. (1994): Representative specimen size for hydraulic conductivity assessment of compacted soil liners, *Hydraulic Conductivity and Waste Contaminant Transport in Soils, ASTM STP 1142*, D.E. Daniel and S.J. Trautwein (eds.), ASTM, pp.3-29.

Benson, C.H., Gunter, J.A., Boutwell, G.P., Trautwein, S.J., and Berzanskis, P.H. (1997): Comparison of four methods to assess hydraulic conductivity, *Journal of Geotechnical and Geoenvironmental Engineering*, ASCE, Vol.123, No.10, pp.929-937.

Benson, C.H., Daniel, D.E., and Boutwell, G.P. (1999): Field performance of compacted clay liners, *Journal of Geotechnical and Geoenvironmental Engineering*, ASCE, Vol. 125, No.5, pp.390-403.

Bowders, J.J. (1988): Termination criteria for clay permeability testing (Discussion), *Journal of Geotechnical Engineering*, ASCE, Vol.114, No.8. pp.947-949.

Bowders, J.J. and Daniel, D.E. (1987): Hydraulic conductivity of compacted clay to dilute organic chemicals, *Journal of Geotechnical Engineering*, Vol.113, No.12, pp.1432-1448.

Committee on Permeability Assessment of Unsaturated Ground (1997): Assessment Methodology on Permeability of Unsaturated Ground, *Research Report of JGS*, 84

Daniel, D.E. (1984): Predicting hydraulic conductivity of compacted clay liners, *Journal of Geotechnical Engineering*, ASCE, Vol.110 No.2, pp.285-300.

Daniel, D.E. (ed.) (1993): *Geotechnical Practice for Waste Disposal*, Chapman & Hall.

Daniel, D.E. (1994): State-of-the-art: Laboratory hydraulic conductivity tests for saturated soils, *Hydraulic Conductivity and Waste Contaminant Transport in Soils, ASTM STP 1142*, D.E. Daniel and S.J. Trautwein (eds.), ASTM, pp.30-78.

Daniel, D.E. (1998): Landfills for solid and liquid wastes, *Environmental Geotechnics*, P.S. Seco e Pinto (ed.), Balkema, pp.1231-1246.

Daniel, D.E. and Benson, C.H. (1990): Water content-density criteria for compacted soil liners, *Journal of Geotechnical Engineering*, ASCE, Vol.116, No.12, pp.1811-1830.

Daniel, D.E. and Koerner, R.M. (1995): *Waste Containment Systems: Guidance for Construction, Quality Assurance, and Quality Control for Liner and Cover Systems*, ASCE Press.

Daniel, D.E., Bowders, J.J., and Gilbert, R.B. (1997): Laboratory hydraulic conductivity testing of GCLs in flexible-wall permeameters, *Testing and Acceptance Criteria for Geosynthetic Clay Liners, ASTM STP 1308*, L.W. Well (ed.), ASTM, pp.3-22.

Eid, H.T., Stark, T.D., Evans, W.D., and Sherry, P.E. (2000): Municipal solid waste slope failure. I: Waste and foundation soil properties, *Journal of Geotechnical and Geoenvironmental Engineering*, ASCE, Vol.126, No.5, pp.397-407.

Fernandez, F. and Quigley, R.M. (1985): Hydraulic conductivity of natural clays permeated with simple liquid hydrocarbons, *Canadian Geotechnical Journal*, Vol.22, pp.205-214.

Fernandez, F. and Quigley, R.M. (1988): Viscosity and dielectric constant controls on the hydraulic conductivity of clayey soils permeated with water-soluble organics, *Canadian Geotechnical Journal*, Vol.25, pp.582-589.

Foose, G.J., Benson, C.H., and Edil, T.B. (1996): Evaluating the effectiveness of landfill liners, *Environmental Geotechnics*, M. Kamon (ed.), Balkema, pp.217-221.

Fox, P.J. (1996): Analysis of hydraulic gradient effects for laboratory hydraulic conductivity testing, *Geotechnical Testing Journal*, ASTM, Vol.19, No.2, pp.181-190.

Giroud, J.P. and Bonaparte, R. (1989): Leakage through liners constructed with geomembranes, *Geotextiles and Geomembranes*, Vol.8, pp.27-67, 71-111.

Giroud, J.P., Badu-Tweneboah, K., and Soderman, K.L. (1994): Evaluation of landfill liners, *Proceedings of the Fifth International Conference on Geotextiles, Geomembranes, and Related Products*, pp.981-986.

Giroud, J.P., Soderman, K.L., Khire, M.V., and Badu-Tweneboah, K. (1998): New development in landfill liner leakage evaluation, Proceedings of the Sixth International Conference on Geosynthetics, IFAI, Vol.1, pp.261-268.

Imamura, S, Sueoka, T., and Kamon, M. (1996): Long term stability of bentonite/sand mixtures at L.L.R.W. storage, *Environmental Geotechnics*, M. Kamon (ed.), Balkema, pp.545-550.

Kamon, M. (1999): Appropriate structural code for waste disposal sites, *Waste Management Research*, JSWME, Vol.10, No.2, pp.147-155 (in Japanese).

Kamon, M. (2000): Solution scenarios of geo-environmental problems, *Proceedings of the Eleventh Asian Regional Conference on Soil Mechanics and Geotechnical Engineering*, Balkema, Vol.2 (in press).

Kamon, M., Katsumi, T., Kanayama, M., Jiang, W., and Morimoto, T. (2000): Stability of solid waste landfills along clay liners, *Proceedings of the Second International Summer Symposium*, JSCE, pp.249-252,

Kamon, M., Katsumi, T., Zhang, H., Sawa, N., and Rajasekaran, G. (2001): Redox effect on hydraulic conductivity and heavy metal leaching from marine clay, *International Symposium on Suction, Swelling, Permeability and Structure of Clay* (in press).

Katsumi, T., Benson, C.H., Foose, G.J., and Kamon, M. (1999a): Evaluation of the performance of landfill liners, *Waste Management Research*, JSWME, Vol.10, No.1, pp.75-85 (in Japanese).

Katsumi, T., Jo, H.-Y., Benson, C.H., and Edil, T.B. (1999b): Hydraulic conductivity performance of geosynthetic clay liners permeated with inorganic chemical solutions, *Geosynthetics Engineering Journal*, Japanese Chapter of IGS, Vol.14, pp.360-369 (in Japanese).

Katsumi, T., Benson, C.H., Foose, G.J., and Kamon, M. (2000): Performance-based design of landfill liners, *Engineering Geology*, Elsevier (in press).

Lin, L., Katsumi, T., Kamon, M., Benson, C.H., Onikata, M., and Kondo, M. (2000): Evaluation of chemical-resistant bentonite for landfill barrier application, *Annuals of Disaster Prevention Research Institute, Kyoto University* (in press).

Lo, I.M.C., Liljestrand, H.M., and Daniel, D.E. (1994): Hydraulic conductivity and adsorption parameters for pollutant transport through montmorillonite and modified montmorillonite clay liner materials, *Hydraulic Conductivity and Waste Contaminant Transport in Soils, ASTM STP 1142*, D.E. Daniel and S.J. Trautwein (eds.), ASTM, pp.422-438.

Lo, I.M.C., Mak, R.K.M., and Lee, S.C.H. (1997): Modified clays for waste containment and pollutant attenuation, *Journal of Environmental Engineering*, ASCE, Vol.123, No.1, pp.25-32.

McBride, M.B. (1994): *Environmental Chemistry of Soils*, Oxford University Press.

Manassero, M., Van Impe, W.F., and Bouazza, A. (1997): Waste disposal and containment, *Environmental Geotechnics*, M. Kamon (ed.), Balkema, pp.1425-1474.

Mesri, G. and Olson, R.E. (1971): Mechanisms controlling the permeability of clays, *Clays and Clay Minerals*, Vol.19, pp.151-158.

Mitchell, J.K. (1993): *Fundamentals of Soil Behavior, Second Edition*, John Wiley & Sons.

Mitchell, J.K. (1997): Geotechics of soil-waste material interactions, *Environmental Geotechnics*, M. Kamon (ed.), Balkema, pp.1311-1328.

Mitchell, J.K. and Madsen, F.T. (1987): Chemical effects on clay hydraulic conductivity, *Geotechnical Practice for Waste Disposal '87*, R.D. Woods (ed.), ASCE, pp.87-116.

Mitchell, J.K., Hooper, D.R., and Campanella, R.G. (1965): Permeability of compacted clay, *Journal of the Soil Mechanics and Foundations Division*, ASCE, Vol.91, No.4, pp.41-65.

Mitchell, J.K., Seed, R.B., and Seed, H.B. (1990): Kettleman Hills waste landfill slope failure. I: Liner-system properties, *Journal of Geotechnical Engineering,* ASCE, Vol.116, No.4, pp.647-668.

Nagaraj, T.S., Pandian, N.S., and Narasimha Raju, P.S.R. (1991): An approach for prediction of compressibility and permeability behaviour of sand-bentonite mixes, *Indian Geotechnical Journal,* Vol.21, No.3, pp.271-282.

Norrish, K. and Quirk, J. (1954): Cristalline swelling of montmorillonite, Use of electrolytes to control swelling, *Nature,* Vol.173, pp.255-257.

Olson, R.E. (1986): State of the art: Consolidation testing, *Consolidation of Soils: Testing and Evaluation, ASTM STP 892,* R.N. Yong and F.C. Townsend (eds.), ASTM, pp.7-70.

Onikata, M., Kondo, M., and Kamon, M. (1996): Development and characterization of a multiswellable bentonite, *Environmental Geotechnics,* M. Kamon (ed.), Balkema, pp.587-590.

Onikata, M., Kondo, M., Hayashi, N., and Yamanaka, S. (1999): Complex formation of cation-exchanged montmorillonites with propylene carbonate: Osmotic swelling in aqueous electrolyte solutions, *Clays and Clay Minerals,* Vol.47, No.5, pp.672-677.

Petrov, R.J., Rowe, R.K., and Quigley, R.M. (1997a): Comparison of laboratory-measured GCL hydraulic conductivity based on three permeameter types, *Geotechnical Testing Journal,* ASTM, Vol.20, No.1, pp.49-62.

Petrov, R.J., Rowe, R.K., and Quigley, R.M. (1997b): Selected factors influencing GCL hydraulic conductivity, *Journal of Geotechnical and Geoenvironmental Engineering,* ASCE, Vol.123, No.8, pp.683-695.

Rad, N.S., Jacobson, B.D., and Bachus, R.C. (1994): Compatibility of geosynthetic clay liners with organic and inorganic permeants, *Proceedings of Fifth International Conference on Geotextiles, Geomembrances and Related products,* pp.1165-1168.

Rowe, R.K. (1998): Geosynthetics and the minimization of contaminant migration through barrier systems beneath solid waste, *Proceeding of the Sixth International Conference on Geosynthetics,* IFAI, pp.27-102.

Rowe, R.K., Quigley, R.M., and Booker, J.R. (1995): *Clayey Barrier Systems for Waste Disposal Facilities,* E&FN Spon (Chapman & Hall).

Seed, R.B., Mitchell, J.K., and Seed, H.B. (1990): Kettleman Hills waste landfill slope failure. II: Stability analyses, *Journal of Geotechnical Engineering,* ASCE, Vol.116, No.4, pp.669-690.

Shackelford, C.D. (1994): Waste-soil interactions that alter hydraulic conductivity, *Hydraulic Conductivity and Waste Contaminant Transport in Soils, ASTM STP 1142,* D.E. Daniel and S.J. Trautwein (eds.), ASTM, pp.111-168.

Shackelford, C.D., Malusis, M.A., Majeski, M.J., and Stern, R.T. (1999): Electrical conductivity breakthrough curves, *Journal of Geotechnical and Geoenvironmental Engineering,* ASCE, Vol.125, No.4, pp.260-270.

Shackelford, C.D., Benson, C.H., Katsumi, T., Edil, T.B., and Lin, L. (2000): Evaluating the hydraulic conductivity of GCLs permeated with non-standard liquids, *Geotextiles and Geomembranes,* Elsevier, Vol.18, Nos.2-3, pp.133-161.

Stark, T.D., Eid, H.T., Evans, W.D., and Sherry, P.E. (2000): Municipal solid waste slope failure. II: Stability analyses, *Journal of Geotechnical and Geoenvironmental Engineering,* ASCE, Vol.126, No.5, pp.408-419.

Wang, X. and Benson, C.H. (1995): Infiltration and saturated hydraulic conductivity of compacted clay, *Journal of Geotechnical Engineering,* ASCE, Vol.121, No.10, pp.713-722.

Clay Science for Engineering, Adachi & Fukue (eds) © 2001 Balkema, Rotterdam, ISBN 90 5809 175 9

Microstructural modelling of transport in smectitic clay buffer

R. Pusch
Geodevelopment AB, Lund, Sweden

L. Moreno & I. Neretnieks
Department of Chemical Engineering and Technology, Royal Technical Institute, Stockholm, Sweden

ABSTRACT: Transport of water through smectite clay is strongly dependent on the microstructural constitution. Conceptual modeling of water percolation has been made and numerical 2D models employed for calculating the hydraulic conductivity have yielded good results. In practice, permeation takes place in 3D and the paper describes an attempt to develop such a model, assuming that flow takes place through capillary-filled channels formed by interconnected porous voids with higher porosity and conductivity than the clay matrix

1 INTRODUCTION

The physical behavior of clay soils is controlled by their microstructural constitution. For clays prepared by compacting bentonite powder it evolves in conjunction with water saturation and alters by percolation of water of different chemical composition. The present paper describes how the microstructure can be modeled with special respect to permeation of water. This matter is of particular importance for the isolating capacity of the clay embedment of canisters with highly radioactive waste in repositories. It is termed buffer material and has a very low hydraulic conductivity at high densities as demonstrated by Figure 1.

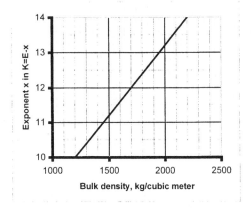

Figure 1. Hydraulic conductivity K of water of MX-80 bentonite as a function of the bulk density for saturation and percolation with distilled water (Pusch 1994).

2 EVOLUTION OF BUFFER MICRO-STRUCTURE

Buffer material is prepared by compaction of air-dry bentonite powder under about 100 Mpa pressure yielding blocks with a density that will be about 2000 kg/m^3 that will be placed around high level waste canisters in holes or tunnels.

The powder grains, which have a size ranging between 0.1 and 2 mm, are ground fragments of the mined natural bentonite beds and hence have a low porosity and high density. The compacted clay material consists of powder grains in close contact with air-filled voids between, and hydration of the clay takes place by capillary suction and adsorption of water molecules in the interlamellar space and on external surfaces of the stacks of lamellae.

Penetration of water into the strongly hydrophilic powder grains causes exfoliation of their shallow parts, which disintegrate and reorganize to form soft gels in the open voids. The gels are consolidated by the expanding dense grains but complete homogeneity is not obtained. Hence, the voids between the expanded grains will be filled with clay gels with a density that is low in the largest voids and high in the smallest voids (Figure 2). The gel-filled voids combine to form continuous channels that are responsible for the bulk hydraulic conductivity.

3 CONCEPTUAL MODEL

Digitalization of transmission electron micrographs of ultrathin sections of suitably prepared, water saturated clay specimens verifies that the microstructure is heterogeneous as illustrated by Figure 3. Figure 4 shows a typical micrograph of smectite clay with a density of 1900 kg/m³ at water saturation.

Grains with interlamellar voids

Clay gels in "external"

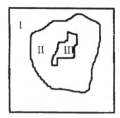

Figure 2. Generalized microstructure of MX-80 clay of compacted powder grains and subsequent hydration. The gel fillings have different densities, lowest for the largest voids and highest for the smallest voids.

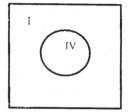

Figure 3. Schematic micrographs. Left: Typical picture of ultrathin section with varying density in soft matrix region (I to III). Right: normalized picture with defined channel cross section and density of the clay gel in the channel (IV) and of the clay matrix (I).

4 NUMERICAL MODELING IN 2D

A practical way of quantifying microstructural heterogeneity (Pusch et al. 1999) is defined in Figure 5, where a represents the clay matrix (I in Figure 3), and b the soft gel and open space in the channels (IV in Figure 3).

F_2 and F_3 can be evaluated from digitalized TEM micrographs with different degrees of greyness representing different densities. Depending on the scale, a or b may dominate. Micrographs with an edge length of at least 30 μ m appear to be representative for the larger part of the clay matrix. F_2 and F_3 are related to the average bulk density of saturated MX-80 clay as shown in Table 1.

Calculation of the bulk hydraulic conductivity

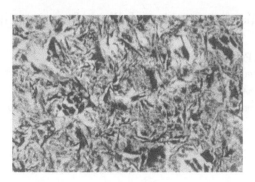

Figure 4. Microstructure of 500 Å ultrathin section of Friedland clay (Type B) with a density of 1900 kg/m³ as observed in a transmission electron microscope. The darkest parts are dense particles of rock-forming minerals, mainly chlorite, medium-gray parts are mainly mixed-layer and montmorillonite clay minerals, while white parts are open voids. The bar is 1 μ m long.

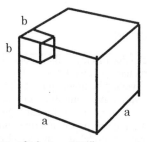

In 2D: $F_2 = b^2/a^2$ $b = a(F_2)^{1/2}$

In 3D: $F_3 = b^3/a^3$
$$F_3 = F_2(F_2)^{1/2} = (F_2)^{3/2}$$
$$\rho_{av} = \rho_b F_3 + \rho_a(1 - F_3)$$

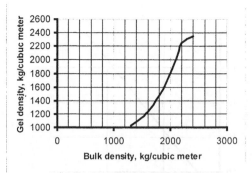

Figure 5. Microstructural parameters is the average bulk density ρ_{av} of the clay and the average density of components a (impermeable clay matrix) i.e. ρ_a, and b (soft gel fillings and open space), i.e. ρ_b. The diagram shows the average gel density versus average bulk density.

Table 1. *F*-parameters as functions of clay density.

Bulk density, kg/m³	Gel density, kg/m³	F_2	F_3
1500	1150	0.85	0.80
1700	1300	0.62	0.50
1800	1500	0.45	0.33
1900	1650	0.32	0.23
2100	2000	0.20	0.10
2200	2200	0.15	0.01

Table 2. Microstructural data and conductivities for MX-80 in Na form, percolation with distilled water (Pusch et al. 1999).

Bulk density, kg/m³	F_2	Gel density, kg/m³	Gel conductivity, kg/m³	Theor. bulk conductivity m/s	Experim. bulk cond. M/s
2130	0.17	2000	7E-14	E-14	2E-14
1850	0.24	1650	2E-12	4E-12	3E-12
1570	0.80	1200	E-10	8E-11	8E-11

k_{11}	k_{12}		k_n
k_{21}			
		K_{ij}	
k_m			K_{mn}

Figure 6. System of elements with different hydraulic conductivity permeated in one direction

of the heterogeneous clay can be made by applying basic flow theory, taking a permeated clay section to consist of a system of elements with different hydraulic conductivity (Figure 6).

Theoretically, the hydraulic conductivity K of a soil with elements of different conductivity can be expressed in the following way:

$$K = 1/m \, \Sigma n \, (\, \Sigma 1/k_{ij}) \qquad (1)$$

where:
K= Average hydraulic conductivity
n= Number of elements normal to flow direction
m= Number of elements in flow direction
k_{ij}= Hydraulic conductivity of respective elements

Using Eq.1 and relevant *F*-parameter values one gets the theoretical hydraulic conductivity in Table 2 for three specific bulk densities. The table also

gives experimental values that appear to be in good agreement with the theoreticallly derived data.

The size and frequency of the channels containing clay gels can be estimated on the following grounds:

- The channels have a circular cross section
- The diameter of the widest channel is 50 μ m in MX-80 clay with a bulk density 1570 kg/m³, 20 μ m in clay with 1850 kg/m3 density, and 5 μ m in clay with 2130 kg/m³ density, as concluded from transmission electron microscopy. Gel-filled voids with a diameter less than μ m.are of no importance to the bulk conductivity
- The size of the channels have normal distribution.

Using these criteria one gets the size distributions in Table 3.

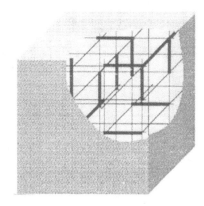

Figure 7.Schematic view of the 3D model concept.

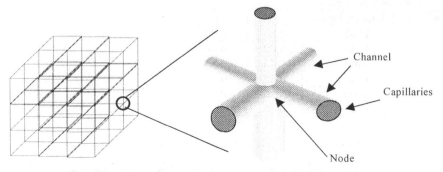

Figure 8.   Channel network mapped as a cubic grid with channels intersecting at a node in the grid.

5   NUMERICAL MODELING IN 3D

Practically all water flow takes place within the three-dimensional network of gel-filled channels with stochastic properties (Figure 7). The channels, which represent the zone termed IV in Figure 3, are characterised by their lengths, widths, apertures and transmissivities. The clay matrix is assumed to be porous but impermeable. The basis of the development of the present model is the code 3Dchan (Neretinieks & Moreno1993).

Calculation of the bulk hydraulic conductivity can be made by assuming that a certain number (commonly 6) of channels intersect at each node of the orthogonal network and that each channel in the network consists of a bundle of N capillaries with circular cross section. Figure 8 shows a small part of the channel network and illustrates that the channels have different diameters.

The number of channels, which are assumed to have the length L, contain bundles of N capillaries with a diameter (d) that is proportional to thechannel width is chosen to match the total

porosity of the clay. After complete grain expansion the voids filled with homogeneous clay gels are assumed to have a normal size distribution with the same intervals as in the 2D model, i.e. 1-5 μ m for the clay with 2130 kg/m³ density, 1-20 μ m for the clay with 1850 kg/m³ density, and 1-50 μ m for the clay with 1570 kg/m³ density. The code generates a certain number of channels for a given volume (Neretnieks & Moreno 1993). Using the Hagen-Poiseuille law the flow rate through the channel network is calculated for given boundary conditions assuming a pressure difference on the opposite sides of the cubic grid and no flow across the other four sides. Table 4 specifies the density, total porosity n, and resulting bulk hydraulic conductivity of the considered clay types.

One finds that the calculated bulk conductivity of MX-80 with the density 1570 kg/m³ at saturation is on the same order of magnitude as typical experimental data, while the model overrates the conductivity of the denser clays. The major reason for the discrepancy is that the model does not account for variations in cross section of the

Table 3. Number of channels with different diameters per 250x250 μ m² cross   section area (Pusch et al. 1990).

Bulk density, kg/m³	Number of 20-50 µm channels	Number of 5-20 µm channels	Number of 1-5 µm channels
2130	0	0	135
1850	0	10	385
1570	2	85	950

Table 4. Hydraulic conductivity (K) of three clay types prepared by compacting MX-80 powder and saturating them with electrolyte-poor water. n is the porosity.

Bulk density of air-dry powder kg/m³	Dry density kg/m³	Density at water saturation kg m⁻³	n	Calculated K, m/s	Experi-mental K, m/s
2000	1800	2130	0.13	3E-12	2E-14
1500	1350	1850	0.20	1.3E-11	3E-12
1000	900	1570	0.47	2.4E-10	8E-11

channels, meaning that the most narrow parts controls the percolation. Taking this effect into consideration, which is part of the ongoing development of the model, the agreement between calculated and recorded hydraulic conductivities is expected to be very good.

6 CONCLUSIONS

The conceptual structural model of clay formed by compacting MX-80 grains with subsequent hydration implies that the clay does not become homogeneous but contains interconnected voids filled with clay gels of lower density than the rest of the clay matrix. Taking micrograph-derived structural parameter data as a basis and applying flow theory for heterogeneous media the calculated bulk hydraulic conductivity is about the same as the experimental. The recently developed 3D capillary-filled channel-type model with theoretically derived flow partly overrates the conductivity, suggesting that the channels have a varying cross section with the most narrow parts controlling the net conductivity.

REFERENCES

Neretnieks I, Moreno L., 1993. Fluid flow and solute transport in a network of channels. *Journal of Contaminant Hydrology*, 14, pp 163-192.

Pusch R, Karnland O, Hökmark, H, 1990. GMM – A general microstructural model for qualitative and quantitative studies of smectite clays. SKB Technical Report TR 90-43, SKB Stockholm.

Pusch R, 1994. Waste Disposal in Rock., Developments in Geotechnical Engineering, 76. Elsevier Publ. Co, ISBN;0-444-89449-7.

Pusch R, Muurinen A, Lehikoinen J, Bors J, Eriksen T, 1999. Microstructural and chemical parameters of bentonite as determinants of waste isolation efficiency. European Commission, Contract No F14W-CT95-0012, Directorate-General Science, Research and Development.

1 Suction and swelling of clays/unsaturated soils

Clay Science for Engineering, Adachi & Fukue (eds) © 2001 Balkema, Rotterdam, ISBN 90 5809 175 9

Predicted and measured suction and volume changes in expansive soil

B. M. El-Garhy & A. A. Youssef
Department of Civil Engineering, Minufiya University, Shebin El-Kom, Egypt

W. K. Wray
Department of Civil Engineering, Ohio University, Athens, Ohio, USA

ABSTRACT: This paper presents the results of a numerical simulation of field measured soil suction and volume changes (shrink/heave) in expansive soil. The model employed was developed by El-Garhy (1999) for predicting soil moisture changes in terms of soil suction and the resulting volume changes in expansive soil under a covered area (e.g. flexible raft foundation or pavements) with respect to time. The model was validated against field data at four test sites in three different countries (i.e. United States, Saudi Arabia, and Australia) with widely varying climatic and soil conditions. Comparison to the field measurements from Al-Ghatt test site, Saudi Arabia over 36 weeks period of time is presented in this paper. Good correlation with field measurements of soil suction and associated soil movements is achieved.

1 INTRODUCTION

Expansive soils are found throughout the world. However, the problems caused by expansive soils are usually associated with areas of arid and semi-arid climate. Because in these areas the climatic conditions change over the year from wet to dry or vice-versa, and these changes in the climatic conditions cause moisture change within expansive soil, and consequently, volume changes (shrink/heave). The performance of shallow foundations is largely dependent on moisture movement and its distribution in the supporting expansive soil. The process of moisture movement through unsaturated soils is very complex and difficult to describe quantitatively due to the change in unsaturated soil properties (e.g. diffusion coefficient) with the change in soil suction. Water flow in saturated soils caused by a driving force resulting from an effective potential or hydraulic gradient, but water flow in unsaturated soils is subject to a free energy, potential, or soil suction likewise constitutes a moving force (Hillel 1971). Most researchers since about 1965 describe moisture movement in unsaturated expansive soil in terms of soil suction (e.g. Richards 1965; Lytton & Kher 1970; Mitchell 1979; Pufahl & Lytton 1992; Fredlund 1997; Wray 1998; El-Garhy et al. 2000). Richards (1974) reported that soil suction could be used to represent the state of the soil water in soils much more effectively than the water content for two reasons. First, soil suction is primarily controlled by the soil environment and not by the soil itself and it tends to be uniform or at least shows continuous trends. The soil suction profile is easier to predict because it tends towards an equilibrium value at a particular depth under particular climatic conditions, while water content is highly sensitive to the soil material variables (e.g. soil type, clay content, density, and soil structure). The water content profile also is difficult to predict because it can show wide variations and discontinuities. Second, the correlation of soil parameters (i.e. permeability, diffusivity, and shear strength) with water content is poor unless other soil properties such as density and clay content are considered, but these parameters can be most conveniently correlated with soil suction.

2 METHODOLOGY AND BASIC FORMULATION

2.1 *Mass flow model*

It has been suggested (Mitchell 1979) that the movement of water through unsaturated soils can be adequately represented by a transient diffusion equation in terms of soil suction expressed in pF units, Eq. (1). With the recent progress in the speed and memory size of digital computers the transient suction diffusion equation can be simply and rapidly solved in three dimensions.

$$\frac{\partial^2 u}{\partial x^2} + \frac{\partial^2 u}{\partial y^2} + \frac{\partial^2 u}{\partial z^2} + \frac{f(x,y,z,t)}{p} = \frac{1}{\alpha}\frac{\partial u}{\partial t} \qquad (1)$$

where:

u =   soil suction expressed in pF units
α =   diffusion coefficient (cm^2 / sec)
p =   unsaturated permeability (cm / sec)
t =   time (sec)
x, y, z =   space coordinates
$f(x, y, z, t)$ =   internal source of water

The diffusion coefficient, α, can be measured in the laboratory (Mitchell 1979) or calculated from empirical equations (McKeen & Johnson 1990; Bratton 1991; Lytton 1994). The different methods used to determine the value of α for expansive soils are discussed by El-Garhy (1999).

2.2 Volume change model

Total and differential vertical volume changes are important parameters in the selection and design of foundations resting on expansive soils. After the determination of suction distribution through the unsaturated expansive soil mass with respect to time, the vertical volume change can be calculated by one of the volume change prediction methods based on the concept of soil suction. In the present work, Wray's model (1997) was used, Eq. (2), to calculate the vertical volume change at each nodal point in the soil mass.

$$\Delta H_{i,j,k} = \Delta z\, \gamma_{h_{i,j,k}} \left[\Delta pF_{i,j,k} - \Delta pP_{i,j,k} \right] \qquad (2)$$

where:

$\Delta H_{i,j,k}$ =   incremental change in elevation (shrink or heave) at grid point (i, j, k) over the increment thickness Δz

Δz =   increment thickness in z-direction over which shrink or heave occur

$\gamma_{h_{i,j,k}}$ =   suction compression index at grid point (i, j, k)

$\Delta pF_{i,j,k}$ =   change of soil suction expressed in pF units at grid point (i, j, k)

$\Delta pP_{i,j,k}$ =   change of soil overburden over increment thickness Δz at grid point (i, j, k)

The different methods used to calculate the value of suction compression index, γ_h, are discussed by El-Garhy (1999). McKeen's chart and equations are the most frequently recommended methods cited in the technical literature to estimate the value of γ_h (e.g. McKeen 1980; Lytton 1994; and Wray 1997).

2.3 Numerical method of analysis

The soil mass is descretized and represented as a grid of points as shown in Figure 1. The finite dif-

ference equation representing the change of soil suction with time at each nodal point takes the form:

$$u_{i,j,k}^{t+1} = R\left[R_x^2\left(u_{i+1,j,k}^t + u_{i-1,j,k}^t\right) + R_y^2\left(u_{i,j+1,k}^t + u_{i,j-1,k}^t\right) \right.$$

$$\left. + \left(u_{i,j,k+1}^t + u_{i,j,k-1}^t\right)\right] + \left[1 - 2R\left(1 + R_x^2 + R_y^2\right)\right]u_{i,j,k}^t \quad (3)$$

where:

$$R_x = \frac{\Delta z}{\Delta x}, \qquad R_y = \frac{\Delta z}{\Delta y}, \text{ and } \quad R = \frac{\alpha \Delta t}{\Delta z^2}$$

For convergence and stability, the ratio R must be:

$$R \le \frac{1}{2\left(1 + R_x^2 + R_y^2\right)} \text{ or } \Delta t \le \frac{\Delta z^2}{2\alpha\left(1 + R_x^2 + R_y^2\right)} \quad (4)$$

where:

Δt =   time interval
Δx =   division length in x-direction
Δy =   division length in y-direction
Δz =   division length in z-direction
α =   diffusion coefficient (cm^2 / sec)
$u_{i,j,k}^t$ =   soil suction at time step, t, at grid point (i, j, k)
$u_{i,j,k}^{t+1}$ =   soil suction at time step, $t + 1$, at grid point (i, j, k)

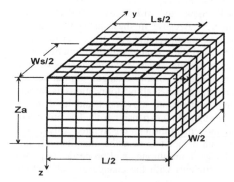

Za   depth of active zone
W   width of soil mass
L   length of soil mass
Ws   width of raft foundation
Ls   length of raft foundation

Figure 1. Idealization of raft foundation and supporting soil mass for moisture movement and soil structure-interaction analysis.

In a mathematical sense, the type of transient suction diffusion equation represented by Eq. (1) is called a parabolic partial differential equation. Two sets of information must be known to solve this equation (Champra & Canale 1998): (1) the initial conditions, i.e. specify the initial value of suction at each node in the soil mass at the initial time, and (2) boundary conditions, i.e. specify the values of suction on the boundaries of the soil mass at each time step.

3 COMPUTER PROGRAM

A FORTRAN computer program was written and named SUCH utilizing the finite difference technique to solve the transient suction diffusion equation in three dimensions. SUCH is able to calculate the distribution of soil suction throughout the unsaturated expansive soil mass with respect to time and the resulting volume changes beneath a covered area when the soil mass is subjected to one or more of seven different effects that commonly cause moisture changes in the supporting expansive soils at the perimeter of the covered area. These perimeter effects are: (1) climate only (i.e. no tree, no ponded water, and no barriers), (2) vertical moisture barrier around the perimeter of the foundation, (3) horizontal moisture barrier around the perimeter of the foundation, (4) both vertical and horizontal moisture barriers together around the perimeter of the foundation, (5) tree roots adjacent to the edge of the foundation, (6) trees and vertical moisture barrier at the edge of the foundation, and (7) ponded water adjacent to the edge of the foundation (no barriers).

4 VALIDATION OF THE MODEL

The model SUCH was used to determine the distribution of soil suction throughout the unsaturated expansive soil mass and the resulting volume changes (shrink/heave) with respect to time at several field sites for the purpose of program validation. The validation process involves comparison of predicted and measured values of either soil suction or surface and subsurface volume changes. Comparison to the field measurements from Al-Ghatt test site, Saudi Arabia is presented in this paper. Other comparisons can be found in El-Garhy (1999).

4.1 *Validation at Al-Ghatt Test Site, Saudi Arabia*

Al-Ghatt test site is selected to represent a two-dimensional problem. Dhowian et al. (1985) established a main field station in Al-Ghatt town in Saudi Arabia for the purposes of monitoring the changes of soil suction and the resulting volume changes (shrink/heave) in expansive soil over a long period of time. The instrumented site covered an area of $12\,m \times 24\,m$ as shown in Figure 2. The site included a saturation system consisting of 4.0-m deep sand drains. The diameter of each sand drain was 10 cm. The sand drains were placed in two rows with 0.5-m

average spacing. A 0.6-m PVC pipe was installed on the top of each sand drain. These pipes were attached to individual water tanks ($0.5\,m \times 0.5\,m \times 0.7\,m$) and sealed to eliminate water leakage. Six instrument units called "Batteries" were installed as illustrated in Figure 2. Each Battery includes a psychrometer stack to measure the change of soil suction with time at different depths, a moisture access tube to measure the change in water content with time, a surface heave plate to measure the changes in surface heave with time, and five deep heave plates at consecutive one meter depth increments to measure the subsurface shrink or heave with time. The instrumentation extended to a depth of 6 m with one psychrometer stack and one access tube was extending to a depth of 8 m.

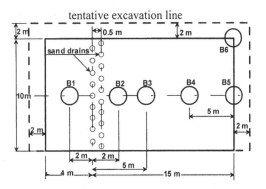

Figure 2. Layout plan of the main field station at Al-Ghatt site (after Dhowian et al. 1985).

The expansive stratum extended to a depth of 8 to 10 m from the ground surface. Soil stratigraphy at the site consists of three strata: (1) a non-expansive soil layer of depth 1.0 m, (2) a silt shale layer of depth 3.0 to 4.0 m, and (3) a clay shale underlying the silt shale layer. The top non-expansive soil stratum was removed from the test area before the installation of the field instruments (Dhowian et al. 1985). Consistency limits and clay content at different depths are presented in Table 1. It has been observed (Dhowian et al. 1985) that the soil suction at 7.5 m depth in Battery B2 was affected by the saturation system. Therefore, the active zone depth is taken as 9.0 m, which is approximately at the depth of the expansive soil a stratum at the site.

Equilibrium soil suction is taken equal to the soil suction corresponding to the plastic limit at the bottom of the active zone depth. The plastic limit at 9.0 m depth is taken as 21% as determined from Table 1. The water content of 21% corresponds to a soil suction of 4.54 pF as determined from the redrawn suction-water content curve (Figure 3). The value of TMI is determined from the Russam and Coleman curve (Figure 4) for the equilibrium soil suction of

4.54 pF to be -30 in./year. The calculated values of diffusion coefficient by three different empirical equations are presented in Table 2.

Based on inspection of the measured data comprising the initial soil suction profile, it is assumed to change linearly from a suction of 4.78 pF at the surface to the equilibrium soil suction of 4.54 pF at bottom of the active zone depth of 9.0 m. Figure 5 shows the initial suction profile assumed along with measured values at the time of site construction in Battery B1 and Battery B2.

Table 1. Soil properties at different depths for Al-Ghatt test site (after Dhowian et al. 1985).

Depth (m)	w_n (%)	LL (%)	PL (%)	Sand (%)	Silt (%)	Clay (%)
1.2	10.8	35	19	22	18	60
2.2	19.2	68	30	3	22	75
3.5	18.5	62	28	4	24	72
7.2	17.7	47	21	8	40	52

w_n = natural water content

Table 2. Calculated values of diffusion coefficient for Al-Ghatt test site.

Method	Diffusion coeff.
McKeen & Johnson (1990) empirical equation	$0.00565 \, cm^2 / sec$
Bratton (1991) empirical equation	$0.00348 \, cm^2 / sec$
Lytton (1994) empirical equation	$0.00373 \, cm^2 / sec$
Average of the above three values	$0.00429 \, cm^2 / sec$

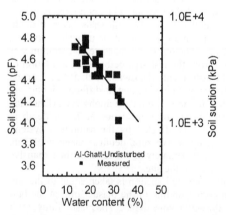

Figure 3. Soil suction-water content relationship for Al-Ghaat test site.

The field measurements of soil suction at a depth of 0.50 m in Battery B1 and Battery B2 are drawn as a function of time to determine a fit equations. These equations can be used to represent the actual change of surface suction as a result of climate change with respect to time. The fit equations are shown in Figure 6.

The measured database of soil suction and the associated surface and subsurface volume changes (shrink/heave) are not available, but the curves which represent these database are available. Therefore, for the purpose of comparison, the available curves were translated and converted to a database for use in this study.

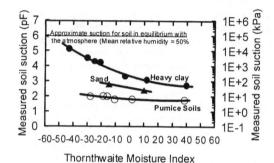

Thornthwaite Moisture Index

Figure 4. Variation of soil suction of road subgrade with Thornthwaite Moisture Index (after Russam and Coleman, 1961).

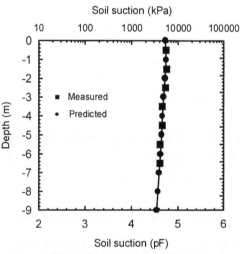

Figure 5. Comparison of assumed initial suction profile and the measured values at the time of site construction at Al-Ghatt test site.

4.1.1 *Analysis of soil suction*

In the first phase of study, an average value of diffusion coefficient of $0.00429 \, cm^2 / sec$ was used in an attempt to match the predicted values of soil suction with that measured in the field in Battery B2. The soil suction distribution after 11 weeks was derived using the average value of diffusion coefficient of $0.00429\ cm^2/sec$ and is illustrated in Figure 7.

Referring to Figure 7, it is observed that the predicted values of soil suction are wetter than the measured values. Therefore, numerous computer runs using different values of diffusion coefficient were made to predict the exact suction distribution profile as measured in the field.

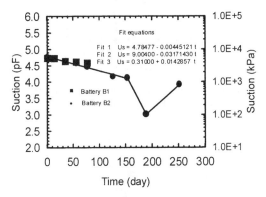

Figure 6. Measured soil suction as a function of time at 0.5-m depth in B1 and B2 at Al-Ghatt test site.

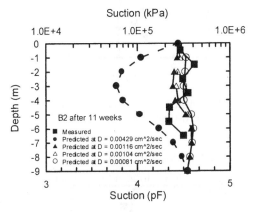

Figure 7. Predicted and measured soil suction profiles for different values of diffusion coefficient at Al-Ghatt test site.

The value of the diffusion coefficient, α, of $0.00081\ cm^2/sec$ seems to be the best for the expansive soil at this site as shown in Figure 7, to a depth of approximately 4m.

The distribution of soil suction throughout the soil and the surface and subsurface vertical volume changes with respect to time were predicted by the moisture diffusion and volume change model SUCH. Comparison between the predicted and measured values of either soil suction or surface and subsur-

face vertical volume changes was carried out through the period of 2 to 36 weeks in Battery B2 and through the period of 18 to 27 weeks in Battery B3. The results are illustrated in Figures 8-18. Referring to these figures it is observed that:

1. In Battery B2, the predicted values of soil suction match very well the measured values through the first 11 weeks as shown in Figures 8-11. However, after the first 11 weeks, the predicted values of soil suction along the sand drains depth of 4 m match very well the measured values, but below the depth of sand drains the measured values are wetter than the predicted values as shown in Figures 12-15. It is believed that the good agreement between predicted and measured values after the first 11 weeks along the depth of the sand drains is probably due to the actual representation of surface boundary condition by fitted equations. The unfavorable agreement below the depth of sand drains is probably due to two reasons. First, the diffusion coefficient of the clay shale is higher than the considered diffusion coefficient in the analysis. Second, the bad representation of the boundary condition at the bottom of the active zone depth where the validation process indicated that the depth of active zone might be greater than the depth of 9.0 m which is considered in the analysis.

2. In Battery B3, the measured values of soil suction is wetter than the predicted values as shown in Figures 16-18. This is probably due to the high horizontal permeability of the shale resulting from horizontal lamination (Dhowian et al. 1985).

4.1.2 *Analysis of volume changes*

The available soil properties are used to calculate the values of suction compression index, SCI, at different depths by using McKeen's chart as shown in Table 3. Comparison of predicted and measured surface and subsurface vertical volume changes (shrink/heave) was carried out through the period of 2 to 36 weeks in Battery B2, and the results are illustrated in Figures 19-27. Referring to these figures, it is observed that:

Through the first 18 weeks, the predicted values of cumulative swell along the top 5.0-m depth are slightly smaller than the measured values as shown in Figures 19-23. However, after the first 18 weeks the predicted values exceeded the measured values as shown in Figures 24-26.

2.The predicted values of vertical volume change at the surface are smaller than the measured values through the first 18 weeks. However, after the first 18 weeks the predicted values are greater than the measured values as shown in Figure 27.

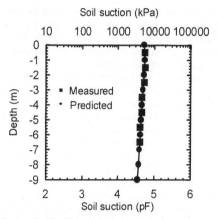

Figure 8. Predicted and measured suction after 2 weeks at B2 at Al-Ghatt test site.

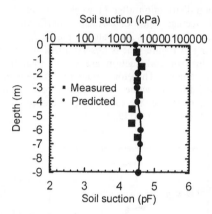

Figure 11. Predicted and measured suction after 11 weeks at B2 at Al-Ghatt test site.

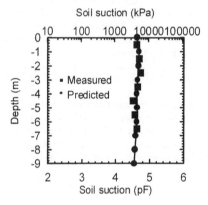

Figure 9. Predicted and measured suction after 5 weeks at B2 at Al-Ghatt test site.

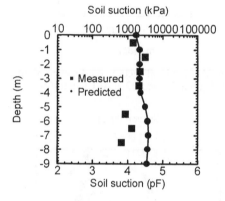

Figure 12. Predicted and measured suction after 18 weeks at B2 at Al-Ghatt test site.

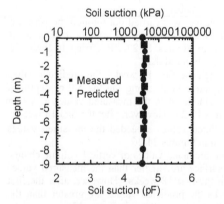

Figure 10. Predicted and measured suction after 8 weeks at B2 at Al-Ghatt test site.

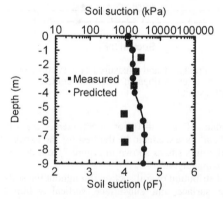

Figure 13. Predicted and measured suction after 22 weeks at B2 at Al-Ghatt test site.

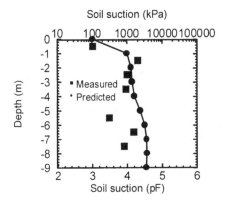

Figure 14. Predicted and measured suction after 27 weeks at B2 at Al-Ghatt test site.

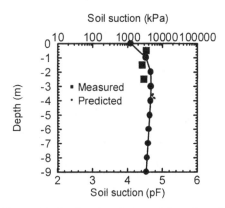

Figure 17. Predicted and measured suction after 22 weeks at B3 at Al-Ghatt test site.

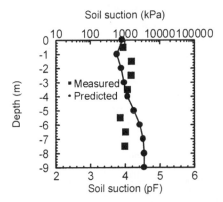

Figure 15. Predicted and measured suction after 36 weeks at B2 at Al-Ghatt test site.

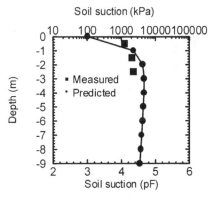

Figure 18. Predicted and measured suction after 27 weeks at B3 at Al-Ghatt test site.

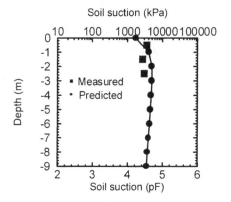

Figure 16. Predicted and measured suction after 18 weeks at B3 at Al-Ghatt test site.

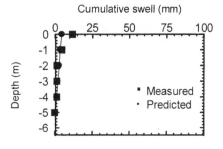

Figure 19. Predicted and measured cumulative swell after 2 weeks at B2 at Al-Ghatt test site.

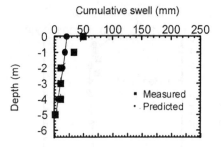

Figure 20. Predicted and measured cumulative swell after 5 weeks at B2 at Al-Ghatt test site.

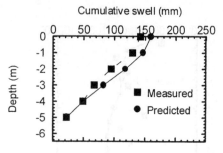

Figure 24. Predicted and measured cumulative swell after 22 weeks at B2 at Al-Ghatt test site.

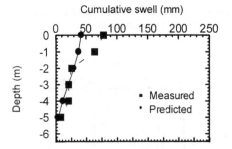

Figure 21. Predicted and measured cumulative swell after 8 weeks at B2 at Al-Ghatt test site.

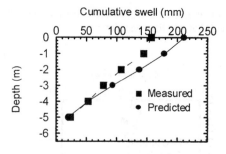

Figure 25. Predicted and measured cumulative swell after 27 weeks at B2 at Al-Ghatt test site.

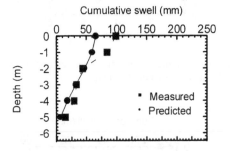

Figure 22. Predicted and measured cumulative swell after 11 weeks at B2 at Al-Ghatt test site.

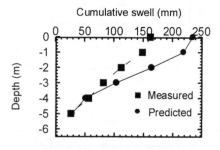

Figure 26. Predicted and measured cumulative swell after 36 weeks at B2 at Al-Ghatt test site.

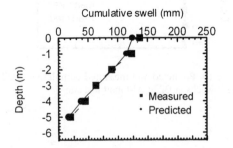

Figure 23. Predicted and measured cumulative swell after 18 weeks at B2 at Al-Ghatt test site.

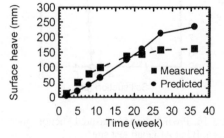

Figure 27. Predicted and measured surface heave with respect to time at B2 at Al-Ghatt test site.

Table 3. Calculated values of SCI by McKeen's Chart (McKeen 1980) for Al-Ghatt test site.

Depth (m)	PI	Ac	CEC	CEAc	SCI
1.2	16	0.27	31.3	0.52	0.0366
2.2	38	0.57	53.5	0.70	0.0720
3.5	34	0.47	49.3	0.68	0.0700
7.2	26	0.50	35.2	0.68	0.0500

Ac = Activity
CEC = Cation exchange capacity
CEAc = Cation exchange activity

5 CONCLUSIONS

This paper presents the results of a numerical simulation of field measured soil suction and volume changes in expansive soil. The model employed was developed by El-Garhy (1999) for predicting soil moisture changes in terms of soil suction and the resulting surface and subsurface volume changes in expansive soil under a covered area (e.g. flexible raft foundation or pavements) with respect to time. The developed model produced good correlation with reported field measurements of soil suction and associated surface and subsurface soil movements over a 36 weeks period of time. If measurement of the diffusion coefficient is not available, the field data can be used by the computer model to predict the magnitude of the field diffusion coefficient of the soil.

The difficulty in applying the developed model to practical problems depends on the quantitative expression of the model parameters (i.e. diffusion coefficient, equilibrium soil suction, active zone depth, and suction compression index) and the initial and final boundary conditions. Therefore, if an accurate representation of the model parameters and boundary conditions can be estimated, the computer model can reasonably predict the moisture movement and the resulting volume changes in unsaturated expansive soils.

REFERENCES

Bratton, W. L., (1991), "Parameters for Predicting Shrink/Heave Beneath Slab-on-Ground Foundations Over Expansive Clays," *Ph.D. Dissertation, Department of Civil Engineering, Texas Tech University, Lubbock, Texas.*

Champra, S. C., & Canale, R. P., (1998), "Numerical Methods for Engineers with Programming and Software Applications," *3rd Edition, WCB - McGraw-Hill, Inc., N. Y.*

Dhowian, A. W., Erol, O., & Youssef, A. F., (1985), "Evaluation of Expansive Soils and Foundation Methodology in the Kingdom of Saudi Arabia," *Research Report, CANCST, No. AT-5-88-3.*

El-Garhy, B. M., (1999), "Soil Suction and Analysis of Raft Foundation Resting on Expansive Soils, *Ph.D. Dissertation, Civil Engineering Department, Minufiya University, Egypt.*

El-Garhy, B. M., Youssef, A. A., & Wray, W. K., (2000), "Using Soil Diffusion to Design Slab-on-Ground Foundations on Expansive Soils," *Asian Conference on Unsaturated Soils from Theory to Practice, 18-19 May, Singapore.*

Fredlund, D. G., (1997), "An Introduction to Unsaturated Soil Mechanics," *Unsaturated Soil Engineering Practice, Geotechnical Special Publication No. 68, ASCE,* pp. 1-37.

Hillel, D., (1971), "Soil and Water Physical Principals and Processes," *Academic Press, Inc., New York*

Lytton, R L., & Kher, R. K. (1970), "Prediction of Moisture Movement in Expansive Clays," *Research Report 118-3, Center for Highway Research, University of Texas at Arlington, Austin, Texas.*

Lytton, R. L., (1994), "Prediction of Movement in Expansive Clay," Vertical and Horizontal Deformations of Foundations and Embankments, *Geotechnical Special Publication No. 40, A. T. Yeung and G. Y. Felio, Editors, ASCE,* Vol. 2, pp. 1827-1845.

McKeen, R. G., (1980), "Field Studies of Airport Pavements on Expansive Clay," *Proc. 4th Int. Conf. on Expansive Soils,* Denver, Colorado, Vol. 1, pp. 242-261.

McKeen, R. G., & Johnson, L. D., (1990), "Climate-Controlled Soil Design Parameters for Mat Foundations," *Journal of Geotechnical Engineering, ASCE,* Vol. 116, No. 7, pp. 1073-1094.

Mitchell, P. W., (1979), "The Structural Analysis of Footings on Expansive Soil," *Research Report No. 1, Kenneth W. G. Smith and Associates, Adelaide South Australia.*

Pufahl, D. E., & Lytton, R. L., (1992), "Temperature and Suction Profiles Beneath Highway Pavements: Computed and Measured," *Transportation Research Record 1307,* pp. 268-276.

Richards, B. G., (1965), "An Analysis of Subgrade Conditions at the Horsham Experimental Road Site Using Two-Dimensional Diffusion Equation on a High-Speed Digital Computer," *Moisture Equilibria and Moisture Changes in Soils Beneath Covered Areas,* Butterworths, Australia, pp. 243-258.

Richards, B. G., (1974), "Behavior of Unsaturated Soils," *Chapter 4, Soil Mechanics-New Horizons, I. K. Lee, Editor, Elsevier, New York,* pp. 112-157.

Russam, K., & Coleman, J. D., (1961), "The Effect of Climatic Factors on Subgrade Moisture Conditions," Geotechnique, Vol. 11, No. 1, pp. 22-28.

Wary, W. K., (1997), "Using Soil Suction to Estimate Differential Soil Shrink or Heave," *Unsaturated Soil Engineering Practice, Geotechnical Special Publication No. 68, ASCE*, pp. 66-87.

Wary, W. K., (1998), "Mass Transfere in Unsaturated Soils: A Review of Theory and Practices," *Proc., 2nd Intl. Conf. on Unsaturated Soils*, Beijing, Vol. 2, pp. 99-155.

Clay Science for Engineering, Adachi & Fukue (eds) © 2001 Balkema, Rotterdam, ISBN 90 5809 175 9

Suction – Water content relationship in swelling clays

M. Yahia-Aissa, P. Delage & Y. J. Cui

Teaching and Research Centre in Soil Mechanics (CERMES), Ecole Nationale des Ponts et Chaussées, Marne La Vallée, France

ABSTRACT: The paper presents a set of experimental results on the water retention properties of a swelling clay, considered as a possible material for engineered barriers, in the confinement of nuclear waste disposal at great depth. Tests were performed on both compacted specimens and clay powder, and suction was controlled using an extended osmotic technique and vapour control technique. Under unconfined swelling condition, similar retention curves were obtained for compacted specimens and clay powder. This suggests that the dry density does not affect the water uptake. This is explained by the fact that the water sorption-desorption phenomena occur within the intra-aggregate porosity, which is mainly controlled by physico-chemical forces exerted by clay mineral. This intra-aggregate porosity is not affected by compaction process, and there is no effect of the density. When swelling is confined, the water retention curves are similar to the previous ones at suctions larger than 10 MPa. For smaller suction, the water uptake is largely reduced due to the disappearance of macropores. This leads to a confined retention curve much different from the unconfined one.

1. INTRODCUTION

The fundamental property for soil water movement, is characterised by the soil water retention curve. It is defined by a soil suction – water content relationship, or a soil suction – degree of saturation relationship. Most investigations have been carried out in order to predict the mechanical and hydraulic behaviour of unsaturated soils (Croney, 1952; Holmes, 1955; Villar et al., 1993; Imbert & Zingarelli, 1993; Wan et al., 1995; Al-Mukhtar, 1994; Tang et al., 1997; Delage et al., 1998).

In sandy and low plastic soils, water retention curves depend mainly on capillarity, in relation to the development of air-water menisci inside the soil microstructure. In this case, soil water can move freely out of, or into the soil structure as the soil suction increases or decreases. In high plastic clays, such as the active FoCa clay being studied for nuclear waste disposal, the effects of physico-chemical clay-water interactions increase with clay plasticity. Moreover, the high compaction level of the clay used as buffer material, induces a very small porosity, and the confined state condition of the buffer material used for engineered barriers could have an additional influence on the soil retention curve.

This paper involves the effects of dry density and wetting conditions on suction – water content relationship. The water retention properties were experimentally determined and compared for both compacted and powder samples under either unconfined or confined swelling conditions.

2. MATERIALS AND METHODS

2.1. Soil material

FoCa clay tested is a highly expansive calcium smectite clay, extracted from Fourges – Cahaignes located in the North - West of France. It contains mainly an interstratified mineral of kaolinite-smectite (80%). Other minerals were identified: kaolinite 7%, quartz 2% and some trails of goethite and gibbsite (Atabek et al., 1991). It has a liquid limit $w_L = 112\%$ and a plasticity index PI = 62%. The specific surface measured using methylene blue test is $S_p = 515$ m²/g. Compacted specimens used in this study were statically compacted to an unit mass $\rho_d = 1.87 \pm 0.02$ Mg/m³ at a water content w = 11% ± 0.2%. After compaction, specimens were put in a plastic film, and stored in an hermetic container. The powder clay samples used, with an initial water content of 11%, were also gathered in the same container.

2.2. Experimental method

For suction values from 0 to 10 MPa, the suction control method used is an extended osmotic technique (Delage et al., 1998), while for higher suctions, vapour control was used. The osmotic control method is shown in Figure 1 (Cui & Delage, 1996). Sample was inserted in a tube-shaped semi-permeable membrane that had been wetted previously. The membrane and sample were immersed in a PEG (Polyethylene Glycol) solution, whose suction value is calibrated as a function of its concentration : the higher the concentration, the higher the suction. The water exchange between the sample and the PEG solution occurs by liquid transfer through the semi-permeable membrane. The weight stabilisation of the sample corresponds to the suction equilibrium between the PEG solution and the specimen.

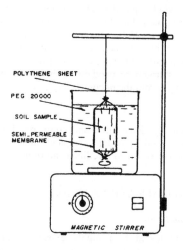

Figure 1: Use of the osmotic technique for the determination of the water retention properties (Cui & Delage, 1996).

The second suction control technique used is a controlled relative humidity technique H_r, namely vapour control. This technique is based on controlling the relative humidity of the air around the sample placed in a desiccator, using a saturated saline solution. In this case, the water exchange occurs by vapour transfer between the sample and the aqueous solution. The equilibrium is reached after about 1 month. This long duration is mainly due to a very low vapour exchange velocity. Saturated saline solutions used in the following were previously calibrated by Delage et al. (1998). Table 1 showed the various salts used and their corresponding suction values. The sensitivity of the relative humidity to temperature variations required

the immersion of all desiccators in a temperature controlled bath at 20° C.

For confined swelling conditions, samples were directly compacted in rigid cylindrical cells of 5 cm in diameter and 1 cm high. In this case, suction controlled technique used is based solely on vapour control. For very low suction values (S < 1 MPa), saturated salt solution is replaced by PEG solution. The suction in the sample is defined in that case by the final concentration of the solution.

Table 1: Salts used and their corresponding suction values

Saturated solution	Suction (MPa)
LiCl	262
$CaCl_2$	139
K_2CO_3	113
$Mg(NO)_3$	82
$NaNO_2$	57
NaCl	38
$(NH_4)_2SO_4$	24.9
$ZnSO_4$	12.6
KNO_3	9
$CuSO_4$	6.1

3. RESULTS

3.1. Unconfined swelling condition

Both compacted and powder samples were used for the measurement of suction-water content relationship. For compacted samples, most results were reported by Delage et al. (1998). Additional investigation using identical samples were done for low suction values, using the osmotic technique as shown in Figure 1. All of the results are gathered in Figure 2, showing an excellent continuity of linearity for a broad suction range, from 0.1 to 262 MPa. Moreover, although compacted samples were submitted to wetting – drying paths, no significant hysteresis was observed, indicating that the capillary phenomenon can be neglected. In general, for a given soil, the hysteresis depends on both the density and the physico-chemical interactions. More is the density, or more are the physico-chemical interactions, less is the hysteresis.

In order to investigate the density effect, powder samples were tested under the same unconfined swelling conditions. Results obtained are presented in Figure 3. As previously, An excellent continuity of linearity over an extended range of suction (0.1 to 82 MPa) was obtained, and moreover, no hysteresis was observed. These results show clearly that for the active FoCa clay, the water retention properties are wholly governed by the physico-chemical interactions.

3.2. Confined swelling condition

Figure 4 shows variation of the water content as a function of suction for compacted specimens, determined using either vapour control or osmotic control. The tests were carried out under confined swelling conditions. The evolution seems to be bilinear in a semi-logarithmic scale.

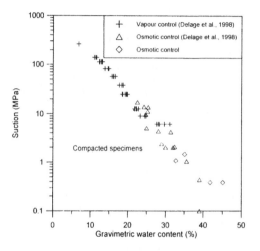

Figure 2: Water retention curve for compacted samples in unconfined swelling condition

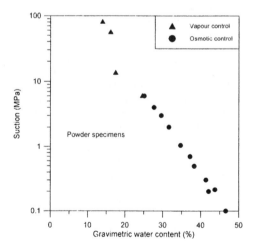

Figure 3: Water retention curve for clay powder samples under unconfined swelling conditions

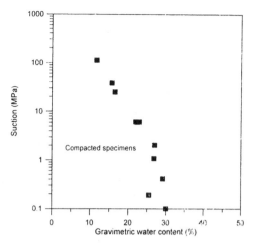

Figure 4: Suction – water content relationship for compacted samples under confined swelling conditions.

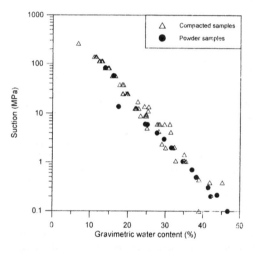

Figure 5: Water retention curve for both compacted and clay powder samples in unconfined swelling condition

4. COMPARISONS AND DISCUSSIONS

Figure 5 shows water retention curve for both compacted and powder samples. The overlapping data illustrated in this figure indicate that the water content – suction relationship for both compacted and powder specimens is identical. The initial dry density did not affect the water retention curve. This suggests that only the physico-chemical interactions are involved in the wetting – drying process for this clay.

Figure 6 shows results carried out on compacted specimens for both unconfined and confined swelling conditions. Little difference is observed at higher suctions (s > 4 MPa) between the two curves.

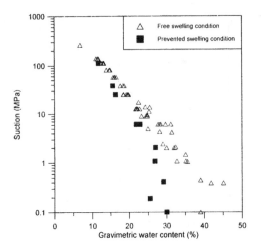

Figure 6: Suction – water content relationship for compacted specimen; comparison between unconfined and confined swelling condition

At lower suctions, the confined swelling condition has a strong influence on the shape of the curve. At the same suction value, less water is adsorbed when swelling is prevented.

This phenomenon could be explained by a microscopic consideration. Observations under the scanning electron microscope showed that the compacted FoCa clay has a microstructure defined by a majority of micropores and a small part of macropores (Yahia-Aïssa 1999). When wetted under confined swelling conditions, clays minerals can swell because of the presence of macropores. This microscopic swelling permits water to go into the soil, increasing the water content. It must be noticed that the microscopic swelling results in also the filling of macropores. When all the macropores fully filled under a certain suction (4 MPa for compacted FoCa clay), no more swelling can be allowed, therefore no more water uptake can be occur.

5. CONCLUSION

Test results presented in this paper show clearly that the dry density does not affect the water uptake, in accordance with the literature data (Villar et al., 1993; Imber & Zingarelli, 1993; Wan et al., 1995). This is explained by the fact that the water sorption – desorption phenomena occur within the intra-aggregate porosity, which is mainly controlled by physico-chemical forces exerted by clay mineral. This intra-aggregate porosity is not affected by the compaction process, and there is no effect of the density. Hence, contrary to sand and low plastic clay, for which water retention curve is mainly controlled by capillary effects, water retention curve of highly plastic clays depends solely on the physico-chemical effects.

Under confined swelling conditions, a suction threshold value is defined. This value represents the beginning of the difference between unconfined swelling conditions and confined swelling conditions. It is associated with the end of the microscopic swelling, or with the disappearance of macro-pores, during suction decrease under confined conditions.

REFERENCES

Al-Mukhtar M. (1994). *Comportement hydromécanique des argiles à faible porosité*. Research report for ANDRA, contract N°315 378BO, CNRS/86.0267.00.

Atabek R. B., Félix B., Robinet J. C. et Lahlou R. (1991). *Rheological behaviour of saturated expansive clays materials*. Workshop on Stress Partionning in Engineered Clay Barriers, Duke University, Durham, N.C.

Croney D. (1952). The movement and distribution of water in soils. Géotechnique, Vol. 1, pp. 1-16.

Cui Y. J. & Delage P. (1996). *Yielding and plastic behaviour of an unsaturated compacted silt*. Géotechnique, Vol. 46, No.2, pp. 291-311.

Delage P., Howat M. D. & Cui Y. J. (1998). *The relationship between suction and swelling properties in a heavily compacted unsaturated clay*. Engineering Geology, No.50, pp. 31-48.

Holmes J. W. (1955). *Water sorption and swelling of clay blocks*. Journal of Soil Science, Vol. 6, No. 2, pp. 200-208.

Imbert C. & Zingarelli V. (1993). *Modelling and testing of hydration of clay backfilling materials*. Annular Progress Report, SCK-CEN, pp. 1-18.

Tang X., Graham J. & Fredlund D. G. (1997-a). *Effects of osmotic suction on strength and stiffness of compacted sand-bentonite*. Proceedings of the 50th Canadian Geotechnical Conference, Ottawa, Vol. 1, pp. 641-648.

Villar M. V., Martin P. L., Cuevas J. & Campos R. (1993). *Modelling and validation of the Thermo-Hydraulic-Mechanical and Geochemical behaviour of clay barrier*. Annular Progress Report, CEC Contract FI2W-CT91-0102 (DOEDO), pp. 1-13.

Wan A. W. L., Gray M. N. & Graham J. (1995). *On the relations of suction, moisture content, and soil structure in compacted clays*. Proceedings of the 1st International Conference on Unsaturated Soils, Paris, Vol. 1, pp. 215-222.

Yahia-Aïssa M. (1999). *Comportement hydromécanique d'une argile gonflante fortement compactée*. Thèse de doctorat de l'Ecole Nationale des Ponts et Chaussées (ENPC), Paris, 241p.

Clay Science for Engineering, Adachi & Fukue (eds) © 2001 Balkema, Rotterdam, ISBN 90 5809 175 9

Influence of stress and suction on volume changes and microtexture of a Ca-smectite

X. Guillot
Ecole Centrale Paris, Châtenay-Malabry & CRMD, Orléans, France

F. Bergaya & M. Al Mukhtar
CRMD, Orléans, France

J.-M. Fleureau
Ecole Centrale Paris, Châtenay-Malabry, France

ABSTRACT: The volume change behavior of clays depends on the hydro-mechanical stress state and physico-chemical properties. This paper deals with the effects of stresses and suctions on volume changes and on parameters that are obtained from X-ray diffraction, such as interlamellar distances. The material studied is a Ca-smectite French clay (FoCa). Experiments have been performed on this clay as powder or paste in initial state. Concerning the mechanical experiments, a large range of suctions and mechanical stresses is covered. Suctions from 0.2 MPa to 155 MPa are applied, requiring different techniques such as osmotic or salt solutions. These techniques are adapted to oedometric cells to perform hydro-mechanical tests allowing to study the behavior under low to high mechanical stresses, from a few kPa to 10 MPa. The history of the material seems to be of major importance on its microscopic and macroscopic behavior under the hydro-mechanical stresses.

1 INTRODUCTION

The future of radioactive wastes is a very important topic in France and their possible burying is studied. In this context, a specific clay, the FoCa clay, that presents interesting characteristics to be used as a manufactured barrier is more especially considered. In addition to this motivation (for its applied side), another more fundamental interest of this work is the connection between macroscopic (soil mechanics) and microscopic (physico-chemical) aspects.

The purpose of this paper is to show the influence of hydro-mechanical experiments on microscopic properties such as those given by X-Ray Diffraction (XRD). Firstly, some general results concerning oedometer tests (macroscopic experiments) in which the clay is initially prepared as a powder or as a paste, are presented. The way of preparing the samples, their history, is dominant in their macroscopic behavior. This point is illustrated all through this paper by means of different comparisons. In particular, the effect of suctions and mechanical stresses, on the interlamellar distance (or basal spacing d_{001}) of the clay layers is highlighted.

Which are the parameters influenced by hydro-mechanical stresses? Are they macroscopic or microscopic? Is the effect of suction predominant in comparison to that of mechanical stresses? These questions are raised in this paper.

2 MATERIAL

The material studied is the French FoCa swelling bentonite, where the main exchangeable cation is Ca. This is confirmed by the XRD results on Figure 1 ($d_{001} = 1.5$ nm). In France, it is taken as a reference in manufactured barriers, for the disposal of highly radioactive wastes. This material was identified as containing 80% of interstratified smectite/kaolinite with small amounts of free kaolinite (6%); the other non clay impurities are mainly quartz, goethite and calcite (Atabek et al. 1991). All these impurities are shown in the previous diffractogram.

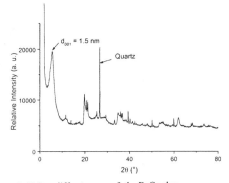

Figure 1. X-Ray diffractogram of the FoCa clay

3 EXPERIMENTAL

Before the oedometer tests,

- the FoCa clay powder is equilibrated to the selected suction with osmotic or salt solutions and this suction is maintained during the oedometer experiments.

- the FoCa clay paste is prepared with an initial water content of 150% and the chosen suction is applied once the paste is consolidated in the oedometer under a stress of 0.21 MPa.

The suctions used: 0.2 and 0.8 MPa are obtained after putting the clay inside a dialysis membrane in contact with osmotic solutions for at least one month. The higher suctions of 2.7 and 155 MPa are obtained by putting the clay in a desiccator containing saturated salt solutions, until equilibrium is reached.

The oedometer device used for the highest suctions and described by Qi et al. (1996) allows us to apply two maximum values of mechanical vertical stress: 0.21 and 10 MPa. With another oedometer device (Fleureau et al 1992) used for the two other lowest suctions the maximum mechanical stress is 0.21 MPa.

All the oedometer tests are conducted at room temperature.

Through an accurate study of diffractograms, XRD provides different kinds of information, on mineralogy (see Material section), on interlamellar distances (d_{001}) and some indications on the material's texture ((hk)bands, midheight width of the peaks). The XRD patterns are performed using a SIEMENS D5000 diffractometer with Cu Kα radiation (0.15406 nm) filtered by a Ni foil. X-Ray Diffraction analyses presented here have been carried out immediately after unloading of the samples submitted to oedometer tests. Each diffractogram is standardised with respect to the peak of quartz at 0.334 nm.

4 RESULTS

Experiments performed on the clay initially prepared as a powder or as a paste are first presented separately, then compared. In order to check if the hydromechanical experiments induce an anisotropy, XRD analyses are made on samples cut both perpendicular and parallel to the vertical stress.

It must be noticed that the way of cutting the sample and its positioning in the diffractometer for the analysis can affect the observed diffractograms.

4.1 *Hydro-mechanical results*

4.1.1 *For the powder*

4.1.1.1 *Effect of suction, under a low mechanical stress (σ_m) of 0.21 MPa*

At σ_m = 0.21 MPa, the results of specific void volume (V_V) changes with suction are presented in Table 1 and Figure 2. At 2.7 MPa this volume change presents a maximum that can be explained by an optimal water content which induces the highest density in the material. Under 2.7 MPa, the volume change decreases as the water content increases. Some interstitial pressures that oppose the mechanical stress may be induced to these water contents.

4.1.1.2 *Effect of stresses (0.21 and 10 MPa), under a suction of 2.7 MPa*

Table 2 shows that the decrease in specific void volume is more important for the powder (from 44 to 86.2%) when the stress increases from 0.21 to 10 MPa. Indeed, the mechanical stress strongly reduces the volume of the sample, from 0.71 to 0.27 cm³/g.

Table 1. Specific void volume evolution with suction

	Suction (MPa)			
	0.2	0.8	2.7	155
$V_{Vinitial}$ (cm³/g)	0.61	0.71	1.27	0.75
V_{Vfinal} (cm³/g)	0.50	0.42	0.71	0.60
ΔV_V (cm³/g)	0.11	0.29	0.56	0.15
$\Delta V_V/V_{Vini.}$ (%)	18	40.1	44	20

4.1.2 *For the paste*

For the material initially prepared as a paste, this reduction is less important, from 0.58 to 0.35 cm³/g (Table 2).

Table2. Specific void volume evolution with mechanical stress

	Mechanical stress (MPa)			
	0.21		10	
	powder	paste	powder	paste
$V_{Vinitial}$ (cm³/g)	1.27	1.5	1.96	1.63
V_{Vfinal} (cm³/g)	0.71	0.58	0.27	0.35
ΔV_V (cm³/g)	0.56	0.92	1.69	1.28
$\Delta V_V/V_{Vini.}$ (%)	44	61.3	86.2	78.5

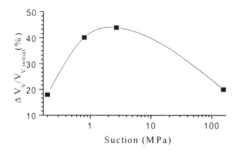

Figure 2. Specific void volume change with suction

4.1.3 *Comparison between powder and paste*

Surprisingly enough, under a high mechanical stress of 10 MPa, the powder becomes more compressible than the paste (Table 2). It can be explained by the fact that the initial density of the material prepared as a powder is very small due to the important porosity created by this suction of 2.7 MPa.

4.2 *XRD results*

4.2.1 *For an initial state as a powder*

4.2.1.1 *Effect of suction, under a low stress of 0.21 MPa*

One must notice, that the sample at the highest suction of 155 MPa appears dry and cannot be trimmed perpendicular or parallel to the stress without crumbling into powder. So it is the powder that is observed in this case.

Samples cut perpendicular to the vertical stress

XRD diffractograms of these samples at different suctions are presented in Figure 3. As suction increases (from 0.2 to 155 MPa), the basal spacing decreases (from 1.85 nm to 1.56 nm). In terms of water interlayers, this corresponds to about 3 and 2 pseudo layers. Although for suctions of 0.2, 0.8 and 155 MPa, the d_{001} peaks are symmetrical, this is not the case for the intermediate suction of 2.7 MPa where the d_{001} peak is broader and dissymmetrical. This could mean that there is an interstratification of 2-3 water layers. The diffractograms for 2θ ranging from $7°$ to $13°$ present irrational higher orders which are extremely difficult to define with accuracy.

By observing the sample perpendicular to the stress, we could expect that the stacking of the layers would result on the diffractograms in some well-defined d_{001} peaks. Indeed, the midheight width of the d_{001} peak gives some information about the size of the particles, using Debye Scherrer formula. Knowing the basal spacing, the number of clay layers per particle can be estimated. This number is higher when the peak is sharper.

The observation of the (hk) bands (Figure 4) which become more visible as suction increases even in the XRD perpendicular held sample, seems to indicate a certain isotropy in the material with increasing suctions. Generally, well-defined (hk) bands are rather expected to be observed in the parallel case. Moreover, the fact that, as suction increases, the intensity of the d_{001} peak decreases relatively to the main peak intensity of the quartz, confirms that the material tends towards an isotropic texture.

Samples cut parallel to the vertical stress

Generally, the tendencies observed on Figure 5 for the samples cut parallel to the vertical stress are similar to those on Figure 4. For these samples, it is normally expected to observe only the (hk) bands and in a lesser extent the d_{001} peaks. It means that the

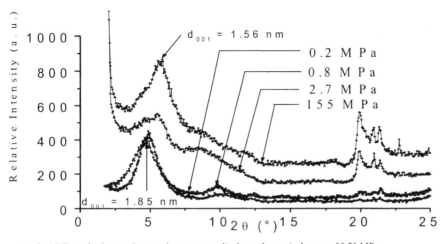

Figure 3. XRD results for powder samples cut perpendicular to the vertical stress of 0.21 MPa

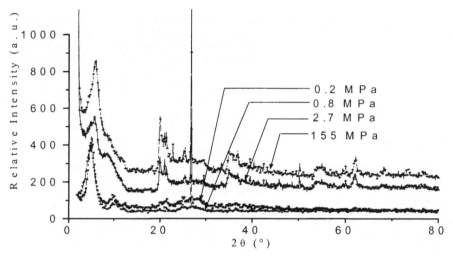

Figure 4. XRD results in wide range of 2θ for powder samples cut perpendicular to the vertical stress of 0.21 MPa

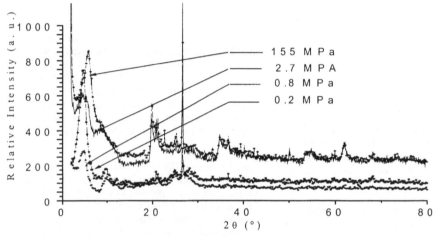

Figure 5. XRD results for powder samples cut parallel to the vertical stress of 0.21 MPa

existence of very well defined d_{001} peaks indicates the presence of layers initially parallel to the stress in the original sample. Moreover, the basal spacing observed in the previous case (1.85 nm) for the lowest suctions (0.2 and 0.8 MPa), is slightly higher in the parallel sample at about 1.91 nm and 1.94 nm respectively. This can be explained by the fact that the mechanical stress has more influence on the interlamellar distance of the initially perpendicular layers.

The (hk) bands, more visible for the highest suctions (Figure 5), confirm the possible isotropy previously evoked for the samples taken perpendicular.

We note that, whether in parallel or in perpendicular positions, the orientated layers exist in these two directions: - since the d_{001} peaks are more visible than the (hk) bands for the lowest suctions,

- the (hk) bands are visible in both cases for the highest suctions.

So, the organisation of the material tends to be isotropic at all suctions. Moreover, when the suction increases from 0.2 to 155 MPa, the number of pseudo water layers decreases as expected from about 3 to 2, going through an interstratification of 2-3 water layers at 2.7 MPa. This particular suction leads to a more complex behavior as shown by the evolution of other physico-chemical properties such as macroporosity (Guillot et al. 1999).

This is one of the reasons why the following data of this paper deal with experiments carried out at this suction.

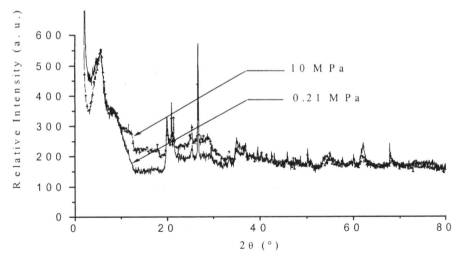

Figure 6. XRD results for powder samples cut perpendicular to the vertical stress for a suction of 2.7 MPa

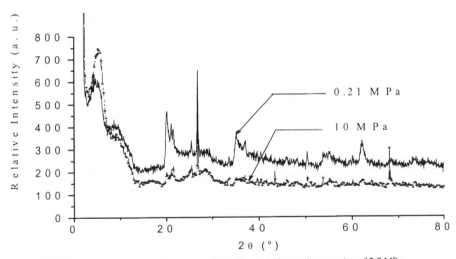

Figure 7. XRD results for powder samples cut parallel to the vertical stress for a suction of 2.7 MPa

4.2.1.2 *Effect of mechanical stresses (0.21 and 10 MPa), under a suction of 2.7 MPa*

Samples cut perpendicular to the vertical stress

The main effect of the highest stress (10 MPa), is the partial elimination of water corresponding to the third interlamellar pseudo water layer. The rather broad and dissymmetrical peak (Figure 6) of the interstratified sample with 2-3 water layers, compacted to 0.21 MPa, tends towards a d_{001} symmetrical peak at approximately 1.6 nm, corresponding to more homogeneous filling of 2 pseudo water layers. This effect is similar to that observed on the sample at the highest suction (155 MPa) under a low stress.

The second effect of the highest stress is probably to orientate the layers perpendicular to it, as shown by the (hk) bands being more distinguishable at the lowest stress. It seems to create a certain anisotropy in the material. However, the comparison between the perpendicular and the parallel diffractograms of the sample compacted to 10 MPa, reveals that there is no difference concerning (hk) bands. Therefore, the material tends towards a certain anisotropy but it still remains basically isotropic under this suction.

This idea is also supported by the fact that, in both cases (0.2 and 10 MPa), the intensity of the d_{001} peak, compared to the main one of quartz, is not so high. So, we can only say that there is a better organization in the material compacted to 10 MPa than in the one compacted to 0.21 MPa.

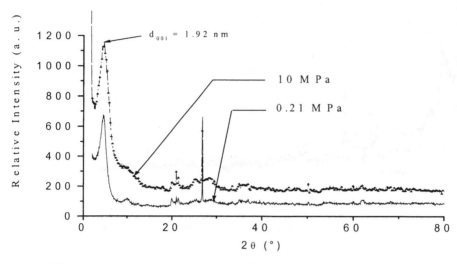

Figure 8. XRD results for paste samples cut perpendicular to the vertical stress for a suction of 2.7 MPa

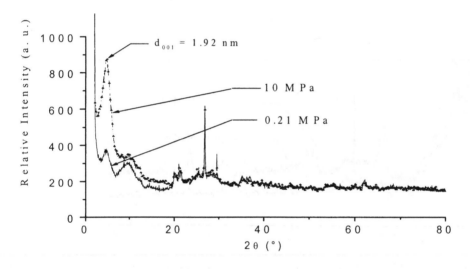

Figure 9. XRD results for paste samples cut parallel to the vertical stress for a suction of 2.7 MPa

Samples cut parallel to the stress

Once again, the same tendencies are observed for the samples cut parallel to the vertical stress (Figure 7) as those on Figure 6, for the samples cut perpendicular. As previously shown, the mechanical stress also has more influence on the interlamellar distance of the perpendicular layers where the basal spacing is 1.60 nm. Indeed, the basal spacing of parallel layers is slightly higher: about 1.72 nm.

The (hk) bands (Figure 7) remain more visible at the lowest mechanical stress of 0.21 MPa confirming a possible isotropy.

Similar results concerning the material initially prepared as a paste are discussed in the following.

4.2.2 For an initial state as a paste
Effect of stresses (0.21 and 10 MPa), under a suction of 2.7 MPa

Samples cut perpendicular and parallel to the stress (Figures 8 and 9)

For the experiments on the FoCa initially prepared as a paste, the highest stress (10 MPa) has no influence on the interlamellar distance (d_{001} = 1.92 nm even at 0.21 MPa). This indicates that the effect of the stress has very little influence on the clay texture. Moreover, it will be seen that there is no preferential orientation of the layers due to the stress. In contrast, the results on the powder show

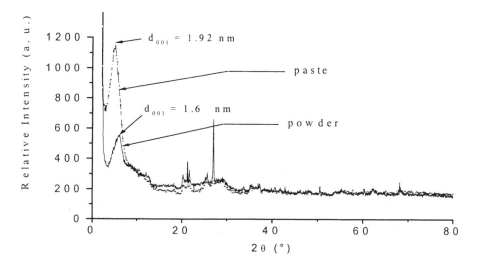

Figure 10. XRD results for samples cut perpendicular to the vertical stress of 10 MPa and for a suction of 2.7 MPa

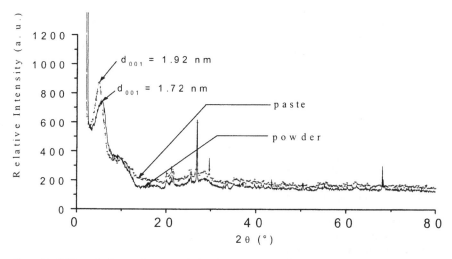

Figure 11. XRD results for samples cut parallel to the vertical stress of 10 MPa and for a suction of 2.7 MPa

the same interlamellar distance in the perpendicular and parallel situations.

4.2.3 Comparison between powder and paste at a mechanical stress of 10 MPa, under a suction of 2.7 MPa

Samples cut perpendicular and parallel to the stress (Figures 10 and 11)

The main difference concerns the basal spacing which is 1.6 nm for the powder and 1.92 nm for the paste in the perpendicular case (1.72 and 1.92 nm respectively in the parallel case). The initial consolidation of the clay paste seems to prevent the evacuation of the third pseudo water layer, even un-

der a suction of 2.7 MPa. The history of the material seems to play a major part in its behavior, and it is always important to specify the way the experiments are done.

5 CONCLUSION

- In all cases, for samples starting as powder or paste and with different applied hydro-mechanical stresses, the number of layers per particle is small, between 2 and 3. So, the texture of the material at a mesoscopic scale seems to consist of an aggregation of small particles orientated in a rather isotropic manner. In each particle, the layers are separated by several pseudo water layers from about 3 to 2, in this range of applied stresses and suctions.

- Mechanical stress plays a major part in the macroscopic volume changes and at the basal spacing microscopic level in a less extent. On the other hand, the suction acts on the interlamellar distance. This is more clearly displayed for the powder than for the paste.
  - For the material initially prepared as a powder, d_{001} decreases as the suction increases with a more complicated behaviour under a suction of 2.7 MPa. Besides, at this suction and at the microscopic level, the effect of a high mechanical stress on the layers that are orientated perpendicular to this stress is only to evacuate the third pseudo water layer. But from a volumic point of view, it tends to reduce the macroporosity. So, the stress affects first this porosity and influences secondarily the distance between layers.
  - For the material initially prepared as a paste, the effect of a high mechanical stress is less important at both scales, micro and macro (no reduction of interlamellar distance; limited volume changes). This means that in this field, suction is preponderant in comparison to the mechanical stress. Obviously, the order in which they are applied is very important.

- At the aggregate scale, the organization tends to be isotropic and the effects of hydromechanical stresses during the experiments are of less importance than the way in which the material has been prepared. These conclusions are deduced from the fact that the differences between samples observed perpendicular or parallel to the stress are not so obvious. But according to our results, suctions and mechanical stresses both act on micro and macro-texture. An overall view of changes is very difficult to assess because this combination of actions affects different scales of the material. This paper which is mainly focused on the particle scale shows the necessity to turn towards a higher scale which is that of the aggregates.

- This work is a first step (most information given by XRD) in trying to understand the textural rearrangement of the FoCa clay when submitted to various hydro-mechanical stresses. A multiscale study needs to do other microscopic analyses, using different techniques such as mercury intrusion, thermogravimetric analysis (TGA), gas adsorption (BET) or electronic microscopy, that should improve our comprehension of the phenomena that occur. Moreover, the range of mechanical stresses applied is being enlarged to a value of 60 MPa in the case of manufactured barriers.

6 REFERENCES

Al Mukhtar, M., Qi, Y., Alcover, J.-F. & Bergaya, F., 1999. Oedometric and water-retention behavior of highly compacted unsaturated smectites. *Can. Geotech. J.* 36: 675-684.

Atabek, R. B., Félix, B., Robinet, J.C. & Lahlou, R. 1991. Rheological behaviour of saturated expansive clay materials. Workshop on Stress Partitioning in Engineered Clay Barriers, Duke University, Durham, NC.

Fleureau, J.-M., Soemitro, R. & Taibi S., 1992. Behavior of an expansive clay related to suction. Proceedings of the 7th International Conference in expansive soils, Dallas, Vol. 1, pp. 173-178.

Guillot, X., Al. Mukhtar, M., Fleureau, J.-M. & Bergaya F., 1999. Poster in Euroclay, sept. 5-9, Krakow, Poland.

Qi, Y, 1996. Comportement hydromécanique des argiles: couplage des propriétés micro-macroscopiques de la laponite et de l'hectorite. Thèse de doctorat, Université d'Orléans.

Clay Science for Engineering, Adachi & Fukue (eds) © 2001 Balkema, Rotterdam, ISBN 90 5809 175 9

Some considerations of clay suction by using thermodynamics theory of water in soil

H. Suzuki

Department of Civil Engineering, University of Tokushima, Japan

ABSTRACT: Soils in general of three phases consist of soil solid, water and air. The water in soil becomes very complicated phenomena, that are related to the repetitions of freezing, liquefying and evaporating. Therefore, it is necessary that unsaturated ground behavior is described by the three phases theory of soil incorporated the changes of the phases and soil suction, a potential energy of water in soil, is considered by using thermodynamics of water in soil. However, in soil mechanics evaluating mainly mechanical phenomena, the definition of soil suction is expressed the difference between pore air pressure u_a and pore water pressure u_w by a few famous results (Bishop & Donald, 1961) triaxial compression tests without such as theoretical approach. It is recognized as a basis of studies of unsaturated soils. In this study, we reconsider the definition of soil suction through the theories of thermodynamics and physical chemistry of water in soil and point out problems that we should solve.

1 INTRODUCTION

Study of thermodynamics of water in soil has been performed since the 1900's in the field of soil science. Schofield (1935) proposed the concept of pF, namely, the logarithm of the specific Gibbs free energy and established the energy concept of water in soil. Edlefsen & Anderson (1943) emphasized the use of the chemical potential in his book entitled "thermodynamics of soil moisture". After that, some researches related to soil moisture movement were carried out using thermodynamics of water in soil. The International Society of Soil Science defined the total potential of water in 1963. Iwata (1972a, 1972b) proposed a more general energy concept of water in soil, which includes the definition of the total potential. In geotechnical engineering, Aitchison (1965) organized the international symposium on the definition of soil suction considering thermodynamics, then soil suction was expressed as a function of vapor pressure. This concept of soil suction was introduced into geotechnical engineering researches, for example, Fredlund and Rahardjo (1988).

In this study, we examine the specialty of the definition of suction by the thermodynamics developed and DLVO theory proposed in colloid science field.

2. CAPILLARY FORCES INVOLVED IN WATER RETENTION

2.1 *The definition of soil suction due to capillary raise in a tube*

Firstly, we explain the basic definition of suction that is often used in a textbook of soil mechanics. Figure 1 shows the meniscus in the cylindrical tube.

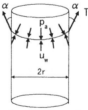

Fig. 1 Schematic diagram for soil suction due to capillary raise in a tube

The vertical balance equation is as shown the following:

$$u_w\left(\pi r^2\right) = p_a\left(\pi r^2\right) - T\cos\alpha(2\pi r)$$

$$u_w = p_a - \frac{2T\cos\alpha}{r} \tag{1}$$

$$s = u_a - u_w = \frac{2T\cos\alpha}{r} = \rho g h_c \qquad (2)$$

where, T is surface tension per unit length and p_a and u_w are atmospheric pressure and pore water pressure, respectively. Substituting u_a, pore air pressure, into p_a in Equation 1, we can obtain an equation expressing suction by capillary:
where, ρ is the density of water, h_c is the height of capillary. This is the basis of matrix suction for soil mechanics.

2.2 The definition of soil suction due to lens water between spheres

We examine soil suction due to capillary mentioned above, but it is only a phenomenon in a tube. To obtain the more actual definition, it is necessary to define suction due to retaining water between spheres. Figure 2 shows a schematic diagram due to

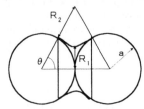

Fig.2 Schematic diagram for soil suction due to capillary raise in a tube

lens water between spheres. The pore water pressure can be calculated from the following Laplace Equation.

$$u_w = p_a + \sigma\left(\frac{1}{R_1} - \frac{1}{R_2}\right) \qquad (3)$$

where, σ is the surface tension of the air-water interface and R_1 and R_2 are the main radii of curvature of the retaining lens water. They can be also calculated by the following equation obtained form the geometrical relations.

$$R_1 = a\tan\vartheta - R_2 \qquad (4)$$

$$R_2 = a(\sec\vartheta - 1) \qquad (5)$$

Deviating similarly Equation 2, soil suction becomes by replacing p_a to u_a.

$$s = u_a - u_w = -\sigma\left(\frac{1}{R_1} - \frac{1}{R_2}\right) = \rho g h_c \qquad (6)$$

In the case of two spheres model, the definition of suction can be expressed by u_a-u_w. Namely, the de-

finition using present studies is consistency with u_a-u_w, if we only consider the influence of capillary.

3. GENERAL FORMULA OF CHEMICAL POTENTIAL

Generally speaking, thermodynamics is applied to wide fields. To simplify the next discussion, we consider using the thermodynamic developed by soil physics. In the study of thermodynamics of water in soil, chemical potential μ_w, which is Gibbs' free energy of unit water, is used as an index describing equilibrium condition. There were some problems

$$d\mu_w = -\underline{S}_w dT + \underline{v}_w dp + \frac{v_w}{4\pi}\left(\frac{1}{\varepsilon} - 1\right)dD$$

$$+ \left(\frac{\partial\psi}{\partial z}\right)dz + gdh + \sum_{j=1}^{k-1}\left(\frac{\partial\mu_w}{\partial C_j}\right)dC_j$$

$$+ \frac{\partial}{\partial r}\left(\frac{2\sigma}{r}\underline{v}_w\right)dr \qquad (7)$$

that the describing of μ_w and its application to unsaturated soils, but we adopt the equation developed by Iwata et al (1995).

where, $\underline{S}_w$ is the partial entropy of water and T is the temperature. $\underline{v}_w$ is the partial volume of water, ε is the partial dielectric constant and D is the electric displacement. ψ is the potential energy of intermolecular interaction between a clay particle and water molecules of unit mass. C_j is the concentration of jth component and k is the number of components, σ is the surface tension of water, p is the total sum of external and internal forces, p_{ex} and p_{in}, respectively. r is radius of curvature of air-water interface, in this case we assume that retaining water simplify as a sphere. The factors of effects from the right side of a equation 7 was shown as follows: 1) temperature, 2) pressure, 3) electric displacement, 4) the distance from the surface of particle 5) gravity field 6) the amount of solution, 7) capillary.

4. GENERAL CONCEPT OF SOIL SUCTION

When we explain the concept of soil suction, a diagram showing forces between particles by retaining water is often used as shown Figure 2. However, it is the part of Equation 7, namely, only considers capillary. Therefore, to discuss more generally, we need the diagram including air-water interaction. Referring to the diagram of soil physics field, we can obtain Figure 3.

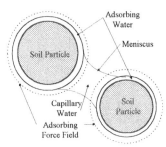

Fig.3 Concept of soil suction

The surface of particles consists of three kinds of water: capillary water existing by meniscus (F_c), electric force acting on surface (F_R), van der Waals Force (F_A). The dashed lines in this figure show the adsorbing forces. F_c acts on the two particles and they pull each other. On the other hand, electric double layer is formed by a charge of particle surface, the two particles have a repulsion (F_R) in case of low concentration of ion. Namely, to discuss generally soil suction, it is necessary to make clear the relation of F_A, F_C and F_R.

5. ESTIMATION OF FORCES ACTING ON PARTICLES

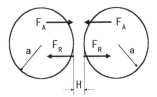

Fig.4 Interface of forces acting on two particles

As we already explained forces due to capillary in the 7th term of Equation 7 by Equations 2 and 6, we consider the 3rd and 4th terms. Forces acting on particles are static electricity interaction forces F_R (the 3rd term) based on overlapping of electric double layer expanding the electric charge of surface and van der Waals F_A (the 4th term). About the former, on the basis of DLVO theory(Derjaguim and Landan, 1941, Verwey and Overbebeek, 1948, Shaw, D. J. 1970, Hiemenz, P. C. 1977) established colloid particles, the repulsion potential by overlapping of electric double layer can be estimated, this theory is gotten for the interaction energy between two plates, but forces acting on between two particles can be obtained by using Derjaguin's approximate equation. This equation can stands up for any kind of force, if H<<a can be satisfied. To simplify the derive of equations, we consider the two spheres which have same radius a as shown Figure 4.

$$F_R(H) = \frac{64\pi a n k T}{\kappa} \gamma^2 \exp(-\kappa H) \quad (N/m^2) \qquad (8)$$

where, n is number concentration of ion in diffuse layer and k is Boltzman constant. T is absolute temperature and κ is given by the following equation.

$$\kappa = \sqrt{\frac{2nz^2 e}{\varepsilon_r \varepsilon_0 k T}} \qquad (9)$$

where, z is and e is elementary electric charge. ε_r is relative dielectric constant of medium and ε_0 is vacuum dielectric constant. $1/\kappa$ is the thickness of diffuse electric double layer. κ^2 is the parameter of Debye-Huckel. In the solution at 25℃ it is If electrolytic concentration c=0.001 mol dm^{-3}, $1/\kappa$

$$\kappa = 3.3 \times 10^9 z\sqrt{c} \quad [m^-] \qquad (10)$$

=10 nm. If c=0.1 mol dm^{-3}, $1/\kappa$ =1 nm. As we have shown above, electric double layer becomes thin depending on c. In Equation 8, γ is

$$\gamma = \tanh\left(\frac{ze\psi_0}{4kT}\right) \qquad (11)$$

$$\psi_0 = \frac{RT}{zF}\ln\frac{c}{c_0}$$

where, ψ_0 is electric potential and R is gas constant. F is Farady constant and c is bulk concentration of ion determining electric potential. c_0 expresses the value of c when ψ_0 is equal to zero. In the solution at 25℃, e ψ/κ T=1 when ψ =25.6mV. Therefore, interaction energy between spheres E_R(H) is

$$E_R(H) = -\int_\infty^H F_R(H)dH$$
$$= \frac{64\pi a n k T}{\kappa^2}\gamma^2 \exp(-\kappa H) \qquad (12)$$

The units of these equations are as follow: a is μm, c is mol dm^{-3}, κ is m^{-3}. The unit of E_R(H) is kT. In the solution at 25℃, we can get the following approximate equation.

$$E_R(H) = \frac{1.21 a c \times 10^{23}}{\kappa^2}\gamma^2 \exp(-\kappa H) \qquad (13)$$

Van der Waals Force (F_A) and its energy E_A(H) are given by the following equation.

$$F_A(H) = -\frac{aA}{12H^2} \qquad (14)$$

$$E_A(H) = -\frac{aA}{12H} \qquad (15)$$

where, A is Hammaker constant which is peculiar constant of material. Therefore, total interaction energy between spheres $E_T(H)$ is

$$E_T(H) = E_R(H) + E_A(H) \qquad (16)$$

6. INTERACTION OF FORCES BETWEEN TWO SPHERE

If we use the equations 8 – 16, interaction forces between spheres shown in Figure 4 can be calculated. However, it is difficult to estimate the values of parameters in these equations. Therefore, we try to obtain relations among F_A, F_B and their resultant forces F_T neglecting the absolute quantity of the calculated. Taking attention to Equation 8 and 14, the former expresses that F_R is in proportion to radius a and reduce exponentially for H and the latter expresses that F_A is also proportion to radius a and inverse proportion to a square of H. To simplify to

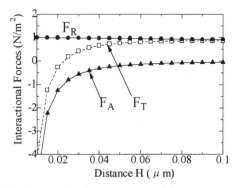

Fig.5 Interaction between two particles

calculate them we assume that these equations are only functions of a and H and the other coefficients are unit amount. Of cause, forces between particles strongly depend on concentration of solution, we adopt unit amount to have no effects. Figure 5 shows the changes of F_A, F_R and F_T according to H assuming a=1 μ m. As shown this figure, in the surface of particle a rather large repulsion and attraction occur, resultant force changes from attraction to repulsion with increase of H. Generally, there are two local minimums, we can not obtain in this calculations. We want to compare with suction expressed by Equation 6 and it is a problem that we must examine. Because the absolute of calculated values have no reliability.

7. CONSOLIDATIONS BY THERMODYNAMICS OF WATER IN SOIL

To correspond to matrix suction we consider the conditions under same temperature and pure water, which neglect the 1st and 6th terms in Equation 7.

$$d\mu_w = \underline{v}_w dp + \frac{\underline{v}_w}{4\pi}\left(\frac{1}{\varepsilon} - 1\right)dD + \left(\frac{\partial \psi}{\partial z}\right)dz$$
$$+ gdh + \frac{\partial}{\partial r}\left(\frac{2\sigma}{r}\underline{v}_w\right) \qquad (17)$$

Equation 17 is a basis of the following considerations. Firstly, we consider the effects of only pressure. In this case pore pressure u becomes equal to $\Delta \mu_w$. Since the gas is governed Boyle's low pv=RT (p: gas pressure) we can obtain the following equation.

$$\Delta\mu_w = RT\log_e \frac{p}{p_0} \qquad (18)$$

where, p is vapor pressure of water in soil and p_0 is vapor pressure of free water. Its ratio (p/p_0) is a relative humility. This Equation expresses a relation between pore pressure and suction. Secondly, we can obtain the following equation considering the effects of surface tension.

$$\log_e \frac{p}{p_0} = \frac{v}{RT}\left[\frac{2\sigma}{r} + (p - p_0)\right] \qquad (19)$$

where, σ is surface tension, r is a radii of curvature of air-water interface. In many cases p can be approximated to p_0 and if r can be equal to capillary radii:

$$\underline{u} = RT\log_e \frac{p}{p_0} = \frac{2\sigma}{r} = \rho g h_c \qquad (20)$$

where, u is the difference of pressure at air-water interface and h is matrix suction that is equivalent to height of water.

8. CONCLUSIONS

1) We insist that soil suction needs the considering not only meniscus but also forces acting on between particles inclusive to static electricity power by overlapping electric double layers and van der Waals Forces.

2) DLVO theory, which is developed by colloid science, applied to obtain forces acting on particles and calculated them. Consequently, it made clear that attractive force and repulsion act on particles at the surface and resultant force changed complicatedly.

3) We verify that chemical potential is approximately consistency with matrix suction under the conditions that can be neglected changes of temperature and solution.

REFERENCE

Aitchison, D. G. 1965., Engineering concepts of moisture equilibria and moisture changes in soils. Statement of the Review Panel, Ed., published in *Moisture Changes in Soils Beneath Covered* pp.7-21. *Areas,* A Symp.-in-print Australia, Butterworths,

Bishop, A. W. and Donald, I. B., 1961, The experimental study of partly saturated soil in the triaxial apparatus, Proc. of the 5th ICSMFE, Vol,1, pp.13-22.

Bocking, K. A. and Fredlund, D. G., 1980. Limitation of the Axis Translation Technique. in *Proc. 4th Int. Conf. Expansive Soils,* Denver, CO, pp117-135.

Dejaguin, B.V. and Landau, L. 1941, *Acta Physicochim. URSS* 14,pp.633-662.

Dineen, K. and Burland, J. B. 1995., A new approach to osmotically controlled oedometer testing, *1st Int. Conf. of Unsaturated* Soils, pp.459-465.

Edlefsen, N. E. and Anderson, B. C., 1943. Thermodynamics of soil moisture. *Hilgardia* 15, pp.31-298.

Fredlund, D. G. and Raharjo, H., 1988. State-of-development in the measurement of soil suction, in *Proc. Int. Conf. Eng. Problems on Regional Soils,* Beijing, China, pp.582-588.

Fredlund, D. G. and Rahardjo, H., 1995. *Soil mechanics for unsaturated soil,* Jhon Wiley & Sons INC, pp.64-106.

Hiemenz, P.C. 1977, Principles Colloid and SurfaceChemistry, *Dekker New York*

Hilf, J. W., 1956. An investigation of pore-water pressure in compacted cohesive soils, Ph. D. dissertation. Tech. Memo. No.654, U. S. *Dep. of the Interior, Bureau of Reclamation, Design and Construction* Div, 654pp.

International Society of Soil Science., 1963. *Soil Physics Terminology, Bull.* 23, p.7.

Iwata, S. 1972a., Thermodynamics of soil water. Vol.1, *Soil Science,* No.113, pp.162-166.

Iwata, S. 1972b., On the definition of soil water potential as proposed by the ISSS in 1963, *Soil Science,* No.114, pp.88-92.

Iwata, S., Tabuchi, T. and Warkentin, B. P. 1995., *Soil-water interactions.* Marcel Dekker, Second Edition, pp.1-38.

Mukhtar, M. A. L., 1995 Coupling phenomena of hydraulic and mechanical stress in low porosity clays, *Int. Symp. of Compression and Consolidation of Clayey Soils,* Vol.1, pp.15-20.

Schofield, R. K., 1935. The pF of the water in soil. *Trans. 3rd Int. Cong. Soil Science* Vol.2, pp.37-48.

Shaw, D. J., 1970. Introduction to Colloid and Surface Chemistry, Butterworth, London.

Verwey, E. J. W. and Overbeek, J. Th. G., 1948. Theory of stability of lyophobic colloid, Elsevier, Amsterdam

Clay Science for Engineering, Adachi & Fukue (eds) © 2001 Balkema, Rotterdam, ISBN 90 5809 175 9

Sample storage time effects on suction and in-situ undrained strength

T. Shogaki, F. Nakamura & N. Sakakibara
Department of Civil Engineering, National Defense Academy, Yokosuka, Japan

ABSTRACT: The sample storage time effects on suction and *in-situ* unconfined compressive strength ($q_{u(l)}$) are examined for natural clay deposits having unconfined compressive strength (q_u) from 50 kPa to 600 kPa. If there is no change in water content, the q_u and suction values of samples do not change in a four month period under atmospheric pressure, which is zero confined pressure. Therefore, there is no effect on the estimation of the $q_{u(l)}$. This means that the unconfined compression test has small sample storage time effect on the undrained shear strength properties and gives a stabilizing undrained shear strength.

1 INTRODUCTION

Unconfined compression test (UCT) is an effective procedure under undrained conditions. It has been considered in general that the undrained shear strength properties obtained from this test are very sensitive to the stress release and mechanical disturbance of samples. Shogaki and Maruyama (1998) proposed a procedure for estimating *in-situ* undrained shear strength using disturbed samples within thin-walled samplers from *in-situ* sampling. The validity of this method was examined using the method for estimating the *in-situ* preconsolidation pressure (Shogaki 1996) and the *in-situ* rate of strength increase measured by a precision triaxial apparatus (Shogaki et al. 1999a).

In this paper, the sample storage time effects on suction and *in-situ* unconfined compressive strength ($q_{u(l)}$) are examined for natural clay deposits having unconfined compressive strengths (q_u) from 50 kPa to 600 kPa. The samples were stored at 17 °C to avoid volume changes caused by temperature changes.

2 SOIL SAMPLES AND TEST PROCEDURES

The undisturbed soil samples used in this study were obtained from Holocene and Pleistocene marine clays located offshore at Kumamoto and the Holocene plain in Nagoya, Japan. Field sampling was performed with a fixed piston sampler having an inner diameter of 75 mm (JGS 1221-1995) for Kumamoto clay and a rotary double-tube sampler (JGS 1222-1995) for Nagoya clays.

The index and mechanical properties of these soils are shown in Table 1. The plasticity index (I_p), q_u and the effective overburden pressure (σ'_{vo}) are in the range of 23 to 46, 83 kPa to 664 kPa and 87 kPa to 438 kPa, respectively.

Ten small specimens (15 mm in diameter and 35 mm in height, called S specimen hereafter) from the 75 mm diameter sample were taken as shown in Figure 1. It had been previously confirmed that the stress strain curves of specimens located too close to the sampling tube wall are similar to those of specimens located in the central site of the plane of the sampling tube for ten S specimens taken from samples 75 mm diameter and 45 mm height (Shogaki et al. 1995). The unconfined compression tests for the three kinds of sample storage time were performed as follows:

Table 1. Indexed values and mechanical properties of Kumamoto and Nagoya clays.

Soil	Kumamoto, T-9	Nagoya, P-2	Nagoya, P-6
Depth, z (m)	20	50.4	62.6
w_n (%)	89	56	36
σ'_{vo} (kPa)	87	355	438
Sand (%)	2	1	25
Silt (%)	45	39	45
Clay (%)	53	60	30
w_L (%)	88	79	47
I_p	46	39	23
q_u (kPa)	83	623	664
σ'_p (kPa)	136	810	995
OCR	1.6	2.3	2.3

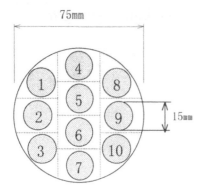

Figure 1. Location of specimens for the plane of the sampling tube.

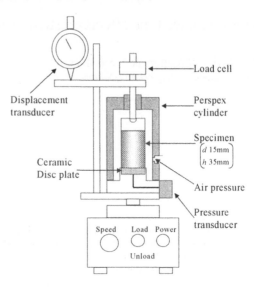

Figure 2. Portable unconfined compression apparatus.

1) Test I : 10 days after sampling for all clays.
2) Test II : 52 days after sampling for Kumamoto, 67
    days after sampling for Nagoya P2 and
    28 days after sampling for Nagoya P6.
3) Test III : 126 days after sampling for Kumamoto
    and 134 days after sampling for Nagoya
    P2 clays.

These samples were stored at 17 ℃ to avoid the volume changes caused by temperature changes. A simple procedure for the wrapping of samples to avoid water content was used. This involved sealing the sample in a plastic pouch after double wrapping with plastic sealing material.

UCT was performed according to the Japanese Industrial Standard for unconfined compression test of soils (JIS A 1216-1993) using S specimen. In an engineering sense, there is no difference in shear strength characteristics between the S specimen and the ordinary specimen (35 mm in diameter and 80 mm in height, called O specimen hereafter), which are examined for soils of $I_p \fallingdotseq 17\sim150$ and $q_u \fallingdotseq 20\sim$ 1000 kPa (Shogaki et al. 1999b).

The portable unconfined compression apparatus as shown in Figure 2 (Shogaki and Maruyama 1998) for measuring the suction (s) and the q_u value is used. This equipment has a height of about 20 cm and a weight of about 8 kg. The load is applied by the linear head and is transmitted though an AC/DC powered motor. Therefore, since the equipment is portable, it is practical for field use. The procedures for preparing the remolded soil samples were as follows:

1) The soil was put into a plastic pouch after the un-
    disturbed specimen was removed. The amount of
    soil used for remolding almost equals the total
    amount of five S specimens.
2) The mouth of the pouch was closed after the air
    was removed by suction.
3) The soil was well kneaded by hand from outside
    the pouch. The q_u value of the Kumamoto and

Nagoya clays did not change after a remolding time longer than 5 min.

The unconfined compression test was performed on S specimens at a strain rate of 1 %/min, after the suction of the specimen was measured using a ceramic disc plate. The air entry value of a ceramic disc is about 200 kPa. The secant modulus (E_{50}) is given by ($q_u/2$)/ε_{50}, in which ε_{50} is the strain at the value of $q_u/2$.

3 SAMPLE STORAGE TIME EFFECT ON
   UNDRAINED SHEAR STRENGTH
   PROPERTIES

The results of the UCT for the distance of sampling points from the cutting edge of the tube (D_s) for Kumamoto clay are shown in Figure 3. Test I and II were done on the specimens at D_s=750 mm and Test III was done on the specimens at D_s=830 mm. The water contents of the specimens D_s=830 mm are about 8 % smaller than those of D_s=750 mm. To examine the effect of sample storage time on the suction, Figure 4 shows the results of suction measurment for three different storage times. The procedure for measuring the suction is the same as that reported in other literature (Shogaki 1995). It is important that the specimen suction can be measured quickly, not only for the shear test using measured specimen suction, but also for the efficiency of the test. The suction measurements of specimens taken in advance were confirmed as follows:

1) The s_o value can be measured quickest when the
    piezometer point indicates the same value as the
    specimen suction.

2) However, the possibility exists that suction greater than that of the piezometer point can not be accurately measured if all air is not removed from the pressure transducer pipe.
3) If the air in the pressure transducer pipe is removed and the piezometer sensitivity is height, the measured s_o values are independent when the specimen is put on the ceramic disc plate for the piezometer values.
4) However, the time in which suction becomes constant is less when the suction goes from a larger value to specimen suction.

In Figure 4, zero time is indicated when the water on the top surface of the ceramic disc plate is wiped with paper. The specimen was put on the ceramic disc plate when the piezometer point indicated $s \fallingdotseq 70$ kPa, which was a larger suction value than expected for the specimen. The suction in the ceramic disc plate reached the expected suction of the specimen a few seconds after the water on the top surface of the ceramic disc plate was wiped with paper. The time in which the suction became constant, namely the suction (s_o) of specimen, was a few minutes for each specimen and this time was independent of sample storage time. The suction of all specimens was in the range of 7 kPa to 14 kPa, independent of sample storage time.

Figure 5 shows the relationship between the pore water pressure (u) measured at the base of the specimen, the stress (σ) and the axial strain (ε_a)for the same specimens as shown in Figure 4. The suction under shear is represented as the pore water pressure in Figure 5 since the suction under shear becomes 0 with a small s_o value. Therefore, the u values at the $\varepsilon_a = 0$ % are s_o values. The s_o, q_u and E_{50} values of each specimen are also given in the inset of Figure 5. In Tests II and III, the specimens having greater s_o values have a tendency toward greater q_u and E_{50} values. However, these tendencies

do not exist for Test I . Therefore the relationships between stress, pore water pressure and axial strain are almost the same independent of sample storage time and in spite of a small change in the water content in test III. The results of the UCT for Nagoya P2 clay are also shown in Figure 6. Test I was done on the specimens at D_s=100 mm and 375 mm, Test III was done on the specimen at D_s=100 mm and Test III was done on the specimen at D_s=215 mm. The water content of these specimens is almost the same and is about 52 %.

Figure 7 shows the results of suction measurment for three different storage times for Nagoya P2 clay. In Figure 7, zero time is also indicated when the water on the top surface of the ceramic disc plate is wiped with paper. The Nagoya clays have a suction greater than atmospheric pressure, since the σ'_{vo} values are in the range of 355 kPa to 438 kPa. Therefore, a back air pressure of 98 kPa to 180 kPa, to balance the suction within the specimen, was applied to the specimen as shown in Figures 7 and 9. It can be judged that the effect of sample storage time

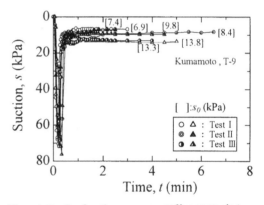

Figure 4. Results of suction measurement (Kumamoto clay).

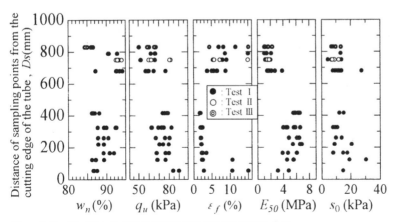

Figure 3. Results of the unconfined compression test (Kumamoto clay).

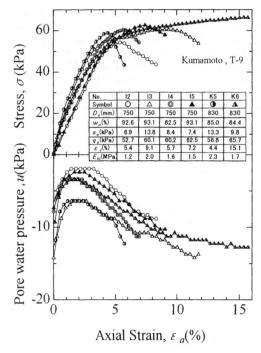

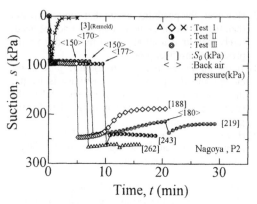

Figure 7. Results of suction measurement (Nagoya P2 clay).

Figure 5. Relationships between stress, pore water pressure and strain (Kumamoto clay).

on the relationship between suction, time and the s_o value are small in spite of the higher σ'_{vo} and the lower I_p values. The relationships between the u, σ and ε_a values for the same specimens as shown in Figure 7 are shown in Figure 8. The w_n, s_o, q_u and E_{50} values of each specimen are also given in the inset of Figure 8. It can be judged that the effect of sample storage time on the water content can be ignored. When the s_o values for E1 and remolded specimens of Test I become small, the q_u values become small

and the ε_f values become large. This may be caused by sample disturbance during sampling and preparing the specimen.

Figure 9 shows the results of measuring suction for Nagoya P6 clay. The time in which suction reaches a constant value is in the range of 8 min to 17 min after back air pressure was applied. The relationships between the u, σ and ε_a values for the same specimens as shown in Figure 9 are shown in Figure 10. The w_n, s_o, q_u and E_{50} values of each specimen are also given in the inset of Figure 10. The w_n, s_o, q_u and E_{50} values are almost the same, independent of Test I and II.

4 SAMPLE STORAGE TIME EFFECT ON IN-SITU UNDRAINED SHEAR STRENGTH

Figure 8 shows the relationships between stress, pore water pressure and axial strain (Nagoya P2 clay). An estimation method for *in-situ* undrained shear strength using disturbed samples within thin-

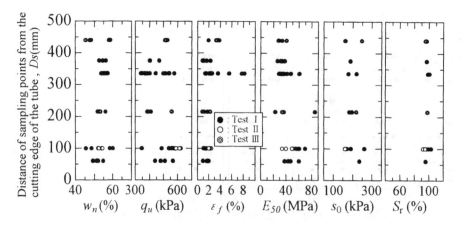

Figure 6. Results of the unconfined compression test (Nagoya P2 clay).

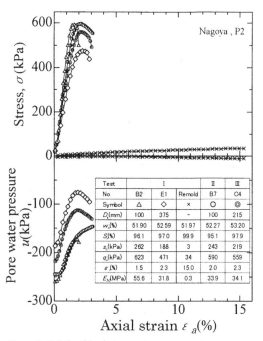

Figure 8. Relationships between stress, pore water pressure and strain (Nagoya P2 clay).

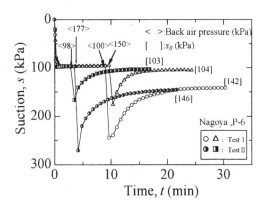

Figure 9. Results of suction measurement (Nagoya P6 clay).

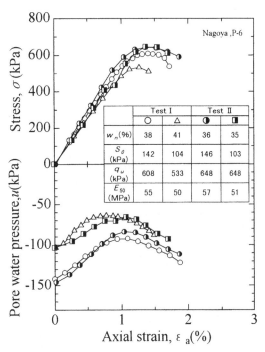

Figure 10. Relationships between stress, pore water pressure and strain (Nagoya P6 clay).

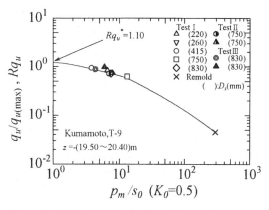

Figure 11. Relationship between Rq_u and p_m/s_o (Kumamoto clay).

walled samplers from *in-situ* sampling has been proposed by Shogaki and Maruyama (1998). To obtain the correcting factor ($Rq_u{}^*$) for estimating in-situ undrained shear strength, the relationships between Rq_u and p_m/s_o for Kumamoto, Nagoya P2 and P6 clays are shown in Figures 11, 12 and 13 respectively, where $Rq_u{}^*$ is the ratio of other samples to that of q_u of the high quality sample within the tube ($q_{u(max)}$) and p_m is the mean consolidation pressure (Shogaki and Maruyama, 1998).

The coefficient of earth pressure at rest (Ko) for Kumamoto clay was assumed to be 0.5. The Ko value for Nagoya clay was assumed to be 0.7 from the

equation ($Ko_{(OC)}/Ko_{(NC)}=OCR^m$) proposed by Ladd et al. (1977). The curves in Figures 11, 12 and 13 are regression curves for plots of Test I. The plots of Test II and III are located close to these regression curves. This means that the storage time does not affect the estimation of the *in-situ* undrained shear strength in Test II and III.

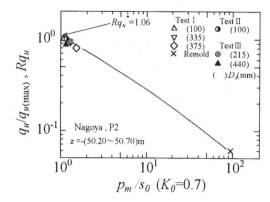

Figure 12. Relationship between Rq_u and p_m/s_o (Nagoya P2 clay).

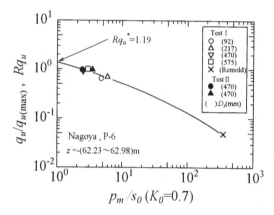

Figure 13. Relationship between Rq_u and p_m/s_o (Nagoya P6 clay).

5 CONCLUSIONS

The sample storage time effects on suction and *in-situ* unconfined compressive strength ($q_{u(l)}$) are examined for natural clay deposits having unconfined compressive strength (q_u) from 50 kPa to 600 kPa.

The main conclusions obtained in this study are summarized as follows:

1) If there is no change in water content, the q_u and suction (s_o) values of samples do not change within a four month period under atmospheric pressure, which is zero confined pressure.
2) Therefore, there is no effect on the estimation of the $q_{u(l)}$. This means that the unconfined compression test has small sample storage time effect on the undrained shear strength properties and gives a stabilizing undrained shear strength.

ACKNOWLEDGEMENT

The authors gratefully acknowledge Dr. Hirofumi Abe of the Chubu-chishitsu Co. Ltd. for his useful suggestions on suction measurement, which made this present study possible.

REFERENCES

Japanese Geotechnical Society 1998. Method for obtaining undisturbed soil samles using a thin-walled sampler with fixed piston (JGS 1221-1995). *Japanese Geotechnacal Society Standards for Soil Sampling*: 1-8.
Japanese Geotechnical Society 1998. Method for obtaining undisturbed soil samles using rotary double-tube sampler (JGS 1222-1995). *Japanese Geotechnacal Society Standards for Soil Sampling*: 8-11.
Japanese Standard Association 1993. Method for unconfined compression test on soils. *JIS A 1216-1993*: 1-11 (in Japanese).
Ladd, C.C., Foott, R., Ishihara, K., Schlosser, F. & Poulous, H.G. 1977. Stress-deformation and strength charactristics. State of the art report. *Proc. of 9th ICSMFE*. 4: Tokyo. 421-494.
Shogaki, T. 1995. Effective stress behavior of clays in unconfined compression tests. *Soils and Foundations*. 35(1): 169-171.
Shogaki, T., Moro, H. & Kogure, K. 1995. Statistical properties of soil data with thin-walled samplers. *Proc. 5th Int. Offshore and Polar Eng. Conf*. Hague. 406-413.
Shogaki, T. 1996. A method for correcting consolidation parameters for sample disturbance using volumetric strain, *Soils and Foundations*, 36(3): 123-131.
Shogaki, T. & Maruyama, Y. 1998. Estimation of *in-situ* undrained shear strength using disturbed samples within thin-walled samplers, *Geotechnical Site Characterization*:Atlanta.419-424 : Balkema.
Shogaki,T., Maruyama,Y. & Shirakawa,S. 1999a. A precision triaxial apparatus using small size specimens and strength properties of soft clay. *Geotechnical Engineering for Transportation Infrastructure*: 1151-1157.Amsterdam: Balkema.
Shogaki, T., Sakamoto, Y. & Sudo, T. 1999b. Effect of specimen size on unconfined compressive strength properties of natural deposits. *54th annual Conf. of JSCE*. 3: 90-91 (in Japanese).

Clay Science for Engineering, Adachi & Fukue (eds) © 2001 Balkema, Rotterdam, ISBN 90 5809 175 9

Behavior of unsaturated silty clay during water infiltration

Tatsuaki Nishigata & Kazuhiko Nishida
Kansai University, Osaka, Japan

ABSTRACT: Natural slopes that are in unsaturated state collapse suddenly during rainfall. Generally, the cause is considered to be dependent on decrease of effective stress due to suction loss at water infiltration. And slopes themselves develop shear stress due to their gravity forces within the ground in unsaturated state, and they suffer due to water immersion. It is important to examine not only the strength of soil but also the deformation characteristic of slope. In this study, a triaxial test apparatus is devised for examining the behavior of the deformation characteristics for an unsaturated silty clay and experiments are conducted with various cases.

1 INTRODUCTION

Natural slopes and embankments are generally in an unsaturated state, and often these slopes collapse suddenly as a result of decreasing of suctions due to infiltration of water during rains. When analyzing the stability of these types of slopes consisting of unsaturated soil, strength and deformation characteristics with regard to unsaturated earth become necessary; this in turn has lead to a recent increase in the popularity of studies into the unsaturated soil mechanics. Among the most famous of them is a study by Bishop (1960), Fledlund et al. (1978) which took up the suctions in unsaturated soil for an evaluation of strength parameters.

Generally, when loose soil on the surface layer of a slope generating a shear stress is subject to water infiltration, there will naturally be some reduction in strength, but deformation of the soil caused by the infiltration is also a very important factor. Even if the slope is not failed, significant deformation may hinder the function of soil structures. Deformation can thus be considered as a factor that should be considered during the actual design. Nevertheless, relatively little research has been devoted to such deformation, and in the case of design in the past, strength parameters were employed from test samples that had been infiltrated with water beforehand. This method did not sufficiently express the actual condition of the slopes, however, and for this reason, attention was drawn to strengths of an unsaturated soil on wetting under shear stress. This research unfortunately did not gain a sufficient information of quantitative deformation characteristics during infiltration, however, and many points remained unresolved.

In particular, it is meaningless to use the strength constants from test samples in which water has been infiltrated previously and a collapse has already occurred, because the collapse due to water infiltration is a discontinuous phenomenon in soil skeleton. Furthermore, the research of Nobari and Dancan (1972) is also questionable in that it expresses the stress-strain relationship for unsaturated soils during water infiltration under shear stress condition as being the same as that of undergoing the water infiltration in advance. Based on the above, this paper aims to gain a clear understanding of the stress-strain relationship, assuming a slope surface undergoing a constant small shear stress.

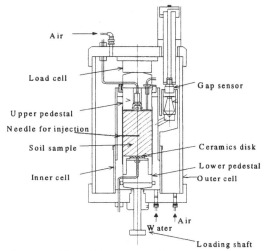

Figure 1. Triaxial compression test apparatus

Table 1. Properties of soil sample

Density of soil particle	2.650g/cm³
Sand fraction	0.1%
Silt fraction	90.4%
Clay fraction	9.5%
Coefficient of uniformity	3.5

Table 2. Test condition

Initial water content	20%
Initial void ratio	0.9, 1.1, 1.3
Initial saturation degree	42, 48, 60%
Dry density	1.26g/cm³
Wet density	1.53g/cm³

2 EXPERIMENTAL FACILITIES

2.1 *Triaxial test apparatus*

The equipment used for this experiment was the low-pressure triaxial compression test unit shown in Fig.1. This equipment features a method by which the axial pressure is applied from bottom of the test sample by a piston; at that time, the weight of the test sample and the weight of the lower pedestal are counterbalanced by the weights. By using an air regulator, it is possible to accurately control the confining pressure with air pressure within a range of 0 to 15 kPa. A test sample with diameter of 8 cm and height of 16 cm is placed inside a double cell, and the volume changes of the sample during the experiment are determined by measuring changes in the water level inside the inner cell using a gap sensor. The pore pressure in the sample is measures through a ceramic disc (A.E.V.=200 kPa) installed in the lower pedestal. This equipment is designed such that both of strain and stress controlled tests can be available.

2.2 *Physical properties of the test sample*

The soil sample used was silty clay called as "DL clay" which is made by grinding kaolinite ore. This sample appears that both strength and deformation relate closely with changing of moisture content.

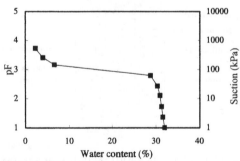

Figure 2. Water retention curve

Physical properties of this sample is summarized in Table 1.

Figure 2 shows the water retention curve for the soil sample. We can see that the suction decreases rapidly when the water content exceeds the 30% level. The DL clay adjusted to a moisture content of 20% was compacted to form a specimen with a height of 16 cm and a diameter of 8 cm. Conditions for the test material used in the triaxial test are shown in Table 2.

3 EXPERIMENTAL METHOD

In many studies in the past, the method for wetting of the specimen in triaxial test was to supply the water to the sample from bottom of sample. When this method is used, however, the wetted area differs from shear deformation area, so it is not possible to determine the actual behavior due to wetting. This study uses a method in which water is injected directly into the shear deformation area, by inserting a hypodermic needle into the center of the specimen. Injection speed was kept to 25cc/min to prevent the side effect of excessive pore pressure when the water was injected. The total volume injected was 50cc.

Confining pressure ranges from 5 kPa to 15 kPa assuming the surface layer of a slope, and the isotropic consolidation was applied to the specimen for 30 minutes before the shear test was carried out. Shear test was conducted for a drained condition while measuring the pore pressure at the lower pedestal. This test program consists Non-infiltrated shear test, Pre-infiltrated shear test and Infiltrating shear test.

a) Non-infiltrated shear test: This is a standard shear test with drainage condition. The test sample is consolidated using an isotropic confining pressure, and axial load is applied with a strain rate of 0.1%/min. Water content remains initial water content.

b) Pre-infiltrated shear test: After sample equilibrium with isotropic consolidation pressure, water of 50cc is injected, and shear stress is applied.

c) Infiltrating shear test: After isotropic consolidation, the shear stress is applied, and water 50cc is injected when a certain level of shear stress is reached. The shear stress level at the time of injecting the infiltration water is defined as:

$$S_L = \frac{(\sigma_1 - \sigma_3)_w}{(\sigma_1 - \sigma_3)_{max}} \times 100 \ (\%) \qquad (1)$$

where $(\sigma_1 - \sigma_3)_w$ = the shear stress at the time of water injected, and $(\sigma_1 - \sigma_3)_{max}$ = the maximum shear strength in the Non-infiltrated shear test.

In this experiment, we used four levels of the shear stress: S_L=20, 35, 50, and 80%. We also made stain controlled test up to the point of the water infiltration, and conducted two types of stain controlled (strain rate = 0.1 and 0.6%/min) and stress controlled tests (constant shear stress) during and after water infiltration.

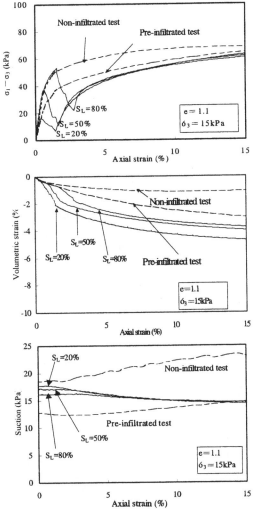

Figure 3. Results of strain controlled test for sample of e=1.1.

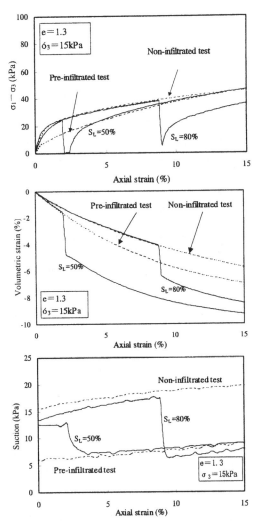

Figure.4 Results of controlled strain test for sample of e=1.3.

4 EXPRIMENTAL RESULTS

4.1 *Strength and deformation in stain controlled tests*

Figure 3 illustrates the changes of shear stress, volumetric strain and suction with axial strain where: e=1.1, σ_3=15kPa, and S_L=20, 50, 80%. This chart clearly indicates that the soil specimen before infiltration displays the same behavior as in the Non-infiltrated test. The specimen after infiltration displays decreased suction and decreased strength, and when infiltration is complete, the shear stress reaches its lowest point. After that, the strength begins to recover, and at around 15% axial strain, it approaches stress-strain curve for the Pre-infiltrated shear test. Nobari et.al.(1972) indicated that the lowest point of the Infiltrating shear test reached on the stress-strain curve for the test of the Pre-infiltrated shear test, but it

should be noted that based on the results of the this experiment, the lowest point of the shear stress is on the lower side of the curve for the Pre-infiltrated shear test. This means that predicting behavior at water infiltration based on the results of the pre-infiltrated shear test will result in an error on the side of danger. The same tendency will exist even with various confining pressures.

Figure 4 shows the results of the stain controlled tests for the specimen with large void ratio (e=1.3). In this case, the shear stress during the water infiltration falls even farther below the curve of the pre-infiltrated shear test.

4.2 *Strength and deformation in stress controlled tests*

Figure 5 illustrates the changes of shear stresses, volume changes and suction under stress controlled

91

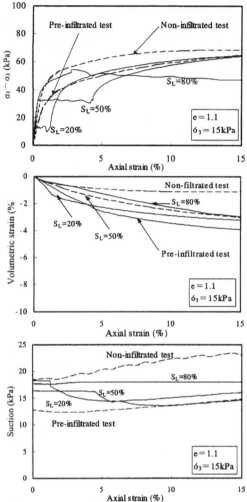

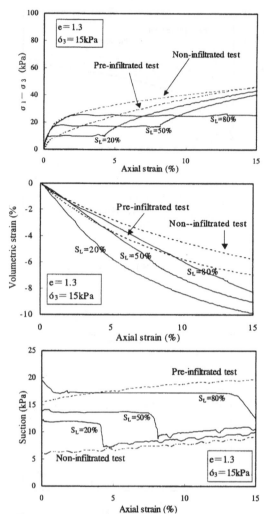

Figure 5. Results of stress controlled test for sample of e=1.1.

Figure 6. Results of stress controlled test for sample of e=1.3.

condition (e=1.1, σ_3=15kPa). From Fig.5, we can see that when the shear stress level at water infiltration is low, axial strain stops at roughly the point where infiltration is complete. We switch to stain controlled test after that, the shear stress increases and approaches to the line of the Pre-infiltrated shear test. On the other hand, as the shear stress level at the time of water infiltration increases, the axial strain after the infiltration also increases. When S_L=80%, the axial strain does not cease, but continues to the point of failure. The volume changes and the suction differ depending on the value of S_L, but both show a tendency to decrease due to water infiltration, and approaching the values for the pre-infiltrated shear test.

Figure 6, similarly, shows the results of stress controlled tests for specimen with large void ratio (e=1.3), as in the case of the stain controlled test, as

the void ratio increases, the effects of water infiltration become more pronounced. Furthermore, in the cases of both Figs 5 and 6, the point that axial strain stops is on the lower side of the stress-strain curve for the pre-infiltrated shear test.

From these results, we have illustrated in Figs.7 and 8 the changes of the amount of shear stress reduction in the stain controlled tests and the amount of total strain in the stress controlled tests under the test conditions of confining pressure of 5 kPa and each e value (e=1.3, 1.1, and 0.9). From these figures, we can see that the amounts of stress reduce and axial strain due to the water infiltration become greater when the level of shear stress at the infiltration is larger.

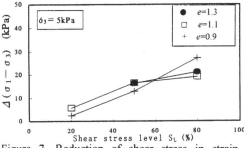

Figure 7. Reduction of shear stress in strain controlled test.

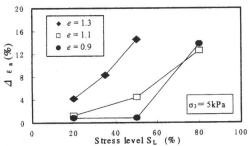

Figure 8. Change of axial strain in stress controlled test.

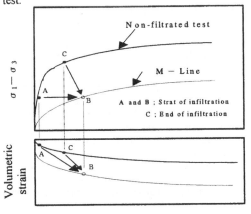

A and B ; Strat of infiltration
C ; End of infiltration

Figure 9. General outline of M-line.

4.3 Characteristics of stress and strain due to infiltration

We can assume that the lowest point of shear stress in a stain controlled test with infiltration and the point where axial strain remains in a stress controlled test represent the deformation characteristics of transient conditions when the sample is subjected to water infiltration under shear stress. If the infiltration volume is equal, it can be considered that all of these points will exist above the same line. Figure 9 gives a general outline of this situation, where the above mentioned line is arbitrarily called the "M-line." Furthermore, from

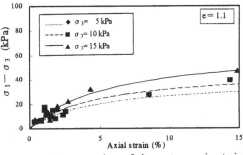

Figure 10. Lowest points of shear stresses in strain controlled tests.

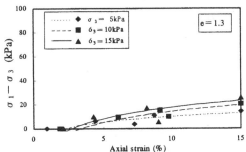

Figure 11. Points where axial strains stop in stress controlled test.

the experimental results already discussed, it is clear that the M-line exists on the lower side of the stress-strain curve for the test of pre-infiltrated shear test.

Figures 9 and 10 show the lowest point of shear stress in a stain controlled test and the point where the axial strain stops in a stress controlled test are plotted on the same graph in groups with confining pressure. We can see some scatter, but it is clear that in each group corresponding to a specific confining pressure, all the points come align to form the M-line, and it could be called the characteristic curve at water infiltration. Furthermore, as confining pressure increases, the M-line is in a higher position. And when the initial void ratio of the specimen increases, the graph is notable in that the M-line begins at a position that is displaced from the origin of the lateral coordinate.

Figure 11 illustrates the various M-lines for the test specimens with differing initial void ratios. According to Fig.11, the effect of the initial void ratios on the M-line is extremely large, and the M-line presents the provisional stress-strain relationship that expresses the transient situation at the water infiltration.

Figure 12 shows the volumetric strain corresponding to the plotted points in Fig.11. Here as well, we can see the formation of a single characteristic curve for each initial void ratio, as in the case of the M-line.

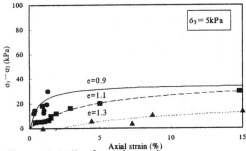

Figure 12. *M*-line for each sample.

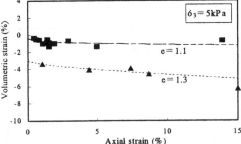

Figure 13. Volumetric strains corresponding to *M*-line.

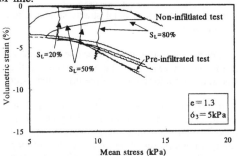

Figure 14. Relationship between volumetric strains and mean stress.

4.4 *Volumetric strain and mean principal stress*

Suction before water infiltration acts in isotropic state, and is thought to decrease uniformly with the infiltration. Figure 13 illustrates the relationship between volumetric strain and mean principle stresswhen the initial void ratio of the specimen is 1.3 under confining pressure of 5 kPa. The line before the sample undergoes infiltration follows roughly the same path as in the case of a Non-infiltrated sample. As the sample undergoes infiltration, it displays a sudden decrease in volumetric strain as a result of reduced suction in both of stain controlled and stress tests. Furthermore, in the case of stain controlled test, mean principle stress drops rapidly by the infiltration. And, as we might have assumed, the lowest point of shear stress in the stain controlled tests and the point where axial strain stops in stress controlled tests attain a position above the same line. This line corresponds to the M-line discussed above.

As above mentioned, we can see that there is no significant difference in volume changes in the case of low confining pressures where the shear stress acting on the sample is small, regardless of whether the sample is under isotropic stress or shear stress condition. This phenomenon is referred to as a "collapse phenomenon." The strength of the soil structure decreases commensurately with the reduction in suction, and this in turn manifests itself as shear deformation.

5 CONCLUSIONS

1. There is a clear difference between the pre-infiltrated shear test and Infiltrating shear test, in the case of loose unsaturated soil under low confining pressure. If we line up the lowest stress point in the stain controlled tests and the point where axial strain ceases in the stress controlled tests, the result is a single curve (M-line). This curve can be considered a characteristic curve that expresses the deformation when unsaturated soil is subjected to infiltration.

2. The position of the M-line noted above is lower than the stress-strain curve for the specimen that underwent the Pre-infiltrated shear test. We can assume that the results of experiments involving the pre-infiltrated shear test will give results that err on the danger side.

3. The position of the M-line will be correspondingly lower when the confining pressure is low and the initial void ratio is large. That is to say, it indicates that a significant deformation is apparent when the soil is subjected to infiltration.

4. The volumetric strain due to water infiltration in stain controlled and stress controlled tests are concentrated on a characteristic curve that corresponds to M-line.

5. It considers that there is no difference between volume changes due to infiltration under isotropic stress nd volume changes due to infiltration under shear stress. From this, we can assume that collapse due to infiltration is equivalent to shear deformation of soil structure.

REFFERENCES

Bishop,A.W. 1960. The measurement of pore pressure in the triaxial test, *Proc. Conf. Pore Pressure and Suction in Soils*, 38-46.

Fredlund,D.G., Morgenstern,N.R. & Widger,R.A. 1978. Shear Strength of Unsaturated Soils, *Canadian Geotechnical Journal*, 313-321.

Nishida,K & Nishigata,T. 1998. Effect of water soaking on deformation of decomposed granite soil under shear stress, *Annual Meeting of The Japanese Geotechnical Society*, 679-680.

Nobari,E.S. & Duncan,J.M. 1972. Effect of reservoir filling on stresses and movements in earth and rockfill dams, *University of California, Berkeley, California*, 71-80.

Residual effective stress in undisturbed samples collected from very large depth

Y. Watabe
Port and Harbour Research Institute, Ministry of Transport, Yokosuka, Japan

ABSTRACT: A series of measurement of residual effective stress and unconfined compression test was carried out to investigate the sample quality of Osaka bay Pleistocene clay collected from very large depths. The suction larger than 98.1 kPa could be measured in the undisturbed samples by applying back air pressure. The disturbance ratio defined by Okumura (1974) ranges from 1.5 to 3, implying that the quality of samples are classified to very good without much strength reduction by mechanical disturbance.

1 INTRODUCTION

In Japan, undrained shear strength for design is often determined by unconfined compression test, which is very sensitive to sample disturbance. Tsuchida & Tanaka (1995) explained that there are two different kinds of disturbances, i.e. one is remolding type and the other is crack type as shown in Figure 1. The crack type disturbance tends to appear in a sample taken from large depths, because this material, which is consolidated under very high pressure and in which structure has developed by aging effect, is very brittle.

Shimizu & Tabuchi (1993), and Mitachi & Kudoh (1996) proposed some methods for predicting in situ undrained shear strength based on unconfined compression strength and residual suction. These correction methods for unconfined compression strength are available for the remolding type disturbance, but are not for the crack type disturbance.

These previous researches were mainly concerned with Holocene clays, which are relatively young and taken from the depths of at most 30 m. However, as the greater scales of structures have been constructed in the coastal area, the Pleistocene clay, which are deep and old deposits, are getting the more important in the stability and the deformation analysis. The typical example is the construction of the Kansai International Airport. In the construction site, the water depth is 18.0 ~ 19.5 m, and the thickness of the Holocene layer is 18 ~ 24 m. For constructing the large-scale artificial island, several boring and undisturbed sampling were carried out up to the depth of 400 m.

A series of unconfined compression test was carried out to investigate the shear strength of Pleistocene clays taken from large depths. Figure 2 shows unconfined compression test results of the bore hole

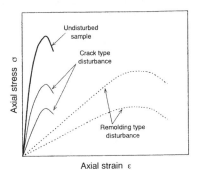

Figure 1. Remolding type and crack type disturbances. (after Tsuchida and Tanaka, 1995)

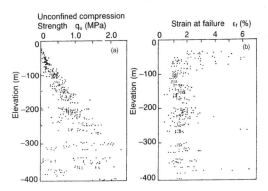

Figure 2. Unconfined compression test results; (a) unconfined compression strength with depth, (b) strain at failure with depth.

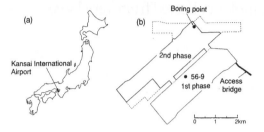

Figure 3. (a) Location of the Kansai International Airport and (b) the point of borehole in this study.

No.56-9, as (a) a variation of unconfined compression shear strength with depth and (b) relationship between elevation and strain at failure. In Figure 2(a), the test results are scattered and it is difficult to determine the profile of shear strength with depth for design work. In Figure 2(b), strains at failure is as small as about 1.5 % for samples collected from a depth larger than 80 m. This fact implies that the sample disturbance was crack type. However, it has not been cleared yet that whether the crack was created before trimming or after.

It has been considered that the unconfined compression test is not to be the appropriate method to get reliable test results for the samples taken from the large depths. This is due to the fact that the negative pressure more than a vacuum does not exist, and it is impossible to maintain residual effective stress larger than 98.1 kPa in a fully saturated sample for the samples taken from the large depths.

In this study, the sample quality of Osaka bay Pleistocene clay collected at depths from 40 to 200 m will be discussed. For this purpose, the residual effective stress in the undisturbed samples was measured and unconfined compression tests were carried out. These two kinds of tests are sensitively influenced by the sample disturbance. Since these depths up to 200 m were mainly considered for prediction of settlement, the quality of these samples are very important.

Figure 3 shows (a) location of the Kansai international airport and (b) the sampling point for this research. In the field investigation for construction of the Kansai international airport, a new wire-line sampling system was developed (Horie et al. 1984), for more efficient sampling.

2 MEASUREMENT OF RESIDUAL EFFECTIVE STRESS

Mitachi & Kudoh (1996) measured suction larger than 98.1 kPa in an artificial Kaolin clay, consolidated under very high pressure of 400 kPa, applying a back air pressure. This technique is coming from research groups studying unsaturated soils who are

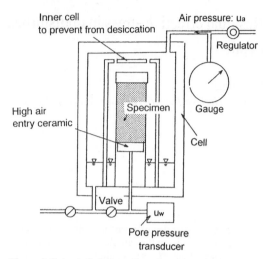

Figure 4. Schematic layout of suction measuring system.

trying to measure suction to evaluate the effective stresses or pore-size distributions (Fredlund & Rahardjo, 1993). Relationships between degree of saturation and suction are well known as soil-water characteristic curves. In unsaturated soil, suction is caused by meniscus between soil particles, and its value is possible to be more than 98.1 kPa. Therefore, very wide-range suctions from 0 to about 10^6 kPa are treated. Since suction S is defined as pore air pressure u_a minus pore water pressure u_w, i.e. $S = u_a - u_w$, it is possible to obtain true suction by giving some high air pressure u_a (back air pressure) and measuring the relative water head ($u_a - u_w$).

In the present study, measurement of residual effective stress larger than 98.1 kPa was attempted by adopting and simplifying a method to measure suction in unsaturated soils mentioned above. This is based on an idea that if a sample is collected from large depths, negative pore pressure becomes larger than the vacuum and fine air foams form innumerable menisci, then residual effective stress, even which is larger than 98.1 kPa, is possibly remained in the apparently saturated sample. The suction measuring system used in this study is schematically illustrated in Figure 4. This system can measure residual effective stress, by placing the specimen under a certain high air pressure u_a (back air pressure). The specimen trimmed 35 mm in diameter and 80 mm in height is placed on the high air entry value ceramic disk (air entry value is 200 kPa), which is preliminarily saturated by de-aired water. The cell is set up, then air pressure is increased up to a certain value u_a, and the valve, between inside of ceramic filter and inside of the cell, is opened for a moment to balance the pressures in the pedestal and in the cell.

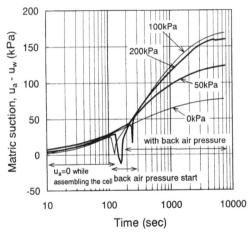

Figure 5. Examples of measured suction on specimens at a depth of 93m.

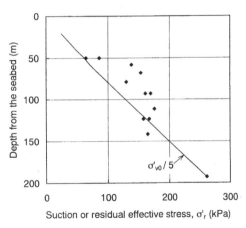

Figure 6. Profile of residual effective stress with depth.

Figure 5 shows an example of data sets measured by suction tests, which were conducted on the specimens from the same depth of 93 m. While assembling the cell at from 0 to about 200 sec, the specimens were under one atmospheric pressure. Then, a back air pressure of 50kPa, 100kPa or 200kPa was applied to the specimen, except for a case noted as "0kPa", which was conducted under the atmospheric pressure. The test result conducted under 0 kPa was approaching to the vacuum (u_a-u_w=98.1kPa under u_a=0kPa; i.e. u_w=-98.1kPa), and then this test result might show only apparent suction. Therefore the true residual effective stress could be larger than this. The similar tendency was observed in the test result, which was conducted under an air pressure u_a of 50 kPa, approaching to u_a-u_w=148.1kPa; i.e. u_w=-98.1kPa. On the other hand, almost the same residual stresses were measured for the tests, which were carried out under 100 kPa and 200 kPa, and the value of residual effective stress does not depend on the back air pressure. This should be the true residual effective stress in the specimen.

3 TEST RESULTS OF RESIDUAL EFFECTIVE STRESS

Figure 6 shows the variation of residual effective stresses with depth for Osaka bay Pleistocene clay, in which it is assumed that the measured suction is directly equal to the residual effective stress. Residual effective stresses in Pleistocene clays collected from large depths were about one fifth of effective overburden stresses.

Since undisturbed sample after extruding from sampling tube loses confined pressure, then it is in isotropic stress condition under one atmospheric

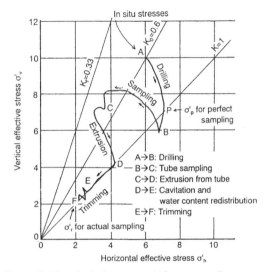

Figure 7. Hypothetical stress path for a normally consolidated clay element during tube sampling. (after Ladd & Lambe, 1963)

pressure. During this procedure, the sample is maintained to be almost undrained condition. Then, if sample was collected perfectly without any physical disturbance, the stress would follow a path of undrained extension from a point on K_0-line until it reaches to isotropic condition. Ladd & Lambe (1963) explained a concept on the stress path during sampling and laboratory testing as shown in Figure 7. The point P of the effective stress σ'_p corresponds to the condition of perfect sampling, in which no disturbance except for the release of in-situ deviator stress has been given to the sample. Mechanical disturbance during sampling and laboratory testing, such as drilling, tube sampling, extrusion from tube

Table 1. Physical Properties of the clays.

	Yokohama clay	Boston blue clay	Osaka bay Pleistocene clay
Liquid limit w_L (%)	93	33	60–100
Plastic limit w_P (%)	42	13	30
Plasticity index I_P	50	20	30–70
Compression index C_c	0.95	0.4	0.7–1.0
Swelling index C_s	0.11	0.08	0.07–0.11
Clay fraction (<2μm) (%)	29	54	20–40

sampler and trimming, are given to the soil sample and the pore water pressure within the specimen builds up. Finally, point F with an effective stress of σ'_r represents the residual effective stress of the specimen. The ideal undisturbed sample on the point P is called "perfect sample" and remaining effective stress is noted as σ'_p. In the present study, σ'_p is evaluated by CAU extension test (Watabe et al. 2000), in which the specimen is anisotropically consolidated under the in situ overburden pressure with K_0 (= 0.5) and σ'_p is defined as an isotropic stress when σ'_1 decreased down to σ'_3, while extension test with undrained condition.

The residual effective stress is possibly a function of air entry value (AEV), plasticity index (I_p) and swelling index (C_s). If I_p is the same among two soils, the larger AEV, the larger residual effective stress can be maintained, and the larger C_s, the specimen is the tougher for disturbance. Tanaka & Locat (1999) reported mercury intrusion test results of Osaka bay Pleistocene clay, and these are shown in Figure 8. According to the following equation (Marshall, 1958):

$$(u_a - u_w) = 2T/R_p \tag{1}$$

wherer $(u_a - u_w)$ is the matric suction, T is the surface tension of water (75 kPa μm) and R_p is the pore radius, it is possible to associate a given matric suction to a pore diameter. Considering eq(1) and Figure 8, it can be understood that the maximum pore radii are in order of 1.0 μm at shallower depths and 0.5 μm at deeper depths, which corresponds to the air entry value in order to 150 kPa and 300 kPa, respectively. This implies that the observed residual effective stresses in Figure 6 are reliable values.

Okumura (1974) defined the disturbance ratio R as a ratio of residual effective stress of specimen σ'_r to the effective stress of perfect sample σ'_p. They discussed relationship between shear strength reduction ratio which is defined as a ratio of shear strength un-

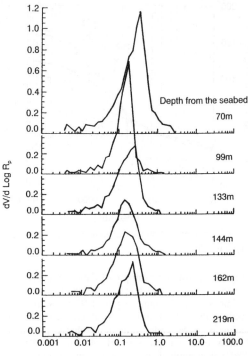

Figure 8. Entrance pore radius distribution versus depth from the seabed as given by mercury porosimetry tests. (after Tanaka & Locat, 1999)

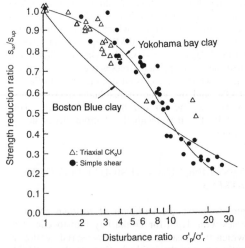

Figure 9. Relationship between strength reduction and disturbance ratio of Yokohama clay. (after Okumura, 1974)

der residual effective stress s_{ur} to shear strength of perfect sample s_{up} and disturbance ratio R for Yokohama clay as shown in Figure 9. According to this figure, shear strength of Yokohama clay is not decreased while the disturbance ratio is less than three. As a reference, the relationship for Boston Blue clay is also drawn in Figure 9. Okumura (1974) indicated that there are two kinds of clays on remolding type disturbance as shown in Figure 9. Taking account of the physical properties shown in Table 1, Osaka bay Pleistocene clay can be considered as a family of Yokohama clay (a family of Japanese marine clay). Figure 10 shows distribution of disturbance ratio R with depth obtained in the present study. The disturbance ratios R for the Osaka bay Pleistocene clay are ranged from 1.5 to 3. From Figure 9, it can be concluded that the quality of these undisturbed samples is very good and shear strength reduction by disturbance is not so serious.

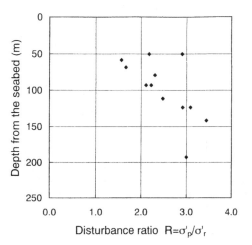

Figure 10. Profile of disturbance ratio with depth.

4 UNCONFINED COMPRESSION TEST

After measuring residual effective stress by suction test, unconfined compression tests were carried out for all of the specimens with a strain rate of 1.0 %/min.

Figure 11 shows the distribution of shear strength determined by unconfined compression tests. In this figure, for comparison, undrained shear strength determined by CAU tests with strain rate is 0.1 %/min, which were carried out by Watabe et al. (2000), are also shown. Generally, it is said that average shear strength (AVE) for compression and extension in CAU test is coincident to unconfined shear strength for Japanese Holocene marine clays. This tendency consistent with Pleistocene clay.

Test results of unconfined compression test are more scattered than test results of triaxial test, however, variance of unconfined compression test results in this study is smaller than that in Figure 2(a) and is almost the same as that for the Holocene clay. In particular, trimming a Pleistocene clay sample is very difficult. For example, even if wire-saw is used to cut into the sample, some part of specimen can be easily chipped off. Therefore, it must be trimmed very carefully, little by little, but rapidly to prevent desiccation. In spite of that unconfined compression test seems to be very simple with easy handling, it requires well skilled technique and concentration when Pleistocene clay is treated. In the case of Figure 11, samples were trimmed very carefully because this test series is designed for the research. Thus it can be considered that the reasons of scattered data in Figure 2(a) are not only reliability of the test method itself and heterogeneity of the specimen, but also trimming technique of a technician. Therefore, it can be concluded that unconfined compression test is not practically applicable to brittle clays like the Pleistocene clay collected from large depth.

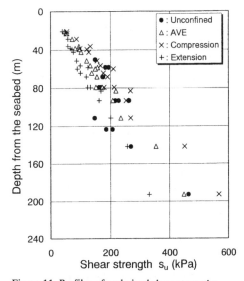

Figure 11. Profiles of undrained shear strengths.

5 CONCLUSIONS

In this study, measurement of residual effective stress and unconfined compression test were carried out for Osaka bay Pleistocene clay which were collected from large depths as geotechnical survey for the construction of the Kansai international airport. The test results are concluded as follows,

1) An attempt to measure residual effective stress, which is larger than 98.1 kPa in undisturbed sample, was conducted by adopting and simplifying a method to measure suction in unsaturated soils. As a result, it can be measured the residual effective stress by placing the specimen under a back air pressure.

2) It was cleared that suction larger than 98.1 kPa can be remained in undisturbed and apparently saturated sample.
3) The disturbance ratio defined by Okumura (1974) was ranged from 1.5 to 3 for Osaka bay Pleistocene clay. According to Okumura's chart (Figure 9), these samples are classified to very good without much strength reduction by physical disturbance.
4) Stress-strain relationships observed in unconfined compression test for Osaka bay Pleistocene clay were very brittle, in which axial strain at failure was only 1% and sudden reduction of the resistance was observed. Thus, it can be said that the trimmed specimen has a tendency to have crack type disturbance.

REFERENCES

Fredlund, D.G. & Rahardjo, H., 1993, "Soil mechanics for unsaturated soils." John Wiley & Sons.

Horie, H., Zen, K., Ishii, I. & Matsumoto, K., 1984, "Engineering properties of marine clay in Osaka bay, (Part-1) Boring and sampling." *Technical Note of The Port and Harbour Research Institute*, Ministry of Transport, Japan, 498, 5-45. (in Japanese)

Ladd, C.C. & Lambe, W., 1963, "The strength of undisturbed clay determined from undrained tests", *ASTM*, STP-361, Laboratory Shear Testing of Soils, 342-371.

Marshall, T.J. 1958. "A relation between permeability and size distribution of pores." *Journal of Soil Science*, 9(1), 1-8.

Mitachi, T. & Kudoh, Y., 1996, "Method for predicting in-situ undrained strength of clays based on the suction value and unconfined compressive strength." *Journal of Geotechnical Engineering*, JSCE, 541(III-35), 147-157. (in Japanese)

Okumura, T., 1974, "Study on the disturbance of clay and improvement of sampling method." *Technical Note of The Port and Harbour Research Institute*, 193. (in Japanese)

Shimizu, M. & Tabuchi, T., 1993, "Effective stress behavior of clays in unconfined compression tests." *Soils and Foundations*, 33(3), 28-39.

Tsuchida, T. & Tanaka, H., 1995, "Evaluation of strength of soft clay deposits. -A review of unconfined compression strength of clay-" *Report of Port and Harbour Research Institute*, Ministry of Transport, Japan, 34(1), 3-37.

Watabe, Y., Tsuchida, T. & Adachi, K., 2000, "Undrained shear strength of Pleistocene clay in Osaka Bay and its effect on the stability of a large scale seawall structure." *Proc. of IS-Yokohama*.

Clay Science for Engineering, Adachi & Fukue (eds) © 2001 Balkema, Rotterdam, ISBN 90 5809 175 9

Swelling and suction properties of six compacted European clays

D.C. Wijeyesekera
Department of Civil Engineering, University of East London, UK

ABSTRACT: For environmental and economic reasons, clay is popularly used as clay barriers or as an essential component in most synthetic landfill liners. Compacted soil consistently produces high swelling pressures. Excessive swelling pressures could further develop as the material became wetter. Higher plasticity soils are prone to greater swelling and shrinkage, combinations of dry density, water content and initial stress state exist that will minimize the development of undesirable swelling pressures.

1 INTRODUCTION

Waste disposal sites need to be designed on the concept of ensuring minimum environmental impact. Clay is inevitably and popularly used as clay barriers or as an essential component in most synthetic landfill liners. Soil compaction improves the material properties in giving it increased shear strength and at the same time it reduces the propensity for settlement and deformation as well as its permeability to water. Compaction in the field must necessarily be controlled and the quality assured in order to avoid failures of engineering works. Compaction cracks observed on the runway construction in Hong Kong's new airport aroused concern(Ground Engineering, April1998). An industry source put it down to a question of compaction quality control. Economic and environmental considerations in the construction process often dictate the necessity to utilize unsuitable clayey fill (or as dug material) in preference to importing suitable fill. A form of compaction is then necessary to improve such ground conditions. Though the mechanical behaviour of clay in contact with water is fairly well established, very little knowledge exists about the effects on the properties of clay when permeated with a leachate. The influence of leachate on the geomechanical properties of Boom clay was investigated by Ourth (1999).

The upper and lower acceptability limits of moisture content are related to the plasticity of the soil as determined in the Atterberg limit test. The upper limit is also influenced by the mobility of compaction plant on the soil. The lower limit of the moisture content is determined by the acceptable percentage of air voids present when the soil is compacted until it reaches the 100% of the maximum dry density line. If a very dry soil is compacted until it reaches the 100% of maximum dry density line, the air voids may be in excess of 10%. Ingress of water into a soil in this state can cause excessive swelling pressures that will lead to a collapse of the soil structure.

Wijeyesekera et al. (1999) outlined the design, construction and performance of a compacted London clay barrier for part of the Heathrow Express Rail Link tunnel. Comparative studies of the compaction and suction characteristics of London Clay with Wadhurst Clay and Brickearth have been reported elsewhere (Wijeyesekera and O'Connor. 1997). Such characteristics for Hamburg Klei, obtained from an harbor extension site in Altenwerder, Germany have also been reported (Wijeyesekera and Von Lewinski. 1998). Collaborative research is under way on the behavior of Boom clay as a confining leachate barrier. This paper presents a synthesis of the characteristics of these clays with reference also being made to data of other European clays.

P wave velocity measurements through samples of compacted clay during drying and wetting cycles are reported. The variations in P wave velocity with changes in air void ratio and dry density are presented and discussed as a possible parameter for assessing the shrinkage and cracking of compacted clayey fill.

2 BACKGROUND

Lambe (1958) suggested that compaction of clay soils results in a distinct microstructure change with the compacting moisture content in that those compacted dry of the optimum moisture content have a flocculated particle pattern and conversely a dispersed particle pattern is achieved with those compacted wet of the optimum. Parsons (1992) showed Proctor curves for heavy clays that show a pronounced depression in the curve at a water content drier than optimum. The pore water pressure within a compacted clayey fill can be negative (a suction) and the way in which these suction pressures change have a significant influence on the long term mechanical behavior of the compacted fill.

Compacted soil consistently produces high swelling pressures. The samples that are compacted dry of the optimum moisture content produce similar stress-strain behavior achieving similar levels of failure stress. However samples with a comparable dry density but with water content above the optimum water content produced different stress-strain responses. At these higher moisture contents the deviator stress at failure was lower, as would be expected, but there was a significant difference in the stiffness response of the soil. The sample prepared by kneading compaction showed that during shearing a constant stiffness was maintained compared with the strain hardening exhibited by the statically compacted samples during shearing. The failure strength was found to be approximately equal in both cases. Seed and Chan (1959) postulated that the difference in stiffness was due to the method of compaction. Samples dry of the optimum water content (prepared by kneading or static compaction) produced a "flocculated" fabric whereas wet of the optimum water content only the statically compacted samples produced a "flocculated" fabric, the kneaded samples producing a "dispersed" fabric. Overcompaction on site can not only be a wasteful effort but should preferably be avoided as if it is not confined by overburden, the compacted soil can readily absorb water, resulting in swelling and cracking that will effect a lower shear strength and greater compressibility.

The changes in pore water suctions have a significant influence on the long term mechanical behavior of the compacted fill. Increases in the water content of well-compacted fill dissipate the pore suctions and can lead to volumetric increases or swelling pressures. Conversely a decreases in water content of the fill increases pore suctions and causes shrinkage and cracking. The swelling pressures developed within compacted clay are shown to be a function of the soil type, dry density, moisture content, initial stress state and degree of confinement.

Although higher plasticity soils are prone to greater swelling and shrinkage, combinations of dry density, water content and initial stress state exist that will minimize the development of swelling pressures (Wijeyesekera and O' Connor 1997). The acceptable upper and lower moisture content limits are related to the plasticity of the soil. This depends mainly on the capacity of the soil to absorb water. A high plasticity clay may have its highest density at an optimum water content of 30% whereas a low plasticity clay may have a similar density at only 15% water content. The lower limit on moisture content for Class 2A - wet cohesive material is taken as PL-4% and the upper limit is taken as 1.2PL. The latter is a moisture content at which a clayey soil has sufficient strength for acceptable trafficability (Parsons 1992).

3 EUROPEAN CLAYS STUDIED

In the UK, the various types of materials for use in backfilling and compaction behind road structures are given in the "Specification for Highway Works", published by the Department of Transport (1991). The specifications are intended to assist the engineer in the selection of soil used for earthworks. Soil properties are required to lie within certain limits to be acceptable for the different methods of compaction. Table 6/1 of the "Specification for Highway Works" lists the acceptable materials for use as backfill. Two basic types of material can be used which are defined in the specification as cohesive and granular. The type of fill relevant to this paper is the cohesive fill that can be used as backfill adjacent to structures. From Table 6/1 this is defined as a Class 7A material. The range of acceptable material properties is:

- Grading: 100% to pass 75 mm sieve and 15 - 100% to pass 63 μm sieve.
- Moisture content: limits set by engineer.
- Moisture condition value: limits set by engineer.
- Undrained shear strength parameters: limits set by engineer.
- Effective angle of internal friction and cohesion: limits set by engineer.
- Liquid limit: not to exceed 45%
- Plasticity index: not to exceed 25%

London Clay is a well known soil and was chosen for its high plasticity and known swelling potential. The Transport Research Laboratory (TRL) also used it in a pilot study investigating the development of swelling pressure (Symons et al, 1989). Brickearth is a silty clay with a plasticity which lies towards the limit of acceptability for clayey backfill. Wadhurst clay is natural clay with a plasticity, which lies between that of Brickearth and London Clay and outside the current UK Department of Transport recommendations.

Boom clay is well-known clay from Mol in Belgium. The samples were obtained in an air-dried condition and then pulverized to a particle size smaller than 0.5 mm.

In the Northern region of Germany, mainly in the vicinity of the brackish-marine environment of the river estuaries, various organic clays are found in layers up to 15m thickness. 'Hamburg Klei' is such a soft organic silty clay, deposited in the Holocene period and forms the subsoil in Altenwerder in Hamburg. This is often used as a compaction material for road works, as clay core in dams or as barriers in landfill design due to its low permeability characteristic.

Swelling soils in Greece belong mainly to the Quaternary clay soils and Neogene Marls (Tsiambaos and Tsaligopoulos, 1995). The Mediterranean climates, where summer months are warm and very dry whilst the winter months have heavy rains, encourage the seasonal volume changes (swell / shrink) or the development of swelling pressures when they are confined.

4 RESEARCH OBSERVATIONS

The aim of laboratory compaction tests is to suggest specifications for field compaction of the soil. The need for quality control in field compaction is emphasized. The method of compaction for this type of material requires the end product to be at least 100% of the maximum dry density or not more than 5% air voids, whichever gives the lower dry density at the field moisture content. This is defined by the Proctor compaction test with a 2.5-kg rammer (BS1377, Part 4; German Standards DIN 18127). The water content-dry density relationship for the soil obtained from laboratory tests form the basis for the assessment of the suitability of a soil and for compaction specifications. Proctor density of the six European soils is compared, in the light of work published by Morin and Todor (1977) in Figure 1. There is close correlation with their suggested variation for the maximum dry density relation given by the bold line shown on the graph.

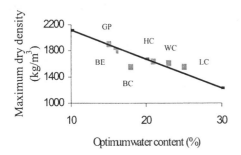

Figure 1. Maximum Proctor dry density – optimum water content variations (LC - London Clay; WC – Wadhurst Clay; HC- Hamburg Clay; BC – Boom Clay; BE – Brickearth; GP- Greek Patras Clay)

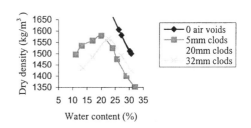

Figure 2. Influence of clod size of London Clay on Proctor compaction.

4.1 Influence of clod size

Laboratory compaction tests provide the basis for control of the degree of compaction. Typical compaction curves for British Standard 'light' compaction for the London clay tested is shown in Figure 2. The compaction curves show a significant shift to higher maximum dry densities at lower optimum moisture contents with decreasing initial maximum clod sizes of the clay.

In the case of the light compaction tests, there is a gain of over 2.5% in the dry density with a nearly 4% decrease in the optimum moisture content when the maximum clod size is reduced to 5 mm from 32 mm. The corresponding values for the heavy compaction are an increase of 4% in the maximum dry density with a reduction of 3% in the optimum water content. A similar effect of clod size at compaction has been shown to have a significant effect on the permeability of the compacted clay by Benson & Daniel (1990). Thus the use of an industrial soil conditioning equipment is encouraged to reduce the clod size to a practicable minimum before compaction.

4.2 Influence of compaction energy level

The British and German Standards define two different energy levels of compaction; light and heavy compaction test. Different levels of energy input result in different degree of compaction. In the laboratory the energy level is defined in terms of mass and height of drop and number of blows or of power input, mass and time of vibration of a vibrating hammer. In the field, the input energy is influenced by the mass and the width of the compacting machine, the vibrating force (if existing) and number of passes of the compactor. However the energy used in the field is more difficult to control than in the laboratory. When the thickness of the layer to be compacted is increased the dry density obtained is generally reduced. The compaction energy applied at the bottom of the layer is less than at the top of the layer. Therefore a non-uniform dry density is achieved through the soil layer with a higher density on top than at the bottom.

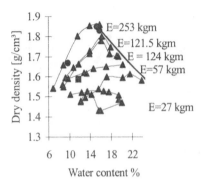

Figure 3. Dry density – water content relationship for Hamburg Clay compacted with different energy levels. E is the level of compaction energy

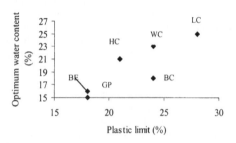

Figure 4. Optimum water content – plastic limit variation

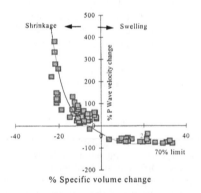

Figure 5. Change in P wave velocity with specific volume

The compaction is quantitatively controlled in terms of the achieved dry density. The value of the dry density is in fact meaningless unless there is information available to assess this value. The in-situ dry density is classed on the basis of the proctor density achieved in the laboratory and is expressed as 'relative compaction' which is the in-situ density expressed as a percentage of the proctor density.

'Proctor curves' for the Hamburg Klei at five different compaction energy levels, E (253, 121.5, 124, 57. And 27 kgm) are presented in Figure 3. In the area of low energy levels (57 and 27 kgm) a clear

Proctor curve cannot be identified. Repeat tests were carried out with the soil mixed to the required water content and left for over 24 hours so that the water is distributed uniformly through the soil before compaction. The test results indicated a 'second peak' in the area of low water content which might be a consequence of the suction developed when the water content is comparatively small.

4.3 Influence of soil type

Different soil types behave differently when compacting energy is applied. This depends mainly of the soil water interaction. A high plasticity clay may have its highest density at the optimum water content of 30% whilst a low plasticity clay may have a similar density at only 15% water content. Morin and Todor (1977) developed correlations between the optimum water content (w_{opt}) the plastic limit (w_p) on the basis of work carried out on South American and African red tropical soils. Figure 4 shows this variation for the European soils described in this paper. They relate as follows;

European soils; $w_{opt} = 0.88\ w_p + 0.24$
South American soils; $w_{opt} = 0.61\ w_p - 0.84$
African soils; $w_{opt} = 0.58\ w_p - 4.54$

4.4 P Wave velocity observations

Material engineers to non-destructively assess the integrity of materials have successfully used Wave propagation techniques. The velocity with which P waves traverse across the axis of 38mm diameter cores samples of compacted London clay extruded from the Proctor compaction moulds was measured using a PUNDIT (Portable Ultrasonic Non-Destructive Index Tester). P wave velocity of the compacted samples was observed to be not only a function of compaction water content but also of the air porosity. It was also observed that with increasing specific volume the P wave velocity decreased, since the compacted clay is a 3 phase medium. The P wave velocities of the samples were measured continuously as the cores were allowed to dry in stages at room temperature (20°C), at 48°C and at 105°C. The P wave velocities were also measured after the samples were wetted and allowed to swell freely. Figure 5 illustrates the changes in P wave velocity and specific volume from the initial compacted state. The best-fit curve suggests that a seventy- percent decrease in P wave velocity will suggest that the soil has fully swollen.

5 SWELLING

The swelling of compacted clay fills involves two processes, which occur simultaneously. Micro-scale swelling involves the hydration of clay mineral platelets, and macro-scale swelling which is concerned with the relief of high suctions in the capillaries during wetting. These cause volume changes resulting from changes in effective stress. Holtz and Gibbs (1954) showed that for a given dry density, a clay compacted dry of optimum will swell more than a sample compacted wet of optimum. This suggests that the degree of saturation or the air porosity is again an important factor in the estimation of the amount of swell.

The free swell (S) of the compacted London clay cores and their results are presented in Figure 6, which indicates a linear trend of S = 7.5812 - $0.251w_c$ where w_c is the compaction water content. According to the prediction by Mc Dowell (1956), the minimum initial water content, w_{min} for London clay is 24.6%. The results of the current study suggest a w_{min} of 30.2%.

5.1 *Swelling pressure*

The shrink / swell character of an unsaturated clayey soil is governed by the equilibrium state of forces. Internal forces governed by clay mineralogy, soil water chemistry and suction are balanced by the external forces of applied stress and capillary tension. This equilibrium is influenced by a number of factors acting within the soil-water system. Any changes from the equilibrium of this system causes volumetric and stress changes. The following influence the internal forces that act on a micro-scale:

Clay mineralogy; The ability of a soil to swell depends upon the amount and type of clay minerals present in the soil and the arrangement and specific surface of the clay particles. Clay particles consist of platelets whose surfaces are negatively charged. To maintain equilibrium, cations in the soil water are attracted to the surface of the platelets. Any changes in the equilibrium of internal and external forces will lead to an adjustment of the particle spacing and consequent volume change. The clays presented in this paper have varying proportions of kaolinite, illite and montmorillonite (smectite).

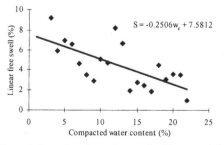

$$S = -0.2506w_c + 7.5812$$

Figure 6. Percentage linear free swell with compacted water content.

Soil water chemistry - The chemistry of the soil water itself can have a large effect on the magnitude of swelling. The type and concentration of cations present affects the absorption of water and consequent swelling e.g. calcium (Ca^{2+}) or magnesium (Mg^{2+}) ions in the soil water result in less swelling than when sodium (Na^{+}) ions are present. The infiltration of a leachate with new chemistry into a landfill barrier can therefore have a significant change. Ourth 1999 presented the chemistry of such leachate.

The mechanical response of soils is dictated by the fabric as well as flaws/ discontinuities that may exist in the material. These flaws can often be in the form of cracks generated during desiccation or shear and serve as stress concentrators.

Many researchers have used the term swelling pressure in many different ways, and the definition is often dependent in some way on the test method. Brackley (1973) stated that swelling pressure could be defined, as the pressure required holding the soil at constant volume when water was added. Johnson (1989) observed that the magnitude of the swelling pressure depended on the degree of confinement of the soil with the greater degree of confinement leading to increased swell pressure. In all the experiments, compacted samples were prepared at different water contents and at different dry densities. They were tested in the various apparatus by allowing inundation and observing any changes with time. The determination of swelling pressure was affected by the degree of confinement. In cases where straining was prevented in all directions as in the modified oedometer where the swelling pressure increased to a fairly constant value. In contrast, when straining was allowed in the 100mm stress path cell, the swelling pressure increased quickly to a peak value then reduced as continued softening and straining was allowed. O'Connor (1994) used a computer controlled apparatus which investigated swelling under boundary conditions of constant vertical stress and zero lateral strain. During swelling, changes in lateral pressure were interpreted as swelling pressure.

A simple classification method for swelling soils based on index properties, such as consistency limits, clay-sized fraction and free swell is useful for the prediction of swelling character. The complementary use of chemical and mineralogical analysis is very often a useful aid in the interpretation and justification of their behavior.

Method specifications state the exact method to be applied for compaction, i.e. the type of compacting plant (weight, dimensions of compacting area), the thickness of the layer of soil and the number of passes that the compacting plant has to pass over the area. This type of specification was based on extensive investigations into compaction performed at the Transport Research Laboratory over many years. It was aimed at allowing contractors to choose from a variety of compaction plant and then by using reference charts it would be possible to understand quickly the thickness of each layer and the number of passes required for a satisfactory compaction of the particular type of soil. Figure 7 compares the swelling pressure data for compacted samples of three test soil types. It can be seen that for a clayey soil (Brickearth) compacted to a state where the percentage of air voids is approximately 10%, the swelling pressure developed on inundation is as much as 50% less than the more densely compacted soil. The degree of compaction of the sample is greater than the 5% air voids given in the current Department of Transport specification.

With regard to plasticity, soil with a higher plasticity index but under the same conditions of density and relative water content produces a higher value of swelling pressure. Under the current criteria higher plasticity soils are excluded from use as fill to structures. Figure 8 suggests that an increase of up to 5% in plasticity, from the plasticity index of 25% (the current recommended limit), would only increase the swelling pressure by about 6% for soils compacted at moisture contents at or above the plastic limit. This relaxation though small would increase the choice of soil for backfilling applications. However, caution must be exercised in the control of the soil placement moisture content because as the moisture content decreases below 0.9PL there is a marked increase in swelling pressure as plasticity index increases.

An alternative method of determining the initial soil condition at which there would not be significant swelling pressure is illustrated in Figure 8. The curves for PL-4 and 1.2PL are again indicated. The assumption is made that a swelling pressure is acceptable if it would be produced by a soil, which is at the limit of the current specification. This swelling pressure is given in Figure 8 at the intersection of the lower limit of water content (PL-4) with the current maximum permitted plasticity index of 25%. It can then be used to indicate the minimum placement moisture content for different plasticity soils. For example, a soil with a plasticity index of 43% placed with water content of 0.9PL will produce the same swelling pressure as the soil currently at the limit of the specification when placed at the minimum water content of PL-4%.

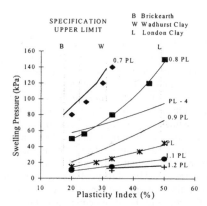

Figure 8. Nett swelling pressure variation with plasticity index

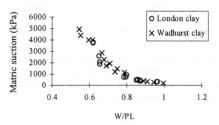

Figure 9. Matric suction variation with ratio of water content expressed as a ratio of plastic limit for London clay and Wadhurst clay.

6 SOIL SUCTION

Soil suction has two main components: osmotic suction and matric suction. Osmotic suction results from the chemical activity and mineralogy of the soil. In contrast matric suction depends on capillary effects of the air-water interfaces that exist in an unsaturated soil. Total suction is a function of both osmotic and matric suctions. Since osmotic forces are generally constant changes in total suction result from changes in matric suction.

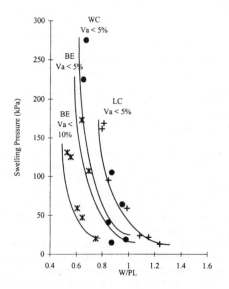

Figure 7. Nett swelling pressure vs. water content / plastic limit

It is difficult to measure forces or stresses at a micro-scale so macro-scale measurements are preferred as primary indications of swelling/shrinking behavior. There are several methods available for the measurement of suctions in soil. The indirect method described by Chandler and Gutierrez (1986) has been adopted in this study. Pieces of Whatman no. 42 filter paper are placed in contact with the soil and the samples sealed in a container. Over a period of between five and seven days the moisture in the filter paper and soil will come into equilibrium. By carefully weighing the pieces of filter paper and calculating their moisture contents, it is possible to estimate the soil suction using their suggested equation.

Figure 9 shows the soil suctions thus measured plotted against the ratio of compaction water content and plastic limit for London clay and Wadhurst clay. The results appear to fit a single curve, suggesting the dependence of suction on the compacted moisture content as a proportion of the plastic limit.

7 CONCLUSIONS

In this paper, the compaction characteristics of six European clay fills are presented and discussed in the context of the factors that influence compaction of soils used as barriers for landfill. Published correlation between Maximum dry density, optimum water content and plastic limit has been examined.

It was shown that for higher plasticity soils, there were combinations of dry density, water content and initial stress level that would minimize the development of swelling pressures.

The upper limit to placement moisture content is linked to mobility of construction vehicles. Based on the minimum shear strength of 50kPa for such acceptable trafficability (Jones and Greenwood, 1993), the results suggest a limit of 1.2PL would be appropriate for London and Wadhurst clays although difficulties could be experienced with brickearth above 1.0PL

On the basis of the study, soils of higher plasticity could be used as fill to landfill structures but the compaction water content must be carefully controlled. The test results indicate that soils with a plasticity index of 26% must have minimum water content of 0.9PL. The minimum compaction water content for soils of intermediate plasticity could be determined by linear interpolation.

REFERENCES

Benson, C.H., & Daniel, D.E., 1985, Influence of clods on hydraulic conductivity of compacted clay., *ASCE Journal of Geotechnical Engineering*, vol 116 (8) , pp 1231-1248.

Brackley J.J.A ,1973, Swell pressure and free swell in a compacted clay. *Proceedings of the Third International Conference on Expansive Clays*. Nr 1. Israel Institute of Technology Haifa.

British Standards Institution, 1990, *BS 1377 Methods of testing of soils for civil engineering purposes*, BSI, London.

Chandler,R.J.,& Gutierrez,C.I., 1986, The filter paper method of suction measurement: *Geotechnique*, vol. 36, pp. 265-268.

Department of Transport, 1986, *Specification for highway works*. 6th Edition. HMSO.

Holtz, W.G.,& Gibbs, H.J., 1954, Engineering properties of expansive clays, *Proc. ASCE*, vol 80, New York.

Johnson L.D,1989, Horizontal and Vertical Swell Pressures from a Triaxial Test: A Feasibility Study. *Geotechnical Testing Journal, GTJODJ*, Vol 12, Nr1, March 1989, pp 87-92.

Jones R.H.,& Greenwood J.R., 1993 Relationship testing for acceptability assessment of cohesive soils. *Proceedings of the conference of Engineered Fills*, New castle Upon Tyne, UK. Thomas Telford Ltd. Pp. 302-311.

Lambe, T.W., 1958, The structure of compacted clay. *Journal of Soil Mechanics and Foundation Engineering, ASCE*, vol.84, no.2, pp. 1-35.

Manual of Contract Documents for Highway Works ,1991, Volume 1, *Specification for Highway Works*, December 1991, HMSO, London.

Morin, W.J., & Todor, P,C., 1977, *Laterites and lateritic soils and other problem soils of the tropics*. United States Agency for International Development, A10/csd 3682.

Morin, W.J., and Todor, P,C., 1977, *Laterites and lateritic soils and other problem soils of the tropics*. United States Agency for International Development, A10/csd 3682.

O'Connor K.J, 1994,. *The swelling pressure of fined grained unsaturated soils*. PhD Thesis, City University, London.

Ourth A-S, & Verbrugge, J-C., 1999, First results about the influence of leachate on the properties of Boom Clay.,*Ground contamination: pollutant management and remediation, Proceedings of the second conference*, British Geotechnical Society and Cardiff School of Engineering, 13-15 September 1999, London.

Parsons A.W. ,1992, *Compaction of soils and granular materials*: A review of research performed at the Transport Research Laboratory. HMSO, London.

Symons I.F., Clayton C.R.I. and Darley P. ,1989, *Earth pressures against an experimental retaining wall backfilled with heavy clay*. Department of Transport, TRRL Research Report 192, Crowthorne.

Tsiambaos, G., & Tsaligopoulos,CH.,1995, A proposed method of estimating the swelling characteristics of soils: some examples from Greece.Bulletinn of the International Association of Engineering Geology, N0 52, pp 109-115.

Wijeyesekera,D.C., & O'Connor K., 1997, Compacted clayey fill: an assessment of suction, swell pressures, shrinkage and cracking characteristics. Proceedings International Conference Ground Improvement Techniques, Macau, 7-8 May 1997, pp. 621-630.

Wijeyesekera, D.C., O'Connor, K., & Salmon, D.E., 1999, Design and performance of a compacted clay barrier for Heathrow Express rail link tunnel., *Ground contamination: pollutant management and remediation, Proceedings of the second conference,* British Geotechnical Society and Cardiff School of Engineering, 13-15 September 1999, London.

Wijeyesekera, D.C., & Von Lewinski,F., 1998, Compaction and suction characteristics of Hamburg Klei, Proceedings 2nd International Conference Ground Improvement Techniques, Singapore, 7-9 October 1998.

Clay Science for Engineering, Adachi & Fukue (eds) © 2001 Balkema, Rotterdam, ISBN 90 5809 175 9

Swelling pressure of a compacted bentonite subjected to high suction

T. Nishimura
Ashikaga Institute of Technology, Japan

ABSTRACT: Engineers are well aware of the severe distress that lightly loaded structures can suffer when placed on swelling soil that is subjected to environmental changes. It is known that extensive structural damages occur in many areas where expansive soils are known to exist. Prediction of swelling pressure is required for geotechnical engineers. Constant volume swelling tests were conducted for a compacted bentonite as an expansive soil. A high suction and varying preconsolidation pressures were applied to a compacted bentonite before swelling. A relative humidity controlling method was used for an application of high suction. A compacted bentonite was desaturated in relative humidity chamber. The relative humidity of 50% was selected in this test program. Swelling properties of desiccated, compacted bentonite are compared with that of compacted bentonite (i.e., no subjecting of high suction). Additionally, this study appears that swelling pressure depends on preconsolidation pressure and dry density.

1 INTRODUCTION

The ground surface deposit can also be subjected to varying and changing environmental conditions. The pore-water pressure of soil above the water table in ground decrease to negative values, while the total stress in the deposit remains essentially constant. The negative pore-water pressures act in all directions (i.e., isotropically), resulting in a tendency for shrinkage and overall desaturation of the upper portion on the ground. A high tension in the water phase (i.e., high matric suction) causes the soil to have a high affinity for water.

Soils exhibiting expansive properties are common throughout most portions of the United States (Krohn and Slosson 1980). Moderately and highly expansive soils together mantle approximately 20 percents of the United States. In the United States, the damage caused by shrinkage and swelling soils amounts to about $9 billion per year, which is greater than the combined damages from natural disasters such as floods, hurricanes, earthquakes, and tornadoes (Jones and Holtz 1987).

Many researchers in many ways use the term, swelling potential, but in generally, it may be taken to include both the percent swell and the swelling pressure of soils. There is generally agreement on the definition of swelling pressure as the pressure required to hold the soil at constant volume, when water is added. Experimental methods to determine the swelling pressure of expansive soils show that the conventional consolidation tests gives an upper bound value, the method of equilibrium void rations different consolidation loads gives the least value, and constant volume method give intermediate values (Sridharan et al. 1986)

2 PURPOSE OF THIS STUDY

Changes in stress state occur during geological deposition, erosion, precipitation, evaporation, and evapotranspiration. An expansive soil near ground surface is generally unsaturated due to desiccation. Under a constant total stress, an unsaturated expansive soil will experience swelling and shrinkage as a result of matric suction variations associated with environmental changes. An unsaturated expansive soil will undergo volume change when the net normal stress or the matric suction variable changes in magnitude. The laboratory information desired by the geotechnical engineer for predicting the amount of heave is an assessment of: 1) the *in situ* state of stress, and 2) the swelling properties with respect to changes in net normal stress and matric suction.

Very high suctions due to evaporation are applying to expansive soils near ground surface boundary. Evaporation is related with change in relative humidity in atmospheric. Example, it is estimated that a relative humidity of 80 % is corresponding to suction of 30,130 kPa based on relative humidity versus suction theory relations.

This study reports the results obtained from (i) constant volume swelling tests using a compacted bentonite subjected to high suction and (ii) constant volume swelling tests using a compacted bentonite with various preconsolidation pressures and dry densities. Mainly application methods of suction are as following; based on pressure plate apparatus, vapor equilibrium method using salt solution in desiccator and relative humidity controlling method (Nishimura and Fredlund 1999). In this test procedure, high suctions were applied to the compacted bentonite using the relative humidity controlling method. This study investigates effect of high suction on swelling pressure of a compacted bentonite. Relationship between swelling pressure and varying preconsolidation pressures applied to bentonite is appeared in this study.

3 LITERATURE

3.1 Swelling properties

"Free-swell" and "Constant volume test" is generally used in the determination of the swelling pressure of the soil for the prediction of heave. In the "free-swell" oedometer test, the specimen is allowed to swell under a token pressure by submerging the specimen in distilled water. After attaining an equilibrium condition, the soil specimen is then loaded and unloaded following the conventional oedometer tests procedure. In the "constant volume" oedometer test, the applied load is increased in order to maintain the specimen at a constant volume after it has been submerged in distilled water. Gilchrist (1963) shows typical results of "free-swell" and "constant volume" oedometer tests on a compacted Regina clay. Gilchrist (1963) states that swelling pressure obtained from "free-swell" oedometer tests is not coincident with swelling pressure obtained from "constant volume" oedometer tests.

The swelling soil behavior depends on the properties of the soil, such as clay content, dry density, water content, soil structure, plasticity index, shrinkage limit, matric suction, and stress state variables. Chen (1988) studies the effect of initial water content, dry density of compacted soils, and surcharge pressure on the amount of total heave using "free-swell" oedometer tests for expansive shales. The results indicate that total heave increases with decreases in the initial water content of compacted specimens compacted at a constant initial dry density. On the other hand, total heave increases with increasing initial dry density for specimens at a constant initial water content. Total heave decreases with an increasing surcharge pressure. Similar observations were reported by Holtz and Gibbs (1956) and Dakshanamurthy (1979). Komornik and David (1969) state that plasticity, initial moisture content and dry density have a significant effect on predic-

tion of the swelling behavior of the clay. Komornik and David (1969) show relationship between free swell and shrinkage limit. However, correlation is poor and the scatter is high.

Fredlund et al. (1980) visualizes the procedures for predicting heave in terms of the stress paths followed on the constitutive surface (Direct model method, Sullivan and McClelland consolidometer method, Richards method, Jennings and Knight double oedometer method). Direct model method is based on "free-swell" consolidation tests on undisturbed samples placed in an oedometer. Specimens are subjected to the overburden pressure and allowed free access to water. Sullian and McClelland consolidometer method is based on a constant volume oedometer test on an undisturbed sample initially subjected to the overburden pressure. Richards method is based on a laboratory suction test, which yield a water content versus matric suction curve. Jennings and Knight double oedometer tests is based on the two oedometer tests; namely a free-swell oedometer test that is initially under a token load of 1 kPa and a natural water content oedometer test.

Fredlund and Rahardjo (1993) show three-dimensional views of the void ratio and water content constitutive surface for the assessment of the in situ stress state and the swelling properties of expansive, unsaturated soils using a one-dimensional oedometer test. The "corrected" swelling pressure obtained from the oedometer tests represents the in situ stress state translated to the net normal stress plane (i.e., the void ratio versus effective stress plot). The "corrected" swelling pressure is equal to the overburden pressure plus the in situ matric suction. The translated in situ matric suction is called the "matric suction equivalent", $(u_a - u_w)_e$. (Yoshida et al. 1983). The magnitude of the matric suctions equivalent, $(u_a - u_w)_e$, will be equal to or lower than the in situ matric suction. The difference between the in situ matric suction and the matric suction equivalent is primarily a function of the degree of saturation of the soil.

Dider et al. (1980) describes the measuring device, which allows the rapid determination of changes in hydraulic conductivity of feebly permeable swelling soils. The sample whether remolded or undisturbed, has a diameter of 7 cm and a maximum thickness of 2.5 cm. They suggest the equation relationship between swelling percent and swelling pressure.

$$(\Delta V/V) = x \log(SP) + y \tag{3.1}$$

where $(\Delta V/V)$ is swelling percent, SP is swelling pressure, x and y are coefficient with depending on the mineralogy nature of the clay.

3.2 Unsaturated soils

An unsaturated soil is considered to have four

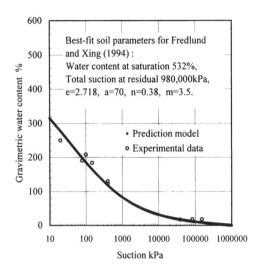

Figure 1. Soil-water characteristics curve for a bentonite.

phases: solid, air, water, and contractile skin (the air-water interface). The contractile skin acts like a rubber membrane as it induces a matric suction in the soil pores. There are significant differences between the behavior of saturated and unsaturated soil. These differences have been found to be related to the role of matric suction as a stress variables. Suction has been shown to affect both strength and volume change characteristics of unsaturated soils. While a saturated soil possesses positive pore-water pressure, an unsaturated soil is characterized by negative pore-water pressure, which exerts a tensile pull at all air-water interfaces in the soil profile. The surface tension on the contractile skin pulls the particles together providing additional strength to an unsaturated soil compared to a saturated soil where, in contrast, the positive pore-water pressure reduces the strength (Wulfsohn et al. 1996).

Two independent stress variables are required to describe the state of stress in an unsaturated soil (Fredlund and Morgenstern 1977). The preferable stress state are $(\sigma - u_a)$ and $(u_a - u_w)$. $(\sigma - u_a)$ is herein called the net normal stress term and $(u_a - u_w)$ is called the matric suction. This combination is most satisfactory since the effect externally applied loads and the effects of environmental changes can readily be separated in terms of stress changes.

4 TEST PROCEDURES

4.1 Soil-water characteristics curve

Expansive soils contain clay minerals that exhibit high volume change upon wetting. Clay minerals fall into three principal categories: kaolinite, illite, and montmorillonite. Although illite and montmorillonite have identical structural units, montrillonite units are separable, making for a large specific surface area. This study used a bentonite as a expansive soil.

The soil-water characteristics curve for the bentonite is measured using a pressure plate apparatus, glass desiccator containing saturated salt solutions (i.e., vapor equilibrium technique) and relative humidity technique. The above mentioned three methods were used to develop the soil-water characteristics curve over the entire soil suction. The air pressure in the pressure plate apparatus was increased to a maximum of 480 kPa. The vapor equilibrium technique and the relative humidity technique were used to impose suction larger than 480 kPa.

The testing program focuses the swelling pressure of an unsaturated, compacted bentonite obtained from "Constant volume" swelling tests. The testing program consists of Testing Program 1 and Testing Program 2.

4.2 Testing program 1

A bentonite of an initial water content of 48% was prepared with weight of 60g. The bentonite was applied precompression load of 200kg in the oedometer ring. A one-dimensional, static compacted bentonite was placed directly into a relative humidity controlled chamber in order to apply a high total suction. The chamber used in this study could control the relative humidity in a range from 20% to 90%. The thermodynamic relationship between soil suction and relative humidity (i.e., relative vapor pressure) can be written as follows:

$$\psi = -135022 \ln\left(u_v \middle/ u_{v0} \right) \tag{4.1}$$

where ψ is soil suction or total suction (kPa), u_v is partial pressure of pore-water vapor (kPa), u_{v0} is saturation pressure of water vapor over a flat surface of pure water at the same temperature (kPa), u_v/u_{v0} and is relative humidity, RH (%).

The testing program selected relative humidity of 50%. A relative humidity of 50% corresponds to a 94000kPa using the equation (4.1). After the bentonite reached equilibrium with the relative humidity of 50%, "Constant volume" swelling tests were conducted in order to measure the swelling pressure of a desiccated compacted bentonite. In the "Constant volume" swelling tests, the unsaturated, compacted bentonite was produced elimination of suction in distilled water. The compacted bentonite at water content of 48% was prepared for comparing the effect of suction on swelling pressure, which no subjected to apply high suction due to relative humidity technique.

4.3 Testing program 2

The powdered bentonite of water content of 0% was used in the Testing Program 2. The powdered bentonite on weight of 20g was applied preconsolidation pressure in oedometer ring with a diameter of 50mm for twenty-four hours. The Testing Program 2 selected the preconsolidation pressure of 98kPa, 980kPa, 490kPa, and 4,900kPa. After preconsolidation pressure applied to bentonite, compacted bentonite was unloaded, and was conducted "Constant volume" swelling tests in the distilled water.

5 TEST RESULTS AND DISCUSSIONS

Figure 1 shows the relationship between soil suction and gravimetric water content for a bentonite. The soil-water characteristics curve is developed using a pressure plate apparatus, vapor pressure technique, and relative humidity technique. The soil-water characteristics curve model can be written as an equation proposed by Fredlund and Xing (1994). Prediction model parameters for the best-fit soil-water characteristic curve are shown in Fig.1. A bentonite begins to desaturate when it is subjected to an increase in soil suction. In general, there are four identifiable stages of desaturation; namely, the boundary effect stage, the primary transition stage, the secondary transition stage, and the residual stage of desaturation. A coarse-grained soil such as a gravel or sand describes clear four stages along the soil-water characteristics curve. However, it becomes to be difficult to distinguish four stages for a soil with a higher percentage of fines.

Gravimetric water content of a bentonite remains to decrease over the entire range of soil suction. The rate of desaturation with suction is great till suction is 10,000kPa. After suction exceeds at 10,000kPa, large increment in suction eventually led to a relatively small change in gravimetric water content. This stage of soil-water characteristic curve is referred to as the residual stage of desaturation in a bentonite.

Figure 2 shows the relationship between swelling pressure and elapsed time. Before swelling test starts, large cracks and shrinkage occurs in compacted bentonite due to applying highly suction (i.e., the void of compacted bentonite reaches equilibrium with the relative humidity of 50%). Desiccated, compacted bentonite had the water content of 12.3%. At beginning of swell in the water, swelling pressure of compacted bentonite subjected to high suction is developed. Compacted bentonite subjected to high suction provides with rapidly growing of swelling pressure. Swelling pressure increases until elapsed time is over one hundreds hours. After time exceeds over one hundreds hours, the rate of increasing of

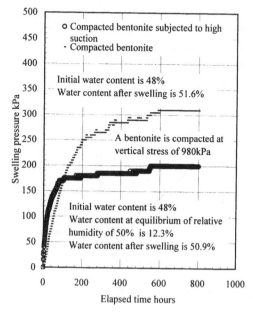

Figure 2. Relationship between swelling pressure of a bentonite and elapsed time.

swelling pressure with time shows the relatively small change. Swelling pressure of desiccated, compacted bentonite reaches 199 kPa at eight hundreds hours. At swelling pressure is attending to equilibrium, the water content of bentonite is measured as 50.9%.

Swelling pressure of compacted bentonite (i.e., no subjecting high suction) is shown in Fig.2. At the beginning of swell of compacted bentonite, swelling pressure is produced similar to that of compacted bentonite subjected to high suction. In case of compacted bentonite, swelling pressure remains to increase after elapsed time exceeds two hundreds hours. After time reaches six hundreds hours, swelling pressure unchanged. Swelling pressure of 309kPa is measured at eight hundreds hours with the water content of 51.6%. Water content of compacted bentonite after swelling is similar with that of compacted bentonite subjected to high suction. Swelling pressure of compacted bentonite is larger than that of compacted bentonite subjected to high suction.

Figure 3 shows relationship between swelling pressure at various preconsolidation pressure and elapsed time. Swelling pressure increases quickly at the beginning of swell in water. Swelling pressure unchanges after elapsed time is over fifty hours with the exception of compacted bentonite applied preconsolidation pressure of 980kPa. "Constant volume" swelling tests were stopped at two hundreds fifty hours. Water content of bentonite and swelling pressure were measured. Water content after swelling increases with decreasing of preconsolidation pressure. Swelling pressure increases with preconsolidation pressure.

Figure 4 shows relationship between swelling pressure and preconsolidation pressure applied to bentonite. Preconsolidation pressure is indicated on logarithm scale. In the range from 490kPa to 4,900kPa in preconsolidation pressure, swelling pressure increases linearly with preconsolidation pressure on logarithm scale. Increment of preconsolidation pressure applied to bentonite relates to develop of dry density in bentonite. Figure 5 shows relationship between swelling pressure and dry density in compacted bentonite. Swelling pressure increases with dry density. The effect of dry density of compacted bentonite on swelling pressure obtained from Testing Program 2 is similar to literature in the past.

6 CONCLUSIONS

This study conducted "Constant volume" swelling tests for a compacted bentonite subjected to high suction and a compacted bentonite with varying preconsolidation pressure. High suction was applied in the relative humidity chamber. This study selected the relative humidity of 50%. The relative humidity of 50% corresponds to total suction of 940,000kPa using the thermodynamic relationship between total suction and relative humidity. The compacted bentonite was desaturated due to high suction. The compacted bentonite was prepared for comparing of swelling properties, which had no subjected to high suction. Swelling pressure increases quickly at the beginning of swell in water. Swelling pressure of desiccated, compacted bentonite is smaller than that of compacted bentonite no subjected to high suction. Swelling pressure increases with increasing of preconsolidation pressure. Water content of bentonite was measured at swelling pressure attending to equilibrium. Water content after swelling decreases with increasing of preconsolidation pressure. As dry density of bentonite is developed due to apply the preconsolidation pressure, swelling pressure increases with increasing of dry density.

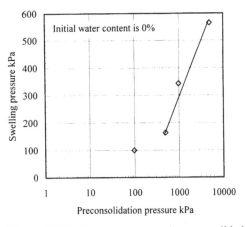

Figure 4. Swelling pressure and preconsolidation pressure.

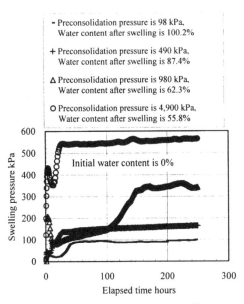

Figure 3. Relationship between swelling pressure and elapsed time.

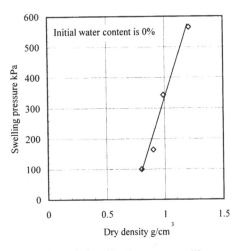

Figure 5. Relationship between swelling pressure and dry density.

REFERENCES

Chen, F.H. 1988. Foundations on expansive soils. *2nd Ed Amsterdam. Elsevior*: 463pp.

Dakshanamurthy, V. 1979. A stress-controlled study of swelling characteristics of compacted expansive clays. *Geotechnical Testing Journal. GTJODJ*. Vol.2. No.1: 57-60.

Didier, G., Kastner, R., & Bourdeau, Y. 1980. New cell for study of swelling soils. *Proceedings of the Fourth International Conference on Expansive Soils. Denver. CO. USA*. Vol.1: 18-93.

Fredlund, D.G. & Morgenstern, N.R. 1977. Stress state variables for unsaturated soils. *ASCE. Journal of the Geotechnical Engineering Division*. 103(G75): 447-466.

Fredlund, D.G., Hasan, J.U., & Filson, H. 1980. The prediction of total heave. *Proceedings of the Fourth International Conference Expansive Soils. Denver. CO*. Vol.1: 1-17.

Fredlund, D.G. & Rahardjo, H. 1993. *Soil Mechanic for Unsaturated Soils. A Wiley-InterScience Publication. JOHN WILEY & SONS. INC*: 517pp.

Fredlund, D.G. & Xing, A. 1994. Equations for the soil-water characteristic curve. *Canadian Geotechnical Journal*. Vol.31: 521-532.

Gilchrist, H.G. 1963. A study volume change of a highly plastic clay. *M.Sc. thesis, University of Saskatchewan. Saskatoon. Saskatchewan. Canada*: 141pp.

Holtz, W.G. and Gibbs, H.J. 1956. *Engineering properties of expansive clays. Trans. ASCE*. Vol.121: 641-663.

Jones, D.E. & Holtz, W.G. 1987. Expansive soils-The Hidden Disaster. *Civil Engineering. ASCE. New York. NY*: 87-89.

Komornik. A. & David, D. 1969. Prediction of swelling of clays. *Journal of the Soil Mechanics and Foundations Division. Proceedings of the American Society of Civil Engineers. SM1. January*: 209-225.

Krohn, J.P. & Slosson, J.E. 1980. Assessment of expansive soils in the United States. *Proceedings of the Fourth International Conference on Expansive Soils. Denver. CO*. Vol.1: 596-608.

Nishimura, T. & Fredlund, D.G. 1999. Unconfined compression shear strength of an unsaturated silty soil subjected to high total suctions. *Proceedings of the International Symposium on Slope Stability Engineering. IS-SHIKOKU'99. Matsuyama Shikoku Japan*: 757-762.

Sridharan, A., Rao, A.S., & Sivapullaiah, P.V. 1986. Swelling pressure of clays. *Geotechnical Testing Journal. GTJODJ*. Vol.9. No.1: 24-33.

Wulfson, D, Adams, B.A., & Fredlund, D.G. 1996. Application of unsaturated soil mechanics for agricultural conditions. *Canadian Agricultural Engineering*. 38(3): 173-181.

Yoshida, R., Fredlund, D.G., & Hamilton, J.J. 1983. The prediction of total heave of a Slab-on-Ground Floor on Regina clay. *Canadian Geotechnical Journal*. Vol.20. No.1: 69-81.

Clay Science for Engineering, Adachi & Fukue (eds) © 2001 Balkema, Rotterdam, ISBN 90 5809 175 9

The swelling of montmorillonite in nonaqueous liquids

M. Onikata
Laboratory of Applied Clay Technology, Hojun Kogyo Company Limited, Annaka, Japan

M. Kondo
Hojun Yoko Company Limited, Tokyo, Japan

ABSTRACT: The swelling properties of montmorillonite tend to decrease by impact of chemicals such as seawater and leachate contaminants. Many researchers attempted to describe an osmotic swelling of montmorillonite with dipole moment and electric permittivity of solvents, but they were of imperfect. Authors were found that formamide - montmorillonite complex behaves as a key reference on the mechanism for swelling of montmorillonite in nonaqueous liquids. Donor numbers and acceptor numbers, which indicate the magnitude of electron donor and acceptor of solvents respectively, affected the swelling behavior of montmorillonite. Polar liquids with smaller donor numbers than that of formamide farther expanded the osmotic swelling of formamide - montmorillonite complex, but the polar liquids with larger donor numbers showed a limited swelling. This behavior indicates that montmorillonite can exhibit the osmotic swelling in polar liquids with bi - functionality of electron donor and acceptor. Similarly, the polar liquids with smaller donor numbers than that of water activated the osmotic swelling of montmorillonite in electrolyte solutions.

1 INTRODUCTION

Since bentonite shows a remarkable swelling in water and restricts the migration of water, it is widely utilized as a barrier material to prevent the leakage of water from reserves and swaps (Konta 1995), or as a compacted clay liners (CCLs) and geosynthetic clay liners (GCLs) in waste containment facilities (Mitchel et al. 1996; Daniel & Bowders 1996). Hydraulic performance is based on the character of montmorillonite, which is a main clay mineral of bentonite.

Montmorillonite consists of silicate layers with negative charge. Exchangeable cations exist in the interlayer in order to compensate for the negative charge of silicate layers (Grim 1953). Montmorillonite in water shows the osmotic swelling with basal spacings of > 4.0nm by hydration of interlayer cations, however, shows the limited swelling (crystalline swelling) in nonaqueous liquids such as polar liquids and electrolyte solutions (Norrish & Quirk 1954; Olejnik et al. 1974). This reduced swelling in nonaqueous liquids restricts the industrial application of montmorillonite.

1.1 Swelling of montmorillonite in polar liquids

Montmorillonite can form complexes with many polar liquids. The complexes indicate the basal spacings of < 2.0nm corresponding to mono - or double - molecular layer of the interlayer polar liquids (Tensmeyer et al. 1960; Mortland 1970; Theng 1974). The driving force for penetration of polar liquids between the interlayer of montmorillonite is mainly due to the solvation of interlayer cations. Polar liquids coordinate to the interlayer cations directly, or through the water molecules strongly bound to the interlayer cations. The dipole moment and electric permittivity of polar liquids, the charge and size of interlayer cations play an important role on the stability of complex (Yamanaka et al. 1974). Ethylene glycol reacts with montmorillonite to form a complex with basal spacing of 1.7nm. This ethylene glycol - montmorillonite complex is widely used for identification of montmorillonite from other phyllosilicate minerals (MacEwan & Wilson 1980).

Thus, montmorillonite shows the limited swelling in a wide variety of polar liquids, however, shows the osmotic swelling in only several polar liquids with large electric permittivity, ε, such as water ($\varepsilon = 78.5$), formamide ($\varepsilon = 109.5$), N - methylformamide ($\varepsilon = 171$) (Olejnik et al. 1974). The mechanism of osmotic swelling in the polar liquids with large electric permittivity can be explained by theory of diffuse electric double layer according to Gouy - Chapman (van Olphen 1977). At large spacing, the interactions between the silicate layers and the cations are much weakened. A repulsive (osmotic) force between the electric double layers on the surface of silicate layer is balanced by attractive van der Waals forces and electrostatic force between the silicate layers bound by the interlayer cations. The thickness of electric double layer, that is, Debye length depends on the electric permittivity of polar liquids, and increases with increasing of the electric permittivity as shown in Equation 1,

$$\frac{1}{\kappa} = \sqrt{\frac{\varepsilon k T}{8\pi n v^2 e^2}} \qquad (1)$$

where $1/\kappa$ = Debye length; ε = electric permittivity; k = constant of Boltzman; T = absolute temperature; n = ion concentration; v = valence of cation; e = elementary charge. For montmorillonite in the polar liquids with large electric permittivity, the repulsive force between the electric double layers is larger than the attractive van der Waals forces. Therefore, these polar liquids can easily penetrate into the interlayer, and montmorillonite will show the osmotic swelling. However, the electric permittivity of many polar liquids is generally small, so that Debye length on the surface of silicate layer come to thin. Therefore, montmorillonite in many polar liquids will show the limited swelling because the attractive van der Waals forces are larger than the repulsive force between the electric double layers.

To understand the swelling behavior of montmorillonite in polar liquids, the swelling index U ε /v^2 was proposed by Norrish (1954). Where U = solvation energy; ε = electric permittivity; v = valence of cation. For a given valence the larger the value of U and ε, there is a greater tendency of the clay to swell. Although this index may explain the osmotic swelling of montmorillonite in water, formamide, and N - methylformamide with large electric permittivity, it does not explain the limited swelling of methyl - substituted molecules such as N - methylacetamide with a very large electric permittivity of 178.9.

1.2 Swelling of montmorillonite in electrolyte solution

Although montmorillonite swells with water infinitely, it shows the limited swelling in electrolyte solution such as seawater (Norrish & Quirk 1954). When the concentration of NaCl, [NaCl] , is increased to 0.3M, the basal spacing of montmorillonite shows 1.9nm, indicating the restriction of penetration of water molecules into the interlayer. At [NaCl] = 0.3M, the basal spacing suddenly jumps from 1.9nm to 4.0nm, and then the spacing increases linearly with $1/\sqrt{}$ [NaCl] . The limited swelling in electrolyte solution can be also interpreted by theory of Gouy - Chapman. Debye length decreases with increasing of ion concentration in electrolyte solution. For example, in pure water at pH 7, Debye length is 960nm. For NaCl

solution, it is 30.4nm at 10^{-4} M, 0.96nm at 0.1M (Israelachvil 1991). For montmorillonite in electrolyte solution, the attractive van der Waals forces between the silicate layers are larger than the repulsive force between the electric double layers. Therefore, montmorillonite in electrolyte solution will show the limited swelling because water molecules cannot penetrate into the interlayer.

Kondo (1996) found that propylene carbonate forms a complex with montmorillonite and activates the osmotic swelling of montmorillonite in electrolyte solution. Onikata et al. (1996) reported an environmental geotechnical application of propylene carbonate - bentonite complexes as impermeable soil materials in areas of coastal brines.

In this study, we have investigated the swelling properties of montmorillonite in nonaqueous liquids, and a possible mechanism of the osmotic swelling is proposed.

2 EXPERIMENTAL

2.1 Materials

Preparation of formamide (FA) - montmorillonite complexes with polar liquids

FA - montmorillonite complex was prepared by simply mixing 1.0g of montmorillonite saturated with Na$^+$ with 3.0g of FA using an agate mortar at room temperature. This FA - montmorillonite complex showed the osmotic swelling with basal spacing of > 4.0nm. Various polar liquids were further mixed at room temperature for 1 min so that the final composition of each mixture was 1/3/3 (montmorillonite/FA/polar liquid) by weight.

Preparation of polar liquid - montmorillonite complexes with electrolyte solution

Polar liquid - montmorillonite complexes were prepared by simply mixing 1.0g of montmorillonite saturated with Na$^+$ with 2.0g of various polar liquids using an agate mortar at room temperature. 0.5M NaCl solution was further mixed at room temperature for 1 min so that the final composition of each mixture was 1/2/1 (montmorillonite/polar liquid/0.5M NaCl solution) by weight.

2.2 Measurement

Powder X - ray diffraction (XRD) patterns were recorded using a Rigaku diffractometer (Model Geiger RAD - IIA) with Ni filtered Cu - Kα radiation. The wet samples were protected from drying by covering with Mylar or polyethylene film.

116

Table 1. Physical properties of polar liquids, and the basal spacings of Na - rich montmorillonite and FA - montmorillonite complexes with polar liquids.

Polar liquid	Dipole moment (Debye)	Electric permittivity	Donor number	Acceptor number	Basal spacing with polar liquids (nm)	
					Na - Mont.[*]	FA - Mont.[**]
Triethylamine	0.66	2.4	61	---	1.4	1.67
Pyridine	2.37	12.4	33.1	14.2	2.06	2.39
N,N - Diethylacetamide	3.69	30.4	32.3	---	2.16	2.33
Ethanol	1.66	24.4	32	37.9	1.73	1.88
Methanol	2.87	32.6	30	41.5	1.73	1.84
N,N - Dimethylformamide	3.86	36.7	26.6	16.0	2.06	1.92
Ethylene glycol	2.28	40.8	20	42.8	1.77	1.73
Formamide	3.37	109.5	24	39.8	Osmotic	Osmotic
Tetrahydrofuran	1.75	7.4	20	8.0	1.50	Osmotic
Water	1.8	78.5	18	54.8	Osmotic	Osmotic
Acetone	2.69	20.6	17	12.5	1.88	Osmotic
Propylene carbonate	4.98	64.9	15.1	18.3	2.01	Osmotic
Sulfolane	4.81	43.3	14.8	19.2	1.52	Osmotic
Acetonitrile	3.44	35.9	14.1	18.9	2.01	Osmotic

*The ratio of montmorillonite/polar liquid = 1/2 by weight.
**The ratio of montmorillonite/FA/polar liquid = 1/3/3 by weight.

3 RESULTS

3.1 Swelling behavior of FA - montmorillonite complexes with polar liquids

Table 1 shows the basal spacings of montmorillonite and FA - montmorillonite complexes with polar liquids and the physical properties of polar liquids used in this study. Donor number (DN) is defined as the negative molar enthalpy of the reaction between a particular solvents and $SbCl_5$, in 1.2 - dichloroethan (Gutmann 1976). Polar liquids with large DN, which have the ability of a solvent to interact with acceptors such as proton, cation and Lewis acids, may solvate strongly to the interlayer cations of montmorillonite. Acceptor number (AN) is based on chemical shift of ^{31}P NMR measurements (Mayer et al. 1975). Polar liquids with large AN, which have the ability of a solvent to interact with donors such as anion and Lewis base, may interact with polar liquids with larger DN and form the structure of hydrogen - bonded network in the interlayer. In Table 1, polar liquids are arranged in decreasing order of DN given by Gutmann.

Montmorillonite showed the osmotic swelling in FA and water among the polar liquids. The other polar liquids showed the limited swelling with basal spacing of 1.4 - 2.0nm, indicating the penetration of mono - or double - molecular layer of interlayer polar liquids. Note that with the exception of alcohols such as ethanol, methanol and ethylene glycol, water and FA have a large AN compared with other polar liquids.

FA - montmorillonite complexes with the polar liquids with larger DN than that of FA (DN = 24) changed from osmotic to limited swelling. In contrast, the XRD peaks observed for FA - montmorillonite complexes with polar liquids with smaller DN than FA became broader which indicates an increase of the osmotic swelling.

3.2 Swelling mechanism of FA - montmorillonite complexes with polar liquids

Figure 1 shows schematic structure model of the swelling behavior of FA - montmorillonite complexes. FA is a bi - functional molecule with an electron donor group, - NH_2 at one end and an acceptor group, - CHO at the other. When FA molecules penetrate into the interlayer spacing of montmorillonite, NH_2 group of FA acted as electron donor solvates with interlayer cations to form primary coordination shell. FA molecules with AN of 39.8 are also an excellent electron acceptor itself. In primary coordination shell formed by FA, FA molecules may act as electron acceptor as electron donor. FA molecules strongly coordinated to the interlayer cations act as an acceptor to other FA molecules to form a larger secondary coordination shell outside the primary coordination shell. Therefore, montmorillonite with FA will show the osmotic swelling (Fig. 1a). These results suggest that bi - functional molecules with electron donor and electron acceptor such as water and FA can expand the interlayer of montmorillonite to an osmotic swelling. In spite of alcohols are bi - functional molecules, which have both a large DN and a large AN, montmorillonite with alcohol showed the limited swelling as shown in Table 1. The alcohols may be exception because of the special formation of OH – O - Si hydrogen bonds between adsorbed alcohol molecules and oxygen ion of surface of silicate layer (Emerson 1957).

If FA - montmorillonite complex is treated using the polar liquids with larger DN than FA, these polar liquids (DN > 24) can penetrate into the primary coordination shell of FA to preferentially coordinate to the interlayer cations and thus, FA molecules are replaced by the polar liquids molecules and is expelled from the interlayer (Fig. 1b).

117

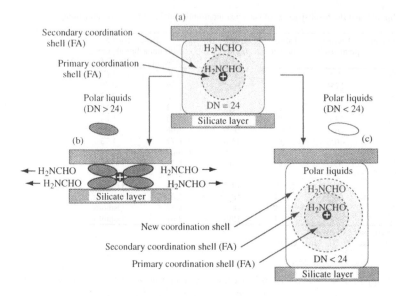

(a)

Secondary coordination shell (FA)

Primary coordination shell (FA)

H₂NCHO

H₂NCHO

DN = 24

Silicate layer

Polar liquids (DN > 24)

Polar liquids (DN < 24)

(b)

← H₂NCHO
← H₂NCHO

H₂NCHO →
H₂NCHO →

Silicate layer

(c)

Polar liquids

H₂NCHO

H₂NCHO

New coordination shell

Secondary coordination shell (FA)

Primary coordination shell (FA)

DN < 24

Silicate layer

Figure 1. Schematic structural model of the post-swelling behavior of FA - montmorillonite complexes: (a) FA - montmorillonite complex with FA (DN = 24) molecules in a strongly bound primary coordination shell and in a loosely bound secondary coordination shell between the silicate layers (osmotic swelling); (b) limited swelling (crystalline swelling) with polar liquids of DN > 24, FA molecules being expelled from the interlayer; (c) osmotic swelling with polar liquids of DN < 24.

On the other hand, if FA - montmorillonite complex is treated using the polar liquids with smaller DN than FA, these polar liquids (DN < 24) cannot replace FA molecules coordinating to the interlayer cations because of weaker electron acceptor than FA. The bi - functional FA molecules remaining coordinated to the cations act as acceptor for the polar liquids with smaller DN than FA, expanding the interlayer spacing to form new coordination shell outside the secondary coordination shell of FA (Fig. 1c).

3.3 Swelling behavior of polar liquid - montmorillonite complexes with electrolyte solution

Table 2 shows the basal spacings of polar liquid - montmorillonite complexes with 0.5M NaCl solution and the DN of polar liquids used in this study. For polar liquid - montmorillonite complexes with 0.5M NaCl solution, the polar liquids with larger DN than water showed the limited swelling. On the other hand, the polar liquids with smaller DN than water showed the osmotic swelling.

Figure 2 shows the basal spacings of propylene carbonate - montmorillonite complex swelled in NaCl solution of different concentrations (Onikata et al. 1999). The basal spacings were compared with those reported by Norrish & Quirk (1954) on montmorillonite in NaCl solutions. The basal spacings of propylene carbonate - montmorillonite complex increased from 1.9 to 2.3nm in [NaCl] = 1 - 2M, and

then suddenly increased from 2.3 to 4.5nm at [NaCl] = 0.75M. In dilute solutions, the basal spacings increased with decreasing NaCl solution. On the other hand, montmorillonite without propylene carbonate also showed a similar steep increase in swelling from 1.9 to 4.0nm at [NaCl] = 0.3M. These results indicate that the basal spacing of 2.0nm is probably a critical value where the electrostatic attractive force between the silicate layer by way of the interlayer cations is sufficiently weak that the limited swelling is transformed into an osmotic swelling with basal spacing of 4.0nm.

Table 2. The basal spacings of polar liquid - montmorillonite complexes* with 0.5M NaCl solution.

Polar liquid	Donor number	Basal spacing (nm)
Triethylamine	61	1.61
Pyridine	33.1	3.15
N,N - Diethylacetamide	32.3	3.05
Ethanol	32	2.52
Methanol	30	2.29
N,N - Dimethylformamide	26.6	2.45
Ethylene glycol	20	1.80
Tetrahydrofuran	20	1.60
Water	18	Osmotic
Acetone	17	Osmotic
Propylene carbonate	15.1	Osmotic
Acetonitrile	14.1	Osmotic

*The ratio of montmorillonite/polar liquid/0.5M NaCl = 1/2/2 in weight.

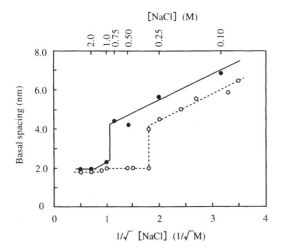

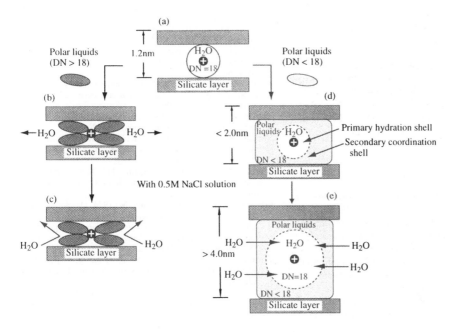

Figure 2. The basal spacings of Na - rich montmorillonite in different concentrations of NaCl solution with (●) and without (○) propylene carbonate. The basal spacings for the system without propylenecarboate are from Norrish & Quirk (1954).

Figure 3. Schematic structural model of the swelling behavior of polar liquid - montmorillonite complexes with 0.5M NaCl solution: (a) montmorillonite with the hydrated interlayer cations; (b) polar liquid (DN > 18) - montmorillonite complexes. Polar liquids (DN > 18) replace water molecules bound to the interlayer cations and coordinate to the interlayer cations directly; (c) limited swelling with 0.5M NaCl solutions; (d) polar liquid (DN < 18) - montmorillonite complexes. Polar liquids (DN < 18) coordinate to the interlayer cations through the water molecules; (e) osmotic swelling with 0.5M NaCl solutions.

3.4 Swelling mechanism of polar liquid - montmorillonite complexes with electrolyte solution

Figure 3 shows schematic structure model of the swelling behavior of polar liquid - montmorillonite complexes with 0.5M NaCl solution. The swelling mechanism of polar liquid - montmorillonite complexes can be interpreted by the bi - functionality of water molecules as electron donor and electron acceptor and the magnitude of the DN of polar liquid relative to that of water similar to swelling mechanism of FA - montmorillonite complexes with polar liquids.

Water molecules with DN of 18 usually bind to the interlayer cations to form the primary hydration shell (Fig. 3a). In the hydration shell, bi - functional water molecules can act as electron donor and electron acceptor. When the montmorillonite is treated using the polar liquids with larger DN than water, these polar liquids (DN > 18) replace water molecules bound to the interlayer cations and directly coordinate to the interlayer cations (Fig. 3b). Therefore, polar liquid (DN > 18) - montmorillonite complexes with 0.5M NaCl solution show the limited swelling because water molecules derived from 0.5M NaCl solution cannot penetrate into the interlayer (Fig. 3c).

On the other hand, when the montmorillonite is treated using the polar liquids with smaller DN than water, the water molecules bound to the interlayer cations can act as acceptors to these polar liquids (DN < 18) to form secondary coordination shell outside the primary hydration shell (Fig. 3d). As shown in Table 1, the basal spacings of polar liquid (DN < 18) - montmorillonite complexes show approximately 2.0nm, which is nearly in the critical range for osmotic swelling. For polar liquid (DN < 18) - montmorillonite complexes with 0.5M NaCl solution, water molecules will be further bound around the interlayer cations to form a large hydration shell within the secondary coordination shell formed by the polar liquids (DN < 18). Therefore, polar liquid (DN < 18) - montmorillonite complexes with 0.5M NaCl solution will show the osmotic swelling (Fig. 3 e).

4 CONCLUSIONS

The swelling mechanism of montmorillonite in nonaqueous liquids such as polar liquids and electrolyte solutions could be interpreted by the bi - functionality with electron donor and electron acceptor and the magnitude of DN of polar liquids.

The polar liquids with large DN preferentially coordinated to the interlayer cations of montmorillonite

Montmorillonite showed the osmotic swelling in bi - functional molecules such as water and FA except for alcohols.

Polar liquids with smaller DN than water activated the osmotic swelling of montmorillonite in electrolyte solutions.

ACKNOWLEDGMENTS

The authors wish to thank Prof. S. Yamanaka of Hiroshima University for his guidance and numerous helpful suggestions on this study.

REFERENCES

Daniel, D.E. & Bowder, J.J. 1996. Waste Containment Systems by Geosynthetics. In M. Kamon (ed.), *Proceedings of Second International Congress on Environmental Geotechnics. Environmental Geotechnics.* 3: 1275-1291. Rotterdam: Balkema.

Emerson, W.W. 1957. Organo-clay complexes. *Nature.* 180: 48-49.

Grim, R.E. 1953. *Clay Mineralogy.* 55-64. New York: McGraw-Hill.

Gutmann, V. 1976. Empirical parameters for donor and acceptor properties of solvents. *Electrochimica Acta.* 21: 661-670.

Israelachvili, J.N. 1991. *Intermolecular and Surface Forces, 2nd edition.* London: Academic Press.

Kondo, M. 1996. *US Patent 5,573,583.*

Konta, J. 1995. Clay and man: Clay raw materials in the service of man. *Applied Clay Science.* 10: 275-335.

MacEwan, D.M.C. & Wilson, M.J. 1980. Interlayer and intercalation complexes of day minerals. In G.W.Brindley & G.Brown (eds), *In Crystal Structures of Clay Minerals and Their X-ray Identification: Mineralogical Society.* 197-248. London.

Mayer, U., Gutmann, V. & Gerger, W. 1975. The acceptor number - A quantitative empirical parameter for the electrophilic properties of solvents. *Monatshefte fur Chemie.* 106: 1235-1257.

Mitchel, J.K. 1996. Geotechnics of soil-waste material interaction. In M. Kamon (ed.), *Proceedings of Second International Congress on Environmental Geotechnics. Environmental Geotechnics.* 3: 1311-1328. Rotterdam: Balkema.

Mortland, M.M. 1970. Clay-organic complexes and interactions. In N.C.Brady (ed.), *In Advances in Agronomy.* 22: 75-117. New York: Academic Press.

Norrish, K. 1954. The swelling of montmorillonite. *Discussion of the Faraday Society.* 18: 120-134.

Norrish, K. & Quirk, J.P. 1954. Crystalline swelling of montmorillonite. Use of electrolytes to control swelling. *Nature.* 173: 255-256.

Olejnik, S., Posner, A.M. & Quirk, J.P. 1974. Swelling of montmorillonite in polar organic liquids. *Clays and Clay Minerals.* 22: 361-365.

Onikata, M., Kondo, M. & Kamon, M. 1996. Development and characterization of a multiswellable bentonite. In M. Kamon (ed.), *Proceedings of Second International Congress on Environmental Geotechnics. Environmental Geotechnics.* 3: 587-590. Rotterdam: Balkema.

Tensmeyer, L.G., Hoffmann, R.W. & Brindley, G.W. 1960. Infrared studies of some complexes between ketones and calcium montmorillonite. Clay-organic studies. Part III. *Journal of Physical Chemistry.* 64: 1655-1662.

Theng, B.K.G. 1974. *The chemistry of day-organic reactions.* London: Adam Hilger.

van Olphen, H. 1977. *An Introduction to Clay Colloid Chemistry. 2nd Edition.* New York: J. Wiley and Sons.

Yamanaka, S., Kanamaru, F. & Koizumi, M. 1974. Role of interlayer cations in the formation of acrylonitrile - montmorillonite complexes. *The Journal of Physical Chemistry.* 78: 42-44.

Clay Science for Engineering, Adachi & Fukue (eds) © 2001 Balkema, Rotterdam, ISBN 90 5809 175 9

Swelling characteristics of fly ash-treated expansive soils

B. R. Phani Kumar & S. Naga Reddayya
Department of Civil Engineering, JNTU College of Engineering, Kakinada, India

ABSTRACT: Expansive soils are highly problematic in nature by virtue of their innate potential for volume changes consequent upon changes in moisture regime. They swell in rainy season and shrink in summer. Because of this alternate swelling and shrinkage many civil engineering structures are severely damaged. Different foundation techniques and soil stabilization techniques have been devised for mitigating this problem. This paper presents the effect of fly ash on swelling characteristics of highly expansive clays. The disposal of fly ash has become a serious problem all over the world. One of the ways of disposing fly ash is utilizing it as a geotechnical material. An experimental programme has been conducted for the determination of different swelling characteristics varying the percentage of fly ash. This paper presents a detailed discussion of test programme and results.

1 INTRODUCTION

The problems posed by expansive soils have been recorded all over the world. They swell or increase in their volume when they imbibe water in rainy season, and shrink or decrease in their volume when water evaporates from them in summer. The reason for this volume change behavior of these soils is the presence of mineral montmorillonite and enormous suction. Because of this alternate swelling and shrinkage, many lightly loaded civil engineering structures like residential buildings, pavements and canal linings etc. are severely damaged. To counteract this dual problem of swelling and shrinkage, many innovative foundation techniques and soil stabilization techniques have been devised by various researchers, some of which are briefly discussed below.

Under-reamed piles developed by Central Building Research Institute, Roorkee, India are bored cast in situ piles with enlarged bases connected at the top by plinth beams. Enormous uplift resistance is mobilized along the surface of the pile and reduces heave. They have been successfully employed in many problematic instances, especially in India.

Granular pile-anchors (Phani Kumar, 1995) are highly effective in arresting heave of foundations in expansive soils. In this technique, the foundation is anchored at the bottom of the granular pile to a mild steel plate with the help of a mild steel rod. Considerable shear resistance will be mobilized along the pile-soil interface and resists the uplift.

In mechanical alteration, the top layers of expansive soil are replaced by a non-expansive material. Sand cushion (Satyanarayana, 1966) and cohesive non-swelling (CNS) layer technique (Katti, 1978) are examples of this technique. The philosophy of the sand cushion is that in monsoon, the saturated sand occupies less volume, accommodating some of the heave of the underlying soil and in summer partially saturated sand bulks and occupies the extra space left by the swell of the soil.

In CNS layer technique, about top 1.0 m to 1.20 m of the expansive soil is replaced by a cohesive non-swelling soil layer. This also has been found to be quite effective in reducing the heave.

In some techniques that bring about a radical change in the nature of the soil are stabilization techniques using chemicals like lime and calcium chloride. Fly ash is an industrial waste, which poses the problem of its disposal, and causes pollution. It is estimated that in India 100 million tonnes of fly ash is produced every year (Sridharan et al. 1996). This calls for a bulk utilization of fly ash through different means. It can be used as a back fill material, construction material in embankments, replacing material in problematic soils like expansive soils, a base course material in pavements and in some other geotechnical applications. Hence, its characteristics and efficacy are to be studied from different angles before actually using it in the field. This paper presents the effectiveness of fly ash as an admixture in expansive soils by conducting a laboratory test programme using highly expansive clays.

Table 1 Index properties of the soil

	S_1	S_2
Specific gravity, G	2.72	2.73
Clay (%)	36	42
Silt (%)	48	44
Sand (%)	16	14
Gravel (%)	0	0
Liquid limit (%)	148	79
Plastic limit (%)	17	26
Plasticity index	131	53
Free swell index (%)	250	190
Indian Standard Classification based on Plasticity	CH	CH
Indian Standard Classification based on Degree of expansion	High	High

Table 2 Effect of fly ash on the swelling characteristics at constant dry unit weight

Flyash(%)	FSI(%)		S(%)		p_s(k Pa)	
	S1	S2	S1	S2	S1	S2
0	250	190	10.80	9	90	80
5	200	150	8.75	7.5	72	65
10	165	110	7.20	6	60	52
15	140	85	6.00	5	50	40
20	125	65	5.50	4	45	32

2 EXPERIMENTAL PROGRAMME

The test programme has been conducted on different expansive soils. The soils used in the experimental programme for studying the efficacy of fly ash in reducing the swelling characteristics have been collected from Warangal and Amalapuram in Andhra Pradesh. They are highly swelling soils having free swell index of 250% and 190%. The index properties of the soils are given in Table 1. The index properties have been determined according to the relevant Indian Standard Codes. The soils are named as S_1 and S_2 respectively.

Flyash has been obtained from a local thermal power station. It has a low specific gravity of 2.10 and is non-plastic in nature. The effect of fly ash on free swell index (FSI), swell potential and swelling pressure has been studied at different percentages of fly ash. The swell potential and swelling pressure has been determined on remolded soil samples by conducting free swell test in one-dimensional consolidometer. In all the tests water content has been kept zero and dry unit weight has been kept constant at 14 kN/m³ for obtaining measurable values of swell potential and swelling pressure. As suction will be maximum in dry clays, swell potential will also be very high. The percentage of fly ash has been varied as 0, 5, 10, 15 and 20 by weight of dry soil. A series of tests has also been performed on flyash-clay mixtures

so that the volume of clay is constant. The results of the tests are presented and discussed in the following section.

3 RESULTS AND DISCUSSION

After complete saturation of the sample the swell potential S(%) is determined as the ratio of increase in thickness to the original thickness of the soil sample expressed as percentage. Swelling pressure p_s is determined as the pressure corresponding to original void ratio obtained from e-log p curve.

Table 2 shows the values of free swell index (FSI), swell potential S(%) and swelling pressure p_s in kPa, when the dry unit weight has been kept constant.

It can be seen from the table that all the swelling characteristics, namely, FSI, swell potential, swelling pressure got reduced considerably with increase in percent fly ash. The trend of variation in FSI, swell potential and swelling pressure with increase in percent flyash is curvilinear as shown in Figs. 1, 2 and 3 respectively. All the above swelling characteristics showed marked reduction at low percentage of fly ash. As the percent fly ash increased, the reduction in the values of these quantities has been found to decrease. It is interesting to find that, when the fly ash content is increased to 20%, all the above swelling characteristics are reduced by about 50%, as can be observed from Table 2.

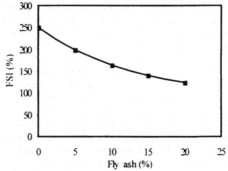

Fig.1 Variation of FSI with percentage fly ash

This reduction in swelling characteristics of expansive clays on addition of fly ash can be attributed to the reduction of amount of suction consequent upon addition of fly ash. The plastic fines contribute most to suction and swelling nature of expansive clays. When they are replaced by non-plastic fines of the material of fly ash the swelling nature is reduced. The interaction between clay particles that is necessary for swelling is reduced quite effectively by the addition of non-plastic fly ash particles. Hence the reduction in FSI, swell potential and swelling pressure.

Table 3 Swelling characteristics of flyash-clay mixture at constant volume of clay

Flayash(%)	S(%)		p,(kpa)	
	S1	S2	S1	S2
0	10.80	9	90	75
5	11.70	10	98	87
10	13.20	11	104	96
15	14.50	12	112	104
20	16.00	13	125	115

Table 3 shows the values of swell potential, S(%) and swelling pressure, p, in kpa when the volume of the clay has been kept constant.

It is interesting to find from Table 3 that the swell potential and swelling pressure increased for both the soils in the second series of tests conducted keeping the volume of the clay constant. As the volume of the clay is kept constant, the dry unit weight of the mixture increases because of increase in % flyash, thus increasing the dry unit weight of the clay also. As the number of clay particles to be compacted in a given volume increases, the interaction between them increases which is quite essential for the increase in swelling. Hence, even though the % of non-plastic fines of fly ash is increased, the net result is that the swell potential and swelling pressure increase when the volume of the clay is kept constant.

4 CONCLUSION

Addition of fly ash to expansive soils reduces the amount of suction and consequently their swelling nature. Free swell index which is an important index property of expansive clays and which reflects their swelling nature in a singular manner can be reduced quite effectively by the addition of fly ash. Swell potential and swelling pressure also decrease satisfactorily as percent fly ash increases. For the type of fly ash and for the type of soil used, 20% fly ash reduced the swelling characteristics by about 50% as determined from the free swell test. This reduction is basically by replacement of plastic fines by non-plastic fines of flyash. However, if the volume of the clay is kept constant with increase in % fly ash thus increasing swelling. Therefore, replacement of expansive soil by fly ash may be recommended for reducing swelling effectively and for solving the problem of disposing fly ash which is a hazardous industrial waste.

5 ACKNOWLEDGEMENTS

The authors gratefully acknowledge the laboratory work done by the following students: R. Jyothi, B. Santhi Sri, H. Sushma. K. Sita and B. Saritha.

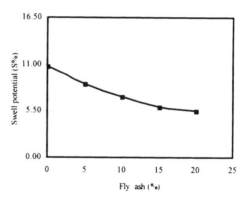

Fig.2 Variation of swell potential with percentage fly ash

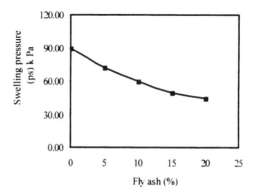

Fig.3 Variation of swelling pressure with percentage fly ash

REFERENCES

Katti. R.K. 1978, "Search for solutions to problems in black cotton soils". *First IGS Annual Lecture*. Indian Geotechnical Journal, Vol.9, No.1, pp.1-80.

Phani Kumar, B.R. 1995, "A study of swelling characteristics of and granular pile- anchors in expansive soils", *Ph. D Thesis submitted to JNT University*, Hyderabad.

Satyanarayana, B. 1966, "Swelling pressure and related mechanical properties of black cotton soils" *Ph.D Thesis, I.I.Sc*, Bangalore.

Sridharan, A. Pandian, N.S. and Rajasekhar, C. 1996, "Geotechnical characterization of pond ash", *Ash Ponds and Ash Disposal Systems*. Raju, V.S. et al (Eds) Narosa Publishing House. New Delhi.

Clay Science for Engineering, Adachi & Fukue (eds) © 2001 Balkema, Rotterdam, ISBN 90 5809 175 9

Stabilization of a compacted swelling clayey soil

B. Wibawa & H. Rahardjo
NTU-PWD Geotechnical Research Centre, School of Civil and Structural Engineering, Nanyang Technological University, Singapore

ABSTRACT: A compacted swelling soil stabilized with cement and lime was investigated. Soil samples were taken from Cikarang, West Java, Indonesia. Stabilization of the compacted soil with cement 5% by weight was found effective to lower swelling pressure and free swell of the swelling soil. Stabilization of the compacted soil with lime 5% by weight was not found to be effective to lower swelling pressure especially at the dry side of the optimum water content. Free swell was found to be lower than that of the original soil.

1 INTRODUCTION

1.1 Background

Expansive soils are found in many parts of the world, particularly in semi-arid areas. An expansive soil is generally unsaturated due to desiccation. Expansive soils also contain clay minerals that exhibit high volume change upon wetting. The large volume change upon wetting causes extensive damage to structures, in particular, light buildings and pavements (Fredlund & Rahardjo 1993). In a review from Chen (1988), expansive soils are detected in Australia, Canada, China, Israel, Jordan, Saudi Arabia, India, south Africa, Sudan, Ethiopia, Spain, the United States. In the United States alone, the damage caused by the shrinking and swelling soils amounts to about US$9 billion per year (Jones & Holtz 1973).

Swelling clayey soils are found in several areas in West Java, Indonesia. West Java is one of the provinces in Indonesia (see Fig. 1). This province consists of mountaineous area in the central and southern part and low land area in the northern part. Locations of the swelling clayey soils are not only in the northern part but also in the central part of West Java (Indonesian Ministry of Mines & Energy 1996).

In West Java, clayey soils are usually utilised as fill materials in industrial areas to save prime cost (Wibawa 1995). The clayey soils are removed from higher lands at the site or surrounding quarries in order to fill lower lands. Cut and fill at the same site area results in a cheaper cost of soil removal compared to that from the surrounding quarries. The swelling clayey soils are compacted layer by layer to obtain a certain degree of dry density

Improving engineering properties of a soil by mixing with admixtures is referred to as soil stabilization. The purposes of adding these admixtures to the soil are to increase strength, reduce deformability, provide volume stability, reduce permeability and increase durability.

Clayey soils can be mixed with cement in order to stabilize it. In United States, the mixing of clayey soils and cement were used in 1917 for highway construction and then in 1932 for low cost and all-weather roads (Catton 1959). Laboratory tests of cement modified soil clearly demonstrated that the grains of cement served as a nucleus to which the fine soil particles adhered (Felt 1955). Therefore, the quantity of clay and silt was significantly reduced as cement was added to produce a coarser grained soil of lower water holding capacity and increased volume stability and strength. Besides, an additional quantity of cement to the clayey soils causes a decrease in liquid limit and plasticity index of the clayey soils.

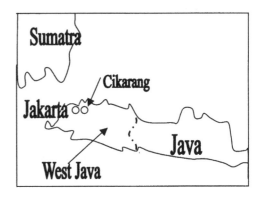

Fig. 1 Map of West Java (not to scale)

Clayey soils can also be stabilized by mixing it with lime. The use of lime stabilization of soils in construction has been in practice over thousands years old. In the United States, treatment of clayey soils with lime started in the 1950's. Many of state highways and major airports were built on lime stabilized clays. In the United Kingdom, lime stabilization techniques were not widely used until the late 1970's (Greaves 1996). The stabilization effect depends on the reaction between lime and the clay minerals. The main effects of this reaction are: an increase in shear strength of the soil, a reduction in swelling and shrinkage, a reduction in the water content and an improvement in workability and compaction characteristics. Either quicklime or hydrated lime can be used for soil stabilization.

1.2 Climate condition

Indonesia including West Java, as a tropical country, has two seasons annually. Generally, dry season is from April to October and rainy season is from October to April.

Figure 2 shows monthly rainfall in Bekasi and Sukamandi observation stations in the year 1997. The two observation stations are the nearest to the location of swelling clayey soils in Cikarang. The monthly rainfall was quite high from January to April, but it was very low from May to October. The rainfall was rather low in November and December. The total rainfall in Bekasi was 1269 mm and it was 2401 mm in Sukamandi in the year 1997.

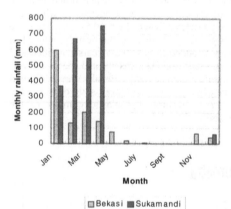

Fig. 2 Monthly rainfall in Bekasi and Sukamandi, West Java in 1997.

1.3 Objective of the paper

The objective of the paper is to investigate swelling pressures and free swells of a swelling clayey soil treated with cement and lime by laboratory testing. Swelling soil samples were taken from Cikarang, West Java.

2 SWELLING SOIL

2.1 Index properties

Index properties of the swelling soil in Cikarang are shown in Table 1. Based on the criteria from the Unified Soil Classification System, the soils from code 1, code 2, code 3 and code 4 are classified as inorganic clay of high plasticity (CH).

Table 1. Index properties of swelling soils in Cikarang

Sample Code	G_s	Liquid limit (%)	Plasticity index (%)	Shrinkage limit (%)	% silt	clay
1	2.65	97.48	72.18	14.92	25	75
2	2.66	95.48	66.08	16.04	35.5	64.5
3	2.64	95.03	70.32	13.50	37	63
4	2.61	97.42	67.82	13.51	53	47

Figure 3 shows grain size distribution of swelling soils in Cikarang. Using MIT classification for the grain size particles, clay fraction is almost predominant in all specimens except for the sample code 4.

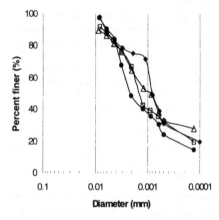

Fig. 3 Grain size distribution curves for soil samples in Cikarang

2.2 Compaction and free swell tests

Compaction tests were carried out to simulate field conditions where compaction was carried out during the filling up process. Laboratory compaction test was performed to prepare soil specimens for free swell test. Compaction test was done in accordance with ASTM D698. Figure 4 shows a relationship between dry density and water content for the sample code 2. The soil has a maximum dry density of 1.5 g/cm^3 and an optimum water content of 24.14%.

Free swell tests using oedometer (Smith 1973) were carried out on compacted specimens of the swelling soil. For each condition of water content and dry density, one specimen was taken for the free swell test. Initial surcharge of 10 kN/m² was applied to the specimens.

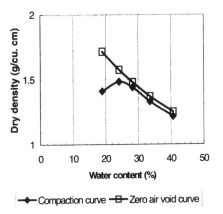

Fig. 4 Compaction and Zero air void curves for the soil sample code 2 in Cikarang

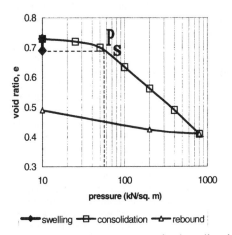

Fig. 5 Void ratio versus pressure for the soil code 2 in Cikarang at the optimum water content (24.14%)

Figure 5 shows a relationship between void ratio (e) and effective pressure (p). The curve is for the soil specimen that has an initial water content corresponding to the optimum water content (w_{opt} = 24.14%). The curve shows the swelling of the specimen as indicated by an increase in void ratio to a value greater than the initial void ratio. This means that water was absorbed by the soil specimen so that volume of the soil became greater than the initial volume. Subsequently the void ratio decreases as water comes out from the specimen due to the in-

creasing pressure. The rebound curve is found to reach a void ratio lower than the initial void ratio. Dotted line from initial void ratio intersects the consolidation curve to get the swelling pressure (P_s).

2.3 Swelling pressure and free swell

Swelling pressure can be defined as the pressure which prevents the specimen from swelling or the pressure which is required to return the specimen back to its original state (void ratio, height) after swelling (Chen 1988). In the above case, the swelling pressure (P_s) of all specimens are found to vary from 17.5 kPa to 55 kPa.

Water absorbed by the specimen results in the swelling of the soils. Figure 6 shows a relationship between the swelling pressures of the compacted soil and their compaction water contents. This figure indicates that the maximum swelling pressure corresponds to the optimum water content. It is also related to the maximum dry density of the compacted soil. This can be attributed to the specific surface that has the greatest value at the maximum dry density condition. Specific surface is defined as total surface area per unit bulk volume of soil (Hillel 1998). Soil absorbs the largest amount of water as compared conditions at other initial water contents.

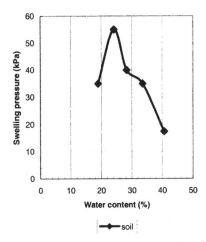

Fig. 6 Swelling pressure versus water content for the soil code 2 in Cikarang

Swelling pressure of the soil in the dry side of the optimum water content was found to be lower than that of the optimum water content. This can be explained by the lower dry density compared to the maximum dry density. Therefore, the specific surface of the soil specimen is smaller than that of the maximum dry density. As a result, the amount of water absorbed by the specimen is also smaller.

127

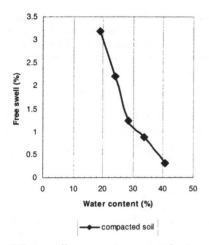

Fig. 7 Free swell versus water content for the soil code 2 in Cikarang

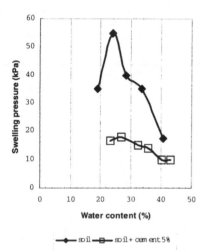

Fig. 8 Swelling pressure versus water content for the natural soil and the soil stabilized with 5% cement (code 2) in Cikarang

The wet side of the optimum water content (w_{opt}) shows a decrease in the swelling pressure from that of w_{opt}. In this case, the swelling pressure drops sharply with increasing water contents. This is due to the dry density at the dry side of optimum was lower than that of the optimum water content.

Figure 7 shows a relationship between free swell and water content for the natural soil code 2 in Cikarang. Free swell is a ratio between a change of volume to its initial volume. Figure 7 indicates that free swell is quite high at the dry side of the optimum water content and it sharply decreases when the initial water content increases. At the dry side of the optimum water content, much water can be absorbed due to a low water content in the soil speci-

men. However, the soil specimen has enough water in its pore in the wet side of the optimum water content. As a result, swelling is not so large. Free swell of the soil specimens is from 0.31% to 3.19%.

3 SOIL STABILIZED WITH CEMENT

Soil can be stabilized by mixing it with a chemical compound such as cement. Stabilization with cement is suitable for rather coarse-grained soils. Stabilization of soils with cement results in a mixture of pulverised soil, cement and water. The compaction of this mixture produces a new material, which due to its strength, favourable deformation characteristics and its resistance to water, is well adaptable as a road pavement, road and building foundation. Hence, stabilization means improving the engineering properties of the soil. In connection with swelling soils, the main concern is its volume stability, i.e., swelling pressure and free swell.

3.1 Compaction and free swell tests

Compaction test of soil and cement mixture was carried out in order to prepare soil specimens for a free swell test. Compaction test was performed in accordance with ASTM D698 and it was done immediately after mixing the soil and cement without waiting period. Delay in compaction, which allows the hydration process to commence and thus builds up the strength of clods, is a major cause of loss in strength because the mixture becomes more difficult to compact and the final density achieved will, therefore, be lower (West 1959). The soil was mixed with 5% cement by weight in a dry condition. Cement used in the test is ordinary Portland cement. The range of water content in the test was from 23.23% to 43%. As a result, the optimum water content was 27% and the maximum dry density was 1.42 g/cm³.

Free swell tests using oedometer were also carried out on compacted specimens of the soil-cement mixture. After finishing the compaction test, specimens for free swell test were prepared immediately from each specimen of the compaction test. Procedure of the free swell test for the mixture was same to that of the natural soil. Hence, swelling, consolidation and rebound stages were also obtained similar to that of the natural soil.

3.2 Swelling pressure

Figure 8 shows a relationship between swelling pressure and water content for the soil-cement mixture specimens as compared to that of the natural soil. This figure indicates that swelling pressure decreases significantly compared to that of the natural soil. Swelling pressure is from 10.2 kPa to 18 kPa. On the dry side of the optimum water content, the swelling pressure was higher than that of the wet side. According to Hausmann (1990), the reaction of cement and water forms cementitious calcium silicate and aluminate hydrates, which bind soil particles together. The hydration releases $Ca(OH)_2$, which in turn may react with components of the soil, such as clay minerals. Hydration occurs immediately upon contact of cement and water, but reactions of $Ca(OH)_2$ and the clay minerals are slower. Therefore, swelling pressure decreases because soil particles can absorb smaller amount of water than that of the natural soil.

3.3 Free swell

Figure 9 shows a relationship between free swell and water content in case of the soil-cement mixture specimens compared to that of the natural soil. This figure indicates that the free swell was from zero on the wet side of optimum water content to 0.95% on the dry side of optimum. The free swell decreases significantly due to the presence of cement mixture in the soil. According to Kezdy (1979), the hydration of cement creates rather strong bonds between the various mineral substances, and forms a mass, which efficiently encloses the non-bonded soil particles. The chemical surface effect of the cement reduces the water affinity of the clay and, thereby, the clay's water-retention capacity. Therefore, the free swell magnitude becomes lower. The use of the cement mixture to this particular swelling soil provides an effective way in keeping its volume stability.

4 SWELLING SOIL STABILIZED WITH LIME

Soils can also be stabilized by mixing with a chemical compound such as lime. The stabilizer may be quick lime or hydrated lime. Stabilization of soils with hydrated lime is broadly similar to cement stabilization in that similar criteria and testing and construction techniques are employed. They differ, however, in two aspects: first, stabilization of soils with lime is applicable to far heavier clayey soils, and is less suitable for granular materials, and, second, it is used more widely in preparing a soil for further treatment. As a temporary measure, such modification or stabilization needs not necessarily be effected to the standards required for permanent construction. Quicklime may also be used for excessively wet or dry soil conditions respectively (Ingles et al. 1972). In connection with swelling soils, the main concern is its volume stability, i.e., swelling pressure and free swell.

4.1 Compaction and free swell tests

Compaction test of the soil-lime mixture was carried out in order to prepare soil specimens for a free swell test. Compaction test was performed in accordance with ASTM D698 and it was done immedi-

ately after mixing the soil and lime without waiting period. Soil was mixed with 5% lime by weight in a dry condition. Lime used in the test was a hydrated lime mixed with a little quick lime. The range of water content in the test was from 13.24% to 47.73%. The results give an optimum water content of 24.95% and a maximum dry density of 1.44 g/cm^3.

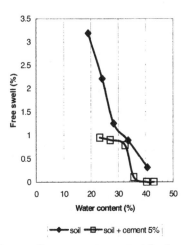

Fig. 9 Free swell versus water content for the natural soil and the soil stabilized with 5% cement (code 2) in Cikarang.

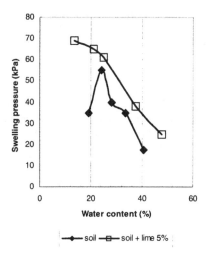

Fig. 10 Swelling pressure versus water content for the natural soil and the soil stabilized with 5% lime (code 2) in Cikarang

Free swell tests using oedometer were also carried out on compacted specimens of the soil-lime mixture. After finishing the compaction test, the specimen for free swell tests were prepared immediately.

129

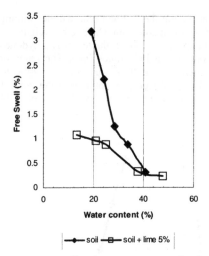

Fig. 11 Free swell versus water content for the natural soil and the soil stabilized with 5% lime (code 2) in Cikarang

Procedure of the free swell test for the mixture was same to that of natural soil. Hence, swelling, consolidation and rebound stages were also obtained similar to that of natural soil.

4.2 Swelling pressure

Figure 10 shows a relationship between swelling pressure and water content in case of the soil and lime mixture specimens compared to that of the natural soil. This figure indicates that swelling pressure increases significantly, especially on the dry side of the optimum water content. Swelling pressure was from 25 kPa to 69 kPa. They were greater than the swelling pressure of the natural soil. When the specimen with the hydrated lime and quick lime was immersed in water during the free swell test, it absorbed water immediately and resulted in the formation of calcium hydrate from the quick lime. As a result, an expansive force was released (Czernin 1980). The presence of quick lime produces an additional swelling pressure in the mixture, so that swelling pressure of the soil-lime mixture was greater than that of the natural soil. On the wet side of the optimum water content, swelling pressure of the soil-lime mixture had no significant difference with swelling pressure of the natural soil due to quite high initial water contents so the specimens can only absorb a small amount of water.

4.3 Free swell

Figure 11 shows a relationship between the free swell and water content in case of the soil and lime mixture specimens compared to that of the natural soil. This figure indicates that free swell was from

0.24% on the wet side of optimum water content to 1.09% on the dry side of optimum. Free swell of the soil-lime mixture decreases significantly than that of the natural soil. However, this phenomenon still needs further research.

5 CONCLUSIONS

Based on the experimental data above, the following conclusions can be drawn for this particular soil:
a. Stabilization of swelling soil with 5% cement by weight is effective to keep a volume stability, i.e., swelling pressure and free swell.
b. Stabilization of swelling soil with 5% hydrated lime and quick lime mixture by weight is not quite effective to decrease swelling pressure, especially on the dry side of the optimum water content. Free swell was found to be lower than that of the natural soil.

REFERENCES

Catton M. D. 1959. Early soil-cement research and development. *Journal of the Highway Division*, ASCE 85: 1 - 14.
Chen, F.W. 1988. *Foundations on expansive soils, Developments in Geotechnical Engineering 54*, New York: Elsevier.
Czernin W. 1980. *Cement chemistry and physics for civil engineers*. London: Crosby Lockwood & Son.
Fredlund D.G. & Rahardjo H. 1993. *Soil Mechanics for Unsaturated Soils*, New York: John Wiley & Sons, Inc.
Hausmann M. R. 1990. *Engineering principles of ground modification*, New York: McGraw Hill.
Hillel D. 1998. *Environmental Soil Physics*. New York: Academic Press.
Indonesian Meteorology & Geophysics. 1997. *Monthly rainfall for Bekasi, Sukamandi and Aljasari*, Jakarta.
Indonesian Ministry of Mines & Energy. 1996. *Engineering Geological Map*. Jakarta.
Ingles O. G. & Metcalf J. B. 1972. *Soil stabilization*. Sydney: Butterworths.
Jones D. E. & Holtz W. G. 1973. Expansive soils-The Hidden Disaster. *Civil Eng. ASCE*: 87-89.
Kezdy A. 1979. *Stabilized earth roads. Developments in Geotechnical Engineering 19*. Amsterdam: Elsevier.
Smith, A. W. 1973. Method for determining the potential vertical rise, PVR. *Proc. Workshop Expansive Clays and shales in Highway design and Construction*, Laramie, 1: 189-205.
West G. 1959. A laboratory investigation into the effects of elapsed time after mixing on the compaction and strength of soil cement. *Geotechnique* 9(1): 22.
Wibawa B. 1995. Cohesive soil at the ground surface as fills material. *Trisakti Science Magazine* 3: 82-92. (in Indonesian)

Clay Science for Engineering, Adachi & Fukue (eds) © 2001 Balkema, Rotterdam, ISBN 90 5809 175 9

Upheaval damage to a wooden house by swelling of mudstone due to weathering

H. Kawakami
Open University of the Air, Nagano Study Center, Japan

H. Abe
Chubu-Chishitsu Corporation, Nagano, Japan

ABSTRACT: A wooden house which was built on hard mudstone was damaged by the swelling of the foundation after three years. A ridge of a small hill was excavated to a depth of 4 meters and the house was constructed on the mudstone. The house was founded on the mudstone through strip footings and small individual footings. The measurements of the house are 11 by 15 meters. Surrounding footing of the house was normal but the central individual footings showed some upheaval movement. The maximum upheaval of the floor has reached ten centimeters. After the upheaval movement of the floor, the surface of the mudstone was changed to dry and fine grained materials. According to the test results of the mudstone in the laboratory, mudstone swelling due to weathering is caused by the repetition of frost -defrost actions and drying-wetting. In one year the range of air temperatures under the floor ranged from 25 to 0 degrees centigrade. The ground temperature at a depth of 40 centimeters below the surface changes from 20 to 10 degrees from the inside to outside. From the measurements of temperature, the results show there is no weathering due to frosting and defrosting or the temperature changes. According to the porewater pressure changes at a depth of 60 centimeters, it is concluded that the consequences of weathering mainly depend on the repetition of drying and wetting in the mudstone layer up to a depth of one meter below the surface.

1 INTRODUCTION

A ridge of a small hill was excavated to a depth of 4 meters and a wooden house was built on the mudstone ground. After three years the floor of the house started to heave and maximum relative upheaval reached 120 mm in five years. There was no upheaval on the surrounding footings of the house. Upheaval movement was observed on the individual footings which are distributed inside the house. Mudstone is very hard and it was not easy to break up the mudstone using a pickaxe during the excavation. However, the surface of the mudstone under the floor changed into dry and granular materials after three years. When the dried surface layer was removed, wet and hard rocks were found. As for the causes of weathering of mudstones, the existence of clayminerals (Chigira 1990; Oyama et al. 1998), temperature changes and freezing and thawing have been identified (Matsukura 1996). The swelling properties of the mudstone accompanying proceeding the weathering action were investigated.

2 UPHEAVAL OF THE HOUSE AND GROUND CONDITIONS

The measurements of the house are 11 by 15 meters as shown in Figure 1. The strip footings were mainly arranged under the surrounding wall and under the inside columns in parts to support structural loads. The individual stone footings were arranged inside the house to support the floor loads. The structures of the footings are shown in Figure 2. As the load applied on the individual footings is less than that of the strip footing, the foundation depth of the individual stone footing is less than that of the strip footing. Therefore the individual footings were easily affected by ground deformation.

The floor heights were surveyed along the centerlines A-A and B-B of the hall. The position of the lines A-A and B-B are shown in Figure 1. Results obtained are shown in Figure 3. The relative upheaval measurement at the center of the hall had reached 120 mm. The strip footing along the surroundings of the house didn't show any upheaval as shown in the results of points 8 and 15. However strip footings arranged inside the house heaved a little as shown in the results of points 3 and 12. The individual stone footing showed a large upheaval as shown in the results of point 4, 5 and 6 etc.. Wooden

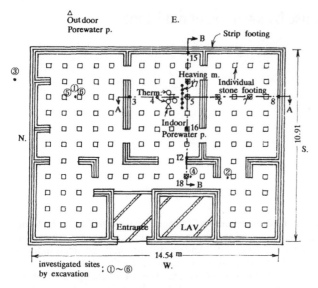

Figure 1   Plan of the wooden house and arrangement of footings

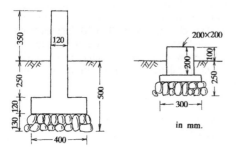

Figure 2   Measurements of a strip footing and an individual stone footing

houses in Japan used to be built on the footings, so that an air space would be kept between the floor and ground surface.

Mudstone properties were investigated by the excavation of thin layers both in outdoor and indoor sites. The investigated sites are shown in Figure 1. Distribution of the void ratios and water contents against the depth of the mudstone are shown in Figure 4. According to the measurements at Site 2, mudstones as far as 10 centimeters from the surface were quite weathered. Their void ratios increased and their water contents decreased with depth.

Dried and finegrained mudstone was observed at the surface of Site 2. The surface of Site 1 also dried and finegrained. Site 3 was outdoor and mudstone was under the saturated condition. Site 4 was under the individual footings and the mudstone was becoming to loose condition. Sites 5 and 6 were different from others on time measured. Their weathering had occurred at a depth as far as 20 centimeters. In order to

show the degree of saturation of the measured results, the relationship between the void ratio and water content is shown in Figure 5. The degrees of saturation S_r had decreased with proceeding weathering.

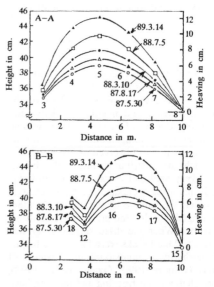

Figure 3   Upheaval of the floor

The test results on consistencies are shown in Table 1. The each natural water content is different, but their consistency indices coincide and show the ordinary properties of low plastic mudstone in the Tertiary period (Abe& Kawakami 1989).

132

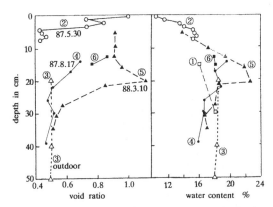

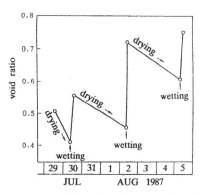

Figure 4 Void ratios and water contents against the depth of mudstones

Figure 7 Repeated test of drying-wetting on the circular column specimen

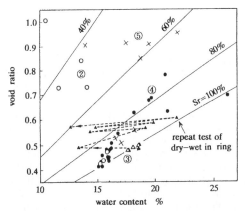

Figure 5 Soil conditions after repeated drying-wetting and in situ

3 BEHAVIOUR UNDER REPEATING OF DRYING AND WETTING

When the mudstone was immersed in the water or dried simply, no remarkable volume change could be observed. However, the mudstone rock showed volume expansion with the repetition of wetting and drying as shown in Figure 6. A rock sample was set in the ring of consolidation test. The sample was soaked with water under the load of 98 kPa and then dried in the air under no external load. The void ratio had changed from 0.4 to 0.5 during 4 repetitions of drying and wetting.

The rock specimen that had formed circular column was also tested by the repetition of drying and wetting as shown in Figure 7. The rock specimen was 6 centimeters in diameter and 7.7 centimeters in height. The specimen was enveloped by a filter paper on which water was sprayed and drying was repeated. After 3 repetitions of drying and wetting, the sample expanded in volume remarkably. It was observed that the volume expansion was also large at the subsequent wetting period when drying time was long.

4 BEHAVIOUR UNDER REPETITION OF FREEZING AND THAWING

The mudstone also showed expansive behaviour under the cycles of freezing and thawing. As shown in

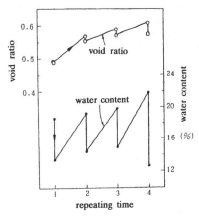

Figure 6 Repeated test of drying-wetting in the ring of consolidometer

Table 1 Consistency of mudstones

Sample	Site ① Depth 22 cm.	Site ③ Depth 50 cm.	Site ③ Depth 70 cm.
Water Content in Situ	22.5	19.4	38.4
Specific Weight	2.72	2.71	2.71
Liquid Limit	66	60	63
Plastic Limit	28	24	29
Plasticity Index	38	36	34

Figure 8, both the water content and the void ratio of rock specimen increased with the cycles of freezing and thawing. The mudstone could maintain a saturated condition during repetitions of freezing and thawing.

5 CONTINUOUS MEASUREMENTS OF HEAVING AND SOIL SUCTION IN SITU

The heaving of floor in the wooden house had been surveyed for five years after completion of the house. The plan view of heaving is shown in Figure 3, showing the heaving along the centerlines A-A and B-B. The heaving behavior over an elapsed time is shown in Figure 9. The measuring Point 5 was set on the individual footing and showed the maximum amount of heaving, and Point 12 was set on the strip footing and showed a relatively small amount of heaving. The details of the heaving in first three years were not clear, but the heaving was almost linear with time. The swelling velocity increased a little after three years. It is considered that the weathering of mudstone might occur from a surface to a deep level.

Pore pressure and temperature were measured continuously at various depths from August in 1988 to February in 1989. The heaving of the ground at

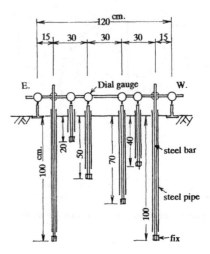

Figure 10   Heaving measuring apparatus of ground at various depths

various depths was measured using the apparatus shown in Figure 10. A level of 100 centimeters in depth was set as the standard. The following surface levels; 20 cm, 40 cm, 50 cm, and 70 cm in depth were measured as the relative heaving with that of 100 cm in depth. After drilling the ground, invar rods of various lengths were installed and the ends of the invar rods were anchored with cement in the ground. Each rod was protected by a steel pipe to cut the friction. The results measured are shown in Figure 11. There are two points that are not realized easily. Firstly, the heaving at a depth 20 cm was larger than that of surface at the west side. Secondly, at a depth 70 cm there was some settlement. These are realized by supposing a depth 100 cm at west side behaved heave a little. There is a possibility that the settlement at the depth of 100 cm showed the greater heave than that at depth 70 cm. Measuring the depth of 100 cm might be not enough to investigate the weathering of mudstone.

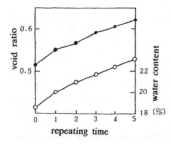

Figure 8   Repeated test of freezing-thawing

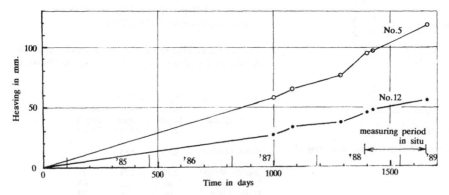

Figure 9   Heaving measurement of the floor after construction

134

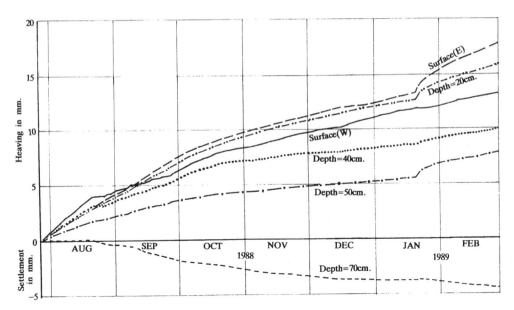

Figure 11   Heaving measurement results at various depths

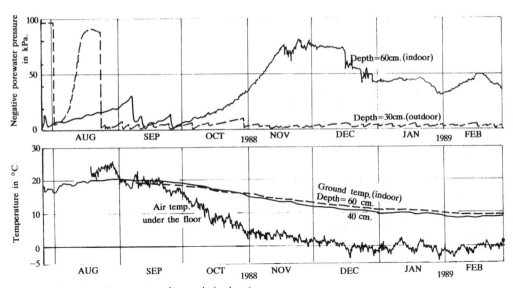

Figure 12   Suction and temperature changes during heaving measurement

Moreover, heaving suddenly increased late in January 1989 in Figure 11. This reason must be considered with other factors measured.

The porewater pressures and temperatures had been measured at the same time as the heaving measurement as shown in Figure 12. The porewater pressures were measured at a depth 30 cm of outdoor and at depth 60 cm of indoor. The temperatures were measured at a depth 40 and 60 cm indoors. Also, the air temperature was measured under the floor. The measured positions of those are shown in Figure 1.

The outdoor suction increased in August and was nearly equal to zero in September. The indoor suction repeatedly rose and fell in September and greatly increase in October and November. The drying of the mudstone might occur in this period. Afterwards, the suction decreased in middle of December and late in January. The ground temperatures was 20 degrees centigrade in summer, and started to decrease in autumn and reach to 10 degrees centigrade in winter. The air temperature under the floor was start to decrease in October and stayed at about zero degree.

135

The reason that we are able to explain the sudden increase of swelling in late January is due to the fact that suction decreased in late January. However, a decrease in suction occurred in the middle of December in same manner. There were not any changes in swelling in middle of December. The clear relation between swelling and suction was not shown. However, it is supposed that the sudden increase in swelling was due to the wetting of dried mudstone.

6 COUNTERMEASURES AGAINST SWELLING OF THE MUDSTONE

It is considered that the swelling of ground was due to the repetition of drying and wetting of the mudstone. The ground surface inside the house was covered by a concrete layer of 10 cm in thickness in order to protect the mudstone from drying. Afterwards, the swelling of the surface was not observed any more.

7 CONCLUSIONS

The wooden house which was built on the hard mudstone had started to heave after three years. This was due to the weathering of the mudstone. This weathering was caused by repetition of drying and wetting. It is necessary to protect the mudstone from drying in order to avoid the weathering of fresh mudstone.

REFERENCES

Abe, H & Kawakami, H. 1989. Ground characteristics and foundations in Nagano Prefecture. *Jour. Kisoko* 44(2): 81-89 (in Japanese)

Chigira, M. 1990. A mechanism of chemical weathering of mudstone in a mountainous area, *Engineering Geology*, 29: 119-138.

Matsukura, Y.1996. Weathering and its rates on rock and rock materials, *Jour. Japanese Geotech. Soc. Tsuchi-to-Kiso* 44(9): 59-64 (in Japanese).

Oyama, T., Chigira, M. Ohmura, N. & Watanabe, Y. 1998. Heave of house Foundation by the chemical weathering of mudstone, *Jour. Japan Soc. Eng. Geol.* 39(3): 261-272 (in Japanese).

Clay Science for Engineering, Adachi & Fukue (eds) © 2001 Balkema, Rotterdam, ISBN 90 5809 175 9

Strain softening of unsaturated swelling clays

Takeshi Ito
Department of Civil and Environmental Engineering, Akita National College of Technology, Japan

ABSTRACT: Triaxial compression tests, direct shear tests and compaction tests are carried out on several blend samples of Toyoura standard sand, kaolinite and montmorillonite to investigate strain hardening and softening characteristics. Through the laboratory test result using above clayey soils, remarkable plastic flow during loading has been recognized especially in montmorillonite content samples, so that the plastic flow phenomena are evaluated based on a constitutive equation proposed by Richard - Abbott. The test specimen's behavior is a new evidence that supports the theory. The relationship between mechanical properties are also investigated with connection of soil physical properties. Several new relationships among parameters derived from different tests will be proposed as a function of strength and physical parameters of unsaturated swelling clays.

1 INTRODUCTION

Fundamental design in expansive soil such as bentonite is complicated and impeded by the fact that they have remarkable swelling behavior. The main reason is that the foundation containing swelling clay minerals especially the montmorillonite. Montmorillonite is contained much in bentonite, and plays an important role in the behavior of such unsaturated swelling clayey soils. Swelling behavior depend mainly on the content of montmorillonite (Ito 1980).

In this study, owing to grasp the strain behavior of unsaturated swelling clayey soils, several blend samples of bentonite at varying proportions mixed with Toyoura standard sand and kaolinite powder were used. Then uniaxial tests and triaxial tests (UU-test) were carried out for these samples. In test results, a typical strain hardening and softening behavior was recognized during loading. To investigate these test results, a statistical method and a non-linear analytical method proposed by Richard - Abbott (1975) were applied. The relationship between engineering properties and physical properties were also investigated with the connection of bentonite contents. It was recognized that soil strength depends highly on the amount of bentonite, that is, in the same water content, a small

quantities of bentonite soil showed larger strength, and the results also indicated that the shear strength parameters, c and ϕ change gradually as the dry density γ_d changes. Thus an analytical equation for swelling clayey soils was conducted by using soil parameters derived from laboratory tests.

In this paper, it was confirmed that unsaturated swelling clayey soil behavior especially the strain softening phenomena were put well by proposed models, and verified as a function of physical and mechanical properties.

2 TEST PROCEDURE AND SAMPLES

Swelling phenomena such as swelling pressure (Ps) depends mainly on the following factors.

$$Ps = f (Ck, Mq, Wp, Sp, Ec, Wi, T, ...) \qquad (1)$$

where, Ck is clay mineral type, Mq is quantity of clay mineral, Wp is composition of ionic content in pore water, Sp is structure or array of clay particle, Ec is degree of compaction or present load, Wi is initial water content, T is temperature.

In this study, we evaluated several above mentioned factors, that is, Ck and Mq are represented by bentonite content, Wp is replaced by distilled water, Ec is given as a constant pre-consolidation load as 3.0 kgf/cm² and 48 hours, Wi is settled optionally, and T is settled at 20 ℃ .

According to our laboratory tests, the strain hardening and softening phenomena were recognized in all samples. These test samples were divided into two types as follows:
(1) blend samples of bentonite and Toyoura standard sand of which bentonite content are 10%, 20%, 30%, 40% and 50%, and named them as Bs_{10}, Bs_{20}, Bs_{25}, Bs_{30}, Bs_{40} and Bs_{50}.
(2) blend samples of bentonite, kaolinite and Toyoura standard sand, of which bentonite content are 5%, 10%, 15%, 20%, 33% and 50%, and defined similarly as Bks_5, Bks_{10}, Bks_{15}, Bks_{20}, Bks_{33} and Bks_{50}.

A bentonite used in this study is Kunigel V3 produced at Kunimine mine, Yamagata prefecture, Japan. It contains montmorillonite more than 70%.

Prior to mechanical analysis, consistency test and compaction test were performed on all samples, then each optimum water content (Wopt) was searched to apply for triaxial compression test. Since it was impossible to measure pore water pressure from such unsaturated bentonite samples, undrained tests such as uniaxial compression test and triaxial UU-test were carried out.

3 TEST RESULTS

3. 1 *Compaction tests*

First of all, all test samples were examined to find out the maximum density ($\gamma_{d \cdot max}$) through the compaction tests.

Figure 1 shows an example of finding γ_d as a function of optional water content, W, to find $\gamma_{d \cdot max}$ for 75% Toyoura standard sand with 25% bentonite (Bs_{25}) and 75% kaolinite with 25% bentonite (Bk_{25}). As is seen from the figure, γ_d - W relationship is quite different between these two, i.e. high maximum density was obtained for Bs_{25} regardless of the same bentonite content.

Figure 2 is also γ_d - W relationships for a series of Bks samples. It was recognized that each maximum density depends highly on the content of bentonite, and a high bentonite including samples showed lower $\gamma_{d \cdot max}$ values. From a series of compaction tests, we found the following relationships:

$$\gamma_{d \cdot max} = 1.802 \exp.(- 4.005 \times 10^{-3} \times Bks) \qquad (2)$$

$$Ip = 16.14 + 5.881 \, (Bks) \qquad (3)$$

$$\gamma_{d \cdot max} = 3.128 - 0.5401 \cdot \ln (W) \qquad (4)$$

in which, Ip is plasticity index.

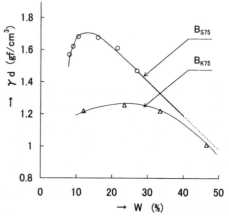

Figure 1 γ_d - W curves of bentonite soils.

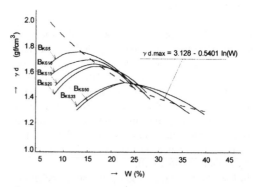

Figure 2 $\gamma_{d \cdot max}$ trend curve of Bks clayey soils.

3.2 *Characteristics of Bks samples*

A series of hexa-diagram shown in Figure 3 (a) to Figure 3 (f) represent physical and mechanical properties of Bks samples. In which, Bks is a bentonite content, C_u is clay content under 2 micron mm. ε_s is 96 hours swelling strain during unloading under the pre-consolidation pressure from 3.2 kgf/cm² to 0.1 kgf/cm². It can be seen that swelling strain is getting larger as the bentonite content increases, while $\gamma_{d \cdot max}$ gradually decreases as is shown in equation (2) and (4).

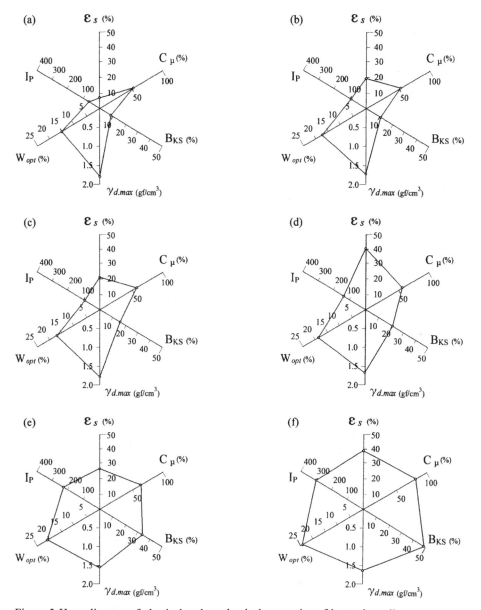

Figure 3 Hexa-diagram of physical and mechanical properties of bentonite soils.

3.3 *Characteristics of uniaxial compression strengths*

Figure 4 shows uniaxial stress strain curves for different initial water content (Wi) and bentonite content (Bs). It is pointed out that stress strain curves change with Bs and Wi, and each curve shows the strain hardening and softening phenomena. A similar tendency was also observed at triaxial compression tests. This is the main reason why these phenomena will be analyzed theoretically later. As is seen in Figure 4, different patterns of peak strength paths were recognized in accordance with initial water content. These particular phenomena can been seen in the relationship between stress and strain curves for such as unsaturated clayey soils. They are considered to be traces of peak strengths based mainly on the both of bentonite content and water

139

content. Their characteristics can be classified into three typical peak strength paths as shown in the Figure 5.

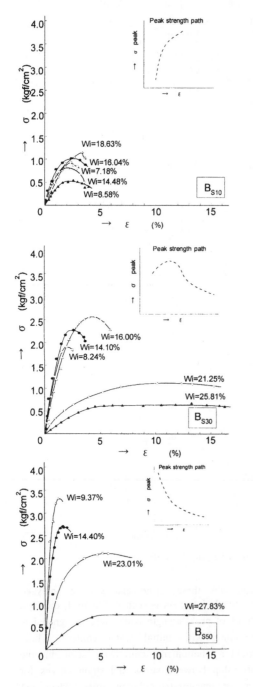

Figure 4   Characteristics of uniaxial peak strength by different bentonite soils.

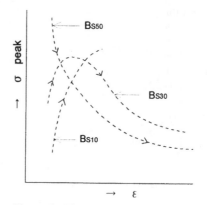

Figure 5   Three patterns of peak strength paths.

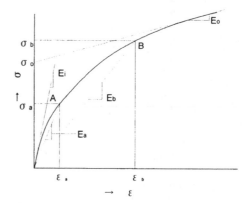

Figure 6   Parameters by Richard-Abbott theory.

4   INVESTIGATIONS ON STRAIN SOFTENING

4.1 *Analytical method*

An unsaturated soil sample including swelling clay minerals shows generally remarkable strain softening phenomena after reached to peak strength. Although there are many theoretical equations that deal with the strain softening phenomena of such soil spacemen, the method of Richard - Abbott is one of widely used theories. This theory is believed to be the best theory to be presented specifically to express the strain softening phenomena. It is characterized that a tangential line parameter (Ep) is introduced for the slope of strain softening part. Parameter m is also introduced optionally in a stress strain curve at point A and B, while several other specified parameters are taken in the Figure 6. The constitutive equation is as follows:

$$\sigma = E_0 \cdot \varepsilon / \{1 + (E_0 \cdot \varepsilon / \sigma_0)^m\}^{1/m} + E_p \cdot \varepsilon \qquad (5)$$

$E_0 = Ei - Ep$ (6)

in which, Ei is the initial elastic coefficient, Ep is the plastic modulus, and it takes usually negative value, m is the shape parameter of stress strain curve. If Ep is zero, then the equation (5) becomes as follows:

$$\sigma = E_0 \cdot \varepsilon / \{1 + (E_0 \cdot \varepsilon / \sigma_0)^m\}^{1/m} \quad (7)$$

In this situation, above equation means elastic-perfectly plastic state. When m = 1, $E_0 = 1/a$ and $\sigma_0 = 1/b$, $\sigma - \varepsilon$ curve becomes like Kondner's hyperbola. Strain softening is expressed in taking account of tangential slope parameter (Ep) after peak strength.

Figure 7 shows a comparison between experimental and theoretical results proposed by Richard -- Abbott. This diagram is a case of UU-test of Bks33 samples for the case of $\sigma_3 = 1.6$, 0.8 and 0.4 kgf/cm². From the figure, peak strength is seen gradually decreasing as σ_3 decreases, and strain softening phenomenon is acceralating.

Figure 8 shows similarly a case of Bs samples with constant $\sigma_3 = 0.6$ kgf/cm². It is recognized that a higher bentonite content sample shows lower strength, and its strain softening phenomena can be described with the Richard-Abbott model that is fairy different from Kondner's hyperbolic model as shown in the Figure 8.

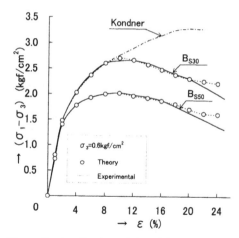

Figure 8   Application of the Richard-Abbott theory for Bs samples.

In this way, unsaturated clayey soils including swelling clay minerals is a new evidence that supports well the Richard-Abbott model.

5 RELATIONSHIP BETWEEN TRIAXIAL COMPRESSION TEST AND DIRECT SHEAR TEST

Let's put the following parameter Ef obtained from triaxial compression test (UU-test) as:

$$(\sigma_1 - \sigma_3)_f / \varepsilon_f = E_f \quad (8)$$

This Ef becomes larger as Bs value decreases. Next equation is to certify that the bentonite content is a

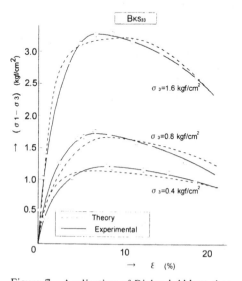

Figure 7   Application of Richard-Abbott theory for bentonite, kaolinite and sand mixture soils.

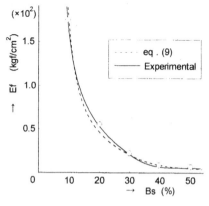

Figure 9   Relationship between bentonite content Bs and strength parameter Ef.

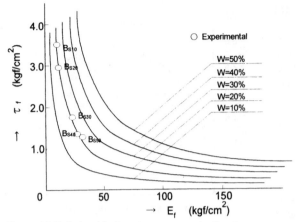

Figure 10 Relationship between τ_f and E_f by proposed theory eq.(13).

key factor to make unsaturated swelling clayey soil strengths as shown in Figure 9.

$$E_f = 1.688 \times 10^4 \times (1/Bs)^2 \qquad (9)$$

Besides above relationship between Bs and E_f, the following equation was also obtained from direct shear tests.

$$\tau_f / W = 0.0281 \times Bs^{1.703} \qquad (10)$$

Equation (9) becomes as follows.

$$Bs = 129.9 \ (1/\sqrt{E_f}) \qquad (11)$$

From above equations, we obtain the following relationship.

$$\tau_f / W = 111.4 \times E_f^{-0.8515} \qquad (12)$$
or
$$\tau_f = 111.4 \times W \ \{ \ (\sigma_1 - \sigma_3)_f / \varepsilon_f \ \}^{-0.8515} \qquad (13)$$

Figure 10 shows the above relationship for various water contents, in which round marks show experimental results obtained at W = 20 %. Thus we found that shear strengths between different tests agreed well with Equation (13). As is seen from the figure, we can estimate direct shearing strengths, τ_f from triaxial compression tests.

6 CONCLUDING REMARKS

Unsaturated swelling clay samples mixed with bentonite, kaolinite and Toyoura standard sand have been used to investigate their strain softening phenomena. From the experimental results, we have studied several remarkable characteristics as follows.

Shear strength was getting lower as increased bentonite content. Strain softening behavior was greatly influenced by water content (more than 10%) and bentonite content (more than 5 %). Thus the compression strengths were found to trace independent 3 peak strength paths accordance with bentonite content.

The behavior of strain softening of unsaturated swelling clayey soils has proved to fit well with the proposed model by Richard - Abbott.

From the results of triaxial and direct shear tests, a constitutive equation among ε_f, W and other strength parameters has proposed in the form of equation (13). The relationship was recognized to have a good expression with experimental results.

REFERENCES

Ito,T.1980. Swelling behavior with respect to geological hazards. *Tsuchi to kiso*. JSSMFE. No.1152:31-38(in Japanese with English abstract).
Richard, R. and Abbott, B. 1975. Versatile elastic-plastic stress-strain formula, *ASCE, EM4*, 511-515.

Clay Science for Engineering, Adachi & Fukue (eds) © 2001 Balkema, Rotterdam, ISBN 90 5809 175 9

Shrinkage strain potentials of fiber reinforced and raw expansive soils

A. J. Puppala
Department of Civil and Environmental Engineering, The University of Texas, Arlington, Tex., USA

C. Musenda
Soiltech Engineering and Testing Incorporated, Fort Worth, Tex., USA

ABSTRACT: This paper presents a fiber reinforcement technique and its effectiveness in reducing the shrinkage strain potentials of two expansive soils from Texas, USA. Two types of fibers and three dosage levels were investigated. Three-dimensional volumetric shrinkage tests were conducted on soil samples. Test results indicate that fibers can reduce the volumetric shrinkage strain potentials of expansive soils. No appreciable difference in the final decrease in shrinkage strain values due to fiber size was noted. However, fiber dosage levels influence the decrease in shrinkage strains.

1 INTRODUCTION

Expansive clays have been found in many parts of the world. The movement of expansive soils is a major concern in many construction projects since it causes significant distress to the infrastructure. Other types of expansive soils are man-made expansive soils that include chemically stabilized sulfate rich soils and compacted clay liners, which exhibit expansive behavior due to pore water chemistry variations. Repair costs of the structures built on the expansive soils run into billions of dollars annually (Wiggins 1978; Chen 1988; Nelson and Miller 1992). In the United States alone, these costs are estimated to be around $10 billion. The knowledge base on expansive soils indicates a need for new and effective expansive soil treatment strategies (Chen 1988; Nelson and Miller 1992; Musenda 1999).

There are a number of treatment methods available for stabilizing expansive soils. These include chemical additives, prewetting, soil replacement with compaction control, moisture control, and surcharge loading. All these methods have limitations and they do not always address the shrinkage movements in dry environments (Musenda 1999). Hence, new methods are needed to solve or reduce volumetric strain movements of expansive soils.

In recent years, polymeric fibers have been increasingly used in various geotechnical design and application areas (Crockford 1993; Gregory and Chill 1998). This is because these materials can be manufactured with desired properties, no leaching problems, or environmental concerns. The fibers are already used in concrete materials to control volumetric shrinkage cracks (Rebeiz et al. 1994; Alhozaimy et al. 1996). Similar advantages can be achieved in the expansive soils in controlling shrinkage strains. However, this needs to be investigated. Hence, a research study was conducted to investigate the influence of fiber reinforcement on the volumetric shrinkage strain potentials of expansive soils. This paper presents a summary of the test results and explains the effectiveness of polypropylene fibers in reducing shrinkage volumetric strains of expansive soils.

2 BACKGROUND

The concept of soil reinforcement has dramatically changed the function of soil as a construction material. The introduction of modern soil reinforcing materials such as polymeric materials has enabled engineers to effectively use even unsuitable in-situ soils as construction materials in a wide range of geotechnical and geoenvironmental applications (Musenda 1999). Reinforcing techniques primarily use continuous inclusions oriented in a preferred direction to achieve strength and stability in a soil mass

(Koerner 1998). Examples are the use of geotextiles or geogrids as reinforcing elements in the soil mass. They are generally placed in a predetermined orientation in alternating layers.

Other reinforcing elements are fibers, which include single monofilament fibers, injected randomly into sands, and oriented and randomly distributed, discrete fibers for effective reinforcement of both sandy and clayey soils (Gregory and Chill 1998). The behavior of continuous, oriented reinforcements has been widely researched and understood. In contrast, there is limited information available on the behavior and field performance of soils reinforced with discrete, randomly distributed fibers. However, in the last few years, there has been a growing interest in the potential use of fibers as a viable soil reinforcing material.

Strength aspects of fiber-reinforced soils are well understood. Maher and Woods (1990) investigated the influence of fiber reinforcement on Kaolinite clays. They conducted laboratory unconfined compressive, split tension, and three point bending tests on fiber reinforced soils which indicated that inclusion of fibers increases the peak compressive strength and ductility of kaolinite. Similar results were also reported on natural clay and sand mixtures (Gregory and Chill 1998; Puppala and Musenda 1998).

Influence of fiber reinforcement on volume change behavior of soils is not reported in the literature. However, it is reasonable to assume that the fibers do reduce shrinkage strains of soils. This is because the recycled polymer fibers were used to reduce shrinkage cracks in concrete materials (Rebeiz et al. 1994; Alhozaimy et al. 1996). The mechanical behavior of concrete and soil materials under shrinkage conditions is similar. Hence, the fibers are assumed to reduce the volumetric movements of the expansive soils in shrinkage environments.

3 EXPERIMENTAL PROGRAM

Two expansive clayey soils sampled from Irving and San Antonio cities in Texas were used as control soils. Atterberg tests were conducted (in accordance with ASTM D4318-93) on both the Irving and San Antonio clays. The results are summarized in Table 1. Results of the Atterberg tests show that both clays are inorganic clays with high plasticity indexes. Both clays are classified as CH as per the Unified Soil Classification System.

The fibers shown in Figure 1 were used in this study. These are fibrillated polypropylene fibers of nominal 25-mm (1-inch) and 50-mm (2-inch) lengths manufactured by synthetic industries. Short fibers are 25.4 mm long and long fibers are 50.8 mm long. The fibers are manufactured from parallel oriented long-chain polypropylene. The fibers have high tensile strength properties ranging from 552 to 758 MPa. This indicates that these materials can provide excellent reinforcement in construction materials. In addition, they offer chemical resistance, are not affected by salts in soils, and survive in high temperature conditions. The melting and ignition points of these materials are 162°C and 593°C, respectively. The polymers are not affected by biological degradation and ultraviolet degradation.

Table 1: Physical characteristics of clays

Property	Irving	San Antonio
Liquid Limit	82	73
Plastic Limit	27	27
PI	55	46
< No. 200	84	80
USCS	CH	CH

Figure 1. Fibers used in this investigation

Volumetric shrinkage tests were then conducted on raw samples of both clays and on soil samples reinforced with short and long fibers. Four levels of fiber-reinforcement were used. These are 0% (control), 0.3%, 0.6% and 0.9% of fibers by dry weight of soil.

Standard Proctor tests were first conducted on both expansive clays to determine the dry unit weight and moisture content relationships. These results, presented in Figure 2, are used to determine the moisture content and dry densities for soil sample preparation.

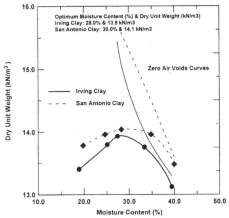

Figure 2. Standard proctor test results

The linear shrinkage test measures the horizontal shrinkage strain of a standard bar of soil paste. The test was based on the Texas Department of Transportation's method, Tex-107-E. The standard bar test and molds of the Tex-107-E method were too small for the present shrinkage tests since the fibers were longer than the width of the molds. This could cause adverse boundary effects on the test results. Hence, standard proctor compaction mold size samples (100 mm diameter X 112 mm height) were prepared (Figure 3) and tested in determining the shrinkage characteristics of soil samples.

The oven-dried soil sample was weighed, proportioned and mixed with water content at the liquid limit state to form slurry. The fiber-reinforcements were added to the soil slurry and mixed uniformly. The mixture was then poured in the molds. Specimens in the molds were then oven-dried at 80°C for one day. After the completion of tests, samples were sliced to assess the presence of fibers at different depth levels. This procedure showed that the slurry and fiber

mixing procedure provided uniform reinforcement levels at various sample cross-sections. The specific gravity of fibers (around 0.91) is less than the specific gravity of soil mix. Hence, the majority of the fibers were still in suspension during the testing period. No significant settlements or segregation of fibers were observed at majority of the fiber dosages. However, at high fiber dosage levels of 0.9%, small amounts of fiber segregation were observed. Therefore, strains were recorded at three different heights in both diametral and vertical directions. These values were then averaged. These average strains were more representative of the total fiber dosage amounts in the soil samples. Total volume changes due to shrinkage were then calculated using the diametrical and vertical shrinkage strains.

Figure 3. Test setup showing molds used for volumetric shrinkage measurements

4 DISCUSSION OF TEST RESULTS

Figure 4 shows the effects of using short fibers on the shrinkage potentials of both Irving and San Antonio clays. The volumetric shrinkage values of raw Irving clay samples averaged around 42%. Samples of the same clay with 0.3% short fibers exhibited shrinkage strains of 36%. This is a 6% reduction in the absolute shrinkage strain value of the raw soil. This also represents a 14% reduction of the original shrinkage strain of raw soil. Similarly, reinforced soils at 0.6% and 0.9% reinforcement exhibited shrinkage strains of 34% and 24% respectively. These values represent

Volumetric Shrinkage for Irving and San Antonio Clay
Fiber Type: Short, 1 inch

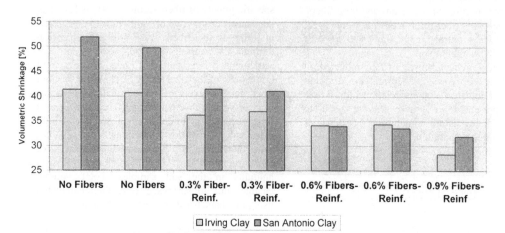

Figure 4. Effects of short fiber reinforcement on the volumetric shrinkage on both Irving and San Antonio clays

Volumetric Shrinkage for Irving and San Antonio Clay
Fiber Type: Long, 2 inches

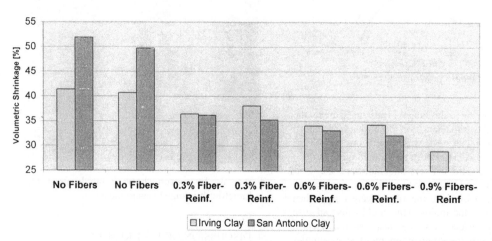

Figure 5. Effects of long fiber reinforcement on the volumetric shrinkage on both Irving and San Antonio clays

absolute reductions of 8% and 12% or percent reductions of 19 and 28 of original volumetric shrinkage strains.

The shrinkage characteristics of San Antonio clay showed similar trends (Figure 4). Raw samples averaged a volumetric shrinkage strain of 50%. Samples with 0.3% short fibers had undergone a volumetric shrinkage strain of 42%. This represents 8% in absolute reduction of shrinkage strain. At 0.6%, the volumetric

shrinkage strain measured was 34% (16% absolute reduction) and at 0.9%, the shrinkage strain was 32% (18% absolute reduction). These results are a clear indication that reinforcement of short fibers is effective in curbing the volumetric shrinkage strains.

Figure 5 presents the shrinkage behavior of both Irving and San Antonio clays reinforced with long fibers. For Irving clay samples, there is a 5% absolute reduction of volumetric shrinkage at

0.3% reinforcement, 8% reduction at 0.6% reinforcement, and a 12% decrease at 0.9% reinforcement, respectively. The absolute reduction in volumetric shrinkage strains for San Antonio clay were 14% at 0.3% reinforcement, 16% at 0.6% reinforcement, and 20% at 0.9% reinforcement. The 20% reduction corresponds to 40% reduction of original shrinkage strains. Such reduction is considered quite significant in producing the modified subgrades that undergo less volume changes in dry environments.

Test results presented in Figures 4 and 5 show that both fibers are effective in reducing the volumetric shrinkage potentials of expansive clays. The increase in the percentage of fiber reinforcement has decreased the shrinkage strain potentials of soils. Fibers provide tensile resistance to shrinkage induced tensile forces within the soil mass. As a result, fiber reinforced soils have higher tensile strength. Therefore, these samples undergo lesser amounts of shrinkage strains than raw expansive soil samples.

A statistical comparison test on both test results of short and long fibers was conducted to evaluate the best fiber size in reducing the shrinkage strains. Statistical t-tests were conducted as a part of this analysis on both test results. A 95% percent confidence level was used. The long fibers decreased shrinkage strains around 2% more than the short fibers. However, this decrease was proven to be small and statistically insignificant. In summary, both fiber sizes provided similar test results.

5 SUMMARY AND CONCLUSIONS

This paper presents a fiber reinforcement methodology, which was shown to decrease the shrinkage potentials of natural expansive clays. Though not tested in this research, the results can also be extended to manmade problematic soils such as clay liners. The following are some of the major conclusions deduced from this research:

- Slurry soil specimen preparation using polypropylene fibers showed smaller amounts of fiber segregation. The method used in this paper provided samples that yielded repeatable shrinkage test results.

- Both short and long fibers decreased the volumetric shrinkage strain potentials of expansive soils. The maximum absolute reduction reported in this study was 20% at 0.9% fiber reinforcement level. This

represents approximately 40% reduction of original volumetric strains.

- Statistical analysis on both fiber test results indicated that both fiber sizes provided similar results.

Future research studies in this area should focus on the matric suction potential measurements of the similar fiber reinforced soils and also on the effects of combined chemical and fiber treatments on expansive soils. Such studies further help in developing effective stabilization methods for natural and manmade expansive soils. Current research at The University of Texas at Arlington is already focusing on these topics.

ACKNOWLEDGEMENTS

This study was partially supported by the Advanced Technology Program (ATP) of Texas Higher Education Coordinating Board, Austin, Texas under the grant no. 1407610-50. The authors would like to acknowledge this support. Also, the authors would like to acknowledge Mr. Gary Gregory, P.E. of Gregory Geotechnical, Inc. and Mr. David Chill, P.E. of Synthetic Industries for providing suggestions and material support for this research.

REFERENCES

Addison, M. B. 1995. Lab Methodology for Optimizing Injected Chemical Modifiers Utilized in Swell Reduction. *Ph.D. Thesis*, The University of Texas at Arlington, Arlington, Texas, 1995.

Alhozaimy, A. M., Soroushian, P., Mirza, F. 1996. Mechanical Properties of Polypropylene Fiber Reinforced Concrete and the Effects of Pozzolanic Materials. *Cement and Concrete*. Vol 18: 1996.

Chen, F.H. 1988. *Foundations on Expansive Soils*. Elsevier, 1988, 463 pages.

Crockford, W. W. 1993. Strength and Life of Stabilized Layers Containing Fibrillated Polypropylene. Transportation Research Record No. 1418, Transportation Research Board. Washington, D.C. 1993.

Gregory, G. H., and Chill, D. S. 1998. Stabilization of Earth Slopes with Fiber Reinforcement. Proceedings, 6th International Conference on Geosynthetics. Atlanta, Georgia, 1998.

Koerner, R. M. 1998. Designing with Geosynthetics. Fourth Edition, Prentice Hall, 1998.

Maher, M. H., and Woods, R. D. 1990. Mechanical Properties of Kaolinite/Fiber Soil Composite. Journal of Geotechnical Engineering, ASCE. Vol. 120: No. 8, 1990.

McGown, A. et. al. 1985. Soil Strengthening Using Randomly Distributed Mesh Elements. Proceedings of the 11th International Conference on Soil Mechanics and Foundation Engineering. San Francisco, Vol. 3, 1985.

Musenda, C. 1999. Investigations on the Effects of Using Discrete Randomly Distributed Fiber Reinforcement in Expansive Foundation Soils. Masters Thesis. The University of Texas at Arlington, 131 pages, 1999.

Nelson, J.D. and Miller, D. J. 1992. Expansive Soils: Problems and Practice in Foundation and Pavement Engineering. John Wiley and Sons, Inc., New York, 1992.

Puppala, A.J. and Musenda, C. 1998. Investigation of Geofiber Reinforcement Method on Strength, Swell and Shrinkage Characteristics of Soils. Fifth International Conference on Composites Engineering. Las Vegas, Nevada, July, 1998.

Rebeiz, K.S., Fowler, D.W., Paul, D.R. 1994. Mechanical Properties of Polymer Concrete Systems Made with Recycled Plastic. ACI Materials Journal. V. 91, Np.1, January-February, 1994, pp. 40-45.

Wiggins, J. H. 1978. Building Losses from Natural Hazards: Yesterday, Today and Tomorrow. National Science Foundation Report, 1978.

Clay Science for Engineering, Adachi & Fukue (eds) © 2001 Balkema, Rotterdam, ISBN 90 5809 175 9

A technique for making unsaturated samples using membrane filters

K. Ando, J. Konishi & T. Toyoda
Department of Architecture and Civil Engineering, Shinshu University, Nagano, Japan

H. Tanahashi
Department of Civil Engineering, Daido Institute of Technology, Nagoya, Japan

ABSTRACT: The most common specimens of element tests for unsaturated soils are unsaturated specimens made by dynamic or static compaction methods and saturated specimens made by a pre-loading method. By dynamic or static compaction methods, it is hard to make highly homogeneous samples which have a clear stress history. By the pre-loading method, it may be easy to make those, but not applicable for low plastic soils. To avoid these difficulties, a method for making unsaturated samples using the pressure plate method in a triaxial cell has been suggested. But this method has some serious difficulties arising from the use of ceramic plates. In this study, we propose the pressure membrane method in which membrane filters are used instead of ceramic plates in the pressure plate method, for the purpose of developing a simple method of making homogeneous unsaturated samples.

1 INTRODUCTION

Adachi & Oka (1981) suggested that the selection of soil types and a method for making specimens are important for study on typical mechanical behaviors of unsaturated soils. An appropriate method for making specimens should therefore be selected to study specific properties of soils. Karube et al. (1986) found that the dynamic compaction method is better than the static compaction method to make specimens which are structually isotropic and easy to treat. Abe (1994) has shown that it is possible to make unsaturated specimens having a clear stress history using the pressure ceramic plate method in a triaxial cell.

In this study, an alternative method is proposed for making unsaturated soil samples by using the pressure membrane method; Membrane filters used in this method are easier to treat than ceramic plates used in the pressure plate method.

2 CHARACTERISTICS OF MEMBRANE FILTER

The membrane filters used in this study, pore size of 0.2μ m and thick of 125μ m, are a cellulose acetate type. Their bubble point, which is the pressure where air bubbles begin continuously to leak through the membrane filter in a completely saturated state, is 0.25MPa. These filters are generally used in eliminating particles and bacteria from liquid.

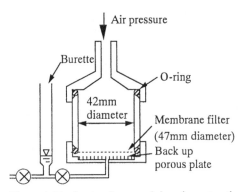

Figure 1. Equipment for examining air permeation.

2.1 *Air permeation behavior of membrane filter*

The equipment for examining air permeation behavior is a filter holder made of stainless steel as shown in Fig.1. A water saturated membrane filter is laid on the porous plate at the bottom of the small chamber and a constant air pressure is then applied on the surface of the membrane filter. When the air permeates through the water-saturated membrane filter, water is gradually squeezed from the bottom of the membrane filter. The water level in the burette rose up as a result of this. The amount of water passed through membrane is equal to the amount of air permeation. Relationships between elapsed time and amount of air permeation is shown in Fig.2. As a result, the amount of air permeation increased

Table 1. Physical properties of soils used in pF tests.

Soil name	Specific gravity	Liquid limit %	Plasticity index %	Initial water content %	Grain size distribution %		
					Clay size	Silt size	Sand size
Toyoura sand	2.65	NP	NP	35	0	0	100
Shirasu	2.49	NP	NP	50	0	2	98
DP50*	2.69	31.5	7.4	60	39	61	0
ASP600**	2.63	70.8	40.1	136	91	9	0

* A mixture of a silt and kaolin clay
** A kind of kaolin clay

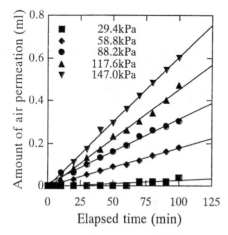

Figure 2. Amount of air permeation.

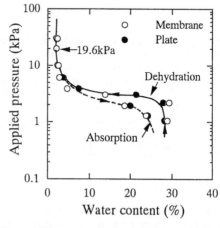

Figure 4. Water retention curves (Toyoura sand).

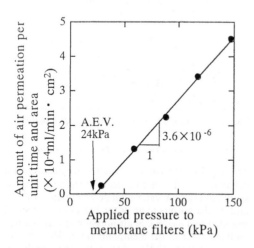

Figure 3. Air entry value of membrane filter.

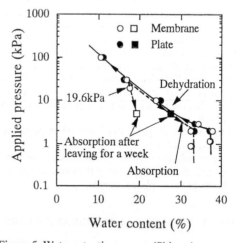

Figure 5. Water retention curves (Shirasu).

proportionaly with the elapsed time and reasonably increased with the applied pressure. The relationship between the amount of air permeated per unit area per minute and the applied pressure is shown in Fig.3. When the amount of air permeation was zero,

the air entry value of this membrane was estimated at about 24kPa. The air entry value of 24kPa in membrane filter is lower than that of ceramic disk with high air entry value used in commonly unsaturated soil tests.

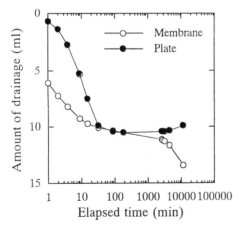

Figure 6. Amount of drainage for a week (Shirasu).

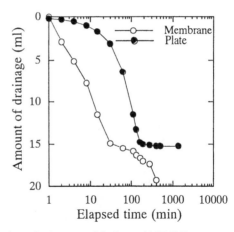

Figure 9. Amount of drainage (ASP600).

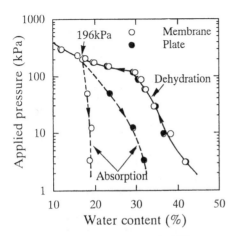

Figure 7. Water retention curves (DP50).

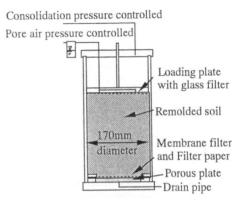

Figure 10. Equipment for making sample.

2.2 pF test with membrane filter

Whether or not water retention curves (i.e., soil-water characteristics curves) obtained by the pressure membrane method coincide with those obtained by the pressure plate method was examined.

The same equipment as shown in Fig.1 was used in pF tests by the pressure membrane method. The water saturated membrane filter was laid on the porous plate at the bottom of the small chamber. The soil sample was put into chamber and a constant air pressure was then applied on the top of the soil sample. Physical properties and initial water content of the soil samples used in this test are summerized in Table 1. Dehydration tests (pF tests with increasing pressure) and absorption tests (pF tests with decreasing pressure) were performed on Toyoura sand, Shirasu and DP50 but only dehydration tests were performed on ASP600.

Figure 4 shows the water retention curves for Toyoura sand. Dehydration curves were obtained by increasing applied pressure to a maximum of

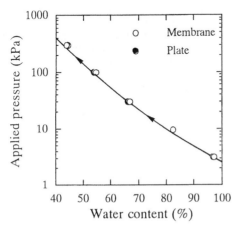

Figure 8. Water retention curves (ASP600).

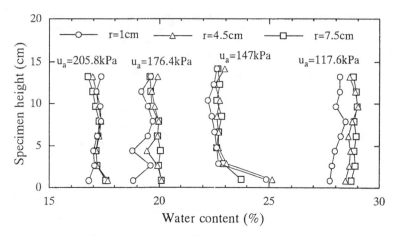

Figure 11. Distribution of water content (DP50).

29.5kPa. Absorption curves show the relationship between applied pressure and amount of water due to reduce applied pressure from 19.6kPa to 1.274 and 1.96kPa. It was found that dehydration and absorption curves using the pressure membrane method agree with those obtained by the pressure plate method.

Figure 5 shows the water retention curves for Shirasu. Dehydration curves were obtained by increasing of applied pressure up to 98kPa. Absorption curves are obtained by reducing applied pressure from 19.6kPa to 0.98, 1.96 and 4.9kPa. It was found that dehydration and absorption curves obtained by the pressure membrane method agree with those obtained by the pressure plate method. To examine the effect of applying pressure for an extended time, pressure was maintained at 19.6kPa for a week, and after that applied pressure was reduced to 4.9kPa in absorption process. The data point obtained by the pressure plate method (■) is indicated on the absorption curve as shown in Fig.5, but that obtained by the pressure membrane method (□) is not shown on the absorption curve.

Figure 6 shows the relationships between the amount of drainage and elapsed time. The drainage curve reached an equilibrium after 30 minutes in the pressure plate method. On the other hand, the drainage curve of the pressure membrane method appeared to reach a constant value. After elapsed time is over 2000 minutes, amount of drainage suddenly increased. This was not the result of the drainage of water but of air permeation through the membrane filter into the drainage. At this stage, absorption was prevented by the air space in the drainage formed with air permeated through the membrane filter.

In Figs.4 and 5 for Toyoura sand and Shirasu, the water retention curve based on pressure membrane method is similar to that obtained from pressure plate method till applied pressure up to the air entry value of the membrane filter. However, if a pressure lower than the air entry value of the membrane filter is applied continuously to the sample for a longer period, the air permeation cause and the water retention curve of the pressure membrane method in Fig.5 separates downward to the point (□). As a result, the pressure membrane method may be suitable for making an unsaturated soil sample with both dehydration and absorption histories in the pressure range lower than the air entry value of the filter and in rather shorter pressure-sustaining periods.

Figure 7 shows the water retention curves for DP50. Dehydration curves were obtained by increasing applied pressure to a maximum of 294kPa. Absorption curves are obtained by reducing applied pressure from 196kPa to 3.33, 12.5 and 49.3kPa. It was found that there was difference between pressure membrane method and pressure plate method in the absorption curves. In the pressure plate method, the maximum applied pressure is always lower than the air entry value of the ceramic plate. Therefore absorption must have taken place in the process. While, in the pressure membrane method, because the maximum applied pressure of 196kPa exceeds the air entry value of 24kPa in the membrane filter, air must permeate through the filter into the drainage line to prevent the return of water to the sample resulting in a small change of water content in Fig.7. This means that even when the applied pressure is released during sample making, only small absorption of water into the sample may occur.

However, both dehydration curves have a strong correlation even in the range of applied pressure which is much higher than the air entry value of the

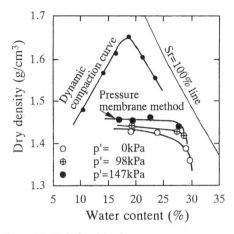

Figure 12. Relationships between water content and dry density.

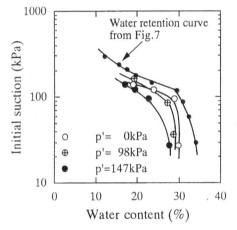

Figure 13. Relationships between water content and initial suction.

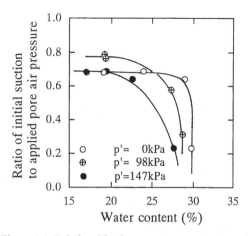

Figure 14. Relationships between water content and ratio of initial suction to applied pore air pressure.

membrane filter. The same result was obtained in the case of ASP600, see Fig.8.

Figure 9 shows relationships between the amount of drainage and elapsed time under the applied pressure of 294kPa. Amount of drainage have no change after 200 minutes in the case of the pressure plate method. On the other hand, the drainage curve of the pressure membrane method appeared to reach a constant value, but suddenly rose up after 100 minutes of time. This is because of the air permeation through the membrane filter into the drainage which is similar to Shirasu, as shown in Fig.6. These results of pF tests for the DP50 and the ASP600 suggest that it is possible to make samples having only the dehydration history in the range of applied pressure which is much higher than the air entry value of the membrane filter.

3 INITIAL STATES OF SAMPLES MADE BY THE PRESSURE MEMBRANE METHOD

Figure 10 shows the equipment for making samples by the pressure membrane method. The procedure is as follows: The remolded soil is put into the equipment after the saturated membrane filter is set up. Then the sample is consolidated by applying the pressure step by step and pore air pressure is applied to the sample when the consolidation is finished. Final equiliburium is certified by the 3t method based on the amount of drainage. Only DP50 samples were used in this test program.

3.1 *Distribution of water content*

A consolidated sample is taken out from the equipment and the cylindrical sample is divided into four pieces immediately. To examine the spacial distribution of water content of the sample, a quarter piece of the cylindrical column is divided into 24 parts, 8 parts in height and 3 parts in radius (r=1 cm, 4.5 cm and 7 cm from the center). The water contents of each part were measured. Figure 11 shows the distributions of water content for the samples in each sustained pore air pressure, u_a of 117.6, 147, 176.4 and 205.8 kPa as well as the same consolidation pressure, p' of 147 kPa.

Differences of water content in radial direction (r=1($\bigcirc$), 4.5($\square$) and 7.5cm($\triangle$)) were less than 2% in each cases. On the other hand, the maximum differences of water content between top and bottom were about 3%. The water content distribution of the specimens will be more homogeneous due to cut both the top and bottom in specimen.

3.2 Dry density

Dry density and water content of the trimmed sample were measured. The trimmed sample had a diameter of 5cm and a height of 12.5cm. Relationships between water content and dry density are shown in Fig.12. Dry density of samples made by the pressure membrane method are smaller than that of samples made by the dynamic compaction method, and the larger the consolidation pressure p', the higher the dry density.

Consequently, it may be suitable to make unsaturated loose samples by the pressure membrane method. However, the pre-loading method is suitable for cohesive soils because samples of cohesive soils can stand without any support or any suction. In conclusion, the method for making samples should be chosen by soil types.

3.3 Initial suction

Initial suction of the trimmed specimens were measured. Figure 13 shows relationships between the water content and the initial suction. All the measured suction of specimens are smaller than the suction of the water retention curve replotted from Fig.7. The reason is, the suction could decrease due to the disturbance during trimming. Near the water content which corresponds to the air entry value of the soil, measured suction rapidly drops. Therefore, initial suction may be lower than applied pore air pressure.

Figure 14 shows the relationships between the water content and ratio of the initial suction of the trimmed specimen to the applied pore air pressure for the making of sample. If a trimmed specimen maintains the applied pore air pressure, the ratio is equal, and if it does not maintain the applied pore air pressure at all, the ratio becomes to zero.

When the consolidation pressure, p' is zero, the ratio of initial suction to applied pore air pressure rapidly decreases till the water content corresponding to the air entry value of the soil. When the consolidation pressure p' is 98 and 147 kPa, the ratio gradually decreases to zero as the water content increases up to the water content corresponding to the air entry value of soil in the p'=0kPa. The maximum ratio is about 0.7, and suction is surely lost at about 30%. It is expected that the water content is about 30 % when the ratio is zero. This water content corresponds to the air entry value of the soil. There is no suction in the specimens if the applied pore air pressure is lower than air entry value of the soil. To obtain samples which can stand without any support, pore air pressure should be more than the air entry value of the soils, and it should be applied for low plastic or non-plastic soils.

4 CONCLUSIONS

1) For a pressure range more than the air entry value of the membrane filter, the water retention curves for dehydration obtained by the pressure membrane method seem to coincide with those obtained by the pressure plate method.
2) In the pressure membrane method, only small absorption occurs when the applied pressure is released. This property may enable us to make samples without absorption of water even when the pressure is released.
3) The pressure membrane method is better than the compaction method for making a looser sample.
4) Initial suction of the trimmed specimens is lower than the applied pore air pressure when the samples are made by the pressure membrane method. To make samples with a suction, higher pore air pressure than the suction should be applied.

ACKNOWLEDGEMENTS

The authors would like to acknowledge Dr. Hirofumi Abe of Chubu Chishitsu Co., for his sincere and very valuable advice and discussions. They are the most grateful to Prof. Ryousuke Kitamura of Kagoshima University for providing the Shirasu samples. They also would like to thank Mr. Ruka Kosugi, Koji Kawamura and Ms Reiko Ichikawa for their help in carrying out the experiments.

REFERENCES

Abe, H 1994. Experimental study on the estimation of mechanical properties for unsaturated soils. Ph.D thesis, Tokyo University, 42-109 (in Japanese).

Adachi, T. & Oka, F 1981. Testing methods and mechanical behaviors of unsaturated soil. Tsuchi to Kiso, Japanese Geotechnical Society, Vol.29, No.6 Ser.No.281, 27-33 (in Japanese).

Karube, D., Kato, S. & Katsuyama, J. 1986. Effective stress and soil constants of unsaturated kaoline. Proceedings of JSCE, III, No.370,179-188 (in Japanese).

Clay Science for Engineering, Adachi & Fukue (eds) © 2001 Balkema, Rotterdam, ISBN 90 5809 175 9

Behavior of unsaturated soil in triaxial test under undrained condition

Katsuyuki Kawai
Department of Civil Engineering, Kobe University, Japan

Daizo Karube & Ryuta Yoshimuta
Graduate School of Science and Technology, Kobe University, Japan

ABSTRACT: In this study, the specimens desaturated from saturated state using the pressure plate method were applied to triaxial compression test under undarined condition, and the suction change and the deformation characteristics were investigated. When the degree of saturation in the specimen was relatively high, the pore water pressure in the specimen increased with compression and decreased with expansion in shear process. However, the pore water pressure in the specimen increased again regardless of the expansion when shear strain became larger. On the contrary, when the degree of saturation was low, the pore water pressure in the specimen increased with shear strain regardless of the volume change. On the failure criterion, it was found that the critical state line obtained from undrained compression test agreed with that obtained from drained compression test.

1 INTRODUCTION

Since the hysteresis exists between the drying and the wetting processes in the water retention curve, the suction and the degree of saturation cannot be uniquely specified. In short, the degrees of saturation may be different even if the suctions in specimens are same. The effects of suction that affects the behavior of unsaturated soil depend on the degree of saturation. Accordingly, we must take the effects of hysteresis in the water retention curve into account to predict the behavior of unsaturated soil. The studies of unsaturated soil have been based on the triaxial test carried out under drained condition. According to these results, some constitutive models have been proposed. However, there are few test data obtained under undrained condition. And it is not confirmed that these knowledge obtained from the test results under drained condition can be similarly adaptable for the undrained condition. Accordingly, to extend our understanding for unsaturated soil, it is needed to accumulate the test data obtained under undrained conditions.

In this study, the undrained compression tests on unsaturated soil were carried out using the triaxial apparatus improved for unsaturated soils. The results were compared with the previous results obtained from the drained compression tests, and the dependency of behavior on the stress path was examined.

2 EFFECTS OF WATER DISTRIBUTION ON BEHAVIOR OF UNSATURATED SOIL

Karube et al. (1996, 1997) and Kato et al. (1996) showed that the behavior of unsaturated soil depends on not only the suction but also the water distribution in the soil mass. The form of pore water in unsaturated soil is classified into two categories as shown in Figure 1. The water occupying the pore of soil (Figure 1(a)) is named "bulk water" and the water clinging to the contact points between soil particles (Figure 1(b)) is named "meniscus water".

The effects of suction depend on these pore waters. The suction acting on the bulk water compresses the void and increases the stiffness of soil mass, whereas the suction acting on the meniscus water affects as a part of cohesion at the contact points between soil particles and increases the stiffness of soil mass. The distribution of these pore waters depends on the degree of saturation. Consequently, the behavior of unsaturated soils differs even if the suctions acting on them are the same values. The effects of these pore waters are evaluated by "bulk stress", p_b and "meniscus stress", p_m. Using two stress components, Karube et al. (1996) proposed the follwing equation to express the stiffness of soil mass of unsaturated soil under isotropic stress state:

$$\frac{\partial e}{\partial p'} = \frac{\lambda}{p' + p_m} \tag{1}$$

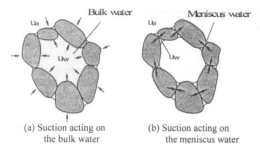

(a) Suction acting on the bulk water

(b) Suction acting on the meniscus water

Figure 1. Effects of suction in unsaturated soil

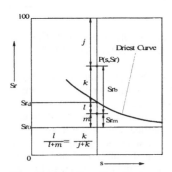

Figure 2. Yield curve defined by Eq. (3) in p' vs. q plane

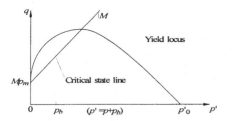

Figure 3. Estimation of S_r fraction

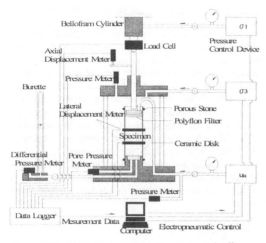

Figure 4. The triaxial test apparatus for unsaturated soil

Table 1. Physical properties of clay used

Gs	w_P	w_L	I_P
2.70	29.6%	43.0%	13.4

Here, e is the void ratio, p' is the skeleton stress and defined as $p' = p + p_b$, and p is the mean net stress defined as $p = p_T - u_a$. p_T is the mean total stress and u_a is the pore air pressure. λ is the compression index.

Equation (1) is based on the idea that the compressibility of soil due to p' is proportional for $p' + p_m$.

Moreover, they proposed following energy equation for the triaxial compression state:

$$q = M\left(p' + p_m\right) - p'\left(\frac{d\varepsilon_v^p}{d\varepsilon_s}\right) \qquad (2)$$

By adopting the associated flow rule, they obtained the following yield function:

$$\frac{q}{M} = -p' \log_e \frac{p'}{p_0'} + p_m\left(1 - \frac{p'}{p_0'}\right) \qquad (3)$$

Here, q is the deviator stress, M is the value of $q/\left(p' + p_m\right)$ at critical state, ε_v^p is the plastic volumetric strain, ε_s is the shear strain and p_0' is the referential value of p' at $q = 0$.

Figure 2 shows an example of yield curve defined by Equation (3) in p' vs. q plane. When we decide the yield curve under some suction condition, we must estimate the bulk stress and the meniscus stress quantitatively. Karube et al. (1996) assumed the "Driest Curve", which means a water retention curve under the condition that the bulk water never exist in soil mass, and which is shown in Figure 3. By using the assumption for the relations between the volumes of the pore waters as shown in Figure 3, these stress component are defined as follows:

$$p_b = \frac{S_{rb}}{100 - S_{r0}} \times s = \frac{S_r - S_{rd}}{100 - S_{rd}} \times s \qquad (4)$$

$$p_m = \frac{S_{rm}}{100 - S_{r0}} \times s = \frac{\left(100 - S_r\right)\left(S_{rd} - S_{r0}\right)}{\left(100 - S_{r0}\right)\left(100 - S_{rd}\right)} \times s \qquad (5)$$

Here, s is the suction, S_r is the degree of saturation, S_{r0} is the residual degree of saturation, S_{rd} is the degree of saturation on "Driest curve", S_{rb} is the volumetric ratio of the bulk water and S_{rm} is the volumetric ratio of the meniscus water. The sum of p_b and p_m represents the suction stress given by the following equation,

$$p_s = p_b + p_m = \frac{S_r - S_{r0}}{100 - S_{r0}} \times s \qquad (6)$$

Based on Equation (6), the following equation is obtained:

$$p' + p_m = p + p_s \qquad (7)$$

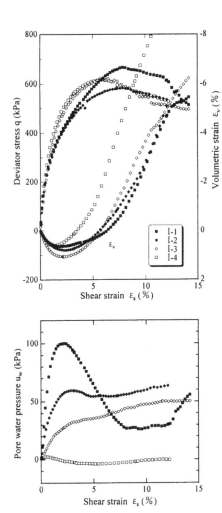

Figure 5. Relations of stress and pore water pressure with shear strain

3 EXPRIMENTAL METHOD

3.1 *Preparation of the Soil specimen*

A powder clay was used for the test. The physical properties of the clay are shown in Table 1. Firstly, we added deaerated water to the powder clay until the clay became slurry state at double of the liquid limit. And then, we preconsolidated it using an oedometer apparatus and obtained a block sample. This block sample was shaved into a specimen of 35mm diameter and 80mm height. The specimen was set in the triaxial compression apparatus improved for unsaturated soil as shown in Figure 4.

3.2 *Experimental procedure*

In the triaxial compression apparatus, the confining pressure, the axial pressure and the pore air pressure were controlled independently by a personal computer. Also, the pore water pressure in the specimen was measured under undrained state during the shear process. The pressure difference between the applied pore air pressure and the measured pore water pressure defines the suction in the specimen.

The suction paths shown in Table 2 were applied to the specimen under 20kPa confining pressure to desaturate the specimen. Then, the confining pressure was increased to the mean net stresses shown in Table 2. These desaturation and compression processes were conducted with small pressure increments, and we confirmed the equilibrium of the state in the specimen before the next incremental pressure was applied. After these desaturation and compression processes, triaxial compression tests were conducted under the constant mean principal net stresses and the constant pore air pressure in undrained conditions. Table 2 also shows the stress conditions and the states of specimens before the triaxial compression tests were started.

4 EXPERIMENTAL RESULTS AND DISCUSSIONS

4.1 *Volume change and pore water pressure in undrained shear process*

Figures 5 (a) and (b) show the relations of the deviator stress and the pore water pressure with the shear strain during undrained shear at $p = 245(kPa)$ and $u_a = 245(kPa)$.

In the specimens of I-1 and I-2 whose degrees of saturation are relatively high, the pore water pressures increased with compression and decreased with expansion. After that, though the specimens kept expanding, the pore water pressures began to increase again regardless of the volume changes after the deviator stress reached at the peak value. This phenomenon is less remarkable when the degree of saturation in the specimen becomes lower.

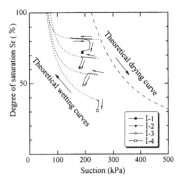

Figure 6. Water retention curve during shear

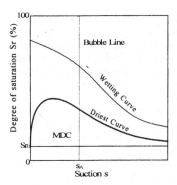

Figure 7. Diagram of modified driest curve

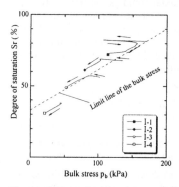

Figure 8. Effects of the expansion on bulk stress

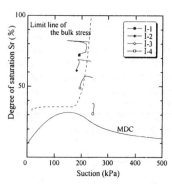

Figure 9. Limit of the bulk stress

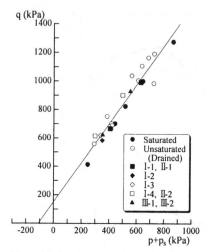

Figure 10. Stress states on the peak point

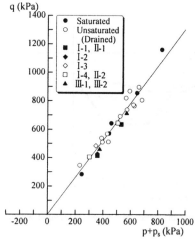

Figure 11. Stress states on the most compressed point

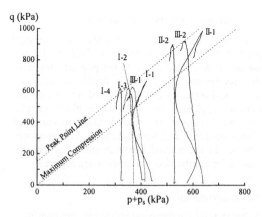

Figure 12. Stress paths of the triaxial tests under undrained condition

The stiffness of void occupied by the bulk water against shear stress is less than that of void sustained by the meniscus water. Consequently, the volume change during shear process mainly occurs in the void occupied by the bulk water, and the behavior of unsaturated specimen with higher degree of saturation is similar to the behavior of saturated specimen. Contrary, the void sustained by the meniscus water is destroyed with the deformation during shear process. The meniscus water is expelled out because it is changed to the bulk water by the shear deformation.

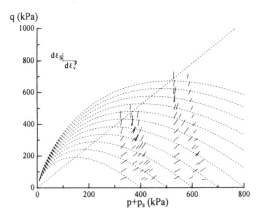

q (kPa)

$\frac{d\varepsilon_s}{d\varepsilon_v^p}$

p+p$_s$ (kPa)

Figure 13. Vectors of the strain increment
on the stress plane

The behavior of pore water in unsaturated soil appears to be affected by both of the behaviors of the bulk and the meniscus waters.

The past test results obtained on unsaturated soil under drained conditions (Karube et al. 1996, 1997) show that the specimen in high degree of saturation compresses greatly and expands little with the increase of shear strain. This behavior corresponds to the prediction of Equation (4), which means that the confining pressure induced by the bulk stress increases with suction and degree of saturation. However, in these test results shown in Figure 5, both the compression and expansion of the specimen with high degree of saturation become unremarkable under undrained shear process. This is because, in the specimen with high degree of saturation, the suction decreases during undrained shear process, and simultaneously the bulk stress shown by Equation (4) decreases. On the expansion, the behavior, observed in Figure 5, dose not correspond to Equation (4). This is because the bulk water cannot keep the shape and the effects of the meniscus stress exceed that of the bulk stress when the expansion begins with shear strain.

4.2 *The water retention curves during undrained shear process*

Figure 6 shows water retention curve obtained during shear process. In this figure, the theoretical drying curve and the theoretical wetting curve proposed by Kawai et al. (1999) are drawn with the dotted line and alternate long and short dash line. These theoretical curves are the water retention curves under the condition that the suction changes under a constant external stress. The theoretical drying curve represents the "Virgin Drying Line" which means the drying curve in the wettest side, and any drying curve will reach and go along this theoretical drying curve with suction increase. It is found that the

some parts of water retention curve obtained during volume compression agree with the theoretical wetting curve. However, the water retention curve obtained during volume expansion deviates from the theoretical water retention curve due to the suction change when the expansion increases. It can be thought that the bulk water resists against the expansion at first, and then the bulk water bubbles at certain expansion when the expansion induced by shear deformation occurs.

Figure 7 shows 'Modified Driest Curve (MDC)' proposed by Karube et al. (1998). The MDC is based on the concept that unsaturated state at $s = 0$ on the wetting curve contains of entrapped air and the meniscus stress is none. The bulk stress is modified with application of the MDC defined by following Equations (8) and (9) for Equation (4):

$$s \geq s_A ; \; \frac{S_{rd} - S_{r0}}{100 - S_{r0}} = \left(\frac{s_w}{s}\right)^{\lambda} \qquad (8)$$

$$s < s_A ; \; \frac{S_{rd} - S_{r0}}{100 - S_{r0}} = \left(\frac{s_w}{s_A}\right)^{\lambda} \left\{ 1 - (\lambda + 1)\left(\frac{s}{s_A}\right)^2 + (\lambda + 2)\left(\frac{s}{s_A}\right) \right\} \quad (9)$$

in which s_A is the air entry value, s_w is the water entry value and λ is the parameter obtained from the application of Brooks and Corey method (1966) for experimental wetting curve.

Figure 8 shows the relationship between the modified bulk stress shown by Equation (4) and the degree of saturation. The bulk stress decreases along the broken line in the figure when the bulk stress reaches certain values though the bulk stress increases with the expansion.

Figure 9 shows the relation between the degree of saturation and the suction during the shear process. In this figure, the dotted line means the limit line of the bulk stress, shown in Figure 8. Thus, we can predict the behavior of water retention curve after it deviates from theoretical curve by means of the limit line.

4.3 *Undrained strength for unsaturated soil*

Figure 10 shows the relations of the peak deviator stresses with confining pressures to which the suction stresses are applied. In this figure, the results from the triaxial compression test on unsaturated soil under drained conditions and on saturated soil are also plotted. In this state, the shear strength includes the effects of dilatancy. Because the dilatancy occurred in the specimen depends on degree of saturation, the stress states at peak point never represent the critical state. Consequently, the solid line shown in Figure 10 does not intersect the origin.

Figure 11 shows the relations of deviator stresses at which the maximum compressive volumetric strain occurred, with the confining pressures to which the suction stresses are applied. The effects of dilatancy on the shear strength are excluded using the stress states at the maximum compression point.

159

Table 2. Stress path and initial condition of specimen

Specimen	Suction Path under p=20 (kPa)	Stress state during shear test		Initial condition before the shear test		
		p (kPa)	u_a (kPa)	Void ratio	Water content (%)	Saturation (%)
I -1	0→245	245	245	0.959	28.84	81.22
I -2	0→294→48→245	245	245	0.919	23.06	67.80
I -3	0→392→48→245	245	245	0.950	19.65	55.87
I -4	0→490→245	245	245	0.975	13.54	37.51
II-1	0→245	441	245	0.905	27.67	82.62
II-2	0→490→245	441	245	0.920	13.92	40.86
III-1	0→294	245	294	0.923	21.03	61.55
III-2	0→490→294	441	294	0.881	20.64	63.24

From these results, it is found that we can estimate the undrained shear strength of unsaturated soil in the same way as the shear strength under drained conditions with application of suction stress p_s.

Figure 12 shows the stress paths expressed on the plane of q versus $p + p_s$. It is found that the stress paths converge along the dotted line means the peak point line.

Figure 13 shows the vectors of strain increment on the plane of q versus $(p + p_s)$. The yield function defined by Equation (3) is based on the assumption that the associated flow rule is applicable. Here, the assumption is proven. The theoretical yield curves expressed as the dotted line in the figure are those which are given by Equation (3) under condition of $p_m = 0$. Though the meniscus stress is not the zero actually, its value is relatively small and can be considered to be nearly equal to zero. Consequently, it is proven that the associated flow rule is applicable for the undrained condition, and that the yield function expands in similar shapes with shear process.

5 CONCLUSIONS

Triaxial compression tests on unsaturated soil were carried out under undrained condition. From the experimental results and discussions, the following conclusions were obtained.

(1) The volume change of unsaturated soil in shear test depended on the drainage condition. The effects of the bulk stress and the meniscus stress can be explained by the difference in the behavior between the bulk water and the meniscus water.

(2) As shown in Figure 6, during the shear process, the relation between the degree of saturation and suction due to the volume change depends on the distribution of the bulk water. The relation due to the volume change during shear process agrees with the water retention curve obtained from tests controlled only the suction in the specimen. However, the relation deviates from the theoretical wetting curve since the bulk water cannot resist the increased expansion.

(3) The stress states expressed as $(p + p_s)$ and q are uniquely arranged at maximum compression point. This result means that undrained shear strength can be estimated in the same way as drained shear strength with application of suction stress.

(4) The vectors of strain increment are plotted on the stress plane expressed as $(p + p_s)$ and q. Though these results never contain the effects of the meniscus stress, it is found that the associated flow rule is applicable if the effects of the meniscus stress can be ignored.

REFERENCE

Karube, D., Kato, S., Hamada, K. and Honda, M. 1996. The relationship between the mechanical behavior and the state of pore water in unsaturated soil. *JSCE Journal*. No.535/III-34 : 83-92 (In Japanese)

Karube, D., Honda, M. Kato, S. and Tsurugasaki, K. 1997. The relationship between shearing characteristics and the compression of pore-water in unsaturated soil. *JSCE Journal*. No.575/III-40 : 49-58 (In Japanese)

Karube, D. Kato, S. Honda M. and Kawai, K. 1998 A constitutive model for unsaturated soil evaluating effects of soil moisture distribution. *Proc.Second International Conference on Unsaturated Soils*. Beijing : 485-490

Kato, S., Karube, D. Honda, M. and Fujiwara, T. 1996. Influence of water distribution on the compression behavior of unsaturated soil. *JSCE Journal*. No.554/III-37 : 57-69 (In Japanese)

Kawai, K., Karube, D. and Seguchi, H. 1999. Factors affecting on water retention characteristic of soils. *Proc. International Symposium on Slope Stability Engineering*. Shikoku. Vol.1 : 381-386

Brooks, R. H. and Corey, A. T. 1966. Properties of porous media affecting fluid flow. *Proc. ASCE92*. IR(92) : 61-88

Clay Science for Engineering, Adachi & Fukue (eds) © 2001 Balkema, Rotterdam, ISBN 90 5809 175 9

Strength characteristics of compacted soils in unconfined compression test

Y. Yoshimura
Department of Civil Engineering, Gifu National College of Technology, Japan

S. Kato & K. Kawai
Department of Civil Engineering, Kobe University, Japan

ABSTRACT: Unconfined compression test, in which the suction in the specimens was measured, was carried out for statically and dynamically compacted specimens of silty clay. The effect of the degree of saturation and the suction on the unconfined compressive strength were examined. It was found that the suction stress affects the unconfined compressive strength, and that the relationship between the suction and the suction stress can be estimated by using the soil-water characteristic curve.

1 INTRODUCTION

Study for unsaturated soil was started as discussion for the property of the Bishop's equation (Bishop, 1959) for the effective stress. After that, Coleman (1962) proposed simple constitutive equations that treated suction as a independent stress parameter. Recently, some of constitutive models (for example, Alonso, et al.1990) which are based on this stand point of view, have been proposed. But in these models, the effects of soil moisture under some suction are not in consideration.

The relation between soil moisture and suction is known as the "water characteristic curve". In recent year, the studies, which relate the water characteristic curve to behavior of unsaturated soil, have been made. Fredlund et al. (1995) proposed one method, which estimates the relation between the increase in shear strength of unsaturated soil caused by suction and the water characteristic curve. Karube et al. (1996) and Kato et al. (1996) proposed a concept of "bulk water" and "meniscus water" which takes the effect of hysteresis in the water characteristic curve to a constitutive equation. And they also proposed a constitutive model for unsaturated soils which stands on this point of view.

In the field of unconfined compression test, studies for unsaturated soils in high degree of saturation have been made. Shimizu&Tabuchi (1993) and Mitachi&Kudoh (1996) showed the test results and the analyses that mean the adaptability of the Terzaghi's effective stress to nearly saturated soils in high degree of saturation. However, in these studies, the suction worked as a negative pore water pressure under a high degree of saturation, and these researchers examined the effect of suction in a different saturation region from that in the present study.

Ridley (1995) took the unconfined compression test as one kind of undrained test, and showed a unique relationship between the "critical water content" and the suction.

The final purpose of this study is to propose a simple method for estimating the relationship between suction and the cohesion (or the suction stress) based on the results of the unconfined compression test in which the suction is measured. The unconfined compression test is a convenient test for undrained conditions. The estimation method proposed, when used with these proposed constitutive models, will contribute to the practical application of predictions for the behavior of unsaturated ground.

In this study, unconfined compression tests, in which the suction was measured, were carried out for statically and dynamically compacted silty clay specimens. The relationships of the suction stress with the unconfined compressive strength and deformation characteristics, and the relationship between suction and the suction stress are examined.

2 SOIL WATER DISTRIBUTION IN SOIL MASS AND DEFINITION OF SUCTION STRESS

The relationship between soil water and suction under equilibrium is called the soil-water characteristic curve. The difference in the soil-water characteristic curve between wetting and drying processes is called hysteresis. Accordingly, there exist different soil moisture states under the same suction value. When soil moisture states are different, distribution patterns of pore water in the voids will consequently change.

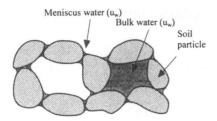

Figure 1 Bulk water and meniscus water in unsaturated soil

Figure 1 shows a key sketch of the bulk water and the meniscus water. The tendency in the soil water distribution patterns is for the bulk water to exist more in the drying process, and the meniscus water to exist more in the wetting process. When we take the influence of suction on the behavior of unsaturated soil into account, we should know the different effects of the bulk water and the meniscus water on the skeleton of the soil mass.

The meniscus water increases the intergranular force that acts between soil particles, and it increases the stiffness of soil skeleton accordingly. The bulk water causes not only an increase of the stiffness of soil skeleton but also a decrease of volume of the soil mass with slippages at contact points.

Karube et al. (1996) named the stress component that arises from the effect of the meniscus water as the meniscus stress, p_m and the stress component that arises from the effect of the bulk water as bulk stress, p_b. They postulated that these two stress components consist of the suction stress (p_s) as shown in the following equation.

$$p_s = p_m + p_b \qquad (1)$$

They also proposed the following equation that defines the suction stress under some suction, s, the residual degree of saturation, S_{r0} of the specimen and degree of saturation, S_r.

$$p_s = p_m + p_b = \frac{S_r - S_{r0}}{100 - S_{r0}} \cdot s \qquad (2)$$

3 TEST PROCEDURE

3.1 *Soil used and sample preparations for specimen*

Silty clay, whose specific gravity of the soil particles, G_s is 2.68 and whose liquid limit is non-plastic, was used. After adding distilled water to the air-dried sample in order to adjust water content, it was stored over a whole day and night in a sealed container.

To prepare the statically compacted specimen, the wetted sample was compacted in a mold of 50mm diameter and 300mm height. The inside of this mold was coated with Teflon. When the wetted sample was compacted in the mold, a spacer disk was used in order to adjust the specimen height to 120mm.

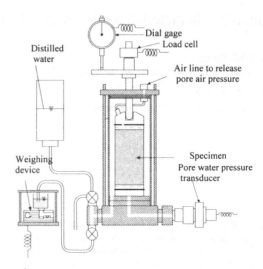

Figure 2 Sketch of the unconfined compression test cell used

The dry densities were set at three cases of 1.250, 1.330 and 1.405g/cm³. The water contents adjusted were about 3~24% at which the preparation of the specimen was possible.

To prepare the dynamically compacted specimen, the wetted sample was compacted with a mold of 10 cm diameter and rammer of 2.5 kg weight. A compacted sample was made from five layers and each layer was compacted with 25 times tamping. The water contents in the unconfined compression tests were 11 to 23%. From the obtained compaction curve, it was found that the optimum water content for this compaction method was about 20%. In this paper, the test results in dry side for the optimum water content are expressed as "Dry side" , and the test results in wet side are expressed as "Wet side". These explanatory notes are often used in the figures shown in later.

After coating the compacted sample with a plastic film and wrapping it with a wetted cloth to prevent the drying of the compacted sample, it was stored for a few days in a closed container, and was taken out before the test. By cutting from the top and bottom of the compacted sample, a specimen of 50mm in diameter and 100mm in height was made. These prepared specimens were used in the unconfined compression test.

3.2 *Unconfined compression test measuring suction*

Figure 2 shows a sketch of the unconfined compression test cell used. After the specimen was set on the pedestal of the cell, the specimen was covered with a rubber membrane and a loading cap, and sealed with O-rings. Before the test was carried out, the cell was filled with water. During the test, the displacement of the cell water was measured with a weighing device that was connected to the inside of the cell. The volume of the specimen was obtained from adjusting the recorded displacement of the cell water for the piston penetration into the cell. A ceramic disk, whose air entry value was 500 kPa, was installed into the pedestal of the cell to measure the pore water pressure in the specimen. In the loading cap, a porous metal plate was installed. An air line was connected to it and released the pore air pressure in the specimen to atmosphere. The suction in the specimen was decided by value of the measured pore water pressure consequently. The rate of strain used was 0.1% per minute.

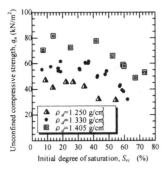

Figure 3 Relationship between unconfined compressive strength and the initial degree of saturation in statically compacted specimen

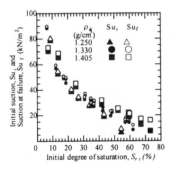

Figure 4 Relationship between initial suction, suction at failure and initial degree of saturation in statically compacted specimen

4 EXPERIMENTAL RESULTS AND DISCUSSIONS

4.1 *Strength and suction change characteristics*

Figure 3 shows the relationship between the unconfined compressive strength, q_u and the initial degree of saturation, S_{ri}, which was measured when the specimen was trimmed, for compacted specimens. The unconfined compressive strength increases as the dry density increases. For about 35, 40 and 45% of the degrees of saturation, the unconfined compressive strengths are constant in each case respectively. But the unconfined compressive strength de-

creases when the saturation increases from these degrees of saturation.

Figure 4 shows the relationship between the initial suction, su_i, and the suction at failure, su_f with the initial degree of saturation, S_{ri} for statically compacted specimens. The su_i and su_f values decrease with an increase in S_{ri}. But the increase of suction (su_f-su_i) is almost constant regardless of S_{ri}, and any effect of the density is hardly observed.

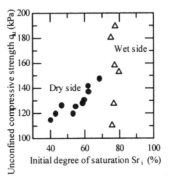

Figure 5 Relationship between unconfined compressive strength and the initial degree of saturation in dynamically compacted specimen

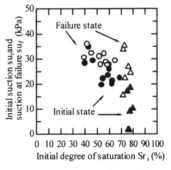

Figure 6 Relationship between initial suction, suction at failure and initial degree of saturation in dynamically compacted specimen

Fig.5 shows the relationship between the unconfined compressive strength, q_u and the initial degree of saturation, S_{ri} for dynamically compacted specimens. The unconfined compressive strength increases as the initial degree of saturation increases to 80% which is near to the optimum water content, and it decreases rapidly about 80% of degree of saturation.

Fig.6 shows the relationship between the initial suction and the suction at failure with the initial degree of saturation for dynamically compacted specimens. The initial suction decreases as the initial degree of saturation increases. The suction at failure decreases gradually, and then increased. The suction in all specimens increases from the initial state to the failure. From Figure 6, it is found that the increase in suction during test is large in the specimen that has high initial degree of saturation.

From the stress-strain relations for statically and dynamically compacted specimens, it was found that all of the specimens expanded during the axial compression process.

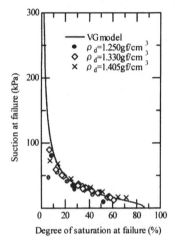

Figure 7 Relation between the suction at failure and the degree of saturation at failure for statically compacted specimen

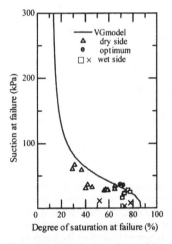

Figure 8 Relation between the suction at failure and the degree of saturation at failure for dynamically compacted specimen

4.2 Estimation of suction stress estimated from water characteristic curve

Various equations have been presented to express the relationship between suction and soil moisture.

In this study, the equation which was proposed by Van Genuchten (1980) (the VG model) is used, because it can be applied to various soil types from sand to clay. The VG model is given by the next equation.

$$Se(\theta) = \left\{1 + (\alpha \cdot s)^n\right\}^{-(1-\frac{1}{n})} \tag{3}$$

Here, α and n are fitting parameters, and $Se(\theta)$ is the relative volumetric water content defined by the next equation.

$$Se(\theta) = \frac{\theta - \theta_r}{\theta_{sat} - \theta_r} \tag{4}$$

where, θ_r is the residual volumetric water content, θ_{sat} is the saturated volumetric water content, and θ is the volumetric water content. In practice, the above parameters of α, n, θ_r and θ_{sat} are determinded by the result of fitting soil-water characteristic curve.

When the volume change of a specimen is very small during the water retention test, the relative volumetric water content can be given by the next equation.

$$Se(\theta) = \frac{Sr - Sr_0}{100 - Sr_0} \tag{5}$$

Figure 7 shows the relation between the suction at failure and the degree of saturation at failure for statically compacted specimens. In this figure, the symbols show the results of unconfined compression test, and the solid line shown in Figure 7 is the approximation result by the VG model to the relation between suction at failure and degree of saturation at failure. The used parameters in the VG model are α=0.06, n=2.3, Sr_{sat}=86.4(%) and Sr_0=0%.

Figure 8 shows the relation between the suction at failure and the degree of saturation at failure for dynamically compacted. In this figure, the dots show the results of unconfined compression tests for

line is the approximation result by the VG model to the water retention curve for the dynamically compacted specimen of 17% water content. The used parameters in the VG model are α=0.0239, n=2.95 and Sr_0=12%. From the Figure 8, it is found that the test results are distributed along the solid line.

The void diameter distributions in the specimens will change when the water contents at compaction are different, because the dry densities of the compacted specimens depend on the water contents at compaction. When the void diameter distributions show different tendencies, the water characteristic curves obtained from each specimen will be different. But the results showed in Figures 8 and 9 mean that the relationships between the suction and the degree of saturation at failure show the same tendency with a water characteristic curve even if the dry densities of the specimens are different.

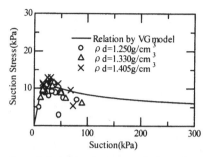

Figure 9 Comparison between prediction and test results of relationships between the suction stress and suction for statically compacted specimen

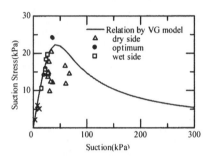

Figure 10 Comparison between prediction and test results of relationships between the suction stress and suction for dynamically compacted specimen

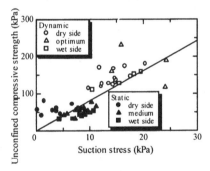

Figure 11 Relationships between the unconfined compressive strength and the suction stress for statically and dynamically compacted specimens

Figure 9 and 10 show comparisons between a prediction and test results of the relationships between the suction stress and the suction for statically and dynamically compacted specimens. In these figures, the solid lines are predictions obtained by the next equation that is derived from Equations (2), (3) and (5).

$$p_s = \frac{Sr - Sr_0}{100 - Sr_0} \cdot s = \left\{ 1 + (\alpha \cdot s)^n \right\}^{-(1-\frac{1}{n})} \cdot s \qquad (6)$$

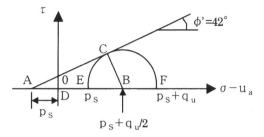

Figure 12 Geometric relationship between the failure criteria and the Mohr's stress circle at failure

The parameters used are the same ones for the solid line shown in Figures 7 and 8. The predictions show a tendency for the suction stress to increase as suction increases to 30~50 kPa, above which it decreases and gradually approaches a fixed value. The symbols shown in these figures are the calculated results obtained by using Eq.(2) for the unconfined compression test results. These calculated results show similar tendencies with the predictions. From these results, it can be stated that the suction stress at failure is known when using the results of the unconfined compression test.

4.3 Relationship between unconfined compression strength and suction stress

Figure 11 shows the relationship between the unconfined compressive strength and the suction stress at failure for statically and dynamically compacted specimens. In this figure, the symbols show test results, and they are distributed around the solid line shown by Eq.(8) mentioned later. From this result, it can be concluded that the suction stress affects the unconfined compressive strength.

The relationship between the unconfined compressive strength and the suction stress can be deduced as follows. Firstly, we postulate that the suction stress will act as a confining pressure for the specimen in an unconfined compression test, and that the Mohr's stress circle at failure is in contact with the failure criteria that corresponds to the suction at failure. Secondly, we postulate that the cohesion of the failure criteria is equal to the suction stress. Based on these assumptions, a geometric relationship between the failure criteria and the Mohr's stress circle at failure is given as shown in Figure 12. In this figure, q_u means the unconfined compressive strength, and p_s means the suction stress at failure. From the geometric relationship concerning for the triangle ABC, an equation that connects the unconfined compression strength to the suction stress at failure is derived as follows.

$$q_u = \frac{4 \sin \phi'}{1 - \sin \phi'} \cdot p_s \qquad (7)$$

where, ϕ' the internal angle of friction is based on

the net stress that is defined by the difference between total stress and pore air pressure. The internal angle of friction is slightly larger than that obtained from a saturated sample with an effective stress (Karube et al. 1986). Abe & Hatakeyama (1997) reported that the internal angle of friction based on effective stress is 42 degrees in saturated specimens that were made under the same condition as the specimens used in the unconfined compression tests. Substituting this value into the Equation (7), the following equation is obtained.

$$q_u = 8.09 \cdot p_s \qquad (8)$$

The solid line shown in Figure 11 represents the relationship obtained from the Equation (8). It is found that the test results distribute around the solid line. This result demonstaretes the appropriateness of the assumptions mentioned above.

5 CONSLUSIONS

Unconfined compression tests, in which the suction was measured, were carried out on statically and dynamically compacted silty clay specimens in order to establish a simple method for estimating the relationship between suction and cohesion. The relations between the suction stress, the unconfined compressive strength and deformation characteristics, and the relation between suction and the suction stress were examined. The results and discussions are summarized as follows.

1. Though the initial degrees of saturation and the dry densities of the specimens were different, the relationship between suction and the degree of saturation at failure was shown by one water retention curve. Also, the relation between the suction stress and suction at failure showed a similar tendency with the prediction by Equation (3) using the same parameters for the water retention curve mentioned above.

2. The unconfined compressive strength is affected by the suction stress, and the suction stress seems to work as a confining pressure for unsaturated specimens. The relation between the suction stress and the unconfined compressive strength showed the similar tendency as the relation shown in Equation (8).

3. These results demonstrate the applicability of the unconfined compression test for estimating the relationship between the suction stress and suction.

REFERENCES

Abe, H. & Hatakeyama, M. 1997, Report of the committee for evaluation of water permeability in unsaturated ground. Japanese Geotechnical Society, pp.82-88. (In Japanese)

Alonso, E.E., Gens, A. and Josa, A. 1990, A Constitutive Model for Partially Saturated Soils, *Geotechnique*, Vol.40, No.3, 405-430.

Bishop, A.W. 1959, The Principle of Effective Stress, *Teknisk Ukeblad*, Vol.3, pp.859-863.

Coleman, J.D. 1962, Stress/Strain Relations for Partly Saturated Soil, Correspondence, *Geotechnique*, Vo.12, No.4, pp. 348-350.

Fredlund, D.G., Vanapalli, S.K., Xing ,A. and Pufahl, D.E. 1995, Predicting the Shear Strength Function for Unsaturated Soils Using the Soil-water Characteristic Curve, *Unsaturated Soils*, Balkema, Vol.1, pp.43-46.

Karube, D., Kato, S. & Katusyama, J. 1986, Effective stress and soil constants of unsaturated kaolin, *Journal of Japanese Society of Civil Engineering*, Vol.370, pp.179-188. (In Japanese)

Karube, D., Kato, S., Hamada, K. & Honda, M. 1996, The relationship between the mechanical behavior and the state of pore water in unsaturated soil, *Journal of Japanese Society of Civil Engineering*, Vol. 535, pp.83-92. (In Japanese)

Kato,S., Yoshimura, Y., Kawai, K. & Sunden, W. 1999, Effects of suction on strength and deformation of unsaturated soil in unconfined compression test, Poster session proceedings of 11th Asian Reginal Conference on Soil mechanics and Geotechnical engineering, Vol.1, pp.17-18.

Mitachi, T. & Kudo, Y. 1996, Method for predicting in-situ undrained strength of clays based on the suction value and unconfined compressive strength *Journal of Japanese Society of Civil Engineering*, No.541/III-35, pp.147-158.

Ridley A. M. 1995, Strength-Suction-Moisture Content Relationships for Kaolin under Normal Atmospheric Conditions, *Unsaturated Soils*, Balkema, Vol.1, 645-651.

Shimizu, M. & Tabuchi, T. 1993, Effective Stress Behavior of Clays in Unconfined Compression Tests, *Soils and Foundations*, Vol.33, No.3, pp.28-39.

Van Genuchten, M. Th.,1980, A closed-form for predicting the hydraulic conductivity of unsaturated soils, *Soil. Sci. Soc. Am. J.*, Vol.44, pp.892-898.

Clay Science for Engineering, Adachi & Fukue (eds) © 2001 Balkema, Rotterdam, ISBN 90 5809 175 9

DEM analysis for effects of intergranular force on behavior of granular material

S. Kato
Department of Civil Engineering, Kobe University, Japan

S. Yamamoto
Obayasi Construction Company Limited, Japan

S. Nonami
Oyotisitsu Company Limited, Japan

ABSTRACT: In this study, with the distinct element method analysis, simulations of the biaxial compression tests for two-dimensional granular material are carried out. In this analysis, the problems encountered in the triaxial compression tests for unsaturated soils are removed. The influence of meniscus water in unsaturated soil is reproduced by introducing the constant intergranular adhesive force that acts perpendicular to the tangential plane at the contact point of disk particles. The influence of the intergranular adhesive force on the stress-strain relation, the strength constants, and the adaptability of the converted normal stress for the analysis of the results are examined. It is found that the internal friction angle and the cohesion of the assembly of the disk particles increases with the intergranular adhesive force, and that there is a limitation for applying the concept of the converted normal stress for arranging the behaviors of cohesive granular materials.

1 INTRODUCTION

In unsaturated soil, meniscus water exists at the contact points of soil particles, and the intergranular adhesive forces, that increase the resistance for the slippage at the contact points, act between soil particles. These forces affect the deformation and strength characteristics of unsaturated soils. Collapse, which is caused by the disappearance of the intergranular adhesive force, is an example of their effects.

The researches of unsaturated soils have been carried out mainly based on the triaxial compression tests by the present (for example, Karube et al. 1986). From these results, it has been understood that the analytical method, that takes the effects of the intergranular adhesive force into account as the cohesion or some kind of stress, is effective. However, in present state, reported data number is not so many, because the triaxial compression test for unsaturated soils is itself difficult, and is including problems that the loading of the suction using the axial translation technique with a ceramic disk is limitative. Accordingly, some parts on the effects of suction are not clarified, and the applicability of arrangement method that takes the effects of suction into account.

In this study, the biaxial compression tests of two-dimensional granular material were simulated by the distinct element method, in which the problems encountered in the triaxial compression test for unsaturated soil are almost removed. Then, the adaptability of the converted normal stress for the analysis of the results are examined.

2 PROBLEMS IN TRIAXIAL TEST FOR UNSATURATED SOIL

In the triaxial compression test for unsaturated soil, the axisial translation technique is often used to apply suction to a specimen. The capacity of the suction applied is limited by the air entry value (AEV) of the ceramic disk used. It is possible to use a ceramic disk of about AEV=1500kPa in practice, but a ceramic disk of about AEV=500kPa is mainly used, since the coefficient of the permeability of the ceramic disk decreases with the increase of AEV. Consequently, as a testing condition, the suction from 0 to about 500kPa is often applied to the specimen.

The intergranular adhesive force acting between particles does not correspond to the suction value applied. It is known that the relationship between suction and the cohesion obtained from the triaxial compression test result is approximated by the hyperbola (Fredlund et al.,1995). According to their result, the cohesion increases proportionally with suction, until the suction reaches the AEV. When the suction exceeds the AEV, the cohesion approaches some constant value with increase in the suction. Generally, 100kPa or less represents the bounds for the value of the AEV in an unsaturated soil. Under a suction of about 500kPa, the rate of increase of the cohesion considerably decreases. This non-linearity means that the intergranular adhesive force acting between particles does not correspond to the suction value applied. Consequently, it is difficult to reproduce the condition in which large intergranular adhesive force acts, with the triaxial compression test apparatus. Accordingly, there exists a limitation in clarifying the effect of the intergranular adhesive force by the triaxial compression test using the axis translation technique void ratio influences the strength and the rigidity of the soil. Accordingly, it is desirable that a soil is tested under the condition in which only the suction differs and the void ratio is the same, for clarifying the effects of suction. However, considerable skill is required to adjust such condition, because void ratio also changes when suction varies.

The triaxial compression test for unsaturated soil using the axis translation technique has the problems mentioned above. Consequently, the triaxial test for unsaturated soil itself is difficult to carry out, and the test results are fewer than those for saturated soil. There are some aspects for which the effect of suction has not been clarified experimentally.

In this study, a biaxial compression test for two-dimensional granular material is simulated by distinct element method analysis. In this analytical method, the volume change of the specimen is clear, and the size of the intergranular adhesive force can also be optionally set. It seems to remove many of the problems in triaxial compression test for unsaturated soil mentioned above.

Based on the analytical results, the effects of the intergranular adhesive force on deformation and strength characteristics of an assembly of disk particles are examined, and they are compared with the tendencies observed in the triaxial compression test results for unsaturated soil.

3 ANALYTICAL METHOD AND CONDITION

3.1 Outline of DEM analysis with intergranular adhesive force

The distinct element method (DEM) is one of the discontinuous corpora analysis method proposed by Cundall(1971).

Unlike continuum analysis of the finite element method or the boundary element method, it is suitable for analyzing the dynamic behavior of the granular material. In the DEM analysis, simple dynamic models (they are generally the Voigt model and Coulomb's friction rule) are introduced at contact points and contacting surfaces between the disk particles with the assumption that the particles are rigid. An independent equation of motion at every element is solved forwardly in the time domain, and interactions between particles and deformation of

particle aggregates are traced. This method has the merit that necessary output data such as stress, strain, and rotation angle of some disk particles, and the setting of boundary conditions are easily treated.

In this study, the analysis is carried out using the granular material distinct element analysis system (GRADIA) programmed by Yamamoto(1995).

The effect of meniscus water observed in unsaturated soil is expressed by introducing an intergranular adhesive force described below.

The intergranular force acting between contacting particle i and j is denoted as P_{ij}. The x-direction component, y-direction component, and moment component of the resultant force of the intergranular adhesive force at time t are given by the following equations considering that the direction of P_{ij} is normal for the line direction of the particle tangent plane.

$$[F_{x_i}^P]_t = \sum_j P_{ij} \cos \alpha_{ij} \tag{1}$$

$$[F_{y_i}^P]_t = \sum_j P_{ij} \sin \alpha_{ij} \tag{2}$$

$$[M_i^P]_t = 0 \tag{3}$$

Where, $[F_{x_i}^P]_t, [F_{y_i}^P]_t, [M_i^P]_t$ are the x-direction component, y-direction component and moment component of the resultant force of the intergranular adhesive force and α_{ij} is the angle between the normal direction of the contact plane and the x-axis.

It is possible to introduce an intergranular adhesive force by deducting these components from each component of the total intergranular force except for the intergranular adhesive force. By integrating these equations of motion for each particle by the Euler method with respect to time, the analysis was carried out. Consequently, the solution becomes stabilized conditionally on the integral time increment.

Cundall(1971) proposed the following equation for the integral time increment.

$$\Delta t < \Delta t_c = 2\sqrt{\frac{m}{k}} \tag{4}$$

Where, m is mass of the disk particle, and k is the spring constant.

It has been found experientially that sufficient stability and accuracy can be ensured in the quasistatic problem at about $1/10\Delta t_c$, though equation (4) is deduced from the equation of motion for a single-degree-of-freedom system. But taking this equation as a standard, the integral time increment must be decided by trial and error. In this study, $\Delta t = 1/10\Delta t_c$ $= 5\times10^{-5}$ sec. is used.

The parameters and the material properties necessary for the analysis are listed in Table 1.

Yamamoto(1995) carried out the simulation, in which disk particles are arranged in a similar position to a biaxial compression test carried out for

Table 1 Parameters and material properties of the sample

	Between particles	Between particle and platen
Normal spring constant (N/m/m)	0.9×10^{10}	1.8×10^{10}
Tangential spring constant (N/m/m)	3.8×10^{8}	6.0×10^{8}
Normal damping constant (N/m/m)	7.9×10^{5}	1.1×10^{5}
Tangential damping constant (N/m/m)	1.4×10^{4}	2.0×10^{4}
Friction coefficient (deg.)	16	10
Material density (kg/m^3)	2700	

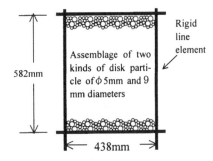

Figure 1 Schematic diagram of the specimen analyzed

582mm

438mm

Rigid line element

Assemblage of two kinds of disk particle of φ 5mm and 9 mm diameters

aluminum bars, using the material property shown in Table 1. And it was confirmed that macroscopic internal frictional angles obtained by the simulations were about 22~24 degrees, and agreed with one observed in the biaxial compression test carried out for aluminum bars.

3.2 Analytical condition

As an analytical model, a specimen in which two kinds of particles of 5mm and 9mm diameters were placed randomly at the proportion of mixing weight ratio of 3:2, was used. These were 6372 5mm particles and 1269 9mm diameter particles. The sum of them was 7641. The shape of the specimen was a rectangle of about 582mm height and about 438mm width as shown in Figure 1. The four edges of the rectangle were surrounded by the rigid line elements that correspond to the loading platens. Both of the side rigid line elements were controlled in order that the lateral stress was constant, and their vertical movements were restricted. The bottom rigid line element was also fixed. When the shear process of the biaxial compression test was simulated, the upper rigid line element was moved down perpendicularly at a constant rate of 10cm/s.

In this study, firstly, without the intergranular adhesive force, the specimen was compressed isotropically under a confining pressure of 49 kPa. Then, the desired intergranular adhesive force was introduced between particle contact points. Afterwards, the shear process of the biaxial compression test was simulated under a confining pressure of 49 kPa, or the specimen was compressed isotropically to σ_3 =490 kPa, and the shear process was simulated under confining pressure of σ_3 =490 kPa. In addition, a isotropic compression test to σ_3=980kPa compressive stress was simulated with an intergranular adhe-

sive force of P=0.0N, 9.8N and 49N in order to observe the effect of the intergranular adhesive force on the compression process.

4 ANALYTICAL RESULTS AND DISCUSSIONS

4.1 Stress and strain relation in shear process under constant intergranular adhesive force

Figures 2 and 3 show $\sigma_1/\sigma_3 \sim \varepsilon_1 \sim \varepsilon_3$ relations and $\varepsilon_1 \sim \varepsilon_v$ relation for the case of σ_3=49kPa. Figures 4 and 5 show a similar relations for the case of σ_3=490kPa. It is found that the $\sigma_1/\sigma_3 \sim \varepsilon_1 \sim \varepsilon_3$ relation shifts at the top, and the shear strength increases when the intergranular adhesive force increases. The characteristics observed in figures 2 and 4 show that the principal stress ratio increases in the initial stage of the shear process when the intergranular adhesive force is higher, and that the change of principal stress ratio decreases after this. From the $\varepsilon_1 \sim \varepsilon_v$ relations shown in Figures 3 and 5, it is found that, when there is no intergranular adhesive force, the volume of the specimen changed from compression to expansion during the shear process, and that, in other cases, expansive volumetric strain occurs with an increase in the intergranular adhesive force. These results correspond to the tendency observed in direct shear tests for sand under constant suction (Shimada,1998). In his results, the volumetric expansion of the specimen during the shear process increased when the suction increased.

4.2 Effects of intergranular adhesive force on strength constants

Figures 6 and 7 show the relationships between the intergranular adhesive force, P and the internal friction angle, φ and the cohesion, c that are decided from Mohr's stress circles for the maximum axial stress and the confining pressure. From Figure 6, it is found that the internal frictional angle, φ increases with the increase in the intergranular adhe-

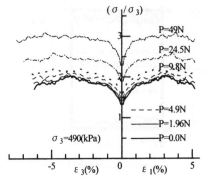

Figure 2 $\sigma_1/\sigma_3 \sim \varepsilon_1 \sim \varepsilon_3$ for the case of σ_3=49kPa

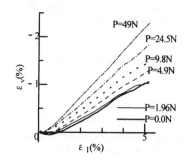

Figure 3 $\varepsilon_1 \sim \varepsilon_v$ relation for the case of σ_3=49kPa

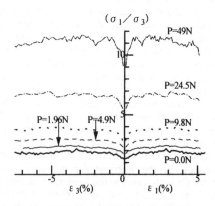

Figure 4 $\sigma_1/\sigma_3 \sim \varepsilon_1 \sim \varepsilon_3$ for the case of σ_3=490kPa

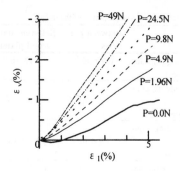

Figure 5 $\varepsilon_1 \sim \varepsilon_v$ relation for the case of σ_3=490kPa

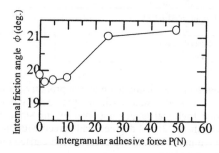

Figures 6 Relationship between the intergranular adhesive force P and the internal friction angle, ϕ

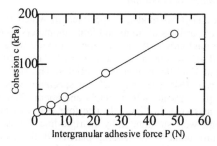

Figures 7 Relationship between the intergranular adhesive force P and the cohesion, c

sive force. From Figure 7, it is found out that the cohesion, c is proportional to the increase in the intergranular adhesive force.

In the research based on triaxial compression test results on unsaturated soil, there are some interpretations for the relationship between suction and strength. According to the proposal by Fredlund et al. (1995), the internal frictional angle is a constant value in spite of the suction. On the other hand, Karube et al. (1986) showed triaxial test results for unsaturated soil, in which the internal frictional angle slightly increases with the increase in suction. From the present analytical result shown in figure 6, it is

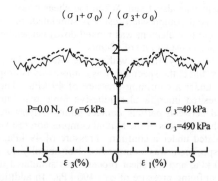

Figure 8 $(\sigma_1+\sigma_0)/(\sigma_3+\sigma_0) \sim \varepsilon_1 \sim \varepsilon_3$ relation for the case of P=0.0N (σ_0=6kPa)

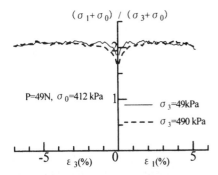

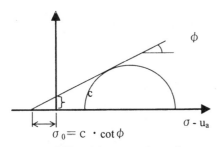

Figure 9 $(\sigma_1+\sigma_0)/(\sigma_3+\sigma_0)\sim\varepsilon_1\sim\varepsilon_3$ relation for the case of P=49N (σ_0=412kPa)

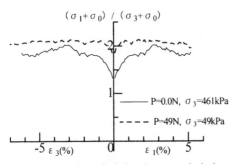

Figure 10 Definition of the converted normal net stress for unsaturated soil

Figure 11 Comparison of relations between principal stress ratio considering the converted normal net stress, maximum principal strain, and minimum principle strain

found out that the internal frictional angle tends to increase with the increase of the intergranular adhesive force. This result corresponds to the test results shown by Karube et al.

4.3 *Adaptability of the converted normal stress*

Kato(1997) conducted triaxial compression tests under constant suctions using compacted cohesive clay specimens, and showed that the $(\sigma_1+\sigma_0)/(\sigma_3+\sigma_0)\sim\varepsilon_1$ $\sim\varepsilon_3$ relations under constant converted normal net stresses agree in spite of the size of the confining pressure.

Figure 8 shows the $(\sigma_1+\sigma_0)/(\sigma_3+\sigma_0)\sim\varepsilon_1\sim\varepsilon_3$ relation for the case of P=0.0N (σ_0=6kPa). Figure 9 shows a similar relation for the case of P=49N (σ_0=412kPa). For the case of P=0.0N, the relation shows a tendency which agrees for the whole area of the shear process. For the case of P=49N, it is found that the difference appears in the initial stage of shear, and that it has come to the upper part, when confining pressure is smaller. When the cases of σ_3=49kPa and 490kPa (shown in figures 8 and 9) are compared, since the void ratios before the shear process are almost equal, it can be considered that there is no effect of void ratio. Consequently, it is supposed that the difference observed in stress-strain relations shown in these figures exist as an effect of the intergranular adhesive force.

Converted normal net stress for unsaturated soil evaluates the cohesion induced by the interparticle adhesive force as an isotropic stress component on the net stress axis, as it is shown in figure 10. (In this figure, the σ-u_a means the net stress that corresponds to the external stress for unsaturated soil, and for saturated soil or some granular materials, the net stress changes to the external stress, σ, and the converted normal net stress becomes the converted normal stress.) That is to say, the effect of the cohesion is converted into the stress component induced by an external force. In the unsaturated soil with the cohesion, the equilibrium condition of stress components by external forces becomes a problem as a balance of the force for the soil element. The cohesion, which induced by the internal force, influences in the constitutive relation that represents a material property. Therefore, the effect of cohesion must be evaluated in some forms, and be introduced into the constitutive relation. It is considered that the converted normal net stress is one of the evaluation methods.

At present, by evaluating the converted normal net stress or the converted normal stress as a stress component, the stress and strain relations in the shearing tests of unsaturated soil (Kato, 1997) and cement improved soil (Matsuoka&Sun, 1994) have been presented. In these test results, the arrangement methods are based on the interpretation that the converted normal net stress and the converted normal stress have the same influence on the behaviors of the materials as the external stress component does. However, in those experimental results, it has not been confirmed whether it always establishes such interpretation in spite of the size of the interparticle adhesive force, because the size of these converted stresses are small.

Using the analytical method in this study, we can examine that, the behavior of granular material under some confining pressure is equal that under some converted normal stress the values of which is the same to the confining pressure.

Figure 11 shows the relations between principal

stress ratio considering the converted normal stress, maximum principal strain, ε_1 and minimum principle strain, ε_3 in cases of confining pressure, σ_3=49kPa with interparticle adhesive force, P=49N and confining pressure σ_3=461kPa with interparticle adhesive force, P=0.0N. In the former case, σ_0 becomes 412 kPa, and the sum total of σ_0 and σ_3 is 461 kPa. In the later case, σ_0 becomes 6 kPa, and the sum total of σ_0 and σ_3 is 467 kPa. From this figure, it is found that two analytical results in which sum totals of σ_0 and σ_3 are almost the same show considerably different behaviors.

This result means that the converted normal stress based on the interparticle adhesive force causes different effect on the stress strain relation from the isotropic stress based on external force, and that there is limit of application in the method for evaluating converted normal net stress in addition to the external stress in the simplicity.

The interparticle adhesive force increases only the perpendicular component of the interparticle force to the particle tangent plane. In the meantime, when the converted normal stress, σ_0 acts as the confining pressure externally, it increases not only the vertical component for the particle tangent plane but also parallel component (the shear component). This microscopic difference is a source of the difference of the macroscopic shearing behavior described in the superscription.

5 CONCLUSIONS

In this study, isotropic compression and shearing tests under the biaxial compression condition in the constant adhesive force were simulated using DEM program which introduced the interparticle adhesive force for two-dimensional granular material. And the effect of the interparticle adhesive force on deformation and strength was clarified. From the results and discussions, conclusions are summarized as follows.

1) The tendencies observed between the principal stress ratio, principal strain and volumetric strain in the shear process differed when the interparticle adhesive forces differs. Under equal confining pressure, when the interparticle adhesive force become bigger, the build-up of the principal stress ratio and principal strain relation became more suddenly, and more expansive volumetric strain is observed.

2) The internal frictional angle decided from Mohr's stress circle at failure showed the tendency that it increases with the increase in the interparticle adhesive force. This result supports the claim by Karube et al. And, the cohesion showed the tendency that it increased in proportion to the interparticle adhesive force.

3) Even if the deformation behavior is arranged by taking the converted normal net stress as a confining pressure, the result is different from that observed under the same increment of confining pressure. This fact shows that there is limit of application in the method for evaluating converted normal net stress in addition to external stress.

4) The interparticle adhesive force increases only the perpendicular component of the interparticle force to the particle tangent plane. In the meantime, when the converted normal net stress, σ_0 acts as the confining stress increment externally, it increases not only the vertical component for the particle tangent plane but also parallel component (the shear component), when it externally worked. It is a source of the difference of the macroscopic shearing behavior.

REFERENCES

Cundall, P.A. 1971, A computer model for simulation progressive, large scale movement in blocky rock system, Symp. ISRM, Nancy, France, Proc.2, pp.129-136.
Fredlund, D.G., Vanapalli, S.K., Xing, A. & Pufahl, D.E. 1995, Predicting the shear strength function for unsaturated soils using the soil-water characteristic curve, *Proc. of 1st Int. Conf. on Unsaturated Soils*, Vol.1, pp.43-46.
Karube, D., Kato,S. & Katsuyama, J. 1986, Effective stress and soil constants in unsaturated kaolin, *Journal of Japanese society of civil engineering*, No.370, pp.179-188. (In Japanese)
Kato, S. 1997, Effects of soil water distribution on behavior of unsaturated soil and its evaluating method, PhD thesis, Kobe University. (In Japanese)
Matsuoka, H. & Sun, D.A. 1994, An elasto-plastic model for frictional and cohesive materials includeing granular materials and metals, *Journal of Japanese society of civil engineering*, No.550, pp.201-210. (In Japanese)
Shimada, K. 1998, Effect of matric suction on shear characteristics of unsaturated fraser river sand, *Jour. of Faculty Environmental Science and Technology*, Okayama Univ., Vol.3, No.1, pp.127-134.
Yamamoto, S. 1995, Fundamental study for behavior of granular material by DEM analysis, PhD thesis, Nagoya Institute of Technology. (In Japanese)

Clay Science for Engineering, Adachi & Fukue (eds) © 2001 Balkema, Rotterdam, ISBN 90 5809 175 9

Interrelationship among consistency limits of clays

J. H. Park
United Graduate School of Agricultural Sciences, Kagoshima University, Japan

T. Koumoto
Faculty of Agriculture, Saga University, Japan

ABSTRACT: The consistency limits (liquid limit w_L, plastic limit w_p, shrinkage limit w_s) are used for the classification and estimation of engineering properties of fine soils. In this paper, the consistency limits and pH values were measured for several types of clays, having a wide range of plasticity. The interrelationship among the consistency limits and the relationship between the consistency limits and pH values were examined.

1 INTRODUCTION

Consistency refers to the resistance of soil offered against forces that tend to deform or rupture the soil aggregate; in other words it means the state of soil is changed depending upon the water content in cohesive soil. The consistency limit which is made up of the liquid limit, the plastic limit, and the shrinkage limit, indicating the soil water content limit for various states of consistency as was suggested by Atterberg (1912). Fig.1 shows the relationship between volume and water content of soil. When a saturated cohesive soil is subjected to being dried out, the curve was obtained. As shown in Fig.1, the liquid limit separates the liquid state from the plastic one, the plastic limit separates the plastic state from the semi-solid one, and the shrinkage limit separates the semi-solid state from the solid one. The physical and mechanical properties of cohesive soils can change depending upon the water content. The consistency limit values are used for the classification and estimation of engineering properties of cohesive soils. So far, many researchers (Terzaghi, 1926; Casagrande, 1932; Skempton, 1944) have studied the relationship between the consistency limits and other properties of soil.

In this paper, the consistency limits and pH values were measured for several types of clays that had a wide range of plasticity. The interrelationships among the consistency limits as well as the relationship between the consistency limits and pH values were examined.

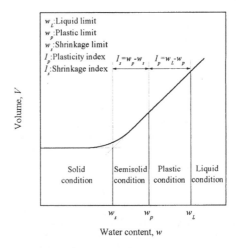

Fig.1 Generalized relationship between volume (V) and water content (w) for soil

2 CONSISTENCY LIMITS TEST AND RESULTS

As shown in Table 1, 13 kinds of samples in which kaolinite, Karatsu clay, 2 kinds of Ariake clay, bentonite, and 8 kinds of mixed clays (M1～M8) were used in this test. The tests of the liquid limit and the plastic limit were carried out by Casagrande's method, which is prescribed by JIS and JGS. The plastic limit values were determined

Table 1. Physical properties of tested clays

No.	Sample	ρ_s (g/cm^3)	pH	Consistency limits (Casagrande method)					
				w_L (%)	w_P (%)	I_P	w_s (%)	I_s	R
1	Kaolinite(K)	2.68	5.4	66.0	33.7	32.3	39.7	-6.0	1.274
2	Karatsu clay	2.62	5.6	50.4	17.6	32.8	16.5	1.1	1.777
3	M1	2.64	6.3	72.1	37.3	34.8	36.6	0.7	1.312
4	M2	2.56	6.9	81.7	42.0	39.7	35.8	6.2	1.314
5	M3	2.62	7.4	86.6	44.3	42.3	36.6	7.7	1.297
6	Ariake clay(A1)	2.54	7.4	78.6	31.1	47.5	20.9	10.2	1.647
7	Ariake clay(A2)	2.63	7.8	138.6	68.6	70.0	41.7	26.9	1.186
8	M4	2.55	8.4	166.2	52.3	113.9	21.9	30.4	1.593
9	M5	2.66	8.7	224.4	45.5	178.9	18.6	26.9	1.682
10	M6	2.68	8.9	279.5	37.8	241.7	15.2	22.6	1.737
11	M7	2.64	9.3	361.9	44.4	317.5	10.8	33.6	1.755
12	M8	2.71	9.6	405.1	40.6	364.5	8.9	31.7	1.848
13	Bentonite(B)	2.67	10.2	479.9	36.2	443.7	5.3	30.9	1.903

Note: A1: Kubota area, A2: Isahaya area, M1 (K:A2=4:1), M2 (K:A2=3:2), M3 (K:A2=2:3), M4 (A2:B=7:3),
M5 (A2:B=6:4), M6 (A2:B=5:5), M7 (A2:B=4:6), M8 (A2:B=2:8)
I_s: Shrinkage index, R: Shrinkage ratio

from the average of three approximate values in the five tests. The test for shrinkage parameters of soil was carried out according to the JGS T 145, but the mass and volume were measured each time for our test. In this test, mercury was used to measure the volume. The pH measurement for the samples were carried out according to the JGS T 211, but the quantity of water added was 2 to 9 times the weight of the oven dried soil for mixed clays and bentonite. Furthermore, the pH value was determined from two out of the three measurement values that were closest to the average value.

Fig.2 shows the typical shrinkage curves of 6 kinds of soils, kaolinite, Karatsu clay, M2, Ariake clay(A2), M5, and bentonite. Also, for each soil, five tests for w_s were conducted and we used the three average values. As shown in Fig.2, kaolinite has a markedly higher shrinkage limit than bentonite, i.e. the kaolinite value is 7 times higher than that of bentonite, and the shrinkage limit of clay having more bentonite is lower than that of kaolinite.

3 RESULTS AND DISCUSSION

3.1 Relationship between w_p and w_L

Fig.3 shows the relationship between w_p and w_L. As shown in the figure, the w_p increases linearly with increasing w_L in the range of $w_L \leqq 120$. The general tendency of test results are very similar to that of the equation, $w_p = 0.30w_L + 15.85$, that was reported by

Koumoto (1990) in the range of $w_L < 120$. But, in the range of $w_L > 120$, the variation of w_p would be independent of w_L and the values of w_p can be obtained in the approximate range from 35 to 70. The plastic limit w_p which is obtained after drying the soil, does not have a large magnitude of difference, but the w_L difference, in the range of $w_L > 120$, is caused by the increase in absorbed water between the plate-like montmorillonite particles.

3.2 Relationship between w_s and w_L

Fig.4 shows the relationship between w_s and w_L. According to Asakawa (1964), the relationship between w_s and w_L shows the linear relation, $w_s = 0.31(w_L + 33.7)$, that has been obtained in the range of $w_L \leqq 100$, but the w_s decreases as w_L increases in Ariake clay and mixed clays (M4, M5, M6, M7, and M8) as shown in Fig.4.

The high w_L for bentonite is caused by its main ingredient, montmorillonite, which has a higher specific surface (800m^2/g) (Ohtsubo, et al. 1987) than other clays, because of the more adsorbed water and the mutual interface action of clay particles. Yong and Warkentin (1975) stated that an increase in diffuse ion layer formation increases the liquid limit in montmorillonitic soils. This result means a relatively easier movement of particles and particle groups during drying are provided due to an oriented particle arrangement (Mitchell, 1976).

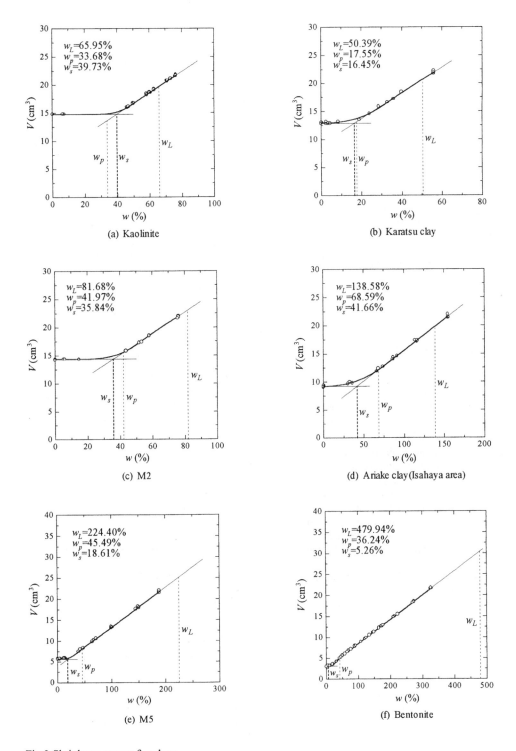

Fig.2 Shrinkage curves for clays

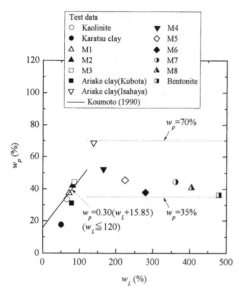

Fig.3 Relationship between w_p and w_L for clays

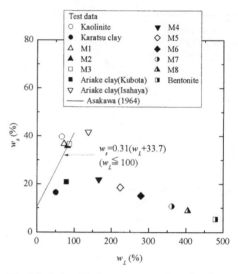

Fig.4 Relationship between w_s and w_L for clays

3.3 *Relationship between w_s and w_p*

Fig.5 shows the relationship between w_s and w_p. As shown in Fig.1 which is the schematic representation of volume-water content relation, the relationship between w_s and w_p generally should be $w_s < w_p$, but in Fig.5 $w_s > w_p$ for only kaolinite.

It was pointed out that the problems from the test results of liquid and plastic limits may have errors due to differences in the remolding time (Ogawa and Kobayashi, 1995) and preparing and adjusting methods for the samples (Yamazaki and Takenaka, 1965). Therefore, as many tests as possible are needed for the same soil sample, so an average value can be obtained (The Clay Science Society of Japan, 1967).

The values of w_s and w_p obtained for kaolinite were opposite to the other clay results in this test. This is due to the fact that the test methods for the plastic limit and the shrinkage parameters of soils are different.

3.4 *Relationship between I_p and w_L*

Fig.6 shows the plasticity chart which is the relationship between I_p ($=w_L - w_p$) and w_L. The plasticity chart was developed by Casagrande (1948) in order to estimate the characteristics of consistency of fine grain soils. As we know, clay and silt are classified by the A-line, and high or low values of compressibility by the B-line.

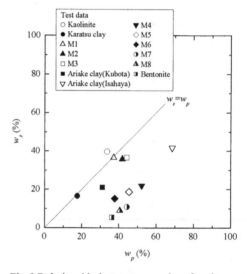

Fig.5 Relationship between w_s and w_p for clays

From the plasticity chart, compressibility increases as the liquid limit increases, and the viscosity increases as the plasticity index increases. Also, the soil that was used in this test shows that I_p increases along the A-line. Active clays like bentonite and mixed clays (M4, M5, M6, M7, and M8) containing a lot of bentonite, generally plot well above the A-line, whereas Ariake clay is silty clay having a high compressibility.

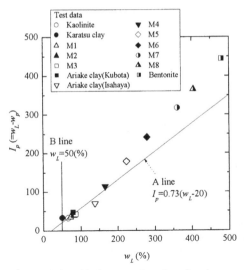

Fig.6 Relationship between I_p and w_L for clays (Plasticity chart)

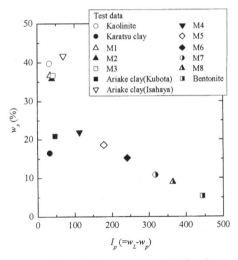

Fig.7 Relationship between w_s and I_p for clays

3.5 Relationship between w_s and I_p

Fig.7 shows the relationship between w_s and I_p. As shown in Fig.7, w_s decreases as I_p increases in the range of $I_p > 100$ and the value of w_s is scattered and is independent of I_p in the range of $I_p < 100$.

Furthermore, this tendency is similar to the results obtained for the relationship between w_s and w_L as shown in Fig.4. Because the value of w_L is 2~13 times higher than w_p, therefore the above relationship can be obtained.

3.6 Relationship between I_s and I_p

Fig.8 shows the relationship between I_s $(=w_p-w_s)$ and I_p. As shown in Fig.8, I_s increases in proportion to I_p in the range of $I_p < 100$, and I_s is a constant ($I_s=30$) being independent of I_p in the range of $I_p > 100$.

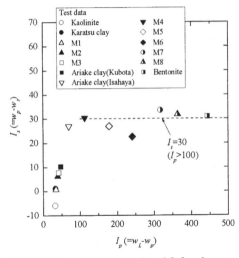

Fig.8 Relationship between I_s and I_p for clays

3.7 Relationship between w_L and pH

Fig.9 shows the relationship between w_L and pH. As reported by Sano, et al. (1998), the consistency limits of soil increases as the pH increases in the small range for w_L and pH. From the figure, our data showed the same tendency. Clay that had a lot of bentonite had high pH values and high w_L values.

Because bentonite is mainly composed of montmorillonite, in which the cohesion of its particles are weak, numerous molecular layers of water are formed, and have the properties of great dispersion and swelling in water.

3.8 Relationship between w_s and pH

Fig.10 shows the relationship between w_s and pH. As shown in Fig.10, with the exception of Karatsu clay, in general, w_s decreases with an increase in pH. The above result was obtained considering that the pH of clay having more bentonite is higher, and its w_s is lower. In addition, Karatsu clay strayed from this tendency, because it's low w_s, in spite of having the same pH value as the kaolinite, due to the Karatsu clay containing lots of sand and silt.

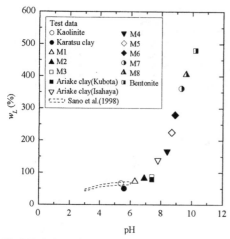

Fig.9 Relationship between w_L and pH for clays

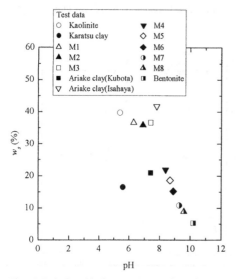

Fig.10 Relationship between w_s and pH for clays

4 CONCLUSIONS

The results on this study may be summarized as follows.

1) A linear relationship was obtained between w_p and w_L when $w_L \leqq 120(\%)$ as reported by Koumoto(1990), but when $w_L > 120(\%)$, w_p was in the range of 35(%) to 70(%).

2) Kaolinite has markedly higher shrinkage limit than that of bentonite, therefore the shrinkage limit of clay having more bentonite is lower than that of kaolinite.

3) w_s is widely distributed regardless of w_L in the range of $w_L < 100(\%)$, but w_s decreases with the increase of w_L in the range of $w_L > 100(\%)$.

4) The shrinkage index I_s ($=w_p$-w_s) increases in proportion to the plasticity index (I_p) in the range of $I_p < 100$ and I_s is a constant (I_s=30) being independent of I_p in the range of $I_p > 100$.

5) w_L increases approximately in proportion to the square of the pH.

REFERENCES

Asakawa. 1964. The test for shrinkage parameters of soil. The Japanese Society of Soil Mechanics and Foundation Engineering.

Casagrande, A. 1948. Classification and identification of soils. Transactions, ASCE. Vol. 113: 901-991.

Clay Science Society of Japan. 1967. Clay handbook: 606.

Koumoto, T. 1990. A consideration of the A-line in the plasticity chart by Casagrande. Soil Physical Conditions and Plant Growth. No.60: 2-5.

Mitchell, J. K. 1976. Fundamentals of Soils Behavior. John Wiley. New York.

Ogawa, F. and Kobayashi, M. 1995. Influence of remolding on the liquid and plastic limits of soil. Soils and foundations. Vol.35, No.4: 115-121.

Ohtsubo, M., Takayama, M. and Egashira, K. 1987. Effects of salt concentration and cation species on the remoulded strength of low swelling smectite marine clays. Soils and foundations. Vol.27, No.2: 85-92.

Sano, H., Yamada, M., Ohta, M., Nozawa, M. and Watanabe, K. 1998. A study on consistency limits of acidified soil by air drying. Journal of JSCE. No.610, III-45: 97-104.

Skempton, A. W. 1944. Notes on the compressibility of clays. Quart. Jour. of the Geol. Soc. of London. 100: 119-135.

Terzaghi, Karl. 1926. Simplified soil tests for subgrades and their physical significance. Public Roads. Vol. 7, October.

Yamazaki, F. and Takenaka, H. 1965. On the influence of air-drying on Atterberg's limit. Transactions, JSIDRE.14: 46-48.

Yong, R. N. and Warkentin, B. P. 1975. Soil Properties and Behavior. Elsevier Scientific Publishing Co., New York.

Clay Science for Engineering, Adachi & Fukue (eds) © 2001 Balkema, Rotterdam, ISBN 90 5809 175 9

A case study on failure of cut slope consisting of swelling clay

M. Mukaitani, E. Ichinose & M. Hori
Takamatsu National College of Technology, Japan

Norio Yagi, Ryuichi Yatabe & Kinutada Yokota
Ehime University, Matsuyama, Japan

ABSTRACT: The failure of the cut slope of weathered serpentine clayey soil occurred. Therefore, the appropriate counter works were executed, but the failure occurred again. In this paper, the mechanism and the cause of this failure are clarified considering the X-ray analysis and laboratory tests. The main reason of the failure is that the difference between the peak and the residual strengths of this serpentine clay whose main minerals are montmorillonite and a chlorite is so large. The results of the stability analysis by the Generalized Limit Equilibrium Method using both strengths of the peak one and the residual one explains the mechanism of this failure.

1 INTRODUCTION

This is a case study of a fail that occurred in a slope due to cutting for the road constraction. The cause of this failure may be due to lack of a survey before cutting operations began. After the part of this slope had failed, some kinds of countermeasure were executed based on the usual design methods. But this slope began to slide again. Then we made detail investigations, that is, finding the mechanism of the failure by carrying out the shear tests, the X-ray analysis of clay mineral and by the stability analysis. Finally, we dicided on appropriate countermeasure, including a suiteble safety factor, by using the results of these investigations.

2 OUTLINE OF THE SITE OF SLOPE FAILURE AND MOVEMENT OF SLOPE

The site is at Shimotsu-cho, Wakayama, Japan. This slope was cut for the road construction. The plan of the site and the section of the slope along which movement was most active, are shown in Figure 1 and 2, respectively. Scale of the failure is as 60m length, 120m width and the deepest position of slip surface of 12m. The geological out line and the positions of boring points are shown Figure 1. This site exists in Sanbagawa belt and consists of several kinds of schist in which extra basic serpentine distributes as shape of lens. The block of serpentine in which parallel to the slope surface has been changed to clayey soil of limited zone. The slope failure oc-

curred along this clayey soil layer which was not found before cutting.

First a small failure occurred due to rainfall of 30 mm/day and of maximum intensity 15 mm/hr on August 24, 1992. Some cracks were observed on August 29 when the slope had cut along A-line in Figure 2. Then counter weight fill work was executed on August 30 and 31. But as cracks grew bigger, prevention pile works were executed along No. 1 line from September 22 to beginning of October. Displacements in the slope were measured by the pipe strain gauges installed in the boreholes on No. 2 line to confirm a slip surface. From the observation of strains, the slip surface as shown in Figure 2 is estimated.

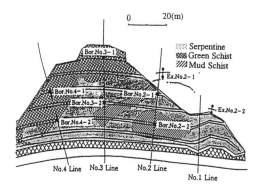

Figure 1 Plane of the cut slope.

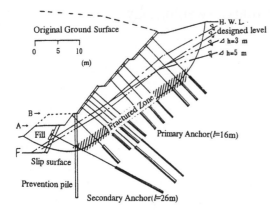

Figure 2 Vertical cross section of cut slope.

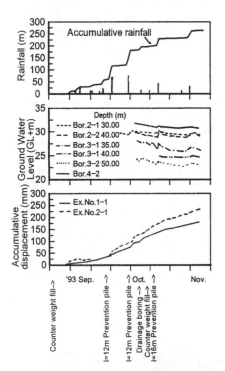

Figure 3 Change of accumulative displacement, ground water level and rainfall.

Furthermore, counter weight work up to level C in Figure 2 from October 19 to 22 and the prevention piles of 16m length, the anchor works and the drainage boring were executed from October 23 as shown in Figure 6.

Amount of rainfall, ground water level and accumulative displacements from the beginning of sliding are shown in Figure 3 in which the dates when counter works were executed are described. The depth of ground water levels was measured from the

cut slope surface. The rate of the displacement is increase by rainfall.

The strains in ground water taken by the pipe strain gauges in the borehole No. 2-2 are shown in Figure 4, in which the clear sliding layer can clearly be seen. The strain in the sliding layer is very large, more than $10000\,\mu$ in a two to three week period.

These countermeasure were designed to provide a suitable safety factor after the planed slope cutting was finished but the displacement of the slope was increased during the cutting process and cracks were generated when cutting reached the final level F in Figure 1, at middle of March 1993.

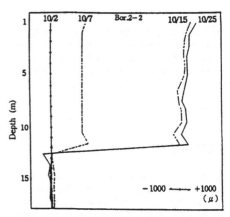

Figure 4 Observations by strain gauge type inclinometer.

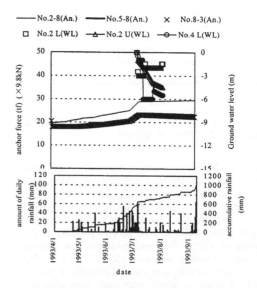

Figure 5 Changes of tensile force of anchor, ground water level and accumulative rainfall.

180

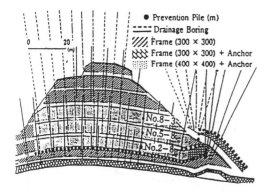

Figure 6 Outline of countermeasure to cut slope.

Therefore, the tension was applied to the anchors and the tensile force of the anchors increased gradually to near the yield point of the anchor due to rainfall in June 1993 as shown in Figure 5. As the slope failure seemed to be occurred, the counter weight work was executed again and the anchor works were added. After the anchor works, the counter weight was removed and the slope stability could be kept.

3 STRENGTH PARAMETERS OF WEATHERED SERPENTINE IN SLIDING LAYER

The strength parameters, which were used in the design of counter works, were obtained by the back analysis of conventional method. But the slope failure occurred after the counter works were executed in spite of keeping a suitable safety factor on calculation. Therefore strength characteristics of the clayey soil in sliding layer was investigated using the some shear testing apparatus in order to make mechanism of the slope failure clear.

3.1 Samples and method of shear tests

The core and the block samples of the sliding layer were gathered from the boreholes and outcrop. The properties of the sample 6 are shown in Table 1.

The shear tests carried out are as follows:

1) The triaxial and the box shear tests were carried out to obtain the peak strength. The undisturbed and the remoulded reconsolidated specimens were tested under the saturated and the unsaturated conditions. The consolidated undrain triaxial test were adopted.

2) The box type and the simple shear type ring shear tests (Mukaitani, et al., 1997) were carried out under the consolidated drain condition to obtain the residual strength. The remoulded and reconsolidated saturated specimens were only used.

3) The box and the simple shear type ring shear tests were carried out for the dried and soaked specimens to investigated strength decrease due to soaking. The air dried sample consisting of grains smaller than 420 μm was used.

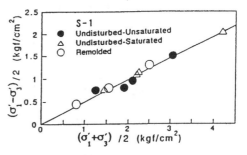

Figure 7 Failure line by triaxial tests.

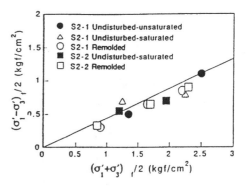

Figure 8 Failure line by triaxial tests.

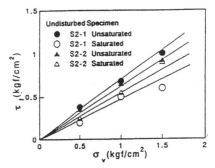

Figure 9 Failure lines by box shear tests

3.2 Test results and discussions

The triaxial test results for the block samples under various states of the specimens are shown in Figures 7 and 8. The box shear test results are shown in Figure 9. The cohesion c' in terms of effective stress is zero for all tests. the internal friction angle

ϕ_p' in terms of effective stress from the triaxial tests is constant for all specimens. The ϕ_p' of the unsaturated specimen is larger than one of the saturated specimen from the box shear tests. The cause of this fact is not clear. The ϕ_p' from the triaxial test is slightly smaller than one from the box shear. This is due to small gravel included in the specimen.

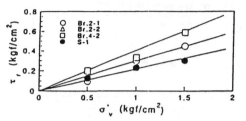

Figure 10 Residual strength lines by box shear

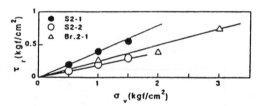

Figure 11 Residual strength lines by simple shear

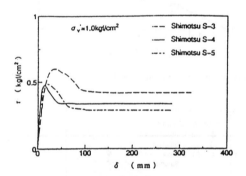

Figure 12 Relation between shear stress and displacement.

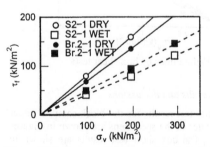

Figure 13 Influence of soak on ϕ by box shear.

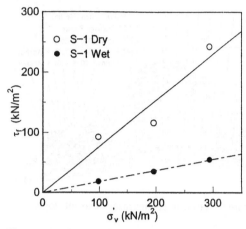

Figure 14 Influence of soak on ϕ by simple shear.

The test results from box type and simple shear type ring shear tests are shown in Figure 10 and 11, respectively. The cohesion for the residual strength is also zero. It is clear that the residual friction angle ϕ_r' for scatter even in the same sliding layer. The ϕ_p' is between $21°$ and $28°$ and the ϕ_r' is between $8°$ and $20°$.

Table 1 Physical properties and grain size distribution.

No.	w_L (%)	Ip	ρs (g/cm³)	Grain size contents (%)			
				Clay	Silt	Sand	Gravel
S-1	28.3	11.9	2.83	5.6	36.0	56.3	2.1
S2-1	28.4	19.0	2.64	13.3	37.8	32.1	16.8
S2-2	32.6	16.7	2.67	13.6	40.7	28.2	17.5
S-3	22.1	9.9	2.93	15.5	35.6	36.3	12.6
S-4	40.1	25.3	2.61	9.3	26.2	61.3	3.2
S-5	31.6	17.0	2.74	11.8	15.0	51.8	21.4
B2-1	39.6	22.7	2.81	8.5	35.4	37.8	18.3
B2-2	42.0	24.0	2.74	4.1	55.4	27.9	12.6
B4-2	32.9	16.4	2.68	2.6	42.9	43.7	10.8

Table 2 ϕ values by various shear tests

No.	ϕ_p' (°)	ϕ_d (°)	ϕ_r' (°)	
			Box	Simple
S-1	27.3		14.2	
S2-1	23.4	24.3 (34.9)	8.4	20.7
S2-2	22.6	28.7 (33.8)		12.4
S-3	21.0		13.2	
S-4	23.2		9.4	
S-5	21.4		8.8	
B2-1	25.6		13.2	
B2-2	24.1		12.7	
B4-2	28.2		20.1	

Table 3 Clay minerals analyzed by X-ray diffraction.

No.	Chrysotile	Antigolite	Chlorite	Talc	Montmorillonite	Felspars	note
S-1			*	—			clay
S2-1	+	—			*		clay
S2-2	—	+			*		clay
S-3		—			*		clay
S-4	—	—	—		*		clay
S-5		*			*		clay
B2-1			*	—			clay
B2-1			*				clay
B4-1			*	—			clay
B4-2			*				clay
S-6	—	*	—				Rock
S-7	—	*					Rock
B3-1		*				—	Gravel
B3-2		*		—		—	Gravel

∗: main content, ＋: secondary, −: little

The cause of the first slope failure is due to relatively small ϕ_p' . The main cause of slope movement after the countermeasure works were executed is due to largely decreasing of the ϕ_r' from the ϕ_p' , because the sliding layer had reached the residual strength state experiencing very large shear strain due to the first sliding. Influence of soaking on the peak, and the residual shear strength from the box shear and the simple shear type ring shear tests is shown in Figures 13 and 14, respectively. The ϕ_d' for drain condition and the ϕ_r' are very large for the dried specimens, but they drop so much (from about 40° to 20° and 16° respectively) due to soaking. Cause of this fact is not clear now.

4 X-RAY ANALYSIS OF MINERAL INCLUDED IN CLAYEY SOIL, OF SLIDING LAYER

Informer chapter it is clarified that the shear strength of the clayey soil is relatively low value. In order to investigate this cause, the X-ray analysis of clay minerals was conducted.

4.1 X-ray diffraction

The conditions of X-ray diffraction were as follows: 30kV and 15mA as electricity condition, Cu as the target and Ni as the filter were used. Scanning speed was as 1° /min. The powder method was adopted. Analysis were conducted on untreated specimens as those heat-treated at 500℃ and ethylene glycol and also by hydrochloric acid.

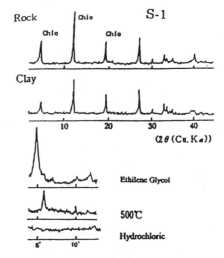

Figure 15 X-ray deffraction pattern.

4.2 Results of analysis

Examples of X-ray diffraction patterns are shown in Figure 15. There are two peaks at 7 Å and 14 Å for the original rock and clay. On the other hand, the sample treated by echylen glycol has also two peaks at the same positions and the position of peak for the sample treated by heat of 500℃ moves from at 14 Å to 13 Å. These facts represent that the rock and the clay contain an expensive chlorite.

From these analysis the clay seems to contains expensive chlorite. Another results are summarized in Table 3. The results of analysis express that the

183

main mineral of the rock and gravel is antigorite with a little chrisotile and chlorite as main mineral. It is considered that relatively small internal friction angles of the weathered serpentine clayey soil are caused by main mineral contained in them.

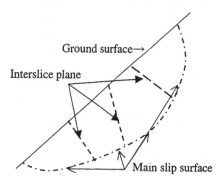

Figure 16 Description of interslice and main slip surface.

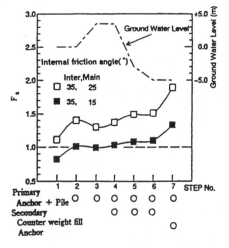

Figure 17 Result of stability analysis.

5 STABILITY ANALYSIS OF THE SLOPE

5.1 *Analysis method and conditions*

The generalized limit equilibrium method (called 'GLEM') was used for stability analysis of the slope (Enoki et al., 1990). In this method, strength parameters on the interslice plane are also needed as same as ones on the main slip surface. The formers mat bigger than the latter because the soil of interslice planes. We determined the main slip surface as shown in Figure 2 through the weathered serpentine layer.

The strength parameters are as follows. The cohesion c' is zero for any case. The internal friction angle ϕ_r' from the ϕ_p' on the main slip surface

are 25° and 15° as average, respectively. The ϕ_p' corresponds to the first sliding and ϕ_r' does to after experiencing very large shear strain due to failure. On the other hand, the ϕ_p' on the interslice plane was determined as 35° from the relation between stress and strain behavior of the triaxial tests.

The stability analysis was conducted as every step of the countermeasure works. The dry unit weight of the ground assumed to be 20 kN/m^3. The sliding soil was mass was divided into the ten blocks by the field was used. The preventing resistance of the pile was calculated according to Brom's method.

5.2 *Results of stability analysis*

The results of stability analysis are shown in Figure 17. The step 1 corresponds to when the first failure occurred due to cutting. The step 3 corresponds to when the ground water level suddenly rose. The period from step 4 to step 6 is during gradual lowering of the ground water level due to the executing drainage borehole.

The safety factors of step 1 calculated using ϕ_r' and ϕ_p', are 1.1 and 0.8, respectively. It is estimated from this calculation that the failure at the step 1 is seemed to be the first sliding of the clayey soil layer. The first sliding means that sliding has not occurred along sliding layer before this failure. After the first countermeasure works, the calculated safety factor is more than 1.3 using 25° of ϕ, but it is around 1.0 using 15° of ϕ. Therefore the result of calculation using the ϕ_r' explains necessity of the second countermeasure works.

6 CONCLUSION

One case of the slope failure due to cutting is described. This slope sliding was occurred during cutting without countermeasure works. But sliding occurred again after the countermeasure works that were designed from results of some kinds of shear tests, of the X-ray analysis for mineral and of the stability analysis.

1) This sliding occurred in the clayey soil layer of weathered serpentine, which exists forming lens-like in the ground. The internal friction angle ϕ_p' in terms of effective stress of this clayey soil is relatively low. This is one of caused of first sliding.

2) It became clear from the results of the triaxial and the ring shear tests that lowering of the ϕ' from ϕ_p' of the peak strength to ϕ_r', of the residual one is very big. This is one of the causes that sliding occurred after executing the countermeasure works, because the sliding layer reached the residual strength state when first sliding occurred.

3) These are two kinds of clayey soil in sliding layer. One consists of montmorillonite and the other does mainly expensive chlorite. These minerals in-

fluence so much on the strength characteristics of the clayey soils.

4) The stability analyses conducted using the ϕ_r' from the ϕ_p', explain well the mechanism of stability of this slope at every step of countermeasure works. Therefore, stability analysis should be conducted considering lowering of strength parameter ϕ' when a slope has slide in the past.

ACKNOWLEDGMENTS

The authors would like to express their gratitude to Mr. Osamu Futagami of technical officer and engineers of the machine industry in Ehime University and former students in Takamatsu National College of Technology for their help in the preparation of this paper.

REFERENCES

Baker D. H. 1986. Embankment of slope stability by vegetation. *Ground engineering.* Vol. 19. No. 3. 11-15.

Endo Y. and Tsuruta T. 1968. The influence of roots on the shear strength of soil. Vol. 1. *Journal of institute of forest research. Hokkaido.* 167-182. (In Japanese)

Enoki Meiketsu, Yagi N. and Yatabe R. 1989. The mechanism of reinforcement in tension in sand. *Memories of Faculty of Engineering. Ehime university.* Vol. 11. No. 4. 441-449. (In Japanese)

Enoki Meiketsu, Yagi N. and Yatabe R. 1990. Generalized Slice Method for Slope Stability Analysis, Soils and Foundations, Vol. 30, No. 2, 1-14.

Enoki Meiketsu, Yagi N. and Yatabe R. 1991. Stability Analysis Method of Natural Slope. *Memories of Faculty of Engineering. Ehime university.* Vol. 11. No. 2. 425-432. (In Japanese)

Gray D. H. 1974. Reinforcement and stabilization of soil by vegetation. ASCE. *Journal of Geotechnical engineering division.* Vol. 100. No. GT6. 695-699.

Gray D. H. and Ohashi H. 1983. Mechanics of fiber reinforcement in sand. ASCE. *Journal of Geotechnical engineering division.* Vol. 109. No. 3. 335-353.

Greenway D. R. 1987. Vegetation and slope stability. Slope stability - *Geotechnical engineering and Geomorphology.* Edited by M. G. Anderson and K. S. Richards. Publicized by John Wiley and Sons.

Hattori T. et al. 1993. Study on engineering characteristics of weathered granite. *Proceedings of 48th Annual Conference. Japanese Society of Civil Engineers* (JSCE). 924-925. (In Japanese)

Iwamoto K. et al. 1985. The shear strength characteristics of soils with roots. Using large shear apparatus. part 1. *Proceedings of 40th Annual Conference of Japanese Society of Civil Engineers* (JSCE). 175-176. (In Japanese)

Ditto. 1985. part 2. 320-321. (In Japanese)

Kobashi S. 1983. The recent findings of influence of roots on stability of slopes. *Journal of technique of greening.* 10 (1). 14-19.

Mukaitani Mitsuhiko 1997. Construction of a dike on the soft ground with super time depending deformation. *Proceedings of 2nd international conference on ground improvement techniques.* 353-360.

Ueno Shouji 1996. The method of application of topographical and geological information for geothechnical engineers, (part 5) Landslide. *Journal of JGS.* Vol. 44. No. 6. 51-56.

Yagi Norio and Yatabe R. 1985. A Microscopic Consideration on Shearing Characteristics of Decomposed Granite Soil. *Journal of JSCE.* No. 364. 131-141. (In Japanese)

Yagi Norio, Enoki M. and Yatabe R. 1992. Reinforcing Mechanism of Sandy Soil by Roots of Plans. 1991. *Proceedings of 27th Annual Conference of Soil Mechanics and Foundation Society of Japan.* 1865-1866. (In Japanese)

Yagi Norio, Yatabe R. and Enoki M. 1994. The Effects of Root Networks on Slope Stability. *Proceedings of Int. Conf. on Landslide*, Slope stability and Safety of Infra-structure. 387-392. Malaysia

Yatabe Ryuichi, Yagi N. and Enoki M. 1991. Consideration from effective stress about strength parameters of slip layer clay of landslide. *Journal of the Japanese Society of Landslide.* Vol. 28. No. 2. 20-26. (In Japanese)

Clay Science for Engineering, Adachi & Fukue (eds) © 2001 Balkema, Rotterdam, ISBN 90 5809 175 9

Methodology for prediction of desiccation of clay liners

G.L.Sivakumar Babu & E.Gartung
Geotechnical Institute (LGA), Nuremberg, Germany

ABSTRACT: Compacted clays are being used as liners in landfills and other waste containment facilities as barriers to water and contaminant flow. Extensive research on long term behaviour of clay liners established that clay liners over a period of time, under various environmental conditions such as leachate soil interactions and temperature tend to desiccate, resulting in cracks in the liner, affecting their performance significantly. Desiccation behaviour of clays is affected primarily by variations in suction pressures induced due to physico-chemical changes and temperature. There remains a need to determine the range of suction pressures developed under different degrees of compaction at temperatures comparable to leachate conditions. In this paper, soil water characteristics of compacted clays represented by van Genuchten parameters available in literature are used and variations in suction pressures are evaluated by a one dimensional coupled transport model of water, water vapour and temperature. Variations in suction pressures and shear strength properties of clay liners are related and a methodology for prediction of onset of desiccation is developed.

1 INTRODUCTION

Compacted clays are being used as liners in landfills and other waste containment facilities as barriers to water and contaminant flow. Extensive research on long term behaviour of clay liners established that clay liners over a period of time, under various environmental conditions such as leachate soil interactions and temperature tend to desiccate, resulting in cracks in the liner, affecting their performance (Melchior, 1993). The compacted liners are essentially partially saturated and have negative pore pressures or suction pressures. Desiccation behaviour of clays is affected primarily by variations in suction pressures induced due to physico-chemical changes and temperature. There remains a need to determine the range of suction pressures developed under different degrees of compaction at temperatures comparable to leachate conditions and examine current guidelines for minimisation of desiccation. It is also known that desiccation is associated with available shear resistance at particle level and corresponding suction pressures and that desiccation is essentially a shear failure at the particle level. Hence, there is a need to link variations in suction pressures and shear strength properties

of clay liners and develop a methodology for prediction of onset of desiccation in terms of variations in suction pressures The following sections present the methodology adopted, test results considered and inferences and finally an approach is presented to predict desiccation behaviour.

2 LITERATURE REVIEW

Studies on partially saturated soils have been extensive and notable contributions have been made by various workers in understanding the behaviour of partially saturated soils and to some extent on desiccation behaviour (Fredlund and Morgenstern, 1977, Morris et. al 1992, Fredlund and Rahardjo, 1993). Fredlund and Morgenstern presented a theoretical stress analysis of a partially saturated soil and identified that the study of soil behaviour in terms of independent stress state variables, viz., total stress relative to pore air pressure ($\sigma - u_a$), suction ($u_a - u_w$) is necessary. Morris et.al (1992) examined the cracking behaviour of soils controlled by suction and developed solutions to estimate crack depths considering tensile and shear behaviour and fracture mechanics concepts.

3 PREDICTION OF SUCTION VARIATIONS

Prediction of suction pressures and corresponding variations in relation to pore fluid environment and temperatures is important for prediction of propensity for cracking. Döll (1997) presented a numerical model of coupled transport of water, vapour and heat in unsaturated soils in terms of suction potentials. The input parameters for model are initial suction profile, thickness of clay liner, porosity, saturated water content, saturated hydraulic conductivity and soil water characteristic curves represented by van Genuchten parameters. Model gives suction profiles corresponding to different time periods and ambient temperatures of leachate. In the present investigation, program developed by Döll is used.

3.1 Soil water characteristic curves

Soil water characteristic curve defines a relationship between water content and suction and is a function of soil type, pore fluid environment, pore size distribution, compactive effort etc. van Genuchten relationship (van Genuchten, 1980) is given by

$$\Theta = \frac{(\theta - \theta_r)}{(\theta_s - \theta_r)} = \left\{ \frac{1}{\left[1 + \left((\alpha\psi)^n\right)\right]} \right\}^m \quad (1)$$

The term Θ is the effective saturation, θ, θ_r and θ_s are the volumetric water content at suction ψ, residual water content and saturation water content respectively, and α, n and m are the fitting parameters for the relationship. Tinjum et. al (1997) presented data on soil water characteristic curves for typical landfill liners for dry state, optimum and wet of optimum water contents corresponding to standard proctor (SP) and modified proctor (MP) efforts. They represented soil water characteristics by van Genuchten parameters α and n. In the present study, some data of the reported results are used. Table 1 gives the index properties such as liquid limit (LL), plasticity index (PI) and compaction characteristics. Table 2 gives van Genuchten parameters for soil water characteristic curves for the soil. Further details are given in Tinjum et. al (1997).

Table 1 Index properties and compaction characteristics (Tinjum et. al 1997)

Soil	LL %	PI%	OMC (MP) %	Max γ_d (MP) kN/m³	OMC (SP) %	Max.γ_d (SP) kN/m³
B	49	26	12.8	19.2	18.5	17.3
C	27	15	9.5	20.7	13.7	18.9

Table 2 van Genuchten parameters for B soil

W/c	Dry density, γ_d	θ_s	n	α (cm⁻¹)
21.8	16.1 (SP)	39.2	1.101	0.0253
29.6	17.2 (SP)	35.2	1.083	0.0032
35.8	16.2 (SP)	38.1	1.120	0.0012
19.3	18.7 (MP)	29.4	1.133	0.0041
26.0	19.2 (MP)	27.5	1.176	0.0024

Computations are performed for a typical 1.5 m thick compacted clay liner of the above properties constructed on a silty clay. The silty clay has saturated volumetric water content (by volume) of 0.44 and saturated hydraulic conductivity of 3.78×10^{-05} cm/s and its soil characteristic curve defined by $n = 1.08$ and $\alpha = 0.2$ cm⁻¹. Water table is assumed to be at 4 m below the bottom of the clay liner. Initially, the temperature is assumed to be $10°C$ everywhere along 5.5 m length. Subsequent temperature as a result of leachate interactions at the top of the clay liner is assumed to be $35°C$. It is also assumed that water table is at $10°C$ through out the simulations for the entire time period. Simulations are performed for five years. Suction pressures are evaluated and examined in terms of proposed methodology for prediction of onset of cracking to understand i) the role of state of compaction (dry, optimum or wet), ii) effect of compaction effort (standard proctor or modified proctor) and iii) influence of temperature. The inferences are discussed in the following sections.

3.2 Effect of state of compaction

Compaction of clay liners to water contents considering different criteria such as shear strength, permeability and compressibility, swell potential or susceptibility for desiccation cracking is a point of discussion in Geotechnical Engineering. Considering all the above criteria, Benson et. al (1997) recommend a zone of water contents at which the clay liner material has all the desirable attributes of a satisfactory liner material. Rowe et. al (1995) indi-

cate that compaction should be done wet of optimum, as it produces a clay barrier of enough self
healing properties. These observations need be examined in terms of suction pressures. Fig. 1 shows
the variation of suction pressures as affected by the
degree of compaction. It can be observed that for
the two soils, of the three conditions, compaction
to wet of optimum and close to optimum leads to
lesser suction pressures.

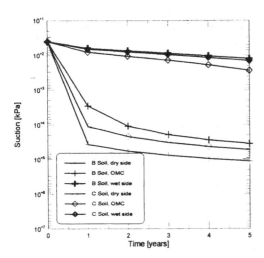

Fig. 1 Effect of state of compaction on suction
pressure development

3.3 Effect of compaction effort

Fig. 2 shows the effect of compaction efforts for
soil C. It can be observed that standard proctor
effort results in lesser suction pressures compared
to modified proctor effort.

3.4 Effect of temperature

Fig. 3 shows the effect of temperature on the suction pressures developed. It can be noticed that i)
as temperatures of the leachate increase, higher
suctions are developed and ii) the effect is more
pronounced in soil state corresponding to dry of
optimum.

4 PREDICTION OF ONSET OF CRACKING

Fredlund and Rahardjo (1993) showed that the
shear strength of a partially saturated soil can be
expressed as

$$\tau_f = c' + (\sigma - u_a) \tan \phi' + (u_a - u_w) \tan \phi^b \quad (2)$$

where, c' is the effective cohesion, τ_f is the shear
strength, $(\sigma - u_a)$ is the normal stress, ϕ' is the friction angle, $(u_a - u_w)$ is the suction and ϕ^b is the
friction angle representing the variation of shear
strength with respect to changes in suction.

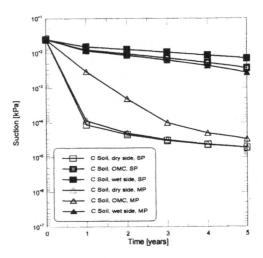

Fig.2. Effect of compaction effort on suction development

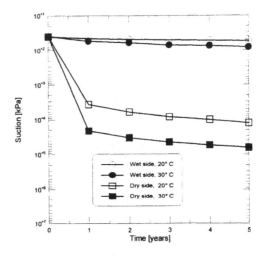

Fig.3. Effect of leachate temperature on suction
development

To determine final suction that induces desiccation
and denoting initial suction as $(u_a - u_w)_0$ and subsequent suction developed as a result of desiccation
as $(u_a - u_w)_1$, a simple expression can be derived as

189

$$(u_a - u_w)_1 = \frac{(u_a - u_w)_0 (1 + \sin \phi^b) + 2c \cos \phi^b}{(1 - \sin \phi^b)} \quad (3)$$

The above equation enables computation of suction pressure $(u_a - u_w)_1$ that could develop as a result of desiccation.. Fredlund and Rahardjo (1993) evaluated the range of this parameter for different types of soils and observed that this value is in the range of 15^0 to 20^0. Hence for a typical value of 15^0 and cohesion of 10 kPa, considering that the initial suction is 40 kPa, using the above equation, a final suction value of 94 kPa is obtained. This is the maximum permissible suction due to desiccation and values equal to this or more than this value induce shear failure leading to desiccation cracks. This value is examined with reference to simulation results and the time corresponding to which this value is likely to be exceeded is noted. As discussed earlier, this value is a function of state of compaction, type of soil, compaction effort and liner may or may not crack depending on the stated conditions. Simulation results obtained for a typical soil compacted wet of optimum (standard proctor) and plotted in terms of extended Mohr-Coulomb envelope in suction $(u_a - u_w)$ versus shear strength plane shown in Fig. 4 makes the point clear. It can be observed that the soil is likely crack at the end of third year.

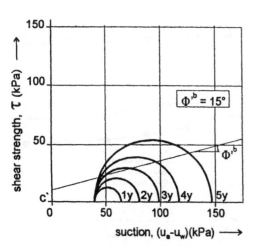

Fig.4 Prediction of onset of cracking

The above relationship is derived considering that the failure envelope in suction plane is linear within the suction range of interest. It is also recognised that the failure plane is non linear and the effect of non linearity is reflected in terms of reduction in ϕ^b.

from a value of friction angle corresponding to saturated state to lesser values subsequently.

5 CONCLUDING REMARKS

The paper evaluates suction pressures developed in typical compacted liner materials at different moisture states and degrees of compaction. It is noted that compaction of soils to wet side results in lesser suction pressures. An approach to evaluate maximum suction pressure that is likely to cause desiccation, knowing initial suction is presented and illustrated with typical example.

6 ACKNOWLEDGEMENTS

First author thanks Alexander von Humholdt Foundation, Bonn for financial support and the authorities of LGA for providing the facilities to carry out the work. The authors thank Mr. H. Sporer and Mr. H. Zanzinger for helpful discussions.

REFERENCES

Benson, C. H, Daniel, Boutwell, G. P (1997) Field performance of compacted clay liners, *Journal of Geotechnical and Geoenvironmental Engineering*, 125 (5), 390- 403.

Döll, P (1997) Desiccation of mineral liners below landfills with heat generation, *Journal of Geotechnical and Geoenvironmental Engineering*, 123(11), 1001- 1019.

Fredlund, D and Rahardjo, H (1993*) Soil mechanics for unsaturated soils*, John Wiley & Sons New York.

Melchior, S (1993) *Water balance and efficiency of multiple surface liners systems for landfills and contaminated sites*, Ph.D. thesis. University of Hamburg, Germany.

Rowe, R.K, Quigley, R. M. and Booker, J. R. (1995) *Clayey barrier systems for waste disposal facilities*, E & FN Spon, pp390.

Tinjum, J. M. Benson, C. H and Boltz. L R (1997) Soil water characteristic curves for compacted clays, *Journal of Geotechnical and Geoenvironmental Engineering*, 123(11), 1060- 1069.

van Genuchten, M (1980) A closed form equation for predicting the hydraulic conductivity of unsaturated soils, *Soil Sci. Soc. Am. Journal*, 44, 892-898.

Clay Science for Engineering, Adachi & Fukue (eds) © 2001 Balkema, Rotterdam, ISBN 90 5809 175 9

Soil water suction and compaction influence on desiccation cracks of mineral liners

S. Wendling & H. Meißner
Department of Soil Mechanics and Foundation Engineering, University of Kaiserslautern, Germany

ABSTRACT: The sealing effect of compacted clay layers which are used as a cover or liner for landfills is directly affected as soon as cracks form. This research work focuses on the relationship between the formation of cracks in clayey soils which results from desiccation and the soil-water suction in relation to the parameters temperature, water content and void ratio. Main topics are the study of tensile strength in clayey soils as well as the analysis of moisture transfer which occurs due to both temperature gradient resulting from decay process or sun rays and moisture gradient in clayey mineral liners. Soil-water characteristics of a kaolinite are experimentally determined. Tensiometers and a vacuum desiccator method have been used to measure the soil-water suction for values of pF between pF0 and approx. pF3 as well as pF5 to pF7. The water transfer coefficients such as the isothermal and thermal diffusivity as well as the unsaturated hydraulic conductivity have been also experimentally determined. Furthermore, a laboratory model is presented to determine the tensile strength of the kaolinite in relation to different parameters such as temperature, water content, radial pressure and void ratio. Tensile stress-strain-curves as well as tensile strengths and strains vs. volumetric water content and void ratio are shown.

1 INTRODUCTION

Desiccation cracks of mineral liners of landfills originate as a result of moisture loss that takes place due to moisture transfer resulting from both temperature and moisture gradients. Therefore, fundamental relations between water content and soil-water suction, i.e. the pF-curve have to be determined. In addition, interrelations between moisture transfer, tensile stresses and shrinkage behaviour of a mineral liner have to be investigated as a function of water content, stress state, temperature and void ratio.

2 DETERMINATION OF THE SOIL-WATER CHARACTERISTICS (pF-CURVE)

In all tests a kaolinite with particles less than to 2μm (clay fraction) of $d_{2\mu m} = 60$ % is used. Two methods were applied to measure the soil water suction in the range of pF0 to approx. pF3 and pF5 to pF7, where pF is the soil water suction expressed in cm of water column on a logarithmic scale.

2.1 Tensiometer method

Tensiometers can measure soil-water suctions in the range of pF0 to pF3. Five tensiometers of the type UMS-T5, (UMS, 1993) were used.

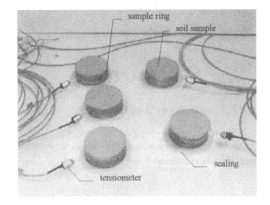

Figure 1. Experiment to determine pF-curve in the range of pF0 to pF3, tensiometer method.

Soil samples were compacted into five sample rings with a diameter of 10 cm and a height of 4cm. An initial void ratio of $e = 1.49$ and a degree of saturation of $S_r \approx 1$ were attained. Tensiometers were then installed in the middle zone of each sample ring, see Fig. 1. The soil samples have open ends and closed bottoms. Evaporation takes place from the open ends and the moisture loss from the soil samples causes an increase of the pF-values that were determined during the test using the tensiometer measurements. Fig. 2 shows the trend of the pF-values as a function of the evaporation time for three soil samples. It can also be seen, that the tensiometers need a few hours until their values stabilize.

Different test completion times are selected in order to reach certain predefined pF-values. For the example shown in Fig. 2, the tests were finished after 24, 32, and 78 hours for which the pF-values were 1.57, 1.77 and 2.44 respectively.

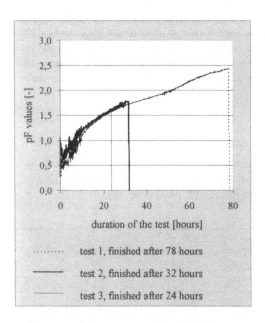

Figure 2. Tensiometer method, test results.

After finishing every test the volumetric water content θ was determined from:

$$\theta = w \cdot \frac{\rho_d}{\rho_w} \quad [cm^3/cm^3] \tag{1}$$

where w is the gravimetric water content of the soil sample determined using the oven drying method, ρ_d is the dry density and ρ_w is the density of water.

The relationship between θ and ψ (pF - curve) is shown in Fig. 7.

2.2 Vacuum desiccator method

2.2.1 Test principle
A soil sample is allowed to reach equilibrium with a known air humidity which can be maintained in a sealed flask using a solution of sulphuric acid (Croney et al. 1952). The Figures 3 and 4 show a diagram and a picture of our modificated vacuum desiccator method respectively.

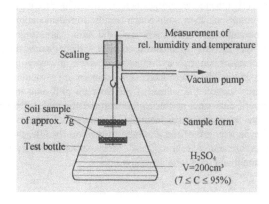

Figure 3. A diagram of the modified apparatus of the vacuum desiccator method.

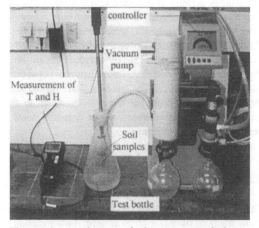

Figure 4. A picture of the vacuum desiccator apparatus used.

2.2.2 *Applied function*

The soil-water suction ψ can be determined (Schofield 1935) as followed:

$$\psi = -\frac{R \cdot T}{M \cdot g} \ln \frac{p}{p_0} = 2{,}303 \frac{R \cdot T}{M \cdot g} \log \frac{100}{H}$$

and the pF-value results in:

$$pF = \log\left(\frac{\psi}{1cm}\right)$$

i.e.:

$$pF = \log\left(2{,}303 \frac{R \cdot T}{M \cdot g}\right) + \log(2 - \log H) \qquad (2)$$

where

pF   is the logarithmic scale of the soil-water suction ψ in cm of water column

R is the universal gas constant and is equal to 8.3143 [J/(mol K)]

T is the absolute temperatur [K]

M is the molecular weight of water and is equal to 18.016 [kg/kmol]

g   is the gravitational acceleration and is equal to 9.81 [m/s²]

$\frac{p}{p_0} = H$ is the relative humidity [%]

p is the vapor pressure [Pa]

p_0 is the saturated vapor pressure [Pa].

In the tests R, M and g are constants and T and H are measured.

The time needed to reach the moisture equilibrium can be radically reduced using a vacuum (Croney et al. 1952).

2.2.3 *Adjustment of the relative humidity*

To obtain a certain value of rel. humidity a sulphric acid solution will be used. An increase of the acid concentration causes a decrease of the rel. humidity, see Table 1, tests 1, 4 and 5 for which a maximum rel. humidity of 50% was reached.

To obtain values of rel. humidity > 50% it was possible to variate the vacuum values that could affect the rel. humidity values for tests with a limit value of acid concentration of approx. 8% by weight, see Table 1. tests 7, 12 and 16.

Fig. 5 shows the trend of relative humidity during several tests. It can be seen that at the beginning of all the tests the relative humidity has a value of

Table 1. Relative humidity as a function of sulphuric acid concentration and the vacuum values, own tests.

Test no.	concentration of H_2SO_4	vacuum	rel. humidity
-	weight - %	mbar	%
5	67.37	16	10.1
1	7.88	16	50.0
4	44.73	17	40.2
7	7.69	19	52.8
12	7.70	20	60.4
16	7.73	28	78.3
13	66.01	936	12.5

approx. 100% which is due to the fact that the sulphuric solution was fresh prepared and therefore high temperatures and relative humidities arise.

After a few hours the sulphuric solution was stabilized and the relative humidities reach their final values. In all the tests the relative humidity was nearly constant in the last 24hours.

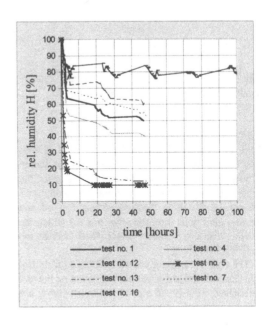

Figure 5. Relative humidity vs. time during some tests.

A comparison with tests that continued 100 hours under the same laboratory conditions shows that the values of rel. humidity between 24 and 48 hours was nearly the same as that at the end of the test, see Fig. 5, test no. 16. Therefore, the moisture equilibrium of the small soil samples of approx. 7g was assumed to be reached after 48 hours.

Measured values of relative humidity and temperature at the end of each test give a pF value, see Eq. 2.

Due to the fact that measured temperatures for all tests lie in the range of $19.3°C \leq T \leq 20.8°C$, the temperature can be considered nearly constant and a relationship between relative humidities and pF values can be made, see Fig. 6.

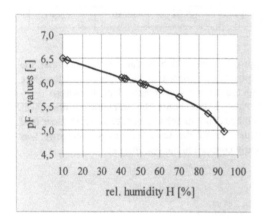

Figure 6. pF-values as a function of the relative humidity, $19.3°C \leq T \leq 20.8°C$, see also Eq. 2.

2.2.4 Soil preparation and test evaluation

Two soil samples of approx. 7g were used in each test. The initial void ratio is $e = 1.49$ and the degree of saturation is $S_r \approx 1$. At the end of the tests the water content w was determined gravimetrically from the one sample using the oven drying method and the dry density ρ_d from the other one for which the volume can be exactly determined using the immersion weighing.

The volumetric water content calculated from Eq. 1 builds with the soil-water suction value of the soil determined from Eq. 2 a point in the pF-curve, see Fig. 7.

2.3 Fitting curve

The measured soil-water characteristics shown in Fig. 7 are fitted with the $\psi(\theta)$-function adopted by van Genuchten (1980). It is given by the following equation:

$$\psi = a\left[\left(\frac{1}{\Theta}\right)^b - 1\right]^c$$

where ψ is the soil water suction in cm of water column, a, b and c are parameters to be estimated from the measured soil-water data and Θ is the dimensionless volumetric water content and can be written as:

$$\Theta = \frac{\theta - \theta_r}{\theta_{sat} - \theta_r}$$

where θ, θ_r, θ_{sat} are the actual, the residual and the saturated water content respectively.

Using the test results of the kaolinite we find:

$$\psi = 1000 \cdot \left[\left(\frac{0.54}{\theta}\right)^4 - 1\right]^{0.71} \tag{3}$$

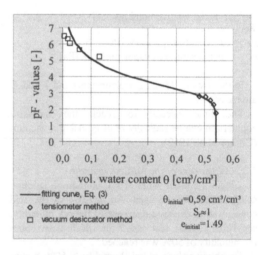

Figure 7. pF-curve, test results and fitting curve.

2.4 Influence of the void ratio on the pF-curve

Results of two another tests with an initial void ratio of $e = 1.87$ and $e = 0.76$ are shown in Fig. 8. Tests in the moist rang (range of tensiometer measurements) were more affected by void ratio than those in the dry range (range of vacuum desiccator measurements).

I.e. the pF-curve (Eq. 3) can only be used if similar initial conditions are available, otherwise a different pF-curve should be determined, particularly in the moist range.

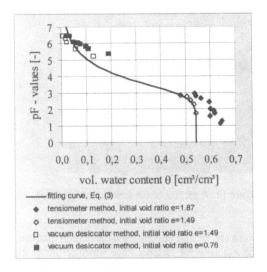

Figure 8. pF-θ-relation for different void ratios.

3 MOISTURE TRANSFER UNDER COMBINED MOISTURE AND TEMPERATURE GRADIENTS

3.1 *Constitutive relation*

The following equation was developed by (Philip & DE Vries 1952). It describes the one dimensional transfer of soil-moisture under combined effects of moisture gradient and a downward oriented temperature gradient:

$$\frac{q}{\rho_w} = -D_\theta \frac{\partial \theta}{\partial z} + D_T \frac{\partial T}{\partial z} + K_u \qquad (4)$$

where

q	total flux density [g·cm²/d]
ρ_w	density of water [g/cm³]
D_θ	isothermal moisture diffusivity [cm²/d]
$\dfrac{\partial \theta}{\partial z}$	moisture gradient [1/cm]
D_T	thermal diffusivity [cm²/d·°C]

$\dfrac{\partial T}{\partial z}$	temperature gradient [°C/cm]
K_u	unsaturated hydraulic conductivity [cm/d].

3.2 *Laboratory model*

Fig. 9 shows the experimental model used to determine the transfer parameters D_θ, D_T and K_u.

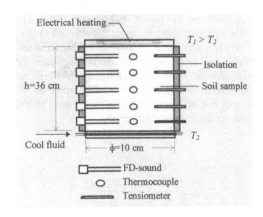

Figure 9. Laboratory model to determine the parameters of moisture transfer, see Eq. 4.

Three variations of test conditions have enabled us to determine the transfer parameters. Test results gave that the moisture diffusivity coefficient D_θ increases with an increase of volumetric water content θ in the range of $0.05 < \theta < 0.32$ cm³/cm³, the thermal diffusivity coefficient D_T becomes greater as the volumetric water content increases in the range of $0.21 < \theta < 0.32$ and the unsaturated hydraulic conductivity K_u increases with the increase of θ in the range of $0.18 < \theta < 0.34$. Missing values that lie outside these ranges are still to be determined.

Because of the length limit it will not be explained in details but it can be refered to the sources (Meißner & Wendling 1999; 2000) and (Wendling & Meißner 2000) that include a detailed description of the experimental methods used and the analytical equations applied.

4 TENSILE TESTS

An apparatus is used for determining the tensile strength in relation to the volumetric water content (soil-water suction) with variation of the parameters temperature, water content (degree of saturation), stresses and void ratio.

The Figures 10 and 11 show a scheme and a picture of the tensile test respectively. The test cylinder consists of an upper and a lower part. Each part has one outer and one inner shell with the dimension (Φ=160mm, h=130mm) and (Φ=110mm, h=130mm) respectively. The shells were placed so that an annular soil sample can be compacted between the outer and the inner shells. The two inner shells are quartered so that a radial pressure σ_r can be applied, see (Wendling & Meißner 2000). The porous material of the shells allows the diffusion of air humidity and the conduction of heat into the soil sample. The crack zone is predefined between the upper and the lower parts of the cylinder. In this zone (middle zone of the test cylinder, see Fig. 10) the soil displacements and strains as well as the tensile stresses can be determined.

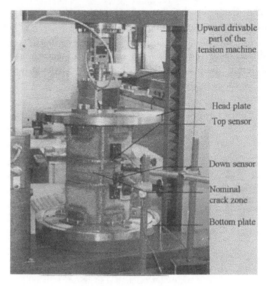

Figure 11. A picture of the tensile test apparatus.

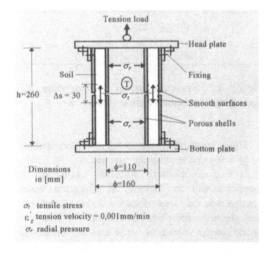

Figure 10. Schematic presentation of the tensile test.

4.1 Determination of tensile stresses

Two openings (b=1cm, h=4cm and t=2,5cm) in the middle zone of the test cylinder are used to observe and measure the crack length so that the net cross-sectional area (A_n) of the soil sample under tension can be obtained and the Tensile stresses (σ_t) result in:

$$\sigma_t = \frac{F_t}{A_n} \tag{5}$$

with $F_t = F_{Tension} - F_{residual}$. $F_{Tension}$ is the tension force

which can be read directly from the tension machine and $F_{residual}$ is the residual force which results at the end of the test when the top and the lower parts of the cylinder are completely separated, see Fig. 12. It is assumed that the crack length observed at the openings is equal to that of the whole sample.

After the peak point was reached the tension force values fall at first rapidly and then slowly until the residual value is reached.

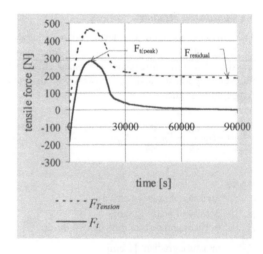

Figure 12. Tension force vs. time during a tension test.

4.2 Determination of tensile strains

Two further openings (b=1cm, h=4cm) are used to measure the soil displacements. The openings are in the middle zone of the test cylinder which has smooth surfaces. Sensors at the top and the down ends of the opening in a distance of $\Delta s \approx 30mm$ were installed, see Fig. 11 and 13. The tensile strain ε results in:

$$\varepsilon = \frac{\Delta h_{top} - \Delta h_{down}}{\Delta s} \quad \% \qquad (6)$$

where Δh_{top} and Δh_{down} are the displacements at the top and the down end of Δs respectively.

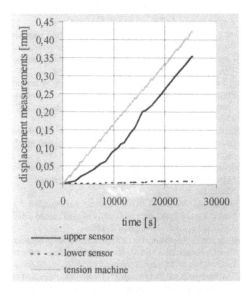

Figure 13. Measured displacements at the top and the bottom of Δs.

4.3 Tensile stress-strain-behaviour

Determined values of tensile stresses, Eq. 5 and tensile strains, Eq. 6 give the stress-strain-diagram shown in Fig. 14.

The diagram shows that in the hardening range, between the start of the test and $F_{t(peak)}$ the tensile stresses increase with the tensile strains. The first crack was observed just after the peak point and a tensile strain of $\varepsilon \approx 0,45\%$ was determined. In the range between the peak point and the first crack a softening of the soil occured.

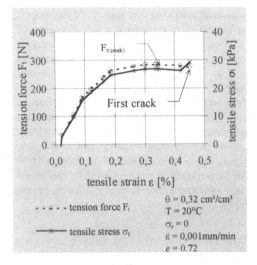

Figure 14. Stress-strain-diagram.

4.4 Tensile stresses as a function of vol. water content and void ratio

Test results are shown in Fig. 15. Tests with an initial void ratio of $e = 0.77$ show an increase of

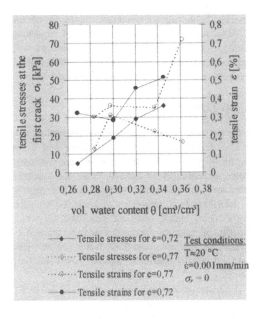

Figure 15. tensile stresses and strains as a function of vol. water content and void ratio.

tensile stresses with an increase of the volumetric water content up to the value $\theta \approx 0.3$. For values of $\theta > 0.3$, the tensile stresses decrease with an increase of θ. However, tensile strains are more and less increasing so long the volumetric water content increases.

Tests with an initial void ratio of $e = 0.72$ show an increase of tensile stresses as long as the volumetric water content rises. Tensile strains increase clearly for values of volumetric water content of $\theta > 0.3$. The tests are to be continued in the range of $\theta > 0.345$. Nevertheless, it can be seen that both tensile stresses and strains are apparently affected by the void ratio.

5 COUPLED PORE WATER DIFFUSION AND STRESS ANALYSIS

The prior objective of the research project is to predict the formation of desiccation cracks in a clay liner. Therefore, a coupled pore water diffusion and stress analysis is in process in an unsaturated clay liner using the finite element method. In reality and in the numerical simulation the mechanical behaviour and the yield function are affected by parameters such as soil water suction, temperature and void ratio. However, the tensile strengths and the shrinkage behaviour are dominant for the formation of desiccation cracks. The studies of moisture transfer, shrinkage behaviour and tensile tests will enable us to develop a constitutive law which will be then implemented in the FE - simulation model.

6 CONCLUSIONS

In the present paper, the soil-water retention curve of a kaolinite soil is determined, where soil water suction of pF0 - pF3 and pF5 - pF7 are measured using a tensiometer method and the vacuum desiccator method. A laboratory model is presented to determine the parameters of moisture transfer. Furthermore, tensile tests are described, in which annular soil samples are used and the variation of parameters such as temperature, water content, void ratio and radial pressure is possible. Test results are shown.

The investigations will be continued. Effects of temperature and radial pressure on the tensile behaviour are to be investigated. A numerical study of a coupled pore water diffusion and stress analysis considering the shrinkage behaviour of the unsaturated soil is in process to describe the formation of cracks in clayey soils.

ACKNOWLEDGEMENTS

The German Research Foundation „Deutsche Forschungsgemeinschaft (DFG)", Bonn - Germany, is gratefully appreciated for its support of the research work presented in this paper.

REFERENCES

Croney, D. & Coleman, J.D. & Bridge, P.M. 1952. *The suction of moisture held in soil and other porous materials.* London: Her Majesty's Stationery Office.

Meißner, H. & Wendling, S. 1998. Crack formation due to desiccation of mineral liners of landfills. *Third International Congress on Environmental Geotechnics*, Lisboa, Portugal.

Meißner, H. & Wendling, S. 1999. Zusammenhang zwischen Rißbildungen und Wasserspannungen in mineralischen Dichtungen unter Berücksichtigung der Parameter Zeit, Temperatur, Verdichtung sowie Sättigungsgrad, Zwischen-bericht DFG-Az.: ME 501/14-1.

Meißner, H. & Wendling, S. 2000. Relationship between desiccation and cracks in Clay liners". *International Conference on Geotechnical & Geological Engineering.* 19 - 24 November 2000. Melbourne, Australia.

Philip, J.R. & DE Vries, D.A. 1957. Moisture Movement in Porous Materials under Temperatue Gradients. *Transactions, American Geophysical Union*, pp. 222-232.

Schofield, R.K. 1935. The pF of the waster in soil. *Third Internationl Congress of soil science Transactions*, 2, pp. 37-48.

UMS GmbH 1993. Bedienungsanleitung zum Miniatur-Druckaufnehmer-Tensiometer T5, Version 1.5, München.

van Genuchten, M. Th. 1980. A closed-form equation for predicting the hydraulic conductiviy of unsaturated soils. *Soil Sci. Soc. Am. J.* 44, pp. 892-898.

Wendling, S. & Meißner, H. (2000). Relationship between Crack Formation and Soil Water Suction due to Desiccation of Mineral Liners of Landfills. *GREEN 3, 3rd International Symposium on Geotechnics Related to the European Environment.* June 21-23, 2000, Berlin, Germany.

Clay Science for Engineering, Adachi & Fukue (eds) © 2001 Balkema, Rotterdam, ISBN 90 5809 175 9

Influence of the distribution of soil suction on crack patterns in farmlands

Shuichiro Yoshida & Kazuhide Adachi
Hokuriku Agricultural Experiment Station, MAFF, Joetsu, Japan

Kyoji Takaki
National Research Institute of Agricultural Engineering, MAFF, Tsukuba, Japan

ABSTRACT: The relation between transpiration from row-planted crops and the location of shrinkage cracks in the soil was studied by means of laboratory experiments and numerical simulations. Experiments were performed with pasty soil with high clay content filled into a rectangular container. Soil moisture was extracted horizontally from opposite directions, which induced cracks in the center. The locations of the observed cracks were analyzed in terms of stress distribution as simulated by a numerical model based on 2-dimensional consolidation theory. Tensile stress was initially distributed in half of the sample, then concentrated into the center. Many cracks initially occurred around the center where tensile stress was estimated to be greater than 0.15 kPa, then part of them widened while others disappeared. These facts show that the peculiar distribution of suction, formed by row-planted crops, determines the region where tensile stresses work, which in turn create cracks in specific places.

1. INTRODUCTION

Cracks in clayey agricultural fields play an important role in water balance affecting such factors as the efficiency of drainage (e.g., Inoue 1993; Maruyama et al. 1997), irrigation (e.g. Mitchell et al. 1993), or the rate of evapo-transpiration (e.g., Adams et al. 1964; Hasegawa et al. 1987). Thus, controlling the cracks' location, width, and depth is of practical concern in various situations.

Cracking patterns in agricultural fields are known to be influenced by both the properties of the soil and the nature of vegetative cover (El Abedine et al. 1971). The effects of soil properties on shrinking or cracking have been widely investigated. But from the standpoint of controlling cracks in agricultural fields, vegetation is a more important factor (Johnson 1962). Johnston et al. (1944) observed that soils cleaved and cracked in the middle of the row in cotton and corn fields. They explained that soil moisture depletion caused by roots led to soil shrinkage and contraction towards the row. Fox (1963) also pointed out that large cracks developed midway between the lines of plants, which were attributed to the rooting system acting as a skeleton that held the soil together. Mitchell et al. (1992) stated that large cracks formed between the plant rows where the soil was wetter, because cleavage planes were produced by shrinkage stress at the point of highest

water content. Yoshida et al. (in prep) investigated the relation between cracking pattern and transpiration from plants in a paddy field, and concluded that not only mechanical reinforcement but also transpiration from roots affect the location and morphology of cracks. From these experimental observations, horizontal extraction of soil moisture from opposite directions may create tensile stresses, which may cause cracking in certain domains. However, theoretical support from mechanics is necessary to elucidate the mechanism that determines the location of cracks.

The objective of this study is to clarify the mechanism which determines the location of cracks in the system where mechanical and hydraulic boundary conditions are controlled. To accomplish this, we performed laboratory experiments and analyses using a numerical model based on the consolidation theory.

2. THEORY AND NUMERICAL FORMULATION OF THE PROBLEM

The shrinkage behavior of soil with high water content can be regarded as consolidation due to suction. Terzaghi's principle of effective stress is known to be applicable to unsaturated conditions only with insular air (isolated from the atmosphere), even if pore water pressure is negative (Kohgo 1995). The consolidation theory for saturated soils is valid for shrinkage phenomena if the soil is considered to be in the normal shrinkage process. However, cracking easily occurs during desiccating when the boundary is constrained, which undermines the validity of the theory. It is necessary to note that the application of this theory is not aimed at simulating crack propagation, but is intended for predicting where cracks appear in an early stage of desiccation.

2.1 Formulation for numerical simulation

Biot's equations of consolidation, represented by equations (1) and (2), are derived from mass balance, momentum balance, principle of effective stress, and Darcy's equation, in which changes in the average soil density and the compressibility of solids and liquids are neglected. Symbols represented by σ', p_w, and δ in equation (1) denote effective stress tensor, pore water pressure, and Kronecker delta, respectively; ε, k, γ_w, and w in equation (2) denote strain tensor, saturated hydraulic conductivity, unit weight of water, and unit vector parallel to gravity, respectively. The commas followed by suffixes represent differentiation in space. The dots above the variables represent derivatives with respect to time. The tensile components are positive in the solid phase while the fluid in the state of compression has a positive pressure.

$$\sigma'_{ij,j} - \delta_{ij}\dot{p}_{w,j} = 0 \tag{1}$$

$$\dot{\varepsilon}_v = \left(k\left(\frac{p_{w,i}}{\gamma_w} + w_i \right) \right)_{,i} \tag{2}$$

Four types of boundary conditions are represented as:

$$\left(\sigma'_{ij} - \delta_{ij}\dot{p}_{wi} \right)n_j = \bar{t} \quad on \; \Gamma_t \tag{3}$$

$$\dot{u}_i = \bar{\dot{u}}_i \quad on \; \Gamma_u \tag{4}$$

$$-k\left(\frac{p_{w,j}}{\gamma_w} + w_j \right)n_j = \bar{q} \quad on \; \Gamma_q \tag{5}$$

$$p_w = \bar{p}_w \quad on \; \Gamma_p \tag{6}$$

where u is the displacement vector, and n is the outward unit vector normal to the boundary. The right-hand term of these equations denote boundary values, where Γ_t, Γ_u, Γ_q, and Γ_p represent traction, displacement, flux, and pressure

boundaries, respectively. The governing equations (1) and (2) combined with the natural boundary conditions (3) and (5) can be transformed into weak forms such as:

$$\int_\Omega \sigma'_{ij}\zeta_{ij}\,dv - \int_\Omega \dot{p}_w\zeta_{ii}\,dv - \int_{\Gamma_t} \bar{t}_i\dot{v}_i\,ds \tag{7}$$

$$\int_\Omega \dot{\varepsilon}_{ii}\eta\,dv + \int_\Omega \left(\frac{k}{\gamma_w}p_{w,i} + kw_i \right)\eta_{,i}\,dv + \int_{\Gamma_q} \bar{q}\eta\,ds = 0 \tag{8}$$

where ζ, v and η are weighting functions. Spatial discretization can be performed by introducing the finite element approximations (9) and (10) into equations (7) and (8), where N_u and N_p are shape functions, and U and P are displacement and pore water pressure at nodal points, respectively.

$$u = N_u U \tag{9}$$

$$p_w = N_p P \tag{10}$$

The weighting functions are replaced by the shape functions by applying the Galerkin method.
Strain tensor and stress tensor can be related to the displacement vector by the following equations, assuming the small strain theory.

$$\dot{\varepsilon}_{ij} = 1/2\left(\dot{u}_{i,j} + \dot{u}_{j,i} \right) \tag{11}$$

$$\dot{\sigma}'_{ij} = D_{ijkl}\dot{\varepsilon}_{kl} \tag{12}$$

Descretization in time is carried out by the generalized trapezoidal method (Lewis & Schrefler 1998). The final algebraic equation is:

$$\begin{bmatrix} -K & C \\ C^T & \Delta t\theta H \end{bmatrix}_{n+\theta} \begin{Bmatrix} U \\ P \end{Bmatrix}_{n+1}$$
$$= \begin{bmatrix} -K & C \\ C^T & -(1-\theta)\Delta t H \end{bmatrix}_{n+\theta} \begin{Bmatrix} U \\ P \end{Bmatrix}_n + \Delta t \begin{Bmatrix} -\dfrac{df^u}{dt} \\ f^p \end{Bmatrix}_{n+\theta} \tag{13}$$

where K is the tangential stiffness matrix, C is the coupling matrix, H is the permeability matrix, f^u is the load vector, f^p is the flow vector, and θ is the time descretization factor. When the elastic modulus and hydraulic conductivity are variable, equation (13) is nonlinear, requiring the repetitions of the calculations to determine the coefficient matrices at time $n + \theta$.

2.2. Introducing variable elastic parameters and variable hydraulic conductivities

The relation between normal stress and the void ratio e is usually linear, whose slope is called compression index Cc, as obtained from consolidation tests. However in our situation, analysis under low stress conditions is important. If we assume a linear relationship, the void ratio separates from the measured value and diverges as stress become close to zero. To avoid this, a quadratic curve shown in equation (14) is applied in place of linear relations.

$$e = a_1 \left(\log_{10} p\right)^2 + a_2 \log_{10} p + a_3 \tag{14}$$

where a_1, a_2 and a_3 are fitting parameters, and p is mean stress. The bulk modulus K_V are related to the void ratio e and mean stress p in the following manner:

$$K_V = \frac{1+e_0}{\left(\dfrac{de}{d \log_{10} p}\right)\left(\dfrac{d \log_{10} p}{dp}\right)} \tag{15}$$

where e_0 is the initial value of the void ratio. From equations (14) and (15), the bulk modulus can be calculated from the void ratio and mean stress using the following equation.

$$K_V = \frac{0.4343 \, p\left(1+e_0\right)}{2a_1 \log_{10} p + a_2} \tag{16}$$

In the plane strain condition, D matrix, appearing in equation (12), can be expressed as:

$$D = \frac{3K_V}{(1+\nu)} \begin{bmatrix} 1-\nu & \nu & 0 \\ \nu & 1-\nu & 0 \\ 0 & 0 & \dfrac{1-2\nu}{2} \end{bmatrix} \tag{17}$$

where ν is the Poison ratio. D matrix is evaluated on each integrating point using equations (16) and (17).

Saturated hydraulic conductivity usually decreases as the void decreases. By using some functional relationships between hydraulic conductivity and the void ratio as described in the literature (e.g., Monte et al. 1976; Takada et al. 1979), we can apply the simplest relation such as:

$$\log k = b_1 e + b_2 \tag{18}$$

where b_1 and b_2 are the fitting parameters.

3. MATERIALS AND METHODS

3.1 Experimental method

The experimental apparatus is shown in Figure 1. The container for the soil sample is made of acrylic resin, having a size of 12cm in length, 20cm in width and 4cm in depth. Both ends of the container are made of a porous filter having a thickness of 7.2 mm and a hydraulic conductivity of 8.0×10^{8} cm/s, which prevents the horizontal gradient of the soil suction from being too steep. The soil sample was prepared by mixing 700g of distilled water and 1200g of air-dried soil, which had passed through a 2mm-mesh sieve, and stirring the mixture for 10 minutes with a spatula. The properties of the soil sample are shown in Table 1. The soil paste was left for four hours and then poured into a container paying attention not to leave large voids in the soil. The initial void ratio, volumetric water content, and the degree of saturation were 1.83, 0.62, and 0.96, respectively. The container was left undisturbed for about 15 hours after that, avoiding loss of water by evaporation. The experiments was initiated by opening taps on the both sides simultaneously. The water pressure on the outer sides of the filters was kept at -10kPa by a hydraulic head. Soil moisture was extracted horizontally from the filters by suction. The drained water was measured using electronic balances. The change of water suction in the soil was measured using porous cups, which were 2mm in diameter, connected to pressure transducers. The formation of cracks was recorded by tracing the surfaces of samples through the transparent cover. An experiment without utilizing the porous sensors was also performed to confirm that the effects of the porous sensors were relatively small.

3.2 Simulation conditions

The simulations were run with 2-dimensional finite elements as shown in Figure 2. The calculations were performed for only half of the experimental domain since the system was symmetrical. The boundary conditions are also shown in the figure. The filters were included in the analyzed domain because their hydraulic conductivities were small compared to that of the soil. An extremely large constant value was set as the elastic modulus of the filter to make the strain of the filters negligible. We applied 9-node isoparametric elements for displacement and 4-node super-parametric elements for pore water pressure. The parameters used for the simulations are shown in Table 2.

Table 1. Properties of the soil

Classification:
  Clayey, montmorillonitic, mesic, typic Epiaquepts
Sampling site:
  Experimental paddy field in Hokuriku Agricultural
  Experiment Station , Niigata Japan.
Texture: LiC
Liquid Limit: 71.2%
Plastic Limit: 34.8%
Plasticity Index: 36.4%

Note: consistency measurements were performed on the
air-dried sample sieved with a 2mm-mesh sieve.

Table 2. Parameters used for the calculations

Elastic Parameters	a_1	-0.0434
	a_2	-0.1799
	a_3	1.6758
	ν	0.2500
Hydraulic Parameters	b_1	1.8816
	b_2	-9.3477
Time Factor	θ	0.6666

Elastic parameters for equation (14) were determined from "hanging water column method (e.g., Jury & Gardner 1991)", which is usually performed to determine the property of soil water retention. In the case of pasty soil with a high water content, the depletion of water results in volumetric shrinkage instead of inducing a decrease in the degree of water-saturation, and the soil can be regarded as water-saturated during the early stage of the shrinkage. Therefore, the water retention curve obtained from this method can be easily transformed to the relation between mean pressure p and the void ratio e (i.e., the e-$logp$ curve), because the suction of water is identical to the mean effective pressure p according to the principle of effective stress, and the void ratio is also easily estimated from the water content by weight. Suction pressures varying from 10 cmH$_2$O (0.98 kPa) to 80 cmH$_2$O (7.8 kPa) were applied in one step to each sample that was filled into a acrylic cylindrical column of 60mm in diameter and 20mm in thickness. Figure 3. shows the relation obtained from the tests, and the result of regression to a quadratic curve. The regression curve is satisfactory in expressing the relation without significant errors.

The hydraulic parameters for equation (18) were determined using hydraulic consolidation tests (Imai & Tsuruya 1984). The tests were performed using an acrylic cylindrical column having a diameter of 50 mm and a height of 100 mm. 5

cmH$_2$O (0.49 kPa) and -40 cmH$_2$O (-3.9 kPa) of hydraulic head were applied to the upper surface and the lower ceramic filter, respectively. The water flowed downward by the difference in the hydraulic head, consolidating the soil. The distribution of pore water pressure was measured by means of five porous sensors inserted into the column. When the drainage rate, measured with an electrical balance, became constant, the column was separated into slices each of 10 mm in thickness in order to measure the distribution of the water content. Figure 4. provides the relation between resultant void ratio and hydraulic conductivity. The regression curve shown by the line corresponded well to the data. The parameters shown in Table 2. were obtained from the linear regression to the results of two repetitions as shown in the figure.

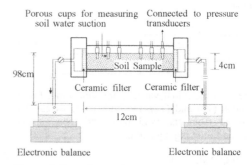

Figure 1. Experimental apparatus

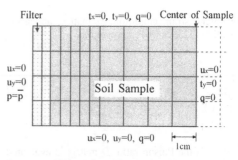

Figure 2. Finite elements and boundary conditions applied to the simulations

4. RESULTS AND DISCUSSION

4.1 Change in suction and cumulative outflow

Figure 5 shows the change in soil water suction as measured and simulated. Measured water suction increased gradually from both ends, forming concavity distributions. The suctions at the central part of the sample did not begin to increase at 120 min when the small fissures were already found. They began to increase at 360 min when a small portion of the fissures began to widen. The simulated results were in fairly close agreement with the experimental data.

Figure 6 shows the change in cumulative outflow through the filters on both ends. The drainage rate did not decrease rapidly since the resistance of the filters controls the flux from the ends of the soil.

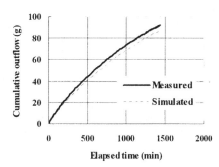

Figure 6. Cumulative outflow from the drains

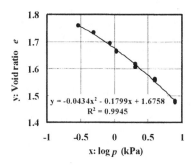

Figure 3. e-log p curve obtained from the "hanging water column method"

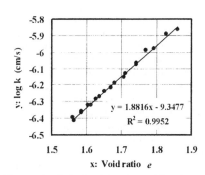

Figure 4. Relation between void ratio e and hydraulic conductivity k

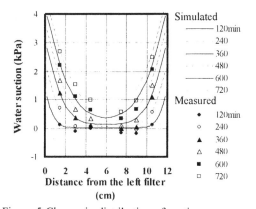

Figure 5 Change in distribution of suction as measured and calculated

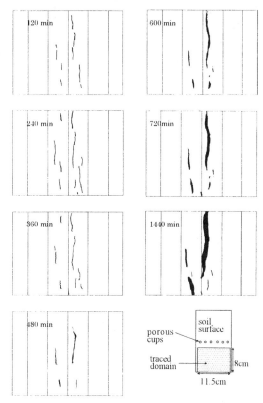

Figure 7. Formation of cracks on the soil surface (The top view)

Though the model estimated less values than the measured values, the difference was so slight that the model is considered to sufficiently express the tendencies as a whole.

4.2 *Development of cracks*

Cracks observed as they developed on the soil surface are shown in Figure 7. Some fissures normal to the water movement were observed in

203

the middle part of the container about 80 min after the drainage started, and then increased in both length and number. But about 360 minutes after the start, many of the fissures disappeared gradually. However some fissures began widening, which together formed a clear wide crack at the very center of the container. It should be noted that the fissures, drawn in Figure 7 as clear lines, observed from 120 min to 360 min were extremely narrow and shallow.

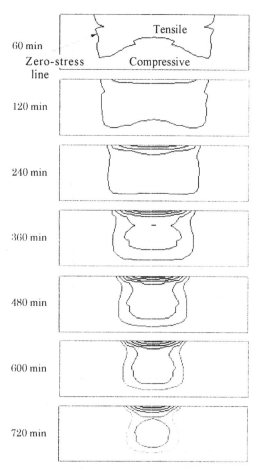

Figure 8. Distribution of horizontal tensile stress simulated by model

4.3 Change in stress distribution and its effects on cracking

In this section, we discuss the distribution of stress based on the simulated results. Figure 8 shows the distribution of horizontal stress calculated from the numerical model. The contour lines are drawn every 0.1 kPa in the area where tensile stress

appears. The outermost contour line surrounding the area with shadowing represent a zero-stress line, which divide the compressive and tensile domain. Small tensile stress was generated at the central half of the sample at 60 minutes, and the magnitude increased gradually. But the area where tensile stress was predicted shrank into the center, which formed a concentrated distribution of tensile stress at the center. These results correspond to the formation of cracks mentioned above: Many fissures developed when and where tensile stress is greater than 0.15 kPa (the simulated value at the cracked points at 80 min), then parts of them widened and the others disappeared when the concentration of tensile stress at the center increased.

From these results, the minimum effective tensile stress for creating new cracks in this soil is deduced to be about 0.15 kPa. The critical conditions in which cracks initially appear have been studied in fields or lysimeters by some researchers (Fujioka et al. 1968; Kohno 1979, 1981). Nemoto et al. (1975), for example, reported that initial cracks appeared when soil water suction was between -2kPa and -8kPa, depending on soil type. As in Nemoto's report, most of the critical conditions shown in previous works are expressed by the state of water, which affects the stress distribution of the soil skeleton. But the fracture conditions would become more theoretical if they were directly expressed by the state of effective stress. Raats (1984) reviewed two crack-slip failure criteria: one the Coulomb/Paul (CP) criterion which introduced the concept of tensile strength into Coulomb's criterion of failure, the other Griffith /Iwin / Orowon (GIO) criterion which relates the tensile strength to the growth of pre-existing minute cracks. According to the former criterion, the condition in which failure occurs is expressed in terms of major and minor principle stresses σ_1, σ_3 as the following.

$$\sigma_3 \le -\sigma_t \quad if \ \sigma_1 \le \sigma_c - \mu\sigma_t \tag{19}$$

$$\sigma_3 \le (\sigma_1 - \sigma_c)/\mu \quad if \ \sigma_1 \ge -\mu\sigma_t \tag{20}$$

where σ_t is called tensile strength. μ is flow value and σ_c is compressive strength, which are defined as equations (21) and (22), respectively.

$$\mu = \tan^2(\pi/4 + \phi/2) \tag{21}$$

$$\sigma_c = 2\sigma_s \tan(\pi/4 + \phi/2) \tag{22}$$

ϕ and σ_s denote the angle of internal friction and cohesion, respectively. This criterion means that if the minor principle stress is negative, not only tensile failure but also shear failure can occur. Figure 9 shows the sets of major and minor principle stresses around the center of the sample. The filled plots in Figure 9 represent the state of stress at cracked points. The initial crack was found at A, where σ_1 is about 0.05 kPa (compressive) and σ_3 is about -0.15 kPa (tensile). If the criterion (19) or (20) is applicable to this case, point A is on the critical line expressed in either (19) or (20). The inclination '$1/\mu$' of the shear critical line expressed by equation (20) is between zero and unit because the ϕ in equation (21) must be a positive angle. The tensile critical line expressed by (19) is parallel to the σ_1 axis. Any combined critical state lines are within the shadowed area in Figure 9. This means that 0.2 kPa is the maximum tensile stress which the soil sample can withstand, regardless of the type of failure.

Once a crack widens and develops more deeply, stress concentrates only at the tip of the cracks; tensile stress around the cracks is released and becomes compressive in order to balance the negative pore pressure. Our numerical model does not strictly cover this stage of cracking, because it assumes the soil sample is an elastic continuum without cracks. But the initial minute cracks are so shallow that only a part of the tensile stress is released. For further discussion, we now assume that the tendency of stress distribution estimated by the model is deduced to continue until some cracks widen. Pre-existing cracks develop on the condition represented in equation (23) according to the GIO model (Raats 1984).

$$\sigma_3 \leq -\sigma_t \quad if \; \sigma_1 \leq 3\sigma_t \qquad (23)$$

Tensile strength σ_t in equation (23) can be estimated with some mechanical and geometrical parameters, which can be different from the tensile strength in the CP criterion. If the concentration of tensile stress at the center of the soil sample progresses, particular cracks among the initial minute ones attained the critical state and begin to widen, while the other cracks around them become narrower due to a release of tensile stress.

Field observations give us much information regarding crack morphology, but the cracking pattern strongly depends on the physical properties of the individual soils, which has prevented us from achieving any general conclusions. This study succeeded in analyzing the location of cracks by simplifying the system and deriving general information about the mechanism. The results of these analyses coupled with the field observations

support the theory that not only mechanical constraint but also transpiration play important roles in determining the location of cracks between rows. But quantitative and comprehensive estimation of cracking behavior by mechanical models is limited due to complications caused by the distribution of roots or heterogeneity in soil properties. Another approach is needed in order to deal with such problems.

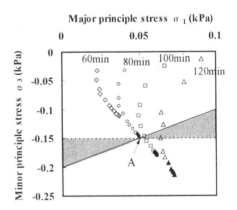

Figure 9. The state of principle stress around the center where cracks occur

5. CONCLUSIONS

1. Cracks can be induced at the center of a rectangular soil block by fixing the faces of both sides and extracting the moisture from them.
2. The location of cracks is determined by the distribution of effective stress induced by the horizontal movement of water.
3. Cracks develop in two stages: one is the formation of minute cracks by small tensile stress or shear stress, and the other is the widening of a particular crack because of the concentration of the tensile stress field.

ACKNOWLEDGEMENTS

We are grateful to Mr. Wakai and Mr. Yokoyama in the Field Management Section of our station for preparing the soil sample and making the experimental apparatus. We also thank Dr. Kohgo and Dr. Haraguchi at the National Research Institute of Agricultural Engineering for their advice and encouragement.

REFERENCES

Adams, J.E. and R.J. Hanks 1964. Evaporation from soil shrinkage cracks. *Soil Sci. Soc. Am. Proc.* 28:281-284.

EL Abedine, A.Z. & G.H. Robinson 1971. A study on cracking in some vertisols of the Sudan. *Geoderma* 5:229-241.

Fox, W.E. 1964. Cracking characteristics and field capacity in a swelling soil. *Soil Science.* 98:413.

Fujioka Y. & K. Sato 1968. On the cracks with drying of the clayey paddy soil (Ⅲ). *Trans. The Japanese Society of Irrigation, Drainage and Reclamation Eng.* No.26:8-14 (in Japanese with English abstract).

Hasegawa, S. & T. Sato 1987. Water uptake by roots in cracks and water movement in clayey subsoil. *Soil Science* 143(5):381-386.

Imai, G. , K. Yano & S. Aoki 1984. Applicability of hydraulic consolidation test for very soft clayey soils. *Soils and Foundations.* 24(2):29-42.

Inoue, H 1993 Lateral water flow in a clayey agricultural field with cracks. *Geoderma* 59:311-325.

Johnson, W.C. 1962. Controlled soil cracking as a possible means of moisture conservation on wheatlands of the southwestern Great Plains. *Agronomy J.* 54:323-325.

Johnston, J.R. & H.O. Hill 1944. A Study of the shrinking and swelling properties of Rendzina soils. *Soil Sci. Soc. Am. Proc.* 9:24-29.

Jury W.A. ,W.R. Gardner & W.H. Gardner 1991. Soil Physics John Willy & Sons, New York

Kohgo, Y. 1995. Study on mechanical properties of unsaturated soils and stability of soil structures. *Bull. National Research Inst. Agricultural Enginerring.* 34:39-162.

Kohno E. 1979. Studies on the engineering properties of soil in view of the shrinkage behavior on the surface soil of paddy field. *Trans. The Japanese Society of Irrigation, Drainage and Reclamation Eng.* No.81:1-8 (in Japanese with English abstract).

Kohno E. 1981. On the shrinkage behavior and crack formation in compacted soil. *Trans. The Japanese Society of Irrigation, Drainage and Reclamation Eng.* No.92:8-15 (in Japanese with English abstract).

Lewis R.W. & B.A. Schrefler 1998. *The finite element method in the static and dynamic deformation and consolidation of porous media.* John Willy & Sons, New York.

Maruyama, T & K.K. Tanji 1997. *Physical and chemical processes of soil related to paddy drainage.* Shinzan-sha Sci. & Tech. Publishing Co. ,Ltd Japan.

Mitchell, A.R. & M.Th.van Genuchten 1992. Shrinkage of bare and cultivated soil. *Soil Sci. Soc. Am. J.* 56:1036-1042.

Mitchell, A.R. & M.Th.van Genuchten 1993. Flood irrigation of a cracked soil. *Soil Sci. Soc. Am. J.* 57:490-497.

Nemoto, S. , K. Kokubu & H. Masujima 1975. Formation of cracks and its factors in paddy fields after releasing ponded water. *Jap. J. Soil Science and Plant Nutrition* 46-6:236-240 (in Japanese).

Raats, P.A.C. 1984. Mechanics of cracking soils. In *ISSS Symposium on water and solute movement in heavy clay soils.* J.Bouma and P.A.C.Raats (eds.). ILRI Publ. No. 37 , Wageningen, the Netherlands: 23-38.

Yoshida, S. , K. Adachi & K. Takaki. (in prep) Effects of cropping and puddling practice on the cracking patterns in paddy fields.

Clay Science for Engineering, Adachi & Fukue (eds) © 2001 Balkema, Rotterdam, ISBN 90 5809 175 9

Viscosity and yield stress of dilute suspension of Na-montmorillonite

Y. Adachi
Institute of Agricultural and Forest Engineering, Tsukuba University, Japan

K. Nakaishi
Faculty of Agriculture, Ibaraki University, Japan

K. Miyahara & M. Ohtsubo
Faculty of Agriculture, Kyusyu University, Japan

ABSTRACT: Rheological properties of dilute suspension of montmorillonite were reviewed to clarify the effectiveness of the approach based on the colloidal interaction. In the electrostatically stable system, the electroviscous effect was found to be significant in addition to shape factor in the determination of viscosity. The functional form of the Gouy-Chapman theory on the interaction of electric double layer was found to be able to explain the relation between yield stress and the volume fraction clay particles which was reported by Fujii et al. The viscosity measured for the coagulated suspension was also analysed using viscosity equation. The data gave the effective volume of flocculated suspension which increases monotonously even under the condition of high ionic strength above the critical coagulation concentration of NaCl. The fractal dimension obtained for coagulated flocs was found to be ≈ 2.0 irrespective of ionic strength, and the increase of intrinsic viscosity was ascribed to an increase of floc strength. The qualitatively same tendency as the result of surface force measurement reported by Israelachvilli et al. implied the role of hydration force.

1 INTRODUCTION

Rheological behavior of montmorillonite suspension is a matter of scientific interest as well as engineering importance and has a long history of study. They are summarized in the well-known monograph of van Olphen (1977) and have motivated plenty of investigations in the field of clay colloid chemistry. However, the interrelation between microscopic colloidal interactions and macroscopically observable rheological properties is poorly understood. The situation can be read from the ambiguios term of so-called cardhouse formation which does not give any quantative information. One of the major reasons for this situation is an unclearness of the element of rheological behavior. In the series of our investigations, we have mainly focused on the dilute suspension with which we can test the analytical framework for the relation between microscopic unit of motion and macroscopic rheological properties.

In the first place, viscosity of dilute suspension of Na-montmorillonite under electrostatically dispersed condition was analysed as a function of electrolyte concentration using data of capillary

viscometer (Adachi et al. 1998). Photon Correlation Spectroscopy (PCS) which yields the Stokes diameter from the diffusion constant was invoked in the analysis of viscosity data. These measurements provided an information on the size and shape of montmorillonite particle suspended individually. The obtained intrinsic viscosity was found to increase with an increase of reciprocal Debye length, κ^{-1}, which demonstrated the evidence of primary electroviscous effect; the viscosity increase due to the interaction between the electric double layer, DL, formed around the particle and the bulk fluid. All data are relatively consistent with the results of analysis on the pair interaction which can be also obtained from the viscosity data. In addtion, the importance of the multiparticle interaction which appeared in the abrupt increment of viscosity against the volume fraction of the clay particle, ϕ, enhanced by the presence of DL. The crossover concentration almost corresponded to the lower detection limit of the yield stress, τ_0, reported recently (Fujii et al., 1996). This correspondence motivated a test of applicability of simple model of τ_0 on the basis of colloidal interaction (Adachi, 1999). Assuming the simple array configuation, Gouy-Chapman theory on the electrostatic repulsive force is expressed as a function of ϕ. This functional form was found to give a good agreement on the reported result of τ_0.

On the other side, in the regime of coagulated suspension, the viscosity data of dilute suspension gave the effective volume of immobile part which is an index of floc size (Miyahara et al., 1999). Analysis of floc structure on the basis of settling velocity did not give any differences of fractal dimension against ionic strength. This result suggests that the increase of floc size is an evidence of strong bond bewteen clay particles. This behavior is qualitatively consistent with the result of microscopic surface force measurement (Israelachvilli and Adams, 1978).

2 ANALYSIS IN THE ELECTROSTATICALLY STABLE REGIME

2.1 Viscosity equation

The viscosity of a colloidal suspension increases monotonously with an increase of ϕ. For a dilute suspension, this behavior can be written in terms of a power series in ϕ;

$$\frac{\eta}{\eta_0} = 1 + K_1\phi + K_2\phi^2 + \mathcal{O}(\phi^3) + \cdots \quad (1)$$

where η and η_0 denotes the viscosity of the suspension and that of medium respectively. The $\mathcal{O}(\phi)$ term represents the disturbance of flow due to the presence of solid particle isolated in a fluid phase. K_1 is generally referred as the intrinsic viscosity. Well-known theory of Enstein gives the value of 2.5 for the case of rigid spherical particle (Einstein, 1905). This value increases with the deviation from spherical shape and with the presence of diffusive electric double layer, DL, around the particle. The $\mathcal{O}(\phi^2)$ term represents the effect of pair interaction between suspended particles. K_2 is denoted as Huggins coefficient. Theoretical value on the basis of two-sphere hydrodynamics is derived as 6.2 (Batchelor, 1977). The increment from this value can be also used as an index of the deviation from the case of solid spheres.

2.2 Experiment

Na-montmorillonite suspension with backgroud pH of ten was prepared from commercially available "Kunipia-F" purchased from Kunimine Co.,Ltd. Viscosity was measured using Ostwald capillary viscometer which enables to detect the difference of relative viscosity with an effective figure of four.

An automatic apparatus for PCS was used to measure Stokes diameter of suspended clay particles.

In Figure 1, reduced viscosity, $\eta/\eta_0 - 1$, is plotted as a function of ϕ. As indicated in this figure, the reduced viscosity increases with an increase of ϕ; however, this behavior is a function of ionic strength. As shown in Figure 2, Stokes diameter of suspended particle is almost constant; therefore, the difference observed in accordance with ionic strength is not due to swelling of individual particle. Intrinsic viscosity was determined from the intercept of reduced viscosity with the ordinate. The obtained values are plotted against κ^{-1} as shown in Figure 3. The linear increment implies the influence of the first electrical viscous effect which simply appears as an increment of effective volume of immobilized part. This is consistent with the result of the analysis of pair interaction (Nakaishi, 1998). These results were used to estimate the shape and size of primary particles. Assuming that suspended particle has a disk shape which is approximated by an ellipsoid with a remarkably large aspect ratio with a thickness of 1.0 nm, we obtained 220 nm as the diameter of the disk from the application of Simha's result with the value of 150 for intrinsic viscosity (Figure 4).

On the other side, the onset concentration of multiparticle interaction is interesting. This appears more remarkably in the system of low ionic strength. This result is an indication of the effect of DL.

2.3 Yield stress of electrostatiacally stable suspension

Recently, Fujii et al.(1996) reported their experimental data on the flow curve of the dilute suspen-

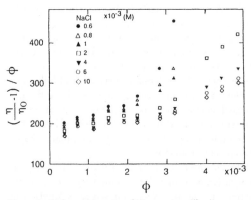

Figure 1. Reduced viscosity of Na-montmorillonite suspension as a function of ϕ.

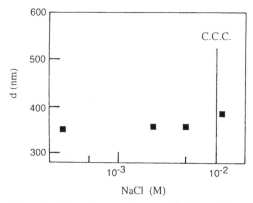

Figure 2. Stokes diameter of montmorillonite particle measured by PCS.

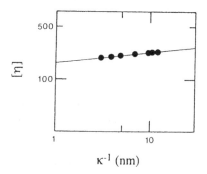

Figure 3. Intrinsic viscosity of Na-montmorillonite suspension against κ^{-1}

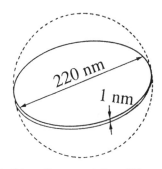

Figure 4. Schematic representation of Na-montmorillonite sheet in suspension.

sion of Na montmorillonite measured using a cone-plate rheometer. With this metod, the yield stress, τ_0[Pa], was expressed as a function of ϕ[-]. The remarkable point of their result is that for the range of measurement ($0.008 < \phi < 0.026$), τ_0 increases more than four order of magnitude. The relation was approximated by the following power law;

$$\tau_0 = 3.59 \times 10^{13} \phi^{7.24} \qquad (2)$$

The power, 7.24, is an important value characterizing the relation. It is interesting to note that similar rapid increase of τ_0 agaist ϕ is found in the literatures published more than three decades ago although the proportional constant reported in the letter is much smaller (Sudo and Yasutomi, 1961 ; Yasutomi and Sudo, 1962).

It should be noted that the minimum concentration of the detection of the yield stress reported by Fujii et al.(1996) almost corresponds to our result on the onset concentration of many body interation described in the previous section. This correspondence suggests the significant role of DL interactions in the appearance of yield stress in more concentrated dispersion. Supposing that the solid-like state is brought about simply by the interference of clay sheets which are geometrically constrained each other in the limitted space as shown in Figure 5, the dominative force acting on each clay sheet is derived as the electrostatic repulsive force with the functional form of,

$$Fr = 64 \times n_0 \times kT \times \Upsilon_0{}^2 \times exp(-\kappa d) \qquad (3)$$

which can be found in the section of Gouy-Chapman theory described in the standard text of colloid science (Hiemens, 1986). Fr, n_0 and kT denote the repulsive force, the number concentration of ions, and Boltzmann factor respectively. Υ_0 is a function of surface potential, ψ_0. With a configuration of simple array, the distance between two plates, d, can be correlated to the volume fraction of solid, ϕ, as:

$$\phi = \frac{\delta}{\delta + d} \qquad (4)$$

where δ denotes the thickness of the sheet. With this relation, the double-layer repulsive force is expressed as a function of ϕ. Substitution of Eq.(4) into Eq.(3) yields:

$$Fr = 64 \times n_0 \times kT \times \Upsilon_0{}^2 \times exp\left(-\kappa\delta(\frac{1}{\phi} - 1)\right) \quad (5)$$

In Figure 6, we compared our calculated result with the experimental data. As indicated in this figure, a good agreement with the prediction on the basis of Eq.(5) was obtained for the data reported by Fujii, et al. The difference observed for other data can be reduced if we introduce the efficiency factor, α_e which represents the ratio of effective volume fraction of solid acting as charged sheets

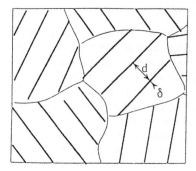

Figure 5. The illustration of the suspension composed of an assemle of lamellar domains.

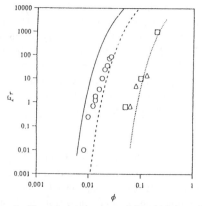

Figure 6. The calculated values of Fr which is equivalent to τ_0 agaist ϕ. Calculations are based on 1 : 1 type ions for $c = 0.001$ M, $\delta = 1$ nm, T=300 K, $\psi_0 = 100$ mv (surface potential) which gives $\Upsilon_0^2 = 0.56$. Plotted data are o Fujii et al., △Sudo et al. and □ Yasutomi et al., respectively. Solid line $\alpha_e = 1$: broken line $\alpha_e = 0.5$: dotted line $\alpha_e = 0.1$

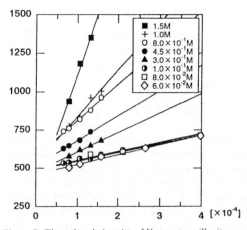

Figure 7. The reduced viscosity of Na-montmorillonite suspension in the coagulated regime.

producing repulsive forces. In this context, Eq.(5) is modified as,

$$Fr = 64 \times n_0 \times kT \times \Upsilon_0^2 \times exp(-\kappa\delta(\frac{1}{\alpha_e\phi} - 1))(6)$$

The calculated result of Fr as a function of ϕ for the case $\alpha_e = 0.5$ and 0.1 shown in the figure demonstrated fairly good agreement with the experimental data. From these correspondences, we confirmed the effectiveness of our method with some speculation of impurity of montmorillonite in the experimental material used by Sudo et al. and Yasutomi et al.

3 THE VISCOSITY OF THE DILUTE SUSPENSION IN THE FLOCCULATED REGIME

3.1 The structure and effective volume of a floc in relation to the hydrodynamic motion

The viscosity equation Eq.(1) can be used to analyse the behavior of coagulated regime. However, in this context, K_1 and K_2 can be used for the evaluation of effective volume of immobilized water trapped inside of floc structure. Assuming suspended flocs are rigid spheres of same size, the viscosity equation can be rewritten as:

$$\frac{\eta}{\eta_0} = 1 + 2.5\alpha\phi + 6.2(\alpha\phi)^2 + \cdots \qquad (7)$$

where α denotes the increase ratio of the effective volume fraction of floc due to the trapped water inside the floc. When floc has a self-similar structure with fractal dimension of D, the value of α can be expressed as:

$$\alpha = (\frac{D_f}{d_0})^{3-D} \qquad (8)$$

where D_f and d_0 are the diameters of floc and primary particle composing the floc respectively. D is the fractal dimension characterizing the self-similar structure of a floc (Adachi and Ooi, 1990). Therefore, we can derive the relation between D_f and D through the value of α obtained from the analysis of the viscosity.

On the other side, the value of D can be derived from the relation between the settling velocity of a floc, v_f, and the diameter of the floc, D_f. This relation follows the following equation.

$$v_f = \frac{g}{18\eta}(\rho_s - \rho_w)d_0^{(3-D)}D_f^{(D-1)} \qquad (9)$$

where ρ_s and ρ_w denote the densitis of clay particle and water respectively.

3.2 Experiment

The reduced viscosity measured for different ionic strength is plotted in Figure 7. As indicated in the figure, it is rather difficult to determine the intrinsic viscosity from the intercept of the line against the ordinate; however, the slope of linear part gives alternative index which is essentially same as intrinsic viscosity. In Figure 8, we plotted the value of α obtained from the slope as a function of ionic concentration. As demonstrated in the figure, α increases monotonously with the increase of ionic strength even in the region of rapid coagulation.

In order to clarify the value of D, we measured the settling velocity of a single floc, v_f, as a function of floc diameter, D_f, using the apparatus illustrated in Figure 9. In this apparatus, the settling movement of a floc in the settling tube is monitored through a CCD camera installed on the microscope. The maximum distance of the projected floc in the monitor is adopted as D_f. The typical result of v_f versus D_f is plotted in a logarithmic paper as shown in Figure 10. As demonstrated in this figure, the obtained value of D is almost 2.0 irrespective of ionic strength.

3.3 Discussion

The increase of α with NaCl concentration should be noted. According to Eq.(8), this increase is due to the increase of D_f or the decrease of D. However, the measurement of the settling velocity demonstrated very little change of D, That is, the experimental result can be interpreted as the increase of the size of flocs with the increase of ionic strength for the region above critical coagulation concentration. The formation of large floc is a sign of the strong bonds which stand for the breakup of flocs (Kobayashi et al., 1999). Usually, the attractive force between colloidal particles is recognized as the van der Waals force which is independent of ionic strength (Verway and Overbeek, 1947). This independency is confirmed for the floc of polystyrene latex particles (Adachi and Ooi, 1999). However, for the case of montmorillonite floc, the prensent result clearly demonstrated the presence of addtional factor which influences the attractive force between colloidal particles. The presence of hydrated cation located adjacent to the colloidal surface can be a reason for this phenomenon (Norrish, 1954). It should be noted that the qualitatively consistent result is reported for the attractive surface force between mica sheets (Israelachvilli and Adams, 1978).

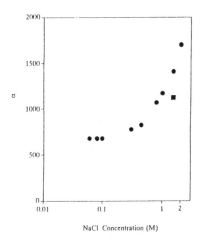

NaCl Concentration (M)

Figure 8. The value of α as a funtion of NaCl concentration.

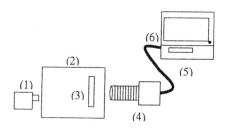

Figure 9. Schematic diagram of experimental set up for the measurement of the settling velocity of a single floc. (1) light source; (2) water bath ; (3) settling tube ; (4) microscope and CCD video camera ; (5) video recorder ; (6) monitor.

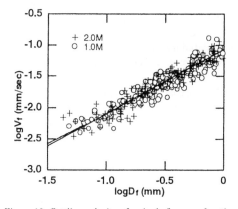

Figure 10. Settling velocity of a single floc as a function of the maximum distance in a floc for different NaCl concentration.

211

4 CONCLUSION

Rheological properties of Na-mountmorillonite suspension was investigated in terms of microscopical colloidal interacition. The two exitreme regimes; electrostatically stable and coagulated under high ionic strength are focused in the analysis. The results of both regimes can be summarized as follows:

(1) In the regime of electrostatically stable suspension, the functional form of yield stress, τ_0, against the volume fraction of clay, ϕ, can be explained by an interactive force of electrical double layers formed arround particles.

(2) In the regime of couagulated suspension, the fractal structure of floc is important. In addition to the van der Waals attractive force, so-called hydration force can be explain the increase of effective viscosity with an increase of ionic strength.

ACKNOWLEDGEMENT

This work is funded partly by a Grant-in-Aid for Scientific Research (No.11460108 and 11896007) from the Japanese Ministry of Education.

REFERENCES

Adachi,Y. (1999): Yield stress of Na-montmorillonite suspension in an electrostatically dispersed state, *Trans. JSIDRE* 200, pp.53-58

Adachi,Y and Ooi,S. (1990): Geometrical structure of a Floc,*J.Colloid and Interface Science* 135, pp.374-384

Adachi,Y and Ooi,S. (1999): Sediment volume of flocculated material studied using polystyrene latex spheres,*J.Chem.Eng.Japan* 32, pp.45-50

Adachi,Y, Nakaishi,K. and Tamaki,M. (1998): Viscosity of a dilute suspension of montmorillonite in an electrostatically stable condition, *J.Colloid and Interface Science* 198, pp.100-105

Batchelor,G.K. (1977): The effect of Brownian motion on the bulk stress in a suspension of spherical particles *J. Fluid Mech.*, 83 pp.97-117

Einstein, A. (1906): *Ann. Phys.*, 19 pp.289-306

Fujii,K., Takahashi,T. and Nakaishi,K. (1996) : Shear flow characteristics and yield stresses of montmorillonite suspensions, *Trans.*

JSIDRE. 195, pp.53-60 (in Japanese with English summary)

Hiemens,P.C.(1986): *Principle of Colloid and Surface Chemistry 2.nd Ed.* , Marcell Dekker

Kobayashi,M., Adachi,Y. and Ooi,S. (1999) : Breakup of fractal flocs in a turbulent flow, *Langmuir* 15, pp.4351-4356

Israelachivili,J.N. and Adams,G.E. (1978) : Measurement of forces between two mica surfaces in aquaous electrolyte solutions in the range 1-100nm, *J. Chem. Soc., Faraday Trans.* I 74, pp.975-1001

Miyahara,K., Adachi,Y. Nakaishi,K. and Ohtsubo, M. (1999) : Viscosity and floc structure of sodium montmorillonite in an alkaline state, *Proc.Rheology in the Mineral Industry*, II, pp.57-68

Nakaishi,K. (1998) : The factor influencing hydrodynamic interaction of dilute sodium montmorillonite suspension under dispersed condition, *Nihon Rheology Gakkaishi.* 26, pp.99-102 (in Japanese with English summary)

Norrish, K. (1954) : Manner of swelling, *Nature.* 173, pp.256-257

Sudo,S. and Yasutomi.R.(1961): Rheology of soil paste (I), *Nougyou Doboku Kenkyu Bessatsu* 2, pp.71-74 (in Japanese with English summary)

Van Olphen (1977): *Clay Colloid Chemistry 2.nd Ed.*, John Wiley

Verwey, E.J.W. and Overbeek, J.Th.G. (1947) : *Theory of the Stability of Lyophobic Colloid*, Elsevier

Yasutomi,R. and Sudo.S.(1962): Rheology of soil paste (II), *Nougyou Doboku Kenkyu Bessatsu* 3, pp.40-45 (in Japanese with English summary)

2 Permeability and contaminant transport

Clay Science for Engineering, Adachi & Fukue (eds) © 2001 Balkema, Rotterdam, ISBN 90 5809 175 9

Predicting the permeability function for unsaturated soils

D.G. Fredlund, M.D. Fredlund & N. Zakerzadeh

Department of Civil Engineering, University of Saskatchewan, Saskatoon, Sask., Canada

ABSTRACT: The permeability function for an unsaturated soil can be estimated by raising the equation for the soil-water characteristic curve to a power and then multiplying the result by the saturated coefficient of permeability. However, when using this procedure there is one variable that must be assumed; namely, the power, q. This paper presents the analysis of several hundred experimental data sets and provides guidance with regard to the most suitable values to use for the exponent, q. The resulting values of q are obtained from best-fit analyses according to various soil types. The overall average q value for all soils analyzed was 3.3 and the standard deviation was 1.4. The results also showed that sandy soils had a lower q value than clayey soils.

1 INTRODUCTION

Several earth structures such as soil covers and soil liners are commonly constructed for the management of various types of wastes. It is important to know the water storage and hydraulic conductivity properties in order to perform an adequate design of these structures.

There are several empirical, but theoretically-based models proposed in the literature for the prediction of the seepage through unsaturated soils. These models are useful for the estimation of the coefficient of permeability with respect to suction. The predictions are commonly made with the use of the soil-water characteristic curve and the saturated coefficient of permeability. The soil-water characteristic curve used in these models is commonly measured in the laboratory using a Pressure Plate apparatus. An estimation of the soil-water characteristic curve can also be obtained through the use of a knowledge-based database system such as SoilVision (Fredlund, 1997).

Fredlund and Xing (1994) proposed an equation for the soil-water characteristic curve that can be used to best-fit data over the entire range of soil suctions from 0 to 1,000,000 kPa. Fredlund (1995) also showed how it was possible to use the saturated coefficient of permeability and the soil-water characteristic curve to obtain a permeability function. The procedure involved integration over the range of the soil-water characteristic curve, starting from saturated soil conditions. The coefficient of permeability over the entire range of soil suctions is defined as the permeability function.

In 1997, Leong and Rahardjo suggested that the permeability function could also be approximated through the use of a single additional parameter applied to the soil-water characteristic curve. The variable was an exponent applied to the equation for the soil-water characteristic curve. When using this procedure, a new soil parameter, q, must be estimated in some manner. The objective of this research paper is to study the nature of the power variable, q, required in order to estimate the permeability function for unsaturated soils.

2 CONCEPTS OF SEEPAGE FOR UNSATURATED SOILS

The constitutive relationship to describe flow through a saturated or unsaturated soil is Darcy's law.

$$v = -k_w \frac{dh}{dy} \qquad (1)$$

where v = flow velocity over the discharge area; and k_w = coefficient of permeability.

The proportionality variable between velocity and hydraulic gradient is assumed to be a constant for saturated soils, k_s, but becomes a permeability function for an unsaturated soil. The coefficient of permeability of an unsaturated soil is a function of the amount of water in the soil which, in turn, can be written in terms of the stress state of the soil (Huang et al. 1998).

$$k_w = func \: [k_s, \: (\sigma - u_a), \: (u_a - u_w)] \qquad (2)$$

It is generally considered sufficient to quantify the amount of water in the soil as a function of soil suction, $(u_a - u_w)$ or ψ. The unsaturated coefficient of permeability can then be written as a function of the saturated coefficient of permeability and the dimensionless water content, Θ, that is equal to $w(\psi)/w_s$. The variable $w(\psi)$ is equal to the water content at any soil suction, ψ, and w_s is the saturated soil water content. It is possible to include an additional fitting parameter, q, to complete the functional relationship.

$$k_w(u_a - u_w) = func \: [(w(\psi)/w_s), \: q, \: k_s] \qquad (3)$$

Numerous analyses have been proposed for the estimation of the permeability function for unsaturated soils (Fredlund et al. 1994; Leong and Rahardjo, 1997). Common to all methods is the existence of a mathematical relationship between the coefficient of permeability and the soil-water characteristic curve.

3 THE SOIL-WATER CHARACTERISTIC CURVE

The soil-water characteristic curve has played a dominant role in understanding the behavior of unsaturated soils in disciplines such as soil science, soil physics, agronomy and agriculture. As a consequence of the long history associated with the use of the soil-water characteristic curve, large amounts of information and experimental data are available from these disciplines. The soil-water characteristic curve is now recognized as one part of the overall water phase constitutive relationship in geotechnical engineering. The soil-water characteristic curve is of great value in predicting unsaturated soil property functions.

The proposed equation defining the soil-water characteristic curve, by Fredlund and Xing (1994) provides a mathematically based function over the entire range of suctions from zero to 1,000,000 kPa. The relationship is empirical, being derived using the assumption that the soil consists of a set of interconnected pores that are randomly distributed. The equation, written in terms of gravimetric water content, w, is as follows:

$$w = \left[1 - \frac{ln\left(1 + \frac{\psi}{\psi_r}\right)}{ln\left(1 + \frac{1,000,000}{\psi_r}\right)}\right] \left[\frac{w_s}{\left(ln\left(e + \left(\frac{\psi}{a}\right)^n\right)\right)^m}\right] \qquad (4)$$

where a = a suction value corresponding to the inflection point on the curve that has physical meaning in its relationship to the air-entry value of the soil; n = soil parameter related to the slope of the soil-water characteristic curve in the transition stage; ψ = soil suction (i.e., matric suction, $(u_a - u_w)$, at low suctions and total suction at high suctions); m = parameter related to the residual water content; w_r = water content at residual conditions; e = natural number, 2.71828.

Equation 4 can be written in a dimensionless water content form, Θ, by dividing both sides of the equation by the saturated water content.

$$\Theta = \left[1 - \frac{ln\left(1 + \frac{\psi}{\psi_r}\right)}{ln\left(1 + \frac{1,000,000}{\psi_r}\right)}\right] \frac{1}{\left(ln\left(e + \left(\frac{\psi}{a}\right)^n\right)\right)^m} \qquad (5)$$

The first term in brockets is a correction factor that ensures that the function goes through zero (at a suction of 1,000,000 kPa). Equation 5 can be used to best-fit the desorption (or adsorption) branches of soil-water characteristic curve data. The fitting parameters (i.e., a, n and m values) can be determined using a non-linear regression procedure such as the one proposed by Fredlund and Xing (1994).

4 COEFFICIENT OF PERMEABILITY (OR HYDRAULIC CONDUCTIVITY) FUNCTION

The shape of the hydraulic conductivity function (or permeability function) bears a relationship to the shape of the soil-water characteristic curve. Figure 1 compares the soil-water characteristic curves and hydraulic conductivity functions for sand and clayey silt. The hydraulic conductivity for both soils remains relatively constant from zero suction up to the air entry value of the soil. The change in the hydraulic conductivity of a soil occurs at approximately the air entry value of the soil. The hydraulic conductivity decreases rapidly beyond the air entry value, for both soils. The decrease in the hydraulic conductivity is due to the reduction in the cross sectional area of flow. The initial hydraulic conductivity, or saturated hydraulic conductivity, k_s, of the sand can be two or more orders of magnitude greater than that of the clayey silt. As the suction increases, it is possible that the hydraulic conductivity of the sand will decrease by more than two orders of magnitude. Under certain conditions, it is possible for the clayey silt to be more permeable than the sand.

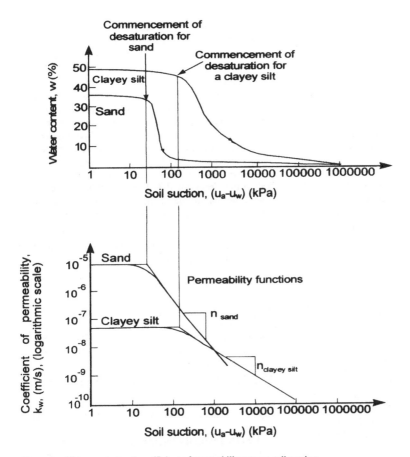

Figure 1. Water content and coefficient of permeability versus soil suction.

5 ESTIMATION OF THE PERMEABILITY FUNCTION FROM THE SOIL-WATER CHARACTERISTIC CURVE

The permeability function of an unsaturated soil can be predicted with sufficient accuracy for many engineering applications with a knowledge of the saturated coefficient of permeability and the soil-water characteristic curve. Several investigators have proposed empirical functions for predicting the permeability function (Huang et al. 1998). The soil-water characteristic curve equation developed by Fredlund et al. (1994) along with the saturated coefficient of permeability, can be used to compute the relationship between hydraulic conductivity and soil suction. Previous studies (Leong and Rahardjo 1997; Benson and Gribb 1997) have shown that proposed integration procedures involving the use of the Fredlund and Xing (1994) soil-water characteristic curve, provide a good estimate of the permeability function.

The calculation of the permeability function is performed by dividing the water content versus suc-

tion relationship into several water content increments. This is equivalent to integrating along the water content axis. The numerical integration procedure can be used to compute data points for a permeability function for the unsaturated soil.

$$k_r(w) = \int_{w_r}^{w} \frac{w-x}{\psi^2(x)} dx \; / \; \int_{w_r}^{w_s} \frac{w_s-x}{\psi^2(x)} dx \qquad (6)$$

where k_r = relative coefficient of permeability; and x = a variable of integration representing water content.

The accuracy of the prediction of the permeability function was shown to improve when the complete soil-water characteristic curve was used. Although the permeability function can be computed down to zero water content, it should be noted that the function may be more indicative of vapor flow in the region beyond the residual stress state. As a result, it may be more reasonable to leave the hydraulic conductivity as a constant, k_{res}, beyond the residual state. When calculating the permeability function, it is convenient to perform the integration along the soil suction axis as shown in the following equation.

$$k_r(\psi) = \int_{\psi}^{\psi_r} \frac{w(y) - w(\psi)}{y^2} w'(y) dy \Bigg/ \int_{\psi_{aev}}^{\psi_r} \frac{w(y) - w_s}{y^2} w'(y) dy$$

(7)

where ψ_{aev} = air-entry value of the soil under consideration; ψ_r = suction corresponding to the residual water content w_r; ψ = a variable of integration representing suction; y = a variable of integration representing the logarithm of suction; and w' = the derivative of the soil-water characteristic curve. Leong and Rahardjo (1997) reported that the "computed coefficient of permeability for this statistical model (i.e., equations 6 and 7 in this paper) showed good agreement with the measured coefficient of permeability."

To avoid the numerical difficulties associated with performing the integration over the entire soil suction range, it is more convenient to perform the integration on a logarithm scale. The proposed models have been found to be most satisfactory for sandy soils whereas agreement with experimental data may prove to be less satisfactory for clayey soils. Equation 8 can be multiplied by the dimensionless water content raised to a power (i.e., Θ^p) in order to provide greater flexibility in computing the permeability function. The additional parameter, p, is assumed to account for tortuosity in the soil pores (Maulem 1986) but it also a parameter whose magnitude must be assumed.

$$k_r(\psi) = \Theta^p(\psi) \frac{\int_{\ln(\psi)}^{b} \frac{w(e^y) - w(\psi)}{e^y} w'(e^y) dy}{\int_{\ln(\psi_{aev})}^{b} \frac{w(e^y) - w_s}{e^y} w'(e^y) dy}$$

(8)

Based on research work by Kunze et al. (1968), the value of the power, p, can be assumed to be 1 unless there is reason to assume otherwise.

Leong and Rahardjo (1997) suggested that rather than performing the above integration, the dimensionless equation for the soil-water characteristic curve, simply be raised to a power, q. Therefore, the permeability function can be written in the following form.

$$k_r(\psi) = k_s [\Theta(\psi)]^q$$

(9)

where $\Theta(\psi)$ = dimensionless form of the soil-water characteristic curve; and q = a new soil parameter.

This form for the permeability function is obviously attractive due to its simplicity. The equation is simple to use and clearly illustrated the relationship between the soil-water characteristic curve and the permeability function.

In order to use Equation 9, it is necessary to know what value to use for the q soil fitting parameter. The primary purpose of the present study is to determine typical values for the new soil parameter, q. The procedure used to evaluate the q parameter, along with the results obtained, are presented below.

In the original study undertaken by Leong and Rahardjo (1997), the data from six soils was used to assess the magnitude of the q soil fitting parameter. The results of their study are shown in Table 1.

Table 1. Summary of typical q parameters from the study by Leong and Rahardjo (1997).

Soil type	a kPa	n	m	q
Beit Netofa clay	389	0.69	1.176	52.12
Rehovot sand	2.25	4.32	1.235	6.04
Touchet silt loam	7.64	7.05	0.506	4.55
Columbia sandy loam	5.81	10.59	0.381	5.79
Superstition sand	2.66	6.86	0.525	6.21
Yolo light clay	2.93	2.11	0.379	9.57

Table 2. Summary of soil types analyzed for permeability function.

USDA Textural Classification	Number of samples
Clay	21
Clay loam	18
Loam	12
Loamy sand	29
Sand	49
Sandy clay loam	17
Sandy loam	30
Silt loam	74
Silty clay	34
Silty clay loam	18

6 PRESENT RESEARCH PROGRAM

The present research study involved the analysis of approximately 300 sets of permeability data. Each data set consisted of: i.) experimental results from the measurement of the soil-water characteristic curve, ii.) experimental results on the measurement of the coefficient of permeability under various applied soil suctions, and iii) the measured saturated coefficient of permeability. All experimental results were extracted from the SoilVision (Software the proprietary property of SoilVision system Ltd.)

knowledge-based system. The soils were divided into a number of categories depending upon their USDA textural classification. The categories, along with the number of data sets in each category, are shown in Table 2.

In addition to analyzing each soil type independently, the combination of all soil types was also analyzed.

Each of the soil-water characteristic curve data sets was first best-fit to determine the a, n, and m parameters associated with the Fredlund and Xing (1994) equation. Then the soil-water characteristic curve and the saturated coefficient of permeability were used to best-fit the permeability data in accordance with the equation proposed by Leong and Rahardjo (1997). The best-fit analysis of the permeability data gave rise to a calculation of the q soil

parameter. All of the above analyses were conducted using the SoilVision software program.

The q soil parameter for all soil types have been statisically analyzed. There is no analysis of the soil parameters associated with the best-fit of the soil-water characteristic curves.

7 PRESENTATION AND ANALYSIS OF THE DATA

A typical best-fit of the Leong and Rahardjo (1997) procedure for a sand soil is shown in Figure 2. A value of q parameter was 2.54. A similar best-fit for a silt loam is shown in Figure 3 and the q parameter was 3.62. Both plots illustrate that the predicted permeability function quite closely fits the experimental data when the q parameter is known.

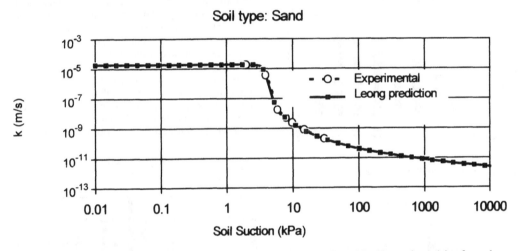

Figure 2. Comparison of permeability function predicted using Leong and Rahardjo (1997) with experimental data for sand.

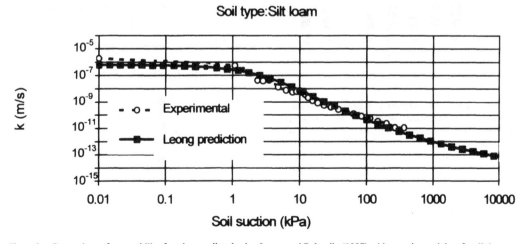

Figure 3. Comparison of permeability function predicted using Leong and Rahardjo (1997) with experimental data for silt loam.

219

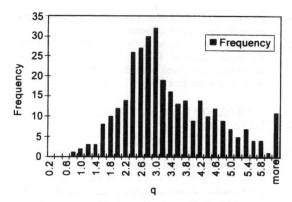

Figure 4. Frequency distribution associated with the q parameter when all soil types are combined.

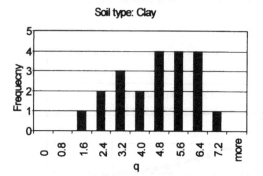

Figure 5. Frequency distribution for clay.

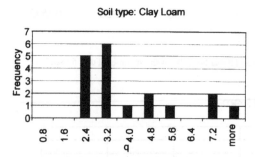

Figure 6. Frequency distribution for clay loam.

Figure 4 shows the frequency distribution associated with the q parameter when all soil types are combined. The statistical analysis of the q parameter is presented in Table 3.

The results show that the overall mean value for q is 3.29 and the standard deviation of q is 1.40. These values yield an overall coefficient of variation of 43%, which is quite high. The overall median value is 2.96 and the mode is 5.61.

Frequency distributions have also been drawn for each of the soil types and are presented in Figures 5 to 14. Some of the frequency distributions are not

close to a normal distribution and this is most likely due to the limited number of samples in each category. Table 4 summarizes the statistical properties associated with the computed q parameters.

The sand showed a mean q value of 2.37 while the clay soils showed a mean q value of 4.34. The sandy loam showed a mean q value of 2.86 while the clay loam showed a mean q value of 3.58. The results indicate that there is a definite trend towards a larger q value for soils with higher plasticity. This trend is quite consistent for all soils categories.

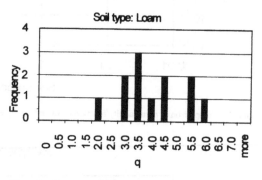

Figure 7. Frequency distribution for loam

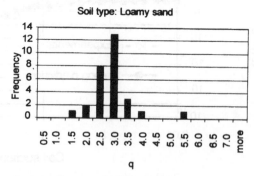

Figure 8. Frequency distribution for loamy sand

220

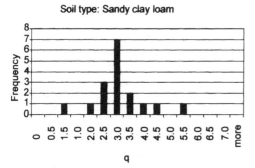

Figure 9. Frequency distribution for sandy clay loam

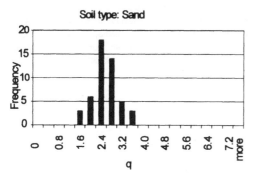

Figure 10. Frequency distribution for sand

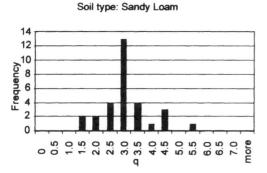

Figure 11. Frequency distribution for sandy loam

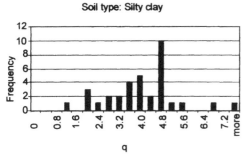

Figure 12. Frequency distribution for silty clay

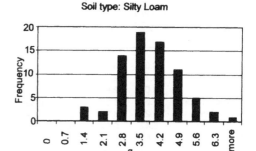

Figure 13. Frequency distribution for silty loam

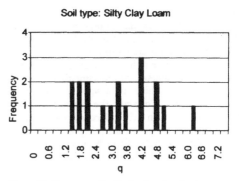

Figure 14. Frequency distribution for silty clay loam

Table 3. Statistical analysis of the q soil parameter

Statistic variable	q soil parameter
Mean	3.29
Standard error	0.08
Median	2.96
Mode	5.61
Standard deviation	1.40
Sample variance	1.96
Kurtosis	14.80
Skewness	2.35
Range	14.39
Minimum	0.64
Maximum	15.03
Number of soils	323

The standard deviation was 0.49 for the sand soil and 1.50 for the clay soil. The standard deviation was 0.84 for the sandy loam and 1.81 for the clay loam. The results indicate that there is less scatter in the fitting parameter q as the soil becomes closer to a sandy material. Also, the fitting parameter, q, moves closer towards 1.0 as the soil becomes sandy.

8 SUMMARY OF FINDINGS

The mean fitting parameters for all soil types ranged from 2.4 to 5.6. The statistical results provide a gen-

Table 4. Statistics of permeability according to Leong function for various soil types

Statistics	Clay	Clay loam	Loam	Sand	Sandy clay Loam	Silty clay	Silty clay loam	Silt loam	Sandy loam	Loamy sand
Mean	4.34	3.58	3.78	2.37	2.80	5.59	3.22	3.52	2.86	2.67
Median	4.71	2.62	3.56	2.36	2.62	4.77	3.18	3.46	2.85	2.59
Mode	3.00	2.80	3.25	2.20	2.75	4.60	4.05	3.15	2.75	4.63
Standard Deviation	1.50	1.81	1.16	0.49	1.00	1.31	1.36	1.09	0.84	0.68
Sample Variance	2.25	3.29	1.34	0.24	0.99	1.73	1.86	1.19	0.71	0.46
Range	5.42	5.92	4.14	2.25	4.71	6.33	4.97	5.99	4.20	4.01
Minimum	1.52	1.84	1.63	1.25	0.64	1.11	1.28	0.83	1.02	1.41
Maximum	6.94	7.76	5.76	3.49	5.35	7.44	6.25	6.82	5.22	5.42
Number of sets	21	18	12	49	17	34	18	74	30	29

eral indication of the range and scatter that can be anticipated when using the Leong and Rahardjo (1997) equation for the estimation of the permeability function for an unsaturated soil.

It would be of value to have a larger data base for analyzing the q parameter. At the same time, it must be recognized that there is always considerable scatter in the results. Certainly, the q parameter will tend to be greater than 1.0 and could vary over a considerable range. The estimation procedure proposed by Leong and Rahardjo (1997) is of value but there does not appear to be unique values for the q parameter.

REFERENCES

Benson, C.H & M. M. Gribb 1997. Measuring unsaturated hydraulic conductivity in the laboratory and field. *Unsaturated Soil Engineering practice, ASCE Geotechnical special publication No. 68*, New York, N.Y: 113 - 168.

Fredlund, D.G. 1995. The prediction of unsaturated soil property functions using the soil-water characteristic curve. *Proc. of the Bengt Broms Symposium in Geotechnical Engineering*, Singapore: 113-133.

Fredlund, M.D. 1997. SoilVision Users Manual. *Version 1.3, SoilVision Systems Ltd.*, Saskatoon, Saskatchewan, Canada.

Fredlund, D.G. & A. Xing 1994. Equations for the soil-water characteristic curve. *Canadian Geotechnical Journal*, 31(3): 521-532.

Fredlund, D.G., A. Xing & S. Huang 1994. Predicting the permeability function for unsaturated soils using the soil-water characteristic curve. *Canadian Geotechnical Journal*, 31(4): 533-546.

Huang, S.Y., S.L. Barbour & D. G. Fredlund 1998. Development and verification of a coefficient of permeability function for a deformable, unsaturated soil. *Canadian Geotechnical Journal*, 35(3): 411-425.

Kunze, R.J., G.Uehara & K.Graham 1968. Factors important in the calculation of hydraulic conductivity. *Proc., Soil Sci. Soc. of Am.*, 32: 760-765.

Leong, E. C. & H. Rahardjo 1997. Permeability functions for unsaturated soils. *Journal of Geotechnical and Geoenvironmental Engineering*, ASCE, 123(12): 1118-1126.

Mualem, Y. 1986. Hydraulic conductivity of unsaturated soils: prediction and formulas. *Methods of soil analysis*, 9(1), Physical and mineralogical methods, Second edition, Agronomy, *Edited by* A. Klute. American Society of Agronomy, Madison, Wisconsin, 799-823.

Clay Science for Engineering, Adachi & Fukue (eds) © 2001 Balkema, Rotterdam, ISBN 90 5809 175 9

Hydraulic properties of high swelling Na/Ca smectites

N. Toride, H. Miyamoto & M. Regea
Department of Agricultural Sciences, Saga University, Japan

ABSTRACT: Clay swelling and clogging of soil pores due to clay dispersion are primary reasons for a decrease in the hydraulic conductivity of clayey soils under sodic conditions. The relative importance and contribution of swelling and dispersion to the hydraulic properties of soils was evaluated by studying the response of high swelling smectites to sodic conditions. We carried out a constant head saturated flow column experiment using the clay-sand mixtures containing 20% clay by weight for various $NaCl$-$CaCl_2$ solutions having different total electrolyte concentration (TEC) and sodium adsorption ratio values (SAR = 0, 10, 20, 30, ∞). Hydraulic head and soil electrical conductivity (EC_a) distributions in a column were monitored using micro-tensiometers and four-electrode salinity probes inserted horizontally at several depths. A 0.5 M solution was applied to equilibrate the soil with a given SAR solution for an entire column. We then reduced the concentration of the influent solution while keeping the SAR value constant. The TEC was successively decreased to 0.05 M, 0.01 M, and 0 M (DW). We determined the conductivity K profile from the hydraulic head distribution as well as the column average conductivity K_{ave} based on the effluent rate. The K_{ave} decreased drastically and finally became negligible for SAR ≥ 20. In case of the complete Na-system (SAR ∞), water flux became negligible immediately after applying a 0.05 M NaCl solution because of clay swelling near the surface. The extremely lower K region at the surface expanded slowly as the concentration below the surface decreased by molecular diffusion. On the other hand, the K decreased gradually throughout the column mainly due to clay dispersion during a DW application for SAR 10. Monitoring pressure and EC_a profiles may prove to be a promising method to distinguish effects of swelling and dispersion on the hydraulic conductivity for various TEC and SAR solutions.

1. INTRODUCTION

The hydraulic conductivity, K, of soils strongly depends on the concentration of the soil solution and the exchangeable cation it contains. Quirk and Schofield (1955) showed that K decreases with increasing exchangeable sodium percentage (ESP) and decreasing TEC. Clay swelling and clogging of soil pores due to clay dispersion are assumed to be primary reasons for a decrease in K under sodic conditions (Quirk and Schofield, 1955; Frankel et al., 1978; Shainberg et al., 1980). Swelling is a reversible process (the reduction in K can be reversed by adding electrolytes or divalent cations to the system). On the other hand, dispersion and subsequent particle movement are essentially irreversible. Clay swelling is not greatly affected by low ESP values less than 10-15, and increases markedly as ESP increases to more than 15 (McNeal et al., 1968; Shainberg and Letey, 1984). Swelling increased continuously and gradually with decreasing TEC. In contrast, clay dispersion occurs throughout the entire ESP range provided TEC is below the flocculation value defined as the minimum electrolyte concentration necessary to flocculate a colloidal suspension (Oster et al., 1980: Shainberg and Letey, 1984). Clay dispersion increases with increasing ESP but dispersion is more sensitive to low ESP values compared to swelling (Regea et al., 1997).

Although most studies to date only considered the average conductivity for an entire soil column K_{ave} based on the effluent flow rate, swelling and dispersion are clay micro-structure changes and occur locally in the soil. The object of this paper is to monitor vertical profiles of the hydraulic head and the bulk soil electrical conductivity (EC_a) in a column packed with high swelling bentonite clay-sand mixtures for a leaching of various $NaCl$-$CaCl_2$ solutions. The response of a high swelling smectite to various TEC and SAR solutions can then be deduced from the vertical distribution of K in the columns.

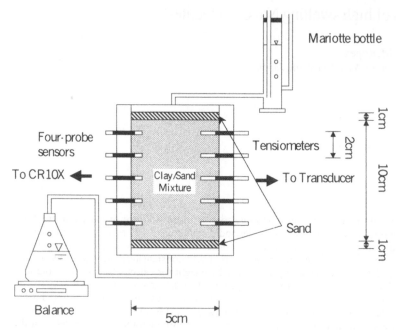

Figure 1 Schematic of constant head setup for saturated hydraulic conductivity experiment

2 MATERIALS AND METHODS

A high swelling smectite of a bentonite clay (Kunimine K.K. Co. Ltd, Yamagata, Japan) was used in this study. The bentonite contained 65 % montmorillonite; CEC was 100 cmol kg^{-1} soil. The bentonite is a typical high swelling smectite. To maintain a sufficiently high flow rate during the experiments and to reduce the resistance against clay particle movement, clay-sand mixtures containing 20% clay by weight were prepared using a coarse dune sand (0.05-0.85 mm diameter). The bulk density was approximately 1.4 kg L^{-1} for the bentonite mixture.

We carried out a constant head saturated flow experiment with clay-sand mixtures packed in a column of 10 cm length and 5 cm diameter (Figure 1). The boundary pressure heads were kept constant throughout the leaching: around 95 cm H$_2$O for the top of the column, and 5 cm H$_2$O for the bottom. A 1 cm layer of coarse sand at the top minimizes the influent disturbance, while a similar 1 cm sand layer at the bottom allows dispersed clay particles to be washed out with the effluent. The column was initially wetted from the bottom with NaCl-CaCl$_2$ solutions (0.5 M Cl^{-1}) having SAR of 0, 10, 20, 30, and ∞, respectively. SAR is defined as c_{Na} / (1/2 c_{Ca})$^{1/2}$ where c_{Na} and c_{Ca} are Na and Ca concentrations (mmol L^{-1}). The corresponding ESP

values are 0 % for SAR 0, 14 % for SAR 10, 24 % for SAR 20, 31 % for SAR 30, and 100 % for SAR ∞. After the clay-sand mixture had been saturated, the 0.5 M solution was applied to the surface. We then reduced the concentration of the influent solution while keeping the SAR value constant. The total electrolyte (Cl^{-1}) concentration, TEC, was successively decreased to 0.05 M, 0.01 M, 0 M (distilled water, DW). Each solution was applied for more than 10 pore volumes of the column. Hence ESP of the clay-sand mixture was in equilibrium with a given SAR solution for an entire column after the leaching.

The effluent rate was measured with an electrical balance connected to a computer. The pressure head, p, was monitored by micro-tensiometers (2 mm in diameter and 10 mm length) with pressure transducers at z = 1, 3, 5, 7, and 9 cm (z = 0 at the bottom of the mixture). The column average conductivity, K_{ave}, and the local K between two tensiometers were calculated based on the water flux and the hydraulic head, $H (= p + z)$.

Total solute concentrations were estimated from the soil electrical conductivity, EC$_a$, using four-electrode salinity probes (Inoue, 1994) inserted horizontally at z = 1, 3, 5, 7, and 9 cm. The probes consisted of four stainless steel pipes (1.6 mm diameter, 5 mm long, 8 mm equal spacing) were connected to a CR10X datalogger with a multiplexer (Campbell Scientific Inc.).

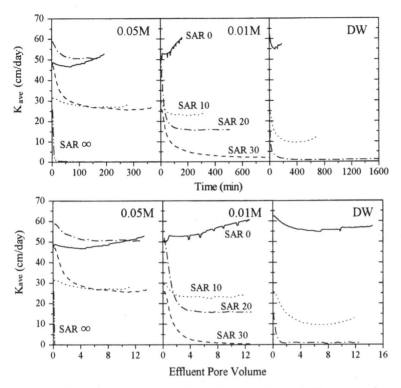

Figure 2 Column average conductivity K_{ave} as a function of time and effluent pore volume.

3 RESULTS AND DISCUSSION

Figure 2 presents the average conductivity K_{ave} based on the pressure heads at the top and the bottom of the clay-sand mixture as a function of time and effluent pore volumes for different SAR values. Note that K_{ave} is proportional to water flux since the boundary pressure heads are constant during the leaching. The pore volume (PV) is defined as the cumulative drainage divided by the amount of water in the mixture. One PV was approximately 86 cm^3 for all cases.

As observed in most of previous conductivity studies (Shainberg and Letey, 1984), K_{ave} decreases with increasing SAR and decreasing TEC. The overall tendency for decreasing TEC was similar as Regea et al. (1998) observed for a high swelling smectite. The response of the bentonite mixture was, however, more sensitive for higher SAR (≥ 20) and less sensitive for lower SAR (≤ 10). Water flux eventually became almost negligible for SAR ≥ 20: during the DW leaching for SAR 20, 0.01 M of SAR 30, and 0.05 M of SAR ∞. On the other hand, the K_{ave} for SAR 10 decreased during the DW application up to around 10 cm d^{-1}, and slightly increased after 10 pore volumes. In case of SAR 0, the K_{ave} did not decrease during the leaching regardless of TEC values. Note that the K_{ave} increased from 50 cm d^{-1} to 60 cm d^{-1} during the application of 0.05M and 0.01M for SAR 0.

The decrease in the average conductivity, K_{ave}, was a result of local K reductions. Figure 3 shows the K profiles based on tensiometer measurements after applying each solution of SAR 30, 20, 10 and 0. Boundary pressure heads were used to determine K values at $z = 0.5$ cm and 9.5 cm. Settlements at the surface were observed during the leaching in some cases. When it was difficult to evaluate the settlement volume at the surface, we could not properly evaluate the K value at the top, e.g. SAR 20. As can be seen from the profiles for 0.5 M in all cases, the K profile was initially not homogeneous. Although we packed the air-dried mixture as uniform as possible from the bottom, the initial K was smaller at the bottom.

Two types of reduction in K was observed in the pressure head p profiles. When decrease in K occurred locally, large fluctuations in p were observed because the p profile was determined according to the K distribution (see Figure 7). On the other hand, when the reduction in K was homogeneous for an entire column, the p profile was almost constant even if the K_{ave} decreased.

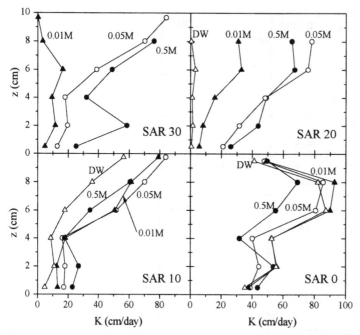

Figure 3 *K* profiles after applying each concentration for SAR 30, 20, 10 and 0.

Figure 4 presents the *K* profiles during the leaching of 0.01 M of SAR 30 and DW for SAR 20. The *K* near the surface firstly become negligible. The lower *K* near the surface then resulted in negligible water flux.

An extreme case is the sodium saturated column (SAR ∞). Figure 5 shows the *K* at *z* = 9.5 cm and 8 cm together with the K_{ave} as a function of log-scaled time during the 0.05 M application. The average conductivity K_{ave} was around 30 cm d^{-1} for 0.5 M NaCl, and became 0.8 cm d^{-1} within 15 min. The rapid drop in *K* just below the surface resulted in the drastic decrease in the water flux.

According to Oster et al., 1980, the flocculation value of Na saturated montmorillonite (SAR ∞) is 0.012 to 0.016 M. The inflow concentration (0.05 M) is well above the flocculation value, but well below the threshold concentration for swelling. Hence clay particles did not disperse, but swell considerably (Shainberg et al., 1971). Local-scale high swelling of clay particles resulted in clogging of conducting pores in a narrow region. As shown in Figure 4, the lower conductivity region at the surface is more distinct and narrower for higher SAR. It is necessary to further investigate flocculation values and swelling threshold concentrations for different SAR and TEC.

Once the water flux decreased, the contribution of convective solute transport decreased compared to molecular diffusion. Figure 6 presents soil electrical conductivity readings, EC$_a$, at 5 depths as a

function of log-scaled time for 0.05 M of SAR ∞. The EC$_a$ at below *z* = 7 cm started to decrease after 1000 min while the EC$_a$ at *z* = 9 cm became 0.5 S m^{-1} within 20 min. The water flux was less than 1.0 cm d^{-1} at *t* = 100 min, and became less than 0.01 cm d^{-1} at *t* = 10^4 min. The concentration in the column, therefore, decreased mainly due to upward diffusion.

Figure 7 shows changes in hydraulic head, *H* (= *p* + *z*) with log-scaled time. The rapid decrease in *H* around at 10 min was a result of the formation of an impermeable layer at the surface. The *H* values at *z* = 9 cm and 7 cm increased again at *t* = 1500 min and 10^5 min, respectively, resulting in higher *H* gradients in the upper region. This implies that the lower *K* region had expanded to the lower part of the column. Note that the *K* at *z* = 9.5 cm and 8 cm in Figure 6 corresponded well to the EC$_a$ decrease at *z* = 9 cm and 7 cm in Figure 6, respectively. Swelling of clay particles occurred just after the reduction in TEC.

When the *K* decreased uniformly throughout the column as observed for 0.05 M of SAR 30, 0.01 M of SAR 20, and DW for SAR 10 (Figure 8), the pressure head, *p*, profiles were constant during the leaching regardless of the decrease in *K*. The K_{ave} after 10 PVs applications was approximately 25 cm d^{-1} for 0.05 M of SAR 30, 15 cm d^{-1} for 0.01 M of SAR 20, and 10 cm d^{-1} in case of DW for SAR 10. The EC$_a$ values became almost constant after 2-3 PVs applications, indicating the solution concentration had decreased to the inflow concentration. Different from the surface local

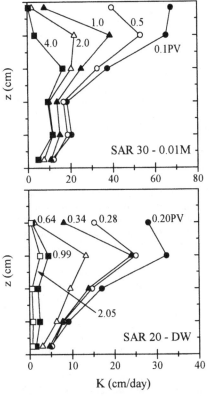

Figure 4 *K* profiles during leaching with 0.01 M of SAR 30 and DW for SAR 20.

case of DW for SAR 10, however, probably because of high viscosity for the bentonite clay. We assume clay dispersion and continuous water flow resulted in the continuous *K* reduction throughout the column. In case of SAR 30 and SAR 20, both swelling and dispersion could operate. While the swelling at the surface was distinct for 0.01 M of SAR 30 and DW for SAR 20, the moderate *K* reduction occurred for higher TEC: 0.05 M of SAR 30 and 0.01 M of SAR 20.

4. CONCLUSIONS

We investigated the response of hydraulic properties for a high swelling smectite to various total electolyte concentration (TEC) and sodium adsorption ratio (SAR). Soil water pressure and electrolyte concentration in a saturated clay-sand mixture column were monitored during the leaching of various NaCl-CaCl$_2$ solutions. We determined the conductivity *K* profile from the hydraulic head distribution as well as the column average conductivity, K_{ave}, from the effluent rate.

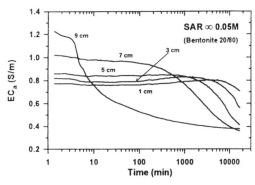

Figure 6 Soil electrical conductivity readings, EC$_a$, at 5 depths as a function of log-scaled time during the 0.05 M application for SAR ∞.

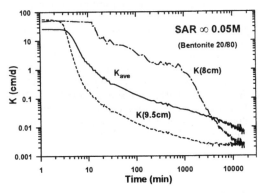

Figure 5 *K* at *z* = 9.5 cm and 8 cm, and K_{ave} as a function of log-scaled time during the 0.05 M application for SAR ∞.

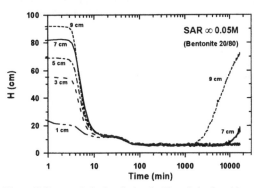

Figure 7 Changes in hydraulic heads, *H*, at 5 depths with log-scaled time during the 0.05 M application for SAR ∞.

reduction due to swelling as shown in Figures 4 and 5, the *K* continued to decrease after the TEC decreased. Since clay swelling will not occur for low ESP such as the case for SAR 10, clay dispersion could be the only mechanism to reduce *K*. Only small amount of clay particles was observed in effluents in

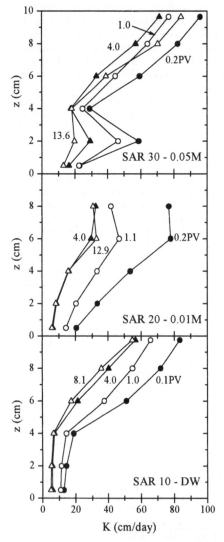

Figure 8 *K* profiles during leaching with 0.05 M of SAR 30, 0.01 M of SAR 20, and DW of SAR 10.

On the other hand, the *K* decreased gradually throughout the column mainly due to clay dispersion during a DW application for SAR 10. When clay dispersion was a dominant mechanism for the reduction in *K*, the *K* continued to decrease even after the TEC became constant. Monitoring pressure head and solute concentration profiles used in this study may prove to be a promising method to improve our understanding of unique response of clayey soils to various TEC and SAR solutions.

REFERENCES

Frankel, H., J. O. Goertzen, and J. D. Rhoades. 1978. Effect of clay type and content, exchangeable sodium percentage, and electrolyte concentration and clay dispersion and soil hydraulic conductivity, *Soil Sci. Soc. Am. J.*, 42: 32-39.

Inoue, M. 1994. Simultaneous movement of salt and water in an unsaturated sand column. *ALRC's Annual Report* 1993-94:13-26. Tottori University, Japan.

McNeal, B. L., and N. T. Coleman. 1966. Effect of solution composition on soil hydraulic conductivity, *Soil Sci. Soc. Am. Proc.*, 30: 308-312.

Oster, J. D., I. Shainberg, and J. D. Wood. 1980. Flocculation value and gel structure of Na/Ca montmorillonite and illite suspension. *Soil Sci. Soc. Am. J.* 43:955-959.

Quirk, J. P. and R. K. Schofield. 1955. The effects of electrolyte concentration on soil permeability. *J. Soil Sci.*, 6: 163-178.

Regea, M., T. Yano, and I. Shainberg. 1997. The response of low and high swelling smectites to sodic conditions, *Soil Sci.*, 162: 299-307.

Shainberg, I., and Letey. 1984. Response of soils to sodic and saline condition. *Hilgardia* 52:1-57.

Shainberg, I., J. D. Rhoades, and R. J. Prather. 1980. Effect of low electrolyte concentration on clay dispersion and hydraulic conductivity of a sodic soil, *Soil Sci. Soc. Am. J.*, 45: 273-277.

The K_{ave} decreased drastically and finally became negligible for SAR ≥ 20. In case of the complete Na-system (SAR ∞), the water flux became negligible immediately after applying a 0.05 M NaCl solution because of swelling near the surface. The extremely lower *K* region at the surface expanded slowly as the concentration below the surface decreased by molecular diffusion. The reduction in *K* at the surface was caused by clay swelling, resulting negligible water flux after forming a impermeable layer at the surface. Clay swelling occurred just after the reduction in TEC.

Clay Science for Engineering, Adachi & Fukue (eds) © 2001 Balkema, Rotterdam, ISBN 90 5809 175 9

Hydraulic properties in compacted bentonite under unsaturated condition

T. Fujita, H. Suzuki, Y. Sugita & H. Sugino
Japan Nuclear Cycle Development Institute (JNC), Tokyo, Japan

M. Nakano
Department of Agriculture, Kobe University, Japan

ABSTRACT: A laboratory test was carried out to investigate the effect of hysteresis and temperature on suction measurements in compacted bentonite using psychrometer. The isothermal water diffusivity and thermal water diffusivity of compacted bentonite were compared with conventional studies. The results were presented as follows: The hysteresis of suction in compacted bentonite was smaller than that of powder bentonite. The suction in bentonite depended on temperature. Conventional models could describe the water retention curve and the water diffusivities under isothermal and thermal conditions in bentonite.

1 INTRODUCTION

High-level radioactive waste management in Japan is based on multibarrier concept. The manufactured components of the multibarrier system constitute an engineered barrier system (EBS), consisting of a stable waste form (vitrified waste), a rigid vessel (overpack) for containment of the waste form, and materials placed between the overpack and the surrounding geological formations during emplacement (backfill or buffer materials). The buffer material is expected to maintain its low water permeability, self-sealing properties, radionuclides adsorption and retardation properties, thermal conductivity, chemical buffering properties, overpack supporting properties, stress buffering properties, etc. over a long period of time (Power Reactor and Nuclear Fuel Development Corporation [PNC] 1992). Natural clay is mentioned as a material that can satisfy above functions (Oscarson and Cheung 1983; Pusch 1983). Among the kinds of natural clay, bentonite when compacted is superior because (i) it has exceptionally low water permeability and properties to control the movement of water in buffer, (ii) it fills void spaces in the buffer and fractures in the host rock as it swells upon water uptake, (iii) it has the ability to exchange cations and to adsorb cationic radioelements. In order to confirm these functions for the purpose of safety assessment, it is necessary to evaluate buffer properties through laboratory tests and engineering-scale tests, and to make assessments based on the ranges in the data obtained. Both water and water vapor are re-distributed through initially unsaturated buffer material during the thermal period due to a rise in temperature rise caused by the radiogenic heating of the vitrified waste after the waste package emplacement, until the buffer eventually reaches a saturated state. A key property of the buffer is its ability to expand (swell) by water uptake, thereby filling voids in the engineered barrier and fractures in the surrounding rock mass. Therefore, water characteristic curve (relationship between the water potential and water content), unsaturated hydraulic conductivity of bentonite in the unsaturated state at elevated temperature must be studied.

In accordance with the recommendations of the 1976 soil physics terminology committee of the ISSS (Bolt, 1976), the potential energy of water in unsaturated soil is defined as follows (Jury et al. 1991).

$$\psi_T = \psi_z + \psi_s + \psi_a + \psi_m \qquad (1)$$

where ψ_T is total potential, ψ_z is gravitational potential, ψ_s is solute potential, ψ_a is air pressure potential, and ψ_m is matric potential. The potential energy of water in soil is negative, because it must be defined relative to a reference or standard state which is customarily defined to be the state of pure, free, water at a reference pressure, temperature, and elevation and is arbitrarity given the value zero (Bolt 1976). In this study, "suction" which delete the sign of potential energy is used. The suction-water content relationship (hereinafter "water retention curve") of a soil shows a hysteresis phenomena in wetting and drying processs (Poulovassilis 1973). This hystereresis is observed in both cohesionless granular soils and cohesive soils (e.g., Jury et al.

1991 : Yong and Warkentin 1975). In clays, hysteresis is attributable partly to the "ink-bottle effect", and partly to changes in the soil fabric and pore size distribution brought about by swelling and shrinking of the soil during the drying and wetting processes (Yong and Warkentin 1975). Suctions in soils are influenced by temperature. Observations show that soil suction generally decreases with increasing temperature. The effect of temperature on soil suction is commonly attributed to the dependence of the surface tension of water on temperature (e.g. Edlefsen and Anderson 1943). Based on the above characteristics and knowledge of suction, this study examines the effect of hysteresis and temperature on suction measurements in compacted bentonite using psychometer.

Water flow in unsaturated soils under the influence of thermal gradients is a complex phenomenon. The thermal gradient of soil is a driving force for movement of both liquid and vapor phases. A common model of water movement under the influence of thermal and hydraulic gradients is the mathematical formulation attributed to Philips and de Vries (1957). This model is based on the concept of viscous flow of liquid water under the influence of gravity, capillary and adsorptive forces, and on the concept of vapor movement by diffusion. For one-dimensional flow, the general differential equations describing the water flow under combined temperature and hydraulic gradients are as follows;

$$\frac{\partial \theta}{\partial t} = \nabla(D_T \nabla T) + \nabla(D_\theta \nabla \theta) + \frac{\partial k}{\partial x} \qquad (2)$$

where θ is volumetric water content, t is time, ∇T is thermal gradient, $\nabla \theta$ is volumetric water content gradient, k is hydraulic conductivity, x is distance, D_T is thermal diffusivity, and D_θ is water diffusivity. D_T and D_θ are divided into following relations;

$$D_T = D_{Tliq} + D_{Tvap} \qquad (3)$$

$$D_\theta = D_{\theta liq} + D_{\theta vap} \qquad (4)$$

where D_{Tliq} is thermal liquid diffusivity, D_{Tvap} is thermal vapor diffusivity, $D_{\theta liq}$ is isothermal liquid diffusivity , and $D_{\theta vap}$ is isothermal vapor diffusivity. These mathematical parameters related to diffusivity are generally found to be dependent on water content and temperature, and are hysteretic (e.g. Radhakrishna et al., 1992). This study examins to estimate the isothermal water diffusivity and thermal water diffusivity from the distributions of water content in compacted bentonite under isothermal and thermal conditions and to compare with conventional studies.

2 MATERIALS

The bentonites examined here are referred to as Kunigel V1, Kunipia F, and MX-80. Kunigel V1 and Kunipia F (Kunimine Industries) are a type of domestic bentonites, produced in Tsukinuno, Yamagata Prefecture. More than 90% of Kunigel V1 or more than 70% of Kunipia F has a smaller grain size than 200 mesh (< 74 µm). MX-80 (American Colloids Company) is considered to be the reference buffer material in Sweden, Switzerland and other countries and is a product of Wyoming, U.S.A. For the silica sand, a mixture of silica sand No. 3 (produced in Seto City, Aichi Prefecture) and No. 5 (produced in West Australia) mixed at a weight ratio of 1 : 1 is used in this study. Table 1 lists mineral compositions of bentonites used in this study (Ito et al. 1993; PNC 1996; Lajudie et al. 1996), Table 2 presents physical properties (Ishikawa et al. 1990; Ito et al. 1993; Lajudie et al. 1996), and Figure 1 shows the particle size distribution curves of bentonites and silica sand (Onofrei and Hnatiw 1997). The chemical compositions of No. 3 and No. 5 silica sands are shown in Table 3. In the following study, the silica sand (hereinafter "sand") used is the same as this mixture (soil grain density = 2.64 Mg m^{-3}), unless otherwise specified.

Table 1. Mineral compositions (%) of the used bentonites

	Kunigel V1[1),2)]	Kunipia F[1),2)]	MX-80[3)]
Montmorillonite	46 – 49	98 – 99	75
Quartz/chalcedony	29 – 38	<1	15.2
Feldspar	2.7 - 5.5		5 – 8
Calcite	2.1 - 2.6	<1	1.4
Dolomite	2.0 - 2.8		
Analcite	3.0 - 3.5		
Pyrite	0.5 - 0.7		0.3
Kaolinite			<1
Mica			<1
Illite			
Gypsum			
Organic matter	0.31 - 0.34		0.4
Other			2

1): Ito et al. 1993 ; 2): PNC 1996 ;3): Lajudie et al. 1996

Table 2. Physical properties of the used bentonites

		Kunigel V1[1),2)]	Kunipia F[1),2)]	MX-80[3)]
True specific gravity (-)		2.7	2.7	2.7
Liquid limit (%)		416	993	400
Plastic limit (%)		21	42	70
Plasticity index (-)		395	951	330
Cation exchange amount (meq/ 100 g)		52	1117	79
Leach cation	Na$^+$	54.6	114.9	56.0
(meq/ 100 g)	K$^+$	1.3	1.1	2.3
	Ca^{2+}	41.9	20.6	30.1
	Mg^{2+}	6.6	2.6	15.6
Na$^+$/Ca^{2+} ratio		1.30	5.58	1.86

1): Ishikawa et al. 1990 ; 2): Ito et al. 1993 ; 3): Lajudie et al. 1996

Table 3. Chemical compositions of No. 3 and No. 5 silica sand

Component [wt%]	Silica sand No. 3	Silica sand No. 5
SiO_2	94.6	99.78
Al_2O_2	3.32	0.05
Fe_2O_3	0.13	0.01
TiO_2	—	0.03
MgO	—	<0.01
Na_2O	—	0.01
K_2O	—	0.01
Ignition loss	—	0.12

3 HYDRAULIC CHARACTERISTICS OF UNSATURATED BENTONITE

3.1 Water retention curve

3.1.1 Method

3.1.1.1 Thermocouple psychrometer

The psychrometers used in this study are Type SC-10A manufactured by Decagon Device Inc. (Figure 2). NT-3 manufactured by Decagon Device Inc. is used to excite the psychrometers and read their outputs. Samples are placed in small, stainless steel cups where the water vapor in the head-space of the sample cup equilibrates with the liquid water in the sample. The humidity of head-space is measured psychrometerically as a wet-bulb temperature depression. Wet-bulb depression is measured by placing a small drop of water on the thermocouple (Richards method). The psychrometer is calibrated prior to use using potassium chloride (KCl) solutions, in order to provide a relationship between suction and instrument output. From a thermodynamic standpoint, the total suction in an unsaturated soil is related to the vapor pressure in the air voids in the soil. Theoretically, the total suction can be calculated using the Kelvin equation (Young and Warkentin, 1975),

$$\psi = \frac{RT}{M} \ln \left(\frac{P}{P_0}\right) \qquad (5)$$

where ψ is the water potential, R is the molar gas constant, T is the Kelvin temperature, M is the molecular mass of water, and P, P_0 are the partial vapor pressure and saturated vapor pressure, respectively, and P/P_0 is the relative humidity.

3.1.1.2 Vapor pressure method

In order to ensure the suction of compacted bentonite, suction is estimated under keeping of the humidity of the surrounding air to a constant, hereinafter "vapor pressure method". This method involves equilibrating soil samples with as atmosphere of known relative humidity. The suction is thermodynamically estimated by a relation to the vapor pressure via the Kelvin equation (Equation 2). Schematic view of vapor pressure method is shown in Figure 3. The humidity is controlled by several easily available neutral or acid salts, summarized in Table 4.

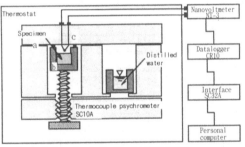

Figure 1. Grain size distribution of several bentonites and silica sand (Onofrei and Hnatiw 1997)

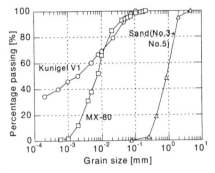

a : Sample chamber, b : Sample cup, c : Thermocouple

Figure 2. Schematic view of suction measurement apparatus including thermocouple psychrometer (SC-10A)

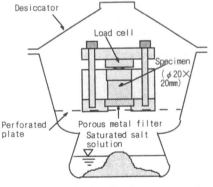

Figure 3. Schematic view of suction measurement apparatus of vapor pressure method

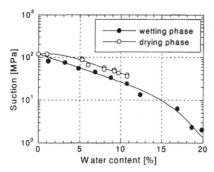

Figure 4. Suction-water content relationships of powder bentonite (Kunigel V1)

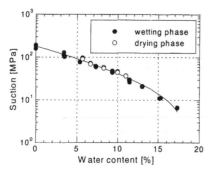

Figure 5. Suction-water content relationships of compacted bentonite (Kunigel V1, dry density = 1.8 Mg m^{-3})

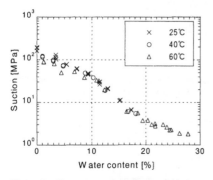

Figure 6. Temperature dependency of suction-water content relationships of powder bentonite (Kunigel V1)

3.1.2 Results and discussion

3.1.2.1 Hysteresis

Figure 4 and 5 show the water retention curves of powder bentonite of Kunigel V1 and its compacted bentonite (dry density =1.8 Mg m^{-3}) controlled to some water content. Each figure compares the water retention curves measured during drying and wetting processs. From Figure 4, it can be seen that the water retention curve of powder bentonite for the drying process differs from that for the wetting process.

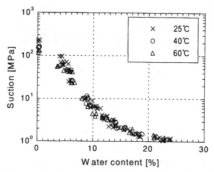

Figure 7. Temperature dependency of suction-water content relationships of powder bentonite-sand mixture (30 wt%)

At a given water content, the suction is higher during drying than wetting. These observations are consistent with those on the behavior of other soil types reported in the literature (e.g., Jury et al. 1991: Yong and Warkentin 1975). On the other hand, it can be seen that the discrepancy of water retention curves of compacted bentonite for the drying process and for the wetting process is not large, as shown in figure 5. This phenomena indicates that hysteresis of compacted bentonite is largely independent of effect due to irregular pore distribution like ink-bottle effects, because pore in highly compacted bentonite is very small. Furthermore, these observations may be taken to imply that suctions in compacted bentonite are largely controlled by the micropores within the peds that are part of the compacted soil structure.

3.1.2.2 Temperature effects

The suctions are measured at 25, 40, and 60°C using thermocouple psychrometer. Figure 6 shows the water retention curve of powder bentonite (Kunigel V1) and Figure 7 the water retention curve of powder bentonite (Kunigel V1) -sand mixture (30 wt%). Qualitatively, it can be seen from the data in the figures that, at a given water content, the suctions in each sample tend to decrease with increasing temperature. However the differences are not large. The influence of temperature on suction of soil is commonly explained by consideration of several factores including thermal expansion of entrapped air, effect of dissolved solutes on the surface tension-temperature relation of the soil water, and increase in the radius of the air-water meniscus associated with thermal expansion and swelling of clay-water structure (Peck 1960; Constantz 1983).

3.1.2.3 Type of bentonite

Figure 8 indicates the water retention curve of Kunipia F, MX-80, and Kunigel V1 at the ambient temperature of 25°C. It can be seen from the data in the figure that, at a given water content, the suctions in Kunipia F has the highest, followed by MX-80, and then Kunigel V1. This discrepancy may be due to the content ratio of montmorillonite. The water content is normalized by the content ratio of montmorillonite, as shown in equation, and the result is shown in Figure 9.

$$\omega_{mon} = \omega/R_{mon} \qquad (6)$$

where ω_{mon} is the water content normalized by content ratio of montmorillonite, ω is the water content, and R_{mon} is the content ratio of montmorillonite.

Figure 9 shows that the differences of suction at a given water content normalized by content ratio of montmorillonite are smaller than that at a given water content in Figure 8. Furthermore, these observations may be taken to imply that suctions are largely controlled by the content ratio of montmorillonite.

3.1.2.4 Modeling of water retention curve of compacted bentonite

A variety of empirical functions have been proposed to relate the water retention curve. One of the S-shape models for water retention curve suggested by van Genuchten (1978), as shown in following equation, is applied to that of compacted bentonite.

$$S_e = [1 + (\alpha\psi)^\beta]^{1/\beta-\beta} \qquad (7)$$

where ψ is the suction, α and β are constants, and S_e is the effective saturation, as follows.

$$S_e = (\theta - \theta_r)/(\theta_s - \theta_r) \qquad (8)$$

where θ is the volumetric water content, θ_r is the residual water content, and θ_s is the saturated water content.

Figure 10 show the water retention curve of compacted and powder bentonites measured by psychrometer and vapor pressure methods with the suctions predicted by the van Genuchten. For the water content range from 3% to 15%, the suctions appear to be independent of measured method and condition of bentonite (powder or compaction). The differences of suction between powder and compacted bentonite for $\omega >15\%$ tend to be large with increasing water content, because water content of compacted bentonite becomes saturation. The van Genuchten model is in fair agreement with measured water retention curve of compacted bentonite. Furthermore, it is necessary to measure suction of compacted bentonite for $\omega<3\%$ and to study its approachability to the van Genuchten model.

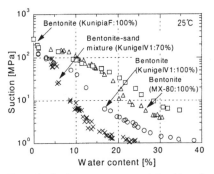

Figure 8. Suction-water content relationships of several kinds of powder bentonite

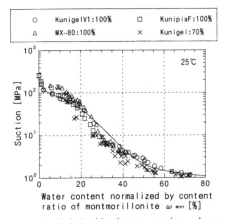

Figure 9. Relationships between suction and water content normalized by content ratio of montmorillonite of powder bentonite

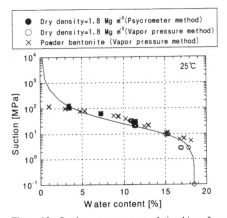

Figure 10. Suction-water content relationships of powder and compacted bentonite measured by several methods (The line presents van Genuchten model ; θ_r=0, θ_s=18.5, α=4.22x10^{-10}, β=2.23)

Salt	Relative humidity (%)	pF	Suction (MPa)
NaCl	75.0	5.60	38.9
NH$_4$Cl	79.2	5.51	31.5
(NH$_4$)$_2$SO$_4$	81.0	5.46	28.5
KHSO$_4$	86.0	5.31	20.4
NH$_4$H$_2$PO$_4$	93.1	5.00	9.7
K$_2$SO$_4$	98.0	4.46	2.7
Distilled water	100.0	-	0.0

pF=log$_{10}$(negative pressure head, in cm)

3.2 Water movement in unsaturated bentonite

3.2.1 Method

3.2.1.1 Isothermal infiltration test
The experiment is designed so that water flow had one-dimensional vertical movement. Figure 11 shows a schematic view of experimental apparatus. Compacted bentonite is installed in a stainless steel cylindrical cell (inner diameter 20mm, height 20mm) and sandwiched between two porous metal filters of stainless steel filter (hole diameter 5 µm) with Teflon filter (hole diameter 2 µm). Distilled water is supplied from the bottom of specimen. The specimen is sliced into 2mm sections after various infiltration periods. The water content of each piece is measured by the oven-drying method (110°C, 24h) and the distribution of volumetric water content in the specimen (hereinafter called the "water distribution curve") is estimated.

3.2.1.2 Infiltration test under thermal gradient
The infiltration tests under thermal gradient condition are carried out in order to understand the behavior of the water movement in unsaturated bentonite. The specimens (diameter 50mm, height 100mm) packaged in a food wrap film are set in a apparatus, named KID-BEN for small sized coupled thermal-hydraulic-mechanical test (Figure 12). The cell of KID-BEN is made of bakelite which is one of thermosetting resin and has air layer in order to enhance thermal-insulating effects. The cooling and heating portions are composed of copper plates and thermostat with circulation system, UNIACE UA-10S manufactured by EYELA. KID-BEN is composed of equipment A for measurement of water content and equipment B for measurement of temperature. In equipment B, the K-type sheathed thermocouples manufactured by TOKO THERMOSCIENCE CORP. are put in the holes drilled in the surface of the specimen side and temperature is continuously monitored. The specimen in equipment A is sliced into 20mm sections after infiltration periods. The water content of each section is measured by oven-drying (110°C, 24h).

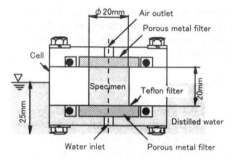

Figure 11. Schematic view of the infiltration test equipment under isothermal condition

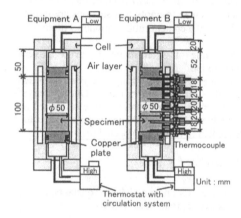

Figure 12. Schematic view of equipment for water movement tests under thermal gradien (Equipment A for water content, Equipment B for temperature)

3.2.2 Results and discussion

3.2.2.1 Water distribution
The distributions of the volumetric water content obtained from the isothermal infiltration tests for the specimens with bentonite (dry density of 1.8 Mg m^{-3} and bentonite-sand mixture (30 wt%, dry density 1.6 Mg m^{-3}) are shown in Figure 13 and 14. The figures indicate that the increases in time and temperature cause an increase in the volumetric water content.

Figure 15 shows the temperature distribution profiles for compacted bentonite (dry density of 1.8 Mg m^{-3} and bentonite-sand mixture (30 wt%, dry density 1.6 Mg m^{-3}) at the steady state. The temperature of the specimen becomes constant after approximately 2-3 hours past. The equipment is dismantled after 96 hours to measure the water content of the specimen. Shrinkage in the direction of the diameter, due to drying, is observed on the high-temperature side of the specimen, while expansion is observed on the low-temperature side. This indicates that moisture is redistributed by evaporation from the hot end and condensation at the low-temperature end. Observa-

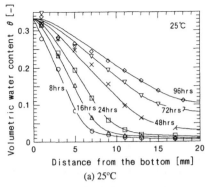

(a) 25°C

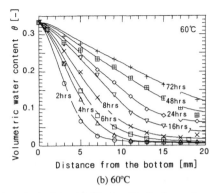

(b) 60°C

Figure 13. Distributions of the volumetric water content obtained from the isothermal infiltration tests for the specimens with bentonite (dry density of 1.8 Mg m^{-3})

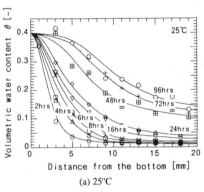

(a) 25°C

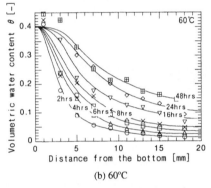

(b) 60°C

Figure 14. Distributions of the volumetric water content obtained from the isothermal infiltration tests for the specimens with bentonite-sand mixture (30 wt%, dry density of 1.6 Mg m^{-3})

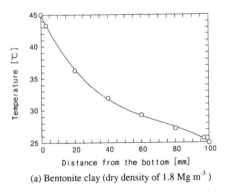

(a) Bentonite clay (dry density of 1.8 Mg m^{-3})

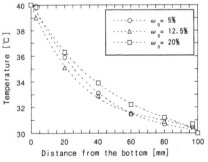

(b) Bentonite-sand mixture (30 wt%, dry density of 1.6 Mg m^{-3})

Figure 15. Distribution of the temperature obtained from the thermal infiltration tests at 96 hours later (ω_0=initial water content)

tion of the entire specimen reveals that the color of the specimen changed gradually from white to black from the high temperature side toward the low temperature side. Figure 16 shows the initial water distribution in the initial state of respective specimens and water-content distribution observed 96 hours later. In either specimen, a decrease in water content is observed on the high-temperature side after 96 hours, while an increase in water content is observed on the low-temperature side. As a result, a water-content gradient is formed in response to a temperature gradient.

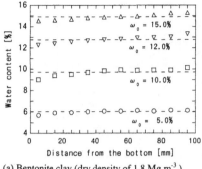

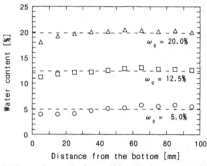

(a) Bentonite clay (dry density of 1.8 Mg m⁻³)

(b) Bentonite-sand mixture (30 wt%, dry density of 1.6 Mg m⁻³)

Figure 16. Distribution of the water content obtained from the thermal infiltration tests at 96 hours later (ω_0=initial water content)

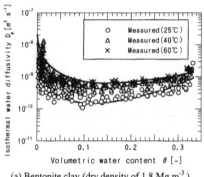

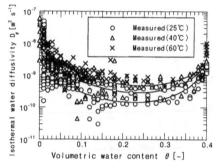

(a) Bentonite clay (dry density of 1.8 Mg m⁻³)

(b) Bentonite-sand mixture (30 wt%, dry density of 1.6 Mg m⁻³)

Figure 17. Isothermal water diffusivity –volumetric water content relations

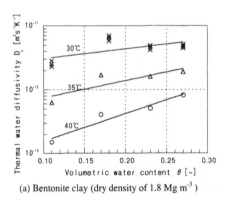

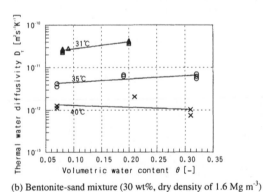

(a) Bentonite clay (dry density of 1.8 Mg m⁻³)

(b) Bentonite-sand mixture (30 wt%, dry density of 1.6 Mg m⁻³)

Figure 18. Thermal water diffusivity –volumetric water content relations

3.2.2.2 Water diffusivity

Water flow in unsaturated soils under the influence of thermal gradients is a complex phenomenon. The thermal gradient of soil is a driving force for movement of both liquid and vapor phases. The water movement in unsaturated bentonite under isothermal and thermal gradient condition is expressed by the following equation.

$$q(\theta,T) = q_\theta(\theta,T) + q_T(\theta,T)$$
$$= -D_\theta \frac{\partial \theta}{\partial x} - D_T \frac{\partial T}{\partial x} \tag{9}$$

where $q(\theta,T)$ [m s⁻¹] is the total water flux, $q_\theta(\theta,T)$ [m s⁻¹] is the isothermal water flux, $q_T(\theta,T)$ [m s⁻¹] is the thermal water flux, D_θ [m² s⁻¹] is the isothermal

236

water diffusivity, D_T [m^{2}s^{-1}K^{-1}] is the thermal water diffusivity, $\partial\theta/\partial x$ is water content gradient, $\partial T/\partial x$ is thermal gradient, and x [m] is the position from bottom of the specimen. Then the total water flux is calculated by the following equation based on the distribution of water content obtained by infiltration test (Nakano 1982).

$$q(\theta,T) = \frac{\displaystyle\iint_{x\ t} \frac{\rho_d}{\rho_w}\frac{d}{dt}\omega(t,x)dtdx}{\displaystyle\int_t dt} \tag{10}$$

where ρ_d [Mg m^{-3}] is the dry density, ρ_w [Mg m^{-3}]is the water density and t [s] is infiltration time. $\partial\theta/\partial x$ and $\partial T/\partial x$ are given by following equation.

$$\frac{\partial\theta}{\partial x} = \frac{1}{2}\sum_{k=t}\left(\frac{\partial\theta}{\partial x}\right)_k \tag{11}$$

$$\frac{\partial T}{\partial x} = \frac{1}{2}\sum_{k=t}\left(\frac{\partial T}{\partial x}\right)_k \tag{12}$$

With above relations, Figure 17(a) shows the relationship between volumetric water content and isothermal water diffusivity of bentonite clay (Kunigel V1, dry density 1.8 Mg m^{-3}) and Figure 17(b) presents the relationship for a bentonite-sand mixture (30 wt%, dry density 1.6 Mg m^{-3}). The isothermal water diffusivity depends on the volumetric water content, and can be expressed as U-shaped curves. And the minimum value of the isothermal water diffusivity increases with the increment of the temperature. The values of isothermal water diffusivities of bentonite clay are larger than those of a bentonite-sand mixture. It may be due to differences of density and suction. Lines in figures show the following equation.

$$D_\theta = D_{\theta_{liq}} + D_{\theta_{vap}} = \frac{a_1(\theta-\theta_s)}{(\theta-b_1)(b_1-\theta_s)} + \frac{a_2\theta}{b_2(\theta-b_2)} \tag{13}$$

where θ_s is the volumetric water content at saturation, and a_1, a_2, b_1, b_2 are constants. Figures indicate that the relationship between volumetric water content and isothermal water diffusivity can be represent as the sum of two hyperbolic functions. The values of constants are summarized in Table 5.

The thermal water diffusivity is calculated using equations (9)-(13) and distribution of water content obtained from the thermal infiltration tests in Figure 16. Figure 18(a) shows the relationship between volumetric water content and thermal water diffusiv-

ity of bentonite clay (Kunigel V1, dry density 1.8 Mg m^{-3}) and Figure 18(b) presents the relationship for a bentonite-sand mixture (30 wt%, dry density 1.6 Mg m^{-3}). The thermal water diffusivity is largely independent of effect due to volumetric water content like isothermal water diffusivity. However, thermal water diffusivity decreases with the increment of the temperature.

Table 5. Values of constants in equation (13)

	Bentonite clay (dry density 1.8 Mg m^{-3})	Bentonite-sand mixture (30 wt%, dry density 1.6 Mg m^{-3})
a_1	$1.76\times10^{-8}T - 3.04\times10^{-7}$	$2.99\times10^{-8}T - 3.74\times10^{-7}$
a_2	$-1.48\times10^{-7}T + 2.98\times10^{-6}$	$-1.50\times10^{-8}T + 1.49\times10^{-7}$
b_1	-3.68×10^{-3}	-2.49×10^{-3}
b_2	$5.22\times10^{-3}T + 2.68\times10^{-1}$	$5.59\times10^{-4}T + 3.93\times10^{-1}$
θ_s	0.333	0.403

4 CONCLUSIONS

This paper describes the hydraulic properties of bentonite as buffer on under unsaturated condition during water migration. A laboratory test was carried out to investigate the effect of hysteresis and temperature on suction measurements in bentonite using psychrometer. The isothermal water diffusivity and thermal water diffusivity estimated from the distributions of water content in bentonite under isothermal and thermal conditions were compared with conventional studies.

The following conclusions are drawn from the studies :
1 The hysteresis of suction in compacted bentonite was smaller than that of powder bentonite.
2 The suction in bentonite depended on temperature and content ratio of montmorillonite.
3 The van Genuchten model was in fair agreement with measured water retention curve of compacted bentonite.
4 The isothermal water diffusivity estimated from water distribution with time and temperature depended on the volumetric water content, and could be expressed as U-shaped curves relations.
5 The relationship between volumetric water content and isothermal water diffusivity could be represent as the sum of two hyperbolic functions.
6 The thermal water diffusivity was largely independent of effect due to volumetric water content and dependent of temperature.

These results are available for evaluating phenomena of re-distribution of water through initially unsaturated buffer material during the thermal period due to a rise in temperature rise caused by the radiogenic heating of the vitrified waste after the waste package emplacement, until the buffer eventually reaches a saturated state.

237

REFERENCES

Bolt, G.H. 1976. Soil physics terminilogy. Bull. Int. Soil Sci. 49:26-36

Constantz, J. 1983. Laboratory analysis of water retention in unsaturated materials at high temperature. in Role of the Unsaturated Zone in Radioactive and Hazardous Waste Disposal, J.W.Mercier et al., (eds.), Ann Arbor Science, Ann Arbor, MI:147-164

Edlefsen, N.E., and Anderson, A.B.C. 1943. Thermodynamics of soil moisture, Hilgardia, 15:31-298.

Ishikawa, H., Amemiya, K., Yusa, Y., and Sasaki, N. 1990. Comparison of Fundamental Properties of Japanease Bentonite as Buffer Material for Waste Disposal, Proc. of the 9th International Conference, Sci. Géol., Mém., 87, pp.107-115 (in Japanese).

Ito, M., Okamoto, M., Shibata, M., Sasaki, Y., Danhara, T., Suzuki, K., and Watanabe, T. 1993. Mineral Composition Analysis of Bentonite, Power Reactor and Nuclear Fuel Development Corporation, PNC TN8430 93-003:41 (in Japanese).

Jury, W.A., Gardner, W.R., and Gardner, W.H. 1991. Soil physics (fifth edition). John wiley & Sons, Inc., New York

Lajudie, A., Raynal, J., Petit, J.-C., and Toulhoat, P. 1996. Clay-based Materials for Engineered Barriers : a Review, Mat. Res. Soc. Symp. Proc., 353:221-229.

Nakano, M., Amemiya, Y., Fujii, K., Ishida, T., and Ishii, Y. 1982. Infiltration and Volumetric Expansion in Unsaturated Clay, Transactions of the Jpn. Soc. of Irrigation, Drainage and Reclamation Eng., 100:8-16 (in Japanese).

Onofrei,C. and Hnatiw,D. 1997. VALUCLAY Database Quality Assurance Reports : Japanese Data Volume II.

Oscarson, D.W. and Cheung, S.C.H. 1983. Evaluation of Phyllosilicate as a Buffer Component in the Disposal of Nuclear Fuel Waste, AECL-7812.

Peck, A.J. 1960. Change of moisture tension with temperature and air pressure: theoretical. Soil Sci., 89:303-310

Philip, J.R., and de Vries, D.A. 1957. Moisture movement in porous materials under tempereture dradient, Trans. Am. Geophys. Un. 38:222- 238.

Poulovassilis, A. 1973. The hysterisis of pore water in presence of non-independent water element, Ecol. Stid. Anal. Synth., 4:161-179

Power Reactor and Nuclear Fuel Development Corporation 1992. Research and Development on Geological Disposal of High-Level Radioactive Waste – First Progress Report -, PNC TN1410 93-059 (in Japanese).

Power Reactor and Nuclear Fuel Development Corporation 1996. Present Status of Research and Development for Geological Disposal, PNC TN1410 96-071 (in Japanese).

Pusch, R. 1983. Use of Clay Buffers in Radioactive Repositories, KBS TR 83-46.

Radhakrishna, H.S., Lau, K.-C., Kjartanson, B.H., and Crawford, A.M. 1992. Numerical modelling of heat and moisture transport through bentonite-sand buffer. Can. Geotech. J., 29:1044-1095

van Genuchten, R. 1978. Calculating the unsaturated hydraulic conductivity with a new closed-form analytical model, Research Report, No.78-WR-08, Princeton Univ.

Young, R.N., and Warkentin, B.P. 1975. Introduction to Soil Behavior, The Macmillan Company, New York

Clay Science for Engineering, Adachi & Fukue (eds) © 2001 Balkema, Rotterdam, ISBN 90 5809 175 9

A study on coupling behavior of seepage and swelling of sand/bentonite mixture soil

Y. Sun
Kiso-Jiban Consultants Company Limited, Tokyo, Japan

M. Nishigaki
Okayama University, Japan

ABSTRACT: In this paper, firstly, a volume strain equation is presented that can calculate the volume strain induced by water seepage and stress change. Secondly, an equation of void ratio(e)-suction(Su)-degree of saturation(Sr) had been established in according to test results. By those equations, the coupling behavior of clayey soil between water seepage and swelling strain active can be calculated. Those equations could be used in design of nuclear waste disposal engineering to estimate the varies of stress, strain and seepage behavior of subsoil of disposal facilities, or a buffer material in complicated environment conditions. Finally, the measurement and calculating results using those equations for a infiltration test by soil column method have been compared. It shows that the equations presented in this paper were useful.

1 INTRODUCTION

In nuclear waste disposal engineering, the mixture soil of sand/bentonite would be generally used as subsoil for disposal facilities, or as a buffer material for the underground storage of the nuclear waste. In high temperature and high pressure active, the behavior of this material will have obvious change. For example, there is swelling of bentonite clay or mixture soil of sand/bentonite during seepage process. In other way, the results of pF test in different void ratio showed that soil-water characteristic curve varied with void ratio change or volume change. It means that there exit the mutual influence between volumetric strain and seepage behavior of bentonite clay or mixture soil of sand/bentonite. Therefore, it is necessary to study the coupling behavior among volumetric strain, water content and suction in design of nuclear waste disposal engineering or other engineering problem.

Fredlund, D.G(1993,1998) had presented a constitute model for describing the stress state and volume change behavior of an unsaturated, expansive soil during the swelling process. And the reasonable of this model had been verified. Alonso.E (1998) had developed a elastic-plastic constitutive model (BBM) for unsaturated soil, it can also be used to predict the swelling behavior of soil. But there are some parameters in those model which need to be determined by tests of controlling suction (Su) and suction is as a stress in the model. As known, suction is difficult to be measured and determined in the field due to hysteresis behavior in

seepage and drain processes, therefore, it will have influence for application of those models.

In this paper, firstly, a constitutive equation for volume strain is presented that can reflect swelling behavior of soil. The constitutive equation is considered as a function of degree of saturation (Sr or suction Su), initial void ratio(e_0)and stress(σ). The volumetric strain of mixture soil included both swelling party induced by water infiltration and compression party induced by stress change. By this constitutive equation, the volume strain induced by water infiltration and stress change can be easily calculated, respectively. Secondly, a equation of void ratio(e)-suction(S_u)-degree of saturation(S_r) had been established in according to test results. By those constitutive equations, the coupling behavior of clayey soil between water seepage and swelling stress active can be calculated. Those equations could be used in design of nuclear waste disposal engineering to estimate the varies of stress, strain and seepage behavior of subsoil of disposal facilities, or a buffer material in complicated environment conditions. Finally, the test results and calculating results using those equations have been compared. It shows that the equations presented in this paper are useful.

2. VOLUME STRAIN CONSTITUTIVE OF SWELL SOIL

2.1 *Volume Strain Equation*

In this paper, the volumetric strain of unsaturated soil is considered as the volume strain sum induced by both stress and water movement. The volumetric strain can be written simply as following equation.

$$d\varepsilon_v = de / (1+e_0) \qquad (1)$$

where:
de is increase value of void ratios; e_0 is initial void ratios of soil. When considering swelling or collapse volume strain, the increase value of soil void ratios can be modified as following equation.

$$d\varepsilon_v = d\varepsilon_{vp} + d\varepsilon_{vw} \qquad (2a)$$

$$de = de_p + de_w \qquad (2b)$$

$$d\varepsilon_v = de_p / (1+e_0) + de_w / (1+e_0) \qquad (3)$$

The volumetric strain caused by stress is considered as a elastic strain and can be calculated by .

$$d\varepsilon_{vp} = (3(1-2\mu)/E)dp \qquad (4)$$

where:
de_p= increase of void ratio caused by stress change;
de_w= increase of void ratio caused by degree of soil saturation change.
μ = Poisson's ratio
E = elastic modulus of soil.
The degree of soil saturation (S_r) is determined by following equation.

$$S_r = (\theta - \theta_r) / (\theta_s - \theta_r) \qquad (5)$$

where:
θ = volumetric water content of soil.
θ_s = volumetric water content of soil in saturated state.
θ_r = volumetric water content of soil in drying state.
The volumetric strain has been divided into two parts, the volumetric strain caused by stress can be calculated as the elastic strain as above-mentioned. A hyperbolic curve relationship between degree of saturation Sr and void increase de$_w$ is suggested to calculate volumetric strain caused by water content change in soil. As known, the swell or collapse behavior of soil is connected with pressure change acted on the soil. Therefore, the parameters of hyperbolic curve are also written as the function of pressure. Figure 1 is a schematic diagram of swelling strain and water content in different pressure. The relationship among degree of saturation, void increase of soil and pressure is showed as following.

$$de_w / (1+e_0) = dS_r / (a'+b'dS_r) \qquad (6a)$$
where:

a' and b' are parameters of hyperbolic curve and function of stress which can be determined by swell

experiment in different pressure. From those test results a' and b' can be described as a line function of stress and determined in reference of hyperbolic curve as shown in Figure 2.

$$a'(\sigma)=A_1P+A_2; \qquad (6b)$$

$$b'(\sigma)=A_3P+A_4 \qquad (6c)$$

where:
P= total normal stress ($\sigma_1+\sigma_2+\sigma_3$)/3.
A_1-A_4 = test parameters

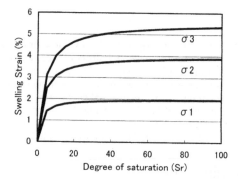

Figure 1. Schematic of volumetric strain (swell & collapse) and water content in different pressures.

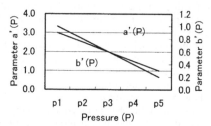

Figure 2. The relationship of swell volumetric parameters a'(σ) and b'(σ) and pressure.

2.2 A Calculating Example on Swelling Strain Change in Different Pressure

The swell volumetric strain can be determined by the variation of stress and volumetric water content. The parameters of volumetric swelling strain constitution can be determined in reference of method of Figure 2 from the results of mixture soil test. For checking those equations, a calculating example had been performed. Test data had been presented by Brackley,I.J.A (1975). The test results and calculation results are shown in Figure 3. From comparing results, it is understand that the swell strain constitutive presented in this paper can reflect the swell volumetric strain behaviors of clayey soil.

The swelling behavior of bentonite clay or the mixture soil of sand/bentonite had been researched by many researchers. Researching results showed that the swelling pressure of mixture soil of sand/bentonite increased with the proportion of bentonite clay increase. In this paper, the mixture soil of sand(85%)/bentonite(15%) is only used as researching material. Its swelling pressure is about in 0.05-0.05 MPa. It will be used in coupling calculation of swelling strain and water infiltration.

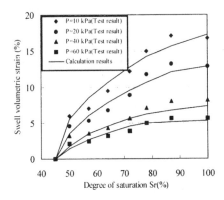

Figure 3. The comparing of test and calculation results about the variation of swell volumetric strain of clayey soil in different pressure (Test data from Brackley, I.J.A 1975).

3. EQUATIONS OF VOID RATIO-SUCTION–DEGREE OF SATURATION

Because the swelling of soil will influence the seepage behavior of soil and increase the danger of seepage failure in soil structure, it is important to establish the relationship between swelling stain and water seepage. As known, the permeability of soil will increase with void ratio increase in swelling process. In other way, when permeability of soil is improved, the swelling strain speed of soil would be increased. Therefore, the interrelation of permeability behavior and volumetric strain variation of soil can be analyzed by the equation of void ratio(e)-suction(S_u)-degree of saturation(Sr).

3.1 Test Results of Water Content Curve in Different Void Ratio

A number of pF tests had been performed in different void ratio (e) for clayey soil and mixture soil. The test results have been showed in Figure 4 and Figure 5. It showed that pF test curve of soil will have variations with respected to different void ratio. When void ratio(e) become large, the suction value in related to saturated water content(θ_s) of pF test curve or air entry value will decrease and the residual volumetric water content(θ_r) of pF curve will also decrease. It means that swelling volumetric strain of clayey soil will increase the permeability of clay. Those test results would be used to set up a equation among volumetric water content(θ), suction(S_u) and void ratio(e).

Compared Figure 4 and Figure 5, it can be found that the variation of pF test curve with respect to void ratio is small in sandy soil (for example granite soil) and large in clayey soil. Because it is difficult to achieve the pF test curve of wetting process for clayey soil, the test results of drying process are used in the paper. When the hysteresis behavior between drying and wetting processes can not be ignored, the pF test curve of the wetting process in different void ratio have to be achieved from pF test. This will be an important problem in the future research.

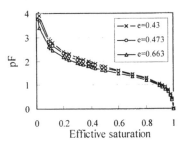

Figure 4 pF test results of a weathered granite soil in different void ratio

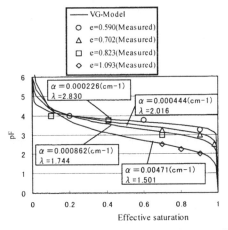

Figure 5 pF test results of a mixture soil of sand/bentonite(15%) in different void ratio (Tata, H 1994)

3.2 Equation of Void Ratio(e) – Suction(Su) – Effective Degree of Saturation(Sr).

The many equations had been presented to describe the pF test curves or soil-water characteristic curves in the past. It had been verified that the equation presented by van Genuchten (1980) is very reasonable for several kinds of soil. Therefor, the van Genuchten model (called as VG model simply) was used to describe the soil-water characteristic curve in this paper. The equations are written as following.

$$S_w = \{1.0 + |\alpha(e)S_u|^{\beta(e)}\}^{-m(e)} \tag{7}$$

where:

m=1-1/β ; α =f(e); β =f(e).

Parameters α , β can be considered as function of void ratio(e). Test results showed that parameter α and β were linear function of void ratio for sandy soil and non-linear function of void ratio(e) for clayey party.

For weathered granite soil, parameters can be expressed as following.

$$\alpha = D_1e+D_2; \quad \beta = D_3e+D_4 \tag{8}$$

where:

D_1, D_2, D_3 and D4 are laboratory experiment parameters. For the mixture soil of sand /bentonite, parameters α and β can be expressed as following.

$$\alpha(e) = D_1 \times 10^{D_2e-D_3} \tag{9}$$

$$\beta(e) = (D_4/(e-D_5)) + D_6 \tag{10}$$

where:

D_1, D_2, D_3, D_4, D_5 and D_6 are laboratory experiment parameters. In other way, the equation among the relative permeability (K_r) and the effective degree of saturation or volumetric water content(θ) is expressed as following.

$$K_r = K/K_S = S_e^{1/2}\ \{1-(1-S_e^{1/m})^m\}^2 \tag{11}$$

$$logKs=D_7e-D_8 \tag{12}$$

where:

K_s is permeability of soil in saturated state. D_7 and D_8 are laboratory experiment parameters from permeability test of saturation state in different void ratio. Those equations are considered in coupling calculation of seepage and deformation.

In this paper, the calculating equations of volumetric strain and soil-water characteristic curve considering the influence of void ratio variation have been combined to analysis the coupling behavior of clayey soil in seepage process. The calculation steps can be concluded as following.

(1) To calculate initial stress field and initial water content of subsoil;

(2) To calculate swelling pressure of soil by the relationship among swelling pressure, water content and clay content in soil;

(3) To calculate the variation of stress and water content of soil due to variation of stress boundary and water boundary as like groundwater table.

(4) To calculate the volumetric strain of soil in according to the volumetric strain equations included degree of saturation (S_r) and pressure.

(5) To calculate the change of permeability of soil by VG model considering the influence of void ratio in according to results of Step (4).

(6) To repeat steps (4) and (5) till calculating results less than the permit error value.

(7) To repeat steps (3) – (6) till end of calculating time.

4. THE COUPLING BEHAVIOR OF CLAYEY SOIL BETWEEN WATER SEEPAGE AND SWELLING STRESS ACTIVE

In this section, a calculating example is given for showing the equations as reasonable and calculating method presented in this paper.

4.1 Seepage Test of Soil Column Method for the Mixture Soil of sand/Bentonite

In nuclear waste disposal engineering, the mixture soil of sand/bentonite would be generally used as subsoil for disposal facilities, or as a buffer material for the underground storage of the nuclear waste. As known, there is an obvious hystersis effect in soil-water characteristic curve. For determining soil-water characteristic curve in infiltration process, it need a infiltration test by soil column method. Some tests has been performed in laboratory (Tata,H 1994). The mixture soil consists of sand (85%) and bentonite (15%), and its dry bulk density in soil column is 15.696 KN/m³. Initial water content is 8% and initial void ratio e_0 is 0.634. Water table level is maintained in 0.0m and let water infiltration from bottom of soil column. Test model and initial condition are showed in Figure 6.

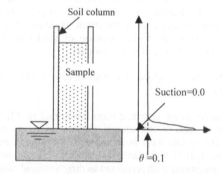

Figure 6. Schematic diagram of soil column seepage test model and the initial and boundary condition.

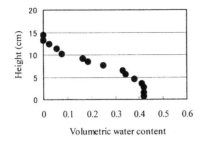

Figure 7. Measurement results of volumetric water content in soil column (after 7 months).

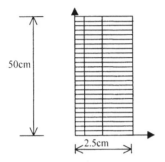

Figure 8   F.E.M model of soil column seepage test

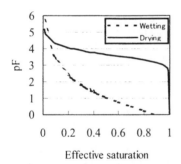

Figure 9 The soil-water characteristic curve in drying and wetting for clayey soil.

Because the permeability of mixture soil is much smaller, it will cost a long time to infiltration from bottom to top layer in soil column. Distribution of volumetric water content after 7 months was measured. The measurement result is showed in Figure 7. In the soil column test, because the pressure born in soil is only dead weight of soil, according to swelling test results of mixture soil in different proportion of bentonite clay, it can be understood that swelling pressure of mixture soil of sand (85%)/bentonite(15%) is larger than 0.05MPa. It means that weight of soil in soil column test is smaller than the swelling pressure of mixture soil in this test case. The mixture soil will produce some swelling strain during infiltration process. Test re-

sults of Figure 7 showed that volumetric water content after 7 months had changed from initial value 0.400 to 0.420 in saturated state due to influence of swelling strain.

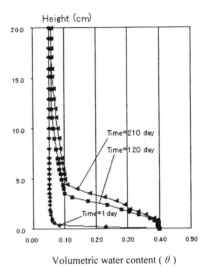

Figure 10. The distribution of volumetric water content (only considering seepage)

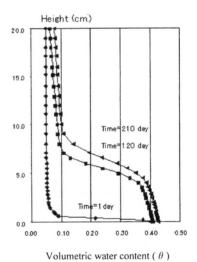

Figure 11. The distribution of volumetric water content (considering coupling effect of swell and seepage)

Table.1 Parameters of calculation

S_r- θ -	D_1	D_2	D_3	D_4	D_5	D_6	D_7	D_8
e-K_r	6.42	2.61	6.0	0.19	0.47	1.21	7.48	11.6
d ε ,	E^*	μ	e_0	A_1	A_2	A_3	A_4	w_0
	500	0.33	0.63	-0.2	3.2	-0.2	1.2	8.6

*MPa

For discussing the coupling effect of seepage and swelling, numerical method is used to simulate infiltration process in soil column test. The F.E.M model of axial symmetry is shown in Figure 8. The initial and boundary conditions of F.E.M. model are set up in reference of Figure 6. The equations presented above sections about volumetric strain and soil water characteristic curve had been used in analysis. The calculation parameters of those equations are collected in Table 1. The difference of drying and wetting process of soil water characteristic curve is increase with the clay content increase. The soil water characteristic curve in wetting side is difficult to achieve for mixture soil. A calculation method of estimating the wetting process of soil water characteristic curve was suggested in the paper (Tata, H 1994). The estimating result about the mixture soil is shown in Figure 9. By this result, the wetting process curve of soil water characteristic curve is applied in seepage analysis.

4.2 Coupling Effect of Seepage and Swell Strain

Two kinds of F.E.M analysis have been performed. One is only saturated-unsaturated seepage analysis in which the variation of permeability of soil with swelling strain have not been considered The other one is a coupling analysis. The swelling volumetric strain and void ratio(e)-suction-volumetric water content(θ) has been included in the constitutive of unsaturated soil presented by author, and a program considering the coupling of seepage and stress has also been developed. The many calculating examples have been performed about coupling behavior of soil between water seepage and swelling stress active in last research (Y.Sun et al 1997).

The calculation results about distribution of volumetric water content in different time are shown in Figure 10 and Figure 11. The results of saturated-unsaturated seepage analysis as shown in Figure 10 are different from the measurement results, since the distribution of volumetric water content in sample of soil column is less than measurement results. The calculation results of considering coupling effect between swell and seepage are shown in Figure 11. Compared Figure 11 with Figure 7, it can be found that the change of volumetric water content in saturated state can be simulated very well after infiltration 7 months. It also reveals that the coupling analysis can produce reasonable result in the case of swelling strain in the soil.

From those results, it can be concluded that although the permeability of mixture soil is very low, the seepage behavior of soil will be increased by the action of swelling strain. It should decrease the effectiveness of mixture soil used as a buffer in nuclear waste disposal facility. Therefore, it is important to consider the coupling effect of swelling and seepage of mixture soil for design of nuclear waste disposal facility.

5.CONCLUSIONS

1. The volumetric strain equation and soil-water characteristic curve equation considered the void ratio change equations presented in this paper could be used to estimate the change of swelling strain and seepage behavior of swelling soil
2. The simulation results about infiltration process of soil column test showed the calculation methods for considering coupling effective of swelling strain and seepage of mixture soil, and they proved to be reasonable.

REFERENCES

Alonso, E.E 1998. Modelling expansive soil behaviour. *Proceeding of the second international conference on unsaturated soils*, Volume 2, pp.37-70. 27-30 Beijing.

Brackley, I.J.A 1975. Swell under load, *Proc. 6th. Reg. Conf, For Africa, SMEF,* Vol.1. pp.65-70, Durban.

Fredlund, D.G. & Rahardjo,H.1993. *Soil mechanics for unsaturated soils*. John Wiley & Sons, Inc.

Tata, H 1994. The research on the seepage behavior of clay. Master thesis of Okayama University (in Japanese*)*.

Research group of soil physics 1979. *Soil physics -soil mechanics base.* Morikita printing co.Ld, (in Japanese)

Shuai,F & Fredlund, D.G. 1998. Theoretical model for the simulation of the swelling process and its application. *Proceeding of the second international conference on unsaturated soils,* Volume 2, pp.509-514, Beijing.

Sun,Y. et al. 1997. Application research on a numerical model of two-phase flow in deformation porous medium. *Proceedings of the ninth international conference on computer methods and advance in geomechanics*. pp.1171-1176.Wuhan.

Van Genuchten,M.T 1980. A closed-form equation for predicting the hydraulic conductivity of unsaturated soils. *Soil Sci. Soc. Am. J.* 44, pp892-898.

Clay Science for Engineering, Adachi & Fukue (eds) © 2001 Balkema, Rotterdam, ISBN 90 5809 175 9

Hydromechanical behavior of a compacted bentonite-silt mixture

Ph. Tabani & F. Masrouri
Laboratoire Environnement, Géomécanique et Ouvrages, Ecole Nationale Supérieure de Géologie, Vandœuvre-lès-Nancy, France

S. Rolland & D. Stemmelen
Laboratoire d'Energétique et de Mécanique Théorique et Appliquée, Ecole Nationale Supérieure d'Electricité et de Mécanique, Vandœuvre-lès-Nancy, France

ABSTRACT: In this study, the results of laboratory infiltration tests performed on unsaturated swelling bentonite-silt specimen, made by static compaction are presented. A non-destructive technique, dual-energy gamma ray attenuation, is used to measure the time variation of water content and the dry bulk density of various sections. The identification of the parameters of a previously selected diffusivity model is performed by inverse method. The influence of swelling and the mechanical boundary conditions on the hydraulic conductivity of the unsaturated soil is clearly highlighted and discussed.

1 INTRODUCTION

Domestic or industrial wastes, generally buried in the natural medium, create contaminants which are likely to infiltrate the soil and pollute the ground water. Two general solutions may be used to avoid groundwater contamination by waste disposal. The first solution is to bury waste in thick deposits of impervious soils or rocks. The second one is to design hydraulic barriers. Horizontal or sloping barriers called liners may be built either with artificial or natural materials such as geomembranes or compacted clays (Auvinet and Hiriart, 1980 ; Day and Daniel, 1985), or soil-bentonite mixtures (Chapuis, 1981). In order to find the bentonite content of these mixtures, effective hydro-mechanical tests (rate and pressure of swelling, permeability) are necessary to determine the appropriate degree of imperviousness.

This paper deals with hydric transfer in swelling unsaturated soils. The influence of swelling on the hydrodynamic properties of the soils and the hydromechanical coupling in these media is discussed.

2 MATERIAL CHARACTERISTICS

The swelling material is a mixture of a bentonite (60 %) and a silt (40 %) in order to respect the limits of our laboratory devices. It is statically compacted to a pressure of 1000 kPa. It has a swelling rate ($\Delta h/h$) of 26,6 %, a swelling pressure of 630 kPa. The optimum Proctor parameters of this mixture correspond to w_{opt}=26% and γ_{dopt}=15 kN/m³.

3 EXPERIMENTAL TECHNIQUE AND TEST METHOD

Various methods can be used to study hydrous and mechanical properties of a porous medium. Factors such as parameters, precision, time, reproducibility, space resolution allow to choose between techniques. The dual-energy gamma ray attenuation system is used in order to study the diffusivity which have an influence on partially saturated swelling porous media.

3.1 *Experimental device*

The understanding of transport processes in swelling soils and materials is hampered by the difficulty of accurately measuring both the volumetric water content and the dry density during the infiltration process. Dual-energy gamma ray attenuation (Nofziger and Swartzendruber, 1974 ; Angulo, 1989) is currently used and it can measure these two parameters in order to characterize the hydro-mechanical properties of an unsaturated soil.

These measurements were made by using two gamma radiation sources : Americium[241] and Cesium[137]. A detector permits to count the number of photons passing through the soil specimen. The radiation attenuation is linearly related to the amount of material within its path. In the case of a water-soil system, a Lambert-Beer type attenuation equation for each gamma energy band was used to determine the attenuation N (photon numbers which go through the soil during 10 minutes) :

$$N = N_0 \exp\left[-\left(\mu_s \rho_s \theta_s + \mu_w \rho_w \theta_w\right)L\right] \tag{1}$$

where the subscripts s et w stand for solid and water phases, respectively, μ is the attenuation constant, θ is the volumetric constant, L is the thickness of the column and N_0 is the attenuation due to the column walls and air between the detectors.

Two independent measurements are needed to study the movement of solid and water particles. They can be obtained from (1) by the radiation attenuation measurements with two different energies: the high-energy attenuation μ_{Cs} and low-energy attenuation μ_{Am}. The volumetric solid and water contents can then determined as :

$$\rho_d = \frac{\mu_w^{Cs} \ln\left(N_0^{Am}/N^{Am}\right) - \mu_w^{Am} \ln\left(N_0^{Cs}/N^{Cs}\right)}{L\left(\mu_s^{Am}\mu_w^{Cs} - \mu_w^{Am}\mu_s^{Cs}\right)} \tag{2}$$

$$\theta_w = \frac{-\mu_s^{Cs} \ln\left(N_0^{Am}/N^{Am}\right) + \mu_s^{Am} \ln\left(N_0^{Cs}/N^{Cs}\right)}{\rho_w L\left(\mu_s^{Am}\mu_w^{Cs} - \mu_w^{Am}\mu_s^{Cs}\right)} \tag{3}$$

where ρ_d is the bulk density (m_s/V_t).

These two attenuation equations can be solved simultaneously if the measured intensities of the two gamma rays are independent.

The lead shield holding the sources was placed by step motor. Count rates were measured by a single detector, and a multi-channel spectrometer (Fig. 1).

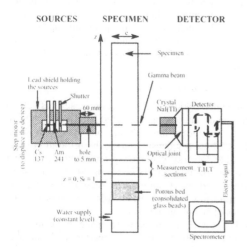

SOURCES   SPECIMEN   DETECTOR

Fig. 1   Experimental device

3.2 *Experimental method*

Infiltration tests were carried out on soil specimen. The soil was compacted in a plexiglass tube which walls were previously lubricated and introduced into an oedometric stainless cell equipped with oblong lights for the gamma beam. A strain gage placed at the top of the specimen allowed to follow the axial

deformations. The specimen was placed on a fixed support, between the source and the detector, on a porous plate to ensure constant saturation at the lower boundary during the experiment. During the test, there was no axial load on the specimen.

3.3 *Diffusivity analysis*

According to the kinematics framework from the continuous media, two modes of description (lagrangian and eulerian type), can be used. In the first case, any physical parameter of a given phase is evaluated by following each particle in relation with its initial position. Therefore the variables are material or of Lagrangian type. In the second case time variation of a particle in relation with a geometrical point are considered; thus the variables are geometrical or of Eulerian type.

The dual-gamma system allows the measurement of eulerian data. In this case, a diffusion equation (assuming that the gravity can be neglected) should be written in terms of volumetric water content :

$$\frac{\partial\theta}{\partial t} = \frac{\partial}{\partial z}\left[D(\theta)\frac{\partial\theta}{\partial z}\right] \tag{4}$$

where θ is the volumetric water content, t is the time, z is the space variable, $D(\theta)$ is the apparent diffusivity of water in relation with the observer reference frame.

Applying the change of variable :

$$\lambda = \frac{z}{\sqrt{t}} \tag{5}$$

λ = Boltzmann variable.

The partial derivative equation is replaced by a second order total derivative equation :

$$\frac{d^2\theta}{d\lambda^2} = \frac{-1}{D}\left[\frac{dD}{d\theta}\left(\frac{d\theta}{d\lambda}\right)^2 + \frac{\lambda}{2}\frac{d\theta}{d\lambda}\right] \tag{6}$$

coupled with the following boundary conditions :

$\theta = \theta_{sat}$ when $\lambda = 0$
$\theta = \theta_{ini}$ when $\lambda \to \infty$

Experimental results showed that this analysis can be used in the present case, since the variation of the volumetric water content at different sections was identical when plotted *versus* the Boltzmann variable. This change of variable allows to calculate the diffusivity.

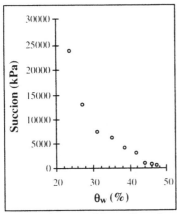

Fig. 2   Water retention curve of the swelling soil

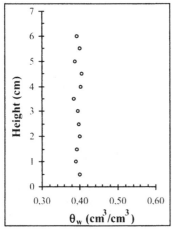

Fig. 3   Initial water content profiles of the soil column

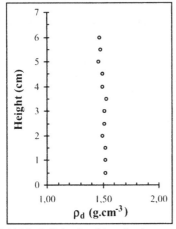

Fig. 4   Initial dry bulk density of the soil column

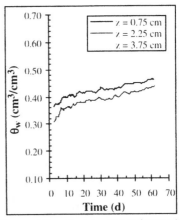

Fig. 5   Time variation of water content at three sections of the column

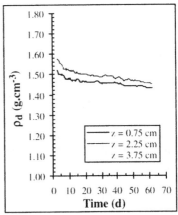

Fig. 6. Time variation of dry bulk density at three sections of the column

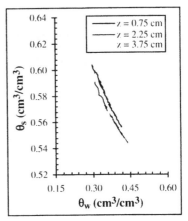

Fig. 7   Solid content versus water content

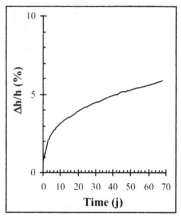

Fig. 8   Time variation of swelling rate of the specimen

In swelling soils, the water flux $q_{w/s}$ given by Darcy's law can be calculated with water $q_{w/o}$ and solid $q_{s/o}$ fluxes, expressed in a fixed (eulerian) co-ordinate system from the following equations (Angulo, 1989):

$$q_{w\,s} = q_{w\,o} - \left(\frac{\theta_w}{\theta_s}\right) q_{s\,o} \qquad (8)$$

Neglecting gravity, the fluxes $q_{s/o}$, $q_{w/o}$ and $q_{w/s}$ are given from Darcy's law by the following equations:

$$q_{s\,o} = -D_{s\,o}\left(\frac{\partial\theta_s}{\partial z}\right) \qquad (9)$$

$$q_{w\,o} = -D_{w\,o}\left(\frac{\partial\theta_w}{\partial z}\right) \qquad (10)$$

$$q_{w\,s} = -D_{w\,s}\left(\frac{\partial\theta_w}{\partial z}\right) \qquad (11)$$

where $D_{s/o}$ and $D_{w/o}$ are the apparent diffusivities of solid particles and water, in the eulerian coordinate system and $D_{w/s}$ is the diffusivity of water associated with the solid phase.

Substituting (9), (10) and (11) into (8), one can obtain the hydraulic diffusivity relative to the solid phase, $D_{w/s}$, as:

$$D_{w\,s} = D_{w\,o} - \left(\frac{\theta_w}{\theta_s}\right)\left(\frac{d\theta_s}{d\theta_w}\right) D_{s\,o} \qquad (12)$$

The apparent diffusivities $D_{s/o}$ and $D_{w/o}$ are calculated by identification of the exponential model parameters. a_s and b_s of $D_{s/o}$ are estimated from Boltzmann curve $\lambda(\theta_s)$, a_w b_w of $D_{w/o}$ are also estimated from Boltzmann curve $\lambda(\theta_w)$ (Table 1). The apparent diffusivity models are validated by solving the second order total derivative equation (6) and by plotting time variation of solid and water content profiles (Figs. 9 and 10).

The hydraulic diffusivity versus water content is presented in figure 11. The hydraulic conductivity versus the explored volumetric water content is shown in Fig. 12.

Various remarks can be made according to these test results:

- Water and solid content variations versus Boltzmann variable for various measurement sections show that the curves are not correctly fitted (Fig. 9 & 10).

This observation could be explained by the lower boundary condition at $\lambda = 0$ of water content $\theta_w = \theta_{sat}$, this parameter can be variable because of the swelling of the soil. However, the resolution of the second order total derivative equation by applying

An exponential model, or Gardner model (1958) which involves two parameters a and b is combined with equation (6):

$$D(\theta) = a\,\exp(b\theta) \qquad (7)$$

The identification of a and b soil parameters is performed by inverse method with a large number of experimental points.

Hydraulic conductivity $k(\theta)$ was determined by the combination of $D(\theta)$ and the water retention curve $s(\theta)$ (Fig. 2).

4 RESULTS AND DISCUSSION

A specimen with a diameter of 6 cm and a height of 7 cm was selected for the test. Initial water content and bulk density profiles are presented in Figs. 3 and 4.

An infiltration experiment was carried out with a swelling soil 24 hours after compaction. Time variation of water content and bulk density for three different sections (z = 0.75 ; 2.25 ; 3.75 cm) are presented in Figs. 5 and 6.

The swelling curve $\theta_s(\theta_w)$, which relates water content to solid content, is required to predict the Darcy's hydraulic diffusivity. This curve is obtained from the dual-gamma spectrometry experiment (Fig. 7). Time variation of swelling rate ($\Delta h/h$) measured with the strain gage are presented in Fig. 8.

Table 1 Diffusivity model parameters

a_s (m²/s)	b_s	a_w (m²/s)	b_w
$4{,}52.10^{-14}$	21,36	$1{,}85.10^{-6}$	-12,59

the variable of Boltzmann together with the identification of the model parameters by inverse method allow to validate the selected model and to plot the diffusivity variations.

- Darcy's hydraulic conductivity versus water content decreases when the volumetric water content increases (Fig. 12). This can be explained by a decrease in interaggregate pore space and connectivity due to soil swelling and a microlevel arrangement of soil aggregates, Tabani (1999). Guerrini and Swartzendruber (1992) and Garnier et al. (1998) also observed a similar phenomenon on swelling soils and on some type of dry nonswelling media.

the parameters of the selected model of diffusivity. Experimental results obtained with a silt-bentonite mixture confirm the potential of the dual-gamma ray technique. This device permits a good understanding of the various mechanisms influencing the flow in partially saturated swelling medium. The obtained results coupled with a very simple exponential model allow to estimate the hydromechanical properties of the swelling soils in infiltration tests.

Complementary experiments on the same type of soil, with the control of volume variations (i.e. triaxial infiltration or free swelling tests) are necessary to complete this study.

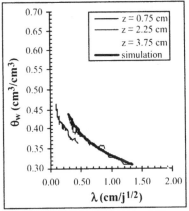

Fig. 9   Water content versus Boltzmann variable : experimental data compared to the fitted curve

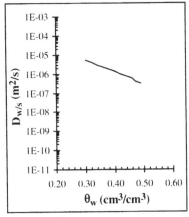

Fig. 11 Hydraulic diffusivity versus water content

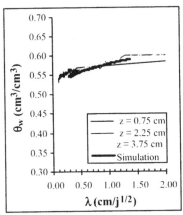

Fig. 10 Solid content versus Boltzmann variable : experimental data compared to the fitted curve

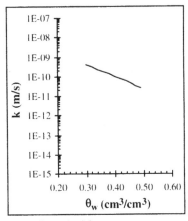

Fig. 12 Hydraulic conductivity versus water content

5 CONCLUSION

The dual-gamma ray attenuation device and an inverse method were used to follow water content and bulk density changes in swelling soils and to identify

BIBLIOGRAPHY

ANGULO R. (1989) : Caractérisation hydrodynamique des sols partiellement saturés. Etude expérimentale à l'aide de la spectrométrie gamma double-sources. Ph. D., dissertation, Inst. Nat. Polytech. de Grenoble, 209 p.

AUVINET G.Y. & HIRIART G. (1980) : An artificial cooling pond for the Rio Escondido coal fired power plant. Proceedings, ASCE Symposium on Surface Water Impoundments, Minneapolis, vol. 2, pp. 1089-1098.

CHAPUIS R.P. (1981) : Permeability testing of soil-bentonite mixtures. Proceedings, 10th International Conference on Soil Mechanics and Foundation Engineering, Stockholm, vol. 4, pp. 744-745.

DAY S.R. & DANIEL D.E. (1985) : Hydraulic conductivity of two prototype clay liners. ASCE Journal of Geotechnical Engineering, 111 : 957-970.

GARDNER W. R. (1958) : Some steady state solutions of the unsaturated moisture flow equation with application to evaporation from a water table. Soil Science, vol. 85, pp.228-232.

GARNIER P., ANGULO R., DICARLO D., BAUTERS T., DARNAULT C. J. G., STEENHUIS T. S., PARLANGE J. Y. & BAVEYE Ph. (1998) : Dual-energy synchrotron X ray measurements of rapid soil density and water content changes in swelling soils during infiltration. Water Resources research, vol. 34, n° 11, pp. 2837-2842.

GUERRINI I.A. & SWARTZENDRUBER D. (1992) : Soil water diffusivity as explicitly dependent on both time and water content, Soil Science Society of America Journal, 56, pp. 335-340.

NOFZIGER D.L. & SWARTZENDRUBER D. (1974) : Material content of binary physical mixtures as measured with dual-energy beam of gamma rays. Journal Applied physics, 45, pp. 5443-5449.

TABANI Ph. (1999) : Transfert hydrique dans des sols déformables. Ph. D., dissertation, Inst. Nat. Polytech. de Lorraine, Ecole Nat. Sup. de Géologie, Nancy, 173 p.

Clay Science for Engineering, Adachi & Fukue (eds) © 2001 Balkema, Rotterdam, ISBN 90 5809 175 9

Numerical determination of coupled flow parameters in unsaturated soil

M. Elzahabi
Department of Civil and Environmental Engineering, University of Carleton, Ottawa, Ont., Canada

R. N. Yong
Geoenvironmental Research Centre, University of Wales, Cardiff, UK

ABSTRACT: A one-dimensional model is developed to analyse and predict the transport of heavy metals in unsaturated clayey soils using the postulates of irreversible thermodynamics. The numerical solution of the governing coupled solute and moisture flow equations were obtained using the implicit finite difference method. The diffusion coefficient was expressed as a function of the volumetric water content and the solute concentration. The diffusion parameters were determined using Powell method for nonlinear optimization technique and were based on the experimental results obtained from laboratory leaching column tests. Results indicated that the presence of the diffusion coefficient is necessary to provide a good agreement between the experimentally measured and theoretically predicted values of contaminant transport through the soil. The numerical results of the coupled solute and moisture equations strongly supported the dependence of the transport coefficients on solute and volumetric water content and showed the reliability of the selected diffusion parameters.

1 INTRODUCTION

Although liner systems are constructed in unsaturated (vadose) soil, most researchers have focused their attention and extensively studied the transport of leachate under saturated conditions while ignoring the effect of the initial insitu soil saturation under acidic conditions. The vadose zone cannot be ignored in the study of contaminant movement because it may be a significant reservoir for capture, storage and release of contaminants to the ground. Where soil is present, the movement of water is largely controlled and may be prevented from continued movement by factors of attraction, the so-called internal gradients (such as capillary and osmotic or adsorption) developed by forces within the soil. A principal characteristic of the vadose zone (water-unsaturated) zone is that the water is held by capillary and adsorption forces and the pore water pressures are generally negative while the water present below the phereatic surface creates a positive hydrostatic pressure (Yong et al. 1992).

Theoretically, water infiltration to the underground is due to gravity effects when pores in the soil are filled with water. Thus, the transport of contaminants will be tied more closely to water movement and the diffusion of contaminants will be seen to be dependent on the water content of the soil and various characteristics and properties that control the internal gradients. Therefore, the degree of water connectivity among the various soil pore classes strongly impacts the hydrologic flux and the mass transfer of contaminants in the system. Contaminant migration from shallow land barrier sites is confined to micropores (matric) regions and most frequently involves unsaturated transport processes (Jardine et al. 1993). During unsaturated conditions, coupled processes of solute and moisture factors control the subsurface transport of contaminants. The extent to which these coupled processes control the movement of contaminants in the vadose zone is largely unknown and questionable in many situations.

Recently, the application of nonequilibrium irreversible thermodynamics to the analysis of coupled flow problems related to transport phenomena in the soil system has been studied by many researchers such as: Taylor & Carry (1962), De Groot & Mazur (1962), Fitts (1962), Olsen (1969,1972), Greenberg et al. (1973), Mitchell et al. (1973), Yong & Samani (1987), Yong et al.(1992), Yeung & Mitchell (1992).

Irreversible thermodynamics are also termed nonequilibrium thermodynamics. It is a theory that can provide a basis for description of a system that is out of equilibrium, (Mitchell 1993). In a system, which is in a non-equilibrium state due to contaminant movement, thermodynamic forces and other forces (advectives) will cause the solutes to move toward the equilibrium state by balancing these forces, (Yong et al. 1992).

However, most existing transport models use an average diffusion coefficient without considering the degree of saturation in order to predict the movement of heavy metals along the soil column. Simplification of the diffusion coefficient as a constant and of the degree of saturation as a linear function cannot be considered a good assumption and may lead to an improper evaluation of the sorption phenomenon in the vadose zone and also to serious errors in predicting contaminant migration. Therefore, 'an adequate characterization of moisture content dependence of the diffusion coefficient is essential for realistic modelling of diffusive mass transport through the unsaturated zone' (Fityus et al. 1999).

In order to minimize the above problem, analysis and prediction of coupled solute and moisture flow through unsaturated clay barriers, based on the dependence of the diffusion and the sorption characteristics on the degree of saturation or the moisture content along the soil depth, is necessary for the design of new disposal facilities. The objective of this study on unsaturated soil-contaminant interaction is, therefore, to develop a method that describes the full coupling effects of solute and moisture on transport coefficients based on experimental evidence and applied to the unsaturated transport theory, with particular attention to the effect of degree of saturation and heavy metals concentrations. In addition, coupled solute and moisture transport parameters of heavy metals in unsaturated clay soil as a function of time and space will be evaluated and predicted based on the proposed method.

2 MODEL DEVELOPMENT

2.1 Assumptions

The following assumptions were considered: 1) one dimensional moisture and solute flow, 2) no volume change, 3) isothermal condition, 4) horizontal flow, 5) neglecting the effect of the gravity potential, 6) validity of Darcy's law (Richards 1931, Klute 1952 and Yong & Warkentin 1975), 7) validity of Fick's (Fick 1855) law (Yong et al.1992), and 8) negligibility of the biological uptake of the minerals.

2.2 Coupled Moisture and Solute Diffusion Equations Development

In a near state of equilibrium, and by applying the second postulate of irreversible thermodynamics (Onsagar 1931), thermodynamic forces and their fluxes can be described by a power or series as follow:

$$J_i \quad L_{ij} X_j \qquad (1)$$

J_i = Rates of Fluxes
X_i = Thermodynamic forces responsible for the fluxes

$$J_\theta = L_{\theta\theta} \frac{\partial \psi_\theta}{\partial x} + L_{\theta C} \frac{\partial \psi_C}{\partial x} \qquad (2)$$

$$J_C = L_{C\theta} \frac{\partial \psi_\theta}{\partial x} + L_{CC} \frac{\partial \psi_C}{\partial x} \qquad (3)$$

L_{ij} = Phenomenological coefficients
$\partial \psi_\theta/\partial x$ = Thermodynamic force due to soil water potential gradient
$\partial \psi_c/\partial x$ = Thermodynamic force due to chemical potential gradient
J_θ = Fluid flux mole/l^2/t
J_c = Solute flux mole/l^2/t
x = Depth of soil column along the direction of flow (x-coordinate), $L_{\theta\theta}$, $L_{c\theta}$ $L_{\theta c}$ L_{cc} are phenomenological coefficients.

The relationship between the chemical potential gradient and concentration gradient is given by (Yong et al. 1992):

$$\frac{\partial \psi_C}{\partial x} = \frac{RT}{C} \frac{\partial(-c)}{\partial x} \qquad (4)$$

R = gas constant, T = absolute temperature, C = solute concentration, t= time.

Also, the relationship between the soil water potential gradient and the volumetric water content gradient is given by (Yong & Warkentin 1975):

$$\frac{\partial \psi_\theta}{\partial x} = - \frac{\partial \psi_\theta}{\partial \theta} \frac{\partial \theta}{\partial x} \qquad (5)$$

Substituting equations 4 and 5 into 2 and 3 to obtain:

$$J_\theta = L_{\theta\theta} \frac{\partial(\psi_\theta)}{\partial \theta} \frac{\partial(-\theta)}{\partial x} + L_{\theta C} \frac{RT}{C} \frac{\partial(-C)}{\partial x} \qquad (6)$$

$$J_C = L_{C\theta} \frac{\partial \psi_\theta}{\partial \theta} \frac{\partial(-\theta)}{\partial x} + L_{CC} c\frac{RT}{C} \frac{\delta(-C)}{\partial x} \qquad (7)$$

The mass conservation equation for the diffusion involving the effect of adsorption reactions can be written as:

$$\frac{\partial \theta}{\partial t} = - \frac{\partial J_\theta}{\partial x} \qquad (8)$$

$$\frac{\partial C}{\partial t} = - \frac{\partial J_c}{\partial x} - \frac{\rho_s}{\theta} \frac{\partial S_c}{\partial t} \qquad (9)$$

where ρ_s = Dry density, S = The adsorbed concentration of solute in the solid phase which is directly proportional to the concentration, θ = Volumetric water content.

Substituting equations 6 and 7 into equations 8 and 9 gives the final one dimensional solute and mass flow equations:

$$\frac{\partial \theta}{\partial t} = \frac{\partial}{\partial x} [L_{\theta\,\theta} \frac{\partial \psi_\theta}{\partial \theta} \frac{\partial \theta}{\partial x} + L_{\theta\,c} \frac{RT}{C} \frac{\partial C}{\partial x}] \qquad (10)$$

$$\frac{\partial C}{\partial t} = \frac{\partial}{\partial x} [L_{C\,\theta} \frac{\partial \psi_\theta}{\partial \theta} \frac{\partial \theta}{\partial x} + L_{C\,c} \frac{RT}{C} \frac{\partial C}{\partial x}] - \frac{\rho_s}{\theta} \frac{\partial S}{\partial t} \qquad (11)$$

Assume :
Moisture diffusivity:

$$D_{\theta\theta} = L_{\theta\theta} \frac{\partial \psi_\theta}{\partial x} \qquad (12)$$

Solute diffusivity:

$$D_{C\,C} = L_{C\,c} \frac{RT}{C} \qquad (13)$$

Solute moisture diffusivity:

$$D_{C\,\theta} = L_{C\,\theta} \frac{\partial \psi_\theta}{\partial \theta} \qquad (14)$$

Moisture solute diffusivity:

$$D_{\theta\,C} = L_{\theta\,c} \frac{RT}{C} \qquad (15)$$

By substituting equations 12 and 15 into equation 10 and 13 and 14 into equation 11, the coupled partial differential equations for one dimensional contaminant transport for unsaturated soil in the boundary layer due to variations in volumetric water content and contaminant concentration can be written as:

$$\frac{\partial \theta}{\partial t} = \frac{\partial}{\partial x} [D_{\theta\theta} \frac{\partial \theta}{\partial x} + D_{\theta C} \frac{\partial C}{\partial x}] \qquad (16)$$

$$\frac{\delta C}{\delta t} = \frac{\delta}{\delta x} [D_{C\,\theta} \frac{\delta\theta}{\delta x} + D_C \frac{\delta C}{\delta x}] - \frac{\rho_s}{\theta} \frac{\delta S}{\delta t} \qquad (17)$$

Furthermore, the relationship between the adsorbed components, S, and concentration, C, can be written as:

$$\frac{\delta C*}{\delta t} = \frac{\delta}{\delta x} [D_{C\,\theta} \frac{\delta\theta}{\delta x} + D_C \frac{\delta C}{\delta x}] \qquad (18)$$

Where C* is the total concentration in mg/100g of soil.

2.3   Finite Difference Formulations

The diffusion coefficient will be calculated for each individual layer in the soil samples and for each pore volume passage of the contaminant by using the implicit finite difference method to solve the one dimensional parabolic second order differential coupled moisture and solute equations (16) and (18). This method is practical because the solution will permit a larger time step and requires less time, which is more economic.

The finite difference formulations are described below as follow:

Dividing the time t and the distance x into j and i intervals, respectively, the left side of equation 16, using the forward difference approximation of order Δt, can be written as follow:

$$\frac{\delta\theta}{\delta t} = \frac{\theta_i - \theta_i}{\Delta t} \qquad (19)$$

Using the second-order central difference approximation of the order of (Δx^2), volumetric water content at time level j+1 can be obtained implicitly from the volumetric water content at time level j as given:

Volumetric water content at time j+1

$$\theta_i^{j+1} = \frac{\Delta t}{2\,\Delta x^2} [\omega[(D_{\theta\,\theta(i+1)} + D_{\theta\,\theta(i)}) (\theta_{(i+1)} - \theta_{(i)})$$
$$- (D_{\theta\,\theta(i)} + D_{\theta\,\theta\,(i-1)}) (\theta_i - \theta_{(i-1)})$$
$$+ (D_{\theta\,C\,(i+1)} + D_{\theta\,C\,(i)}) (C_{(i+1)} - C_{(i)})$$
$$- (D_{\theta\,C\,(i)} + D_{\theta\,C\,(i-1)}) (C_{(i)} - C_{(i-1)})]^j$$
$$+ (1-\omega)[(D_{\theta\,\theta(i+1)} + D_{\theta\,\theta(i)}) (\theta_{(i+1)} - \theta_{(i)})$$
$$- (D_{\theta\,\theta(i)} + D_{\theta\,\theta\,(i-1)}) (\theta_i - \theta_{(i-1)})$$
$$+ (D_{\theta\,C\,(i+1)} + D_{\theta\,C\,(i)}) (C_{(i+1)} - C_{(i)})$$
$$- (D_{\theta\,C\,(i)} + D_{\theta\,C\,(i-1)}) (C_{(i)} - C_{(i-1)})]^{j+1}] + \theta_i^j \qquad (20)$$

and the implicite finite difference form of solute concentration at time j+1 can be written as follows:

$$C_i^{j+1} = \frac{\Delta t}{2\,\Delta x^2} [\omega[(D_{C\,C(i+1)} + D_{C\,C(i)})\,(C_{(i+1)} - C_{(i)})$$
$$- (D_{C\,C(i)} + D_{C\,C(i-1)})\,(C_i - C_{(i-1)})$$
$$+ (D_{C\,\theta\,(i+1)} + D_{C\,\theta\,(i)})\,(\theta_{(i+1)} - \theta_{(i)})$$
$$- (D_{C\,\theta\,(i)} + D_{C\,\theta\,(i-1)})\,(\theta_{(i)} - \theta_{(i-1)})\,]^j$$
$$+ (1-\omega)[(D_{C\,C(i+1)} + D_{C\,C(i)})\,(C_{(i+1)} - C_{(i)})$$
$$- (D_{C\,C(i)} + D_{C\,C(i-1)})\,(C_i - C_{(i-1)})$$
$$+ (D_{C\,\theta\,(i+1)} + D_{C\,\theta\,(i)})\,(\theta_{(i+1)} - \theta_{(i)})$$
$$- (D_{C\,\theta\,(i)} + D_{C\,\theta\,(i-1)})\,(\theta_{(i)} - \theta_{(i-1)})\,]^{j+1}\,] + C_i^j \quad (21)$$

where : $x = i\,\Delta x$ for $i = 1$, I; $t = j\,\Delta t$ for $j = 1$, J and ω is considered 1/2, yielding the Crank-Nicolson implicit method, where the method is considered unconditionally stable for $1/2 \leq \omega \leq 1$.

Equations 20 and 21 are the final differential equations in the finite difference format and are used with the optimization technique to obtain the diffusion parameters.

2.4 Determination of Diffusion Parameters and Unkown Material Coefficients

Prediction of the moisture and concentration profiles at time j+1 can be found numerically once the concentration and moisture profile at time j are measured experimentally and the diffusion function is assumed. As mentioned in the introduction, most existing transport models use an average diffusion coefficient without considering the coupling effects of solute and moisture flows along the soil column. In the unsaturated soil, simplification of the diffusion coefficient and of the degree of saturation as a linear functions cannot be considered a good assumption and may lead to an improper evaluation of the sorption phenomenan. Thus, based on experimental results and for realistic evaluation of diffusive mass transport through the unsaturated clayey soils, the diffusion parameters are expressed as an exponential function of volumetric water content and solute concentration and are described as follow:

$$D_{CC} = a_9 + a_1\,\exp^{-a_{13}C_{N_I}} + a_5\,\exp^{-a_{17}\,\theta_{N_I}} \quad (22)$$

$$D_{C\theta} = a_{11} + a_3\,\exp^{-a_{15}C_{N_I}} + a_7\,\exp^{-a_{19}\theta_{N_I}} \quad (23)$$

$$D_{\theta\theta} = a_{10} + a_2\,\exp^{-a_{14}\,C_{N_I}} + a_6\,\exp^{-a_{18}\,\theta_{N_I}} \quad (24)$$

$$D_{\theta C} = a_{12} + a_4\,\exp^{-a_{16}C_{N_I}} + a_8\exp^{-a_{20}\,\theta_{N_I}} \quad (25)$$

where a_1 to a_{20} are constant material parameters and can be obtained from the optimization procedures. At time j, volumetric water content and concentrations at different depths and times were measured and under the same conditions the volumetric water content and concentrations were predicted numerically at time j+1 using the model described above.

If $C_{exp}(x,t)$ and $C_{cal}(x,t)$ are the measured and the calculated concentrations respectively, and $\theta_{exp}(x,t)$ and $\theta_{cal}(x,t)$ are the measured and calculated volumetric water content, respectively, then the best choice for these material parameters coefficients (a_1 to a_{20}) are those which minimize the following functions:

- For solute concentration:

$$\sigma_1 = \sum_{}^{m} |\,C_{exp}(x,t) - C_{cal}(x,t)\,| \quad (26)$$

- For volumetric water content:

$$\sigma_2 = \sum_{}^{m} |\,\theta_{exp}(x,t) - \theta_{cal}(x,t)\,| \quad (27)$$

where m represents the number of measured and calculated concentrations C and volumetric water content θ, and σ is a function of the unknown material coefficients.

The best way to obtain the minimum of the function σ is to use Powell's conjugate directions method of non-linear optimization (Powell, 1964).

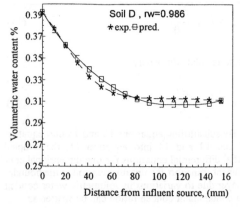

Figure 1. Volumetric water content model calibration.

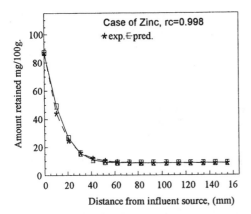

Figure 2. Model calibration for zinc.

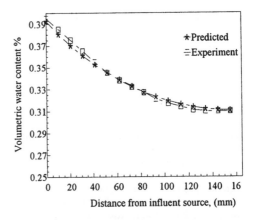

Figure 5. Comparison between predicted and measured volumetric water content after 50 days.

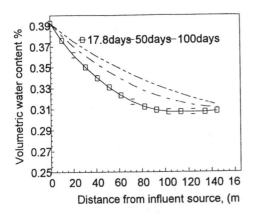

Figure 3. Predicted volumetric water content.

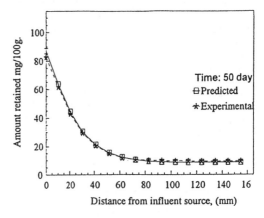

Figure 6. Comparison between predicted and measured concentration profiles.

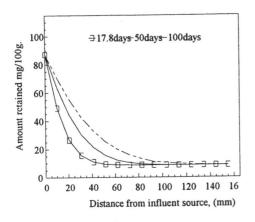

Figure 4. Predicted zinc concentration

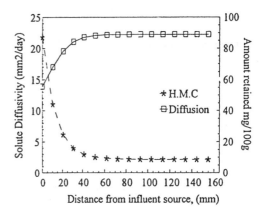

Figure 7 Solute diffusivity and Zinc concentration variations with distance.

So, the derivation of σ with respect to a specific unknown coefficient can be determined in a simple way. This makes Powell's method more useful because it does not require derivatives of the objective function.

2.5   Model Calibration

Laboratory evaluation of heavy metal migration using one dimensional solute (of 5000mg/l of zinc nitrate) and moisture flow (leaching column) tests were conducted on an unsaturated illitic soil at pH 3.5. The migration profiles of heavy metals along the soil column obtained from experiment for each individual layer were used for the calibration of the proposed model. A computer code called the Coupled Moisture and Solute Diffusion Parameter Technique (CMSDPT) has been developed in conjunction with the application of the theory of irreversible thermodynamic, Fick's law, Darcy's law and equilibrium mass transfer principles. Then, the diffusion parameter is calculated for each individual layer in the soil samples using Powell's optimization technique and the implicit finite difference method to solve the coupled diffusion equations.

To calibrate the model, experimentally measured moisture and concentrations were correlated with the corresponding calculated moisture concentrations through the following equations:

For volumetric water content:

$$r_1 = \sqrt{1 - \frac{\sum\limits_{i=1}^{m} [\,\theta_{(Exp.i)} - \theta_{(Calc.i)}\,]^2}{\sum\limits_{i=1}^{m} [\,\theta_{(Exp.i)} - \theta_{Avr.}\,]^2}} \qquad (28)$$

For solute concentration:

$$r_2 = \sqrt{1 - \frac{\sum\limits_{i=1}^{m} [\,C_{(Exp.i)} - C_{(Calc.i)}\,]^2}{\sum\limits_{i=1}^{m} [\,C_{(Exp.i)} - C_{Avr.}\,]^2}} \qquad (29)$$

Where r represents the correlation coefficients, $C_{avr.}$ is the average experimental concentration and $\theta_{avr.}$ is the average experimental volumetric water content. Experimental results from volumetric water content and total heavy metal concentration profiles of zinc after 52 hrs and 17.8 days for illitic soil at pH 3.5

were used to calibrate the model, (for metal partitioning analysis of heavy metals in the liquid (soluble ions) and the solid phases (extractable ions), refer to Elzahabi &Yong 2000). The experimental and the calibrated results calculated by the model for the volumetric water content and heavy metals variations with depth are shown in Figure 1& Figure 2. Experimental moisture and solute data used to calibrate the model, and the initial and boundary conditions are listed below.

The following initial and boundary conditions were used:

- Initial conditions:
  C= 5000mg/l   for $0 \le x \le 165$mm
  $\theta = 0.265$ for $0 \le x \le 165$mm

- Boundary conditions:
  $C = C_0$ for $0 \le x \le 165$mm   t=0
  $\theta = \theta_0$ for $0 \le x \le 165$mm   t=0

  $C = C(x,t)$ for $0 < x < 165$mm   t>0
  $\theta = \theta(x,t)$ for $0 < x < 165$mm   t>0

The following equations were used to calibrate the model between 52 hrs and 17.8 days as listed:

a- Time t=52 hrs.
$\theta_i = 0.289395401 + 0.096475303\ e^{\,(-x\,/\,37.61637976)}$

$C_i = 8.668151957 + 87.32740871\ e^{\,(-x\,/\,3.641068278)}$

b- Time t=17.8 days.
$C_f = 8.618209996 + 78.43227136\ e^{\,-x\,/\,12.9868636}$

$\theta_f = 0.392412593 - 0.00592318\ x + 8.95157e\text{-}05(x^2 /(1-0.01170626x)) + 0.000217252\ x^2 + 2.21355e\text{-}07\ x^3$

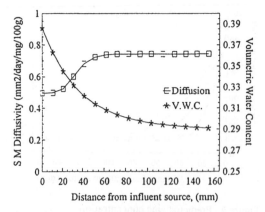

Figure 8 Solute moisture diffusivity and volumetric water content variations with distance.

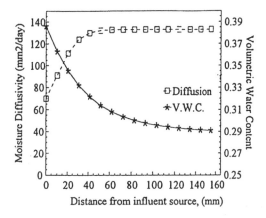

Figure 9 Moisture diffusivity and volumetric water content variations with distance.

2.6 *Validation*

Then, using the computer code CMSDPT, with the known material coefficients and the calculated moisture, solute, solute moisture and moisture solute diffusivity coefficients, and by applying the Implicit finite difference method combined with Powell's optimization technique as mentioned before, one can predict volumetric water content and solute variations along the soil column for different time duration as a function of space and time as shown in Figure 3 & Figure 4.

In order to validate the calculated diffusion parameters, experimentally measured concentration and volumetric water content were compared with the predicted volumetric water content and the concentration under the same conditions to check the capability of the model to predict heavy metals movement for different time duration-- as shown in Figure 5 & Figure 6. Comparison between the measured and the calculated moisture and solute profiles shows a good agreement and increases the level of confidence in predicting coupled solute and moisture profiles in the vadose zone. The calculated solute, solute moisture, and moisture diffusivity parameters as a function of distance, volumetric water content and heavy metal concentration for illitic soil at pH 3.5 are shown in Figure 7, Figure 8 and Figure 9 respectively. Results show that the diffusion coefficient cannot be constant, it is an exponential function of heavy metal concentration and volumetric water content and varies with time, depth of soil, volumetric water content or degree of saturation and heavy metal influent concentration. Simplification of the diffusion coefficient as a constant and of the degree of saturation as a linear function cannot be considered a good assumption and may lead to an improper evaluation of the sorption phenomenon in the vadose zone and also to serious errors in predicting contaminant migration. Therefore, an adequate characterization of moisture content dependence on the diffusion coefficient is essential for realistic modelling of diffusive mass transport through the unsaturated zone. Further examination of the same numerical model was also performed on more representative landfill leachate to better simulate the problem of contaminant transport in the unsaturated (vadose) zone. Results will be presented in a later publication.

3 CONCLUSIONS

1. A numerical technique is developed to evaluate the diffusivity parameters and predict the transport of heavy metals in unsaturated clayey soils.
2. The diffusion coefficient is an exponential function of heavy metal concentration and volumetric water content.
3. Powell's optimization technique is appropriate to calculate the material parameters that govern the diffusion process.

REFERENCES

De Groot, S.R. & Mazur, P., 1962. NonEquilibrium Thermo-dynamics", *North Holland Pub. Co.*, Amsterdam.

Elzahabi M. & Yong, R.N., 2000. PH influence on sorption characteristics of heavy metal in the vadose zone, *Engineering Geology Journal*, (in press).

Fick, Adolf, Uber Diffusion, Annalen der Physik und Chemie 94, ed. In Berlin by Poggendorff, J. C., (a.k.a. Poggendorffer Annalen), Leipzig, 1855.

Fitts, D.D.,1962. NonEquilibrium Thermodynamics. McGraw Hill, New York.

Fityus, S.G., Smith, D.W. & Booker, J.R.,1999. Contaminant Transport Through an Unsaturated Soil Liner Beneath a Landfill. *Canadian Geotechnical Journal*, Vol. 36:330-354

Greenberg, J.A., Mitchell, J.K. & Witherspoon, P.A., 1973. Coupled Salt and Water Flows in a Groundwater Basin", *Journal of Geophysical Research*, Vol.78, No.27, pp. 6341-6353.

Jardine, P.M., Jacobs, G. K. and Wilson, G.V., 1993. Unsaturated Transport Processes in Undisturbed Heterogeneous Porous Media:I-Inorganic Contaminants. *Soil Sci. Soc. Am. J.* 57:945-953.

Klute, A.1952. A Numerical Method for Solving the Flow Equation for Water in Unsaturated Material *Soil Sci.*, pp. 73:105-116.

Mitchell, J.K.1993. Fundamentals of Soil Behavior. John Wiley and Sons Inc Toronto, 437pp.

Mitchell, J.K., Greenburg, J.A. & Witherspoon, P.A. 1973. Chemico -Osmotic Effects in Fine-Grained Soils, *Journal of the Soil Mechanics and Foundation Division*, A.S.C.E , Vol. 99, No. SM 4, pp. 307-322.

Onsager, L. 1931. Reciprocal Relations in Irreversible Processes. *Physical Review*, Vol. 37, pp. 405-426 and vol.38, pp. 2265-2279.

Powell, M.J.D., 1964. An Efficient Method for Finding the Minimum of a Function of Several Variables Calculating Deriviatives, *Computer Journal* Vol. 7, pp. 155-162.

Richards, L.A., 1931. Capillary Conduction of Liquid Through Porous Medium, *Physics*, pp.1:318-333.

Taylor, S.A. & Cary, J.W., 1960. Analysis of Simultaneous Flow of Water and Heat or Electricity with the Thermo dynamics of Irreversible Processes, *Seventh International Congress of Soil Science Transactions*, Vol. 1,pp 80-90.

Yong R.N., Mohamed, A. M. O. & Warkentin, B. P., 1992. Principles of Contaminant Transport in Soils.Elsevier, Amsterdam, pp.211.

Yong., R.N. and Samani, H.M.V., 1987. Modelling of Contaminant Transport in Clays Via Irreversible Thermodynamics, Proceedings of Geotechnical Practice for Waste Disposal, '87/GT Div. ASCE, pp. 846-860

Yong, R. N. and Warkentin, B. P., 1975. Soil Properties and Behaviour, Elsevier, New York.

Yong, R.N., Xu, D.M. & Mohamed, A.M.O., 1992. An Analytical Technique for Evaluation of Coupled Heat and Mass Flow Coefficients in Unsaturated Soil, *International journal for Numerical and Analytical Methods in Geomechanics*. Vol.12 pp.283-299.

Clay Science for Engineering, Adachi & Fukue (eds) © 2001 Balkema, Rotterdam, ISBN 90 5809 175 9

Variation of the intrinsic permeability of expansive clays upon saturation

M.V. Villar

Environmental Impact of Energy Department, CIEMAT – Centro de Investigaciones Energéticas, Medioambientales y Tecnológicas, Madrid, Spain

A. Lloret

Geotechnical Engineering and Geosciences Department, UPC – Technical University of Catalonia, Barcelona, Spain

ABSTRACT: The intrinsic permeability of compacted bentonite has been determined on samples saturated and unsaturated, using different techniques but, in all the cases, keeping the volume of the specimen constant. The saturated hydraulic conductivity of compacted bentonite has been measured for a variety of void ratios, which has allowed the determination of an exponential relationship between permeability and void ratio. In parallel, the permeability to nitrogen gas has also been measured in specimens of compacted bentonite with different void ratio and degree of saturation. It has been observed that, for a given void ratio, the value of intrinsic permeability obtained when the measurement is done with water in the saturated sample, is much lower than that obtained when the measurement is performed with gas in the unsaturated sample. This behaviour is a consequence of the variation of the structure of the bentonite as a function of the degree of saturation. As the degree of saturation increases, the clay particles swell and fill the big pores among aggregates, thus causing the decrease of the average pore size, and consequently, the decrease of the intrinsic permeability, although the total porosity remains the same.

1 INTRODUCTION

This investigation has been performed in the framework of the FEBEX project, a study of the near field for a repository of high-level radioactive waste in crystalline rock (ENRESA 2000). This repository concept considers the excavation, in stable host rocks, of deep galleries in which the waste containers are surrounded by compacted bentonite blocks (ENRESA 1995). The clay blocks are initially unsaturated, and an important percentage of their pores (40-50 %) will be filled with air. Thermal and hydraulic gradients will be generated in the repository, due to the heat emitted by the wastes and to the resaturation of the near field. Consequently, a combined movement of gas and water will take place in the clay barrier, that must be considered as a closed system in which no volume change is allowed.

In a repository of radioactive waste, gas generation, temperature and the changes in the accessible pore volume due to humidity variations, generate gas pressure gradients, which control the advective movement of gas species.

The flow of air requires the continuity of the gas phase. Water-filled pores block gas flow, consequently, gas permeability decreases as soil water content increases. It is considered that, for degrees of saturation greater than 90 %, the gas movement takes place mainly by advective flux of gas species dissolved in water and by diffusion through water (Matyas 1967). Conversely, the permeability to water drops sharply when air invades the system, and reaches a very small value while the saturation is still considerably greater than zero (Corey 1957).

Water and gas advective fluxes ($\mathbf{q}_w$, $\mathbf{q}_g$) follow Darcy's law with relative permeabilities (k_{rw}, k_{rg}) dependent on liquid degree of saturation:

$$\mathbf{q}_w = -\frac{\mathbf{k}_i k_{rw}}{\mu_w}\left(\nabla P_w - \rho_w \mathbf{g}\right) = -\mathbf{k}_w k_{rw}\left(\frac{\nabla P_w}{\rho_w \mathbf{g}} + \mathbf{i}_z\right) \quad (1)$$

$$\mathbf{q}_g = -\frac{\mathbf{k}_i k_{rg}}{\mu_g}\left(\nabla P_g - \rho_g \mathbf{g}\right) = -\mathbf{k}_{g0} k_{rg}\left(\frac{\nabla P_g}{\rho_g \mathbf{g}} + \mathbf{i}_z\right) \quad (2)$$

where $\mathbf{k}_i$ is the intrinsic permeability tensor, P_w, P_g, ρ_w, ρ_g, μ_w and μ_g are the pressure, density and dynamic viscosity of water and gas phases, $\mathbf{g}$ the gravity vector, z the vertical co-ordinate and $\mathbf{i}_z$ the unit vector in the z direction. $\mathbf{k}_w$ is the hydraulic conductivity in saturated conditions and $\mathbf{k}_{g0}$ the gas permeability in dry conditions.

The value of the intrinsic permeability is generally associated to pore diameter and pore size distribution (Romero et al. 1999). Qualitatively, the effect of pore size and total porosity on intrinsic permeability may be evaluated through Poiseuille's equation (Kovács 1981):

$$k_i = \frac{1}{2} \frac{n^3}{(1-n)^2} \left(\frac{D_h}{\alpha}\right)^2 \qquad (3)$$

where n is the porosity, D_h is the effective grain diameter and α is a grain shape coefficient.

Consequently, as air-filled porosity varies with soil water content and compaction, these factors have a major effect on the rate of gas and water exchange in soils.

Ideally, intrinsic permeability (k_i) depends only on soil structure and has the same value for gas and liquid flow. This value can be obtained either from gas flow tests in the totally dry soil, or from water flow tests in the saturated soil, and should be the same regardless the fluid employed in its determination. However, if fluid-media interactions alter the medium structure, the intrinsic permeability can be greatly altered. This is the case of expansive soils, in which water reacts with clay minerals causing the swelling of the clay lattice, thereby reducing the pore space available for flow (Tindall & Kunkel 1999).

2 MATERIALS AND METHODS

2.1 Material

The tests have been performed with a bentonite from the Cortijo de Archidona deposit (Almería, Spain), which has been selected by ENRESA, the Spanish agency for radioactive waste management, as suitable material for the backfilling and sealing of high-level radioactive waste repositories. The conditioning of the bentonite in the quarry, and later in the factory, was strictly mechanical (homogenisation, removal of rock fragments, drying to temperatures lower than 60 °C and crumbling of clods). Finally, the clay was sieved by a 5 mm mesh to obtain the granulated product to be used in the laboratory.

The bentonite has a content of dioctahedric smectite of the montmorillonite type higher than 90 %. Besides, it contains variable quantities of quartz, plagioclase, cristobalite, potassium feldspar, calcite and trydimite.

The cation exchange capacity is of 111 ± 9 meq/100g, and the exchangeable cations are Ca (38 %), Mg (28 %), Na (23 %) and K (2 %).

The liquid limit of the bentonite is 102 ± 4 % and the density of solid particles is 2.70 g/cm³. The hygroscopic water content of the clay at laboratory conditions (relative humidity 50 ± 10 %, temperature 21 ± 3 °C) is about 13.7 ± 1.3 %.

The swelling pressure $(P_s$, MPa) of samples compacted with its hygroscopic water content can be related to the dry density $(\rho_d$, g/cm³) through the following equation:

$$\ln P_s = 6.77 \, \rho_d - 9.07 \qquad (4)$$

The structure of the clay, compacted with its hygroscopic water content and freeze-dried, has been analysed by means of mercury intrusion porosimetry, which allows the quantification of pores with diameter greater than 0.006 μm. The percentage of porosity intruded by mercury is, on average, 33 % of the total porosity, what means that most of the pores have a diameter smaller than 0.006 μm, i.e. most of the porosity lies in the micropore range, according to the classification given by Sing et al. (1985). Figure 1 shows the pore size distribution obtained from mercury intrusion porosimetry.

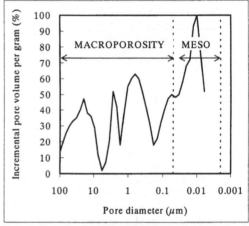

Figure 1. Pore size distribution of the compacted bentonite obtained by mercury intrusion

2.2 Hydraulic conductivity measurements

The method used to measure the hydraulic conductivity in expansive materials has been developed at CIEMAT. The permeability of these types of materials cannot be determined in standard permeameters, due to the important changes in volume undergone by the clay during saturation.

The method is based in the constant-head permeameter technique and consists on the measurement, as a function of time, of the water volume that traverses the saturated specimen, while a constant hydraulic gradient is imposed between top and bottom. The specimen is confined in a cylindrical rigid cell that prevents any change of the clay volume. The swelling of the saturated clay against the cell wall guarantees a perfect contact between clay and cell, avoiding a preferential pathway.

A sketch of the test arrangement is shown in Figure 2. The cell is made of stainless steel and has an inner section (S) of 19.63 cm^2 and a height of 2.5 cm. The granulated clay, with its hygroscopic water content, is uniaxially compacted to the desired dry density directly inside the permeability cell. Porous stones are placed on its top and bottom. The sample is saturated by injecting distilled water through the porous stones at a pressure of 0.6 MPa. This pressure is applied by means of a mercury pressure system. The time needed for full saturation depends on the dry density of the clay, and ranges between 1 and 4 weeks. An automatic volume change apparatus measures the water intake.

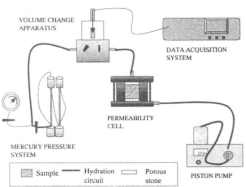

Figure 2. Schematic representation of the set-up to measure hydraulic conductivity

Once the sample is saturated, the lower inlet of the cell is switched to a piston pump. The injection pressure in the lower part of the cell is then increased, while the back-pressure on top is kept constant and equal to 0.6 MPa. Hydraulic gradients (∇h) between 1100 and 10500 m/m have been applied, depending on the dry density of the clay. The automatic volume change apparatus, connected to a data acquisition system, measures the water outflow through the upper outlet of the cell.

When the volume of water that crosses the specimen (ΔV) for a given time (Δt) is constant, the hydraulic conductivity (k_w) is calculated applying the Darcy's law.

$$k_w = \frac{\Delta V}{S \times \Delta t \times \nabla h} \qquad (5)$$

From the values of the saturated hydraulic conductivity (k_w), the intrinsic permeability evaluated from water flux (k_{iw}), can be calculated by the following expression:

$$k_{iw} = \frac{k_w \times \mu_w}{\rho_w \times g} \qquad (6)$$

Values of 10^{-3} Pa·s and 10^3 kg/m^3 have been taken for the viscosity and the density of water, respectively.

The determinations have been made at room temperature, between 20 and 25 °C.

2.3 Gas permeability measurements

The method for gas permeability measurement used in this work is a non-steady-state method whose basic principle was developed by Kirkham (1946). It consists on the pressurisation of a gas tank, and the release of this pressure to the atmosphere through the soil column of unknown permeability. The gas permeability is calculated from the rate of decrease of the tank pressure.

A schematic representation of the experimental set-up is shown in Figure 3. The cylindrical clay sample is placed in a triaxial cell, covered by two latex membranes with silicone paste between both, and porous stones placed on top and bottom. A confining pressure of 1.6 MPa is applied to the cell, in order to assure a perfect adherence of the membranes to the sample surface. This avoids any transport of gas between the sample and the membranes and guarantees that the gas flows exclusively through the soil. The bottom of the sample is connected to a hermetic tank of known volume, in which nitrogen gas has been previously injected at a pressure slightly higher than the atmospheric one. A pressure transducer connected to a data acquisition system measures the gas pressure in the tank. The upper outlet of the cell, connected to the sample top, is open to the atmosphere. During the test, the gas in the tank is allowed to flow through the sample, while the pressure decrease in the tank is monitored as a function of time.

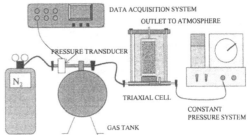

Figure 3. Schematic representation of the set-up to measure gas permeability

The gas permeability is calculated according to the following expression (Yoshimi & Osterberg 1963):

$$k_g = 2.3 \times \frac{V \times L \times \rho_g \times g}{A \times \left(P_{atm} + \frac{P_0}{4} \right)} \times \frac{-\mathrm{Log}_{10} \left(\frac{P(t)}{P_0} \right)}{t - t_0} \quad (7)$$

where k_g is the gas permeability (m/s), V is the volume of the deposit (m³), L is the height of the sample (m), A is the section of the sample (m²), ρ_g is the gas density (kg/m³), P_{atm} is the atmospheric pressure (N/m²), P_0 is the excess pressure in the tank over the atmospheric one for time t_0 (s), and $P(t)$ is the excess over the atmospheric pressure for time t. This equation has been derived in the same way as the water permeability expression for a falling head permeameter, taking into account the gas compressibility to solve the continuity equation. For its derivation, it has been assumed that the initial pressures (P_0) are only slightly higher than the atmospheric one.

The volume of the deposit is $2.21 \cdot 10^{-2}$ m³, the nominal height of the sample is 8.0 cm and its section 11.40 cm². All the tests have been performed with nitrogen gas, whose density has been taken as 0.04 mol/L (Lide 1995). The pressure in the deposit at the beginning of the tests is around 0.101 MPa, in order to preserve the properties of the gas constant throughout the test, and avoid sample disturbance.

Gas permeability (k_g) is equal to the product of the gas permeability in dry conditions (k_{g0}) by the relative permeability to gas (k_{rg}). Taking into account the dynamic viscosity of nitrogen (μ_g, $1.79 \cdot 10^{-5}$ Pa·s), the following relationship between gas permeability (k_g, m/s) and the product of the intrinsic permeability measured with nitrogen gas (k_{ig}, m²) by the relative permeability to gas (k_{rg}), can be established:

$$k_g = \frac{\rho_g \times g}{\mu_g} \times k_{ig} \times k_{rg} = 6.2 \cdot 10^5 \times k_{ig} \times k_{rg} \quad (8)$$

The specimens have been prepared by uniaxial compaction of the granulated clay with different pressures in order to obtain different dry densities. Both the clay with its hygroscopic water content and with added water have been used. In the latter case, the clay is mixed with the appropriate quantity of distilled water, according to the target water content, and the mixture is allowed to stabilise for several days in closed plastic bags, in order to facilitate a homogeneous distribution of humidity. To manufacture specimens with water contents lower than the hygroscopic one, the granulated clay has been oven-desiccated for short periods of time before compaction.

The tests have been performed at room temperature, between 20 and 25 °C.

3 RESULTS

3.1 Hydraulic conductivity

The saturated permeability to distilled water of clay compacted to dry densities between 1.30 and 1.84 g/cm³ has been determined according to the procedure described above. The values obtained are shown in Figure 4.

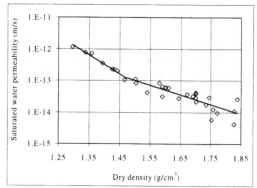

Figure 4. Water permeability of the compacted clay as a function of dry density and fittings obtained

There are important variations of the permeability values with the dry density; the hydraulic conductivity clearly diminishing as dry density increases.

The hydraulic conductivity values (k_w, m/s) can be related to the dry density (ρ_d, g/cm³), through an exponential law, which is different according to the dry density interval:

$$\log k_w = -6.00 \, \rho_d - 4.09; \quad \rho_d < 1.47 \text{ g/cm}^3 \quad (9)$$

$$\log k_w = -2.96 \, \rho_d - 8.57; \quad \rho_d > 1.47 \text{ g/cm}^3 \quad (10)$$

This threshold value (1.47 g/cm³), separates two density intervals in which the hydraulic behaviour of the clay is somewhat different. For densities lower than this value, the permeability increase as dry density decreases is sharper. This different behaviour is probably a consequence of the sharp increase of the swelling pressure for densities higher than this value, what, at constant volume, induces an important reduction of effective porosity and, consequently, of permeability. On the contrary, for densities lower than this value, the swelling pressure is not enough to counteract the greater porosity, and therefore, the hydraulic conductivity increases in a sharper way, giving rise to the breaking of the slope of the correlation line (Villar & Rivas 1994).

The relationship between hydraulic conductivity and porosity is exponential for a wide variety of geological materials. Geneste et al. (1990) observed a relationship between hydraulic conductivity and density, for a French calcic smectite, similar to those presented in this work. For sodic bentonite, Pusch (1979) points out the existence of two density intervals in which the increase of hydraulic conductivity with dry density reduction is different.

3.2 Gas permeability

The gas permeability has been measured in samples with nominal dry density between 1.50 and 1.90 g/cm^3, and degrees of saturation between 25 and 92 %. The method employed does not allow to test clay with lower degree of saturation, as no cohesive compacted specimens can be obtained with very low water contents. On the other hand, for high degrees of saturation, the gas in the pores of the sample becomes occluded and no flow occurs under the low gradients applied in the determinations. As the dry density of the specimen is higher, the water content for which no gas flow is observed, is lower. At the same time, for each water content, no gas flow occurs for dry densities higher than a given value.

Figure 5 shows the evolution of the pressure in the gas tank during different tests, which were performed on specimens compacted to nominal dry density 1.70 g/cm^3 with different water contents. The gas flow becomes slower as the degree of saturation is higher, which causes lower permeability values (the slower flow of the sample with water content 13.8 % compared to that of the sample with water content 14.0 % is probably due to a dry density slightly above the nominal one for the first sample).

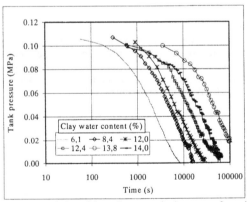

Figure 5. Evolution of the tank pressure in the tests performed on clay compacted to dry density 1.70 g/cm^3 with different water contents

The gas permeability values obtained in the tests can be plotted as a function of the dry density of the clay for different water contents, as shown in Figure 6.

For each water content, the gas permeability decreases as dry density increases, following an exponential law.

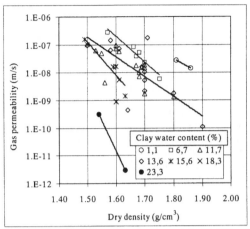

Figure 6. Permeability to gas of clay compacted with different water contents as a function of dry density

The variation of the permeability to gas (k_g, m/s) is shown in Figure 7 as a function of the accessible void ratio, which indicates the ratio between gas accessible pore volume and particle volume and can be expressed as the product of the void ratio (e) by the unit minus the degree of saturation ($1-S_r$). This factor, $e (1-S_r)$, shows in fact a higher correlation with the permeability values than the dry density or the water content do. The correlation found, plotted in Figure 7, can be expressed as:

$$k_g = 2.29 \cdot 10^{-6} \left(e \left(1-S_r\right)\right)^{4.17} \qquad (11)$$

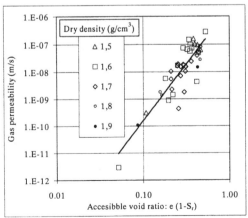

Figure 7. Relationship between the gas permeability and the volume of pores accessible to gas

4 DISCUSSION

In Figure 8, the values of the product of intrinsic permeability by relative permeability obtained via gas flow are plotted versus the degree of saturation for different dry densities. The values of intrinsic permeability obtained via water flow, which correspond to the saturated clay, are also included. No measurements of water flow have been performed on the unsaturated clay, but even lower values would have been obtained.

It can be observed that the intrinsic gas permeability for a given degree of saturation depends on the dry density of the sample.

From Figure 8 it becomes also apparent that the decrease of gas permeability for degrees of water saturation higher than a threshold value (between 65 and 80 %, depending on the dry density) is very sharp, which indicates the discontinuity of the gas phase.

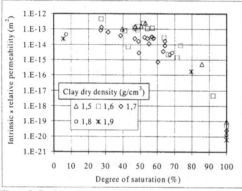

Figure 8. Product of relative by intrinsic permeability obtained via gas flow on compacted samples of different dry density and water content (the values for degree of saturation 100 % come from the hydraulic conductivity tests)

An estimation of the intrinsic permeability derived from gas flow tests can be performed assuming a value of degree of saturation equal to zero in equation (11), in order to obtain the gas permeability in dry bentonite (k_{g0}). Afterwards, from equation (8), the intrinsic permeability can be evaluated, resulting the values plotted in Figure 9. This figure shows also the variation of the values of intrinsic permeability obtained from water flow with the void ratio (a joint correlation for the two density intervals pointed out in section 3.1 is shown). For the same volume of pores, differences of about eight orders of magnitude in the intrinsic permeability values, for dry or saturated clay, can be observed.

These observations suggest that a fundamental difference in the microstructural arrangement of the saturated and the unsaturated sample exists, due to

the swelling of the clay as it saturates, which would not happen in non-expansive materials. The hydration of clay particles under constant volume reduces the size of pores between aggregates of clay. In dry conditions, the diameter of macropores accessible to gas flow may be up to more than 10 μm (Fig. 1). Hydrated clay in confined conditions presents the same global volume of pores, but the spaces between aggregates have been reduced or eliminated due to the swelling of the clay particles. That is to say, during the hydration of the clay at constant volume, the volume occupied by small mesopores and micropores increases, while the volume occupied by macropores decreases.

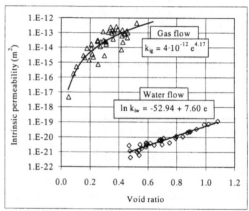

Figure 9. Intrinsic permeability of compacted bentonite obtained from saturated water flow and from unsaturated gas flow tests

In saturated conditions, accessible mean pore diameter is in the range of meso and micropores, i.e. more than three orders of magnitude smaller than the big interaggregate pores. This change in the mean pore diameter available to flow explains the big differences between the values of measured intrinsic permeability for dry and saturated clay.

Changes in pore size can be observed in the study of the compacted clay by environmental scanning electron microscopy (ESEM). This technique allows the observation of the microstructure of the material under different relative humidity conditions and consequently, different degrees of saturation. Figure 10 shows the aspect of the clay aggregates in a sample compacted to dry density 1.70 g/cm^3 with the hygroscopic water content and observed under a relative humidity of 50 % (similar to that of the laboratory). This sample was progressively hydrated during 5 hours in the microscope chamber, by increasing the relative humidity while keeping the temperature constant. The volume of the sample tested is very small, ensuring its quick saturation.

Figure 11 shows the final aspect of the same sample under a relative humidity of 100 %. Despite the absence of confinement, a diminution of the size of some voids can be observed.

Figure 10. Environmental scanning microscope image, taken under relative humidity 50 %, of the bentonite compacted to dry density 1.70 g/cm^3

Figure 11. Environmental scanning microscope image, taken under relative humidity 100 %, of the bentonite compacted to dry density 1.70 g/cm^3

Once saturated, the very low hydraulic conductivity of the bentonite is justified because, at high densities, the water content needed to saturate a compacted specimen is very low. This means that the interlamellar water film will have a thickness of only 3 to 5 Å. The water molecules constituting this film will be strongly adsorbed to the clay mineral surfaces, what leaves only narrow, tortous interparticle passages for water and ion transportation. Moreover, in the calcic montmorillonite compacted to dry densities higher than 1.60 g/cm^3, more than 90 % of the total pore volume is occupied by interlamellar water of restricted mobility (Pusch et al. 1990). Both aspects difficult the transport of water in compacted bentonite.

5 CONCLUSIONS

During the hydration of expansive clays at constant volume, the pore size distribution changes, due to the swelling of the clay particles. In the saturated clay, a greater pore volume fraction is occupied by the interlamellar water, which has a limited mobility and does not contribute to the effective porosity. Thus, although the porosity of a confined saturated and unsaturated specimen is the same, the size of the pores in the saturated sample is lower, and consequently the intrinsic permeability is also lower, since both are directly related.

ACKNOWLEDGEMENTS

This research work has been carried out in the context of the FEBEX Project, financed by ENRESA (Spanish National Agency for Waste Management) and the European Commission. R. Campos, J. Almendrote and J. Aroz have performed the laboratory work. P.L. Martín has supported the gas permeability measurements. The ESEM analyses have been performed in the Department of Materials Science and Metallurgic Engineering of the Technical University of Catalonia (UPC, Barcelona), in collaboration with Dr. J.M. Manero. Dr. B. de la Cruz made the linguistical revision.

REFERENCES

COREY, A.T. 1957. Measurement of air and water permeability in unsaturated soil. *Proc. Soil Sci. Soc. Amer.* 21(1): 7-10.

ENRESA 1995. Almacenamiento geológico profundo de residuos radiactivos de alta actividad (AGP). Diseños conceptuales genéricos. *Publicación Técnica* 11/95. Madrid. 105 pp.

ENRESA 2000. Full-scale engineered barriers experiment for a deep geological repository for high-level radioactive waste in crystalline host rock (FEBEX project). *EUR* 19147. *Nuclear science and technology series.* Luxembourg. 362 pp.

LIDE, D.R.(ed.) 1995. *CRC Handbook of Chemistry and Physics.* 75th Edition. Bocaraton: CRC Press.

GENESTE, P.H.; RAYNAL, M.; ATABEK, R; DARDAINE, M. & OLIVER, J. 1990. Characterization of a French clay barrier and outline of the experimental progamme. *Engineering Geology* 28: 443-454.

KIRKHAM, D. 1946. Field method for determination of air permeability of soil in its undisturbed state. *Soil Sci. Soc. Am. Proc.* 11:93-99.

KOVÁCS, G. 1981. *Seepage Hydraulics*. Developments in Water Science 10. Amsterdam: Elsevier. 730 pp.

MATYAS, E.L. 1967. Air and water permeability of compacted soils. In: *Permeability and Capillary of Soils. ASTM STP* 417. 160-175.

PUSCH, R. 1979. Highly compacted sodium bentonite for isolating rock-deposited radioactive waste products. *Nuclear Technology* 45: 153-157.

PUSCH, R.; KARNLAND, O. & HÖKMARK, H. 1990. GMM -A general microstructural model for qualitative and quantitative studies on smectite clays. *SKB Technical Report* 90-43. 94 pp.

ROMERO, E., GENS, A. & LLORET, A. 1999. Water permeability, water retention and microestructure of unsaturated compacted Boom clay. *Engineering Geology* 54: 117-127.

SING, K.S.W.; EVERETT, D.H.; HAUL, R.A.W.; MOSCOU, L.; PIEROTTI, R.A.; ROUQUÉROL, J. & SIEMIENIEWSKA, T. 1985. Reporting physisorption data for gas/solid systems with special reference to the determination of surface area and porosity. *Pure & Appl. Chem.* 57(4): 603-619. IUPAC.

TINDALL, J.A. & KUNKEL, J.R. 1999. Unsaturated Zone Hydrology for Scientists and Engineers. Upper Saddle River: Prentice Hall. 624 pp.

VILLAR, M.V. & RIVAS, P. 1994. Hydraulic properties of montmorillonite-quartz and saponite-quartz mixtures. *Applied Clay Science* 9: 1-9.

YOSHIMI, Y. & OSTERBERG, J.O. 1963. Compression of partially saturated cohesive soils. *J. Soil Mechanics and Foundations Division*. ASCE 89, SM 4: 1-24.

Clay Science for Engineering, Adachi & Fukue (eds) © 2001 Balkema, Rotterdam, ISBN 90 5809 175 9

Evaluation of permeability of bentonite mud cake

M. Minase
Hojun Kogyo Company Limited, Annaka, Japan

M. Kondo
Hojun Yoko Company Limited, Tokyo, Japan

M. Kamon
Disaster Prevention Research Institute, Kyoto University, Uji, Japan

ABSTRACT: Permeabilities of various swelling clay suspensions were evaluated by performing constant-pressure filtration tests in accordance with Ruth's equation and Darcy's law at several pressures ranging from a low hydraulic head, which simulated typical magnitudes of hydraulic gradients at waste landfills, up to pressures used in conventional filtration tests for bentonite muds. Values of filtration rate coefficient, specific resistance of wet cake and hydraulic conductivity of Na-bentonite, Na-exchanged bentonite, and kaolin-bentonite mixture (1:1) were examined. Ca-bentonite and kaolin were used as references for the low or non-swelling samples. Mud cakes from suspensions containing bentonite exhibited significant compressibility. Consequently, void ratio and specific resistance of wet cake were used as indications of permeability and compressibility of clay samples.

1. INTRODUCTION

Bentonite mud has been used as a drilling fluid for boreholes in petroleum exploration, and as clay suspensions for diaphragm wall construction and cast-in-place pile construction in civil engineering. Bentonite is also applied as an impervious natural clay constituent of compacted clay liners (CCLs) and geosynthetic clay liners (GCLs). Recently, there has been frequent use of composite liners fabricated with geomembrane, CCL and GCL in waste landfills. In reference to this, bentonite mud is proposed for use as an element of a detection and restorative system for leaks in the components of the liner systems which occur through their service lives (Ichikawa et al. 1999). Both applications are based on an ability of natural sodium bentonite particles, which impedes water transport due to sorption of a large quantity of water and a high swelling potential.

Up to this time, performance of bentonite in drilling fluids or clay suspensions has been conducted by evaluating the filtration rate of a constant-pressure filtration test of the mud (e.g. API Spec 13A) and this remains the same. Whereas in seepage liner applications, performance is exclusively based on evaluation of the hydraulic conductivity of the specimen containing the bentonite in a solid state (e.g. ASTM D 5084, D 5887). The "Filtrate Volume" test of API Spec 13A is a practically modified method based on Ruth's constant-pressure filtration equation (Ruth et al. 1933), which is the solution deduced originally from Darcy's law for flow of water through porous soils.

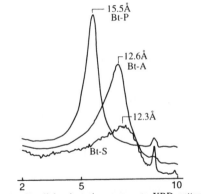

Figure 1. Parallel orientation aggregate XRD patterns of d(001) of Bt-P, Bt-A and Bt-S in RH40% (atmosphere).

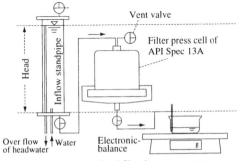

Figure 2. Low constant-head filtration test apparatus.

Table 1. The probability distribution of monovalent and divalent ion of montmorillonite layer.

Specimen	Bt-S	Bt-A	Bt-P
W1*	0.5~0.6	0.75	< 0.1
P11**	0.5~0.6	0.75	0.20
P22***	0.5~0.4	0.35	0.95

* W1:The probability of occurrence of unit layer of M^+
** P11:The junction probability of unit layer of M^+
***P22:The junction probability of unit layer of M^{2+}

In this study, laboratory constant-pressure tests for suspensions incorporated a high or low swelling clay or non-swelling clay were conducted to relate the coefficient of filtration rate and the hydraulic conductivity. The tests were executed in a range over extended filtration pressures from a very low pressure (head), where a leachate head acts on a liner giving a practical hydraulic gradient at landfills, up to pressures applied in the conventional drilling mud test. From the results, the permeability of the mud cake and its structure were described.

2. EXPERIMENT METHODS

2.1 Clay samples

Four types of clays containing primarily a single cation and one type of mixed clay were used in this study. They are as follows: as two swelling clays, an API grade bentonite (Bt-S) from Wyoming, USA, and an activated bentonite (Bt-A) modified to be swellable by sodium-exchanging Ca-bentonite from Alabama, USA; as a low swelling clay, the same Ca-bentonite mentioned above (Bt-P); as a non-swelling clay, Georgia kaolin (Kl: kaolinite having a mean particle size of 0.68 μ m); and as an additional sample, a mixed clay (Kl-Bt) prepared by mixing a non-swelling clay (Kl) with a swelling clay (Bt-S) equally. Three bentonite samples were observed by X-ray powder diffraction under a relative humidity (atmosphere) of 40% (Fig. 1) and probability distribution of M^+-layer and M^{2+}-layer in montmorillonite was analyzed in accordance with Iwasaki & Watanabe (1988). The result shown in Table 1 explains that Bt-A is exchanged with Na^+.

2.2 Conventional constant-pressure filtration tests

Except Bt-P, all clay suspensions were prepared by adding 22.5g (as 8.0% of moisture) of clay into 350ml of distilled water, then the suspensions were filtered under stated pressures with N2 gas in accordance with API Spec 13A. To improve accuracy for measurement of filtrate, a device was installed at the outlet of the filter press cell, whereby formation of a drop of filtrate due to surface tension was prevented and continuous falling of liquid could occur. Measurement of quantity of filtrate was gravimetrically conducted by using an electronic-balance, then converted to volume (cm³). Elapsed time was measured at intervals of 2.0g of filtrate and

continued from immediately after pressurizing until the cumulative quantity of filtrate reached to 20.0g. Next, at once the quantity of filtrate reached to 20.0g the filtration was terminated and the mass of wet cake was measured, then the cake was dried (110℃, 18hr) and weighed. The mass of wet cake resulting from the filtration was calculated from the final quantities of wet cake and filtrate, since the quantity of wet cake produced is directly proportional to the quantity of filtrate (Gray & Darley 1980; Kondo 1998) in filtration of a stable suspension. The experimental procedure for Bt-P was the same as mentioned above except for the concentration of suspension. In order to make a stable Bt-P mud, a suspension was prepared at a high concentration by adding 67.5g (as 8.0% of moisture) of the clay into 350ml of distilled water.

2.3 Low constant-head filtration tests

A pressure gas inlet on the top of the filter press cell was connected with a headwater standpipe and the filter press was assembled, whereby the headwater in the standpipe worked directly on the mud in the cell. Schematic drawings of this filter press assembled is shown in Figure 2. This low constant- head filtration test was carried out only for Bt-S mud.

2.4 Calculation

The coefficient of filtration rate of suspension, K, the specific resistance of wet cake, α', and the hydraulic conductivity of mud cake, k, were calculated in accordance with the following Equations 1 - 7 (Ruth et al. 1933; Kondo 1998). Two coefficients were obtained from the differential equation and the integral equation respectively, and the mean value of them was adopted in calculating the coefficient of filtration rate. The specific resistance of wet cake represents a resistance of filtrate through a unit thickness of wet cake, which was deduced from the specific resistance of cake.

$$K = \frac{(V + V_0)^2}{\theta + \theta_0} = \frac{K_i + K_d}{2} \qquad (1)$$

$$\alpha' = \frac{\alpha \times \omega}{l} \qquad (2)$$

$$\alpha = \frac{2A^2 P g_c}{K \mu \kappa} \qquad (3)$$

$$\kappa = \frac{\rho_w s}{1-ms} \qquad (4)$$

$$k = \frac{\Delta q / \Delta \theta}{i} \qquad (5)$$

$$K_d = 2 \times V \times \frac{\Delta V}{\Delta \theta} \qquad (6)$$

$$K_i = \frac{V^2}{\theta} \qquad (7)$$

where K= coefficient of filtration rate(cm⁶/s)
  K_d= differential form of K
  K_i= integral form of K

$V=$ filtrate volume(cm^3)

$V_0=$ filtrate volume equivalent to resistant at point V=0 (cm^3)

θ = filtering time(sec), to secure volume V

$\theta_0=$ hypothetical filtering time (sec), to secure volume V_0

α = specific resistance of filter cake (cm/g)

α' = specific resistance of wet cake (i.e. specific resistance of a unit thickness of wet cake (cm^{-2}))

ω = mass of dry cake produced /area (g/cm^2)

$l=$ thickness of wet cake (cm)

$A=$ area of filtration (cm^2)

$P=$ pressure drop through filter cake (g/cm^2)

μ = viscosity of filtrate (g/cm $\cdot$ s)

κ = mass of filter cake produced per unit volume of filtrate (g/cm^3)

$g_c=$ conversion coefficient of gravity (980 g $\cdot$ cm/gf $\cdot$ s^2)

$\rho_w=$ density of filtrate (g/cm^3)

$m=$ mass ratio of wet cake / dry cake (-)

$s=$ fraction of solid content (-)

$k=$ hydraulic conductivity (cm/s)

$\Delta q / \Delta \theta$ = flow rate of filtrate through cake (cm/s)

$\Delta V / \Delta \theta$ = volume of filtrate per unit elapsed time (cm^3/s)

$i=$ hydraulic gradient (-)

In addition to the above, Gray & Darley (1980) introduced another solution mentioned below for hydraulic conductivity of drilling mud

$$k = Q_w Q_c \frac{\mu}{2 \theta P A^2} \qquad (8)$$

where Q_w is filtrate volume, and Q_c is cake volume. However, Equation 8 is deduced in Equation 5.

3. EXPERIMENT RESULTS AND DISCUSSION

Experiment results are shown in Tables 2-4.

3.1 *Conventional constant-pressure filtration tests*

In the filtration test (API Spec 13A) of bentonite mud for oil well drilling, 690kPa is applied as a filtration pressure, whereas 294kPa is conventionally used for clay suspension test in civil engineering. Relationships between the coefficient of filtration rate and the pressure obtained from the conventional constant-pressure filtration tests, which contain those pressures of filtration, are shown in Figure 3. Bt-S and Bt-A exhibit the smallest coefficient of filtration rate and the differences between them are of little significance. On the other hand, Kl, a non-swelling clay, shows the largest coefficient of filtration rate, which is about 100 times that of Bt-S or Bt-A. However, the coefficient of filtration rate of Kl-Bt, a mix of non-swelling clay with swelling clay, is of a significantly smaller magnitude - at most 2.0 times of Bt-S. Except for Bt-P, every suspension increases its coefficient of filtration rate with increase of the pressure.

Between 138 - 345 kPa, the coefficient of filtration rates of Bt-S, Bt-A or Kl-Bt increase 1.3 to 1.4 times, whereas Kl increases as much as 1.8 times. Consequently Kl's coefficient of filtration rate is more sensitive to the pressure. Regarding the Bt-P suspension, it can not compare instantly with the others since the concentration is high, the coefficient is of middle magnitude between Bt-S and Kl in simple comparison. For the matters, discussion will be further done with specific resistance and hydraulic conductivity of mud cake, since concentration effect does not intervene in these parameters.

Hydraulic conductivity of mud cake compared to pressure of filtration is shown in Figure 4. Bt-S, Bt-A and Kl-Bt form a group having the smallest hydraulic conductivity of 10^{-9}cm/s order. On the other hand, Bt-P and Kl form a group having hydraulic conductivity of 10^{-8} - 10^{-7}cm/s that are one or two orders of magnitude larger than the former. Gates & Bowie (1942) measured hydraulic conductivity of 20 field muds and 40 laboratory muds, and obtained values from 3×10^{-10} to over 2.4×10^{-7}cm/s at 690kPa. The present results are quite similar to their values. Plots of Bt-S and Kl-Bt are overlapped and it is understood that reasonably small values of hydraulic conductivity are obtained by replacing 50% of Kl with Bt-S. Each volume of soil particle of Kl and Bt-S is nearly equal respectively in Kl-Bt. However, the gel volume of Bt-S is estimated to be about three times as large as that of Kl, since the each hydrated gel volume is obtained from its void ratio of the wet cake. Thus Kl-Bt gel may form a structure of hydrated Kl particles dispersed in a continuous matrix of Bt-S gel. The more this dispersion becomes homogeneous, the more the hydraulic conductivity of Kl-Bt approaches that of Bt-S. Hydraulic conductivity of Bt-A (artificially Na-exchanged) is the smallest and the magnitude is 60 – 70% for that of Bt-S. It is considered that such a difference of hydraulic conductivity between Bt-A and Bt-S would appear based on native character of montmorillonite which affects the interaction between clay particles and pore water constructing mud cake.

Filtration rate depends on specific resistance of mud cake formed during filtration. Variation of the specific resistance of wet cake, α', relating to pressure of filtration is shown in Figure 5. Bt-S, Bt-A, and Kl-Bt exhibit a high magnitude of the specific resistance, whereas Bt-P and Kl exhibit the magnitude of about two orders smaller than those. Every specific resistance of clay increases with the pressure of filtration. It is also noted that every ascending line belonging to the bentonite and the mixed clay has a similar slope, but the slope of Kl is very small. Gray & Darley (1980) described compressibility of mud cake as follows: the mud filter cakes are to a greater or lesser extent compressible, so that the permeability is not constant, but decreases with increase in pressure. Thus, from Darcy's law, the relation of between filtrate volume and pressure :

269

$$Q_w \propto P^x \qquad (9)$$

where the exponent, x, varies from mud to mud, but is always $x < 0.5$. The value of x depends largely on the size and shape of the particles composing the cake.

Table 2. Filtration test results of Bt-S and Bt-A suspension at conventional pressures.

Sample and symbol	Bt-S; O					Bt-A; Δ		
Pressure of filtration (psi)	20	50	75	100	125	20	50	100
Pressure of filtration (kPa)	138	345	517	690	862	138	345	690
Head (cm)	1410	3526	5289	7052	8815	1410	3526	7052
Water content of wet cake (%)	595.6	512.6	471.5	424.4	420.9	488.7	346.0	344.1
Void ratio (-)	14.69	12.64	11.63	10.47	10.38	13.34	9.44	9.39
Thickness of wet cake produced per unit filtrate (10^{-2} cm/cm^3)	1.4471	1.2508	1.1521	1.1307	1.1239	0.9779	0.8082	0.7566
Coefficient of filtration rate of suspension (K: cm^6/sec)	0.0776	0.0984	0.1089	0.1105	0.1133	0.0748	0.1055	0.1248
Specific resistance of wet cake (α': 10^{13} cm^{-2})	1.37	3.35	5.07	7.62	9.39	1.89	5.15	8.75
Hydraulic conductivity of wet cake (k: 10^{-9} cm/sec)	9.31	4.29	3.09	2.32	1.90	5.84	2.74	1.55
Hydraulic gradient (-)	4859	14052	22883	31091	38867	7190	21748	46463

Table 3. Filtration test results of Kl and Kl-Bt suspensions at conventional pressures.

Sample and symbol	Bt-P; ×			Kl-Bt; □		Kl; ◇	
Pressure of filtration (psi)	20	50	100	20	50	20	50
Pressure of filtration (kPa)	138	345	690	138	345	138	345
Head (cm)	1410	3526	7052	1410	3526	1410	3526
Water content of wet cake (%)	204.02	217.29	218.53	268.0	228.7	124.4	128.6
Void ratio (-)	5.568	5.9303	5.9641	6.77	5.78	3.22	3.33
Thickness of wet cake produced per unit filtrate (10^{-2} cm/cm^3)	1.5619	1.7129	1.7301	0.7227	0.6252	0.2646	0.2704
Coefficient of filtration rate of suspension (K: cm^6/sec)	2.1141	1.6898	1.9062	0.1575	0.2193	11.1863	20.3713
Specific resistance of wet cake (α': 10^{13} cm^{-2})	0.044	0.127	0.223	1.81	3.84	0.053	0.071
Hydraulic conductivity of wet cake (k: 10^{-9} cm/sec)	254	93.3	54.0	9.01	4.54	162.46	118.84
Hydraulic gradient (-)	4501	10262	20319	9728	28114	26574	64999

Table 4. Filtration test results of Bt-S suspensions at low head pressures.

Sample and symbol	Bt-S; O		
Pressure of filtration (psi)	0.7090	0.9926	1.276
Pressure of filtration (kPa)	4.889	6.844	8.800
Head (cm)	50.0	70.0	90.0
Water content of wet cake (%)	851.8	814.4	783.7
Void ratio (-)	21.01	20.09	19.33
Thickness of wet cake produced per unit filtrate (10^{-2} cm/cm^3)	3.1900	2.8796	2.7338
Coefficient of filtration rate of suspension (K: cm^6/sec)	0.0071	0.0107	0.0120
Specific resistance of wet cake (α': 10^{13} cm^{-2})	0.291	0.295	0.364
Hydraulic conductivity of wet cake (k: 10^{-9} cm/sec)	58.84	53.36	46.73
Hydraulic gradient (-)	78	121	164

Bentonite cakes, for example, are so compressible that x is zero, and they show data that Q_w is constant with respect to pressure in a range over 690 kPa to 69,000 kPa. The reason for this behavior is that bentonite is almost entirely composed of finely-divided platelets of montmorillonite, which tend to align more nearly parallel to the substrate with increase in pressure. It is of interest that the graphs, the hydraulic conductivity vs. pressure (Fig. 4) and the specific resistance of cake (Fig. 5), reveal that such compressibility of bentonite cake does not only belong to high swelling Na-bentonite (Bt-S and Bt-A) but also to low swelling bentonite (Bt-P). However, as compared with the high compressibility of bentonite, that of kaolin is considerably small. It is considered that the small compressibility of kaolin is due to development of the interlayer hydrogen bond in kaolinite crystallites and the rigid platelets.

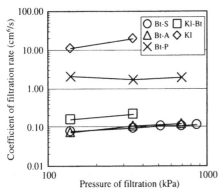

Figure 3. Coefficient of filtration rate vs. pressure of filtration.

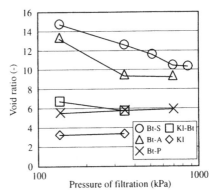

Figure 6. Void ratio of mud cake vs. pressure of filtration.

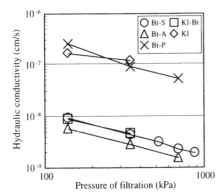

Figure 4. Hydraulic conductivity vs. pressure of filtration.

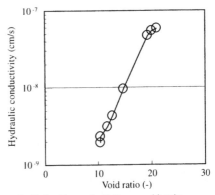

Figure 7. Hydraulic conductivity vs. void ratio.

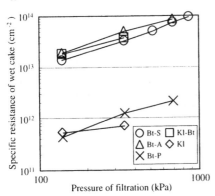

Figure 5. Specific resistance of wet cake vs. pressure of filtration.

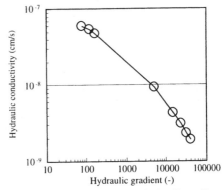

Figure 8. Hydraulic conductivity vs. hydraulic gradient.

The amount of water retained in mud cakes with different test specimens depends on the swelling properties of the clay minerals contained. For this reason, water content of mud filter cake is regarded as an excellent measure for swelling potential of the major clay (Gray & Darley 1980). In this study, the void ratio represents a swelling capacity per unit volume of the clay particles since the mud cake is in saturate.

Variation of void ratios depending on pressure of filtration is shown in Figure 6. In the comparison of Bt-A to Bt-S, either clay decreases void ratio with an increase of pressure, that demonstrates the fact that swelling volume depends on pressure (Takahashi et al. 1996). However, Bt-S reaches a nearly constant void ratio at 690kPa, while Bt-A attains it at 345kPa with significantly smaller void ratio than the former. This difference reveals that the stacking structure in the montmorillonite platelets of Bt-A grows relatively well. Because, as shown from X-ray diffraction patterns of three bentonite samples in Figure 1, 001 reflection of Bt-A is sharp due to a structural habit of original Ca-montmorillonite in contrast to very broad 001 reflection of Bt-S. That is, it appears that montmorillonite platelets of Bt-A tend to align more parallel than Bt-S, and a close gel structure is formed as a result. The void ratio of Kl-Bt mixed clay is not more than 74% for an additive value from the constituent fractions, Kl and Bt-S. This may be explained as a filling effect by entering Kl particles into the space between the domains of the swelled montmorillonite fractions. Each void ratio of Bt-P and Kl-Bt is of similar value, but the coefficient of filtration rate and the hydraulic conductivity of Kl-Bt are considerably smaller. This demonstrates that the water transport impedance potential of Bt-S is dominant.

3.2 Low constant-head filtration tests

In the low constant-head filtration test, the experiment was executed for only Bt-S under low pressures (head), and compared to simulated magnitude of practical hydraulic gradient, since bentonite mud is expected as an elemental fluid of a leak detection and restorative system for defects in the components of the liner systems at waste landfills. As shown in Table 4, results exhibit one order larger magnitude of hydraulic conductivity than the results of the conventional constant-pressure test (Table 2), but 10^{-8} cm/s order of reasonable magnitude of hydraulic conductivity. Relationships between hydraulic conductivity versus void ratio and hydraulic gradient over the whole range of the pressures containing the conventional constant-pressure test are shown in Figures 7 and 8. It is obvious that the mud cake of Bt-S is highly compressible. Calculation of hydraulic conductivity may be related to large hydraulic gradient even under low head pressures since the thickness of the mud filter cake is very thin. For low hydraulic gradient, the pressure of filtration may be at a too low a level due to only static head works of headwater in a standpipe. Consolidation effects may naturally act in the practical fields by working an effective stress from solid waste adding to the leachate head. For this situation, less magnitude of hydraulic conductivity will be expected with additional consolidation effects.

4. SUMMARY

Filtration rate of the suspensions, and permeability and structure of the mud cakes were examined and the clay-water interaction was discussed. The clay samples were as follows: (Bt-S) natural Na-bentonite, swelling; (Bt-A) artificially sodium-exchanged Ca-bentonite, swelling; (Bt-P) Ca-bentonite, low swelling; (Kl)

kaolin, non-swelling; and (Kl-Bt) Kl and Bt-S equal mixture, swelling.

Coefficient of filtration rate of the suspensions and hydraulic conductivity of the mud cakes of two swelling clays (Bt-S, Bt-A) were the smallest values together among the clay samples, 10^{-1} cm^6/s and 10^{-9}cm/s orders of magnitude, respectively. Either of the mud cakes composed with both bentonite samples, also, showed to be obviously compressible that hydraulic conductivity decreased significantly with increase of the pressures, and void ratio as a measure of swelling capacity also considerably decreased. Bt-A exhibited the properties that approach closely to those of Bt-S. This was inspected from X-ray diffraction patterns identified with a synthetic Na-bentonite formed by sodium-exchanging.

Kl showed about two orders larger magnitude of coefficient of filtration rate of the suspension and hydraulic conductivity of the mud cake together compared with Bt-S and Bt-A, as the result Kl exhibited low compressibility for the pressures.

The Kl-Bt mixture showed permeability near to each alone of both swelling bentonite samples. This was described that bentonite gel hydrated and swelled in ten times or more volume occupies a relatively large volume fraction in the mud cake and forms a continuous matrix, whereby dominates permeability of the mud cake. Bt-P showed permeability near to Kl, but larger compressibility than kaolin.

As for observation of filterability of suspension or permeability of mud cake, the void ratio and the specific resistance of wet cake, α', which represents a resistance of the filtrate through unit thickness of the wet filter cake, provided useful information.

5. ACKNOWLEDGMENTS

We would like to thank Dr.T. Iwasaki of Tohoku National Industrial Research Institute for the valuable help given in X-ray diffraction works.

REFERENCES

Gates, G.L. & C.P.Bowie 1942. Correlation of Certain Properties of Oil-Well Drilling Fluids with Particle Size Distribution. *U.S.Bureau of Mines Report of Investigations*, No.3645.

Gray, G.R. & H.C.H.Darley 1980. *Composition and Properties of Oil-Well Drilling Fluids, 4th Ed* : 277-312. Houston, Texas: Gulf Publishing Company.

Ichikawa, T., T.Toki, T.Hongo, S.Narusima, H.Kamano, T.Kawahara, K.Akaboshi, F.Satou, T.Nakamura, K.Mizuno & M.Koga 1999. Development of automatic pressuring system for wastes landfill. *Proceeding of 34th Japan National Conference on GE* : 2241-2242 (in Japanese).

Iwasaki, T. & T.Watanabe 1988. Distribution of Ca and Na Ions in Dioctahedral Smectites and Interstratified Dioctahedral Mica/Smectites. *Clays and Clay Minerals.* 36, No.1: 73-82.

Kondo, M.1998. Poiseuille and Darcy and Ruth. *SMECTITE*. 8, No.1: 44-47 (in Japanese).

Ruth, B.F., G.H.Montillon & R.E.Montonna 1933. Studies in Filtration Ⅱ. Fundametal Axiom of Constant-Pressure Filtration, *Industrial and Engineering Chemistry*. 25, No.2: 153-161.

Takahashi, S., M.Kondo & M.Kamon 1996. Evaluation of swelling behavior and permeability of a geosynthetic clay liner, In M.Kamon (ed.), *Environmental Geotechnics*: 609-614. Rotterdam: Balkema.

Clay Science for Engineering, Adachi & Fukue (eds) © 2001 Balkema, Rotterdam, ISBN 90 5809 175 9

The effects of density and hydraulic gradient on clay permeability

Steven C.H.Cheung
Geotechnical and Tunnelling Department, Atkins China Limited, Hong Kong, SAR, People's Republic of China

Ringo Yu
Fraser Construction Company Limited

ABSTRACT : The Permeability of compacted clay-based materials and the mechanisms controlling this property have been investigated based on the Kozeny-Carmen relationship and the diffuse double-layer theory together with the available data. The study showed that permeability depends on a number of parameters such as density, soil type and structure, water salinity, hydraulic gradient and temperature. The existence of a threshold gradient particularly in active clay mineral is shown to depend on soil fabric, density and water salinity. This information is important in the design of clay-based hydraulic barriers.

1 INTRODUCTION

Groundwater contamination is an issue of great concern these days. Potential contaminants include agricultural, municipal, industrial and radioactive wastes. Natural and engineered clay barriers are often used to limit the movement of contaminants into the groundwater supply system.

Soil permeability is one of the critical parameters required to predict the migration rate of contaminant in clay soil. The permeability has been shown to depend on soil type, pore water chemistry, density, and temperature. Among these factors, there are conflicting literature results on the effects of hydraulic gradient and density. For example, some researchers show that a threshold gradient (a gradient below which no water flow will occur in soil) exists in the clay soils. However, others have shown that such a threshold does not exist in similar soils. Indeed, the permeability of the clay soil has often been found to be lower in clay with lower density as opposed to the expected decrease of permeability with increasing density. The reasons for such conflicting relationships are not clear and further study is required for the design of hydraulic clay barriers.

In this paper, a theoretical model based on the Kozeny-Carmen relationship and diffuse-ion layer theory is used to qualitatively describe the mechanism of permeability. The model will then be compared with the available data to investigate the effects of density and the existence of threshold gradient on clay permeability.

2 THE EFFECTS OF HYDRAULIC GRADIENT, SOIL DENSITY AND FABRIC ON CLAY PERMEABILITY

The rate of water flow through a water saturated porous medium is commonly described by Darcy's Law

$$q = Ki \qquad (1)$$

where q is the rate of water flow per unit area, i is the hydraulic gradient across the material and K is the hydraulic conductivity.

$$K = k\gamma_w g / \eta \qquad (2)$$

where γ_w, is the fluid density, g is the acceleration due to gravity, η is the fluid viscosity and k is the permeability.

Using n_w as the fractional volume of liquid, m as the pore shape factor, t as the tortuosity, S_o as the specific surface area of the solids per unit volume, and n_s as the fractional volume of solids (= 1 - n_w), Kozeny and Carmen suggested that

$$k = n_w^3 / m t^2 S_o^2 n_s^2 \qquad (3)$$

where m and t are approximately 2.5 and $2^{1/2}$, respectively (Yong and Warkentin, 1975).

The Kozeny-Carmen equation is generally satisfactory for sand. However, in clay soils, not all the water can be mobilized for flow : this is related to interparticle forces arising from the surface charge of clay particles. Figure 1 shows the distribution of ions from a clay plate from which the osmotic pressure can be derived (Yong & Warkentin 1975 and Mitchell 1976). The fractional volume of potentially mobile water, termed here as effective porosity, n_w' has to be substituted for n_w in equation (3).

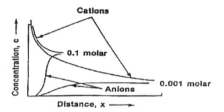

Figure 1 Theoretical distribution of ions (after Yong and Warkentin, 1975).

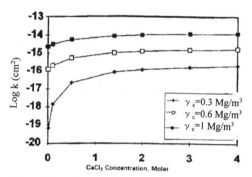

Figure 2a Model results of permeability k vs hydraulic gradient for smectite clay (i=2000 and T =20℃).

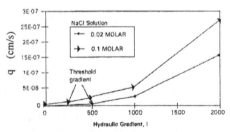

Figure 2b Model results of flow rate, q vs hydraulic gradient for smectite clay (γ_c=0.3 Mg/m³ and T = 20℃).

The osmotic pressure, P_c, developed between two parallel plates can be calculated using the following equations (Yong & Warkentin, 1975),

$$P_c = RT (C_c - 2C_o) \qquad (4)$$

$$C_c = \pi^2 / z^2 \ B(d+x_o)^2 \ 10^{-16} \qquad (5)$$

$$B = 8\pi e^2 / \epsilon k_b T \qquad (6)$$

$$x_o = \epsilon kT / 2\pi z e \sigma \qquad (7)$$

where R is the gas constant, T is the absolute temperature, C_o is the electrolyte concentration in the free solution (water not influenced by the particle forces), C_c is the electrolyte concentration at the midpoint between two parallel plates, z is the valence of the exchangeable cations, d is the half distance between two parallel plates, e is the electronic charge, ϵ is the dielectric constant, k_b is the Boltzmann constant, σ is the surface charge density.

From equation (4) and (5), P_c, is predicted to increase with increasing T or decreasing d and C_o. Good agreement between experimental and theoretical values of P_c, particularly for bentonite systems, has been found (Yong & Warkentin, 1975).

The forces holding water to the clay plates (P_c) will influence flow; the force driving flow (Hydraulic Gradient) must exceed hydraulic pressure P_h, before water can be mobilized. If the hydraulic pressure, P_h, per unit length of the soil is less than or equal to P_c, all the water is immobilized, and no flow will occur. When P_h exceeds P_c, previously bound water flows, the rate of flow increasing with increasing P_h. The thickness of this immobilized layer of water (bound water) depends on the interparticle force (P_c) and the hydraulic gradient. This provides a theoretical explanation for the non-linear relationship of flow rate versus hydraulic gradient and the existence of a threshold hydraulic gradient reported by some investigators (Mouradian and Cheung, 1993).

The thickness of the immobilized water layer, di, adjacent to each clay plate can be calculated from equations (4) and (5) by substituting d = d_i and P_c = P_h. The effective porosity, n_w', can now be calculated for parallel arrangement of clay plates with uniform particle spacing using the following equations, (Yong & Warkentin, 1975)

$$n_w' = \gamma_c S \ d' / 10^6 \qquad (8)$$

$$d' = d - d_i \qquad (9)$$

$$d = (1 / \gamma_c - 1/G_c) \ 10^6 /S \qquad (10)$$

where γ_c, is the clay dry density, S is the specific surface area per unit mass of clay in m²/kg and G_c, is the specific gravity of clay (Yong & Warkentin, 1975 and Gray et al., 1984).

The model suggests that the permeability k depends on n_w', that in turn, depends on T, C_o, S, i, and γ_c. For soils with relatively limited interparticle force, P_c, such as kaolinite, k is principally dependent on γ_c. For soils with relatively high P_c, such as bentonite or smectite, k may be expected to decrease with increasing γ_c, or decreasing C_c. (see Figure 2a).

The occurrence of threshold gradient is also

shown to be smaller with higher NaCl solution (see Figure 2b).

This explains why threshold hydraulic gradients are generally found to exist in soils with high P_c. A parallel arrangement of particles is unlikely in nature and some water will exist in large pores, and not be significantly affected by interparticle forces (see Figure 3).

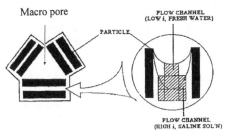

Figure 3 Schematic showing flow channel between parallel clay plates in fresh and saline water and macro pore.

This water is referred to as free water and can easily be mobilized for flow. The amount of this water is expected to depend not only on γ_c, but also on soil fabric. The model has been developed for pure clay systems, but can be applied to sand-clay mixtures. Like pure clay, the permeability of sand-clay mixtures is controlled by the relative proportions of free and bound water. Compared with clays, sand has a very small specific surface area and surface charge density. Thus, in sand-clay mixtures containing kaolinite, there is little bound water on the clay, $n_w' = n_w$ and, n_w is directly related to γ_d. For mixtures containing clays such as smectite, calculations show that n_w' is controlled by the clay component and is therefore related to γ_c (Gray et al., 1984)

The stern layer, x_o, is a layer of water molecules and associated cations that are very tightly bound to the clay surface, and would not, in practice, be influenced by i (or P_h). This is important for clays with high surface charge density and high specific surface area. For example, for smectite clays, the stern water layer is estimated to be 0.4nm. Parallel plates < 0.8 nm apart will not allow the passage of water. In material with S = 800 m^2/g such as smectite, G_c, = 2.75 and a clay density (γ_c) of approximately $\geq$ 1.25 Mg/m^3 platelets have a calculated particle separation of 0.8 nm. In such a system, a k of zero would be expected. Experimental k values are reported in the literature for γ_c > 1.25 Mg/m^3. This is due to the soil fabric where clay platelets form clusters with interparticle distance less than 0.8 nm, thus, allowing the passage of water through the soil around the clusters. The

number of particles and the interparticle distance in a cluster is very important and depends upon the soil system (dispersed, flocculated or aggregated) and the clay dry density γ_c. In this case d is calculated in the following way,

$$d = (1/\gamma_c - 1/\gamma_{cc})\, 10^6/S_c \qquad (11)$$

where S_c, is the effective cluster specific surface area per unit mass of clay in m^2/kg and γ_{cc} is the dry density of the cluster and is calculated as follows,

$$d_c = (1/\gamma_{cc} - 1/G_c)\, 10^6/S \qquad (12)$$

d_c depends upon the number of hydrated water layers present between clay platelets, which depends upon the dry density, γ_c. For three layers of hydrated water γ_{cc} is calculated to be 1.45 Mg/m^3. From literature data of n_w' and γ_c, γ_{cc} and S_c could be calculated and consequently the permeability k.

The above model describes the hydraulic properties of different clay minerals in terms of S with a range of densities, groundwater ionic concentration and hydraulic gradients for both parallel and non-parallel clay systems. This model is particularly useful in the design of hydraulic barriers undergoing environmental change. (i.e. variation in confining pressure, water salinity or temperature)

3 LITERATURE & MODEL DATA DISCUSSION

3.1 Clay Type & Density

The effects of clay type and density on permeability from model and literature results are shown in Figure 4. The permeability of clay generally decreases with increasing density (assuming parallel clay arrangement). At the same dry density (as shown on Figure 4b), mixtures containing smectite (S = 800 m^2/g) possess much lower permeabilities than similar mixtures containing kaolinite (S = 10m^2/g) of illite (S = 80m^2/g). The discrepancy between model and literature data for kaolinite and illitic soils (see Figure 4a) is due to the lower content of illite in the experimental illitic soil resulting in similar specific surface area as that of kaolinite. This tends to support the significance of the surface activity of the clay and the lack of mobility of bound water.

3.2 Hydraulic Gradient Effect

Table 1 shows the values of k for selected mixtures containing smectite or illite, measured under hydraulic gradients ranging from 30 to 10000. For most mixtures containing illite as the predominant clay mineral, k increased as i was increased to approximately 300. Increasing i above 300 had little effect on k. In mixtures containing smectite as the predominant clay mineral, with clay densities less than approximately 1.25 Mg/m³, k increased as i was increased.

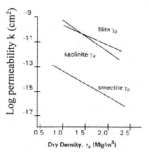

Figure 4a   The relationship between γ_d and k for illitic, kaolinitic and smectitic materials (after Fransham et al., 1981).

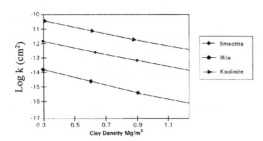

Figure 4b   Model results of permeability vs clay density.

The observations in literature suggest that a gradient of less than 400 for illite and 3000 for smectite systems is insufficient to mobilize all the water in the pore space available for flow. At these gradients, in illite systems of lower densities, the bonds between the water and the clay are stronger than the shear stress causing the flow. The model predicts this phenomenon by considering the effects of diffuse double layer (see Figure 1).

The phenomenon on increasing k with increasing i have been shown by Yong and Warkentin (1975). The existence of a threshold hydraulic gradient for mobilizing the bound water layers in smectite has been shown previously. In smectite systems with γ_c > 1.25Mg/m³ it is dominated by strongly bound water, stern layer. It is important to note that, at this density, whilst n_w' (Fractional volume occupied by free water) clearly tends to zero, the fractional volume occupied by Stern water is still greater than 50% of the volume occupied by the clay (i.e. > n_s).

For illite, Table 1 indicates that k increased as the total dissolved solids in the water were increased from 0 to 70 g/l. As predicted by the diffuse double-layer theory, the interparticle forces appear to be active in relatively dense illite systems and will influence k. However, the effects of water salinity on k are very small compared with those of density. For kaolinite such effect is found to be even smaller from model prediction and experiments.

For smectite, i > 3000, it is shown in Table 1 that most of the bound water (except the Stern layer) was almost fully mobilized in systems with γ_c>1.2 Mg/m³ and is apparently unaffected by i. At γ_c >1.2Mg/m³, the permeability is shown to be dictated mainly by soil fabric, i.e. cluster formation and macropores development in view of the limited development of interlayer water.

3.3 *Soil Fabric Effect*

For the mixtures containing smectite, n_w approached very low values and tends to zero at high γ_c. There is very little further decrease, as γ_c is increased above 1.25 Mg/m³. However, the results show non-zero values of k. This is due to soil fabric, where clay platelets start to form clusters and water flows through the macropores. Under this condition, threshold gradients do not exist any more and interparticle forces start losing their effect. Below 1.25 Mg/m³, a threshold gradient exists in smectite clay with full development of a diffuse double layer. However, it has been shown that development of the interlayer water in smectites can be limited. Under such conditions, clusters and large macropores may be formed and a threshold gradient may be absent or undetected if it is less than the experimentally used one.

4   CONCLUSION

A model based on the Kozeny–Carmen relationship and diffuse ion layer theory has been developed and utilized to study the existence of threshold gradient and the effect of density on clay permeability. The results show that clay permeability is largely dependent on density for less active clay soils such as kaolin. For active clay, soil fabric plays an important role in clay permeability and the presence of threshold gradient as well. Such presence is more prominent in active clay mineral such as smectite at density <1.25Mg/m³. At higher density, the threshold gradient is largely dictated by the extent of interparticle forces in large macropores. It can not be easily detected if the interparticle force is small. For design of bentonite hydraulic barrier, the threshold gradient can be considered in loose systems when needed. However, in dense bentonite systems, consideration of threshold hydraulic gradient is unnecessary as the permeability is expected to be adequately low.

Table 1. The Influence of Density and Hydraulic Gradient on Permeability of Illite and Na-Smectite Systems (After Mouradian and Cheung 1993)

Density (Mg/m^3)	Pore fluid	Hydraulic Gradient (i)	Permeability (cm^2)
Illite (γd)			
1.42	DDW[t]	159	1.1×10^{-13}
	DDW	319	7.6×10^{-13}
1.52	Saline	60	1.6×10^{-12}
	DDW	320	8.3×10^{-13}
1.77	DDW	720	1.0×10^{-13}
	DDW	1 444	1.1×10^{-13}
2.07*	DDW	30	3.5×10^{-13}
	DDW	269	5.6×10^{-13}
Smectite (γc)			
0.6	DDW	3 000	6.1×10^{-13}
	Saline[tt]	3 000	7.0×10^{-12}
Smectite (γc)			
1.02	DDW	460	2.6×10^{-16}
		1 700	6.3×10^{-16}
1.12	DDW	1 550	4.5×10^{-17}
		3 100	1.6×10^{-15}
		>3 100	3.1×10^{-16}
1.24	Saline	1 560	3.1×10^{-16}
	DDW	1 560	5.0×10^{-16}
	DDW	3 120	3.5×10^{-16}
	DDW	> 3 120	3.2×10^{-16}
1.42	DDW	2 000	1.9×10^{-16}
	DDW	5 000 to 10 000	2.1×10^{-16}
1.43	DDW	1 700 to 3 400	1.8×10^{-16}
1.46	Saline	1 600 to 8 000	1.8×10^{-16}
1.50	DDW	2 000	1.9×10^{-16}
1.80	DDW	2 000	9.0×10^{-17}
		5 000 to 10 000	1.1×10^{-16}

t   Distilled Deionized Water
tt   Saline solution 70g/l TDS

5 ACKNOWLEDGEMENT

The authors would like to thank Anthony Hardingham , head of Geotechnical & Tunnelling Department, Atkins China Ltd., for reviewing the manuscript.

REFERENCES

Fransham, P.B., Reesor, S., LaHay, C., Roubanis, D. and Hann, F. 1981. Mechanical and Hydraulic Properties of Two Potential Buffer Materials, Vol. 1, Waterloo Research Institute, Final Report to Atomic Energy of Canada Limited., Project 002-08031, Waterloo Ontario.

Gray, M.N., Cheung, S.C.H. and Dixon, D.A. 1984. Swelling Pressures of Compacted Bentonite/Sand Mixtures. Proc.44th Mat. Res. Soc. Conference, Boston, Vol. 44, 1984.

Mitchell, J.K. 1976. Fundamentals of Soil Behavior, J. Wiley & Sons, Toronto, p.422.

Mouradian, A and Cheung, S. C. H. 1993 Permeability of Clay Soil. Canadian Society for Civil Engineering Annual Conference in Fredericton, June 1993 p.277-286.

Yong, R.N. and Warkentin, B.P. 1975. Soil Properties and Behavior, Elsevior, Amsterdam, p.449.

Clay Science for Engineering, Adachi & Fukue (eds) © 2001 Balkema, Rotterdam, ISBN 90 5809 175 9

Permeability, swelling and microstructure of Pliocene clays from Warsaw

R. Kaczynski

Institute of Hydrogeology and Engineering Geology, University of Warsaw, Poland

ABSTRACT :   The paper presents the results of laboratory and field tests of Pliocene clays in the Warsaw area. The tested clays are overconsolidated and expansive soils containing about 35-80% of clay fraction (< 2 µm) and large quantities of beidellite. The following tests were carried out on the clays: quantitative analysis of microstructures (scanning electron microscope - SEM), swell pressure and permeability tests with the aid of seven different apparatuses. The functional relation between the swelling pressure and the water content (and the degree of saturation) was characterized. The relationship between the various methods of permeability coefficient tests was also determined.

1 INTRODUCTION

The evaluation of permeability, swelling and suction force of cohesive soils is of particular importance for the direct foundation of various engineering objects. It gains special attention in the problem of transport of pollution connected with the increasing number of waste dumps. The soil structure - microstructure - is of fundamental importance for the estimation of soil permeability and swelling.

The above mentioned conditions occur in the Warsaw area. Pliocene clays are often the direct subsoil of engineering objects. The currently built underground in Warsaw is partly situated within Pliocene clays. This paper presents the results of tests conducted in the Śródmieście and Mokotów area (Warsaw districts). All the tests were carried out in the Hydrogeology and Engineering Geology Institute of the Warsaw University.

2 GEOLOGICAL CONDITIONS OF PLIOCENE CLAYS IN THE WARSAW AREA

Warsaw is situated on NE flank of the rim synclinorium. During the Tertiary, the Cretaceous depression (the Warsaw basin) was filled with sandy-clay deposits. The youngest part of the Tertiary, Pliocene, is represented by deposits of a shallow, periodically drying up basin. Pliocene deposits in the Warsaw area are strongly deformed - many folds of different amplitudes and discontinuous deformations occur here. The topmost part of Pliocene deposits is situated at various levels; the difference exceeds sometimes 100 m. A clear, longitudinal upwarp (NNW-SSE), several kilometres long and 0.5-2.0 km wide is observed in the region. The top of the discussed deposits within this upwarp exceeds 100m a.s.l., and   approaches the land surface. Several elevations and depressions have been distinguished within it. The thickness of Pliocene deposits reaches a few dozens of metres, sometimes even 100-150m, 50 m on average. The age and origin of Pliocene deformations has not yet been precisely determined. They probably originated in stages, during several glaciation periods. The Pliocene clays in the Warsaw area are covered with Quaternary deposits, including anthropogenic, of a variable thickness.

3 LITHOLOGICAL AND MINERALOGICAL CHARACTERISTICS OF PLIOCENE CLAYS

Several sedimentation cycles, from sands to clays, have been determined in the Pliocene clays succession. The whole complex of Pliocene clays consists of clays (60-70%), silts (10-20%) and sands (10-20%). Pliocene clays are characterised by a variable colour - the older clays (lowermost horizons) are typically grey, occasionally greenish, green and blue colours prevail in the middle part, with frequent yellow or red spots, while the topmost parts are characterised by vivid colours with frequent red, brown and reddish spots (flame coloured clays).

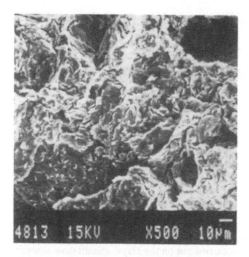

Figure 1. Microstructural surfaces of the Pliocene clays from Warsaw. Undisturbed structure.

Clay minerals and quartz constitute the main mineral content of Pliocene clays. Subordinately feldspars and mica also occur, along with pyrite, marcasite, gethite and hematite. Trivalent iron is present in variegated clays, while grey and green clays are characterised by the occurrence of divalent and trivalent iron in variable proportions. Calcium carbonate composes up to 5% of the bulk mass of clays. Mixed-layer minerals from the beidellite-illite group, with a large quantity of kaolinite represent clay minerals. Calcium and magnesium are the most frequently occurring exchange cations.

Derivatograms of clays from the underground construction site and from Mokotów reveals the occurrence of montmorillonite group minerals: beidellite-B, kaolinite-K and illite-I in the clay fraction, in the following proportion:

$$B^{50-85} >> K^{15-30} >> I^{10-30}$$

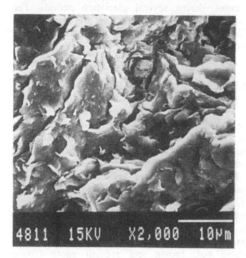

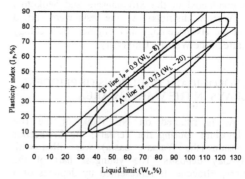

Figure 2. Location of Pliocene clays from Warsaw against a background of Casagrande chart

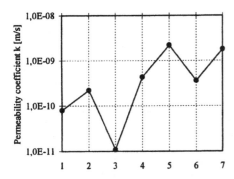

Figure 3. Average value of permeability coefficients
for 7 methods of testing.
1 - BAT, 2 - sucoedometer, 3 - Geonor permeameter,
4 - conventional edometer, 5 - consolidometer, 6 - triaxial
apparatus, 7 - SEM/STIMAN

Occasionally, exception can be observed i.e. in
sandy-silty intercalations the content of kaolinite
reaches 60-90% and of beidellite 10-40%. The
organic matter content of kaolinite reaches 60-90%
and of beidellite 10-40%. The organic matter content
typically does not exceed 0.5%.

4 PHYSICAL AND MECHANICAL
PROPERTIES OF CLAYS

Pliocene clays reveal a considerable spatial
variability of parameters. This results from such
factors as:
- differentiated granulometric content depending
  on sedimentation conditions;
- presence of glaciotectonic deformations causing
  the decrease of strength parameters;
- hydrogeological conditions causing a variable
  degree of saturation of clays with water.

The basic parameters of clays in the investigated
areas are as follows:

Clay sizes, $D < 2\mu m$,   30-75 %
Density, ρ_s,   2.68-2.75 Mg/m³
Bulk density, ρ,   1.85-2.20 Mg/m³
Porosity, n,   30-45 %
Liquid limit, W_L,   40-95 %
Plastic limit, W_P,   20-40 %
Plasticity index, I_P,   20-50 %
Activity, A,   0.50-1.25 %
Swell pressure, σ_{sp},   50-250, somet. to 400 kPa
Overconsolidated ratio, OCR, 1-3, sometimes to 10
Cohesion, c_u,   30-120 kPa
Angle of internal friction, ϕ_u, 10-20, somet. to 25°
Modul. of compressibility, M_o, 5-20, somet. to 40
MPa.

5 MICROSTRUCTURAL PARAMETERS OF
POROUS SPACE

Quantitative and qualitative microstructural
investigations of the Pliocene clays were carried out
with the use of Stiman software connected to a
SEM. The software, based on Fourier and Walsh
analysis was developed by prof. V. Sokolov (1990)
from the Moscow State University. The results of
the analysis enabled to determine microstructural
parameters of the porous space in the tested soils:
porosity, number of pores, total pore area, total pore
perimeter, average diameter, average perimeter of
pores. The morphology of pores is characterised by
the form index equalling to the inverted relation of
two extreme linear diameters of the pore's area. By
using the analysis of the distribution of density
probability function for the diameters of pores in
clay samples it was possible to distinguish ultra-,
micro- and mesopores. The assessment of the degree
of image intensity enabled to determine the degree
of orientation of structural elements – the
quantitative parameter anisotropy coefficient.

The results of the investigations are presented
below and on Figure 1.

Microstructural parameters of the porous space in
Pliocene clays:

Type of microstructure: matrix-turbulent
Porosity,   30-45 %
Number of pores,   2000-19000
Total pore area,   2500-142000 μm^2
Total pore perimeter,   4500-189000 μm
Average diameter,   1.18-2.47 μm
Average area,   3.9-13.9 μm^2
Average perimeter,   9.7-21.0 μm
Average form index of pores,   0.45-0.51
Filtration coefficient,   0.01-0.29 mD
Anisotropy coefficient,   3.8-27.0 %
Number of tests,   10

Presented data show that the analysed clays are
characterised by a matrix-turbulent microstructural
type, and the FF type (face-face) prevails in
microaggregate contacts. The microstructure is
usually medium oriented. Anisometric pores prevail,
with an admixture of fissure pores.

6 CLAY SWELLING

The swelling pressure was analysed by an apparatus
produced by a Norwegian company Geonor. The
swelling pressure σ_{sp} (pressure caused by water -
moisture of soil, at a constant sample volume) was
determined on samples with a natural moisture and
undisturbed structure as well as on samples with
artificially changes moisture. For the natural
moisture of clays the swelling pressure varies

between 50 and 250 kPa, at a saturation degree of $S_r > 0.95$. An incomplete saturation of clays cases a considerable increase of swelling pressure. The following results were obtained:

$\sigma_{sp} = -63.1w^3 + 78.8w^2 - 32.5w + 4.5$
$R^2 = 0.646$
$w = 0.05 - 0.35$
and
$\sigma_{sp} = 5.7S_r^2 - 14S_r + 8.5$
$R^2 = 0.588$
$S_r = 0.4 - 1.0$
where:
σ_{sp} - swell pressure (MPa)
R - correlation coefficient
S_r - degree of saturation
w - water content.

Location of Pliocene clays from Warsaw against a background of Casagrande chart is shown on Figure 2.

7 PERMEABILITY OF CLAYS

The permeability coefficient of the analysed clays was tested with the use of different methods and the following average values (m/s) were obtained:
1. BAT equipment $\quad$ 8.0×10^{-11}
2. sucoedometer $\quad$ 2.2×10^{-10}
3. Geonor permeameter $\quad$ 1.1×10^{-11}
4. conventional edometer $\quad$ 4.3×10^{-10}
5. consolidometer $\quad$ 2.2×10^{-9}
6. triaxial compression apparatus TRX $\quad$ 3.6×10^{-10}
7. SEM/STIMAN microstructural tests $\quad$ 1.8×10^{-9}.

All analyses were carried out on samples with undisturbed structure and with natural moisture. At least 5 analyses were conducted with each method. Additionally, the permeability coefficient was also determined indirectly from microstructural analyses.

The coefficient of variation within each method varied between 51 and 62%. The results reveal the indirect methods (6 and 7) considerably differ from direct methods (2 and 5). The BAT method, despite different analysis conditions (analysis in situ), gives results similar to direct laboratory methods. Graphically average value of permeability coefficient for 7 methods is shown on Figure 3.

8 CONCLUSIONS

1. The analysed Pliocene clays from the Warsaw area often constitute the direct soil foundation and can act as isolation layers against water as well as barriers retaining the migration of pollution.
2. Quantitative microstructural parameters defining pore space can in the future (after conducting more analyses) be used to determine functional relations between microstructural parameters and physical and mechanical characteristics of clays.

3. Analysis of permeability coefficient carried out by seven methods reveal that results obtained by indirect methods considerably vary from results obtained by direct methods. The BAT method, despite different analysis conditions, gives results comparable to those obtained in direct methods.
4. The following functional relations have been obtained in clay swelling analyses:
   - between swelling pressure and moisture
   - between swelling pressure and degree of saturation.
5. Pliocene clays contain mixed-layer minerals from the beidellite (B) - illite (I) group, with considerable admixtures of kaolinite (K): $B^{50-85} >> K^{15-30} >> I^{10-30}$

REFERENCES:

Fredlund, D.G. & Rahardjo, H. 1993. Soil mechanics for unsaturated soils. A. Wiley - Interscience Publication. New York: J. Wiley & Sons Inc.

Garbulewski, K. & Żakowicz, S. 1995. Suction as an indicator of soil expansive potential. Proc. 1st Int. Conf.: On Unsaturated Soils. Paris. Vol. 2: 593-599. Rotterdam: Balkema.

Garbulewski, K., Żakowicz, S., Wolski, W. 1995. Laboratory methods for testing of the swelling soils permeability. Proc. 10th Europ. Conf.: Soil Mechanics and Foundation Engineering. Copenhagen.

Grabowska-Olszewska, B., Osipov, V.I., Sokolov, V.N. 1984. Atlas of microstructure of clay soils. Warszawa: PWN.

Grabowska-Olszewska, B. (Ed.) 1998. *Applied Geology. Properties of unsaturated soils* (in Polish). PWN, Warsaw.

Head, K.H. 1992. Manual of soil laboratory testing. London.

Kaczyński, R. & Grabowska-Olszewska, B. 1997. Soil mechanics of the potentially expansive clays in Poland. *Applied Clay Science* 11: 337-355.

Sokolov, V.N., Osipov, V.I., Tolkacheva, M.D. 1980. The electron microscopic studies of pore space of soils by method consignate surfaces. *Journ. of Micr.* 120: 363-366.

Sokolov, V.N. 1990. Engineering-geological classification of clay microstructure. Proc. 6th Cong. IAEG 3: 753-760. Balkema.

Torstensson, B.A. 1984. A new system for groundwater monitoring. *Ground Water Monitoring Review* IV: 131-138. Stockholm.

Clay Science for Engineering, Adachi & Fukue (eds) © 2001 Balkema, Rotterdam, ISBN 90 5809 175 9

Viscous fingering of solution in bentonite slurries

M. Mizoguchi

Department of Biological and Environmental Engineering, University of Tokyo, Japan

ABSTRACT: Fingering flow in soil that causes contamination of groundwater is one of important phenomena in leaking from hazardous waste landfill. Viscous fingering patterns have been observed when solution is injected into Na-bentonite slurries through a pinhole at the center of Hele-Shaw cell. Morphological patterns were obtained according to water/clay ratio, injection pressure and concentration of solution for NaCl, and $CaCl_2$. As the concentration of the solution increased, the fingers became narrower and the patterns displayed a more ramified morphology. In addition, the fingers produced by $CaCl_2$ solution were more stinging than those by NaCl solution. These results may be attributed to the interaction between clay particles and cations at the tip of a moving.

1 INTRODUCTION

Hazardous waste landfills must be designed and safe managed in our society. For the requirement a double-liner system is normally used to stop the flow of liquids, called leachate, from entering the soil and groundwater beneath the hazardous wastes site. In the double-liner system, the upper and the lower liners must be a flexible-membrane lining made of sheets of plastic or rubber, but recompacted clay is often used beneath the lower liner (Masters 1991). Since the clay has low-permeabilty and swells, it is useful as a material to seal water from the landfill. However, attention must be paid to the interaction between the clay and soluble salt as well as water because the leachate is normally high concentrated solution.

Water often flows heterogeneously through actual soil; one of such water flows is fingering flow. The fingering flow has been studied intensively and some models have been proposed in unsaturated soil (Nieber 1996). However, only a few studies have been made in saturated soil. Van Damme et al. (1986) have first observed that viscous fingering occurs when water flows through clay slurries in a radial Hele-Show cell and that the pattern has a fractal morphology.

The viscous fingering have been investigated using various Newtonian fluids and the computer simulation has predicted some morphological patterns with anisotropy and the linear stability (Ben-Jacob & Garik 1990). Due to such studies the viscous fingering for Newtonian fluids is now well-understood whereas our understanding is not satisfactory for non-Newtonian fluids such as polymer solutions and clay suspensions.

For non-Newtonian fluids, Deccord et al. (1986) have conducted the fingering experiments of aqueous polymer solutions in a radial Hele-Show cell and observed more ramified fingering patterns than Newtonian fluids. Zhao & Maher (1993) injected water into a radial Hele-Show cell, which contains aqueous solutions with different concentration, and found a transition from fingering to fracturing patterns beyond a threshold value of the injection rate. Kawaguchi et al. (1997) compared viscous fingering patterns in polymer and Newtonian solutions and concluded the transition in the pattern would be related to changes in both elastic properties and surface tension.

Clay has different elastic properties according to its environment such as water content, soluble salt, and pressure. Therefore, it is more difficult to predict the fingering pattern in clay suspensions than polymer solution. Considering clay is normally used in an actual hazardous wastes site, it is quite important to know the fingering pattern in clay layer on the assumption that leachate would leak from a hazardous wastes landfill. Lamaire et al. (1991) performed experiments using smectite clays and showed the transition between a viscous fingering (VF) regime characterized by fractal patterns of "fingers" and a viscoelastic fracturing (VEF) regime characterized by fractal patterns of "cracks". However, no experiment has been made focusing on the interaction between clay particles and ions.

If the fingering in clay is generated by only hydraulic instabilities at the water-clay interface as Van Damme et al. (1986) pointed out, the fingering pattern will change dramatically when solution is injected into clay slurries because the interaction between clay particles and ions changes at the tip of a finger. In this study, I observed fingering patterns of solution injected into clay slurries as a function of injection pressure, species of cation and solute concentration using a radial Hele-Show cell.

2 EXPERIMENT

2.1 *Material*

The clay used in this study was Na-bentonite (Kunimine Koukakougyou, Kunigel-V1). The dry powder of the clay was mixed with distilled water to water/clay ratios of 8.5 - 12.5 by weight.

2.2 *Solution*

Distilled water, NaCl solutions (0.1, 0.5, 1.0 N) and $CaCl_2$ solutions (0.1, 0.5, 1.0 N) were injected into the clay slurries.

2.3 *Experimental Apparatus*

Figure 1 shows experimental apparatus to observe the fingering patterns. The experiments were performed in radial geometry, in a square Hele-Show cell made by a pair of acrylic plates (A: $1 \times 30 \times 30$ cm) with a spacer of 0.4 mm clamped in between the plates at four symmetrical positions. The cell, which

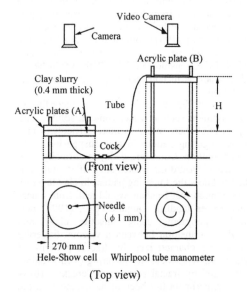

Figure 1. Experimental apparatus

had 60 small holes to keep the pressure of the slurries atmospheric at the circumference with the diameter of 27 cm, was set horizontally over a black rubber sheet to distinguish fingers from the clay slurry. A 1-mm pinhole made by a syringe needle was set at the center of the lower plate and connected to whirlpool tube. Injection rate was measured by reading a moving meniscus in the whirlpool tube placed on a horizontal acrylic plate (B). Injection pressure was given by controlling the elevation of the acrylic plate, H, from 10 to 50 cm.

2.4 *Experimental Procedure*

A homogenized clay slurry was poured on the lower plate of Hele-Show cell and the upper plate was covered over the slurry, removing trapped air. Before solution was injected, injection pressure was kept to be H cmH_2O. Then solution was injected through a pinhole at the center of the cell immediately after a cock, which was equipped near the entrance of the pinhole, was open. Fingering patterns were recorded by a camera over the Hele-Show cell (A), and a meniscus in the tube was also recorded by a video camera over the acrylic plate (B).

3 RESULTS AND DISCUSSION

Figure 2 shows the fingering patterns obtained by injecting 1.0 N NaCl solution into the clay slurry with the water/clay ratio of 9.0 under the pressure of 20 cmH_2O. Radial fingers evolved from the center of the cell like tree branches. The characteristics are in good agreement with the previous studies on viscous fingering patterns obtained by Van Damme et al. (1986). This result indicates that the water/clay ratio of 9.0 is high enough to observe the viscous fingering in my experiment. Once a finger reached the circumference at about 60 sec, the solution flew mainly out of the finger, and the evolution rate of the other fingers decreased. Most of the previous studies just described the fingering patterns into clay slurries confined in a Hele-Show cell, but did not refer to the patterns after a finger reached the boundary with the atmospheric pressure. My experiment is different from the previous studies with respect to this point.

Figure 3 shows the fingering patterns obtained by injecting distilled water, 0.5 N and 1.0 N NaCl solutions into the clay slurry with the water/clay ratio of 11.0 under the pressure of 20 cmH_2O. For the 1.0 N NaCl solution, more and narrower fingers evolved in the clay slurry with the water/clay ratio of 11.0 than 9.0 shown in Figure 2(c). This result may be attributed to the viscosity of the clay slurry; solution will limit to invade into the clay slurry with low water/clay ratio because the viscosity is high. Moreover, as the concentration of NaCl solution increased , the fingers was narrower and the number of the fin-

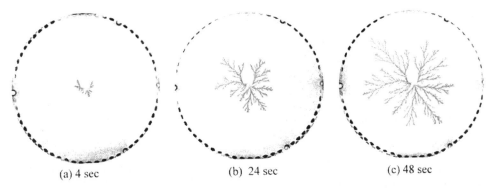

| (a) 4 sec | (b) 24 sec | (c) 48 sec |

Figure 2. Evolution of finger in clay slurry. Fingering patterns obtained by injecting 1.0 N NaCl solution into the clay slurry with the water/clay ratio of 9.0 under the pressure of 20 cmH$_2$O.

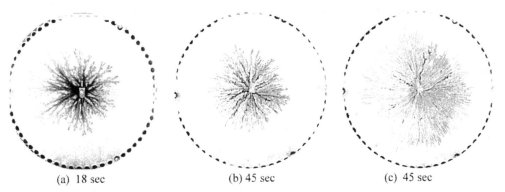

| (a) 18 sec | (b) 45 sec | (c) 45 sec |

Figure 3. Influence of solute concentration. Fingering patterns obtained by injecting distilled water (a), 0.5 N (b) and 1.0 N NaCl (c) solutions into the clay slurry with the water/clay ratio of 11.0 under the pressure of 20 cmH$_2$O.

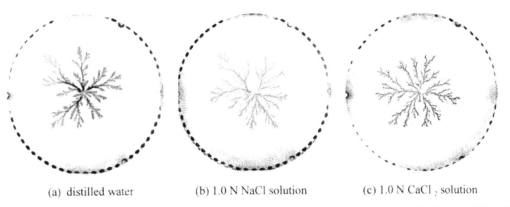

| (a) distilled water | (b) 1.0 N NaCl solution | (c) 1.0 N CaCl$_2$ solution |

Figure 4. Influence of cation in the solution on fingering patterns into a clay slurry with the water/clay ratio of 9.0 under the pressure of 20 cmH$_2$O after 36 sec.

ger increased. This result suggests Na ions will interact with clay particles at the tip of a finger to develop a flow path of the solution.

Figure 4 shows influence of cation in the solution on fingering patterns into a clay slurry with the water/clay ratio of 9.0. The fingers of the solution were displayed a more ramified morphology than those of water. The fingers produced by CaCl$_2$ solution were more stinging than those by NaCl solution. The differences in the patterns may be attributed to the interaction between clay particles and cations: the aggregation of the clay particles caused by the ion exchange from Na$^+$ to Ca^{2+} at the tip of a moving finger; when CaCl$_2$ solution invade into Na-saturated clay slurry, Na-ions on clay particles is exchanged by Ca-ions and the attractive force between clay particles increases, then causing the aggregation of clay particles (Figure 5). Due to such a mechanism of the aggregation at the tip of a finger the fingering patterns of the CaCl$_2$ solution will be more stinging than those by NaCl solution. Although the fingering patterns in clay have been observed as a function of water/clay ratio, injection pressure or injection rate of water, and spacer thickness in the previous experiments (Daccord G & J. Nittmann 1986, Lemaire et al. 1991), no study has been made on the fingering patterns with respect to the ion exchange except my experiment. Therefore, the mechanism that I proposed here must be investigated in details from the viewpoint of colloid chemistry.

Figure 6 shows injection rate of water as a function of injection pressure and water/clay ratio. Solid circles(●) and open squares (□) denote the injection pressure for the water/clay ratio of 10.5 and the water/clay ratio under the injection pressure of 30 cmH$_2$O, respectively. As the injection pressure increased, the injection rate increased linearly. The injection rate of water also increased with the water/clay ratio. These results indicate that the fingering flow of water in the clay slurry with the water/clay ratio of 10.5 obeys Darcy's law:

$$v = - (b^2/12 \eta) \operatorname{grad} p$$

where v is the injection rate, p the pressure, η the viscosity, and b is a constant related to the thickness of the Hele-Show cell.

Figure 7 shows injection rate of solution as a function of solute concentration in clay slurry with the water/clay ratio of 10.5 under the injection pressure of 10 cmH$_2$O. The injection rate decreased as the concentration increased. This result indicates that the fingering flow rate of solution will be smaller than that of water in the slurry with the higher water/clay ratio. Lemaire et al. (1991) observed that smectite clays had a morphological transition from viscous "fingers" to viscoelastic "cracks" at about the water/clay of 10.0. According to their classification, all fingering patterns obtained in my experiment belong to the viscous fingering. However, the recompacted clay beneath the hazardous wastes site will have more viscoelastic property with lower water content. Therefore it is quite dangerous to conclude that concentrated leachte will flow slower than water in clay layer only from the result shown in Figure 7. We must examine the influence of concentration on the flow rate in clay with the lower water/clay ratios.

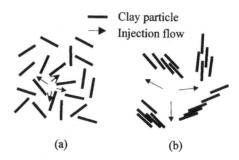

Figure 5. Concept of injection flow in (a) dispersed Na - bentonite and (b) aggregated Ca - bentonite.

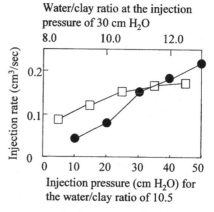

Figure 6. Injection rate as a function of injection pressure (●) and water/clay ratio(□).

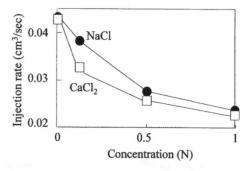

Figure 7. Injection flow rate as a function of solute concentration in clay slurry with the water/clay ratio of 10.5 and the injection pressure of 10 cm H$_2$O.

4 CONCLUSIONS

I observed viscous fingering patterns when solution flows through clay slurries in a radial Hele-Show cell. As a result, the fingers of solution were displayed a more ramified morphology than those of water. The fingers produced by $CaCl_2$ solution were more stinging than those by NaCl solution. In addition, the injection rate of solution decreased as the concentration increased. These results can be attributed to the interaction between clay particles and cations at the tip of a moving finger. We must examine the influence of concentration on the flow rate in clay with the lower water/clay ratios.

The author wishes to acknowledge the assistance of Mr. J. Nishizawa and Mr. T. Ezaki, who were my students at Mie University, for the experiment.

REFERENCES

Ben-Jacob, E. & P. Garik 1990. The formation of patterns in nonequilibrium growth. Nature, 343: (6258) 523-530.

Daccord G & J. Nittmann 1986. Radial viscous fingers and diffusion-limited aggregation - fractal dimension and growth sites. Physical Review Letters, 56: (4) 336-339.

Kawaguchi, M., K. Makino & T. Kato 1997. Comparison of viscous fingering patterns in polymer and Newtonian solutions. Physica D. 105: (1-3) 121-129.

Lemaire E, P. Levitz, G. Daccord & H. Van Damme 1991. From viscous fingering to viscoelastic fracturing in colloidal fluids. Physical Review Letters, 67: (15) 2009-2012.

Masters, G. M. 1991. Introduction to environmental engineering and science. Prentice Hall, 266.

Nieber, J. L. 1996. Modeling finger development and persistence in initially dry porous media. Geoderma 70: 207-229.

Van Dame, H, F. Obrecht, P. Levitz, L. Gatineau & C. Laroche 1986. Fractal viscous fingering in clay slurries, Nature. 320:731-733.

Zhao H & J. V. Maher 1993. Associating-polymer effects in a Hele-Shaw experiment. Physical Review E, 47: (6) 4278-4283.

Clay Science for Engineering, Adachi & Fukue (eds) © 2001 Balkema, Rotterdam, ISBN 90 5809 175 9

Critical coagulation concentration and permeability of the Fukaya clayey soils

T. Nishimura & M. Kato
Tokyo University of Agriculture and Technology, Japan

K. Nakano
National Institute of Agro-Environmental Sciences, Tsukuba, Japan

T. Miyazaki
School of Agricultural and Life Sciences, Tokyo University, Japan

ABSTRACT: Permeability depends on soil structure and the structure is affected by behavior of soil clays. As clay has charge so that behavior of the clay is affected by species and concentration of electrolytes in soil solution as well as clay mineralogy. The presenting study aimed to clarify the relation between Critical Coagulation Concentration (CCC), soil structural stability, and permeability. Two clayey soils in Fukaya, Saitama, Japan were used in this study. Vermiculite and kaolinite were the dominant clay mineralogy of the soils and considerable amount of smectites and mica were also detected by the standard X ray diffraction technique. Texture, bulk density, saturated hydraulic conductivity of 100cc core samples, cation exchange capacity, total carbon as well as clay mineralogy of the soils was almost identical. CCCs of the soil with $CaCl_2$ solution were similar but the soil having unstable structure (water logged) showed greater CCC than the more stable Fukaya-A, well drained soil with the NaCl solution. Soils were packed into an 8.5cm in inner diameter and 15cm length acrylic plastic cylinder and various concentration of NaCl and $CaCl_2$ solutions were applied on to the soil. The unstable water logged soil was sensitive to concentration of soil solutions while the stable well drained soil showed steady permeability with soil solutions having concentrations lower than the CCC. However, effect of coagulation feature of clay on soil structure was almost neglectable under simulated rainfall.

1 INTRODUCTION

Permeability depends on soil texture (Mualem & Assouline 1989). As well as bulk density and soil hardness clay mineralogy, texture, cations and anions in the soil, and organic matters are the factors affecting soil structural stability and then may affect permeability (Quirk (1994), Bradford et al. (1987), Russo and Bresler (1977), Shainberg and Letey (1984), Suarez et al. (1984)).

Recently, some upland fields in Fukaya area, a northern suburb of Tokyo, have started to show water logging while others are still having high permeability. Poor permeability of upland fields may cause loss of production, divergence of agricultural chemicals like excess fertility, pesticide, and herbicide as well as soil particles with runoff. This results pollution of rivers and ponds. Nakano et al. (1998) reported there was not significant difference between soil properties such as, texture, cations and anions, saturated hydraulic conductivity of core samples, and organic matters of waterlogged and well-drained fields.

Purpose of presenting study is to explain the mechanism of recent water logging in Fukaya fields. The presenting study focused on effect of clay mineralogy, dispersion of soil and disintegration of soil aggregates on soil structural stability and permeability. It is difficult to control destructive force onto soil particles quantitatively. In presenting study we assumed minimum energy exerted to the soil by flow through the soil column, and medium by a wet sieving and greatest destructive action was caused by simulated raindrops.

2 METHODS and MATERIALS

2.1 Soil

Two soils from a waterlogged and a well-drained upland field were used in this study. Soil was passed through 3 mm opening screen. A half of the soils were kept in a plastic bag to keep moisture content, moist soil, and rest of the soils were air dried. The moisture content of the moist soil was 27% by weight that was the same as the water content at the sampling site at the field. Some of soil physical and chemical properties are shown Table 1. Clay mineralogy of the soils were determined by standard X-ray diffraction technique and Lim and Jackson (1986) reported, that employed glycerol adsorption following to heating of Li saturated clay.

Table 1 Properties of Soil Material

	Well-drained soil	Waterlogged soil
Texture	CL	L~CL
Bulk density (g/cm³)	0.9~1.3	0.9~1.3
Hydraulic conductivity (cm/sec)*	1.6×10^{-4}	7×10^{-4}
CEC (CEC after DBC) (cmol+/kg)	16.4 (29.8)	13.4 (27.3)
Exchangeable Sodium Percentage (%)	0.36	0.40
pH (1:5=soil:water ratio)	7.2	6.3
Total-Nitrogen (g/kg)	0.69	0.82
Total-Carbon (g/kg)	6.5	7.4
CCC (NaCl) (mmol_c/L)	75	120
CCC (CaCl₂) (mmol_c/L)	1.5	1.7

*: undisturbed, sample size was 5cm in diameter and 5cm in length.

2.2 *Critical Coagulation Concentration (CCC) of the soil clay*

Fractionate the clay particles smaller than 2 μm by settlement. Clays were saturated with 1 mol_c/L NaCl or CaCl₂ solution and dialysised until outer solution reached electrolyte free. Clay suspensions were mixed with a NaCl or CaCl₂ solutions to attain 0 to 500 mmol_c/L with NaCl and 0 to 6 mmol_c/L with CaCl₂ solutions. Clay concentration of the suspension was 3 g/L. The suspensions were allowed to settle for 2 hours, after which time 3 ml of suspension was sampled at a depth of 2cm. The suspension was transfer to a cuvette and light transmittance of 700mn wavelength light was measured by spectrophotometer (Shimazu Co.). Results of the CCC of the soils is also shown in Table 1.

2.3 *Percolation experiment*

Moist soil samples were packed into a plastic acrylic cylinder of 8.4 cm inner diameter. A 10cm thick of the sample soil was packed onto a 5cm thick glass beads, 1 mm in diameter, layer. Bulk density of the soil was 1.15 g/cm³. Solutions were supplied onto the center of the soil surface by using tube pump. The supply rate was 0.0047 cm/sec which was larger

than saturated hydraulic conductivity of the soils at the beginning of the experiment and the rate was controlled to keep shallow ponding on the soil surface. Effluent from the end of the column was sampled to determine flux density through the column and electrical conductivity (EC) of the effluent. Initially, electrolyte solution of the highest concentration was supplied, then the concentration decreased gradually when a cumulative drainage reached 2 pore volumes or EC of the effluent reached the EC of the input solution. Based on the result of CCC, a 500, 100 and 10 mmol_c/L were selected for the NaCl solution, and a 6, 2, and 0 (distilled water) mmol_c/L CaCl₂ solutions were chosen in this study. pH of the solution was between 6.3 to 6.9.

2.4 *Water stable aggregate (wet sieving)*

Wet sieving was conducted by the method similar to Kemper and Rosenau (1987) except electrolyte solutions instead of water were used in presenting study. NaCl and CaCl₂ solutions were employed in this study. Most of the case pH of the solutions was between 6 to 7.

2.5 *Rainfall experiment*

Rainfall experiment has conducted in the Arid Dome of the Arid Research Center of Tottori University, Japan. Raindrops were generated from nozzles of 12m high, and drops were expected to reach terminal velocity. An 85% of drops were smaller than 2mm. Mean rainfall intensity in presenting study was 40 mm/hr.

A plastic acrylic box of 50cm×30cm×10cm was used in this study. Soil was packed to a 5cm deep over a 5cm thick sand and gravel layer. The box has drains at the bottom and effluent was collected. A flume was connected to lower end of the soil surface, and runoff was collected at five to 10 minutes interval. EC and pH of runoff were measured as well as weight of runoff. 2.5 t/ha of gypsum was spread to the soil surface prior to the rainfall to manipulate the electrolyte concentration of runoff.

3 RESULTS

3.1 *Critical Coagulation Concentration (CCC)*

It was assumed that a 40% light transmittance of 700nm wave length light was the threshold of CCC where more than 85% of clay particles settled. The CCC of the both soil clays were sensitive to the cation type. The CCC of the well-drained soil was 75 mmol_c/L with NaCl and 1.5 mmol_c/L with CaCl₂,

292

while waterlogged soil showed 120 mmol$_c$/L and 1.7 mmol$_c$/L , respectively.

3.2 Changes in permeability due to species and concentration of electrolyte solution

The terminal velocity of raindrops having 2 mm in diameter is 6 to 7 m/sec (Hudson 1995), and reciprocation of wet sieving is 0.034 m/sec or slower. As mean water flow velocity through a pore of the soil column was 10^{-4} m/sec or smaller, the permeability experiment assumed to have smallest disturbing action onto soil aggregates.

Decrease in flux density through the soil column may indicate decrease in permeability of the soil. This has happened due to changes in soil structure, clogging pores by dispersed particles or sealing at the surface. Waterlogged soil showed decrease in flux when the electrolyte concentration was similar to or lower than the CCC of the soil clay, 100 mmol$_c$/L with NaCl and 0 mmol$_c$/L following to CaCl$_2$ solution. The well drained soil maintained the flux density when solutions had concentrations lower than CCC, 10 mmol$_c$/L with NaCl and distilled water following to CaCl$_2$ solutions (Figure 1).

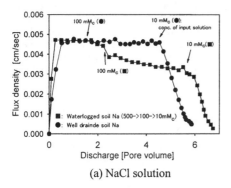

(a) NaCl solution

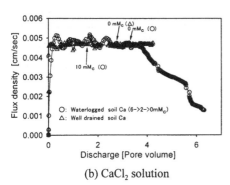

(b) CaCl$_2$ solution

Figure 1 Changes in water flux through soil column due to concentration and species of electrolyte solutions

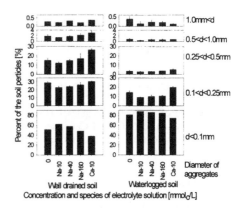

Figure 2 Effect of concentration and species of electrolyte solutions on water stable aggregates

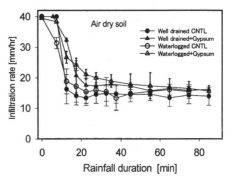

Figure 3 Gypsum application and infiltration of airdry Fukaya soils

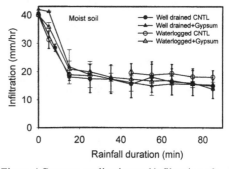

Figure 4 Gypsum application and infiltration of moist Fukaya soils

3.3 Water stable aggregates

Wet sieving assumed to have medium destructive energy to soil aggregates. Waterlogged soil consisted of more finer water stable particles than the well-drained soil (Figure 2). Difference between waterlogged and well drained soils were clear for fractions of particles smaller than 1.0mm. The well drained soil contained more 0.1 mm <d< 1.0 mm particles, while waterlogged soil had finer, d<0.1 mm, particles. This suggested disintegration of larger aggregates into finer ones had happened during wet sieving.

Concentration and type of cation had a little effect on water stable aggregates. ESP of soils were quite low (Table 1), thus dilute NaCl solution could promote degrade of coarser aggregates. However, greater concentration of NaCl and CaCl₂ seemed to stabilize coarser aggregates. This was more apparent for the well drained soil.

3.4 Rainfall experiment

Permeability of the soils were greater than rainfall intensity thus decrease in infiltration and/or beginning of runoff may suggest the change in soil structure, i.e. surface sealing, and decrease in permeability. The well drained and water logged soils showed decrease in infiltration within 20 min from the beginning of the rainfall. Effect of gypsum application on soil structural stability and prevention of sealing was observed only first 20 min from the beginning of the rainfall Surface soil was disintegrated by the rainfall and surface seal has formed with 20 to 30 min of rainfall regardless to gypsum treatment. The seal was the reason of decrease in infiltration rate during a rainfall. Additionally, final infiltration rate of the soils were the same regardless to gypsum application (Figure 3 and 4).

Figure 5 shows electric conductivity (EC) of runoff during a rainfall. EC of runoff was 1 dS/m or greater when gypsum was applied prior to the rainfall. EC of runoff suggested electrolyte concentration of the runoff was greater than 10 mmol$_c$/L for the gypsum treatment and around 1 mmol$_c$/L without the gypsum (Nishimura 1996). The CCC of the soils showed 10 mmol$_c$/L of the Ca^{2+} was high enough to flocculate the clay of the soils.

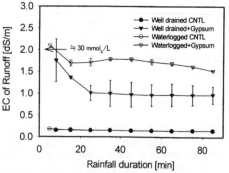

Figure 5 Electric conductivity of runoff during a simulated rainfall

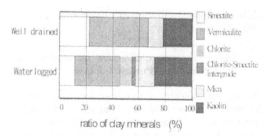

Figure 6 Clay minerals of Fukaya soils

3.5 Detail of clay mineralogy of the soils

Figure 6 shows clay mineralogy of Fukaya soils. In Kanto area, soils that do not have alluvial histo·y may generally contain amorphous clay minerals such as allophane and imogolite. The sample soils are alluvial soil that consisted of sediments have been brought from upstream of Tonegawa river and Arakawa river so that the soils contained significant amount of layered clay minerals and were less affected by amorphous clay minerals.

The well drained soil contained more smectites than water logged soil. However, after a method reported by Lim and Jackson (1986), that was employed glycerol adsorption following to heating of Li saturated clay, 2/3 of the smectites in the well drained soil was low swelling beidellite while a half of the smectites in waterlogged soils was montmorillonite. Thus, in terms of montmorillonite the water logged and well drained soils contained similar amount.

4 DISCUSSION

It is often assumed CCC as an index of soil structural stability. Many studies reported that permeability of soils decreased when electrolyte concentration of flowing solution became lower than CCC (Quirk and Schofield (1955), Russo and Bresler (1977), Regea et al. (1997)). Miller (1987) also reported effect of gypsum application on infiltration of rainfall.

In presenting study, only the result of the permeability experiment, which exerted smallest disturbing energy onto soil aggregates, showed clear agreement with the CCC and the soil structural stability. Water stable aggregates by wet sieving showed slight difference with electrolyte concentration and cation type. It implied destructive action of wet sieving procedure could be equilibrated or exceed contribution of flocculation-dispersion feature of the soil clay on stability of soil structure, and this made effect of electrolyte concentration and cation species on water stable aggregates less. In the rainfall experiment, EC of runoff showed electrolyte concentration of runoff from gypsum treated soil was higher than the CCC. This suggested clays in the soil were flocculative condition, however, effect of gypsum application on infiltration of rainfall was small and limited at the beginning of the rainfall. It may be concluded qualitatively that colloidal behavior of soil has an important role in soil structural stability and hydraulic properties however as the action applied to the soil becomes greater the contribution of the CCC, effects of soil colloidal feature, on soil structural stability would be neglectable. It requires furhter study on the threshold in extent of disturbance within that coagulation feature of clay particles dominates stability of soil structure.

The water logged and well drained soils showed different feature in the permeability experiment. There was no distinct difference in soil properties (Table 1) that could affect structure of water logged and well drained soils. Moreover, smectite, which generally makes the soil structure unstable, was more in the well drained soil after the result of standard X-ray diffraction. However, using a method reported Lim and Jackson (1986) suggested well drained soil contained more beidelite. Beidelite is a smectite which has isomorhpic substitution in a Si-tetrahedra layer, and expected to behave rather like vermiculite than montmorillonite. More fraction of beidelite could be a reason that well drained soil was more flocculation though it contained more smectite than the waterlogged soil.

5 CONCLUSIONS

Many studies reported flocculation-coagulation feature of soil clay affects permeability of the soil. In presenting study, critical coagulation concentration (CCC) could explain changes in permeability due to concentration and species of flowing electrolyte solution. However, presenting study suggested CCC could not describe water stable aggregates, stability of soil structure and permeability under simulated rainfall suitable. It suggested that threshold in external disturbance, excess which, effect of coagulation features of clay on soil structure and permeability becomes relatively small would be expected.

ACKNOWLEDGEMENT

The author would like to thank to Ms. Akiko. Kimura for her help to conduct some of experiments in presenting stugy. Rainfall experiment of this stugy was carried out under the Cooperative Research Program of ALRC, Tottori University (A-IV,1999).

REFERENCES

Bradford, J.M., J.E.Ferris, and P.A. Remley 1987. Interrill soil erosion processes: I. Effects of surface sealing on infiltration, runoff, and soil splash detachment. Soil Sci. Soc. Am J. 51:1566-1571

Hudson N. 1995. Chapter 3 In Soil Conservation. 55-68, Iowa State Univ. Press, Ames, Iowa, USA

Kemper, W.D and R.C. Rosenau 1987. Aggregate stability and size distribution. In A. Klute et ^l. (ed.) Methods of Soil Analysis, Part I. Agro:425-442. Am. Soc. Agronomy Inc. Madison Wisconsin USA

Lim C.H. and M.L. Jackson 1986. Clays and Clay Mierals. 34(3):346-352

Miller, W.P. 1987. Infiltration and soil loss of three gypsum-amended Ultisols under simulated rainfall. Soil Sci. Soc. Am. J., 51:1314-1320

Mualem, Y. and S. Assouline 1989. Modeling soil seal as a non-uniform layer. Water Resour. Res. 25:2101-2108

Nakano, K, M. Nakano, T. Miyazaki 1998. Physical fertility of soils and the drainage efficiencies of upland field. Trans. of Japan Soc. Irrig. Drain. and Reclamation Engg. 195:112-121

Nishimura, T. 1996 Changes in infiltration and ion exchange subsequent to gypsum application to a Japanese acid soil. Trans. of Japan Soc. Irrig. Drain. and Reclamation Engg. 184: 167-173

Quirk, J.P. 1994 Interparticle forces: A basis for the interpretation of soil physical behavior, Adv. in Agronomy, 53:121-183

Quirk, J.P. and R.K. Schofield 1955. The effect of electrolyte concentration of soil permeability. J. of Soil Sci. 6:163-178

Russo, D. and E. Bresler 1977. Effect of mixed Na-Ca solutions on the hydraulic properties of unsaturated soils. Soil Sci. Soc. Am. J. 41; 713-717

Regea, M., T. Yano, and I. Shainberg 1997 The response of law and high swelling smectites to sodic conditions. Soil Sci. 162(4): 299-307

Russo,D., and E.Bresler: 1977 Effect of mixed Na-Ca solutions on the hydraulic properties of unsaturated soils, Soil Sci. Soc. Am. J., 41: 713~717

Shainberg, I. and J. Letey 1984 Response of soils to sodic and saline conditions, Hilgardia 52:1-57

Suarez, D.L., J.D. Rhoades, R. Lavado, and C.M. Grieve 1984 Effect of pH on saturated hydraulic conductivity and soil dispersion, Soil Sci. Soc. Am. J., 48: 50-55

Clay Science for Engineering, Adachi & Fukue (eds) © 2001 Balkema, Rotterdam, ISBN 90 5809 175 9

Soft X-ray radiography of structure-induced macropore flow in clayey soils

Y. Mori, I. Takeda & A. Fukushima
Life and Environmental Science, Shimane University, Matsue, Japan

ABSTRACT: Soft X-ray radiography was performed in combination with saturated hydraulic conductivity and outflow experiments to characterize structure-induced macropore flow in agriculture field soils. Image analysis showed that isolated cylindrical macropore paths were predominant in paddy soils and tortuous networked macropore flow paths were observed in upland field soils. Interaggregate macropore flow paths were also found in upland field soils. Numerical analysis revealed the presence of discontinuity in hydraulic conductivity near saturation, which is small in loamy soils and large in clayey soils. Interaggregate macropore might lessen the discontinuity at near saturation. Therefore, the domain was divided into two (cylindrical macropore and matrix) in clayey paddy soils and into three (cylindrical macropore, interaggregate macropore, and matrix) in loamy upland field soils. Macropore flow in clayey paddy soils might conduct water and solutes to a lower profile with few interactions with bulk soils.

1 INTRODUCTION

Preferential flow (also called bypass flow) is often observed in structured soils, whereby free water and its constituents move through preferred pathways in a porous medium. This flow occurs, in part, via continuous wormholes, root channels, or interaggregate pores, known as macropores. Paddy field is one of the major land-uses in Asia, where 66 percent of soils in Tropical Asia fall into clayey classes (Kawaguchi & Kyuma 1977). Surface-applied chemicals such as fertilizers and pesticides might move rapidly down to groundwater along a small crack or single root channel in clayey soils. Characterization of this flow is one of the major challenges for soil scientists these days, because conventional approaches based on averaged, laminar, and convection-dispersion flow theories might fail to describe it properly. Moreover, recent experimental evidence suggests that preferential flow is more the rule than the exception under field conditions (Flury 1996). In order to predict the flow and solute transport through these channels, this flow should be properly characterized through non-destructive observation.

Most of the non-destructive techniques employed in soil science use strong radioactive rays, such as 200kVp of X-ray computed tomography (CT) and 662keV of ^{137}Cs gamma ray. Because X-ray attenuation in this range has linear relationship to water content or bulk density, and is not different by material, quantitative analysis was successfully conducted for these indexes (Aylmore 1993). Three-dimensional distribution of water or solute was obtained, although the resolution of about 1mm is insufficient to observe macropore flow in root channels, with a diameter of less than 1mm (Vepraskas et al. 1991; Mori et al. 1999a). On the other hand, low energy X-rays attenuate differently according to the material, making high-contrast images when contrast media are used properly. Sometimes simple radiography is sufficient to observe the morphological feature of the flow paths. Thus, soft X-ray radiography utilizes the difference in attenuation effectively. Tokunaga et al. (1984) and Tokunaga (1988) developed a technique of high-resolution soft X-ray radiography for soils. The image resolution was several ten micrometers, which was suitable for studying morphological properties of root channels. Narioka (1987) developed the tree-dimensional reconstruction technique for root channels. If the dynamics of macropore flow and interaction with soil matrix are properly captured, the results might be further utilized for characterizing the fate of surface applied chemicals.

The objectives of this study were to characterize macropore flow and successive drainage in paddy field soils, where the flow was visualized using soft X-rays. Loamy upland field soils were also examined to compare the effect of soil texture. Saturated hydraulic conductivity experiments and then outflow experiments were conducted with a contrast medium as a tracer of the flow. The patterns of medium movement were photographed on an X-ray film to discuss the geometrical properties of the flow paths. Macropore flow was captured with a soft X-ray TV camera to examine the hydraulic properties such as flow velocity and flow regime.

2 MATERIALS AND METHODS

2.1 Soils

Undisturbed soil cores were collected from a paddy field and upland field to determine the effect of differences in soil texture. Paddy field soils (Gray Lowland soils) and upland field soils (Para-Kuroboku soils) were considered as clayey and loamy soils, respectively. The physical properties of these soils are shown in Table 1. Undisturbed soil samples were taken using duralumin square sample holders of 5 x 5 x 5 cm with a 0.2 mm wall thickness. Light material, duralumin, was used not to cancel the effect caused by the contrast media.

2.2 Soft X-ray radiography of macropore flow followed by drainage

Soft X-ray radiography was performed according to the procedure of Mori et al. (1999a) using a side-beam soft X-ray apparatus (SOFTEX Inc., SV-100 AW extra). Saturated hydraulic conductivity experiments followed by outflow experiments were conducted on the soil samples after setting the apparatus (Fig. 1) in the soft X-ray system. A constant hydraulic gradient was maintained during saturated hydraulic conductivity experiment using a mariotte tank: four for paddy soils; two for upland field soils. A contrast medium was introduced dropwise from the constant-head injector, allowing the medium to move with the water flow. In order to obtain high-contrast images, we selected CH_2I_2 as the medium. It is heavy, yet has a dynamic viscosity similar to and surface tension smaller than water (Table 2). Hydrophobic CH_2I_2 moved as small drops with water flow. The fraction of CH_2I_2 to water was kept

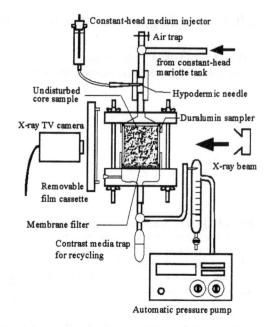

Figure 1. Soft X-ray radiography of macropore flow and successive drainage process

Table 2. Physical properties of CH_2I_2 and water

Material	CH_2I_2	H_2O
Specific gravity (Mg m^{-3})	3.32	0.998
Kinematic viscosity (10^{-6} m^2 s^1)	0.818	1.004
Viscosity (10^{-3} Pa s)	2.716	1.002
Surface tension (10^{-3} N m^{-1})	3.572	7.275

less than 0.1 to avoid lowering the velocity and to maintain relative hydraulic conductivity to water at almost one (Mori et al. 1999a). A hypodermic needle (27 G, ϕ 0.40 x 19-mm) was used for injection of small drops of the medium. De-aerated 0.005 M $CaSO_4$ solution was used to minimize clay particle dispersion.

After saturated hydraulic conductivity experiment, outflow experiments in combination with soft X-ray radiography (Mori et al. 1999b) were performed to examine the drainage process. About 1-cm^3 of CH_2I_2 was dropped onto the saturated soil surface to just cover its surface, and then suction was applied at the bottom of the sample. Since CH_2I_2 has a lower surface tension than water, the contrast medium was easy to follow the water movement. Therefore, the resulting tracing pattern was considered to represent the movement of the drainage front. One-step (Kool et al. 1985, Parker et al. 1985), or multi-step outflow experiments (Eching & Hopmans 1993b) were performed to estimate the unsaturated hydraulic conductivity of these soils. One step of 10 kPa was applied for upland field soils, whereas multi-step of 10, 20, and 31.6 kPa for paddy field soils.

Table 1. Physical properties of undisturbed core samples

Soil	Clayey soils		Loamy soils	
Land use	Paddy field		Upland field	
Sampling depth (cm)	30	50	30	50
Soil texture*	CL	CL	SL	LS
Ks** (10^{-5} m s^{-1})	0.455	0.557	4.05	5.85
Porosity (m^3 m^{-3})	0.554	0.552	0.720	0.672
Macroporosity (m^3 m^{-3})	0.020	0.021	0.097	0.085
Bulk density(Mg m^{-3})	1.38	1.33	0.64	0.74

*; Category standard of Int. Soc. of Soil Sci.
**; Saturated hydraulic conductivity

The model used for the estimation is the combination of van Genuchten's θ (h) model (1980),

$$S_e = \left(1 + |\alpha h|^n\right)^{-m} \tag{1}$$

$$S_e = (\theta - \theta_r)/(\theta_s - \theta_r) \tag{2}$$

with the pore-size distribution model of Mualem (1976) to yield (van Genuchten, 1980) :

$$K(\theta) = K_s S_e^l \left[1 - \left(1 - S_e^{1/m}\right)^m\right]^2, \tag{3}$$

where S_e is the effective saturation ($0 \leq S_e \leq 1$); θ_r (m³ m⁻³) is the residual water content; θ_s (m³ m⁻³) is the saturated water contents; K_s (m s⁻¹) is the saturated hydraulic conductivity; α (cm⁻¹), n, m (m=1-1/n), and l (assumed to be 0.5) are empirical parameters. MLSTPM code (Eching & Hopmans 1993a) was used to estimate these parameters.

Movements of the contrast medium were monitored using a one-inch soft X-ray camera, PbO vidicon (HAMAMATSU PHOTONICS), and recorded with a videocassette recorder (VCR), while flow paths were radiographed on X-ray films. Video images were used for dynamic analysis, and still images for structural analysis. The imaging conditions for soft X-ray radiography were 85kVp for Gray Lowland paddy soil, and 80kVp for Para-Kuroboku upland field soil. Exposure time of 75 s with 1 mA was needed for imaging the flow paths onto the X-ray film, IX-150 (FUJI PHOTO FILM CO., LTD.).

2.3 Image analysis

2.3.1 Saturated water flow

The saturated water flow was traced by CH_2I_2 and recorded on a VCR. The velocity of CH_2I_2 (V_{CH2I2}) was measured and the velocity of water (V_{water}) was estimated. Because the relative hydraulic conductivity was one, these two values were considered as identical. Then, individual macropore flow was examined for cylindrical macropore flow. Samples with cylindrical macropore flow paths were selected, then the flow velocities and the Reynolds number were calculated for V_{water}. Interaggregate macropore flow was set aside since the same measurement could not be applied.

The Reynolds number (Re) expresses the ratio of inertial to viscous forces acting on the fluid:

$$Re = \frac{vd}{\nu} \propto \frac{(inertial\_term)}{(viscous\_term)}, \tag{4}$$

where v is the velocity of the flow, d is the macropore diameter, and ν is the kinematic viscosity.

2.3.2 Drainage process

For outflow experiments, the time sequential imaging of drainage patterns (512 by 512 pixels) was imported to a personal computer. The computer will then "subtract" a mask image of the area obtained before the arrival of contrast medium from an image obtained after the arrival of contrast medium, leaving only an image of the contrast medium movement. A series of this subtraction show the time sequence of drainage patterns. Because attenuations by soil particles were wiped out, high-contrasted image of drainage patterns would be expected. The technique is known as Digital Subtraction (DS) in medical diagnostics and used for visualizing vessel effectively.

Macropores appeared as clearly delineated high-contrast regions, whereas the soil matrix appeared faint in soft X-ray images. Since these regions were clearly distinguishable, the areal fraction of each pore in the total sample area was calculated by pixel counting.

3 RESULTS AND DISCUSSION

3.1 Geometrical properties of flow patterns

Saturated flow paths and drainage patterns (Fig. 2) are shown as positive images. Therefore, darker areas correspond to the areas of medium movement; namely macropore flow paths and successive drainage front movements. The numerical values in the figures show the average volumetric water content calculated using cumulative outflow volume data.

Flow paths showed different shapes depending on soil texture. For paddy field soils, cylindrical macropores were predominant and flow paths were restricted to several cylindrical macropores. The drainage pattern showed that cylindrical macropores drained first, followed by the soil matrix, namely two-domain flow. Mori et al. (1994) revealed that these channels were made by paddy rice roots. Primary roots of rice grow downward in response to gravity and they grow straight downward without interference with gravel in clayey and an inundated field condition. Bulk density in Table 1 showed that clayey soils are strongly packed. Under such conditions, root channels could be conserved well and macropores would channel the water flow without any interaction with bulk soil. This means that solutes might bypass the surrounding soil and go down to a lower profile easily.

In upland field soil, flow paths are connected with each other, creating channel networks. These macropores were considered to have been made by plant roots (Narioka 1990; Tokunaga et al. 1992). Under a condition where aggregates develop in soils, non-

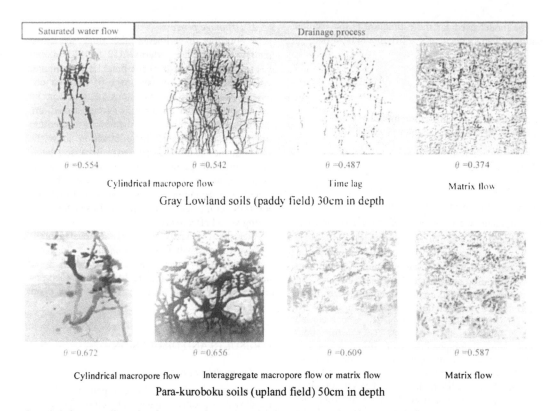

Saturated water flow	Drainage process

θ =0.554　　　　θ =0.542　　　　　　　θ =0.487　　　　　　θ =0.374

Cylindrical macropore flow　　　　Time lag　　　Matrix flow

Gray Lowland soils (paddy field) 30cm in depth

θ =0.672　　　　θ =0.656　　　　　　θ =0.609　　　　　θ =0.587

Cylindrical macropore flow　　Interaggregate macropore flow or matrix flow　　　Matrix flow

Para-kuroboku soils (upland field) 50cm in depth

Figure 2. Soft X-ray radiography of macropore flow paths and drainage patterns of paddy and upland field soils

uniformity in bulk density would be expected, which results in a tortuous macropore flow paths. Sometimes, interaggregate macropore flow paths were found in upland field soils. Drainage originated first from the cylindrical macropores followed by interaggregate macropores, and then matrix drained. However, the order of drainage was not distinct from that in paddy field soils, especially between interaggregate macropore and matrix.

We examined the readiness of exchange between macropores and matrix by drawing the time sequential areal fraction of macropore/matrix drainage in Figure 3. Suction was estimated from the water content calculated by the outflow volume. The matrix drainage in upland field soil started at the suction of around 1.5kPa, while at 4.0kPa in paddy soil. Although image analysis showed similar drainage patterns between the two soils, macropore-matrix exchange was different. Surface applied chemicals would interact with or be caught by soil matrix easily in upland field, whereas with difficulty in paddy field soil. Furthermore, a lag was observed in paddy field soil at Figure 2, between macropore flow and matrix flow, which was not found in upland field soil. Lack of interaggregate macropores might cause

the lag. The effect of the boundary condition between macropores and matrix should be considered at convection and dispersion in a process-based modeling.

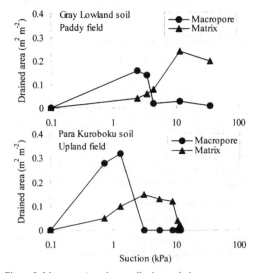

Figure 3. Macropore/matrix contribution to drainage process

300

3.2 Hydraulic properties of the flow

Because the flow area was restricted to a small number of macropores, the movement of the contrast medium showed a Reynolds number ranging from 18.8 to 40.8 (Table 3). When inertial force becomes larger than viscous force, Re in Equation (4) becomes greater than one, and the flow gradually becomes turbulent as inertial force increases. At a Reynolds number from 10 to 100, the flow was considered to be in transition between laminar and turbulent flow (Bear, 1979). Under actual conditions of cylindrical macropore flow, laminar flow was not realized and the flow regime was in the transition region to turbulent flow. Therefore, the traditional approach for a two-domain system may not be strictly valid for the cylindrical macropore system at relatively high velocities.

Figure 4 shows the unsaturated hydraulic conductivity of the examined soils. Some soils showed non-uniqueness problems when saturated hydraulic conductivity was applied as fixed values. Since the image analysis showed the evidence of two- or more-domain flow, we anticipated a discontinuity in the hydraulic conductivity curve near saturation. Therefore, the saturated hydraulic conductivity was assumed to be variable in the parameter estimation procedure. In Figure 4, the unsaturated hydraulic conductivity curve was drawn using the fitted saturated hydraulic conductivity, whereas the actual measured saturated hydraulic conductivity is shown as a single data point. Although the Mualem-van Genuchten equation does not account for multi-

Table 3. Flow regimes in cylindrical macropore

Sample	Flow No.	Pore Diameter	Velocity	Reynolds number
Units		10^{-4} m	10^{-2} m s^{-1}	-
Paddy field soil (30 cm)	P1	9.56	2.11	20.1
	P2	8.24	2.29	18.8
	P3	8.89	2.78	24.6
	P4	9.75	2.77	26.9
Upland field soil (50 cm)	U1	13.5	2.95	39.7
	U2	15.4	2.66	40.8
	U3	11.6	2.28	26.3
	U4	12.1	3.28	39.5

domain flow, discontinuity near saturation implied that multi-domain flow has been observed also in numerical analysis. Paddy field soils showed differences of more than one order of magnitude, with differences for upland field soils being less than one order of magnitude. This could be expected also from Figure 2, where paddy soils showed distinct order from macropore to matrix. Different discontinuity in near saturation even with fast cylindrical macropore flow in both cases needs more explanation other than two-domain flow.

3.3 Three-domain flow in agricultural field soils

In order to explain this discontinuity more precisely, we measured pore-distribution. Figure 5 shows a bimodal shape of pore distribution, macropore and matrix with few pores between the suction of 3 and 10kPa in paddy field soils. Upland field soils also showed bi-modal distribution, but the porosity decreased gradually from 2 to 10kPa. The difference in pore-distribution explains the discontinuity in hydraulic conductivity near saturation. Namely, abrupt changes in pore distribution caused discontinuity in near saturation for paddy field soils, whereas gradual changes in pore distribution caused less discontinuity for upland field soils. Interaggregate macropore found in Figure 2 were due to this gradual change in pore distribution. Most of the studies on dual-porosity or domain concept referred to only the macropore-matrix system. However, we should divide

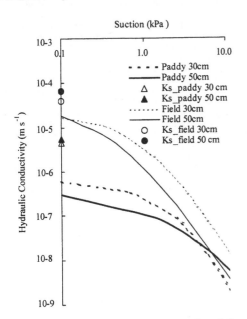

Figure 4. Unsaturated hydraulic conductivity estimated from outflow experiment

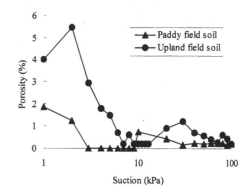

Figure 5. Pore-distribution of examined soils

macropore system into two types: cylindrical macropore and interaggregate macropore. Cylindrical macropore needs the description other than Darcian flow. Interaggregate macropore flow is rather predictable. This concept explains the flow process in agriculture field, where root channel is predominant pore type, much more precisely. Furthermore, distinct dual-porosity of paddy field (clayey) soil would channel the water faster than expected with few interactions with bulk soil.

4 CONCLUSIONS

Macropore flow in undisturbed soils was characterized by both dynamic and structural analyses. The flow was captured with a soft X-ray TV camera and the flow paths were photographed on X-ray film, when a contrast medium, CH_2I_2 was used. The following conclusions were obtained on macropore flow and successive drainage.

1. The cylindrical macropore flow of the root channel conducted water faster than averaged Darcian flow. Reynolds number exceeded one, making abrupt changes in hydraulic conductivity near saturation.

2. The interaggregate macropore flow also bypassed the surrounding bulk soil, however, continuous pore distribution makes dynamics of flow predictable. A small discontinuity was found in hydraulic conductivity near saturation.

3. Matrix flow occurs in tightly packed micro-pores. Therefore, the conventional flow theory may be adequate.

Loamy upland field soils consists of triple-porosity, namely, cylindrical macropore, interaggregate macropore and matrix. On the other hand, clayey paddy soils consists of dual-porosity of cylindrical macropore and matrix. Paddy soils showed unique flow processes because of their clayey and inundated condition.

Soft X-ray radiography revealed that water flow is affected by soil texture and land management. Although they are two-dimensional images, they were sufficient to characterize the flow process in undisturbed soils. This study provided soil hydraulic properties which could be utilized for further modeling of water and solute transport.

ACKNOWLEDGEMENT

We are grateful to Dr. Toshisuke Maruyama, President of Ishikawa Agricultural College, Honorary Professor of Kyoto University and Dr. Toru Mitsuno, Kyoto University for making available their X-ray apparatus and their valuable comments during the development of this study.

REFERENCES

Aylmore, L. A. G. 1993. Use of computer-assisted tomography in studying water movement around plant roots. Advances in Agronomy, Vol. 49:1-54.

Bear, J. 1979. Hydraulics of ground water. McGrau-Hill, Inc.: 65.

Eching, S.O. & J.W. Hopmans. 1993a. Inverse solution of unsaturated soil hydraulic functions from transient outflow and soil water pressure data. Land, air and water resources paper 100021. Univ. of California, Davis.

Eching, S.O. & J.W. Hopmans. 1993b. Optimization of hydraulic functions from transient outflow and soil water pressure data. Soil Sci. Soc. Am. J. 57:1167-1175.

Flury, M. 1996. Experimental evidence of transport of pesticides through field soils- A review, J. of Environ. Qual. 25: 25-45.

Kawaguchi, K. & K. Kysuma. 1977. Paddy soils in tropical Asia. The center for Southeast Asian Studies, Kyoto University. Kyoto: 95.

Kool, J.B., J.C. Parker & M. Th. van Genuchten. 1985. Determining soil hydraulic properties from one-step outflow experiments by parameter estimation: I . Theory and numerical studies. Soil Sci. Soc. Am. J. 49: 1348-1354.

Mori, Y., T.Watanabe & T.Maruyama. 1994. Formation of morphology of root created macropores in paddy field. Trans. JSIDRE 171:13-20 (in Japanese).

Mori, Y., K. Iwama, T. Maruayma & T. Mitsuno. 1999a. Discriminating the influence of soil texture and management-induced changes in macropore flow using soft X-rays. Soil Science,164: 467-482.

Mori, Y., T. Maruyama & T. Mitsuno. 1999b. Soft X-ray radiography of drainage patterns of structured soils. Soil Sci. Soc. Am. J. 63: 733-740.

Mualem, Y. 1976. A new model for predicting the hydraulic conductivity of unsaturated porous media. Water Resour. Res. 12: 513-522.

Narioka, H. 1987. Three-dimensional measurement of soil macropore using soft X-rays. Journal of the JSIDRE 55: 29-35 (in Japanese).

Narioka, H. 1990. Study on physical functions of soil macropores and the instrumentation for their measurement. Bulletin of NODAI Research Institute of Tokyo University of Agriculture (in Japanese).

Parker, J.C., J.B. Kool & M. Th. van Genuchten. 1985. Determining soil hydraulic properties from one-step outflow experiments by parameter estimation: II. Experimental studies. Soil Sci. Soc. Am. J. 49: 1348-1354.

Tokunaga, K. 1988. X-ray radiographs using new contrast media on soil macropores. Soil Science 146:199-207.

Tokunaga, K., H. Narioka & T. Fukaya. 1984. Development of heavy liquid infiltration method and consideration on the soil void images photographed by soft X-ray projection with the above method. Trans. JSIDRE 114:61-68 (in Japanese).

Tokunaga, K., T. Ishida, N. Yano & T. Sase. 1992. Approach to analysis of subterranean structure under plant communities by X-ray stereoradiographs imaged by new contrast medium on soil macropores –proposal for new method for studies on plant-soil system-. Jpn. J. Ecol. 42:249-262 (in Japanese).

van Genuchten, M. Th. 1980. A closed-form equation for predicting the hydraulic conductivity of unsaturated soils. Soil Sci. Soc. Am. J. 44:892-898.

Vepraskas, M.J., A.G.Jongmans, M.T.Hoover & J.Bouma. 1991 Hydraulic conductivity of saprolite as determined by channels and porous groundmass. Soil Sci. Soc. Am. J. 55:932-938.

Clay Science for Engineering, Adachi & Fukue (eds) © 2001 Balkema, Rotterdam, ISBN 90 5809 175 9

Permeability test of unsaturated soil approximating moisture distribution

T. Sugii & K. Yamada
Department of Civil Engineering, Chubu University, Kasugai, Japan

Y. Umeda
Umeda Geo-Research, Gifu, Japan

M. Uemura
Institute of Space Create, Aichi, Japan

ABSTRACT: An experiment method has been developed in Chubu University to estimate the permeability of unsaturated soils. It is an improved type of the original instantaneous profile method, and is able to measure the permeability of many kinds of soils. The new method is characterized by the unsteady condition of water profile and linear approximation of moisture distribution. This improved instantaneous profile is applicable to clayey soils or sand. There is a considerable volume change of soil draining in this test. Volume change of the specimen affects the measured permeability in this method, therefore, the authors correct the volume of the specimen by the water retention curves considering bulk density. In this paper, the proposed method is investigated using an indirect method with function models of the water retention curve and the inverse problem method.

1 INTRODUCTION

Methods of estimating the permeability of unsaturated soils are classified as steady or unsteady methods. Although the former type can easily estimate the stable permeability of unsaturated soils, it is time-consuming. The latter method is often easier to perform and requires less time. Watson (1966) proposed an instantaneous profile method as an unsteady method for measuring the hydraulic conductivity of unsaturated soils. In almost every unsteady method, diffusivity is determined from instantaneous outflow at first, then the permeability is calculated from the diffusivity and the water retention curve. Therefore, the accuracy of the water retention curve affects the hydraulic conductivity of unsaturated soils. On the other hand, the instantaneous profile method measures the transient flow and the hydraulic gradient during which conditions are approximately steady. As the permeability is determined directly, the water retention curve is needed. We improved the original experiment apparatus of the instantaneous profile method, which consisted of several tensiometers and water content sensors. The improved method has only a water content sensor (ADR: Amplitude Domain Reflectometry), a tensiometer and an electronic balance substituting for many measuring implements. Therefore, the apparatus is smaller. A sand was measured using this method, and the applicability of it was obtained (Sugii et al. 2000a).

When the permeability of fine-grained soils is measured, higher air pressure is needed. Hence, the volume of the specimen will be found to decrease in volume during draining in this test. In order to calculate the volumetric water content using height of the specimen in this method, the accuracy of measuring the height of the specimen affects the results. Sugii et al. (2000b) drew the water characteristic face of the same soil using many water retention curves. The volume change of the specimen during the test is corrected by this relationship among suction, volumetric water content and dry density. The applicability of the corrected permeability is investigated using an indirect method with the function models of water retention curve and the inverse problem method (multi-step method (Eching et al. (1993))).

2 WATER RETENTION CURVES CONSIDERING DRY DENSITY

Using the axis translation method (Olson et al. 1965), the water retentively of soils is measured. The volume of specimen is measured after it has reached moisture equilibrium, and water retention curves are obtained with data of the same dry densities. Figure 1 shows a estimated water characteristic face by several of such curves (Sugii et al. 2000).

If the volume of soils is known, the total volume of soils can be estimated from the relationship among suction, volumetric water content and dry density.

This relationship is follows:

$$\log S = 1.67 + \left(\frac{\ln(V - V_s + V_w)}{0.783 V_w} + 1.06\left(\frac{m_s}{V} - 1\right)\right) \quad (1)$$

where S=Suction (kPa), V=total volume of soil, V_s=volume of soils, V_w=volume of water and m_s=mass of dry soils.

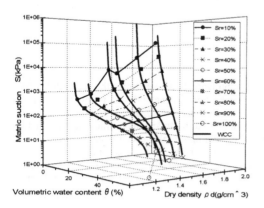

Figure 1 Water retention curves (The fine grained soil)

3 PRINCIPLES OF MEASUREMENT

3.1 Darcy's law

Darcy's law for one-dimensional vertical flow is given by

$$\int_0^z \frac{\partial \theta}{\partial t} dz = K(\theta)\left(\frac{\partial \varphi}{\partial z} + 1\right) \quad (2)$$

where θ =volumetric water content; t=time; $K(\theta)$=permeability; φ =pore water pressure; z=elevation of specimen.

Rearranging equation (2), permeability at time t, and elevation of specimen are given by

$$K(\theta) = \frac{\left(\int_0^z \frac{\partial \theta}{\partial t} dz\right)_{z,t}}{\left(\frac{\partial \varphi}{\partial z} + 1\right)_{z,t}} \quad (3)$$

where the numerator shows the velocity at elevation, and the denominator the hydraulic gradient.

The permeability of unsaturated soils against the volumetric water content corresponding to the point at the elevation can be obtained by measuring the velocity and hydraulic gradient at time t. The permeability of variable degree of saturation can be obtained as time proceeds.

3.2 Calculation of velocity

In the instantaneous profile method, volumetric water content is measured at several points. However, in our method, volumetric water content is measured at the top of the specimen only. The amount of drainage from the specimen is measured over time. After test is finished, the water content of the bottom of the specimen is measured using the oven-dried method. If the vertical distribution of moisture is assumed to be linear, the volumetric water content at the bottom of the specimen can be estimated as follows.

The amount of drainage through the specimen, q, between time, t_{n-1} and time, t_n is given by the hatched area in Figure 2.

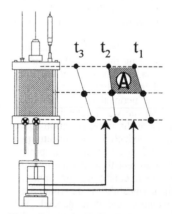

Figure 2 Distribution of volumetric water content in specimen over time.

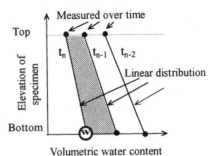

Figure 3. Amount of flow through the specimen.

The volumetric water content at point W is determined from the water content measured after the test is finished. Since the volumetric water content at the top of the specimen is measured during test, the volumetric water content at the bottom in the previous time-step can be calculated from the area of the trapezoid section (the amount of drainage per unit area of specimen).

The amount of flow from t_1 to t_2 through the middle of the specimen is calculated as shown in Figure 3. The numerator of equation (3) is calculated from following equation:

$$\left[\int_0^{z_m} \frac{\partial \theta}{\partial t} dz\right]_{z_m, \frac{t_1+t_2}{2}} = \frac{A}{t_2 - t_1} \qquad (4)$$

where z_m = elevation of middle of specimen, A= hatched area in Figure 3.

This experimental method is characterized by linear approximation of the vertical distribution of moisture in the specimen. In this paper, this method is called "Approximated Profile Method" (A.P.M.).

3.3 Calculation of hydraulic gradient

The change of pore water pressure over time is measured as in shown Figure 4.

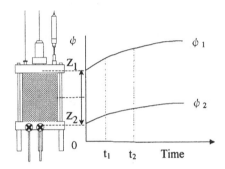

Figure 4 Change of pore water pressure over time.

The hydraulic gradient at the middle of the specimen is given by

$$i_{z_m} = \left(\frac{\partial \psi}{\partial z} + 1\right)_{z_m, \frac{t_1+t_2}{2}} = \left(\frac{\psi_2 - \psi_1}{z_2 - z_1}\right)_{\frac{t_1+t_2}{2}} + 1 \qquad (5)$$

The pore water pressure is controlled by elevation of the drainage tank or supplied air pressure. As there is filter on the bottom of the specimen, the pore water pressure is corrected using the equation:

$$p_w = p_{wf} + \frac{dq}{AK_f} \qquad (6)$$

where p_w = pore water pressure at bottom of specmen, p_{wf} = calculated pressure from elevation of drainage tank or supplied air pressure, d = thickness of filter, q = amount of flow per unit time, A = area of specimen section, K_f = permeability of filter.

4 EXPERIMENT APPARATUS

In the original instantaneous profile method, many sensors are used to measure volumetric water content and water pore pressure. Watson (1966) used RI sensors (using gamma-ray), to measure water content. Instead of RI sensor, recently, FDR (Frequency Domain Reflectometry (Topp et al. 1988)), ADR (Amplitude Domain Reflectometry (Gaskin et al. 1996)) are used for moisture measurement. Their sensors can accurately measure soil moisture easily and safely. We uses an ADR sensor to measure the moisture of soils, and a micro tensiometer to measure pore water pressure. To measure the amount of drainage over time, a load cell is used. In the case of

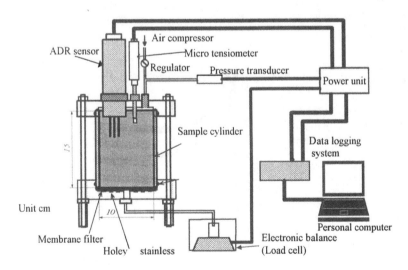

Figure 5 Experiment apparatus

sandy soils, pore water pressure is controlled by changing the elevation of the drainage tank (suction method). Figure 5 shows the experiment apparatus in this study, in which a sample cell unit that includes container filter is attached to the bottom of the sample specimen. The load cell can measure to an accuracy of 0.1g. When air pressure is necessary, air pressure is supplied from the top of the sample cell. The data logger and personal computer record the measured data of pore water pressure, volumetric water content, supplied pressure and amount of drainage over time. In this study a membrane filter were used.

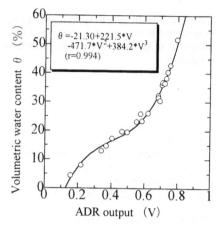

Figure 6 Calibration curve of ADR for sample soils

5 MATERIALS AND METHOD

5.1 *Description of materials*

The fine grained soils are applied as shown in Table 1. The sample is classified as fine grained soil or silt.

Table 1. Sample soils

Property		Value
Grain size	D	~75 μ m
Density of soil particles	ρ_s	2.676 g/cm³
Liquid limit	w_L	36.4 %
Plastic limit	w_P	28.2 %
Plasticity index	I_P	8.2
Permeability of saturated Soils	k_{sat}	3.4E-04 cm/s

5.2 *Soil specific calibration of ADR sensor*

The relationship between the output voltage of ADR and the square root of dielectric constant can be represented as a polynomial approximation:

$$\sqrt{\varepsilon} = 1.07 + 6.4V - 6.4V^2 + 4.7V^3$$
$$(0 \le \theta \le 100\%) \tag{7a}$$
$$or \quad \sqrt{\varepsilon} = 1.1 + 4.44V \quad (\theta \le 50\%) \tag{7b}$$

where ε = dielectric constant, V= output voltage.

The linear relationship between the square root of dielectric constant and volumetric water content was discussed by Whalley (1993):

$$\theta / 100 = \frac{\sqrt{\varepsilon} - a_0}{a_1} \tag{8}$$

where a_0 and a_1 are constants for a specific soil type having typical values of 8.1 and 1.6 respectively. As the moisture of fine grained soils is more than 50%, from equations (7b) and (8) the calibration curve becomes:

$$\theta/100 = \frac{\left[1.07 \div 6.4V - 6.4V^2 + 4.7V^3\right] - a_0}{a_1} \tag{9}$$

Values of a_0 and a_1 are about 1.4 and 1.3 respectively, from Figure 6.

5.3 *Properties of filter and boundary conditions*

A membrane filter is placed under the specimen to supply high air pressure. The lower the permeability of the soil to be measured is, the higher necessary air entry value of the filter is. The vertical distribution of moisture in the specimen is assumed to be linear, therefore, a filter which does not disturb the seepage flow should be used. Accordingly, the permeability of the filter should be close to that of the saturated soil. The properties of the used filter are shown in Table 2.

Table 2 Properties of filter

Thickness	Opening diameter	Air entry value	Permeability
0.05mm	0.2 μ m	189 kPa	2.2E-6

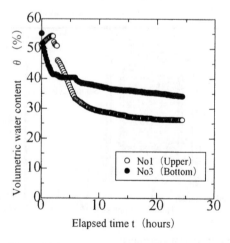

Figure 7 Volumetric water content overtime

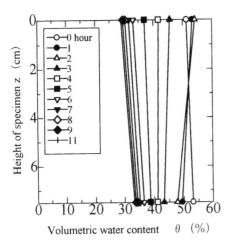

Figure 8 Vertical distribution of moisture overtime

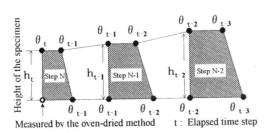

Figure 9 Calculation of height of specimen

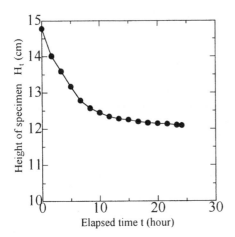

Figure 10 Height of specimen over time

The pressure of the bottom and the top of the specimen are kept 0 kPa (atmosphere) and 211 kPa, respectively.

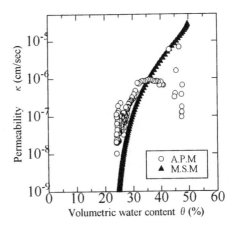

Figure 11 Comparison of A.P.M with M.S.M

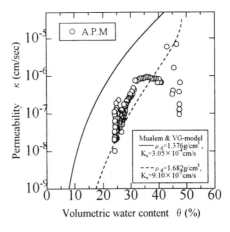

Figure 12 Comparison of A.P.M. with unsaturated soil hydraulic function

5.4 *Preparing and installing the specimen*

Fill the inside of the apparatus with distilled water. Divide the dried sample into five containers. Supply distilled water until the sample is not in a slurried state. After assembly into the cell, compact each layer with a compacting tool. Initial dry density was 1.36 g/cm^3. Insert the ADR and the tensiometers into the top of the specimen and seal the cell. Control the required air pressure and level the drain tank with the bottom of the specimen. Open the valves at the same time and start the measurement.

6. DISCUSSION OF RESULTS

6.1 *Volumetric water content over time*

Volumetric water content over time is shown in Figure 7. The volumetric water content of the bottom is less than that of the top up to three hours.

This is because the volumetric water content of the bottom is estimated from the moisture distribution the last time. Thus, the early moisture distribution is considered not to be linear. The same result was obtained using the pressure method with sand (Sugii et al. 2000a). In our method, it is necessary to investigate the measurement method in the range of high saturation. Figure 8 shows the

vertical distribution of volumetric water content over time. The volumetric water content of the top fractionally increases from the initial moisture. It seems that the pore water flowed around the circumference of the rods of the ADR sensor, therefore, the water content increased.

6.2 *Volume change over time*

Using the measuring data of suction, volumetric water content and equation (1), volume change is

calculated as shown in Figure 9. The area section of the specimen and height of the specimen in unit time is constant. The calculated height of the specimen over time is shown in Figure 10.

6.3 *Comparison of the experimental results with other results*

There are no verifiable results of other experiments, therefore, the experimental results were investigated by an indirect method with function models of the water retention curve and the inverse problem method (multi-step method (M.S.M.)).

Figure 11 shows the comparison of the approximated profile method with the multi-step method (Eching et al. 1993). The multi-step method is an inverse solution method using transient outflow and pore water pressure over time. The measured data of the approximated profile method are employed as input data for the numerical inverse problem. Mualem & van-Genucthen's model is used as unsaturated soil hydraulic function. Although volume change is not considered in the multi-step method,

the results of A.P.M are roughly the same as those of M.S.M. in the range of low saturation and low density. The lower permeability in the range higher water content is the result of approximating the initial moisture distribution as linear in the same as that described for Figure 7.

Figure 12 shows A.P.M. and the indirect method with the unsaturated soil hydraulic function using equations (10). The solid line is loose density and the dashed line is high density.

$$k(\theta) = ks \cdot Se^{0.5}\left[1 - \left(1 - se^{1/m}\right)^m\right]^2 \qquad (10)$$

where k(θ)=permeability of unsaturated soil, ks=permeability of saturated soil, Se=effective saturation, m=1-1/n , n=parameter of van Genuchten's closed-form expressions (van Genuchten ,1980).

There is good correlation between the solid and dashed lines. In addition this, the experimental result is close to dashed line (ρ_d=1.682 g/cm^3). Taking account of volume change (Figure 10), the experimental result is judged to be appropriate.

These comparisons show that the approximated profile method is valid for estimating the permeability of unsaturated soils.

7. CONCLUSIONS

The approximated profile method was developed to estimate the permeability of unsaturated soils. Though this method is an improved version of the instantaneous profile method (unsteady method), it is not necessary to use many expensive measuring sensors nor a large amount of soils.

Taking account of the considerable volume change during the test, a correcting method of A.P.M. is proposed.

The experimental result was investigated by an indirect method with function models of the water retention curve and multi-step method. As a result, the approximated profile method was shown to be valid for estimating the permeability of unsaturated soils directly.

It is considered that the early moisture distribution is not linear, therefore, it is necessary to investigate a measurement method in the range of high saturation.

ACKNOWLEDGMENTS

The authors wish to thank former graduate students of Chubu University Mr. T. Okumura and Mr. T. Saso for their support in carrying out the experiments. This study received support from a Grant-in-Aid for Scientific Research (C) titled "Evaluation of hydraulic properties of unsaturated soils considering bulk density" (No.11650515).

REFERENCES

Eching,S.O. & Hopmans,J.W. 1993. Inverse Solution of Unsaturated soil hydraulic functions from transient outflow and soil water pressure data. LAWR-Hydrologic Science. 1993.
Gaskin,G.J. & Miller,J.D. 1996. Measurement of soil water content using a simplified impedance measuring technique. J. Agric. Engng Res., Vol.63, 153-160.
van Genuchten, M.TH. 1980. A Closed equation for predicting the hydraulic conductivity of unsaturated soils. Soil Sci. Soc. Am. J. Vol.44, 892-898.
Olson, R.E. & Langfelder, L.J. 1965. Pore-water pressures in unsaturated soils. J. Soil Mech. Found Div., Proc.of ASCE, Vol.91, SM4, 127-160.

Sugii,T. Yamada,K. & Uemura,M. 2000a. Measuring hydraulic properties for unsaturated soils with unsteady method, *Proc. Asian Conf. on Unsaturated Soils* "UNSAT-ASIA2000", 439-444.

Sugii,T. Yamada,K. & Kondo,T. 2000b. Study on water characteristic curves considering a bulk density, Proc.of the thirty-fifth Japan national Conf. on Geotechnical Engineering, (be in the press) ,(in Japanese).

Topp,G.C, Yanuka,M., Zebchuk,W.D. & Zegelin,S. 1998. Determination of electrical conductivity using time domain reflectometry: soil and water experiments in coaxial lines, *Water Resources Research*, Vol.24, No.7, 945-952.

Uno,T., Sugii,T. & Tsuge,H. 1991. On the flow of pore water and air in sandy soil, *Proc. national symp. on Multiphase Flow*, 147-150 (in Japanese).

Uno,T. Sato,T. & Sugii,T. 1995. Laboratory permeability measurement of partially saturated soil: Alonso & Delage (ed.), *Unsaturated Soils, Proc. intern. symp., Paris, 6-8 September*, 573-578.

Watson,K.K. 1966. An instantaneous profile method for determining the hydraulic conductivity of unsaturated porous materials, *Water Resources Research*, Vol.2, 709-715.

Whalley,W.R. 1993. Considerations on the use of time-domain reflectometry (TDR) for measuring soil water content, *J. Soil Science*, Vol.44, 1-9.

Clay Science for Engineering, Adachi & Fukue (eds) © 2001 Balkema, Rotterdam, ISBN 90 5809 175 9

Diffusion and sorption coupled transport in mineral sealing cut-off walls

R. Hermanns Stengele, G. Kahr & X. Wittmann
Institute of Geotechnical Engineering, Swiss Federal Institute of Technology (ETH), Zürich, Switzerland

ABSTRACT: A simple technique has been developed for measuring the ionic penetration into the surface mineral sealing composite specimens exposed to a salt solution, from which the apparent diffusion coefficient for chloride, sulphate, sodium and potassium are derived. Results obtained have indicated that for a given composite, the depth of the ion diffusion is dependent upon the age of the specimens. If the sealing composite is older than about 2 years, i.e. the changes in its chemical compositions and microstructures are negligible to the rate of penetration of the ions. The apparent diffusion coefficient determined from diffusion equation according to second Fick's law can realistically characterise the diffusion and sorption coupled transport of ions into the mineral sealing composite materials. The diffusion of the ions is depending on the hydratation of cement, the chemical compositions and microstructure of the sealing materials.

1 INTRODUCTION

Since most of the sealing composites used in the construction of deposit barrier systems are porous, the sorption and desorption of contaminated ground water and its transport have important consequences in the design and the durability of the barrier systems. So far many investigations with respect to the diffusion of ions, both anion and cation, into clays such as bentonite, illite, and kaolinite have been carried out (Hasenpatt 1988; Gray & Weber 1989; Madsen & Kahr 1991). For the mineral sealing materials used in the two-phase cut-off wall systems, the diffusivity was calculated using the difference in concentration in the solution phases on either side of the specimen (Hermanns 1993). The typical value of chloride diffusivity D_{CL} for specimens in contact with NaCl solution is 8.6 x 10^{-11} m^2/s, with ZnCl solution 3.2 x 10^{-11} m^2/s, the sodium diffusivity D_{Na} is smaller than 2 x 10^{-12} m^2/s. The practical importance of the diffusivity lies in that the 'break through time' for certain ions can be estimated once the concentration of the contact solution and the thickness of the cut-off wall are known. Care must be taken as the diffusivity obtained here for the medium concerned has not the same meaning as is derived from Fick's second law of diffusion. According to the basic hypothesis of Fick's theory, diffusion is the process by which matter is transported from one part of a system to another as a result of random molecular motions. Because of the further hydration of the cement mixed in the sealing

material, the chemical compositions and microstructure, and thus the electrical charges of the inner surfaces are changing as the diffusing substances are transported through the solid. As a consequence some ions behave in a more mobile way than the others in the pore solution, thus the transport of cation and anion through the sealing materials is actually a process of diffusion coupled with sorption of both chemical and physical nature. The mechanism of ion movement is very complex and there are several factors affecting the diffusion coefficients (Schneider & Göttner 1991; Nye 1979). In practice, the transport process can still be simply characterised by the apparent diffusivity D_{app}. This operational definition of the diffusion coefficient in the sealing material has the advantage that it can be determined directly by a relative simple diffusion test to be described later, and it makes no assumptions about the mechanism of diffusion or the mobilities of the diffusing ions in their various states in the sealing material.

This paper focuses on the transport of ionic contaminants from contaminated sites through the mineral sealing material of a cut-off wall as an application of diffusion theory. We assume the ingress of anions and cations into the sealing material to be an isothermal diffusion and sorption coupled process generated by concentration gradients. The material after a certain period of immersion into the solution will be characterised by an apparent diffusion coefficient.

The aim of this study is to try to find out the applicability of the diffusion theory to the transport of contaminated ground water into the sealing material, taking the time-dependent ion content in the pore solution as the driving force. For a given immersion time, the content of anions such as Cl^- and SO_4^{-2}, cations Na^+ and K^+ within the element are determined with the help of Ion Chromatograph (IC). By fitting the ion content distribution to a real seepage water from an old hazardous site, the apparent diffusion coefficients are evaluated. Finally the results of anion content and cation content with immersion time from theoretical predictions and the experimental measurements are compared. It will be shown that the apparent diffusion coefficient based on the diffusion equation is not a constant for the sealing material investigated. It decreases with the age of the specimen. If the sealing composite is older than about 2 years, i.e. the changes in its chemical compositions and microstructures are negligible compared to the rate of penetration of the ions, the diffusivity can be used as an important parameter for the transport of contaminated solution in the barrier systems.

2 THEORY

The diffusion equations can be expressed in terms of the nomenclature of vector analysis as

$$\partial C/\partial t = \text{div} (D \text{ grad } C) \qquad (1)$$

where C is the concentration of diffusing substance, t the time during which diffusion has been taking place, D the diffusion coefficient (Crank 1975).

In the case of the diffusion test we performed, i.e. one-dimensional diffusion in a semi-infinite medium, $X > 0$, the boundary is kept at a constant concentration C_0, the initial concentration being zero throughout the medium. From equation (1) we get

$$\partial C/\partial t = D \, \partial^2 C/\partial X^2 \qquad (2)$$

satisfying the boundary condition

$$C = C_0, \quad X = 0, \quad t \geq 0, \qquad (3)$$

and the initial condition

$$C = 0, \quad X > 0, \quad t = 0. \qquad (4)$$

By applying the Laplace transform, equation (2) can be reduced to

$$C = C_0 \, \text{erfc} (X/(4Dt)^{0.5}) \qquad (5)$$

Where erfc is the complementary error-function. In a given diffusing time t, by fitting concentration

profiles $C = C(X, t)$, the apparent diffusivity D_{app} can be obtained.

Thus diffusion may take place within the pores of the medium which can absorb some of the diffusing substances or react chemically. In the simplest case, the concentration S of immobilised ion is directly proportional to the C of ion free to diffuse, i.e.

$$S = R \times C \qquad (6)$$

If the diffusion is accompanied by absorption, equation (2) becomes

$$\partial C/\partial t = D \, \partial^2 C/\partial X^2 - \partial S/\partial t \qquad (7)$$

For a constant R, we have

$$\partial C/\partial t = [D/(R+1)] \, \partial^2 C/\partial X^2 \qquad (8)$$

Obviously the overall process of diffusion with absorption is slower than the simple process governed by diffusion alone. For transport of ions in concrete, equation (7) was found to be a realistic description (Pereira & Hegedus 1984). In our study, the modified diffusion coefficient $D/(R + 1)$ was expressed as the apparent diffusion coefficient D_{app}. The apparent concentration C_{app} which includes the free and immobilised diffusing ions can be calculated according to:

$$C_{app} = C_{0ext.} \, \text{erfc} (X/(4D_a t)^{0.5}) \qquad (9).$$

3 MATERIAL AND METHOD

The composition of the investigated sealing material was chosen according to Hermanns (1993). The sealing mixture contains the commercial clay "Opalit" produced in Switzerland (30% in mass), sulphate-resisting cement (22%), flyash (7%) and water (41%). Opalit contens about 50% of clay, about 25% of quartz and about 11% of carbonate. The rest of about 10% consists of feldspar, pyrite and organics. The fraction of clay is composed of illite, mixed-layers (illite/smectite), kaolinite and chloride. Its cation exchange capacity is 13 meq/100 g. The sulphate-resisting cement contents less than 3% tricalciumaluminate. The flyash used here contains about 50% quartz, 30% aluminiumoxide and the rest is iron oxide.

The fresh mixture was filled in cylinder specimens of diameter 55 mm and length 100 mm and were allowed to hydrate under water for 37 and 750 days respectively. In order to deal with one-dimensional diffusion problem, the circumferential surfaces were sealed with impermeable polymers, while the two end-faces remained free for penetration.

The specimens in the state of water saturation were immersed into a salt solution at laboratory temperature (20°C). The concentrations of the main diffusing ions were simulated as in the contaminated ground water with Cl- 8220 mg/l, SO_4^{-2} 1500 mg/l, Na^+ 3500 mg/l and Ca^{+2} 350 mg/l (pH 7 and conductivity of 25 mS/cm). The transport behaviour of chloride, sulphate, sodium and potassium into the specimen were of major interest. After a given immersion time, specimens were dried at 105°C and then milled step by step in specimen depth of 0.5 mm. The milled powder from each step was carefully collected and dissolved into water solutions for anion analysis and 0.1 M formic acid solutions for cation analysis. The contents of the diffusing ions were measured by ion chromatography (IC). The concentration of the diffusing ions in the pore solutions was obtained through the total water content and the porosity of the specimen.

Though in the solution chloride content is kept constant $C_0 = 8220$ mg/l (immersion time = 30 days) the measured content in the surface zones X = 0.5 mm was C = 12290 mg/l. With regard to the ingress of other ions, similar tendencies are observed. So for practical solution of equation (5), instead of boundary condition (3), C_0 is taken from extrapolated value corresponding to the concentration profile as X = 0.

Results carried out from very similar experimental procedures for hardened cement past (hcp) and brick have demonstrated that if the diffusing substance is reactive with the medium, as in the case of Cl- in hcp, the concentration in its pore solution C in the contact surface zone is nearly double of that in test solution C_0. With the penetration depth proceeds, say X = 12 mm, C/C_0 gradually decreases to 1, and for X = 40 mm $C/C_0 \rightarrow$ 0. For brick specimens, this sharp 'border effect' was not observed.

4 RESULTS

Specimens at the age of 37 days and 750 days were immersed into the test solution. In a given diffusing time, the ionic contents Cl-, SO_4^{-2}, Na^+ and K^+ were measured for each defined penetration depth. Table 1 summarises the apparent diffusion coefficients D_{app} determined according to equation (9) for immersion time t = 30 days and t = 60 days respectively.

Results obtained are in good agreement with previous published data (Hermanns 1993). The higher diffusivity for young specimen can be explained from two points: 1) Consumption of water due to further hydration. This means specimens of sealing material at early ages which were placed into

test solution undergoes a process involving simultaneous intrinsic chemical reaction and diffusion. The hydration-consumed water would be supplied from the test solution. This extra up-take of cations and anions in the solution were not driven by the concentration gradient.

Table 1. Apparent diffusion coefficients for specimens at different ages.

Age of the specimens	D_{app} [m^2 / s]			
	Cl-	SO_4^{-2}	Na^+	K^+
37 days (t = 30 days) [*]	2×10^{-11}	2×10^{-12}	3×10^{-12}	5×10^{-12}
37 days (t = 60 days)	2×10^{-11}	6×10^{-13}	9×10^{-13}	1×10^{-12}
750 days (t = 30, 60 days)	8×10^{-12}	8×10^{-13}	4×10^{-13}	9×10^{-13}

[*] t = immersion time

2) Porosity: Measuring results on its porosity, permeability and compressive strength at the age of 37 and 750 days are given in Table 2. It is known that higher porosity corresponds to larger diffusivity.

Table 2. Sealing material parameters.

Age (days)	porosity	Water content w	Permeability k	Strength σ_c
37	61 %	59 %	$\cong 8 \times 10^{-9}$ m/s	4 N/m^2
750	57 %	54 %	$< 1 \times 10^{-13}$ m/s	9 N/m^2

The apparent diffusion coefficients obtained for anions are generally larger of one order of magnitude than that of cations. The reason lies in that the negatively charged surfaces of the interlayer in clay adsorb preferentially positive ions (Rogner 1993). Thus the cation diffusing process is retarded. Studies on electrical double layer in hardened cement paste have supported the hypotheses for accelerated anion transport (Chatterji & Kawamura 1992). Apart from the effect of electrical charges on the surface of solid particles, the continuing chemical reactions within the composite play also an important part. Though the sulphate-resisting cement was used, there is enough Al_2O_3 in flyash and clay available for the formation of ettringite which has the constitutional formula $[Ca_3Al(OH)_6 \cdot 12H_2O]_2(SO_4)_3 \cdot 2H_2O$, once the composite is in contact with sulphate solution. Because of the increase in solid volume the mineral sealing composite as a whole becomes denser (Taylor 1997; Ggollop & Taylor 1995). As a consequence the diffusion process slows down. On the other hand, the formation of enttringite reduces the concentration of SO_4^{-2}, which in turn promotes the ingress of diffusing ions. It is difficult to identify which factor at early ages is dominant. From our test results, it shows the D_{app} for SO_4^{-2} varies with the age of the material irregular and is smaller than diffusivity for Cl^-.

5 CONCLUSIONS

The transport of anions and cations in the contaminated ground water into the mineral sealing composite can be realistically characterised by the diffusion equation according to second Fick's law. The ingress of the diffusing ions is a diffusion and sorption coupled process. If the sealing composite is older than about 2 years, i.e. the changes in its chemical compositions and microstructures are negligible compared to the rate of penetration of the ions, the apparent diffusion coefficients can be used to predict the ion content profiles. The apparent diffusion coefficient obtained from diffusion test decreases with the age of the specimen, i.e. due to the hydration process.

Much more work is needed in order to specify to what extent diffusion or sorption contributes to the transportation process. Also the influence of the age on the apparent diffusion coefficient should be studied in detail.

6 REFERENCES

CHATTERJI, S. & KAWAMURA, M. 1992. A critical reappraisal of ion diffusion through cement based materials. Part I: Sample preparation, measurement technique and interpretation of results. *Cement and Concrete Research.* 22:525-530.

CHATTERJI, S. & KAWAMURA, M. 1992. Electrical double layer, ion transport and reactions in hardened cement paste. *Cement and Concrete Research.* 122:774-782.

CRANK, J. 1975. *The mathematics of diffusion.* Second edition. Oxford: University Press.

GOLLOP, R.S. & TAYLOR, F.H.W. 1995. Microstructural and microanalytical studies of sulfate attack III Sulfate-resisting portland cement: Reactions with sodium and magnesium sulfate solutions. *Cement and Concrete Research.* 25:1581-1590.

GRAY, D.H. & WEBER, W.J. 1984. Diffusional transport of hazardous wasteleachate across clay barriers. *7th Annual Madison Waste Conf. Municipal and Industrial Waste.* 11-12:373-389.

HASENPATT, R. 1988. Bodenmechanische Veränderungen reiner Tone durch Adsorption chemischer Verbindungen (Batch-und Diffusionsversuche). *Mitteilungen des Institutes für Grundbau und Bodenmechanik.* Eidgenössische Technische Hochschule Zürich ETHZ. 134.

HERMANNS, R. 1993. *Sicherung von Altlasten mit vertikalen mineralischen Barrierensystemen im Zweiphasen-Schlitzwandverfahren.* Institut für Geotechnik. ETH Zürich. 204. Zürich: vdf.

MADSEN, F.T. & KAHR, G. 1993. Diffusion of Ions in Compacted Bentonite. In D. Alexandre et al. (eds), *Proceedings of the International Conference on Nuclear Waste Management and Environmental Remediation.* 10354A:239-246. Prag.

NYE, P.H. 1979. Diffusion of ions and uncharged solutes in soils and soil clays. *Advances in Agronomy.* 31:225-272.

PEREIRA, C.J. & HEGEDUS, L.L. 1984. Diffusion and reaction of chloride ions in porous concrete. *8th international Symp. Chem. Reaction Engineering.* 87:427-438.

ROGNER, J. 1993. Modelle zur Beständigkeitsbewertung von Dichtwandmassen auf der Basis von Lagerungsversuchen. *Mitteilungen des Institutes für Grundbau und Bodebmechanik.* Universität Hannover. 37. Hannover: Eigenverlag.

SCHNEIDER, W. & GÖTTNER, J-J. 1991. Schadstofftransport in mineralischen Deponieabdichtungen und natürlichen Tonschichten. *Geologisches Jahrbuch.* 58. Hannover.

TAYLOR, F.H.W. 1997. *Cement Chemistry.* 2nd edition. London: Thomas Telford Publishing.

Clay Science for Engineering, Adachi & Fukue (eds) © 2001 Balkema, Rotterdam, ISBN 90 5809 175 9

Hydraulic conductivity of an Andisol leached with mixture of dilute acids

M. Ishiguro – *Faculty of Environmental Science and Technology, Okayama University, Japan*

T. Nakajima – *Raito Kogyo Company Limited, Tokyo, Japan*

K. Nakaishi – *Faculty of Agriculture, Ibaraki University, Japan*

T. Makino – *National Institute of Agro-Environmental Sciences, Tsukuba, Japan*

ABSTRACT: Recently, adverse effects of acid rain and chemicals on soils are a serious problem. Saturated hydraulic conductivity (K) changes of allophanic Andisol (volcanic ash soil) , which has much amount of pH-dependent charges, during dilute acid leaching and their causes were examined in this study. When nitric acid at pH 3 or 4 was leached, K decreased rapidly. On the other hand, decrease in K became smaller as the ratio of sulfate in the dilute acid increased. From the results of zeta potential measurement and calculation of repulsive potential energy between the clay particles, we concluded that the decrease in K for nitric acid was due to swelling and dispersion of the soil induced by electrostatic repulsive force. By using Stern-Gouy-Chapman model and batch adsorption experiment, we clarified that most of sulfate adsorbed on the clay surface chemically and electrostatically. These surface adsorption maintained flocculation of the soil and inhibited decrease in K.

1 INTRODUCTION

Hydraulic conductivity (K) decreases when swelling and clay dispersion occur. Swelling and clay dispersion occur due to the increase of electric repulsive force among soil particles. The repulsive force increases with increase of an absolute value of the surface potential of the clay, or decrease of ion concentrations or valency of the counterion (Iwata, 1995). Therefore, charge characteristics of soil are very important when K is considered.

Recently, acid rain has become a serious problem throughout the world. The adverse effects of acid rain on the hydraulic conductivity of soils should be assessed. Because soil charge affects its K and soils have pH-dependent charges, pH is an important factor for K. However, few studies have focused on the influence of pH on K.

Allophanic Andisol contains substantial pH-dependent charge and its K is therefore strongly affected by pH. Nakagawa & Ishiguro (1994) found that K for allophanic Andisol (Hapludand) decreased during leaching with HCl solution at pH 3 and NaOH solution at pH 11. Kakubo et al. (1995) showed that the decrease in K for surface soil of a volcanic ash soil during HNO_3 leaching was less than for subsoil due to the presence of rich organic matter. Matsukawa et al. (1998) noted K for allophanic Andisol to remain constant during H_2SO_4 leaching. Ishiguro & Nakajima (2000) clarified the

relationships among K, pH and clay dispersion in the soil profile experimentally.

Allophanic Andisol is a typical volcanic ash soil in Japan. About 15% of the land is covered with volcanic ash soils. Because the soils are well aggregated, about 50% of the non rice producing crop areas are located in regions of these soils. The objective of this study was to determine the mechanism of change in K for the soil during dilute acid leaching. Because the main contaminants of acid rain are HNO_3 and H_2SO_4, mixtures of these dilute solutions were used in the experiment. For this purpose, zeta potentials, anion adsorption amounts, and dispersion characteristics of the soil were measured. Electrical repulsive potential energies of the soil and surface adsorption amounts of sulfate were also evaluated.

Table 1. Physical and chemical characteristics of the soil.

Soil characteristics			
Coarse sand		1.7	%
Fine sand		11.2	%
Silt		37.5	%
Clay		49.5	%
Amorphous material		41.4	%
Organic Carbon		1.16	%
Texture class		Heavy clay	
Porosity		82.4	%
Cation exchange capacity	pH5	5.7	ceq/kg
	pH7	10.6	ceq/kg
Anion exchange capacity	pH5	10.3	ceq/kg
	pH7	0.6	ceq/kg

2 MATERIALS AND METHODS

2.1 *Soil*

Allophanic andisol (Typic Hydrudand) was obtained from a field at the National Institute of Agro-Environmental Sciences in Tsukuba, from the 4Bw1 horizon. Its physical and chemical properties measured by the National Institute of Agricultural Sciences are listed in Table 1.

2.2 *Hydraulic conductivity studies*

In order to clarify the effects of mixtures of dilute nitric acid and sulfuric acid, K was determined at a constant hydraulic gradient of 5 m m^{-1}. To control the constant hydraulic gradient, a mariotte bottle and a tube connected to the bottom of the soil column through a small water-filled head space and filter net were used. The mariotte bottle was used to maintain a constant water pressure on the top of the soil column. The pressure at the bottom of the soil column was maintained at almost zero as the outlet of the tube was placed at the same height as the bottom surface of the soil. Influent solutions were mixtures of nitric acid and sulfuric acid at pH 3 or pH 4. Mixed ratios of NO_3 concentration (mol$_c$ m^{-3}) to NO_3+SO_4 concentration(mol$_c$ m^{-3}) in the solutions were 0, 50, 70, 90 and 100 %. The solutions at pH 3 and pH4 were equivalent to electrolyte concentrations of 1.0 and 0.1 mol$_c$ m^{-3} respectively. The experimental procedure was as follows:
1 A column (3.2 cm in diam. by 3 cm in height) was packed with <2-mm sieved field-moist soil to a bulk density of 510 kg m^{-3} (same as under field conditions).
2 The soil column was saturated by capillary rise with 1000 mol$_c$ m^{-3} solution of the mixture of $NaNO_3$ and Na_2SO_4 for one day.
3 100 cm^3 (about 5 pore volume) of the solution was percolated through the column to saturate the clay charges with the solutes. Then, 1 mol$_c$ m^{-3} mixture solution was percolated through the column sufficiently until the outflow solution concentration became the same as the inflow solution concentration. The mixed ratio of the solution was the same as that of the 1000 mol$_c$ m^{-3} solution used for the saturation.
4 Finally, a solution with the same mixed ratio at pH 3 or pH4 was percolated through the column. The flow rate was measured during acid leaching and K was calculated following Darcy's law. Each solution was changed quickly using 2 mariotte bottles and a three-way stopcock.

2.3 *Dispersion studies*

Light transmittance of soil suspensions was measured to examine the relation between K and dispersion. Mixed ratios of NO_3 concentration (mol$_c$ m^{-3})

to NO_3+SO_4 concentration(mol$_c$ m^{-3}) in the suspended solutions (NO_3 ratio) were also 0, 50, 70, 90 and 100 %. The procedure was as follows:
1 <2-mm sieved, field-moist soil (0.02 g by dry soil) was equilibrated with 1000 mol$_c$ m^{-3} mixture solution of $NaNO_3$ and Na_2SO_4.
2 The soil was equilibrated with ($NaNO_3$ + HNO_3 + Na_2SO_4 + H_2SO_4) solution at electrolyte concentration of 0.1 mol$_c$ m^{-3} at specified pH from 4 to 5.4, or at electrolyte concentration of 1 mol$_c$ m^{-3} at specified pH from 3 to 5.2. The mixed ratio of the equilibrated solution was the same as that of the 1000 mol$_c$ m^{-3} mixture solution in procedure 1.
3 30 cm^3 of the soil/water mixture, contained in a 50 cm^3 tube, was shaken for 1 min. After 18 h of settling, 15 cm^3 were sampled from the upper portion (2.5 cm in depth) of the mixture, and light transmittance was measured, using visible light. Transmittance intensities were 0 % through a shaded plate and 100 % through pure water in a cell.

2.4 *Measurement of zeta potentials*

In order to evaluate the electrostatic repulsive potential energy in the soil, the zeta potentials of the clay in the soil were measured. The zeta potentials were derived using electrophoretic light scattering (ELS-800; Otsuka electronics) under the solution conditions similar to the influent solutions in the hydraulic conductivity studies.

2.5 *Anion adsorption experiments*

Amounts of NO_3 and SO_4 adsorptions were measured by the batch method of Wada and Okamura (1980) to evaluate the electric charge densities of the soil and the chemical adsorption amount of SO_4. The soil solution conditions were also similar to the influent solutions in the hydraulic conductivity studies. The procedure was as follows;
1 About 2g of soil sample was equilibrated with 1000 mol$_c$ m^{-3} of the mixture of $NaNO_3$ and Na_2SO_4 over night. The soil solution pH was roughly adjusted at pH 3 or 4 with HNO_3 or H_2SO_4.
2 The soil was shaken with 200 cm^3 of the mixture of HNO_3 and H_2SO_4 at 1.0 or 0.1 mol$_c$ m^{-3} for 1 hour. Centrifuged and discarded the supernatant. Repeated this 6 times. The final supernatant was kept to analyze the concentrations.
3 The soil was shaken with 60 cm^3 of 1000 mol$_c$ m^{-3} KCl for 15 min. Centrifuged and gathered the supernatant. Repeated this 3 times.
4 The soil was shaken with 200 cm^3 of 10 mol$_c$ m^{-3} NaOH for 15 min. Centrifuged and gathered the supernatant. Repeated this twice.

5 NO₃ concentration of the supernatants was measured by steam distillation method and those SO₄ concentration was measured by ion chromatography. Al and H concentration of the supernatant at procedure 2 were measured by ICP and pH meter, respectively. The NO₃ and SO₄ adsorption amounts were calculated with these values.

3 THEORY

3.1 *Electrical repulsive potential energy*

Electrical repulsive potential energies between the soil clay particles were estimated in order to clarify the mechanism of the decrease in K. These energies were derived from the measured zeta potentials.

In the case of low potentials, a potential at the distance x from the charged clay surface, Ψ, can be approximated as follows:

$$\Psi = \Psi_s \exp(-\kappa x) \qquad (1)$$

where, Ψ_s = potential of the clay surface. κ is defined as follows:

$$\kappa^2 = \frac{F^2 \Sigma z_i^2 C_{o,i}}{\varepsilon RT} \qquad (2)$$

where, F = Faraday const.; z_i = valence of ion i; $C_{o,i}$ = concentration of ion i in the bulk solution; ε = permittivity of the liquid; R = gas constant; T = absolute temperature. If a zeta potential, ζ, is used instead of Ψ_s, we can get the potential distribution from the plane of the zeta potential to the bulk solution.

We consider that two clay particles approach each other. Then, the potential at the midpoint between the clays, Ψ_m, is,

$$\Psi_m = 2\Psi \qquad (3)$$

where, Ψ = potential at the point when the clay exists alone. The concentration of ion i at the midpoint, C_i, is given as

$$C_i(x) = C_{o,i} \exp\left[-\frac{z_i F \Psi_m}{RT}\right] \qquad (4)$$

The electrical repulsive force between the clays is the osmotic pressure at the midpoint. The osmotic pressure, Π, is,

$$\Pi = RT \Sigma (C_i - C_{o,i}) \qquad (5)$$

where, unit of the concentrations is mol m⁻³. We get the repulsive potential energy, V_ζ, when the two clay particles approach to the plane of the zeta potential as follows:

$$V_\zeta = -2 \int_b^0 \Pi dD \qquad (6)$$

where, D = the distance from the plane of the zeta potential to the midpoint; b = the distance from the zeta potential plane to the boundary of the bulk solution. The point of the repulsive potential energy= 0 is given in the bulk solution.

3.2 *Stern-Gouy-Chapman model*

In order to estimate the amount of SO₄ adsorbed on the clay surface, Stern-Gouy Chapman model (SGC model) was adopted in the soil clay. To simplify the complicated shapes of the soil clay surfaces, the surface is assumed to be flat plane. The charge density of the clay was calculated from the measured amount of anion adsorption. As the predominant clay minerals are allophane and imogolite, the specific surface was given to 1000 m² g⁻¹.

If we suppose that all ions take the Boltsmann's distribution in the diffuse double layer (Gouy-Chapman model; GC model), the relationship between the surface potential of the clay, Ψ_s, and its surface charge density, σ, is given as follows:

$$\sigma^2 = 2 \varepsilon RT \left[[NO_3]\left\{\exp\left[\frac{F\Psi_s}{RT}\right]-1\right\} \right.$$
$$+ [SO_4]\left\{\exp\left[\frac{2F\Psi_s}{RT}\right]-1\right\} + [H]\left\{\exp\left[-\frac{F\Psi_s}{RT}\right]-1\right\}$$
$$\left. + [Al]\left\{\exp\left[-\frac{3F\Psi_s}{RT}\right]-1\right\} \right] \qquad (7)$$

where, [ ] = ion concentration in the bulk solution. Ψ_s is calculated from Eq.7.

The potential distribution in the diffuse double layer is given by an approximation calculation with the following equation.

$$\left[\frac{d\Psi}{dx}\right]^2 = \frac{2RT}{\varepsilon} \left[[NO_3]\left\{\exp\left[\frac{F\Psi}{RT}\right]-1\right\} \right.$$
$$+ [SO_4]\left\{\exp\left[\frac{2F\Psi}{RT}\right]-1\right\} + [H]\left\{\exp\left[-\frac{F\Psi}{RT}\right]-1\right\}$$
$$\left. + [Al]\left\{\exp\left[-\frac{3F\Psi}{RT}\right]-1\right\} \right] \qquad (8)$$

Then, we get the ion concentration distributions.

As much amount of SO₄ is supposed to be adsorbed on the clay surface, the ion distributions are different from that of GC model. Therefore, we adopt SGC model. Assuming that all the adsorbed NO₃ takes the Boltzmann's distribution in the diffuse double layer. The measured amount of adsorbed NO₃ per unit surface area, q(NO₃), is

$$q(NO_3) = \int_c^d \{C_N - C_{N,o}\}\ dy \qquad (9)$$

where, y = distance from the clay surface in GC model; c = distance from the clay surface in GC model to that in SGC model, d = distance from the clay surface in GC model to the boundary of the bulk solution, C_N = NO_3 concentration at y; $C_{N,o}$ = NO_3 concentration at the bulk solution. Then, we can calculate the amount of adsorbed SO_4 in the diffuse double layer, $q_D(SO_4)$.

$$q_D(SO_4) = \int_c^d \{C_S - C_{S,o}\}\ dy \qquad (10)$$

where, C_S = SO_4 concentration at y; $C_{S,o}$ = SO_4 concentration at the bulk solution. The amount of SO_4 adsorbed in the stern layer, $q_S(SO_4)$, is

$$q_S(SO_4) = q(SO_4) - q_D(SO_4) \qquad (11)$$

where, $q(SO_4)$ = measured amount of adsorbed SO_4 per unit surface area.

4 RESULTS

Data for changes in relative K when leaching with dilute acid are given in Figure1. The abscissa shows the percolated amount of the dilute acid solution (m) as the height of solution when collected into a cylinder with the same sectional area as the soil column. The relative K is defined as the ratio of K during leaching to initial K in inflow solution at 1 mol_c m^{-3}. Initial Ks ranged from 6.9×10^{-6} to 1.1×10^{-4} m s^{-1}. The changes in Ks differed with the NO_3 ratio in the inflow solution. Ks decreased more rapidly as the NO_3 ratio increased as shown in Figure 1. For pH 3 solution, K of 70 % NO_3 initially decreased, but finally recovered to initial value. K of 50 % NO_3 maintained the initial value. K of 0 % NO_3 rather increased slightly after 2.5 m percolation. For pH 4 solution, K of 70 % NO_3 showed a similar tendency of K of 90 % NO_3. It gradually decreased. Ks of 50 % and 0 % NO_3 slightly increased. The decrease in K at pH 3 for higher NO_3 ratio was much steeper than that at pH 4.

Light transmittances in soil suspensions are shown in Figure 2. The transmittance decreased as the NO_3 ratio increased. NO_3 kept stable dispersion condition. The pH influence was not clear for the suspensions at 1 mol_c m^{-3}. The transmittance for 0.1 mol_c m^{-3} solution decreased as pH decreased. The transmittance for 0.1 mol_c m^{-3} solution was relatively lower than that for 1 mol_c m^{-3}.

The zeta potentials are given in Figure 3. It increased as the NO_3 ratio increased. Especially, it showed steeper increase near 100 % NO_3 ratio. The potentials at pH 3 were larger than those at pH4.

The calculated repulsive potential energies are shown in Figure 4. It also increased as the NO_3 ratio increased. Especially, it showed much steeper increase near 100 % NO_3 ratio. The potential energy

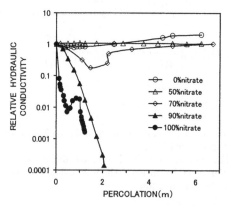

a. pH 3.

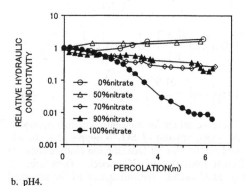

b. pH4.

Figure 1 Relative hydraulic conductivity change during dilute acid leaching.

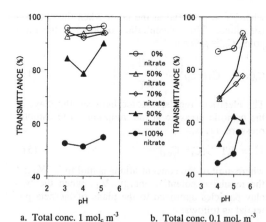

a. Total conc. 1 mol_c m^{-3} b. Total conc. 0.1 mol_c m^{-3}

Figure 2 Light transmittance of the soil suspensions.

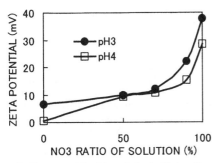

Figure 3   Zeta potentials of the soil clay in dilute acids.

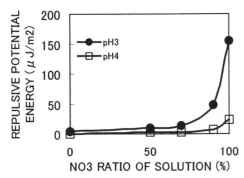

Figure 4   Calculated repulsive potential energies in dilute acids.

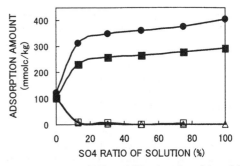

Figure 5   Amounts of NO_3 and SO_4 adsorption of the soil in dilute acids. ●=Total adsorption amount; ■=AEC; □=anion adsorption amount in diffuse double layer; △= NO_3 adsorption amount.

diffuse double layer ‐ the NO_3 adsorption amount in Figure. 5. They were only 0.5 to 1.1 % of the total SO_4 adsorption amount. In this model, 72 to 73 % of the total SO_4 adsorption amount were the exchangeable SO_4 on the clay surface.

became smallest at 0 % NO_3 ratio. The repulsive potential energies at pH 3 were much larger than those at pH 4.

The measured NO_3 and SO_4 adsorption amounts at pH 3 are shown in Figure 5. Those at pH 4 showed a similar trends. The NO_3 adsorption amounts became very small when SO_4 coexisted. The SO_4 adsorption amounts exchanged with KCl (exchangeable SO_4) equal to the AEC ‐ the NO_3 adsorption amount in Figure 5. The SO_4 adsorption amounts extracted with NaOH (strongly adsorbed SO_4 amount) equal to the total adsorption amount ‐ the AEC in Figure 5. The exchangeable SO_4 amount at 100 % SO_4 ratio was 294 $mmol_c$ kg^{-1}. The NO_3 adsorption amount at 100 % NO_3 ratio was 103 $mmol_c$ kg^{-1}, that was less than half of the exchangeable SO_4 amount. The exchangeable SO_4 amounts were larger than the strongly adsorbed SO_4 amount. The latter ranged from 25.9 to 27.5 % of the total SO_4 adsorption amount.

When we adopted Gouy-Chapman model in the soil, SO_4 adsorption amount was underestimated and NO_3 adsorption amount was overestimated. Therefore, SGC model was introduced. The calculated amounts of adsorbed SO_4 in the diffuse double layer were equal to the anion adsorption amount in the

5   DISCUSSION

Based on the change in K, soil structure observations, dispersion studies, repulsive potential energy calculations, and adsorption studies, K was clearly shown to decrease during higher NO_3 ratio acid solution leaching mainly as a result of swelling and dispersion of soil particles. Charge of allophanic andisol is pH-dependent and thus positive charge predominates at low pH (Iimura, 1966; Okamura and Wada, 1983; Ishiguro et al., 1992) as shown in Table 1. At low pH, when NO_3 ratio of the solution is higher, the positive charge generates electric repulsive forces among soil particles as shown in Figure 4. Then, swelling and dispersion (Fig.2) occurs. Larger water-conducting pores will consequently narrow.

On the other hand, when lower NO_3 ratio solution was leached, the repulsive potential energy was much smaller (Fig.4) and the soil particles maintained flocculated condition (Fig.2). Therefore, the K decrease was restricted (Fig.1). According to the SGC model, most of adsorbed SO_4 was on the clay surface. Moreover, NO_3 adsorption amount became much smaller under the existence of SO_4 in the solution (Fig.5). Because the repulsive force generates when diffuse double layers overlap, it reduces due to SO_4 adsorption on the clay surface. SO_4 adsorption on the clay surface keeps the soil under flocculated condition. Therefore, at lower NO_3 ratio, K did not decrease. It rather increased slightly.

The repulsive potential energies at pH 3 at higher NO_3 ratio were higher than those at pH 4 as shown in Figure 4. Because the soil swelled and dispersed well with these higher repulsive potential energies, the Ks at pH 3 at 90 and 100 % NO_3 ratio decreased more rapidly than those at pH 4. Although the repulsive potential energy at pH 3 at 70 % NO_3 ratio was larger than those at pH 4 at 70 and 90 % NO_3 ratio, the K at pH 3 at 70 % NO_3 ratio recovered to the initial value and the Ks at pH 4 at 70 and 90 % NO_3 ratio decreased gradually. The repulsive potential energy could not explain among these K differences. However, these calculated results showed that the decrease in K at higher NO_3 ratio was due to the higher repulsive potential energy among the soil particles.

The results of dispersion studies showed the clear tendency of flocculation at lower NO_3 ratio and dispersion at higher NO_3 ratio. They corresponded with the change in K. However, though the transmittances at pH 4 at higher NO_3 ratio were lower than those at pH 3, the decrease in K at pH 4 was slower than that at pH 3. The dispersion studies gives rough estimation for decrease in K. But they did not give the exact prediction.

Three SO_4 adsorption types were found with the adsorption experiments and the SGC model; adsorption in diffuse double layer, adsorption on the clay surface with electrostatic charge, and that extracted with NaOH. The strong adsorption extracted with NaOH is supposed to be a chemical adsorption. Most of the SO_4 adsorption (about 98.9 to 99.5 %) were on the clay surface. These surface adsorption is supposed to flocculate the soil and increase the K slightly.

6 CONCLUSIONS

Because allophanic andisol has a large number of pH-dependent charges, the positive charge increased and the negative charge decreased with dilute acid leaching. The counterion, NO_3, is monovalent and the electrostatic repulsive force between soil particles is stronger than that for SO_4. Increasing repulsive forces induced swelling and dispersion which resulted in a decrease in K during the leaching of dilute acid with high NO_3 ratio. On the other hand, flocculation resulted in a slight increase in K during the leaching of dilute acid with high SO_4 ratio. The counterion, SO_4, is divalent and, moreover, has strong affinity for soils with pH-dependent charge. Most of them (about 98.9 to 99.5 %) adsorbed on the clay surface and about 25.9 to 27.5 % of the adsorbed SO_4 were supposed to be chemical adsorption at pH 3. Therefore, the attractive force is stronger than the electrostatic repulsive force between soil particles at higher SO_4 ratio.

When acid rain consists mainly of HNO_3 and the pH is lower than 4, there is a high likelihood that soil permeability will decrease in the region of an al-

lophanic andisol with few organic compounds. Because the buffer action of the soil is strong, the pH does not easily decrease. However, the permeability can be lowered by collapse of aggregates in a thin surface layer. The other possible cause of dilute acid leaching is artificial or accidental addition of HNO_3. One possible means for maintaining a ponded condition or reservoir on allophanic andisol soil is addition of HNO_3 solution, as HNO_3 appears to destroy the naturally high K of this soil.

REFERENCES

Iimura, K. 1966. Acidic properties and cation exchange of allophane and volcanic ash soils. (in Japanese with English summary) Bull. Natl. Inst. Agric. Sci. B17:101-157.

Ishiguro, M., K-C. Song, and K. Yuita. 1992. Ion transport in an allophanic andisol under the influence of variable charge. Soil Sci. Soc. Am. J. 56:1789-1793.

Ishiguro, M. & T. Nakajima. 2000. Hydraulic conductivity of an allophanic andisol leached with dilute acid solutions. Soil Sci. Soc. Am. J. 64. (in press)

Iwata, S. 1995. Interaction between particles through water. P.154-228. In S. Iwata, T. Tabuchi & B.P. Warkentin (ed.) Soil-water interactions: 154-228. New York: Marcel Dekker.

Kakubo, A., S. Matsukawa, and H. Katoh. 1995. Effects of nitrate solution on soil permeability and salts leaching. (in Japanese with English summary) Trans. Japanese Soc. Irrigation, Drainage and Reclamation Eng. 180:103-110.

Matsukawa, S., H. Tomita, S. Suzuki, and H. Katoh. 1998. Relation between soil dispersion, hydraulic conductivity and pH of soil water for allophanic soils during acid solution leaching. (in Japanese with English summary) Soil Physical Conditions and Plant Growth Japan. 77:3-9.

Nakagawa, T., and M. Ishiguro. 1994. Hydraulic conductivity of an allophanic andisol as affected by solution pH. J. Environ. Qual. 23:208-210.

Okamura, Y., and K. Wada. 1983. Electric charge characteristics of horizons of Ando (B) and Red-Yellow B soils and weathered pumices. J. Soil Sci. 34:287-295.

Wada, K. & Y. Okamura. 1980. Electric charge characteristics of Ando A1 and buried A1 horizon soils. J. Soil Sci. 31:307-314.

Clay Science for Engineering, Adachi & Fukue (eds) © 2001 Balkema, Rotterdam, ISBN 90 5809 175 9

Role of clay in preventing arsenic dispersion in the Ganges delta of Bangladesh

A.A. Khan, S.M.M. Alam & S.H. Akhter
Department of Geology, University of Dhaka, Bangladesh

ABSTRACT: A clay (carbonate mud) layer of about 110 m thick occurs immediate below the arsenic contaminated aquifer zones and the geological evidences suggest for the clay to occur almost throughout the delta plain at variable depths. The clay is illitic in nature having three layers clay crystals like smectite. The dominant elemental constituents are Si, O, Al, and Ti with subordinate As, Fe, P, Ca, Mg, Na, Mn, S and C. The framework grains are highly coated with iron oxyhydroxide and to lesser extent with manganese oxyhydroxide together with arsenic. Sodium concentration shows a steady increase with depth in the clay layer. Carbon dioxide ranges between 3.9 to 7.6 percent. The formation of soda ash (Na_2CO_3) probably has forced the clay to be dispersed. The clay is very sticky, hard in drying and also calcareous in nature. The framework grain coatings have significantly deteriorated the pore geometry and subsequently reduced the vertical seepage potentiality. An extremely high resistivity signature for this clay supports the deterioration of pore-geometry. The expanding nature of clay resulted in diffusion of arsenic from pores in the overlying aquifers failing to compete with phosphate at bonding sites for adsorption. The clay has been interpreted as a "dispersed clay" and should act as a potential barrier to arsenic transport vertically. The frequent inter-fingering between the channel-fills and overbank-deposits in the arsenic contaminated regions also indicate that the clay is a potential barrier to lateral dispersal of arsenic.

1 INTRODUCTION

Ganges Delta Plains of Bangladesh and West Bengal, India constitute one of the largest deltas in the world. The major rivers viz., the Ganges, the Brahmaputra-Jamuna, the Meghna and their tributaries and distributaries have been the primary transporting media of the sediments in the depositional domain. This large delta can be divided into three broad flood-plain regions i.e., the Ganges flood-plain, the Jamuna flood-plain and the Meghna flood-plain. These flood-plain regions are also bounded by three well defined Quaternary basinal depressions Ganges-Mahananda depression, Jamuna depression, and Meghna depression (Figure 1). The Quaternary aquifers in the Ganges Delta Plains of Bangladesh are largely contaminated by arsenic. Arsenic contamination has so far been observed in the aquifers within the Quaternary deposits situated in all the three flood-plain regions bounded by the basinal depressions.

The sub-surface geology of the region bounded by the Ganges floodplain suggests that all the geological formations have a regional gradient towards south and southeast, indicating that the sedimentation progressed in these directions throughout (Figures 2 and 3). The Quaternary deposits of the floodplain regions are typically characterised by finning upward sequence of channel migration. This has resulted in intercalation of channel-fill deposits with over-bank deposits. The sandy silt facies within overbank association are interpreted as levee deposits formed by deposition of suspended sandy sediments near the channel banks during over-bank floods (Imam et al. 1998). In addition to intercalated finning upward sequence of channel migration, the Quaternary deposits below finning upward sequence are composed mostly of clay, peaty clay, silty clay, clayey peat, peat and silty sand at variable depths. The occurrence of unusually thick clay which is basically a carbonate mud has been the focal point of study. The physico-chemical studies of the clay have been made to understand the fate of arsenic transport from the overlying contaminated aquifers to underlying aquifers.

2 GENESIS OF CLAY

A group of detrital sedimentary rocks are commonly composed of clays, shales, mudstones, siltstones and marls. Two grades of particle size are recognised, the silt grade, with the particles ranging in size from 1/16 to 1/256 mm, and the clay grade, with particles of less than 1/256 mm. A clay has been defined as a natural plastic earth (though some clays are nonplastic), composed of hydrous aluminum silicates (the "clay minerals"). Claystone is indurated clay. To those claystones which are neither fissile nor laminated but are blocky or massive, the term mudstone is applied. Shale is a laminated or fissile claystone or siltstone.

Clays and shales can be classified as residual or transported. Residual clays are the products of deposits are dependent upon climate, drainage, and parent-rock materials. The transported clays derived their constituents from three sources (Figure 4, Pettijohn 1957). They are composed in varying proportions of: a) the products of abrasion (mainly silt), b) the end products of weathering (residual clays), and c) chemical and biochemical additions. These chemical additions are either materials precipitated from solution and then deposited concurrently with the accumulating clays, such as calcium carbonate or they are materials added by reaction or exchange with the surrounding medium, usually sea water, such as potassium and/or magnesium. Based on clay crystal, three major types of clays may be identified viz., two-layer nonexpanding lattice type kaolinitic clay, three-layer expanding lattice type smectite clay, and three-layer with a nonexpanding lattice type illitic clay. Three-layer clay crystals have two outside layers, made of silicon and oxygen and the middle layer made of aluminum and oxygen. The structural formula of illite is:

$(OH)_4 K y (Al_4 . Fe_4 . Mg_4 . Mg_6) (Si_8 - y . Al y) O_{20}$ where y varies from 1 to 1.5 for aluminum in the illite unit cell (Grim, 1942).

Genesis of the carbonate mud is attributed to the deposition in a lagoon environment due to marine transgression.

3 DISTRIBUTION OF CLAY

A 110 m thick carbonate mud occurs 40 m below ground level. The entire column of mud occurs within the Quaternary deposits and is regionally distributed within the Quaternary depressions (Figure 2). The major marine transgression progressed along Meghna and Jamuna depressions and the depressions within the Ganges delta plain (Figure 1). During transgressive phase, the supra-tidal zone was extended as much as 300 km land interior resulting in the deposition of clay in a lagoon environment. High stand sea level of about 5 m above the present one prevailed about 5500 yr. B.P. (Fairbridge 1966). The carbonate mud has been originated from carbonate sediments in clear, warm, sunlit waters of tropical shallow sea by organically secreted solids formed within the basin of deposition and remained at their original sites of formation.

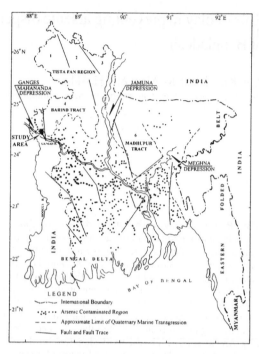

Figure 1. Map showing the limit of Holocene transgression and arsenic contaminated region.

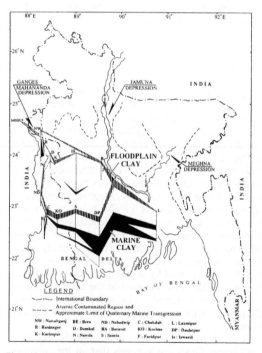

Figure 2. Arsenic contaminated region and the possible limit of the Quaternary marine transgression in the Bengal Delta and sub-surface panel diagram.

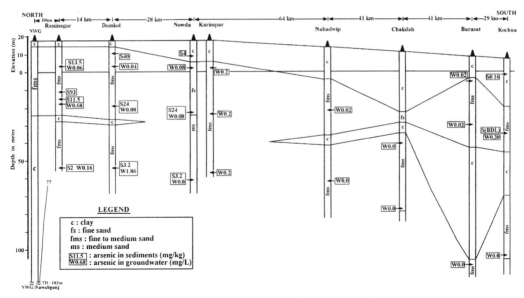

Figure 3. North-South subsurface cross-section across Quaternary deposits.

PARENT ROCK

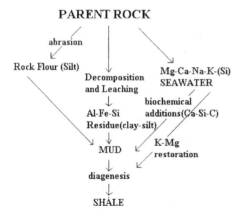

Figure 4. Provenance of shale.

Table 1. Concentration of OC and CO_2 with depths (DW#1).

Sampling depth (m)	Organic Carbon (%)	Carbon Dioxide (%)
3.1	1.9	4.2
9.1	1.9	3.9
24.4	1.9	6.0
33.5	0.7	5.6
39.6	1.7	7.3
——— Boundary between aquifer sand and lower clay ———		
64.0	1.7	7.5
94.5	1.4	6.9
134	1.2	7.6

4 PHYSICO-CHEMICAL PROPERTIES OF CLAY

The physical properties of the clay are very sticky, hard in drying and calcareous in nature. As high as 30% calcite grains, readily soluble in cold dilute acid, occur in clay. The crystal habit of the grains ranges from anhedral to rhombohedral. The dominant elemental constituents as determined by ammonium-oxalate extracts are Si, Al, As, Fe, P, Mg, Na, K, Mn, and S (Figure 5). The SEMEDX analysis of clay also supports the presence of the above elements including O, C, Ca and Ti (Figure 6). The framework grains are highly coated with iron oxyhydroxide and to lesser extent with manganese oxyhydroxide together with arsenic in adsorption. The XRD analysis (Figure 7) exhibits the characteristics of illitic clay. Organic carbon and carbonate contents of clay have been determined by ignition loss method (Table 1). It is envisaged that the organic carbon content is relatively higher in the overlying aquifer sediments than in the underlying clay. On the otherhand, the carbonate content is higher in the underlying clay than in the overlying aquifer sediments.

The framework grain-coatings (Khan et al. 1998) and an unusually high electrical resistivity (Figure 8) has been observed in the clay.

5 PROPERTIES OF AQUIFERS AND ARSENIC CONTAMINATION

Most of the arsenic contaminated aquifers occur within 40 - 70 m below the ground level underlain by clay layer of varying thickness. The aquifers are unconfined in nature. The aquifer sediments are dominantly channel-fill of finning upward sequence (Imam et al. 1998). Intercalations of clay occur due to frequent channel migration that has resulted in the deposition as over-bank deposits over the channel-fill deposits. After the deposition of underlying clay, the channel network started to develop when sea began to retreat. As a result, the underlying clay was cut away by the newly developed channels. The channel-fill deposits got mixed-up by the eroded bank sediments (clay). As the sedimentation progressed, the intercalations of channel-fill and over-bank deposits continued with the transformation of over-bank deposit towards more flood-plain clay at the upper part of the aquifer zones.

It is envisaged that an extensive development of grain coatings of iron oxyhydroxide (FeOOH) and manganese oxyhydroxide (MnOOH) and the presence of arsenic in the coatings suggest that the adsorption of arsenic have been derived from arsenic rich aqueous solution (Alam 1999). The larger proportion of concentration of arsenic in the finer facies (clay) zone indicates that the concentration progressed under calm and stagnant low energy regime.

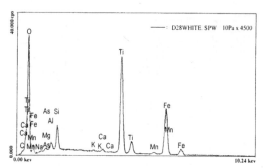

Figure 6. SEMEDX identification of elements in clay (49 m depth).

An extensive NW dispersal of "Toba Ash" (volcanic ash) and its occurrence in the Quaternary sediments (Acharyya & Basu 1993), normally rich in arsenic, possibly has enriched the transgressing sea-water in arsenic and subsequently arsenic has been adsorbed in the sediment grains.

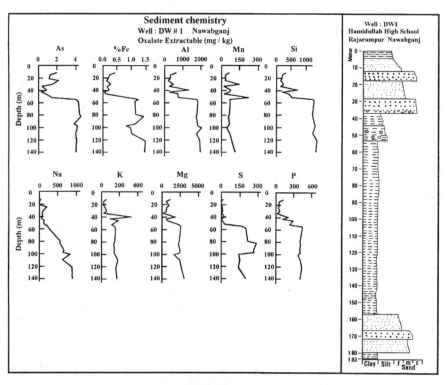

Figure 5. Various elemental concentrations versus depths and vertical columnar section.

6 USE OF CLAY FOR PREVENTING DISPERSION

It has been revealed from XRD analysis that the clay under study is illite that has been formed under marine and alkaline environment. The presence of Ca, Mg and K in clay is of significance. Although illite has three-layer lattice (2:1 of Si : Al), the presence of potassium signify a non-expanding lattice and an intermediate behavior on exchange of ions in solution. Nevertheless, the decrease in concentration of Na at the upper part of the clay signifies the order of replacibility of ions namely, Na < K < Mg < Ca < H.

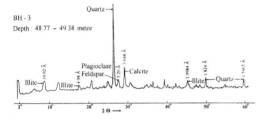

Figure 7. XRD signature of illite.

The occurrence of illite further signifies that the Ganges delta has been experienced an alkaline or soda lake environment along well defined linear zones of depressions bounded by faults viz., Ganges-Mahananda depression, Jamuna depression and Meghna depression. The presence of Na and CO_2 in appreciable proportion in clay signifies the existence of soda lake where the environment might have prevailed for forming Na_2CO_3 (Soda Ash). Soda ash can force clay to be dispersed or deflocculated so that the clay subsequently can act as a seal for seepage loss (Reginato et al. 1973). Also, in Na rich clay, the repulsive electrostatic forces between the negatively charged particles exceed the attractive van der Waals forces, causing the clay particles to exist as separate particles in a dispersed or deflocculated condition having tendency to seal in low permeability (Bouwer 1978). The sticky, amorphous and hard upon drying nature of the clay also signify its dispersed or deflocculated character.

Pore geometry has a direct bearing on electrical current flow efficiency (Herrick & Kennedy 1994) since electrical efficiency is a function of pore geometry, and is not inherently dependent on porosity. An extremely high resistivity values for the clay and

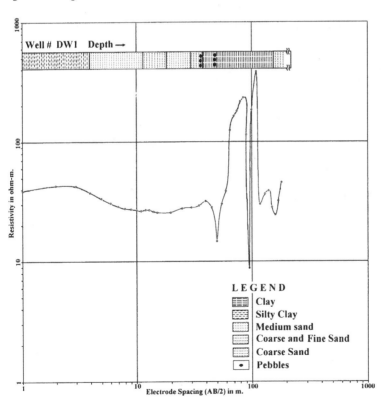

Figure 8. High resistivity signature against clay in well # DW1.

the thick grain coatings of FeOOH and MnOOH signify that water from the pores have been diffused deteriorating the pore geometry.

7 CONCLUSIONS

1. Thick clay layers occur regionally beneath the arsenic contaminated aquifers.
2. The type of clay is illite having being dispersed or deflocculated which prevents vertical seepage from overlying aquifer system. Thick grain coatings possibly have deteriorated pore geometry significantly. This is indicated by high electrical resistivity of the clay.
3. Although the inter-figuring between channel-fill and over-bank fill can reduce lateral seepage, the lateral transport of arsenic may occur where the contaminated aquifers are somehow linked to uncontaminated aquifers. However, the contamination could be prevented by applying this clay for grouting at the down flow directions.

8 ACKNOWLEDGEMENT

Dr. Pauline L. Smedley of British Geological Survey, Wallingford, UK is gratefully acknowledged for assistance in providing analytical data on oxalate extractable elements of the sediment samples. Dr. Muhammad Ali of Bangladesh Atomic Energy Commission is duly acknowledged for the help in providing SEMEDX analysis of sediment samples. Dr. Sohel Kabir of Bangladesh Geological Survey is sincerely acknowledged for conducting XRD analysis of sediment samples.

REFERENCES

Acharyya, S. K. & P. K. Basu 1993. Toba Ash on the Indian Subcontinent and its implications for correlation of Late Pleistocene Allivium. *Quaternary Research,* 40:10-19.

Alam, S.M.M 1999. Physico-chemical status of arsenic contaminated aquifers in Nawabganj Sadar and its surroundings, Nawabganj district. *Unpublished M.Sc Thesis, Department of Geology, University of Dhaka, Dhaka, Bangladesh.*

Bouwer, H. 1978. *Groundwater Hydrology.* New York: McGraw-Hill.

Fairbridge, R.W. 1966. Mean sea level changes, long-term-eustatic and other. In R. W. Fairbridge (Ed.), *Encyclopedia of Oceanography*: 479-485. New York: Reinhold.

Grim, R. E. 1942. Modern concepts of clay materials. *Geol. J.* 50: 225-275.

Herrick, D. C & W. D. Kennedy 1994. Electrical efficiency - A pore geometry theory for interpretating the electrical properties of reservoir rocks. *Geophysics.* 59(6): 918-927.

Imam, B., S.H. Akhter, A. A. Khan, M. A. Hasan & K.M. Ahmed 1998. Sedimentological and mineralogical studies on aquifer sediments within Bangladesh. *Report Groundwater Circle, BWDB, Dhaka, Bangladesh.*

Khan, A.A., B. Imam, S.H. Akhter, M.A. Hasan & K.M.U. Ahmed 1998. Subsurface investigation of arsenic contaminated areas of Rajarampur, Chanlai and Baragharia of Nawabganj district. *Research report Geohazard Research Group, Department of Geology, University of Dhaka, Dhaka, Bangladesh.*

Pettijohn, F.J 1957. *Sedimentary Rocks*, 2nd Edition, New York: Harper & Brothers.

Reginato, R. J., F. S. Nakayama & J. B. Miller 1973. Reducing seepage from stock tanks with uncompacted sodium-treated soils. *Jour. Soil Water Conserv.*, 28: 214-215.

Clay Science for Engineering, Adachi & Fukue (eds) © 2001 Balkema, Rotterdam, ISBN 90 5809 175 9

Transport of aqueous species through aggregated soil

T. Sato, Y. Usui & K. Kawaberi
Department of Civil Engineering, Gifu University, Japan

H. Tanahashi
Department of Construction Engineering, Daido Institute of Technology, Nagoya, Japan

ABSTRACT: A volcanic soil called as the Kanuma soil was applied to laboratory column tests for studying the role of inter-aggregate and intra-aggregate pores on transport of soluble chemical compounds. Break-through curve was analyzed by the mobile-immobile transport model (M-IM model) which can take into account different two types of pore water, one is easy to move and the other stagnant. The paper highlights characterization of breakthrough curve for aggregated soil on the basis of M-IM model. The discussions conclude that (1)volume fraction rate of mobile water is not constant but decreases with the increase of water content, (2)mass transfer coefficient depends on pore water velocity, (3)dispersion coefficient strongly depends on average pore water velocity, and (4)retardation factor is constant at more than 70% of degree of water saturation.

1. BACKGROUNDS

The most common way to describe solute transport in porous media is the classical advection-dispersion model (A-D model). The one-dimensional A-D model is written by

$$R\frac{\partial(\theta c)}{\partial t} = \frac{\partial}{\partial}\left(\theta D\frac{\partial c}{\partial x}\right) - \frac{\partial(qc)}{\partial x} \qquad (1)$$

where c is solute concentration, D is hydrodynamic dispersion coefficient, θ is volumetric water content, q is Darcy velocity, x is distance, t is time and R is retardation factor.

The A-D model has been used successfully to describe transport in most saturated media. However, some solute transport studies have shown that the A-D model may be inadequate to predict the asymmetric nature of vadose zone transport (DeSmedt and Wierenga, 1984) and preferential flow accompanying the tailing in biporous structured soil (Zurmühl and Durner, 1996).

To account for non-ideal transport, the A-D model has been altered to mobile-immobile transport model (M-IM model). The most popular one was given by vanGenuchten and Wierenga (1976). The fundamental equations are

$$\left(\theta_m + f\rho_d k_d\right)\frac{\partial c_m}{\partial t} + \left[\theta_{im} + (1-f)\rho_d k_d\right]\frac{\partial c_{im}}{\partial t} =$$
$$\theta_m D_m \frac{\partial^2 c_m}{\partial z^2} - \theta_m v_m \frac{\partial c_m}{\partial z} \qquad (2)$$

$$\left[\theta_{im} + (1-f)\rho_d k_d\right]\frac{\partial c_{im}}{\partial t} = \alpha(c_m - c_{im}) \qquad (3)$$

where the subscripts m and im represent the mobile and immobile regions respectively, D_m is the mobile zone dispersion coefficient, ρ_d is dry density, K_d is solid-liquid isotherm distribution coefficient, f is fraction of adsorption site in mobile region, α is mass transfer coefficient.

A schematic diagram of the M-IM model is shown in Figure 1. In order to apply this model to actual soils, involved parameters have to be fixed. Pore systems of natural soils may not consist of two distinct fractions that can be termed mobile and immobile in the model. Therefore the parameters are usually determined by optimization procedures for

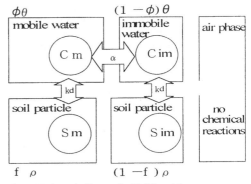

Figure 1. Schematic diagram for M-IM model.

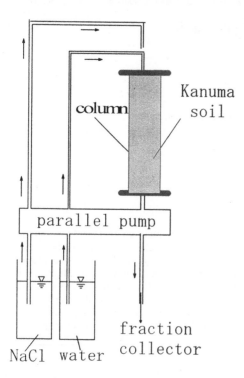

Kanuma
soil

column

parallel pump

NaCl water

fraction
collector

Figure 2.Laboratory column test.

Table 1. Series of column tests.

test cases	concentration (g/l)	darcy velocity (cm/min)	degree of saturation (%)
Run 4	1.648	0.37	61.6
Run 7	1.648	0.63	71.3
Run 8	1.648	1.17	87.4
Run 9	1.648	0.31	71.8
Run 10	1.648	0.07	76.1
Run 11	1.648	1.10	80.1
Run 13	0.164	0.21	76.8
Run 14	1.648	1.04	71.7
Run 15	0.164	0.61	83.6
Run 17	16.480	0.63	86.7
Run 18	16.480	0.20	82.0

the transport model best describing the breakthrough curve of a laboratory column test. Most frequently, a simulated breakthrough curve is fitted to experiments from a laboratory column test, where solute moves under steady state water flow. In field situations, where the flow regime is transient, the question arises as to how to treat the parameters when water content and pore water velocity change.

The purpose of this study is to make clear relationship between the transport parameters of M-IM model and experimental conditions. The previous works indicated the need for an exhaustive, systematic study focussing on the transport parameter changes with different water content, flow rates, flow regimes and soil types (DeSmedt and Wierenga 1984). The goal of this research is to develop and validate the relationships for the M-IM model parameters estimated from laboratory column test at different conditions of flow rates and water content under a steady state flow.

2. LABORATORY COLUMN TEST

A column with 5cm inner diameter and 20cm length was used in the tests to characterize a solute transport for specifying the M-IM model parameters. The laboratory column setup was described in Figure 2.

The Kanuma soil, one of volcanic clay soil, was packed with $1.22g/cm^3$ of dry density into the column. The density of soil particle is $2.65g/cm^3$ and the size of aggregation varies from 7 to 10mm.

Chloride sodium solution with 0.028mole/L was supplied to the top of the soil by the use of a parallel pump after completing steady flow condition by drinking water. The effluent was collected at the bottom to survey the features of solute transport on breakthrough curve.

Test series were described in Table 1. They were determined so as to specify the dependency on pore water velocity, degree of saturation and solute concentration.

3.PARAMETERS ESTIMATION

The equations (2) and (3) in M-IM model were converted into simple forms for convenience of parameters estimations in column tests. The four transport parameters were used in the analytical solutions for one-dimensional solute transport with linear adsorption as follows; ω: Stanton number, β: ratio of retardation in dynamic region with reference to that in stagnant, Pe: Peclet number and R: average retardation factor (vanGenuchten and Wierenga 1976). These quantities are explained in the succeeding section.

The parameters were determined from the choice of the best model for describing the experimental breakthrough curve on the basis of the modified Marquardt method. The initial guess of R, which is determined from the area bounded by the breakthrough curve and the horizontal line at 1.0 of relative concentration, takes a constant in the parameters identification. The breakthrough curve in Run 17 (see Table 1)was described in Figure 3. Simulated breakthrough was displayed as a solid line which parameters were given in this figure. It is well known that relative concentration becomes 0.5 at one pore volume in A-D model with none of reactive chemical constituent. Even for chemical reaction with soil particle, there are no way describing the breakthrough curve passing through 0.5 of relative concentration at less than one pore volume. The experiments show two typical aspects of solute transport through aggregated soil, preferential flow and tailing. Preferential flow cannot be well described by the A-D model (M-model). To describe this phenomena in accurate, the soil is often divided into two (or even more) sub-domains (M-IM model). A simple modeling proposed by van Genuchten and Wierenga (1976) was applied to the study. Simulated breakthrough curve shows a good similarity with the experiments as shown in Figure 3.

4. ESTIMATED PARAMETERS AND EXPERIMENTAL CONDITIONS

Simulated values were fitted to the experiments and the pair of the parameters showing the best describing breakthrough of each test was determined. Estimated values are described in accordance with the test cases in Table 2. If a field scale, predictive model is to be developed, conceptual models such as M-IM model must be well understood and rigorously validated. The model can only be physically validated when all model parameters can be determined

independently of breakthrough curve. It is necessary to develop an accurate way to estimate the transport parameters. Mechanistic relationship between the parameters and the experimental conditions offers well understanding to the transport structure in the model and accurate manner to specify the model parameters in a field-scale. The parameters were checked with the experimental conditions as follows.

4.1 Stanton number(ω)

The amount of ω is the ratio of mass transfer rate across the mobile and immobile interface to the rate of advection. Based on the work of vanGenuchten and Wierenga (1976), the following equation for the Stanton number was described

$$\omega = \frac{\alpha L}{q} \qquad (4)$$

where α is mass transfer coefficient, L is length of column, q is specific discharge or flow rate divided by the cross-sectional area of the soil.

Mass transfer coefficient of α is the parameter to be considered. The experimental conditions described in Table 1 were inserted into the Eq.(3) to specify the relationship with the flow conditions in each experiment. The relationship was given in Figures 4(a) and (b). The parameter of α is not dependent on the degree of saturation but on the average pore water velocity. Linear increase was expected in the relationship between α and v. Mass transfer co-

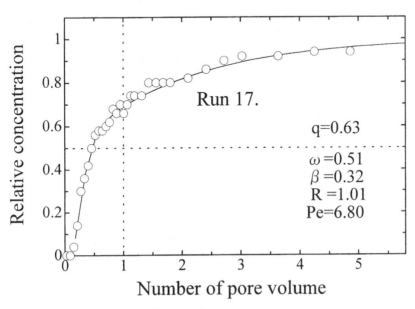

Figure 3. Experimental and computed breakthrough curves.

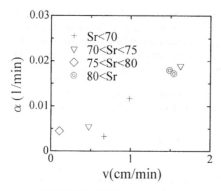

Figure 4(a). Relationship between α and v.

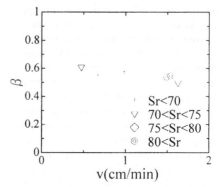

Figure 5(a). Relationship between β and v.

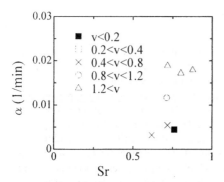

Figure 4(b). Relationship between α and Sr.

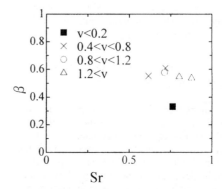

Figure 5(b). Relationship between β and Sr.

Table 2. Estimated parameters for each case.

test cases	ω (stauton number)	β (*)	R (retardation factor)	Pe (peclet number)
Run 4	0.17	0.55	1.54	4.7
Run 7	0.37	0.58	1.21	10.2
Run 8	0.31	0.54	1.04	6.6
Run 9	0.36	0.61	1.06	7.7
Run 10	1.23	0.33	1.03	26.5
Run 11	0.31	0.55	1.07	9.8
Run 13	0.13	0.62	1.40	3.5
Run 14	0.36	0.50	1.25	8.3
Run 15	0.38	0.45	1.45	7.2
Run 17	0.51	0.32	1.01	6.8
Run 18	0.34	0.41	1.52	2.5

*:ratio of retardation factor dynamic region
to the averaged one in the whole.

efficient is defined as

$$\alpha = \frac{D_{molecule}a}{\delta} \qquad (5)$$

where δ is the thickness of the stagnant film, a is the specific interfacial area and $D_{molecule}$ is the molecular diffusion coefficient. The area of interface will increase with the increase of pore water velocity.

4.2 Ratio of retardation in mobile region to the averaged one in the whole (β)

The most complicated, difficult to understand parameter is β which is defined by

$$\beta = \frac{\theta_m + \rho_d f k_d}{\theta + \rho_d k_d} \qquad (6)$$

where ρ_d is dry density of the soil, f is mass fraction of soil particle contacting with mobile water, k_d is distribution coefficient of liquid-solid face, θ is volumetric water content and θ_m is volumetric water content in dynamic region. Two unknowns, θ_m and f, are involved in the equation. Figures 5(a) and (b), however, show that the parameter of β is almost constant independently on the experimental conditions.

The equation (6) can be converted to

$$\beta = \frac{\phi + \dfrac{\rho_d f K_d}{\theta}}{R} \qquad (7)$$

where ϕ is volumetric fraction of pore water in dynamic region and R is averaged retardation factor in the whole region. The density of the soil particle is 1.10g/cm^3, the specific surface area of aggregation is $145 \text{m}^2/\text{g}$, and the average radius of aggregation is 1cm. The specific surface area of aggregation without intra-aggregate pore amounts to $5 \text{cm}^2/\text{g}$ that is much less than the specific surface area containing intra-aggregation. This implies that the amount of f is almost zero for the Kanuma soil. Insert of f=0 into Eq.(7) yields

$$\phi = \beta R \qquad (8)$$

The relationship between ϕ and degree of saturation is described in Figure 6. The volumetric fraction of dynamic region tends to decrease dependently of water content. This tendency is not the same as the linear expression of

$$\theta_m = 0.853\theta \qquad (9)$$

which was proposed by DeSmedt and Wierenga (1984).

Pore volume required for 0.5 of relative concentration ($PV_{0.5}$) was estimated in breakthrough curve to survey preferential flow. Small amount of pore volume is consumed for breakthrough curve in large amount of water content (Figure 7). This implies preferential flow is conspicuous in a soil with higher water content. The tendency is correspondent to the results of Figure 6 showing the decrease of fraction rate in mobile region (ϕ) with the increase of degree of saturation.

4.3 Peclet number (Pe)

The Peclet number is defined as

$$Pe = \frac{v_m L}{D_m} \qquad (10)$$

The mobile zone dispersion coefficient was displayed in accordance with average pore water velocity in Figure 8. A linear relationship is expected between *Dm* and *v*, which is the same in the classical one region model (Bear 1972). The estimated values were much larger than the empirical relationship proposed by DeSmedt and Wierenga (1984) who tested glass beads.

4.4 Retardation factor (R)

The average retardation factor was directly estimated

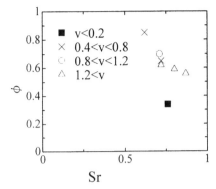

Figure 6. Relationship between ϕ and v.

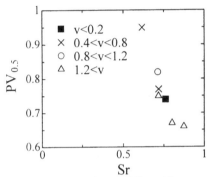

Figure 7. Relationship between $PV_{0.5}$ and Sr.

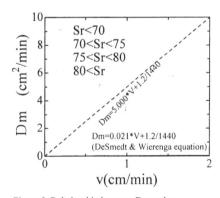

Figure 8. Relationship between Dm and v.

from the experimental breakthrough curve. The area bounded by the breakthrough curve and the horizontal line at 1.0 of the relative concentration is mass of solute absorbed in the column. Therefore the solid-liquid distribution coefficient is estimated from the absorbed solute mass and the dry density of the soil.

The retardation factor is almost constant independently of average pore water velocity (Figure 9(a)). A linear increase of R is expected with the decrease of degree of saturation below 70% while it keeps constant at 1.0 more than that of degree of saturation (Figure 9(b)).

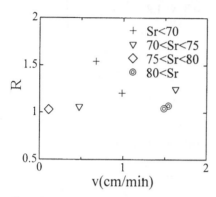

Figure 9(a). Relationship between R and v.

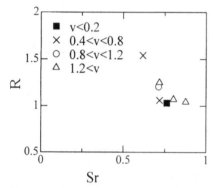

Figure 9(b). Relationship between R and Sr.

5. CONCLUSIONS

The study concludes that
(1) Mass transfer coefficient is not dependent on degree of saturation but depends on average pore water velocity.
(2) Volume fraction rate of dynamic region decreases with the increase of degree of water saturation. Water content in dynamic region is almost constant independently of degree of saturation.
(3) Dispersion coefficient in dynamic region increases linearly with the increase of pore water velocity.
(4) Retardation factor is almost constant above 70% of degree of saturation while it increases below it.

REFERENCES

Bear, J. 1972. Dynamic of fluids in porous media. *American Elsevier*.

DeSmedt, F. & P.J. Wierenga 1984. Solute transfer through columns of glass beads. *Water Resources Research.* 20(2):225-232.

Van Genuchten, M. Th. & P.J. Wierenga 1976. Mass transfer studies in sorbing porous media: 1. Analytical solutions. *Soil Science Society of American Journal.* 40(4):473-480.

Zurmühl, T. & W. Durner 1996. Modeling transient water and solute transport in a biporous soil. *Water Resources Research.* 32(4):819-829.

Clay Science for Engineering, Adachi & Fukue (eds) © 2001 Balkema, Rotterdam, ISBN 90 5809 175 9

BEM solutions of contaminant transport with time-dependent sorption

C.J. Leo
School of Civic Engineering and Environment, University of Western Sydney, N.S.W., Australia

ABSTRACT: A boundary element method (BEM) is formulated to solve the well known advection-dispersion equation governing the fate of contaminant transport with time-dependent sorption in the porous media. Laplace transform is introduced to eliminate the time variable which simplifies the governing equation and from which a fundamental solution for the BEM can be obtained. The time-dependent sorption, considered herein as a linear hereditary process, is efficiently dealt with using the proposed solution approach. After the BEM solutions are found in transform space, the contaminant concentrations in real time are obtained by numerical inversion. This method can be used in a number of contaminant related applications in clays and soils, such as to assist in the design of clay liners for a waste repository

1 INTRODUCTION

The study of contamination in soils receives much research interest after it was realized that many of the poorly attempted methods of disposing waste could endanger the environment and human health in the long term. As the fate of contaminants in soil is well hidden beneath the ground surface and poses a health hazard for an extremely long period of time, the challenge of neutralizing the effects of contamination is an onerous one.

As a part of the strategy to reduce the impact of contamination, computer models are widely used to assist in predicting soil contamination and planning remediation strategies. These models commonly rely on such domain approaches as the FEM or FDM (e.g. Pinder, 1973; Neumann, 1981; Sudicky, 1989) to solve the advection-dispersion equation governing the fate of contaminant transport in soils. A different approach was employed by Leo and Booker (1999a,b), using instead the boundary element method (BEM) to solve the governing equation whereby the Laplace transform technique was used to eliminate the time variable. It is well known that one of the main advantages of using a boundary integral approach is the possibility of reducing the dimensionality of the problem by one and in the number of algebraic equations that must be solved. This approach also has the advantage of dealing efficiently with an infinite boundary and so would result in savings in computational effort.

In the earlier papers of Leo and Booker (1999a,b), only time equilibrium (instantaneous) sorption of the contaminants onto soil particles were considered. It is shown in this paper that the BEM technique is easily generalized to consider a time dependent sorption without loss of computational efficiency due to the employment of the Laplace transform technique. The BEM could be used generally for contaminant transport related applications, such as to assist in the design of clay liners for waste landfill.

2 DEVELOPMENT OF BOUNDARY ELEMENT METHOD FOR 2D PROBLEMS

It is possible in many practical situations to reduce a 3D physical problem in the real world to a more idealized 2D problem without undue loss of realism. The attention in this paper will be given to developing a boundary element method for plane analysis only, which is particularly relevant for studying the effects of prismatic waste sources.

2.1 *Governing Equation of Contaminant Transport*

For a homogeneous soil where the groundwater advection is temporally steady and spatially uniform, the equation of contaminant transport in the local coordinate (x, z) axes is given by (Leo and Booker, 1999a),

$$\int_0^t \{\nabla \cdot (D_a \nabla c) - V_a \cdot \nabla c\} d\tau = nc + n\gamma \int_0^t c(x, z, \tau) d\tau + q \quad (1)$$

where,

$$\nabla = (\partial/\partial x, \partial/\partial z)^T$$

D_a = tensor of 'effective' hydrodynamic dispersion

V_a = vector of the components of the Darcy velocity,

n = porosity of the soil,

γ = sum of the first order radioactive and biodegradation constants,

q = the source/sink term due to sorption of the contaminants onto the soil particles,

t = time variable

In this paper, the 'effective' hydrodynamic dispersion D_a is preferred to D the hydrodynamic dispersion used elsewhere in literature (e.g. Bear, 1979). The two tensors are related as: $D_a = nD$.

It may be noted that in equation (1), the contaminant concentration is assumed to be zero everywhere initially, i.e. $c(x,z,t=0) = 0$. Following Bear (1979), the components of the hydrodynamic dispersion tensor are given as,

$$D_{aij} = nD_0 + \alpha_T V_a \delta_{ij} + (\alpha_L - \alpha_T)\frac{V_{ai}V_{aj}}{V_a} \quad (2)$$

where D_0 is the coefficient of molecular diffusion, α_L, α_T are the coefficients of longitudinal and transverse dispersivities respectively, δ_{ij} denotes the Kronecker delta, V_a is the magnitude of the Darcy velocity, V_{ai} are the components of the Darcy velocity and the indices i,j span the index set (x, z).

The quantity q represents the sorption of the contaminant onto the soil particles. For time-dependent sorption, the model proposed here is a linear hereditary one given by (Smith and Booker, 1993) where,

$$q = \int_0^t \frac{\partial c(x, z, \tau)}{\partial \tau} \rho K_d(t - \tau) d\tau \quad (3)$$

where,

ρ = dry density of the soil,

$K_d(t)$ = time-dependent distribution coefficient.

It is noted that when $K_d(t)$ = constant, equation (3) reduces to an equilibrium-controlled sorption i.e.

$$q = \rho K_d(c - c_0) \quad (4)$$

The distribution coefficient $K_d(t)$ may take any general form to be determined from laboratory measurements. However, a simple non-equilibrium sorption arising from a first-order rate-limited sorption reaction is given as,

$$K_d(t) = K_{d\infty} + (K_{d0} - K_{d\infty})e^{-at} \quad (5)$$

where K_{d0} is the initial distribution coefficient, $K_{d\infty}$ is the ultimate distribution coefficient and a is a decay constant.

2.2 Laplace transform

The contaminant transport problem is a time dependent one whereby different techniques have been used to treat the time variable in Equation (1). A common approach (e.g. Taigbenu and Liggett, 1986) is to use a finite difference time marching scheme to advance the solutions from one time level to the next. It will be observed that this technique requires the evaluation of a domain integral in the boundary element method which is deemed unattractive. In this paper, the Laplace transform technique is proposed which would avoid the undesirable domain integral. Now, taking the Laplace transform defined as:

$$\bar{c} = \int_0^\infty e^{-st} c(x, y, z, t) dt \quad (6)$$

of equation (1), it is found that,

$$\nabla \cdot (D_a \nabla \bar{c}) - V_a \cdot \nabla \bar{c} = n(s + \gamma)\bar{c} + s\bar{q} \quad (7)$$

where the Laplace transform of equation (3) shows,

$$\bar{q} = \rho s \bar{c} \bar{K}_d \quad (8)$$

Combining equations (7) and (8) leads to,

$$\nabla \cdot (D_a \nabla \bar{c}) - V_a \cdot \nabla \bar{c} = \Theta \bar{c} \quad (9)$$

where,

$$\Theta = \left(n + \rho s \bar{K}_d^*\right)(s + \gamma), \quad \bar{K}_d^* = (s + \gamma_0)\bar{K}_d/(s + \gamma).$$

It is noted that the decay constant in the sorbed phase defined as γ_0 may or may not be identical to γ. Further, assuming that the co-ordinate system is chosen so that the axes are parallel to the principal directions of D_a, then equation (9) reduces to,

$$D_{axx}\frac{\partial^2 \bar{c}}{\partial x^2} + D_{azz}\frac{\partial^2 \bar{c}}{\partial z^2} - V_{ax}\frac{\partial \bar{c}}{\partial x} - V_{az}\frac{\partial \bar{c}}{\partial z} = \Theta \bar{c} \quad (10)$$

Equation (10) simplifies to the more mathematically convenient modified Helmholtz equation by using a second series of co-ordinate transform developed by Leo and Booker (1999a) as follows:

$$x = uX \quad (11a)$$

$$z = wZ \tag{11b}$$

$$V_{aX} = wV_{ax} \tag{11c}$$

$$V_{aZ} = uV_{az} \tag{11d}$$

$$\bar{f}_N = \left(wl_x L_X + ul_z L_Z\right)\bar{f}_n \tag{11e}$$

where $\bar{f}_n, \bar{f}_N$ are the normal contaminant fluxes, l_x, l_z and L_X, L_Z are the direction cosines of the normals, on the problem boundary in the (x, z) and the transformed (X, Z) spaces respectively, and

$$u = \left(\frac{D_{axx}}{D_a}\right)^{\frac{1}{2}} \tag{12}$$

$$w = \left(\frac{D_{azz}}{D_a}\right)^{\frac{1}{2}} \tag{13}$$

$$D_a = \left(D_{axx}D_{azz}\right)^{\frac{1}{2}} \tag{14}$$

Furthermore, introducing the change of variable:

$$\bar{c} = \bar{c}^* e^{(\varpi X + \lambda Z)} \tag{15}$$

where,

$$\omega = \frac{V_{aX}}{2D_a} \tag{16a}$$

$$\lambda = \frac{V_{aZ}}{2D_a} \tag{16b}$$

it is found that equation (9) can be reduced to the familiar modified Helmholtz equation,

$$D_a \nabla^2 \bar{c}^* = S\bar{c}^* \tag{17a}$$

where,

$$S = \Theta + D_a\left(\omega^2 + \lambda^2\right) \tag{17b}$$

$$\bar{f}_N = \bar{f}_N^* e^{(\varpi X + \lambda Z)} \tag{17c}$$

$$\bar{f}_N^* = \frac{V_{aN}}{2}\bar{c}^* - D_a\frac{\partial \bar{c}^*}{\partial N} \tag{17d}$$

2.3 Boundary integral equation

The boundary integral equation derived from equation (17a) gives,

$$\varepsilon(r_0)\bar{c}^*(r_0) = \int_\Gamma D_a\left(\bar{c}^{\#}\frac{\partial \bar{c}^*}{\partial N} - \bar{c}^*\frac{\partial \bar{c}^{\#}}{\partial N}\right)d\Gamma \tag{18a}$$

where,

$\varepsilon(r_0) =$

$\begin{cases} 1 \text{ if } r_0 \text{ is within the domain of problem} \\ 0 \text{ if } r_0 \text{ is outside the domain of problem} \\ \frac{1}{2} \text{ if } r_0 \text{ lies on a smooth boundary of problem or its value is} \\ \quad \text{the subtended angle} \div 2 \text{ if the boundary is not smooth} \end{cases}$

r_0 = position vector of the point of disturbance,

Γ = transformed boundary of problem in the transformed (X, Z) space,

$\bar{c}^{\#} = \dfrac{1}{2\pi D_a}K_0\left(\sqrt{S/D_a}\,R^{\#}\right)$, the fundamental solution of equation(17a),

K_0 = modified Bessel function of the second kind of order zero,

$$R^{\#} = \left[\left(X - X_0\right)^2 + \left(Z - Z_0\right)^2\right]^{\frac{1}{2}}$$

X_0, Z_0 = co-ordinates of the point of disturbance.

Using equation (18a), an equivalent boundary equation based on the transformed normal flux is given by,

$$\varepsilon(r_0)\bar{c}^*(r_0) = \int_\Gamma\left(\bar{c}^*\bar{f}_N^{\#} - \bar{c}^{\#}\bar{f}_N^*\right)d\Gamma \tag{18b}$$

where,

$$\bar{f}_N^{\#} = \frac{V_{aN}}{2}\bar{c}^{\#} - D_a\frac{\partial \bar{c}^{\#}}{\partial N}$$

V_{aN} = the normal component of the Darcy velocity on the boundary of problem in (X, Z) space.

2.4 Boundary element approximation

Thus proceeding formally in the manner outlined in Booker and Leo (1999a,b), a boundary element approximation of the boundary integral equation (18a) can be formulated using conventional techniques widely described in literature (e.g. Leo and Booker, 1999a,b) and takes the form of,

$$H^*\bar{c}^* = G^*\bar{f}_N^* \tag{19}$$

where H^*, G^* are the fully populated influence matrices, $\bar{c}^*, \bar{f}_N^*$ are the vectors of nodal values on the boundary elements of problem. For 'constant-valued' boundary elements, the coefficients of the influence matrices H^*, G^* are given as:

$$h_{ij}^* = -\int_\Gamma \bar{f}_N^* (r_j - r_{0i}) d\Gamma \qquad \text{if } i \neq j \qquad (20a)$$

$$h_{ii}^* = -\int_\Gamma \bar{f}_N^* (r_j - r_{0i}) d\Gamma + \tfrac{1}{2} \quad \text{if } i = j \qquad (20b)$$

$$g_{ij}^* = -\int_\Gamma \Delta \bar{c}^* (r_j - r_{0i}) d\Gamma \qquad (20c)$$

where $r_j - r_{0i}$ is denotes the position vector of a point on element j from the point of disturbance at node 'i'. If the co-ordinate transformations in equations (11) and the change in variables given in equation (15) were recovered, it is found that equation (19) can be expressed in natural co-ordinates and in terms of the nodal values of the variables $\bar{c}$, $\bar{f}_n$ as,

$$H\bar{c} = G\bar{f}_n \qquad (21)$$

where H, G are the global influence matrices and vector. Equation (21) is used to solve for the unknown concentration and normal mass flux on the external boundaries of the problem domain. Once these values are known the value of the concentration can be solved at any internal point in the domain using equation (18b).

It should however be observed that the solution found by applying these equations will be in the Laplace transform domain and thus has to be inverted to obtain the solution in the time domain. This is done by performing a numerical inversion of the Laplace transform using the algorithm by Talbot (1979).

3 APPLICATION

3.1 Verification

The results of a test problem are given here verifying the correctness of the boundary element codes. The problem consists of a long 4m radius cylindrical source deeply buried below ground and contaminating the surrounding soil where contaminant mass is sorbed onto the soil grains in a time-dependent fashion. Semi-analytical solutions are found using the approach developed by Rahman and Booker (1989) and Leo and Booker (1999c), and modified for time-dependent sorption. A flow occurs in the z-direction, the magnitude of the Darcy velocity $D_{az} = 0.02$ m/year. The properties of the contaminant and soil medium are given in Table 1. The source itself has an initial concentration of 1000 mg/L. The time-dependent relationship in Equation (5) is used.

Solutions were found in the boundary element method using a total of 80 constant-valued boundary elements. Figure 1 shows the results at elapsed time

of 10, 20 and 100 years. The boundary element results are shown as the triangle symbols while the semi-analytical solutions are represented as solid lines. It is observed that agreement between the semi-analytical solutions and the boundary element solutions are generally very good.

Table 1. Verification problem: properties of the contaminant and soil.

Properties	Values
D_0 (m²/yr)	0.02
α_L (m)	0.0
α_T (m)	0.0
n	0.4
ρ (kg/m³)	1900
K_{d0} (m³/kg)	1.05×10^{-3}
$K_{d\infty}$ (m³/kg)	0.0
a	0.5

Table 2. Illustrative problem: properties of the contaminant and soil.

Properties	Values
D_0 (m²/yr)	0.05
α_L (m)	0.1
α_T (m)	0.05
n	0.4
ρ (kg/m³)	1900
K_{d0} (m³/kg)	1.05×10^{-3}
$K_{d\infty}$ (m³/kg)	0.0
a	0.5

3.2 Illustrative Problem

To further illustrative the use of the boundary element solutions, a simple single cell landfill is analysed using 320 boundary elements. The landfill is idealized as a 20m x 10 m rectangle, although this in no way is a limitation of the boundary element method which can model a source of any shape by breaking the source boundary into small elements. A schematic of the illustrative problem is given in Figure 2. The initial concentration of the source is 1000 mg/L and a Darcy groundwater flow in the downward direction of magnitude 0.02 m/year is present. The properties of the soil with the parameters for the time-dependent sorption are given in Table 2. However, for the purpose of providing a basis for comparison, the cases of no sorption and for time-independent (or instantaneous) sorption where K_{d0} =1.05 x 10^{-3} m³/kg, $K_{d\infty} = 0$, $a = 0$ are also analysed.

The boundary conditions applied for this problem are as follows: the concentration is assumed to vanish everywhere along the boundary except the boundary at the top of the rectangular source where the normal contaminant flux to the boundary is taken as zero. For this problem, results for the instantaneous sorption are clearly quite different in comparison to the other two cases (Figures 3 and 4). An observation of the time-dependent relationship Equation (5) would suggest that its result would probably lie between the no sorption and instantaneous sorption cases.

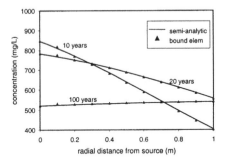

Figure 1. Comparision of semi-analytical and boundary element solutions.

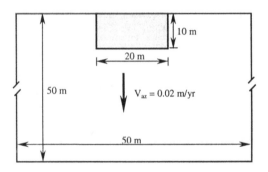

Figure 2. Schematic of boundary element illustrative problem.

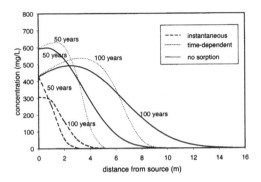

Figure 3. contaminant profiles at 50 and 100 years for different levels and types of sorption.

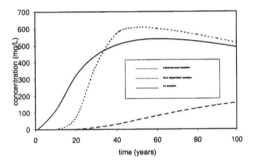

Figure 4. Variation of concentration with time at a point directly 2 m below the source for different types of sorption.

4 CONCLUSIONS

A boundary element method has been presented for analysing contaminant transport in the porous media where time-dependent sorption is present. The solution method applies a Laplace transform and a series of co-ordinate transforms to facilitate the formulation of a boundary integral equation. A system of algebraic boundary element equations can then be developed. As an illustrative example, the boundary element method developed has been used to help assess the contamination from an embedded rectangular source.

REFERENCES

Bear, J. (1979), 'Hydraulics of groundwater flow', *McGraw-Hill*, New York.

Brebbia, C.A. and Walker, S. (1980), 'Boundary element techniques in engineering', *Newnes-Butterworths*, London.

Leo, C.J. and Booker, J.R. (1997), 'A note on co-ordinate transform and fundamental solutions of the equation of contaminant migration in porous media', *Personal Communications*.

Leo, C J. and Booker, J.R. (1999a), 'A boundary element method for analysis of contaminant transport in porous media I: Homogeneous porous media', *International Journal for Numerical and Analytical Methods in Geomechanics*. V23, 14, pp1679-1680.

Leo, C J. and Booker, J.R. (1999b), 'A boundary element method for analysis of contaminant transport in porous media II: Non homogeneous porous media', *International Journal for Numerical and Analytical Methods in Geomechanics*. V23, 14, pp1701-1715.

Leo, C.J. and Booker, J.R. (1999c), 'Semi-analytic solutions of contaminant transport from deeply buried cylindrical repository surrounded by zoned media', *International Journal for Numerical and Analytical Methods in Geomechanics*. V23, 14, pp1797-1815.

Neumann, S.P. (1981), 'A Eulerian-Lagrangian numerical scheme for the dispersion-convection equation using conjugate space-time grids', Journal of Computational Physics, V41(2), pp270-294.

Pinder (1973), 'A Galerkin finite element simulation of groundwater contamination in Long Island, New York', *Water Resources Research*, Vol. 9(6), pp1657-1669.

Rahman, M.S. and Booker, J.R., (1989), 'Pollutant migration from deeply buried repositories', *International Journal for Numerical and Analytical Methods in Geomechanics,* Vol. 13, pp37-51.

Smith, D.W., Rowe, R.K. and Booker, J.R. (1993), 'The analysis of pollutant migration through soil with linear hereditary time-dependent sorption', Vol.17, pp255-274.

Sudicky, E.A. (1989), 'The Laplace transform Galerkin technique: A time-continuous finite element theory and application to mass transport in groundwater', *Water Resources Research*, V25(8), pp1833-1846.

Taigbenu, A. and Liggett, J.A. (1986), 'An integral solution for the diffusion-advection equation', *Water Resources Research*, 22. Pp1237-1246.

Talbot, A. (1979), 'The accurate numerical inversion of Laplace transform', *J. Inst. Math. Applic.* 23, pp97-120.

APPENDIX

The relationships between the parameters in the real and transformed spaces used in this paper are shown derived in this Appendix.

The starting point is the Laplace transform governing equation of contaminant transport in 3D giving,

$$\frac{\partial \bar{f}_x}{\partial x} + \frac{\partial \bar{f}_y}{\partial y} + \frac{\partial \bar{f}_z}{\partial z} + \Theta \bar{c} = 0 \tag{A.1}$$

where,

$$\bar{f}_x = V_{ax}\bar{c} - D_{axx}\frac{\partial \bar{c}}{\partial x} \tag{A.2a}$$

$$\bar{f}_y = V_{ay}\bar{c} - D_{ayy}\frac{\partial \bar{c}}{\partial y} \tag{A.2b}$$

$$\bar{f}_z = V_{az}\bar{c} - D_{azz}\frac{\partial \bar{c}}{\partial z} \tag{A.2c}$$

Substitution of equation (A.2) into (A.1) yields the general form similar to equation (10) used in this paper:

$$D_{axx}\frac{\partial^2 \bar{c}}{\partial x^2} + D_{ayy}\frac{\partial^2 \bar{c}}{\partial y^2} + D_{azz}\frac{\partial^2 \bar{c}}{\partial z^2} - \\ V_{ax}\frac{\partial \bar{c}}{\partial x} - V_{ay}\frac{\partial \bar{c}}{\partial y} - V_{az}\frac{\partial \bar{c}}{\partial z} = \overline{\Theta c} \tag{A.3}$$

where in general the values of $D_{axx}, D_{ayy}, D_{azz}$ are not identical and so equation (A1) has an 'anisotropic' form with respect to the coefficients of hydrodynamic dispersion. The formulation of the boundary integral equation will be facilitated if equation (A.1) is rendered 'isotropic', this may be achieved by implementing a coordinate transformation as discussed below.

Suppose that the following coordinate transformation is introduced,

$$x = uX \tag{A.4a}$$

$$y = vY \tag{A.4b}$$

$$z = wZ \tag{A.4c}$$

then it follows from equation (A.1) that

$$\frac{1}{u}\frac{\partial \bar{f}_x}{\partial X} + \frac{1}{v}\frac{\partial \bar{f}_y}{\partial Y} + \frac{1}{w}\frac{\partial \bar{f}_z}{\partial Z} + \Theta \bar{c} = 0 \tag{A.5}$$

It will be convenient to define the components of flux in the transformed space by the relationship,

$$f_N d\Sigma = f_n d\sigma \tag{A.6}$$

where $d\sigma$ represents an elemental area in the original space and f_n denotes the normal component of the flux passing through that elemental area, and $d\Sigma$ represents the image of that elemental area in the transformed space and f_N denotes the normal component of flux passing through the area.

Suppose that the equation of the surface being considered is given by,

$$x = x(\xi, \eta) \tag{A.7a}$$

$$y = y(\xi, \eta) \tag{A.7b}$$

$$z = z(\xi, \eta) \tag{A.7c}$$

and that the elemental area is bounded by, $\xi = \xi_0$, $\xi = \xi_0 + d\xi$, $\eta = \eta_0$, $\eta = \eta_0 + d\eta$ then

$$d\sigma = \left(j_x^2 + j_y^2 + j_z^2\right)^{\frac{1}{2}} d\xi d\eta \tag{A.8}$$

where,

$$j_x = \frac{\partial(y, z)}{\partial(\xi, \eta)} \tag{A.9a}$$

$$j_y = \frac{\partial(z, x)}{\partial(\xi, \eta)} \tag{A.9b}$$

$$j_z = \frac{\partial(x, y)}{\partial(\xi, \eta)} \tag{A.9c}$$

and the normal l can be shown to be,

$$l = \left(j_x, j_y, j_z\right)^T \frac{d\xi d\eta}{d\sigma} \tag{A.10}$$

In the transformed space, the equation of the surface becomes

$$X = \frac{x(\xi,\eta)}{u} \qquad \text{(A.11a)}$$

$$Y = \frac{y(\xi,\eta)}{v} \qquad \text{(A.11b)}$$

$$Z = \frac{z(\xi,\eta)}{w} \qquad \text{(A.11c)}$$

so that the transformed elemental area is

$$d\Sigma = \left(\frac{j_x^2}{v^2 w^2} + \frac{j_y^2}{w^2 u^2} + \frac{j_z^2}{u^2 v^2} \right) d\xi d\eta \qquad \text{(A.12)}$$

and the normal to the transformed surface is

$$L = \left(\frac{j_x}{vw} + \frac{j_y}{wu} + \frac{j_z}{uv} \right)^T \frac{d\xi d\eta}{d\Sigma} \qquad \text{(A.13)}$$

The relationship (A.6) may now be written as

$$\bar{f}_x j_x + \bar{f}_y j_y + \bar{f}_z j_z = \bar{f}_x \frac{j_x}{vw} + \bar{f}_Y \frac{j_y}{wu} + \bar{f}_Z \frac{j_z}{uv} \quad \text{(A.14)}$$

This must be true for any surface within the body so that,

$$\bar{f}_X = vw \bar{f}_x \qquad \text{(A.15a)}$$

$$\bar{f}_Y = wu \bar{f}_y \qquad \text{(A.15b)}$$

$$\bar{f}_Z = uv \bar{f}_z \qquad \text{(A.15c)}$$

Thus using equations (A.2a), (A.2b), (A.2c) the components of the transformed mass flux may be represented as,

$$\bar{f}_X = vw V_{ax} \bar{c} - D_{axx} \frac{vw}{u} \frac{\partial \bar{c}}{\partial X} \qquad \text{(A.16a)}$$

$$\bar{f}_Y = wu V_{ay} \bar{c} - D_{ayy} \frac{wu}{v} \frac{\partial \bar{c}}{\partial Y} \qquad \text{(A.16b)}$$

$$\bar{f}_Z = uv V_{az} \bar{c} - D_{azz} \frac{uv}{w} \frac{\partial \bar{c}}{\partial Z} \qquad \text{(A.16c)}$$

where,

$$V_{aX} = vw V_{ax} \qquad \text{(A.17a)}$$

$$V_{aY} = wu V_{ay} \qquad \text{(A.17b)}$$

$$V_{aZ} = uv V_{az} \qquad \text{(A.17c)}$$

$$D_{aXX} = \frac{vw}{u} D_{axx} \qquad \text{(A.17d)}$$

$$D_{aYY} = \frac{wu}{v} D_{ayy} \qquad \text{(A.17e)}$$

$$D_{aZZ} = \frac{uv}{w} D_{azz} \qquad \text{(A.17f)}$$

In passing, it is noted that a similar relation to equation (A.6) holds,

$$V_n d\sigma = V_N d\Sigma \qquad \text{(A.18)}$$

This implies that both the diffusive and advective flux transform similarly. Equations (A.16) may be reduced to the equivalent expressions for an isotropic case by choosing D_a so that,

$$D_{axx} \frac{vw}{u} = D_a \qquad \text{(A.19a)}$$

$$D_{ayy} \frac{wu}{v} = D_a \qquad \text{(A.19b)}$$

$$D_{azz} \frac{uv}{w} = D_a \qquad \text{(A.19c)}$$

yielding,

$$D_a^3 = D_{axx} D_{ayy} D_{azz} uvw \qquad \text{(A.20)}$$

and following from equation (A.5),

$$\frac{1}{uvw} \left(\frac{\partial \bar{f}_X}{\partial X} + \frac{\partial \bar{f}_Y}{\partial Y} + \frac{\partial \bar{f}_Z}{\partial Z} \right) + \Theta \bar{c} = 0 \qquad \text{(A.21)}$$

It is now convenient to choose,

$$uvw = 1 \qquad \text{(A.22)}$$

which preserves the form of equation (A.1) in the transformed space. Thus the isotropic form of equation (A.3) gives,

$$D_a \frac{\partial^2 \bar{c}}{\partial X^2} + D_a \frac{\partial^2 \bar{c}}{\partial Y^2} + D_a \frac{\partial^2 \bar{c}}{\partial Z^2} - \\ V_{aX} \frac{\partial \bar{c}}{\partial X} - V_{aY} \frac{\partial \bar{c}}{\partial Y} - V_{aZ} \frac{\partial \bar{c}}{\partial Z} = \overline{\Theta} \bar{c} \qquad \text{(A.23)}$$

It thus follows that in 3D space,

$$D_a = \left(D_{axx} D_{ayy} D_{azz} \right)^{\frac{1}{3}} \qquad \text{(A.24a)}$$

$$u = \left(\frac{D_{axx}}{D_a} \right)^{\frac{1}{2}} \qquad \text{(A.24b)}$$

$$v = \left(\frac{D_{ayy}}{D_a} \right)^{\frac{1}{2}} \qquad \text{(A.24c)}$$

$$w = \left(\frac{D_{azz}}{D_a} \right)^{\frac{1}{2}} \qquad \text{(A.24d)}$$

The relationship between the normal l and the transformed L may be established using equations (A.10) and (A.13) and it is found that,

$$l^T YL = \frac{1}{uvw}\left(j_x^2 + j_y^2 + j_z^2\right)\frac{(d\xi d\eta)^2}{d\sigma d\Sigma} \qquad (A.25)$$

where

$$Y = \text{Diag}\left[\frac{1}{u}, \frac{1}{v}, \frac{1}{w}\right]$$

Using equation (A.10) leads to,

$$l^T YL = \left(\frac{d\sigma}{d\xi d\eta}\right)^2 \frac{1}{uvw} \frac{(d\xi d\eta)^2}{d\sigma d\Sigma} \qquad (A.26a)$$

and so,

$$l^T YL = \frac{d\sigma}{d\Sigma} \qquad (A.26b)$$

Now, if the components of the unit normal in the natural and transformed space are respectively given by,

$$l = \left(l_x, l_y, l_z\right)^T \qquad (A.27a)$$

$$L = \left(L_x, L_y, L_z\right)^T \qquad (A.27b)$$

then it follows that,

$$\frac{d\sigma}{d\Sigma} = \frac{l_x L_X}{u} + \frac{l_y L_Y}{v} + \frac{l_z L_Z}{w} =$$
$$vwl_x L_X + wul_y L_Y + uvl_z L_Z \qquad (A.28)$$

so that,

$$\bar{f}_n = \left(ul_x L_X + vl_y L_Y + wl_z L_Z\right)\bar{f}_N \qquad (A.29a)$$

$$\bar{f}_N = \left(vwl_x L_X + wul_y L_Y + uvl_z L_Z\right)\bar{f}_n \qquad (A.29b)$$

The derivation for two spatial dimensions (in x-z plane) is similar and is easily recovered by setting $v = 1$ so that,

$$D_{ayy} = D_a \qquad (A.30a)$$

hence,

$$D_a^{\frac{2}{3}} = \left(D_{axx}D_{azz}\right)^{\frac{1}{3}} \qquad (A.30b)$$

$$D_a = \left(D_{axx}D_{azz}\right)^{\frac{1}{2}} \qquad (A.30c)$$

$$u = \left(\frac{D_{axx}}{D_a}\right)^{\frac{1}{2}} \qquad (A.30d)$$

$$w = \left(\frac{D_{azz}}{D_a}\right)^{\frac{1}{2}} \qquad (A.30e)$$

and it thus follows that,

$$\bar{f}_X = w\bar{f}_x \qquad (A.30f)$$

$$\bar{f}_Z = u\bar{f}_z \qquad (A.30g)$$

$$\bar{f}_n = \left(ul_x L_X + wl_z L_Z\right)\bar{f}_N \qquad (A.30h)$$

$$\bar{f}_N = \left(wl_x L_X + ul_z L_Z\right)\bar{f}_n \qquad (A.30i)$$

Clay Science for Engineering, Adachi & Fukue (eds) © 2001 Balkema, Rotterdam, ISBN 90 5809 175 9

Modelling the transport of lead in partly saturated soil

H. R. Thomas, R. N. Yong & A. A. Hashm
Geoenvironmental Research Centre, Cardiff School of Engineering, University of Wales, UK

ABSTRACT: Results of research on the modelling of lead transport in partly saturated soil are presented. A general approach to contaminant transport in partly saturated soil is first developed embracing pore water movement, chemical solute movement and chemical interaction phenomena. A numerical solution algorithm is then proposed, which extends previous modelling capabilities, to more fully take account of contaminant behaviour. Pore water and chemical solute transport equations are solved simultaneously in a coupled fashion. Chemical interaction equations are linked to the transport equation set in a sequential iterative manner. Comparisons of numerical results with experimental observations are presented, for a range of soil leaching experiments, covering a variety of conditions. Good agreement is achieved, lending confidence to the approach proposed. Further research is however acknowledged to be required and in that sense, the results presented are seen as a first step in that task.

1 INTRODUCTION

Contamination of groundwater from various sources, has become a significant problem within the last decade. Investigations of contaminant migration through the soil have therefore received increased attention (Mangold & Tsang 1991). Given the varied nature of the chemical composition of a contaminant leachate, and the varied nature of soil, it follows that theory needs to be supported with physical evidence if rational application to field studies and prediction is to be sought.

An improved understanding of contaminant transport in partly saturated soil is necessary since i) in many parts of the world the depth to the water table is so great that the total region of interest is partly saturated and ii) where a water table exist near the surface, the partly saturated zone forms the boundary region through which pollutants may have to travel to reach an aquifer. Contaminant transport mechanisms in partly saturated soils can differ significantly from those in fully saturated conditions. This is because the forces of contaminant-soil interaction are more dominant as transport control processes because of the reduced amount of water present within the system (Yong et al, 1992).

Heavy metals are commonly found in several kinds of wastes including sludge and landfill leachates. They are highly toxic to humans, animals, and aquatic life (Yong et al, 1992). As such the fate and transport of such pollutants is of considerable interest. It is known that the sorption of heavy metals is a selective processes. In other words some heavy metals are preferred over other species in adsorption by a soil.

The present study is directed towards the migration of lead in partly saturated soil The transport of a number of heavy metals has been investigated, as part of a larger program of research (Hashm, 1999). Results for the movement of lead are selected here for presentation.

Contaminants interact hydrologically, physically, and chemically with both the pore water and the solid matrix. The major physical processes of interaction include advection and dispersion. The chemical reactions include aqueous complexation, sorption and precipitation. Any of the above physical and chemical processes can contribute to the distribution and redistribution of chemical components after they are introduced into the subsurface system.

In this paper the transport is described by a set of partial differential equations (Mangold & Tsang, 1991) and the chemical reactions, under the assumption of equilibrium, are described by a set of nonlinear algebraic equations(Westall et al, 1976). Comparison of numerical results with experimental data are presented.

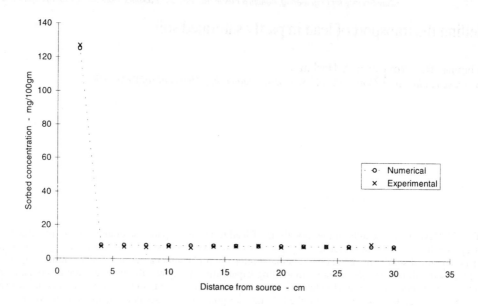

Figure 1. Numerical and experimental profiles of sorbed concentrations, experiment 1

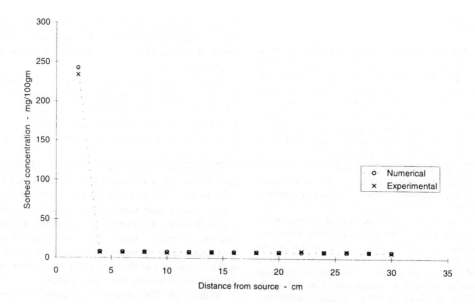

Figure 2. Numerical and experimental profiles of sorbed concentrations, experiment 4

2 GOVERNING PHYSICO-CHEMICAL EQUATIONS

2.1 Pore Water Transport Equations

Pore water movement in the soil is assumed to be defined by a generalised *Darcy's* law (Bear and Verruijt, 1987) according to:

$$\mathbf{V}_w^l = -K_w \left(\nabla \left(\frac{u_w}{\gamma_w} \right) + \nabla z \right) \tag{1}$$

where $\mathbf{V}_w^l$ is the pore water velocity, u_w is the pore water pressure, γ_w is the unit weight of water, K_w is the hydraulic conductivity and z is the elevation.

From considerations of mass conservation, the governing differential equation for the problem, can be written as (Bear, 1972)

$$\frac{\partial (n S_w)}{\partial t} + \nabla \cdot \left(\mathbf{V}_w^l + \mathbf{V}_w^{c_s} \right) = 0 \tag{2}$$

where $\mathbf{V}_w^{c_s}$ is the flow velocity due to the chemical gradient, n is the porosity and S_w is the degree of saturation

2.2 Chemical Solute Transport Equation

The transport of solutes is a process governed by the principle of conservation of mass energy and charge, and the processes of advection, molecular diffusion, mechanical dispersion and sorption. A solute may be defined as any chemical constituent dissolved in the fluid phase, whether as ions or electrically neutral complexes.

Considering pore water flow as a homogeneous liquid moving through a porous medium containing a given mass of chemical solute, the chemical solute conservation equation, in the absence of adsorption, sources and sinks, is given by (Wiest, 1969 ; Bear and Verruijt 1987] as

$$\frac{\partial (n S_w c_s)}{\partial t} + \nabla \cdot \left[c_s \left(\mathbf{V}_w^l + \mathbf{V}_w^{c_s} \right) \right] - \nabla \cdot \left[\mathbf{D}_h \nabla (c_s n S_w) \right] = 0$$

$$\tag{3}$$

where c_s is the chemical solute concentration. $\mathbf{D}_h$ is the coefficient of hydrodynamic dispersion which can be expressed as:

$$\mathbf{D}_h = \mathbf{D}_d + \mathbf{D} \tag{4}$$

where $\mathbf{D}$ is the coefficient of mechanical dispersion and $\mathbf{D}_d$ is the molecular diffusion coefficient.

Transport equation (3), for reacting solutes may be written with the sink term expanded, to account for mass accumulation from reactions due to the sorption process (Bear and Verruijt 1987).

$$\frac{\partial (n S_w c_s)}{\partial t} + \frac{\partial (n S_w S)}{\partial t} + \nabla \cdot \left[c_s \left(\mathbf{V}_w^l + \mathbf{V}_w^{c_s} \right) \right] - \nabla \cdot \left[n S_w \mathbf{D}_h \nabla (c_s) \right] - \nabla \cdot \left[c_s \mathbf{D}_h \nabla (n S_w) \right] = 0 \tag{5}$$

where S is the sorbed chemical concentration .

2.3 Chemical Interaction equations

The chemical reactions that are considered significant for chemical transport are ion complexation in the aqueous solution (liquid phase), sorption on solid surfaces (interphase boundaries) and precipitation/dissolution of solids (solid phase) (Mitchell, 1993).

The sorption process is represented mathematically by some form of the law of mass action, relating the thermodynamic equilibrium constant to the activities(the thermodynamically effective concentration) of the reactants and products (Westall et al. 1976).

$$K_L^{act} = \frac{[SOH.M]}{\gamma_i [M][SOH]} \tag{6}$$

where *[SOH.M]* represents the concentration of the adsorption sites occupied by an ion M or surface-bound metal, *[M]* is the equilibrium dissolved concentration, *[SOH]* represents the concentration of unbound adsorption surfaces K_L^{act} is the thermodynamic equilibrium constant at the given temperature, and the activity coefficient γ_i is treated as a function of ionic strength in dilute solutions. The ionic strength is defined as(Westall et al, 1976):

$$I = \frac{1}{2} \sum m_i z_i^2 \tag{7}$$

where m_i is the molality (mol/M) of species i and z_i is the valence or charge of that species, summed over all the ions in solution, both positive and negative. It is clear that the law of mass action gives rise to a set of non-linear algebraic equations for each of the sorption processes.

Chemical equilibrium and sorption are considered by incorporating the geochemical model MINTEQA2. Seven options are available in MINTEQA2 for modelling surface reactions.

3 NUMERICAL SOLUTION

A solution of the set of equations presented above can only be achieved via numerical techniques. The approach adopted herein is to consider the two transport equations as a coupled set of partial differential equations and solve them simultaneously. The finite element method is employed for the spatial discretisation and a finite difference scheme for the time variation of the variables. A backward difference mid-interval scheme was found to yield excellent results (Thomas & He, 1995).

A solution of the chemical interaction equations was achieved via a Newton-Raphson approximation method (Westall et al, 1976).

Coupling of the two sections of the overall solution process, the transport equations and the chemical interaction equations was accomplished via a sequential iteration approach.

The whole of the above procedure was achieved within the framework of a software package developed at Cardiff, called COMPASS (Code for Modelling Partly Saturated Soil) (Thomas et al, 1998)). COMPASS has been developed over recent years and has been extensively applied to issues related to the Thermo/Hydraulic/Mechanical (THM) behaviour of engineered barriers(Thomas et al, 1996)). Such problems are of importance in the disposal of high level nuclear waste (Thomas et al, 1996). Numerous verification and validation exercises have been performed in this context (Thomas et al, 1998). Initial extension of COMPASS to accommodate chemical solute transfer have been presented recently (Thomas & Cleall, 1999). This work presents the further extension of the modelling approach to couple COMPASS with a chemical interaction model, albeit in a sequential iterative manner.

4 MATERIAL CONSIDERED

Soil samples were obtained from selected landfill sites in South wales. A 100 kg of soil was prepared by mixing 30% by weight of the air dried clayey soils, ground to pass through a 2mm sieve with 70% of sand.

To acquire soil samples of different pH, 10 kg of this soil was washed several times with 10% Nitric acid over a long period of time to ensure complete soil pH equilibrium (Elzahabi & Yong, 1997). The soil was then air dried and ground to pass through a 2-mm sieve.

A series of batch equilibrium tests were carried out to evaluate the retention properties of the soil. Physical and chemical analyses were carried out to determine the properties of the soil (Thomas et al 1999).

5 EXPERIMENTAL TESTS

6 horizontal leaching tests were performed on the soil samples. Table 1 describes the set of experiments.

Before testing, the dry soil was mixed with different volumes of distilled water to achieve the three different degrees of saturation. The samples were then placed in environmentally controlled conditions for at least 24 hours to prevent moisture losses and to allow for uniform moisture distribution. The soil samples were then compacted directly into the leaching cells to a dry density of 1.69 Mg/m^3.

Each cell was then leached, at room temperature, with either 500 ppm or 1000 ppm lead solution at constant atmospheric pressure under zero head until the cell reached full saturation. The test sample was then removed and sectioned horizontally into 15 sections.

Concentration distributions in the liquid and solid phases of the soil were determined for each section. The total heavy metal concentration was determined using the X-Ray Fluorescence Spectrometry (XRF) method with calibration by the acid digestion method. Concentration of heavy metal in the solution was determined using Induced Couple Plasma Emission Spectroscopy (ICP).

Table 1. Testing Schedule

Experiment	Initial Soil pH	Initial S_w	c_s ppm
1	7.6	0.2	500
2	7.6	0.4	500
3	7.6	0.6	500
4	7.6	0.2	1000
5	4.5	0.2	500
6	3.5	0.2	500

6 MATERIAL PARAMETER DETERMINATION

The relationship between suction and degree of saturation for the soil was determined experimentally by pressure plate extraction equipment

In order to obtain the variation of unsaturated hydraulic conductivity with volumetric moisture content, the application of the pore-size interaction model to the suction/water content relationship, as suggested by Green and Corey (1971), was used.

The porosity of the soil was 0.33.

Trials were carried out to choose the best sorption model that fitted the data from the batch equilibrium tests. This resulted in the selection of the Langmuir model. Sorption constants were then found for the sorption of Lead, at three different soil pH of 7.6, 4.5 and 3.5.

7 NUMERICAL SIMULATION

At the start of the simulation work, a series of numerical experiments were performed to establish the optimum element and timestep size to be used. This yielded a mesh of 60, 5 mm thick elements, to represent the 300 mm length of each cell, and a timestep size of 1200 seconds.

For the simulation of experiments 1 to 3, uniform initial conditions for pore water pressure, u_w, chemical solute concentration c_s, and sorbents concentration were employed. The values adopted are given in Table 2.

Table 2. Initial conditions for different degrees of saturation.

S_w	u_w kPa	c_s mol/m^3	Sorbents mol/l soil
0.2	-2560	0.161	0.66
0.4	-800	0.080	0.66
0.6	-400	0.054	0.66

At the leachate entry boundary, Dirichlet boundary conditions were employed as follows

$u_w = 0.0$ Pa

$c_s = 2.41$ mol/m^3

A sorption reaction constant of 19.7 was employed

To simulate the transport of lead at different soil pH, ie. experiments 1, 5 and 6, uniform initial conditions are again assumed for pore water pressure, chemical solute concentration and sorbents concentration. The values used in this case are presented in Table 3

Table 3. Initial conditions for different soil pH.

pH	u_w kPa	c_s mol/m^3	Sorbents mol/l soil
7.6	-2560	0.161	0.66
4.5	-2560	0.087	0.016
3.5	-2560	0.148	0.015

It can be seen that although the soil samples were originally from the same soil, there are differences in the initial properties of these samples. These can be attributed to the washing process used to modify the soil pH.

The Dirichlet boundary conditions now employed were:

$u_w = 0.0$ Pa

$c_s = 2.41$ mol/m^3

Sorption reaction constants of 19.7, 2018, and 1958 were employed for pH values of 7.6, 4.5, and 3.5 respectively.

8 RESULTS AND DISCUSSION

Numerical and experimental results for each of the three experiments at different degrees of saturation, namely experiments 1,2 and 3, have been compared. However, because of the similarity in behaviour, only the results for one experiment, experiment 1, are presented (Figure 1). The numerical results for sorbed concentrations were calculated at 20 mm intervals throughout the soil. The simulated values represent therefore the average over 4 elements.

Good agreement between numerical and experimental results can be seen to have been achieved. The experimental results show the retention of the lead in the first 20 mm of the sample. This has been reproduced in the numerical work.

The simulated dissolved concentration profiles, for the three different degrees of saturation, are not graphically presented as it was found that they have zero, or near zero values, along the soil sample. These simulated dissolved concentration results were found to have lower values than the corresponding experimental results. However, the experimental values of the dissolved concentration were very small, having a maximum value of 0.36 mg/100gm, and can be considered to be negligible compared with the sorbed concentration.

Numerical and experimental results for experiment 4 are presented in Figure 2. Good agreement can once again be seen to have been achieved. Again the retention pattern observed experimentally has been reproduced numerically.

Finally, numerical and experimental results for each of the three experiments at different soil pH values have been compared, namely experiments 1, 5 and 6. Again because of similarity in behaviour, only the results for one experiment are presented, experiment 6 (Figure 3). Similar experimental trends are observed and once again, the numerical simulations produce results in good agreement with experimental observations.

9 CONCLUSIONS

It can be seen that the modelling work performed has produced numerical results that are in good agreement with experimental observations. Confidence in the approach proposed is thus developed. It is however acknowledged that considerable further work is required to fully explore and develop the model's capabilities. The work presented here is thus seen as a first step in this task. Predictions of long term fate of pollutants and environmental mobility may then in due course be addressed.

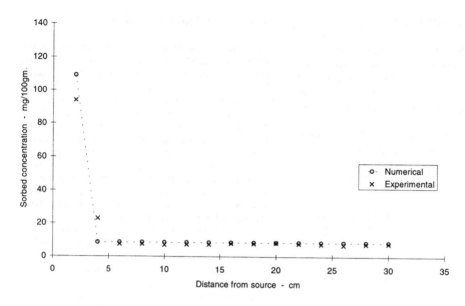

Figure 3. Numerical and experimental profiles of sorbed concentrations, experiment 6

REFERENCES

Bear, J. 1972. Dynamics of fluids in porous media. New York: Elsevier.

Bear, J. & A Verruijt 1987. Modeling groundwater flow and pollution. Dordrecht, D. Reidel Publishing Company.

Elzahabi, M. & R N Yong 1997. Vadose zone transport of heavy metals. Proc. 1st Int. BGS Conf. on Geoenvironmental Engineering., 173-180, Thomas Telford.

Green, R E. & J C Corey 1971. Calculation of hydraulic conductivity: A further evaluation of some predictive methods. Soil Sci. Soc. Am. J 35(3-8).

Hashm, A A 1999. A study of the transport of a selection of heavy metals in unsaturated soil. PhD Thesis, School of Engineering, Geoenvironmental Research Centre, University of Wales, Cardiff.

Mangold, D C. & C F Tsang. 1991. A summary of subsurface hydrological and hydrochemical models. Review of Geophysics, 29:1, 51-79.

Mitchell, J K. 1993. Fundamentals of soil behaviour. 2nd Ed. John Wiley Pub. New york.

Thomas, H R & Y. He. 1995, "Analysis of coupled heat, moisture and air transfer in a deformable unsaturated soil". Geotechnique, 45(4) 677-689.

Thomas, H R, Rees, S W, Kjartanson, B H, Wan A W L. & N A Chandler. 1996, "Modelling in-situ water uptake in a bentonite - sand barrier". Geological Society Engineering Geology Special Publication No 11 on "The Engineering Geology of Waste Storage and Disposal", 215-222.

Thomas, H R, He, Y. & C. Onofrei. 1998, "An examination of the validation of a model of the hydro/thermo/mechanical behaviour of engineered clay barriers". Int J for Num and Anal Meth in Geomechanics, 22, 49-71.

Thomas, H R & P J. Cleall. 1999, "Inclusion of expansive clay behaviour in coupled thermo-hydraulic-mechanical models". Engineering Geology Special Issue – "Microstructural modelling with special emphasis on the use of clays for waste isolation". 54, Nos 1-2, 93-108.

Thomas, H R, Yong, R N. & A A Hashm 1999. Transport of Lead in partially saturated soil. Proc. 2nd Int. BGS Conf. on Geoenvironmental Engineering, 310-317, Thomas Telford.

Westall, J C, Zachary, J C. & F M Morel 1976. MINQEQL: A computer programme for the calculation of chemical equilibrium composition of aqueous systems. Cambridge, Massachuesetts, Department of Civil Engineering, Massachusetts Institute of Technology. Tech. Note 18.

Wiest, R.J.M.(ed.). 1969. Flow through porous medium. New York, Academic Press, Inc.

Yong, R N, Mohammad, A M. & B P Warkentin 1992. Principles of contaminant transport in soils. Elsevier Pub. Amsterdam.

Clay Science for Engineering, Adachi & Fukue (eds) © 2001 Balkema, Rotterdam, ISBN 90 5809 175 9

Evaluation of groundwater discharge through nuclear waste repository

T. Shiraishi & M. Horita
Institute of Technology, Shimizu Corporation, Tokyo, Japan

T. Sasaki
Japan Nuclear Fuel Limited, Japan

ABSTRACT: The facilities for low-level nuclear waste disposal are plan to be composite barrier system consisted in natural barrier and engineering barrier. The facility must be evaluated in order to constraint leakage of nuclear material through the repository for a long period. The dose of nuclide transported through the facility is a very important parameter for safety evaluation in the point for nuclear exposure. There are two types of transportation of nuclide in groundwater. One is in groundwater movement and the other is in defusion, which is independent from ground water movement.

This report proposes a simple method to evaluate the dose of nuclide transported through the repository with ground water movement in a case of the nuclear waste facility surrounded with bentonite barrier. This method evaluates the groundwater discharge through a region surrounded with impermeable material with certain permeability under certain hydraulic gradient in the basis of Darcy's law. This method is applied to the different shape of facilities and shows good correspondence with the results by a finite element analysis in certain conditions. This paper shows a correction method for different shape of facility. This method can be effectively used to determine parameters in safety evaluation because it can be applied to the three dimensional evaluation.

1 INTRODUCTION

The task of nuclear waste disposal in Japan is charged in The Japan Nuclear Fuel Limited and the low-level waste disposal has been started at Rokka-sho village in Aomori prefecture. The facilities for higher level waste disposal in part of the low-level waste are examined in their design and safety evaluation.

The composite barrier system consisted in natural barrier and engineering barrier is evaluated in order to constraint leakage of nuclear material through the repository for a long period. The engineer barrier system has cement materials and bentonite materials. The cement materials are placed inside of waste containers and inside of storage and the bentonite materials are placed around storage. The bentonite material is expected to constraint seepage of groundwater and to adhere to nuclide.

The dose of nuclide transported through the facility is a very important parameter for safety evaluation in the point for nuclear exposure. There are two types of transportation of nuclide in groundwater. One is in groundwater movement and the other is in defusion, which is independent from ground water movement.

This paper proposes a simple method to evaluate the dose of nuclide transported through the repository with ground water movement in a case of the nuclear waste facility surrounded with bentonite barrier. This method evaluates the groundwater discharge through a region surrounded with impermeable material with certain permeability under certain hydraulic gradient in the basis of Darcy's law. This method shows good correspondence with the results by a finite element analysis in certain conditions. For different shape of facility, this paper shows a correction method based upon a parameter study. This method is very simple to evaluate the discharge through the facility and can help to determine design parameters for facilities in the consideration of the discharge through the facility. This paper also proposes a extended method to apply for three dimensional cases. Different shape of facility can also be examined by this method.

2 EVALUATION METHOD FOR DISCHARGE THROUGH FACILITY

In the process of dose evaluation on radioactive waste, the evaluation methods for discarge through fasility are considered as follows.

(1) Evaluated from the velocity of groundwater before the construction of facility

(2) Evaluated with groundwater FEM analysis

The former method gives a conservative discarge because the permeability of facility is equevalent or less than that of surrounding rock. However, the method does not consider the permeability of engineering barrier and can be improved.

The latter method can consider the condition of natural barrier and engineering barrier in FEM modelling and it is considered more reliable than the former method. However, FEM analysis takes time for modeling and analysis in design process.

The following is a proposal of a simple evaluation method for discharge through a facility considering the size and permeability of engineering barrier.

3 FUNDAMENTAL CONCEPT

First introduces the result of groundwater analysis on a low-level waste repository, one pit with 20m wide and 6m high, under operation. The FE model has engineering barrier of impermeable material around the facility by 2m width. Fig.1 shows the FE model and the condition, and Fig.2 shows the hydraulic potential distribution after the analysis. In the FE model, the facility itself is assumed very permeable and the impermeable engineering barrier arround the facility acts for seeling material. Under the analytical condition, Fig.2 shows that the hydraulic gradient in the permeable facility is very small and that in engineering barrier is large.

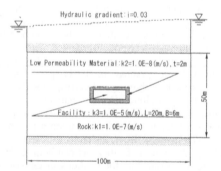

Fig.1 2D FEM Model Condition

Fig.3 shows the hydraulic potential distribution along a horizontal cross section and it shows that the potential lines are concentrated in the engineering barrier at the both side of the facility.

Considering the above result, a simplified evaluation method for the in-flow and out-flow discharge of a facility shown in Fig.4 for two-dimensional problem is derived.

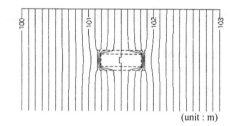

(unit : m)

Fig.2 Contour Lines for Total Head

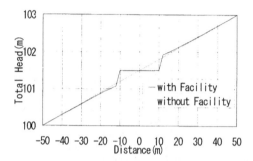

Fig.3 Total head distribution along center line of facility

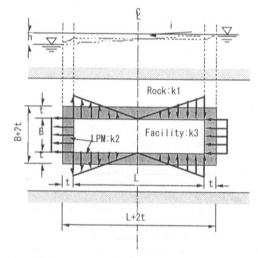

Fig.4 Calculation model of discharge from facility

The hydraulic gradient in the inpermeable engineering barrier, i', is expressed as,

$$i' = h / 2t = i (L + 2t) / 2t \tag{1}$$

where h is the head difference between upstream and downstream, t is the thickness of engineering barrier, i is the hydraulic gradient of original ground, and L is the length of facility as shown in Fig.4. Therefore, the maximum velocity of groundwater in the engineering barrier, v' , is expressed as follows,

348

$$v' = k_2 \, i' = k_2 \, i \, (\, L + 2 \, t \,) \, / \, 2 \, t \qquad (2)$$

where k_2 is the coefficient of permeability of the barrier material. Assuming that the distribution of inflow and out-flow discharge along the facility is triangular shape as shown in Fig.4, the total in-flow discharge, Q, is expressed as follows,

$$Q = B \, v' + 2 \, (0.5 \, v' \, L \, / \, 2)$$
$$= k_2 \, i \, (\, L + 2 \, t \,) \, (\, B + L \, / \, 2 \,) \, / \, 2 \, t \qquad (3)$$

where B is the height of the facility and this Eq.(3) is the evaluation equation for the total discharge through the engineering barrier in the consideration of high hydraulic gradient and permeability of engineering barrier.

The characteristics of the evaluation equation are as follows,
  (1) to be able to consider the permeability, k_2, and thickness of the engineering barrier which surrounds a facility,
  (2) to be able to consider the size of facility, L in length and B in height,
  (3) to be able to consider the hydraulic gradient of original ground,

  (4) to be unable to consider the permeability of natural ground, k_1, and the permeability of facility, k_3, and
  (5) to assume the facility facing against the stream lines perpendicularly.

Comparing with the result of FEM analysis under the condition shown in Fig.1, the FEM result gives 2.98E-8 (m³/s/m) for inflow discharge in unit depth of the facility and Eq.(3) gives 2.88E-8 (m³/s/m). The difference for the evaluation of discharge is about 3%.

4 EVALUATION ON TWO-DIMENSIONAL CASE

On the applicability of the evaluation equation (3), the assumption of the hydraulic gradient in the engineering barrier should be checked. The followings are main points for discussion.
  (1) The effect of the permeability of facility, engineering barrier, and original ground, and
  (2) The effect of size of facility.

4.1 *Effect of permeability of facility, engineering barrier, and original ground*

The hydraulic gradient in each material is depend upon the ratio of their permeability. In order to evaluate the proposed method, Eq.(3) is compared with the result of FEM analysis in the several per-

meability ratios between materials. The results are shown in Figs.5 and 6.

Fig.5 shows the comparison in the discharge evaluated by Eq.(3) and the FEM result in the different combination of permeability of facility, k_1, and engineering barrier, k_2. Eq.(3) matches the result of FEM where the ratio of k_1/k_2 is more than 100. On the permeability of ground, k_3, and engineering barrier, the ratio of evaluated discharge by Eq.(3) and by FEM is about 0.7 where the ratio of k_3/k_2 is larger than 100.

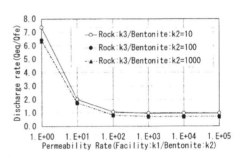

Fig.5 Discharge Rate by Eq.(3) and FEM(Facility vs Bentonite)

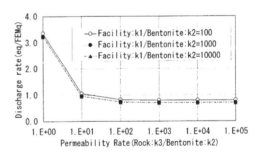

Fig.6 Discharge Rate by Eq.(3) and FEM(Rock vs Bentonite)

Fig.6 shows the comparison in the discharge evaluated by Eq.(3) and the FEM result in the different combination of permeability of ground, k_3, and engineering barrier, k_2. As mentioned before, the ration of evaluated discharge by Eq.(3) and by FEM is about 0.7 where the ratio of k_3/k_2 is larger than 100.

The summary in this section is as follows,
  (1) The proposed Eq.(3) for evaluation of the discharge through facility matches FEM where the permeability of facility is more than 100 times of that of engineering material. The evaluated discharge by Eq.(3) is multiplied by 1.4 to gain better match with the FEM results.
  (2) Eq.(3) can be used when the permeability of ground is more than 10 times of that of engineering barrier. For better evaluation, the evaluated discharge is multiplied by 1.4.

4.2 *Effect of facility size*

The resistance to groundwater stream depends upon the shape of facility. In order to evaluate the applicability of evaluation equation Eq.(3), comparison between Eq.(3) and FEM analysis was performed on the cases in Fig.7.

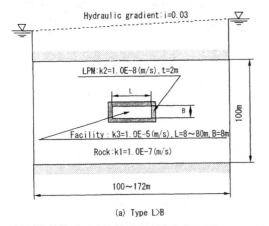

(a) Type L>B

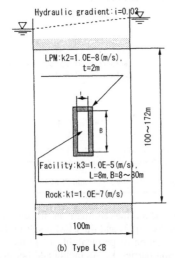

(b) Type L<B

Fig.7 Calculation Model for Facility Dimensional Effect

Fig.8 shows the comparison on the parameters B and L, where B and L are the height and length of facility, respectively. Fig.8 shows that the larger the parameter L/B is, the larger the discharge evaluated by Eq.(3). The evaluated discharge by Eq.(3) mostly matches the discharge obtained by FEM at L/B=3. Fig.9 shows the total head distribution along the centerline through the facility. Fig.9(a) shows that the hydraulic gradient assumed in Eq.(3) is larger than that in FEM where L/B is relatively large. On the contrary, Fig.9(b) shows that Eq.(3) give the gradient smaller than FEM result when L/B is relatively small.

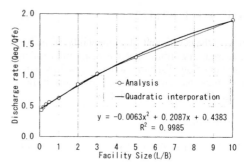

Fig.8 Effect of Facility Size

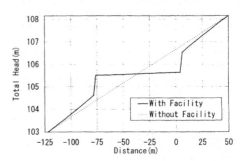

Fig.9 (a) Size Effect (L/B=10)

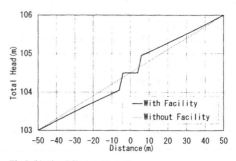

Fig.9 (b) Size Effect (L/B=0.1)

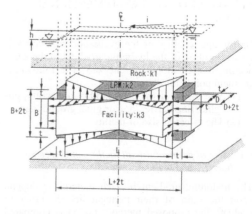

Fig.10 3D Calculation Model of Discharge from Facility

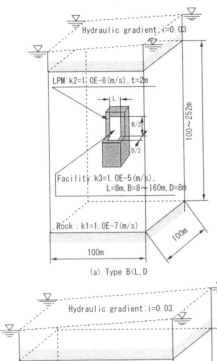

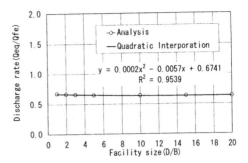

Fig.12 Size Effect in Fig.11 (a) Type

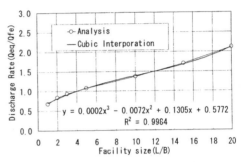

Fig.13 Size Effect in Fig.11 (b) Type

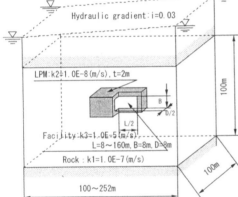

(a) Type B<L, D

Fig.11 3D FEM Model Conditions

Therefore the discharge evaluated by Eq.(3) should be modified with the interpolation factor in Fig.8 when the shape effect of facility is considered.

4.3 Discharge volume through facility in two-dimensional case

On the basis of above-mentioned discussion, the discharge volume through a facility is expressed as follows,

$$Q = R\, k_2\, i\, (L + 2t)\, (B + L/2)/2t \qquad (4)$$

where, Q is the discharge volume through a facility per unit depth, R is a shape factor mentioned below, k_2 is the permeability of impermeable material, t is the thickness of engineering barrier, i is the hydraulic gradient of the ground, L is the width of facility in Fig.4, and B is the height of the facility in Fig.4.

The application range of the evaluation equation is as follows,

$k_2 \leqq 0.01 k_1$ where k_1 is the permeability of facility, and

$k_2 \leqq 0.1 k_3$ where k_3 is the permeability of ground.

The shape factor R is express as follows,

$R = R_1 \times R_2 \times R_3$

where R_1 is the shape factor on the shape of facility as follows,

$R_1 = 1/\{-0.0063(L/B)2 + 0.2087(L/B) + 0.4383\}$

R_2 is the shape factor on the permeability of facility as follows,

$R_2 = 1.00$ at $k_2 = 0.1\, k_3$

$R_2 = 1/0.7$ at $k_2 < 0.1\, k_3$

R_3 is the shape factor on the permeability of ground as follows,

$R_3 = 1.00$ at $k_2 = 0.01\, k_1$ and $k_2 = 0.1\, k_3$

$R_3 = 1/0.8$ at $k_2 = 0.01\, k_1$ and $k_2 < 0.1\, k_3$

$R_3 = 1/0.8$ at $k_2 < 0.01\ k_1$ and $k_2 < 0.1\ k_3$.

5 EVALUATION ON THREE-DIMENSIONAL CASE

In the same manner with two-dimensional case, the discharge volume in three-dimensional case is expressed as follows,

$$Q = k_2\ i\ B_2\ (L + 2t)\ (BD + LB/2 + LD/2)\ /\ 2t \quad (5)$$

where, Q is the discharge volume through a facility, k_2 is the permeability of impermeable material, t is the thickness of engineering barrier, i is the hydraulic gradient of the ground, L, B, and D are the width, the height, and the depth of facility in Fig.10 respectively.

In the same manner as 2D model, Eq.(5) is evaluated for its applicability on different shape of facility. Fig.11 shows the shape of facility and two directions of flow, a) perpendicular to the long axis of facility and b) along the long axis. The comparison between Eq.(5) and FEM analysis is shown in Figs.12 and 13, respectively.

Fig.12 shows that the discharge ratio between Eq.(5) and FEM is almost constant with the flow perpendicular to the long axis of facility. This case is similar condition with the 2D model and the 2D evaluation, Eq.(3), can be used for this condition.

Fig.13 shows that discharge ratio changes with the length of facility in the case with the flow along the long axis of facility. The longer the length of facility is, the larger the discharge ratio is. It implies that the hydraulic gradient assumed larger in Eq.(5) than in FEM. This tendency is recognized in the previous 2D model, Eq.(3). In order to compensate the shape effect, the interpolation function in Fig.13 should be taken into account.

6 CONCLUSION

This report proposes a simple method to evaluate the groundwater discharge through a region surrounded with impermeable material with certain permeability under certain hydraulic gradient in the basis of Darcy's law. The summary of comparison on the evaluated discharge by proposed method to the result by FEM as follows,
(1) The proposed method matches FEM where the permeability of facility is more than 100 times of that of engineering material and the permeability of surrounding rock is more than 10 times of that of engineering material.
(2) Judging from the comparison in 2D models,the evaluated discharge by Eq.(3) is multiplied by 1.4 to gain better match with FEM result where the above-mentioned condition is satisfied.

(3) The estimated discharge from facility varies due to the shape and direction of the facility. Therefore, it is important to consider the proper correction in proposed evaluation method.

Clay Science for Engineering, Adachi & Fukue (eds) © 2001 Balkema, Rotterdam, ISBN 90 5809 175 9

Modelling and analysis of fully coupled thermo-hydro-diffusion phenomena

Ö. Aydan

Department of Marine Civil Engineering, Tokai University, Shimizu, Japan

ABSTRACT: Radioactive nuclear waste disposal recently receives great attention in geo-engineering. Many experimental, theoretical and numerical studies have been undertaken and most of these studies have been concerned with thermo-hydro-mechanical aspects of the phenomenon and diffusion phenomenon of the radioactive substances is mostly omitted. Since the diffusion phenomenon should be quite important in long term, a mechanical model based on the mixture theory is described to couple thermal, hydrological and diffusion fields. In this model, Duffour and Soret effects, which are mostly neglected in previous studies, are also considered to couple the thermal and diffusion fields. Then a finite element formulation of the derived theoretical model is given and a series of numerical analyses carried out on the simulation of laboratory tests is presented. Furthermore, some parameteric studies are performed to investigate the coupling effects of Duffour and Soret effects on thermal and diffusion fields.

1 INTRODUCTION

Radioactive nuclear waste disposal and geothermal energy extraction are some typical examples of thermo-hydro-diffusion phenomena in geo-engineering. Particulary, the radioactive nuclear waste disposal receives great attention and one may find many studies in literature. Most of these studies on radioactive nuclear waste disposal are based on thermo-hydro-mechanical concepts and it is very rare to see any study which includes diffusion phenomena of radiocnive substances. Therefore, a mechanical model, which includes 4 different aspects, should be the most appropriate approach for radio-active waste disposal. Although the variation of stress field around the disposal site may be of great concern for cavity stability in very near field, its effect is expected to be quite limited after back-filling the excavated space in long term. In this article, the mechanical effect in this coupling model presented herein is omitted and a theoretical formulation, based on the mixture theory is described for the thermo-hydro-diffusion phenomena. In the theoretical formulation, Duffour and Soret effects are considered for coupling the thermal and diffusion fields with each other. Then, a finite element formulation of the derived coupled model is presented and it is used for numerical analysis of some laboratory tests and compared with experimental results. Furthermore, a series of parametric numerical analyses are performed to investigate the coupling effects of Duffour and Soret effects on thermal and diffusion fields.

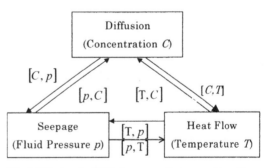

[C,p]:Effect of Diffusion on Seepage (density variation)

[p,C]:Effect of Seepage on Diffusion (Advection of Concentration due to fluid flow)

[T,p]:Effect of Heat Flow on Seepage (Density variation due to Temperature variation)

[p,T]:Effect of Seepage on Heat flow(Advection of Temperature due to fluid flow)

[T,C]:Effect of Heat Flow on Concentration(Duffour effect))

[C,T]:Effect of Diffusion on thermal field(Soret effect))

Figure 1 Coupling model

2 MECHANICAL MODELLING

2.1 Fundamental Equations

The mechanical modelling of thermo-hydro-diffusion phenomena is based on the mixture theory (Trusdell and Toupin 1960, Eringen and Ingram 1965). Figure 1 illustrates how three fields are coupled.

Following the principles of the mixture theory, one can easily derive that the fundamental equations for each field are as follows:

Seepage field

$$\frac{1-n}{\rho_s}\frac{\partial \rho_s}{\partial t}+\frac{n}{\rho_f}\frac{\partial \rho_f}{\partial t}+\frac{1}{\rho_s}\nabla\cdot\{\mathbf{q}_s\}+\frac{1}{\rho_f}\nabla\cdot\{\mathbf{q}_f\}=0 \tag{1}$$

where $\mathbf{q}_s = (1-n)\rho_s\mathbf{v}_s$ and $\mathbf{q}_f = n\rho_f\mathbf{v}_f$.

Diffusion field

$$\frac{\partial C}{\partial t}+\mathbf{v}_s\cdot\nabla\{(1-n)C_s\}+\mathbf{v}_f\cdot\nabla\left(nC_f\right)=-\nabla\cdot\{\mathbf{f}\} \tag{2}$$

where $C = (1-n)C_s + nC_f$ and $\mathbf{f} = (1-n)\mathbf{f}_s + n\mathbf{f}_f$.

Thermal field

$$(1-n)\rho_s\frac{d_sU_s}{dt}+n\rho_f\frac{d_fU_f}{dt}=\nabla\cdot\mathbf{h}+Q \tag{3}$$

where $\mathbf{h} = (1-n)\mathbf{h}_s + n\mathbf{h}_f$ and $Q = (1-n)Q_s + nQ_f$

2.2 Constitutive Laws

Seepage field

D'Arcy law is usually used to relate the relative average fluid velocity $\mathbf{v}_r$ to pressure in the following form

$$\mathbf{v}_r = n\left(\mathbf{v}_f - \mathbf{v}_s\right) = -\frac{k}{\eta}\nabla p \tag{4}$$

where k is permeability and η is viscosity coefficient of fluid. If fluid density is assumed to be a function of p, T, C, its variation may be written in the following form

$$\frac{1}{\rho_f}\frac{\partial \rho_f}{\partial t}=\frac{1}{K_f}\frac{\partial p}{\partial t}-\beta_{fT}\frac{\partial T}{\partial t}-\beta_{fC}\frac{\partial C_f}{\partial t} \tag{5}$$

where K_f is fluid compressibility, β_{fT} is thermal expansion coefficient and β_{fC} is diffusive expansion coefficient.

Diffusion field

Fick's law is often employed as a constitutive law for diffusion problems. This law is extended by associating concentration flux $\mathbf{f}_s$ of solid phase with gradients of temperature and concentration as given below:

$$\mathbf{f}_s = -D_{sT}\nabla T_s - D_{sC}\nabla C_s \tag{6}$$

where D_{sT} is thermal diffusivity of solid phase and D_{sC} is Dufour's coefficient (Bear 1988). Similarly, one may also write the following relation for fluid phase.

$$\mathbf{f}_f = -D_{fT}\nabla T_f - D_{fC}\nabla C_f \tag{7}$$

If $T_s = T_f = T$ and $C_s = C_f = C$, the average concentration flux $\mathbf{f}$ takes the following form

$$\mathbf{f} = -D_T\nabla T - D_C\nabla C \tag{8}$$

where
$D_T = (1-n)D_{sT} + nD_{fT}$ and $D_C = (1-n)D_{sC} + nD_{fC}$.
D_C is average diffusion coefficient, D_{sC} and D_{fC} are the diffusion coefficient of solid phase and fluid phase, respectively.

Thermal field

Fourier's law is well known as a constitutive law for associating heat flux to temperature gradient. This law is also expanded by associating heat flux $\mathbf{h}_s$ with temperature and concentration gradients as

$$\mathbf{h}_s = -\lambda_{sT}\nabla T_s - \lambda_{sC}\nabla C_s \tag{9}$$

where λ_{sT} is thermal conductivity coefficient and λ_{sC} is Soret's coefficient of solid phase. Similarly the following relation can be written for fluid phase.

$$\mathbf{h}_f = -\lambda_{fT}\nabla T_f - \lambda_{fC}\nabla C_f \tag{10}$$

where λ_{fT} is thermal conductivity coefficient and λ_{fC} is Soret's coefficient of fluid phase. If $T_s = T_f = T$ and $C_s = C_f = C$, then average heat flux $\mathbf{h}$ takes the following form.

$$\mathbf{h} = -\lambda_T \nabla T - \lambda_C \nabla C \qquad (11)$$

where

$\lambda_T = (1-n)\lambda_{sT} + n\lambda_{fT}$ and $\lambda_C = (1-n)\lambda_{sC} + n\lambda_{fC}$. λ_T is average conductivity coefficient. λ_{sT} and λ_{fT} are conductivity of solid and fluid phases, respectively.

2.3 Simplified form of Fundamental Equations

Seepage field

If the density variation of solid phase and the porosity variation of skelton are negligible, Eq. (1) may be re-written together with D'Arcy's law as:

$$n\left(\frac{1}{K_f}\frac{\partial p}{\partial t} - \beta_{fT}\frac{\partial T}{\partial t} - \beta_{fC}\frac{\partial C}{\partial t}\right) = -\nabla \cdot \left(\frac{k}{\eta}\nabla p\right) \qquad (12)$$

Diffusion field

Eq. (2) may take the following form together with the use of Eq. (8)

$$\frac{\partial C}{\partial t} = \nabla \cdot (D_C \nabla C + D_T \nabla T) - \mathbf{v}_r \cdot \nabla C \qquad (13)$$

The second term on the right-hand side is the advective term.

Thermal field

Internal energies of solid and fluid phases may be related to temperature field with the use of specific heat coefficients c_s and c_f as

Solid Phase
$$\frac{\partial U_s}{\partial t} = \frac{\partial U_s}{\partial T}\frac{\partial T}{\partial t} = c_s \frac{\partial T}{\partial t} \qquad (14)$$

Fluid Phase
$$\frac{\partial U_f}{\partial t} = \frac{\partial U_f}{\partial T}\frac{\partial T}{\partial t} = c_f \frac{\partial T}{\partial t} \qquad (15)$$

With the use of equations above, Eq. (11) and $\mathbf{v}_s = \mathbf{0}$, Eq. (3) becomes

$$\rho c \frac{\partial T}{\partial t} = \nabla \cdot (\lambda_T \nabla T + \lambda_C \nabla C) - \rho_f c_f \mathbf{v}_r \cdot \nabla T \qquad (16)$$

where $\rho c = (1-n)\rho_s c_s + n\rho_f c_f$.

3 FINITE ELEMENT FORMULATION

3.1 *Weak Forms of Fundamental Equations*

Seepage field

The governing equation of seepage field is assumed to be subjected to the following boundary conditions
Pressure boundary condition
$$p = p_0 \text{ on } S_p \qquad (17)$$

Fluid flux boundary
$$-(n\rho_f \frac{k}{\eta}\nabla p) \cdot \mathbf{n} = \hat{q} \text{ on } S_q \qquad (18)$$

Taking a variation on pressure field δp and integrating by parts, one gets the weak form of Eq. (12) as

$$n\left(\int_V \delta p \frac{n}{K_f}\frac{\partial p}{\partial t}dV - \int_V \delta p \beta_{fT}\frac{\partial T}{\partial t}dV - \int_V \delta p \beta_{fC}\frac{\partial C}{\partial t}dV\right)$$
$$+ \int_V \nabla(\delta p) \cdot \frac{k}{\eta}\nabla p dV = \int_{S_q} \delta p \hat{q} dS \qquad (19)$$

Diffusion field

The diffusion equation is assumed to be subjected to the following boundary conditions
Concentration boundary
$$C = C_0 \text{ on } S_C \qquad (20)$$

Concentration flux boundary
$$-(D_C \nabla C + D_{CT}\nabla T) \cdot \mathbf{n} = \hat{f} \text{ on } S_f \qquad (21)$$

Taking a variation δC and applying the integration by parts to Eq. (15), we have the weak form of Eq.(13) as

$$\int_V \delta C \frac{\partial C}{\partial t}dV + \int_V \nabla \delta C \cdot D_C \nabla C dV + \int_V \nabla \delta C \cdot D_C \nabla T dV$$
$$+ \int_V \delta C \mathbf{v}_r \nabla C dV = \int_{S_q} \delta C \hat{f} dS \qquad (22)$$

Thermal field

Eq. (16) is assumed to be subjected to the following boundary conditions

Temperature boundary
$$T = T_0 \text{ on } S_T \tag{23}$$

Heat flux boundary
$$-(\lambda_T \nabla T + \lambda_{TC} \nabla C) \cdot \mathbf{n} = \hat{h} \text{ on } S_h \tag{24}$$

Taking a variation δT on temperature field and applying the integration by parts to the first term of Eq. (16), one easily gets its weak form as:

$$\int \delta T \rho c \frac{\partial T}{\partial t} dV + \int \nabla \delta T \cdot \lambda_T \nabla T dV + \int \nabla \delta T \cdot \lambda_{TC} \nabla C dV$$
$$+ \int \delta T \rho_f c_f \mathbf{v}_r \cdot \nabla T dV = \int_{S_q} \delta T \hat{h} dS \tag{25}$$

3.2 Discretization of Weak Forms

3.2.1 Discretization in Physical Space
Pressure, concentration and temperature variables are interpolated in a typical finite element with the use of shape functions as given below:

$$p = [N]\{P\}, \; C = [N]\{\phi\}, \; T = [N]\{\chi\} \tag{26}$$

Inserting these relations into each respective weak form and after some manipulations, one easily gets the following equation system for each field as

Seepage field

$$[M]_{PP} \{\dot{P}\} + [M]_{PC} \{\dot{\phi}\} + [M]_{PT} \{\dot{\chi}\} + [K]_{PP} \{P\} = \{Q\}_P \tag{27}$$

where

$$[M]_{PP} = \int \frac{n}{K_f} [N]' [N] dV \; ;$$

$$[M]_{PC} = -\int n\beta_{fc} [N]' [N] dV \; ;$$

$$[M]_{PT} = -\int n\beta_{fT} [N]' [N] dV \; ;$$

$$[K]_{PP} = -\int \frac{k}{\eta} [B]' [B] dV \; ; \{Q\}_P = \int_{S_q} N^T \hat{q} dS$$

Diffusion field

$$[M]_{CC} \{\dot{\phi}\} + [K]_{CC} \{\phi\} + [K]_{CT} \{\chi\} = \{Q\}_C \tag{28}$$

where

$$[M]_{CC} = \int [N]' [N] dV \; ;$$

$$[K]_{CC} = \int D_C [B]' [B] dV - \int \frac{k}{\eta} [N]' ([B]\{P\})' \{B\} dV$$

$$[K]_{CT} = \int D_{CT} [B]' [B] dV \; ; \{Q\}_C = \int_{S_f} N^T \hat{f} dS$$

Thermal field

$$[M]_{TT} \{\dot{\chi}\} + [K]_{TT} \{\chi\} + [K]_{TC} \{\phi\} = \{Q\}_T \tag{29}$$

where

$$[M]_{TT} = \int \rho c [N]' [N] dV$$

$$[K]_{TT} = \int \lambda_T [B]' [B] dV -$$
$$\int \rho_f c_f \frac{k}{\eta} [N]' ([B]\{P\})' [B] dV$$

$$[K]_{TC} = \int \lambda_{TC} [B]' [B] dV \; ; \{Q\}_T = \int_{S_h} N^T \hat{h} dS$$

Finally the following simultaneous equation system is obtained for whole domain as

$$[M]\{\dot{X}\} + [K]\{X\} = \{Y\} \tag{30}$$

where

$$[M] = \begin{bmatrix} [M]_{PP} & [M]_{PC} & [M]_{PT} \\ [0] & [M]_{CC} & [0] \\ [0] & [0] & [M]_{TT} \end{bmatrix} ;$$

$$[K] = \begin{bmatrix} [K]_{PP} & [0] & [0] \\ [0] & [K]_{CC} & [K]_{CT} \\ [0] & [K]_{TC} & [K]_{TT} \end{bmatrix}$$

$$\{\dot{X}\} = \begin{Bmatrix} \dot{P} \\ \dot{\phi} \\ \dot{\chi} \end{Bmatrix} ; \{X\} = \begin{Bmatrix} P \\ \phi \\ \chi \end{Bmatrix} ; \{Y\} = \begin{Bmatrix} \{Q\}_P \\ \{Q\}_C \\ \{Q\}_T \end{Bmatrix}$$

3.2.2 Discretization in Time Domain
Although there are several techniques for discretization in time domain, θ-method is chosen in this study. Since Eq. (30) holds at any time, one may easily write the following relation for a time step ($m + \theta$) as

$$[M]\{\dot{X}\}_{(m+\theta)} + [K]\{X\}_{(m+\theta)} = \{Y\}_{(m+\theta)} \tag{31}$$

With the use of Taylor expansions of variables and after some manipulations, Eq. (31) takes the following form:

$$[C^*]\{X\}_{m+1} = \{Y^*\}_{m+1} \tag{32}$$

where

$$[C^*] = \left[\frac{1}{\Delta t} [C] + \theta[K] \right]$$

$$\{Y^*\}_{m+1} = \left[\frac{1}{\Delta t} [C] - (1-\theta)[K] \right] \{X\}_m +$$
$$\theta \{Y\}_{m+1} + (1-\theta)\{Y\}_m$$

It should be noted that matrices $[K]_{CC}$ and $[K]_{TT}$ contains unknown vector variable $\{P\}$. Therefore, the resulting equation system is non-linear. However, if time step is sufficiently small, it can be linearized with the use of variable $\{P\}$ of the previous time step.

4 ANALYSES AND DISCUSSIONS

The first example is concerned with the simulation of a diffusion-seepage test on a sandstone sample under isothermal condition by Igarashi & Tanaka (1998). Figure 2 illustrates the test conditions. The pressure gradient in the sample was set to 260MPa/m. Figure 3 shows pressure distribution through the sample at different time steps. Since the pressure distribution attains the steady state in a short period of time, this implies that the diffusion takes place under constant fluid velocity through the sample. Material properties used in the analyses are given in Table 1. The computations were carried out for two different situations, specifically, 1) Advection + Conduction and 2) Conduction only.

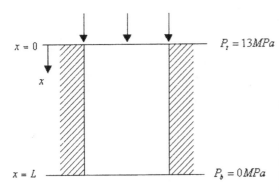

Figure 2 Conceptional illustration of a diffusion-seepage test on sandstone sample

Table 1 Material properties used in analyses

Fick's diffusion coefficient of solid phase (m²/day)	5.1×10^{-5}
Fick's diffusion coefficient of fluid phase (m²/day)	2.1×10^{-4}
Porosity (%)	16
Permeability (m²)	3.1×10^{-12}
Sample length (mm)	50
Sample diameter (mm)	50

Figure 4 compares the breakdown curves for computed two situations with the experimental one.

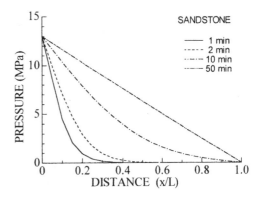

Figure 3 Pressure distribution in a sandstone sample

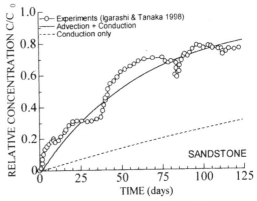

Figure 4 Comparison of computed breakdown curves with the experimental curve

In the analyses, the effect of diffusion on seepage is neglected. As seen from the figure, the best fit to the experimental response was obtained from the computation for the advection+conduction situation while much longer period of time is required for conductive diffusion. This fact simply implies that if the diffusion coefficient, which is obtained from a conductive diffusion model under different pressure gradients, will differ from each other.

The next example is concerned with moisture transport through a buffer material for radioactive waste disposal under different ambient temperatures, reported by Kanno et al. (1999). Recently this problem was theoretically solved by Basha and Selvadurai (1998) as a spherically symmetric problem. Kanno et al. (1999) used bentonite as a buffer material in their tests and the ambient temperature was varied from 25° to 60° C. They determined Fick's diffusion coefficient of bentonite for moisture transport using a conductive diffusion model. Figure 5 shows a simulation of moisture concentration distributions through the sample at different time steps, carried out at an ambient temperature of $25\,^{\circ}\mathrm{C}$. They found that the diffusion coefficients obtained for different ambient temperatures differ from each other.

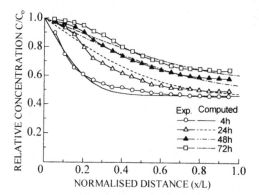

Figure 5 Comparison of computed moisture concentration distributions through the sample with experimental distributions at different time steps

With the experimental finding of Kanno et al. (1999) in mind, a series of parametric studies were carried out. Specifically the following four CASES were considered in parametric studies:

CASE 1:Dufour coefficient and Soret coefficient are nil (thermal and diffusion fields are uncoupled)

CASE 2:Dufour coefficient is not nil, but Soret coefficient is nil

CASE 3:Dufour coefficient is nil, but Soret coefficient is not nil

CASE 4:Dufour coefficient and Soret coefficient are not nil (fully coupled).

Material properties used in analyses are given in Table 2 and computed results are shown in Figures 6-11. For all cases, temperature distributions through the sample become uniform and the effect of diffusion field on temperature distributions are not observed. However, the effect of thermal field on diffusion field are clearly observed in computed results for CASE 2, CASE 3 and CASE 4. Particularly, the effect of thermal field on results for CASE 2 is of great interest. If diffusion coefficient is determined from the conductive diffusion model, the value of the diffusion coefficient should be larger since the concentration reaches to its injection level in a shorter period of time. On the other hand, there is no remarkable difference in computed results for CASE 3. Nevertheless, if slicing method is employed for determining the diffusion coefficient, there may be an apparent scattering among the computed diffusion coefficients.

CASE 4 is probably the most close situation to actual situations. Particularly, the effects of Duffour coefficient and Soret coefficient are quite remarkable. However, there are almost no experimental results reported on these coefficients to confirm the findings from computations. Therefore, it may be stated that

Table 2 Material properties used in numerical analyses

Parameter	CASE 1	CASE 2	CASE 3	CASE 4
Solid's thermal conductivity ($Wm^{-1}K^{-1}$)	0.277	0.277	0.277	0.277
Fluid's thermal conductivity ($Wm^{-1}K^{-1}$)	0.600	0.600	0.600	0.600
Dufour's coefficient of solid ($Wcm^{-1}kg^{-1}$)	0	72	0	72
Dufour's coefficient of fluid ($Wcm^{-1}kg^{-1}$)	0	72	0	72
Soret's coefficient of solid ($cm^2kg\ s^{-1}K^{-1}$)	0	0	3.6×10^{-5}	3.6×10^{-5}
Soret's coefficient of solid ($cm^2kg\ s^{-1}K^{-1}$)	0	0	3.6×10^{-5}	3.6×10^{-5}
Solid's density ($kg\ m^{-3}$)	1600	1600	1600	1600
Fluid's density ($kg\ m^{-3}$)	1000	1000	1000	1000
Solid's specific heat ($kJ\ kg^{-1}K^{-1}$)	5.0	0.5	5.0	5.0
Fluid's specific heat ($kJ\ kg^{-1}K^{-1}$)	1.0	1.0	1.0	1.0
Fick's diffusion coefficient of solid (cm^2/s)	6.1×10^{-3}	6.1×10^{-3}	6.1×10^{-3}	6.1×10^{-3}
Porosity (%)	46.3	46.3	46.3	46.3

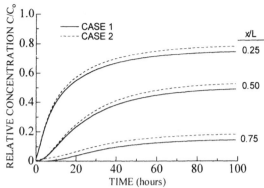

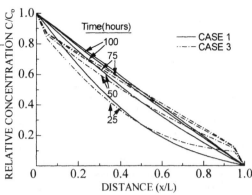

Figure 6 Comparison of computed relative concentration versus time curves for CASE 1 and CASE 2

Figure 9 Comparison of computed relative concentration distributions at different time steps for CASE 1 and CASE 3

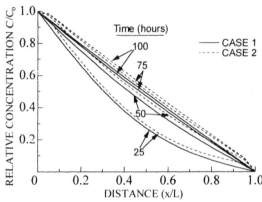

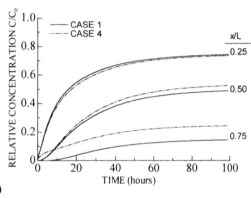

Figure 7 Comparison of computed relative concentration distributions at different time steps for CASE 1 and CASE 2

Figure 10 Comparison of computed relative concentration versus time curves for CASE 1 and CASE 4

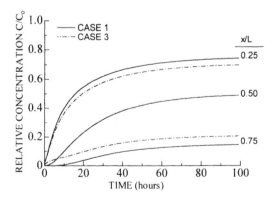

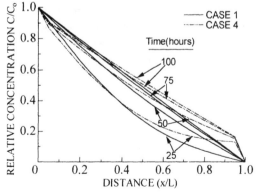

Figure 8 Comparison of computed relative concentration versus time curves for CASE 1 and CASE 3

Figure 11 Comparison of computed relative concentration distributions at different time steps for CASE 1 and CASE 4

359

re-assessment of existing experiments with the consideration of Duffour coefficient and Soret coefficient is urgently necessary.

5 CONCLUSIONS

In this article, a new mechanical model for fully coupled thermo-hydro-diffusion phenomena on the basis of mixture theory and its finite element representation are presented. The validity of this model is checked through some experiments, which correspond to its special forms. From these comparisons, some shortcomings of determination of Fickian diffusion coefficients from conductive diffusion models are pointed out. Parametric studies on the effect of Duffour and Soret coefficients which are generally neglected in diffusion models indicated that these coefficients must be taken into account in numerical analyses. Such considerations can explain the dependency of Fickian diffusion coefficient on ambient temperature and fluid velocity. Furthermore, there is an urgent experimental necessity to obtain the actual values of Duffour and Soret coefficients for a meaningfull assessment of fully coupled thermo-hydro-diffusion phenomena.

ACKNOWLEDGEMENTS

The author would like to thank Dr. H. Ito and Dr. S. Igarashi for providing reports and articles on laboratory and in-situ tests performed at the Central Research Institute of Electric Power Companies, Japan.

REFERENCES

Basha, H.A. and Selvadurai, A.P.S. (1998): Heat induced moisture transport in the vicinity of a spherical heat source. *International Journal for Numerical and Analytical Methods in Geomechanics*, 22, 969-981.

Bear, J. (1988): *Dynamics of Fluids in Porous Media*. Elsevier, New York.

Eringen, A.C. and Ingram, J.D. 1965. A continuum theory of chemically reacting media-I. *Int. J. Eng. Sci.*, 3, 197-212.

Igarashi, S. and Tanaka, Y. (1998): The effect of a discontinuity on solute transport through sandstone. *10th National Rock Mechanics Symposium*, Tokyo, JRMS, 359-364.

Kanno, T., Fujita, T., Takeuchi, S., Ishikawa, H., Hara, K. and Nakano, M. (1999): Coupled thermo-hdro-mechanical modelling of bentonite buffer material. *International Journal for Numerical and Analytical Methods in Geomechanics*, 23, 1281-1307.

Trusdell, C.A. and Toupin, R.A. 1960. The classical field theories. In S. Flügge (Ed.) *Handbuch der Physik III/1*, Berlin, Springer Verlag, 226-793.

Clay Science for Engineering, Adachi & Fukue (eds) © 2001 Balkema, Rotterdam, ISBN 90 5809 175 9

Application of a connected tubes model to gas permeability of a silty soil

J.-M. Couvreur & A. Mertens de Wilmars
Laboratoire de Génie Civil, Université Catholique de Louvain, Louvain-la-Neuve, Belgium

A. Monjoie
Laboratoire de Géologie de l'Ingénieur et d'Hydrogéologie, Université de Liège, Belgium

ABSTRACT: A Proctor cell was modified in order to perform gas permeability tests. This study focuses on permeability k of dry samples of a silty soil. The use of Kozeny-Carman equation to determine the soil permeability k from its porosity ε proved to be unsatisfactory. The plot of experimental k values shows that permeability becomes zero for a positive value of porosity (about 35%). The connected tubes model was developed by Freijer for gas diffusion. This model, which takes into account tubes of different radii, was adapted by the authors for gas advection. The application of this adapted model to a silty soil shows very good results and is able to predict a zero permeability value for a porosity value of 32%. Further studies should focus on the determination of the empirical parameters used in this model.

1 THEORETICAL MODELS

1.1 Kozeny-Carman equation

In Kozeny-Carman model, the soil grains texture is modelized by series of tortuous circular pores. These tubes are characterized by one unique diameter d_c and a tortuosity coefficient $\tau = L_c/L$ (L being the sample length and L_c the pore length)

By using Poiseuille's law, Kozeny and Carman obtained the following relationship:

$$k = \frac{1}{h_k U_o^{\,2}} \cdot \frac{\varepsilon^3}{(1-\varepsilon)^2} \qquad (1)$$

where k is the soil permeability (only depending on the soil texture), h_k is a factor depending on both tortuosity and shape of the tubes (usually taken equal to 5), U_o is the specific soil particles surface ($U_o = S_s/V_s$), ε is the total soil porosity ($\varepsilon = V_v/V$).

As it will be later discussed in this paper, this model is unable to represent pore blockage due to capillary water bonding. It is however suitable for quite homogeneous sandy soils in which the hypothesis of one unique tube diameter is acceptable (Bousmar et al, 1995).

1.2 Connected tubes model

Freijer (1994) developed a model considering the porous media as cylindrical tubes isolated one from another and with changing radius along the tube. Freijer's equations are actually written considering only diffusive transport. The authors adapted this model to gas movement due to a pressure gradient (pure advection).

The equations we will obtain here are slightly different to those from Freijer but the thought process is similar.

Figure 1 shows the porous media representation: one channel consisting of two connected cylindrical tubes with different radius. In the case of a compressible fluid flowing through a cylindrical tube (radius r and length l), mass flow (and therefore the product pq) is a constant along the whole tube. Poiseuille's law can be written as:

$$pq = \frac{\pi . r^4}{8\mu} \cdot \frac{\Delta(p^2)}{l} \qquad (2)$$

with $\quad \Delta(p^2) = p_{inlet}^2 - p_{outlet}^2$

μ is the fluid dynamic viscosity.

Expression (2) can be rewritten for the first (radius r_1 and length l_1) and the second tubes (radius r_2 and length l_2) shown in Fig. 1:

$$pq = \frac{\pi . r_1^4}{8\mu} \cdot \frac{\Delta(p_{a-b}^2)}{2l_1} = \frac{\pi . r_2^4}{8\mu} \cdot \frac{\Delta(p_{b-c}^2)}{2l_2} \qquad (3)$$

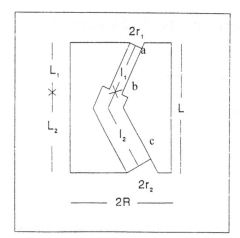

Figure 1. Connected tubes model.

a, b and c are the inlet and outlet points of the single tubes (see Fig. 1) and:

$$\Delta(p_{a-b}^2)=p_a^2-p_b^2$$
$$\Delta(p_{b-c}^2)=p_b^2-p_c^2$$

Actually in a sample of radius R, there is not only one channel but a number n (obviously depending on the sample size). Then expression (3) becomes:

$$pq = n\frac{\pi.r_1^4}{8\mu}.\frac{\Delta(p_{a-b}^2)}{2l_1} = n\frac{\pi.r_2^4}{8\mu}.\frac{\Delta(p_{b-c}^2)}{2l_2} \qquad (4)$$

By integrating Darcy's law on the whole sample length, we get the following expression :

$$pq = \frac{k}{\mu}A\frac{\Delta(p_{a-c}^2)}{2L} \qquad (5)$$

Which is equal to the previous one. Seen that:
$\Delta(p_{a-c}^2)=\Delta(p_{a-b}^2)+\Delta(p_{b-c}^2)$ and $A = \pi.R^2$,
we obtain the following relationship for k:

$$k = \frac{nL}{8R^2}.\frac{1}{\dfrac{l_1}{r_1^4}+\dfrac{l_2}{r_2^4}} \qquad (6)$$

By defining a tortuosity value for each tube:

$$\tau_1 = \frac{l_1}{L_1} \text{ and } \tau_2 = \frac{l_2}{L_2}$$

We obtain another expression of k :

$$k = \frac{nL}{8R^2}.\frac{1}{\dfrac{L_1\tau_1}{r_1^4}+\dfrac{L_2\tau_2}{r_2^4}} \qquad (7)$$

In this model we can also calculate the value of porosity ε :

$$\varepsilon = \frac{n(r_1^2 L_1\tau_1 + r_2^2 L_2\tau_2)}{R^2 L} \qquad (8)$$

The channels should actually be represented by more than two cylindrical tubes. More generalized expressions are then :

$$k = \frac{n}{8R^2}.\frac{L}{\sum_i \dfrac{L_i\tau_i}{r_i^4}} \qquad (9)$$

$$\varepsilon = \frac{n\sum_i r_i^2 L_i\tau_i}{R^2} \qquad (10)$$

From equation (9), it can be seen that, provided only one value of radius r_i is zero, then permeability k will be zero for a (non-zero) positive value of porosity ε. All the channels are blocked and inefficient for gas flow because of that single radius r_i. This situation is similar to the soil texture: pores are blocked due to capillary water bonding between grains. This model can also be compared to a stratified media: only one single impermeable layer induces the impermeability of the whole media.

Expressions (9) and (10) imply the knowledge of characteristic values r_i, L_i, τ_i for each single tube. The porous media can actually be represented by only two different tubes, by taking into account the larger and smaller radius values (Van Brakel et al, 1974). Moreover, the same length ($l_1=l_2=L/2$) and the same tortuosity τ can be considered for each tube. We get the following expressions:

$$k = \frac{n}{4R^2\tau}.\frac{1}{\dfrac{1}{r_1^4}+\dfrac{1}{r_2^4}} \qquad (11)$$

$$\varepsilon = \frac{n\tau}{2R^2}.(r_1^2 + r_2^2) \qquad (12)$$

which depend on the sample size. We can assume that, for a homogeneous media, permeability and porosity are independent of the sample size unless the sample dimensions are smaller than those from the « Elementary Representative Volume » (Jacquin, 1989).

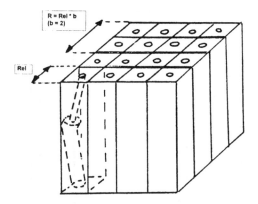

R = Rel * b
(b = 2)

Rel

Figure 2. Isolated parallelepipeds.

We isolate the factor n/R^2 that will be further referred to as C_1. It is the channels number per cross surface unit (perpendicular to the gas flow). It can be seen as a channel density by considering each channel isolated within a single squared base parallelepiped (side R_{el}). The situation is represented in Fig. 2.

Considering only one « elementary » parallelepiped, the channels density is :

$$C_1 = \frac{1}{R_{el}^2} \qquad (13)$$

Whereas if we consider a larger parallelepiped with $R = bR_{el}$, it contains b^2 channels and therefore:

$$C_1 = \frac{n}{R^2} = \frac{b^2}{(bR_{el})^2} = \frac{1}{R_{el}^2} \qquad (14)$$

The value of C_1 is the same in both expressions (13) and (14) which shows that the channel density is constant for a homogeneous media whatever the sample size is.

Equations (11) and (12) may be rewritten as:

$$k = \frac{C_1}{4\tau} \cdot \frac{1}{\frac{1}{r_1^4} + \frac{1}{r_2^4}} \qquad (15)$$

$$\varepsilon = \frac{C_1\tau}{2} \cdot (r_1^2 + r_2^2) \qquad (16)$$

Still we have to evaluate r_1 and r_2. By adjusting his diffusion model, Freijer (1994) found the following empirical relationship between r_1 and r_2:

$$r_2 = R.(\frac{R}{a} + \frac{R}{r_1})^{-m} \qquad (17)$$

where a and m are empirical parameters with: $m \leq 1$ and $a < R$
This implies $r_1 < r_2 < R$.

To get simpler expressions, we define the following dimensionless parameters :

$$r_1' = \frac{r_1}{R}; r_2' = \frac{r_2}{R}; a' = \frac{a}{R} \Rightarrow r_1' < r_2' < 1 \text{ and } a' < 1$$

Then: $r_2' = (\frac{1}{a'} + \frac{1}{r_1'})^{-m}$

Equations (15) and (16) may be rewritten as :

$$k = \frac{nR^2}{4\tau} \cdot \frac{1}{\frac{1}{(r_1')^4} + (\frac{1}{a'} + \frac{1}{r_1'})^{4m}} \qquad (18)$$

$$\varepsilon = \frac{n\tau}{2} \cdot ((r_1')^2 + (\frac{1}{a'} + \frac{1}{r_1'})^{-2m}) \qquad (19)$$

Considering only one elementary parallelepiped, R is replaced by R_{el} and the final expressions are:

$$k = \frac{R_{el}^2}{4\tau} \cdot \frac{1}{\frac{1}{(r_1')^4} + (\frac{1}{a'} + \frac{1}{r_1'})^{4m}} \qquad (20)$$

$$\varepsilon = \frac{\tau}{2} \cdot ((r_1')^2 + (\frac{1}{a'} + \frac{1}{r_1'})^{-2m}) \qquad (21)$$

These allow us to draw curves of permeability k versus porosity ε. To do so, we have to choose the values of a', m, R_{el} and τ. We consider then a range of values for r_1' in order to provide different values of porosity and the corresponding permeability values.

If the so obtained curve does not match the experimental results, another curve has to be drawn with another set of a', m, R_{el} and τ values.

As initially observed by Freijer for relative diffusion coefficient Q, we observe the following influence of the empirical parameters on k vs. ε curves:

- Parameter m influences the porosity value corresponding to permeability equal to zero.
- The curve shape (concave or convex) depends on parameter a.
- R_{el} influences the range of k values. This can be seen from the factor R_{el}^2 in (20). If R_{el} is expressed in 0.001 mm, k will straight be expressed in Darcy units.
- Tortuosity τ affects both values of k and ε.

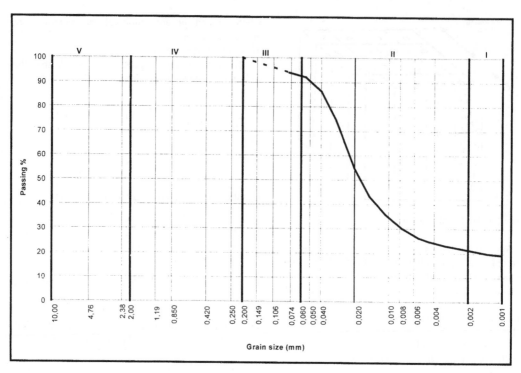

Figure 3. Silt grain size distribution curve.

2 EXPERIMENTAL RESULTS

2.1 Characteristics of the silty soil

The soil samples originate from the shallow quaternary layers (depth between 0.5 and 1 m) present in the area of Louvain-la-Neuve (30 km south of Brussels, Belgium). The grain size distribution curve is shown on Figure 3.

The following values of the investigated soil were obtained by the Laboratory of Civil Engineering Department at Louvain-la-Neuve:

- soil particles density: ρ_s=2694 kg/m³
- Standard Proctor test: w_{opt}=15.53% and $\gamma_{d,opt}$=17.56kN/m³
- Atterberg limits : w_p=21.5%; w_l=33.4%; I_p=11.9%
- Blaine value (ASTM C204-96a) : $\frac{S_s}{M_s}$=1654 cm^2/g. From this value, it is possible to calculate the specific soil particles surface :

$$U_o = \frac{S_s}{M_s} \cdot \frac{M_s}{V_s} = \frac{S_s}{M_s} \cdot \rho_s = 445,000 \ m^{-1}$$

In Casagrande's Soils Classification (1948), this soil is classified as CL (Lean Clay). According to GBMS (Groupement Belge de Mécanique des Sols) method (1994), this soil is described as a silt.

2.2 Experimental procedure

After being dried at a temperature of 105 °C during at least 12 hours (which does not affect the capillary water content), the soil sample is cooled down in a drier in order to avoid any water condensation. The sample is then reduced to a homogeneous grain powder using a hand roller.

The sample is compacted in the gas permeameter cell in different layers, each of them being compacted using a Proctor dam. Using different compaction energies enables to achieve different values of porosity.

Once the cell is filled, its weight allows determination of soil dry density and therefore porosity:

$$\varepsilon = \frac{V_v}{V} = 1 - \frac{V_s}{V} = 1 - \frac{\rho_d}{\rho_s}$$

The gas permeameter allows to measure the gas flow, the pressure at the inlet and the outlet, and the temperature. It is therefore equipped with a set of two ball flow meters, two Torricelli barometers (tubes filled with water) and a thermocouple.

Darcy's law for a compressible fluid through a porous media is:

$$q_m = \frac{k}{\mu} A \frac{\Delta p}{l} \qquad (22)$$

where :
- q_m is the gas flow (m³/s) expressed for the mean pressure in the sample and measured by means of a ball flow meter.
- k is the soil permeability (m² or expressed in Darcy with $1 Darcy = 0.987.10^{-12} m^2 \cong 10^{-12} m^2$)
- μ is the gas dynamic viscosity (Pa.s), depending on the gas temperature.
- A is the sectional area (m²)
- Δp is the pressure difference between the inlet and outlet of the cell (Pa)
- l is the sample length (m) in the flow direction.

During the gas permeability test, different pressure gradients (at least 5) are applied to the sample and the k value is obtained from linear regression between q_m and Δp. Regression coefficient r^2 is also determined.

The permeability value is therefore:

$$k = \mu (\frac{q_m}{\Delta p})_{regression} \cdot \frac{l}{A}$$

2.3 Experimental results

A total of 13 tests were performed with different values of compaction energy. Results are summarized in Table 1.

2.4 Application of Kozeny-Carman equation

In order to discuss the validity of Kozeny-Carman equation, values of k vs. factor $\varepsilon^3/(1-\varepsilon)^2$ are plotted on Fig. 4. It is possible to draw a linear regression with a quite good regression coefficient r² (equal to 0.97). The regression equation is as follows:

$$k = 1350 \frac{\varepsilon^3}{(1-\varepsilon)^2} - 215 \quad (\text{in mDarcy})$$

In opposition to Kozeny-Carman equation, this equation contains a constant term. Permeability becomes equal to zero for a positive value of porosity. This value can be calculated from the regression and is equal to:

$$\frac{\varepsilon^3}{(1-\varepsilon)^2} = 0.159 \Rightarrow \varepsilon = 0.390$$

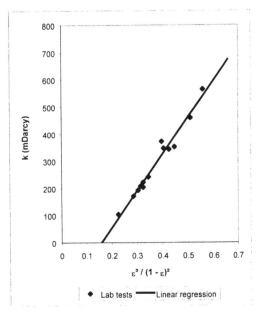

Figure 4. Comparison between experimental results and Kozeny-Carman equation.

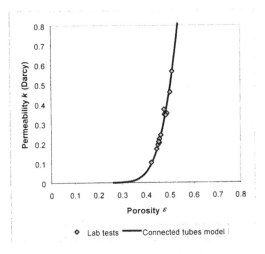

Figure 5. Comparison between experimental results and connected tubes model.

Kozeny-Carman equation is however not sufficient.

2.5 Application of connected tubes model

After different trials, the following set of empirical parameters provided a satisfactory matching with the experimental results:
- $\tau = 1.4$ which corresponds to a 45° angle between the channels and the flow direction.

This value is recommended by Carman.
- $a' = 0.55$
- $m = 0.16$
- $R_{el} = 0.015$mm

Comparison between the experimental values and the curve provided by the model is shown on Fig. 5. As expected, the model is able to predict that the permeability becomes zero for a positive value of porosity ε. Figure 5 shows that k approaches zero for a porosity value about 32 %.

A theoretical model is thus able to explain the experimental results. Taking into account two different tube radii allows getting closer to the true soil texture. One single narrow section along the channel path can greatly reduce the overall permeability.

Some values r_1, r_2, k and ε used to plot the curve are listed in Table 2. It can be seen that, unlike r_2, r_1 undergoes great changes. From this table, we can evaluate the pores dimensions.

The grain size distribution curve shows that d_{50} is about 0.018 mm and we know from Terzaghi's criteria that, for a uniform soil characterized by spherical grains with one unique diameter d, the pore opening is about $d/4..d/5$, which yields in this case 0.003 to 0.005 mm, thus confirming the r_1 range.

3 CONCLUSIONS

The Kozeny-Carman equation is inapplicable for silty soils. Pore blockage arises due to capillary water bonding between grains. The specific yield (percentage of void volume able to convey the gas flow) is greatly reduced when comparing silty to sandy soils. This can be seen from Eckis diagram on Fig. 6 (Monjoie, 1994).

The connected tubes model was initially developed by Freijer for gas diffusion. The authors adapted it to gas flow caused by a pressure gradient (advection). In this model, pores are represented by two connected tubes with different radii. Permeability vs. porosity curves can then be drawn and enable to predict that permeability is equal to zero for a positive value of porosity (in this case 32 %).

The drawn curves, however, depend on the empirical relationship between r_1 and r_2 that was found by Freijer. Further studies should then focus on that relationship and assess its physical meaning. It would then be possible to determine the model parameters a', m, R_{el} and τ from soil characteristic values like Atterberg limits and grain size distribution curve.

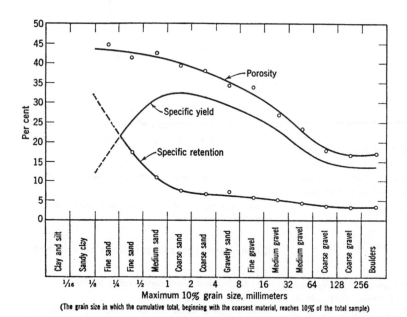

Figure 6. Porosity, specific yield and specific retention variations with grain size.

Table 1. Summary of gas permeameter tests.

Test	$r^2 (q_m, \Delta p)$	porosity ε	k (Darcy)	k (m²)
01.01.01	0.998	0.451	0.193	1.90E-13
01.01.02	0.998	0.511	0.565	5.58E-13
01.01.03	0.999	0.477	0.372	3.67E-13
01.01.05	0.999	0.457	0.204	2.01E-13
01.01.06	0.999	0.457	0.222	2.19E-13
01.01.07	0.999	0.445	0.171	1.69E-13
01.01.08	0.999	0.489	0.352	3.47E-13
01.01.09	0.998	0.479	0.347	3.42E-13
01.01.10	0.999	0.484	0.343	3.39E-13
01.01.11	1.000	0.502	0.460	4.54E-13
01.01.12	1.000	0.454	0.208	2.05E-13
01.01.13	0.996	0.422	0.103	1.02E-13
01.03.01	1.000	0.463	0.242	2.39E-13

Table 2: Numerical values corresponding to the connected tubes curve plotted on Figure 5

r_1 (0,001 mm)	r_2 (0,001 mm)	ε	k (Darcy)
0.8	9.2	0.263	0.000
1.5	10.1	0.325	0.004
3.0	11.0	0.407	0.064
3.8	11.3	0.442	0.155
4.5	11.5	0.477	0.318
5.3	11.7	0.513	0.580
7.5	12.1	0.631	2.189
8.3	12.2	0.675	3.041
9.0	12.3	0.721	4.042

ACKNOWLEDGMENTS

The authors wish to thank the staff of the Laboratory of Civil Engineering Department, especially Mr E. Bouchonville.

NOTATIONS

The units refer to SI.

A	sample cross section	(m²)
a, a', m	empirical parameters	(m,-,-)
d_{50}	mean grains diameter	(m)
h_k	Kozeny's constant	(-)
I_p	plasticity index	(%)
k	permeability	(m² or Darcy)
L, l	sample length	(m)
L_c	pore length	(m)
l_1, l_2	single tube length	(m)
M_s	soil particles mass	(kg)
n	channels number	(-)
p	pressure	(Pa)
q	volumetric flow	(m³.s⁻¹)

q_m	volumetric flow expressed for the mean pressure	(m³.s⁻¹)
R	sample radius	(m)
R_{el}	elementary parallelepiped side	(0.001 mm)
r^2	statistical linear regression coefficient	(-)
r_1, r_2	single tube radius	(m)
r_1', r_2'	single tube dimensionless radius	(-)
S_s	soil particles surface	(m²)
U_o	specific soil particles surface	(m⁻¹)
V	total volume	(m³)
V_s	soil particles volume	(m³)
V_v	void volume	(m³)
w_{opt}	Proctor water content	(%)
w_l	liquid limit	(%)
w_p	plastic limit	(%)
Δp	pressure gradient	(Pa)
ε	porosity	(%)
$\gamma_{d,opt}$	Proctor dry density	(kN/m³)
μ	fluid dynamic viscosity	(Pa.s)
ρ_d	density of dry soil	(kg.m⁻³)
ρ_s	density of solid particles	(kg.m⁻³)
τ	channel tortuosity	(-)

REFERENCES

Bousmar, D. and Moreau, Chr. 1995. *Contribution à l'étude des écoulements gazeux à travers un sol pulvérulent : approche théorique de la perméabilité, conception et expérimentation d'un perméamètre à gaz pour les sables*, Graduate thesis in Civil Engineering. Louvain-la-Neuve: Université Catholique de Louvain, 151 pages.

Casagrande, A. 1948. Classification and Identification of Soils. *Trans. ASCE* 113: 901.

Couvreur, J.-M. 1996. *Perméabilité des sols aux gaz : confrontation entre approches théorique et expérimentale appliquées aux cas de sols cohérents*, Graduate thesis in Civil Engineering. Louvain-la-Neuve: Université Catholique de Louvain, 129 pages.

Freijer, J. I. 1994. *Mineralization of Hydrocarbons and Gas Dynamics in Oil-contaminated Soils: Experiments and Modeling*, Thesis. Universiteit van Amsterdam.

GBMS/BGGF 1994. *Proposition de modification de la dénomination géotechnique des sols basée sur la classification de l'Institut Géotechnique de l'Etat*.

Jacquin, Ch. 1989. Structures des réseaux poreux et propriétés pétrophysiques des roches. *Revue Française de Géotechnique* 49 : 25-42.

Monjoie, A. 1994. *Hydrogéologie: Notes de cours*. Louvain-la-Neuve : SICI.

Van Brakel, J. and Heertjes, P.M. 1974. Analysis of Diffusion in Macroporous Media in terms of a Porosity, a Tortuosity and a Constrictivity factor. *Int. J. Heat Mass Transfer*.

3 Adsorption and desorption in clays

Clay Science for Engineering, Adachi & Fukue (eds) © 2001 Balkema, Rotterdam, ISBN 90 5809 175 9

Role of laboratory procedure on determination of heavy metal retention

R. N. Yong
Geoenvironmental Research Centre, University of Wales, Cardiff, UK

ABSTRACT: This study addresses the problem of laboratory determinations of the distribution of heavy metal pollutants retained in clay soils through various partitioning processes. The three principal laboratory methods examined are: selective sequential extractions (SSE), selective addition of soil fractions, and selective removal of soil fractions. The relevance of the results obtained are discussed.

1 INTRODUCTION

The distribution of heavy metals (HMs) retained in a soil mass, is a reflection of the sorption performance of each kind of soil fraction that comprises the total soil, i.e. the distributed HMs amongst the soil fractions is a direct function of the different sorption capabilities of the various soil fractions. A knowledge of the distribution of partitioned HMs is useful since it allows us to assess or determine: (a) the sorption potential of candidate soils based, (b) the fate and mobility of the sorbed HMs, and (c) the potential for removal of the sorbed HMs.

Common procedures available for determination of the distribution of sorbed HMs include: (a) SSE, selective sequential extraction techniques which can selectively remove the HMs from specific soil fractions, and (b) techniques relying on systematic inclusion or removal of soil fractions in heavy metal-soil interaction studies. This systematic inclusion technique can be identified as the SSFA (selective sequential fraction addition) technique, and the systematic removal of soil fraction technique is identified as the SSFR technique.

The nature of the tests is such that accountability for the role of natural soil structure on the retention properties of the soil is not always included in the results obtained. This is because the SSE test is generally conducted using procedures similar to those for batch equilibrium studies. While the SSFA test is a laboratory test which systematically "constructs" a soil sample to study its retention characteristics, assessment of the retention

characteristics of the individual soil fractions can be made in a fashion similar to the SSE procedure, or in a "sample-leaching" procedure. This will be discussed in detail in the next Section. The SSFR test may be seen to be a compromise between the two insofar as accountability for soil structural contribution to retention performance is concerned.

2 LABORATORY STUDY OF HM RETENTION

The basic idea of each of the tests (SSE, SSFA, SSFR) is: (a) to determine the extent of retention of heavy metals by each type of soil fraction, (b) to infer from the extraction procedures the "strength" of the bonds between the metals and soil fractions. Fig. 1 shows the basic elements of these tests.

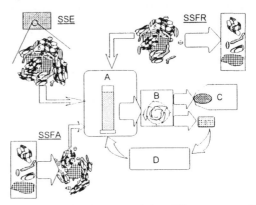

Fig. 1 - Basic elements of the different types of sequential extraction tests. (Adapted from Yong, 2000)

In the general technique for sequential extraction, the extractant is mixed with the soil sample and the soil solution formed will be buffered to the desired pH (item A in Fig. 1). Following equilibrium, the solution is centrifuged (item B) and the supernatant analyzed for heavy metals (item C), whilst the residue is washed with distilled water to ensure a "clean sample" is obtained for the next extractant treatment (item D). The cycle is repeated until all the soil fractions are examined for their individual retention characteristics.

It is the source of the original sample that distinguishes between the SSE, SSFR and SSFA types of test. The common thread throughout the extraction process begins with the soil solution and ends with the re-use of the cleaned residue for the next extraction step. The SSE technique seeks to determine the retention characteristics of individual soil fractions from selective destruction of the bonds between the heavy metals and the target soil fraction. The SSFR technique selectively removes target soil fractions before proceeding with the extraction tests, and finally, the SSFA procedure selectively adds soil fractions for contamination and then proceeds forthwith with the extraction procedure.

All tests are essentially conditioned to determine the optimum retention performance of the various soil fractions – a condition obtained by default because of the use of soil solutions. The choice of whether to use dispersants to obtain a fully dispersed state of the soil is a laboratory decision. Without the use of dispersants, all the available reactive surfaces of the soil fractions are involved in the reactions with the reagents used as extractants.

A considerable amount of insight into soil structure contribution and its influence on the retention characteristics of a soil can be obtained through systematic testing in a parallel scheme arrangement – as will be shown in the next set of discussions. The preceding having been said, it must be emphasized that the results obtained from all the tests are "operationally defined", and that certain laboratory-manufactured conditions are assumed to be insignificant in contributing to the results obtained. In particular, the use of the different extractants at various stages of the extraction procedure assumes that the reactive surfaces of the soil fractions are not altered or affected, i.e. retention of the unextracted HM by the other soil fractions remains "intact". The assumption that nothing changes is perhaps the most tenous of the assumptions.

3 COMPARISON OF RESULTS OBTAINED

Before embarking on the discussion concerning extraction of heavy metals from compact soil samples, it is useful to compare results obtained with the classical soil solution extraction procedure shown in Fig. 1. Available data for direct comparison is relatively scarce. The studies reported by Yong and MacDonald (1998) provide a direct comparison between the SSE and the SSFR types of results. This is shown in Fig. 2.

The procedure used by Yong and MacDonald (1998) for sequential removal of carbonates and amorphous oxides from the illitic soil tested consisted of the following:

(a) Extraction of soil carbonates using 1M NaOAc at pH 5 (adjusted with HOAc) at a soil:solution ratio of 40:1.

(B) Extraction of amorphous oxides using 0.1 mol/L oxalic acid, buffered to pH 3 by ammonium oxalate and mixed in the dark at the same soil:solution ratio.

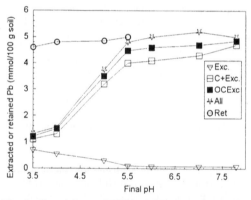

Fig. 2 - Comparison of SSE and SSFR results. Note C = carbonates, O = amorphous oxides, and "Ret" = retained by SSFR sample. (Adapted from Yong and MacDonald, 1998)

The results shown in Fig. 2 are very instructive. Considering the "Ret" curve first, it is noted that this is the SSFR sample where both the carbonates and amorphous oxides were removed using the methods described in the above. The Pb retained, or conversely, the Pb that can be extracted (if such were the case) is seen to be considerably larger at the lower pH values.

The standard SSE type of test results shown in Fig. 2 show that the amount of exchangeable ions extracted (identified as "Exc" in the graph) decreases from the high value at pH 3 to their lowest value at about pH 5.5. It appears that the largest amount of Pb is held by the carbonate soil fractions. The significant point of note is the large difference in Pb retained by the SSFR sample between the pH 3 and pH 5.5 range which is not accounted for in the SSE test.

The large difference seen in Pb retention in the SSFR results or extracted (SSE results), especially at the lower pH range where Pb exists in solution below its precipitation pH, raises some very interesting questions. Recognizing that both these methods of data gathering (SSFR and SSE) depend very highly on operational procedures and reagents used, the differences are nevertheless highly significant. Thus whilst the quantitative values obtained for each test are operationally defined, the value of data as seen from a comparative qualitative sense cannot be denied. To seek a better appreciation as to the possible cause of the differences, the SSFR results of Xing *et al.*, (1995) for three heavy metals are shown in Fig. 3.

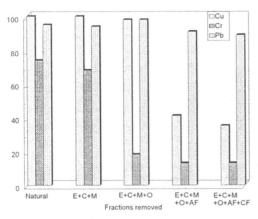

Fig. 3 - SSFR procedure used to determine heavy metals held by the various soil fractions. (Adapted from Xing *et al.*, 1995).

The legend used in Fig. 3 shows E = exchangeable, C = carbonates, M = manganese oxide, O = soil organics, AF = amorphous Fe and CF = structural Fe. The procedure used by Xing *et al.*, (1995) for removal of the various soil fractions was somewhat similar to that used by Yong and MacDonald (1998). For example, removal of carbonates was achieved by using 1 M NH_4AC-HAC at pH 5. To distinguish between the amorphous Fe and the structural Fe (in the layer lattice), 0.2M $(NH_4)_2C_2O_4$ + 0.2 M $H_2C_2O_4$ was

used to extract the amorphous Fe whilst structural Fe was removed by the same extractant but with 0.1 M ascorbic added to the reagent. Extraction of the Fe was conducted in the dark.

3.1 *Assessment of soil structure influence*

The results in Fig. 3 confirm the SSFR results obtained by Yong and MacDonald (1998) as shown in Fig 2, i.e. there appears to be very little differences in the retention of Pb with the both carbonates and amorphous oxides removed. However, the results shown by Xing et al., (1995) in respect to Cu and Cr show distinct differences when carbonates and oxides were removed. The significance of these results, and the ones portrayed in Fig. 2 has yet to be fully determined.

Several hypotheses can be advanced to explain the differences obtained in Figs. 2 and 3. To a very large extent, it is assumed that in reconstructing the soil after removal of a specific soil fraction, the resultant SSFR structure will be conditioned by "what is left" in the soil. Amorphous oxides and carbonates are known to provide coatings on soil particles and "bridges" between particles – dependent on the concentration of these soil fractions and on the pH environment. The assumption is made that in the absence of dispersants used to produce the soil solution, the different soil structures obtained in the SSFR constituted compact soil samples will be somewhat reflected in the soil solutions produced for the extraction procedure. The CEC (cation exchange capacity) and SSA (specific surface area) of the SSFR reconstituted soils reinforce the above speculation. Bearing in mind that the CEC and SSA tests are conducted on fully dispersed samples, a comparison of the values for natural and SSFR sample with carbonates and amorphous oxides removed show CEC values of 23 and 10 meq/100g soil for natural and SSFR soil, and SSA values of 131 and 116m^2/g for the natural and SSFR soils respectively.

Both SSE and SSFR types of extraction tests can be conducted on actual soil samples retrieved from contaminated sites. In that sense, one could argue that some of the original soil structure could participate in the production of test results. Unfortunately, there is no means for direct determination of residual soil structure and its relationship to measured results.

Unlike the SSE and SSRF types of tests, the SSFA test constructs its soil sample from individual soil fractions, e.g. specified clay minerals, laboratory-produced amorphous oxides and carbonates, etc. In that sense, the SSFA technique is more suitable for purposes of study of the contribution of specific soil fractions to the retention characteristics of soil. SSFA tests could also be used as calibration or benchmark tests.

4 THE CASE FOR COLUMN STUDIES

From the preceding, it is apparent that the role of soil structure in retention of heavy metal contaminants in soils cannot be directly determined from the extraction tests designated as SSE, SSFR and SSFA types of tests. Since the core of the problem appears to be the use of treatments involving soil solutions at one stage or another, it would appear that a method of test with soil columns using intact soil samples would be required.

The sorption characteristics as determined by batch equilibrium tests and by leaching column tests can be compared to provide one with some indication of the contribution of soil structure to the retention of contaminants. Fig. 4 shows the basic elements of the proposition which says that if equilibrium sorption of the contaminant is obtained in a leaching column test, this can be plotted on the same diagram with the adsorption isotherm of the same soil.

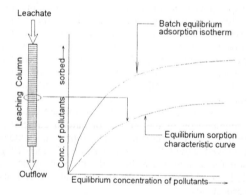

Fig. 4 - Sorption characteristic curves obtained from batch equilibrium and equilibrium leaching column tests.

At any one position in the leaching soil column, equilibrium sorption can be obtained when the sorption capacity of the soil is reached. At that time, any further transport of contaminants or pollutants in the leachate through the soil will not be partitioned, i.e. the contaminants or pollutants will remain in the leachate as it is transported through the soil. At that time, the "equilibrium sorption characteristic curve" shown in Fig. 4 will be obtained.

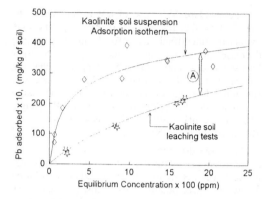

Fig. 5 - Pb sorption characteristic curves for kaolinite soil.

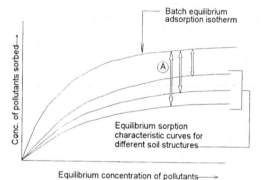

Fig. 6 - Influence of different soil structures on retention characteristics.

The adsorption isotherm obtained for the same soil and the same set of contaminants using batch equilibrium tests can also be represented on the same graph – as shown in Fig. 4. Since the adsorption isotherm represents sorption performance with a dispersed soil in a soil solution, we would expect a maximum sorption performance. However, since intact soil samples are used in the leaching column test, not all the reactive soil surfaces are available. Masking and changes of surface reactive forces due to coatings by oxides, and formations of peds and other microstructural features contribute to the significant changes in available reactive surfaces. The result is the "equilibrium sorption characteristic curve" obtained as shown in Fig. 4.

Actual test results for a kaolinite soil are shown in Fig. 5. The double-arrow shown next to "A" is drawn at the beginning of the region where the adsorption isotherm curve appears to be closely parallel to the leaching column sorption characteristic curve. This value "A" appears to be constant as one proceeds further to the right of the graph – as seen in the diagram.

One could argue that if "A" is reasonably constant, or if one could extend the tests so that the two curves would provide for a reasonably constant "A" value, this would represent the Pb retention "modification" due to the influence of soil structure. Thus if different soil structures for the same soil are obtained, as for example demonstrated as different densities, the results obtained would be shown as those represented in Fig. 6.

5 SOIL STRUCTURE AND SSA CHANGE

The results shown in Fig. 5 suggest very strongly that when equilibrium sorption is reached in column leaching tests, the reduction in sorption of Pb appears to be a constant proportion of the Pb sorbed in the batch equilibrium sorption tests. This presumes that an asymptotic performance in the adsorption isotherm is obtained. If such is the case, a strong case can be made for interpretation of the constant "A" value as a proportional reduction in the surface area presented to (i.e. interacting with) the Pb. In this particular instance, the results shown in Fig. 5 appears to suggest a reduction of about 25% of the surface area presented by the batch equilibrium tests.

If one assumes that the batch equilibrium test procedure provides total particle dispersion and therefore ALL surfaces exposed to interaction with Pb, this can be construed to mean that the surface area presented might perhaps be represented by the specific surface area SSA measured for example by the EGME method prescribed by Eltantaway and Arnold, (1973). If this assumption is carried further, it can be argued that the leaching sample used to provide the equilibrium sorption results shown in Fig. 5 would have a specific surface area roughly equal to 75% of the SSA . The effect of different soil structures obtained as a result of different soil densities (for the same soil), as represented in Fig. 6, can thus be evaluated in terms of the changes in the surface areas presented to an intruding contaminant or pollutant carried in a leachate stream.

6 CONCLUDING REMARKS

The realization that soil structure "must" somehow have some influence on the pollution-retention (i.e. contaminant-retention) characteristics of a soil has always been ever present. However, direct simple methods for determination of the influence of soil structure on sorption characteristics have not been readily available.

The use of extraction tests to provide information on retention of contaminants by the various soil fractions do not satisfy the requirements for analyses

of contributions from soil structure. The proposition that soil column leaching tests using intact soil samples can provide information on soil structure influences, is captivating. On the assumption that equilibrium sorption can be achieved in leaching column tests, i.e. sorption capacity is reached. it is argued that this will demonstrate an asymptotic behaviour similar to batch equilibrium tests that also reach total sorption capacity. At that time, the differences in sorption performance can be interpreted in respect to the influence of the soil structure of the intact soil sample in the leaching column.

Recognizing that all test results in batch equilibrium tests and in column leaching tests are operationally influenced and conditioned, it is still possible to use the results in a qualitative sense. The ramifications arising therefrom can be articulated as follows: (a) A means for "rough" determination of the reduction in SSA can be obtained , and (b) the role of soil structure in control of partitioning of contaminants and pollutants in leachate streams being transported through soil can be assessed.

Determination of the amount of contaminants held by soil fractions is fraught with difficulty because of the problems associated with "removal" of the contaminants from the soil fractions. At the one extreme, we have total digestion of a contaminated soil as a "final" means for release of the contaminants. Other less aggressive methods which seek to "de-bond" or "de-sorb" the contaminants from soil fractions face problems which range from insufficient or inadequate removal of the contaminants, to transformation and alteration of the surfaces of the soil fractions.

In addition to the above, the use of "removal agents", – i.e. extractants, leaching agents, solvents, surfactants, etc. – to obtain release of the contaminants from the soil fractions brings with it (them) the question of effectiveness of the removal agent to perform its function. It is almost virtually impossible to determine if complete removal has been obtained – without resorting to sample digestion.

One concludes therefore that most, if not all, of the laboratory methods designed to obtain an understanding or evaluation of the retention capacity or capability of soil fractions for contaminants must be used with the understanding that:

1. The test results obtained are a direct function of the test technique and procedures used, i.e. "operationally defined";

2. The results must not be used quantitatively. At best, the results can serve to provide some insight into the contaminant holding capability.

3. Qualitative comparisons made using the same technique for determination of contaminant holding capacity must also recognize that available soil particles' surfaces in laboratory tests are not necessarily reflective of those of the soil in the contraminated site.

REFERENCES

Eltantaway, I.N., & P.W. Arnold 1973. Reappraisal of ethylene glycol mono-ethylether (EGME) method for surface area estimation of clays. *J. Soil Sci.* 24:232-238.

Xing, G.X., Xu, L.Y. & W.H.Hou 1995. Role of amorphous Fe oxides in controlling retention of heavy metal elements in soils. In P.M.Huang, J.Berthelin, J.-M.Bollag, W.B.MCgill &A.L.Page (eds.) *Environmental impact of soil component interactions:* 2:63-73. Lewis Publishers.

Yong, R.N., & E.M. MacDonald 1998. Influence of pH, metal concentration, and soil component removal on retention of Pb and Cu by an illitic soil. In E.A.Jenne (ed) *Adsorption of metals by geomedia:* 229-253.Academic Press.

Yong, R.N. 2000. *Contaminated soils: Pollutant fate and remediation.* Lewis Publishers.

Clay Science for Engineering, Adachi & Fukue (eds) © 2001 Balkema, Rotterdam, ISBN 90 5809 175 9

The use of selective extraction procedures for soil remediation

C. N. Mulligan
Concordia University, Montreal, Que., Canada

R. N. Yong
Geoenvironmental Research Centre, University of Wales, Cardiff, UK

B. F. Gibbs
Bivan Consultants, Montreal, Que., Canada

ABSTRACT: The interaction of contaminants with soil is a very complex phenomenon and means are required to understand this matter more thoroughly. To determine the speciation of metals in soils (the distribution of elements among chemical forms or species), specific extracts are used in a process called sequential selective extraction. This method can be used to determine if heavy metals are able to be removed by remediation techniques or to predict removal efficiencies. In this paper, we show that sequential extraction can be employed for the evaluation of the most appropriate soil remediation technology and for monitoring remediation procedures. Exchangeable, carbonate, reducible oxide and organic fractions are amenable to soil washing techniques and that residually bound contaminants are not economical or feasible to remove. This information is important in designing the most appropriate conditions for soil washing.

1 INTRODUCTION

The term speciation is related to the distribution of an element among chemical forms or species. Heavy metals can occur in several forms in water and soils. Interest has increased in these techniques to relate the degree of mobility with risk assessment, (i.e., the more mobile the metal is, the more risk associated with it (Bourg 1995)). Not only is total metal concentration of interest, but it is now accepted that understanding the environmental behaviour by determining its speciation is of paramount importance.

To determine the speciation of metals in soils, specific extractants are used. By sequentially extracting with solutions of increasing strengths, a more precise evaluation of the different fractions can be obtained (Tessier et al. 1984). A soil or sediment sample is shaken over time with an extractant, centrifuged and the supernatant is removed by decantation. The residue is washed in water and the supernatant removed and combined with the previous supernatant. A sequence of reagents are used following the same procedure until, finally, mineral acid is used to extract the residual fraction. Heavy metal concentrations are then determined in the various extracts by atomic absorption or other means. Numerous techniques and reagents have been developed and have been applied to soils (Shuman 1985), sediments (Lum & Edgar 1983),

and sludge-treated soils (Petrozelli et al. 1983). These methods are not standardized and even the results can vary with the same reagents, pH, temperature, extractant strength and solid to volume of extractant ratio. None of the extractions is completely specific, however the extractants are chosen in an attempt to minimize solubilization of other fractions.

To extract the exchangeable fraction, ammonium acetate, barium chloride or magnesium chloride at pH 7.0 is generally used (Lake 1987). They cause the displacement of the ions in the soil or sediment matrix bound by electrostatic attraction. Pickering (1986) showed that magnesium chloride leached low quantities of other sulfides, organic matter, and ions of aluminum and silica.

The carbonate phase (calcite and dolomite) is extracted at pH. 5.0 with sodium acetate acidified with acetic acid. This solubilizes the carbonates, releasing carbonate-entrapped metals. Organic matter, oxides or clay components are not solubilized.

The reducible phase (iron and manganese oxides) is extracted with hydroxylamine hydrochloride with acetic acid at pH 2.0. The hydroxylamine hydrochloride reduces the ferrous and manganese hydroxides to soluble forms. Other components such as organic matter and clay components are not solubilized to any great extent (Tessier et al. 1979).

Hot hydrogen peroxide in nitric acid is used to oxidize the organic matter. The oxidized organic matter then releases metals that are complexed, adsorbed and chelated. These agents are used so that the silicates are not affected by this treatment. In the final step, the silicates and other materials are dissolved by strong acids at high temperatures. This fraction is usually used to complete the mass balances for the metals.

Kabata-Pendias (1992) demonstrated that the speciation of trace metals in natural soils depends on the physical and chemical characteristics of the soil. Soil pH, redox, organic, carbonate, clay and oxide contents all influence metal speciation and mobility. Simple and complex cations are the most mobile, exchangeable cations in organic and inorganic complexes are of medium mobility and, chelated cations are slightly mobile. Metals in organic or mineral particles are only mobile after decomposition or weathering and precipitated metals are mobile under dissolution conditions (e.g., change in pH). Kabata-Pendias (1992) also showed the speciation of trace metals such as zinc, copper, cadmium and lead. Zinc and cadmium are mostly organically bound, exchangeable and water soluble. Copper is mainly organically bound and exchangeable, whereas, lead is slightly mobile and bound to the residual fraction. Chlopecka (1992) showed, however, that the cadmium and zinc speciation of the soils depended significantly on the application of sewage sludge on the soil due to the metal speciation in the sludge. Water and air pollution can also effect speciation.

Recently, sequential extraction techniques have been studied as a tool in various applications. Yong et al. (1993) examined sequential extraction to obtain a better appreciation of the ability of clay soil barriers to contain contaminants in landfill barriers. The effect of soil pH, constituents and heavy metal types were evaluated. In a study by Ramos et al. (1994), sequential extraction techniques were used to evaluate the mobility of cadmium, zinc, lead and copper in contaminated soil in a national park. Cadmium was found to be the most mobile and would likely be the most bioavailable.

Ravishankar et al. (1994) evaluated several sludges for Al, Cu, Fe, Mn and Zn speciation to predict leaching processes. They concluded that more stabilized sludges contained higher contents of organically bound metals and that sludges vary considerably making leaching prediction difficult.

A potential method to evaluate if heavy metals can be removed by remediation techniques or predict removal efficiencies is to determine speciation with selective extractive techniques. It is believed that exchangeable, carbonate and reducible oxide fraction may be amenable to soil washing techniques (Li et al. 1995). Removal of organically and residually bound fractions may not be economical to recover or necessary due to lack of

bioavailability. Gombert et al. (1994) used sequential extraction to determine if cesium, cobalt and chromium could be removed by soil washing. Since less than 20% was extracted after dissolving 20% of the soil mass, soil washing was abandoned as an option. Clearly, more work is needed in this field to correlate the use of sequential extraction of the soil with the appropriate remediation techniques.

In this paper, we have examined the use of sequential extraction procedures for soil contaminated with oil and heavy metals to initially characterize the soil. The sequential extractions were then performed after each wash up to five washes with three types of microbial surfactants. These surfactants were chosen since they are anionic, capable of removing metals and hydrocarbons (Mulligan et al. 1999), and are less toxic and more biodegradable than most synthetic ones available. We have shown that sequential extraction on the washed soils enables determination of the types of metal removed during the washing step. Different washing conditions and the results of the sequential extractions under the various conditions are presented in this paper. This information can then be used to design washing treatments based on the metal speciation in the soil.

2 EXPERIMENTAL PROCEDURES

2.1 Contaminated soil

A hydrocarbon and metal contaminated soil was obtained from a harbour area which contained numerous leaking underground soil storage tanks from refineries, oil terminals, coal storage and processing industries which released fly ash with elevated levels of lead, zinc and copper. It was characterized to determine oil and grease and heavy metal contents (APHA 1995), cation exchange capacity (Chhabra et al. 1975), grain size distribution (ASTM 1970), amorphous content (Segalen 1968) and carbonate content (Hesse 1971). X-ray diffraction was also performed according to the procedures of Starkey et al. (1984). The grain size distribution of the soil indicated a sandy soil. Approximately 74.5% was finer than 2 mm and only 5% passed through a 0.075 mm sieve (10% silt and 90% sand) X-ray analysis indicated the presence of quartz (30%), feldspar (36%), illite (2%), kaolinite (27%), chlorite (3%) and carbonate (0.5%). Other results are summarized in Table 1.

2.2 Biosurfactants

Three types of biological surfactants were evaluated including surfactin from *Bacillus subtilis* ATCC 21332, rhamnolipids from *Pseudomonas aeruginosa* ATCC 9027 and sophorolipids from *Torulopsis bombicola* ATCC 2214 as described in Mulligan et al. (1999).

Table 1. Characterization of soil sample[1]

Parameter	Conc. (mg/kg) unless specified
Moisture content (%)	6.6
Inorganic carbon as carbonate (%)	0.5
Al_2O_3 (%)	1.8
Boron	12
Cadmium	3
Calcium (%)	9.6
Copper	414
Fe_2O_3 (%)	9.0
Magnesium	8000
Manganese	534
Molybdenum	8
Nickel	71
Lead	320
Potassium	1650
SiO_2 (%)	25.1
Sodium	5900
Zinc	870
Cation exchange capacity	7.1 meq/100g (pH 7)
Total oil and grease	12.6%

[1] Mulligan et al. (1999)

2.3 Soil washing procedures

Batch soil washing procedures were performed in centrifuge tubes containing 10:1 wt/wt solution to soil ratios and varying surfactant concentrations and pH values (Ellis et al. 1988). The three biosurfactants were evaluated to determine their capabilities in removing heavy metals from the soil. Controls included the same additives as for the biosurfactant studies without the presence of the biosurfactant. The series of washings was performed by washing the soil for 24 h, removing the supernatants (5,000 x g, 10 min) and adding fresh surfactant solution. The procedure was repeated five times. The supernatants were analyzed for metal concentration with an atomic absorption spectrophotometer according to standard methods (APHA, 1995). The percentage metal removal was determined.

2.4 Sequential extraction procedures

The procedure used for the sequential extraction was similar to that of Yong et al. (1993). Soil samples (1.5 g) were washed with the surfactant solutions and controls and then subsequently dried prior to sequential extraction. Each of the fractions were collected and the concentrations of heavy metals were determined in each of the fractions by atomic absorption spectrometry. The amounts of metals extracted from each of the extractants were then calculated.

3 RESULTS

3.1 Soil washing

A series of washings was performed on the soil using 0.25% surfactin and 1% NaOH by removing the washing solution each day and replacing it with a fresh solution. Copper (Figure 1) and zinc (Figure 2) were followed each day for 5 days. The control was 1% NaOH. For copper, the control showed a final cumulative removal of 20% while approximately 70% was removed by the surfactin. For zinc, removal rates were lower than for copper, 10% for the control and 25% for surfactin. For both metals, the rates of removal were the highest up to three days. Therefore, multiple washings with surfactin (with 1% NaOH) were highly beneficial for copper, with less success for zinc.

A series of washes on the soil was performed using 0.1% rhamnolipid and 1% NaOH, in a similar manner to the surfactin experiments. The control consisted of 1% NaOH. Copper (Figure 1), and zinc (Figure 2) removal were followed for five days. Approximately, 38% of the copper was removed after the five washes compared to 20% for the control. Total zinc removal after the five washes was 15% compared to the 11% for the control. In all cases, the levels of removal continued to increase with each wash. This would indicate that more heavy metals could be removed after further washes with the rhamnolipid and surfactin.

Figures 1 and 2 show the removal of copper, and zinc after a series of washes of soil for 5 days. For copper, the combination of 4% sophorolipid with 0.7% HCl showed total removal of 100%. In addition, clearly the acid alone showed inferior results. The total copper removal for the acid was 40% but this did not occur until the fifth wash. By this time enough acid had contacted the soil to destroy the structure of the soil and solubilize copper. Zinc removal was highest for the acid with sophorolipid combination. The latter reached 100% removal after 4 washes compared to 5 washes for the acid alone where 80% removal was obtained.

Table 2. Heavy metal affinity for various fractions in the soil

Metal	Affinity for each fraction
Cu	Organic>residual>oxide>carbonate>exchangeable
Zn	Oxide>carbonate>organic>residual>exchangeable

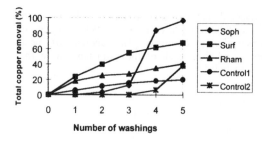

Figure 1. Cumulative copper removal after five washings using different surfactants and controls. Control 1: 1% NaOH ; 0.1% surfactin/1% NaOH; rham: 0.1% rhamnolipid/1% NaOH; control 2: 0.7% HCl; soph: 4% sophorolipid/0.7% HCl.

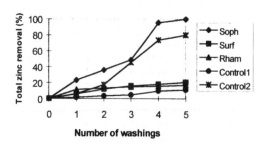

Figure 2. Cumulative zinc removal after five washings using different surfactants and controls. Control 1: 1%; NaOH: 0.1% surfactin/1% NaOH; rham: 0.1% rhamnolipid/1% NaOH; control 2: 0.7% HCl; soph: 4% sophorolipid/0.7% HCl.

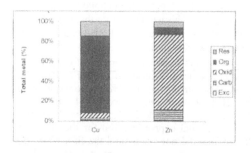

Figure 3. Sequential extraction of the soil. The following are represented: Exc stands for exchangeable, Carb for carbonate fraction, Oxid for oxide fraction, Org for organic fraction and Res for residual fraction.

3.2 Sequential extraction of the soil

Sequential extractions were then performed on the soil without washing or pH adjustment prior to the procedure. As can be seen from Figure 3 for the soil, the exchangeable fractions of copper and zinc were negligible. The carbonate fraction was significant

only for zinc, whereas the oxide fractions accounted for over 50 and 70% of the lead and zinc present in the soil, respectively. The organic fraction constituted over 70% of the copper. Weight losses after the washes to remove the oxide, carbonate and organic fractions correlated within 5% of the oxide, carbonate and organic contents as determined in Table 1. Residual amounts made up about 20% of the copper and approximately 10% of the zinc. The affinities for each of the fraction for copper and zinc are shown in Table 2.

3.3 Sequential extraction in combination with soil washing

Sequential extraction experiments were performed following soil washing of selected samples to determine which fractions were removed by the surfactants and other additives. The pH of the soil was not adjusted prior to sequential extraction. The fraction removed by the surfactant or control was designated as the soluble fraction. Soil residues from several soil washing experiments that showed significant metal removal results were chosen for study.

Experiments that were performed with the soil were used for sequential extraction after washing with surfactin (Figs 4 and 5). Figure 4 shows the sequential extraction of copper with 2% surfactin with 1% NaOH after one washing and for the 0.25% surfactin with 1% NaOH after one, three and five washes. The controls (1% NaOH) are also shown after one and five washes. The oxide, carbonate and exchangeable fractions of copper were fairly consistent until the fifth washing where the oxide fraction was finally removed. The organic fraction , however, decreases with increasing copper removal. For zinc (Fig. 5), the carbonate and organic fractions decreased insignificantly. Removal was mainly from the oxide fraction.

Other sequential extractions were performed with the residue after a single washing with 4% rhamnolipid (with and without 1% NaOH) and multiple washings with 0.1% rhamnolipid (with NaOH) using the soil (Figs. 6 and 7). Figure 6 shows the removal extraction of copper from the various fractions. The combination of 4% rhamnolipid decreased the copper in the organic phase more than the rhamnolipid or NaOH alone. There was also some reduction in the carbonate fraction. After washing five times with the NaOH alone or 0.1% rhamnolipid with NaOH, there was a substantial reduction in the copper in the organic fraction. In the presence of the biosurfactant, additional removal was seen in the carbonate and oxide fractions.

For zinc (Fig. 7), significant solubilization occurred for the single washing with 0.1 % rhamnolipid and 0.1% rhamnolipid with 1% NaOH. Some reduction occurred for the zinc in the organic fraction with rhamnolipid and NaOH. The zinc oxide fraction decreased in all cases. Reduction of the zinc carbonate fraction was also noted for the 0.1% rhamnolipid with NaOH and the NaOH alone after the five washings. Copper removal by the sophorolipid was not examined since washing with this surfactant did not give significant results.

Figure 8 shows the results of the sequential extraction after one and five washings with 0.7% HCl or 4% sophorolipid with 0.7% HCl . For the single wash, the addition of the sophorolipid removed more zinc from the carbonate fraction than the acid. The oxide fraction decreased by the same amount in both cases from the initial 70% to 60%. Whereas washing five times with the acid decreased the carbonate and oxide fractions, the combination of the surfactant and the acid almost completely eliminated both of these fractions.

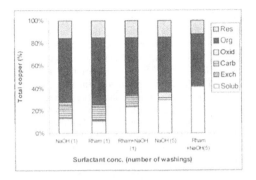

Figure 6. Sequential extraction of copper from soil after rhamnolipid washing. The following are represented: Sol for the soluble fraction, Exc stands for exchangeable, Carb for carbonate fraction, Oxid for oxide fraction, Org for organic fraction and Res for residual fraction.

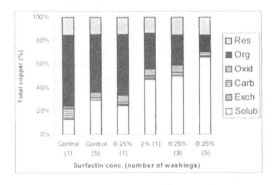

Figure 4. Sequential extraction of copper after surfactin washing of soil. The following are represented: Sol for the soluble fraction, Exc stands for exchangeable, Carb for carbonate fraction, Oxid for oxide fraction, Org for organic fraction and Res for residual fraction.

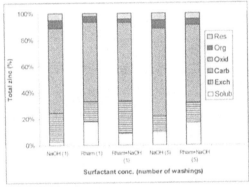

Figure 7. Sequential extraction of the soil for zinc after rhamnolipid washing. The following are represented: Sol for the soluble fraction, Exc stands for exchangeable, Carb for carbonate fraction, Oxid for oxide fraction, Org for organic fraction and Res for residual fraction.

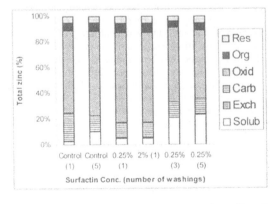

Figure 5. Sequential extraction of zinc after surfactin washing of the soil. The following are represented: Sol for the soluble fraction, Exc stands for exchangeable, Carb for carbonate fraction, Oxid for oxide fraction, Org for organic fraction and Res for residual fraction.

The purpose of the sequential extraction studies was to determine where the metals are present in the soil among the exchangeable, oxide, carbonate, organic and residual fractions. With this information it will then be possible to determine how difficult it will be to remove the metals. The exchangeable metal forms are the easiest to remove while the residual are the most difficult. Very strong acids are required to remove the latter metals.

Few studies have been performed on the sequential extraction of soils and sediments contaminated with copper and zinc. Cesium, cobalt and uranium fractions were examined by Gombertb II (1994). Others such as the study by Yong et al. (1993) have determined the distribution of zinc, lead and copper in kaolinite, illite, montmorillonite and clay soils which have organic contents of less than 2%. Despite the high organic contents (greater than 10%) of the soil used in this study comparisons can be made with illite and natural clay soil (pH 7) for zinc since the organic phase does not play a large role in the retention of these metals. For the soil, the results for zinc, were similar, indicating preference for the oxide phase followed by the carbonate. Both the natural clay and the illite, showed equal contributions of the oxide and carbonate fractions but a higher exchangeable fraction. In this study, the carbonate content of the soil was much lower (0.5%) compared to the contents of 15% and 10% for the illite and natural clay soil, respectively.

In a study by Ramos et al. (1994), copper, lead, cadmium and zinc were sequentially extracted from soils with organic contents of approximately 15%. Similar to this research, they found that copper was associated with the organic matter or residual fractions, and zinc was primarily found in the oxide fraction.

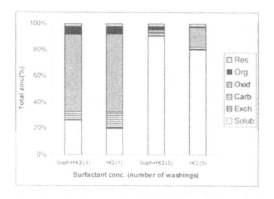

Figure 8. Sequential extraction of the soil for zinc after sophorolipid washing. The following are represented: Sol for the soluble fraction Exc stands for exchangeable, Carb for carbonate fraction, Oxid for oxide fraction, Org for organic fraction and Res for residual fraction.

Organic matter is an important factor in the sorption of metals (Ellis 1973, Hodgson 1963). Humic acid, one of the components of organic matter, is soluble in alkali only, precipitates in acids and is darker and of higher molecular weight than the other components. It has been used for its surfactant ability and is a known chelator. Chelation (Keeney& Wildung 1977) is a type of complexation that occurs between a metal ion and an organic ligand resulting in the formation of two or more bonds between the two in a heterocyclic ring. Humic acids have high charge density due to their acidic functional groups (carboxyl, hydroxyl, phenolic, carbonyl and enolic). This characteristic leads to a strong affinity for cations. They account for most of

the metal immobilization in soil organic matter. With the present sequential extraction procedure, however, it is impossible to distinguish between the contribution of organic matter and petroleum components to metal retention in soils since oxidation is used to destroy both these fractions.

The purpose of our sequential extraction studies on soils that have been previously washed with the surfactants was to determine from what fraction these surfactants are removing the metals. This information can then be used to determine if soil washing is useful and if surfactant use is potentially effective and what surfactants might be useful and under what conditions. For example, if the metals are retained mainly in the residual fraction, it may be very difficult, if not impossible to use any of these surfactants to remove these metals.

Few attempts have been made to correlate sequential extraction results with soil washing results. Gombert II (1994) examined the sequential extraction of cesium, cobalt and chromium to determine if soil washing could be used to treat the soil. It was determined that any fraction other than the residual could be removed by chemically enhanced soil washing. Therefore, cesium could not be removed since it occurred mainly in the residual phase (80%). However, no detail is given for what types of additives can be used for effective soil washing from the different phases.

In this research, only the residual fraction was untouched during soil washing procedures. All fractions under different conditions could be decreased. For example, under acidic conditions, the oxide phase containing the zinc would be released. However, under basic conditions (co-addition of the surfactants with NaOH), copper could be released since it was found mainly in the organic phase.

Both surfactin and rhamnolipid were able to remove copper more easily than the other heavy metals. As seen from the sequential extraction studies, the copper was found mainly in the organic fraction in the soil and sediment samples. Usually oxidative conditions are required to release metals in this form. Copper has been found to bind strongly to organic matter by forming complexes (Cameron 1992). The use of a surfactant and hydroxide is a combination that has been used for enhanced oil recovery known as alkaline flooding. The presence of acidic components in the petroleum or organic matter such as carboxylic acid, phenolics and asphaltene can hydrolyze under basic conditions to form surfactant products (de Zabala et al. 1982, Peru & Lorenz 1990). These surfactants are thus negatively charged and have been postulated to be responsible for enhanced oil recovery and probably contribute to metal removal.

The combination of the added surfactants and the base (whether it is NaOH or sodium bicarbonate) in this study has been shown to enhance copper removal and is therefore a good strategy to remediate soils containing copper in the organic fraction. Zinc also forms complexes with organic matter (Cameron 1992) and some removal from the organic fraction was also noted using the biosurfactants with NaOH. The initial content of zinc in the organic matter in the soil was significantly less (10 and 15% for zinc compared to 80 and 85% for copper in soil and sediments, respectively).

Zinc was associated with carbonates in the soil. Illite, a component shown in the X-ray diffractograms of the soil contains carbonates (Yong et al. 1993) that retain zinc in this fraction. As the pH decreases from 7, the amount in the carbonate phase is dissolved (particularly at pH below 5). In this study, addition of 0.7% HCl enhanced the dissolution since the pH was lowered to 5.5. Addition of sophorolipid with acid removed more zinc from the carbonate occurred even though there was not a significant difference in pH. The sophorolipid most likely enhanced the solubility of this fraction. Multiple washings were required to totally remove the zinc associated with this fraction which indicates that this fraction is hard to remove. Stover et al. (1976) indicated that zinc carbonates are difficult to dissolve. Surfactin (pH 9) and rhamnolipid (pH 6.5) seem to enhance the solubility of the carbonate phase of zinc from the sediments. High pH conditions after multiple washings can also decrease copper in the carbonate fraction.

The oxide fraction was particularly important for zinc retention by the soil. Cameron (1992) indicated that zinc is often associated with iron and manganese oxides. The sequential extraction procedure, however, cannot distinguish between oxides, oxyhydroxides or hydroxides but the latter two forms are the most probable species (Felmy et al. 1983). Yong et al. (1993) indicated that zinc was retained in the oxide fraction at pH above 5. Minimum solubilities for zinc are at pH 9.0 (Radha Krishnan 1993). This is why high pH conditions removed small amounts of zinc even after five washes.

Another study by Li et al. (1995) examined the use of the sequential extraction procedure with EDTA extraction of metals. They concluded that the organic phase was very stable since metals in this phase could not be removed by the EDTA and should not be considered in soil washing processes since it would be uneconomical to treat. As seen here, however, the organic phase associated metals can be easily removed using the appropriate conditions.

Metals within the residual fractions are generally considered stable and can only be removed with very harsh conditions such as strong acids at high temperatures. This fraction should be considered as background values when developing criteria for cleanup as it is not affected by any of the treatments used in this project. In the soil, the residual fraction included 79 mg/kg copper and 44 mg/kg zinc. These levels are comparable to normal levels of copper and zinc which are 20 mg/kg, and 10 to 300 mg/kg, respectively (Cameron 1992).

4 CONCLUSIONS

These results show that the organic phase-associated metals can be removed by either surfactin or rhamnolipid with sodium hydroxide. Acidic conditions with sophorolipid addition were effective for removing the zinc in the oxide and carbonate phases. Multiple washes are effective since they can remove the easier to remove phases first and then the more difficult (such as the oxide and then carbonate fractions). Residual fractions, the most difficult to remove, were not affected during the surfactant washing studies. This information is important in designing the appropriate conditions for soil washing.

REFERENCES

American Public Health Association (APHA) American Water Work Association (AWWA) Water Pollution Control Federation (WPCF) 1995. *Standard Methods for the Examination of Water and Wastewater.* 17th Edition.

American Society of Testing Materials. 1970. Special procedure for testing soil and rocks for engineering purposes. (D422-54). In *1970 Annual Book of ASTM Standards.*101-103, Philadelphia: ASTM.

Bourg, A.C.M. 1995. Speciation of heavy metals and implications for their mobility. *Heavy Metals:* 19 -32 Berlin: Springer.

Cameron, R.E. 1992. *Guide to Site and Soil Description for Hazardous Waste Site Characterization. Volume 1: Metals.* Environmental Protection Agency EPA/600/4-91/029.

Chhabra, R., J. Pleysier. & A. Cremers. 1975. The measurement of the cation exchange capacity and exchangeable cations in soil: A new method. *Proceedings of the International Clay Conference,* 439-448. Illinois: Applied Publishing Ltd., Illinois, USA.

Chlopecka. A. 1993. Forms of trace metals from inorganic sources in soils and amounts found in spring barley. *Water, Air and Soil Pollution,* 69: 127-133.

de Zabala, E.,J. Vislocky, E. Rubin. & C. Radke. 1982. A chemical theory for linear alkaline flooding. *Society of Petroleum Engineering Journal 5 :* 245-258.

Ellis, B.G. 1973. The soil as a chemical filter, In W.E. Sopper and L.T. Kardos (eds), *Recycling treated municipal wastewater and sludge through forest and cropland:* 46-70, University Park, PA: Penn State Press.

Ellis, W.D., J.R. Payne. & G.D. McNabb. 1985. *Treatment of Contaminated Soils with Aqueous Surfactants,* U.S. EPA No. EPA/600/2-85/129 .

Felmy, A.R., D. Girvan & A. Jenne. 1983. *MINTEQ: A computer program for calculating aqueous equilibrium. Report,* USEPA. Washington.

Gombert II, D. 1994. Treatability testing to evaluate what can work, *Nuclear Technology.* 108: 90-99.

Hesse, P.R. 1971. *A Textbook of Soil Chemical Analysis.* 519, London: William Clows and Sons.

Hodgson, J.F. 1963. Chemistry of the micronutrient elements in soils. *Advances in Agronomy.* 15: 119-159.

Kabata-Pendias, A. 1992. Trace metals in soils in Poland - occurence and behaviour. *Trace Substances in Environ. Health.* 25: 53-70.

Keeney, D.R., & R.E. Wildung. 1977.Chemical properties of soils, In *Soils for Management of Organic Wastes and Wastewaters,* ASA, CSSA, SSSA, Madison, WI.

Lake, D.L. 1987. Chemical speciation of heavy metals in sewage sludge and related matrices. In J.N. Lester. (ed.) *Heavy Metals in Wastewater and Sludge Treatment Processes. Volume I: Sources, Analysis and Legislation* 125-153, Boca Raton: CRC Press.

Li, W., R.W. Peters, M.D. Brewster, & G.A. Miller. 1995. Sequential extraction evaluation of heavy-metal contaminated soil: How clean is clean? *Proceedings of the Air and Waste Management Association., 88th Annual Meeting and Exhibition,* San Antonio, Texas, June 18-23.

Lum, K.R. & D.C. Edgar. 1983. Determination of chemical forms of cadmium and silver in sediments by Zeeman effect flame atomic-absorption spectrometry. *Analyst* 108: 918-924.

Mulligan, C.N., R.N. Yong & B.F Gibbs.1999. On the use of biosurfactants for the removal of heavy metals from oil-contaminated soil. *Environmental Progress,* 18: 50-54.

Pickering, W.F. 1986. Metal ion speciation - soils and sediments (a review). *Ore Geology Review.* 1: 83-146.

Petrozelli, G., G. Giudi, & L. Lubrano. 1983. *Proc. Int. Conf. Heavy Metals in the Environment,* 475.

Peru, D. & P. Lorenz. 1990. Surfactant-enhanced low-pH alkaline flooding, *SPE Reservoir Engineering* 5(3): 327-332.

Radha Krishnan, E. 1993. *Recovery of Metals From Sludges and Wastewaters.* Park Ridge, NJ.: Noyes Data Corp.

Ramos, L., L.M. Hernandez, & M.J. Gonzalez. 1994. Sequential fractionation of copper, lead, cadmium and zinc in soils from or near Donana National Park. *Journal Environmental Quality.* 23: 50-57.

Ravishankar, B.R., J.-C. Auclair, & R.D.Tyagi. 1994, Partitioning of heavy metals in some Quebec municipal sludges. *Water Pollution Research Journal of Canada.* 29: 457-470.

Richard, F.C.& A.C. Bourg, 1991. Aqueous geochemistry of chromium: A review. *Water Research Journal.* 25: 807-816.

Segalen, P. 1986. Note sur une methode de determination des produits mineraux amorphes dans certains sols a hydroxides tropicaux. *Cahier ORDSTOM. Serie Pedologie.* 6: 105-126.

Shuman, L.M. 1985. Fractionation method for soil microelements. *Soil Science.* 140: 11-22.

Starkey, H.C., P.D. Blackmon & P.L. Hauff. 1984. The routing minerological analysis of clay bearing samples. *United States Geological Survey Bulletin 1563:* 2-18.

Tessier, A., P.G.C. Campbell & M. Bisson. 1979. Sequential extraction procedure for the speciation of particulate trace metals. *Analytical Chemistry.* 51: 844-851.

Yong, R.N., R. Galvez-Cloutier & Y. Phadingchewit. 1993. Selective extraction analysis of heavy-metal retention in soil, *Canadian Geotechnical Journal* 30: 834-847.

Clay Science for Engineering, Adachi & Fukue (eds) © 2001 Balkema, Rotterdam, ISBN 90 5809 175 9

Surface complexes of Cd adsorbed on montmorillonite

R. Takamatsu – *Faculty of Veterinary Medicine and Animal Sciences, Kitasato University, Towada, Japan*

T. Miyazaki – *Graduate School of Agricultural and Life Sciences, University of Tokyo, Japan*

K. Asakura – *Catalysis Research Center, Hokkaido University, Japan*

M. Nakano – *Department of Agriculture, Kobe University, Japan*

ABSTRACT: It is important to know the behavior of Cd adsorption on clay-water interfaces in order to understand its pollution mechanism in soils. We investigated the pH dependence of Cd surface complex formation by combining an adsorption experiment with EXAFS. In the adsorption experiment, we found several index pH values which exhibit different structure of Cd surface complexes on montmorillonite. Cd K-edge EXAFS spectra at representative pH values proved that the local atomic structures surrounding adsorbed Cd are unique. For the Fourier transformation of EXAFS oscillation for clay samples at pH3.2 and pH4.8, no atom appeared in the second neighbor atom position as occurs in a Cd solution of $Cd(NO_3)_2$. The local atomic information at pH7.1 and 10.2 is similar to Cd precipitation. The results of regression analysis on the EXAFS spectra of clay samples could numerically indicate the coexistence of outer- and inner-sphere complexes.

1 INTRODUCTION

To solve soil pollution problem by heavy metals, the removal of heavy metals is required and, therefore, the knowledge of their adsorption reaction on clay-water interfaces is vitally important. This removal is especially important for people who eat rice, since Cd moves from soil particles and clay minerals and accumulates in rice plants causing a toxic reaction in human bodies. Many studies on the selectivity of adsorption of Cd as well as Cu, Co and Ni and their ion exchange reactions have been reported. A special emphasis has been placed on the pH dependence of adsorption behavior. Schulthness & Huang (1990), Naidu et al. (1994) and Spark et al. (1995) explained adsorption behaviors of Cd at differing pH values was due to the various adsorption sites on different parts of clay structure.

For several decades, surface complexation models have been applied to describe the metal adsorption reactions quantitatively. These models are based on the concept in which metal ions form both inner-sphere and outer-sphere complexes with surface functional groups. Comprehensive discussion of surface complexation formation can be found in the books edited by Sposito(1984) and Stumm(1987).

Katze & Hayes (1995) could predict adsorption behavior using the surface complexation model over range of moderate surface coverage; however, the models were inadequate to predict the amount of adsorption for data collected at a high surface coverage and wide pH range because the construction of the models mentioned above are based only on macroscopic data.

To obtain molecular level information about surface complexes, surface spectroscopic analysis is needed (Johnston & Sposito (1987); Gordon & Brown (1990)). Recent studies using spectroscopic analysis are showing how heavy metals such as Co, Ni and Pb are bound to surface functional groups of clay minerals and oxides. EXAFS (extend X-ray absorption fine structure) spectroscopy is a useful spectroscopic technique to obtain important information about local atomic structure of adsorbed ions, such as the type of atoms surrounding the central atom, the coordination number (N) and inter-atomic distances (R).

In this paper, we find the representative pH values which exhibit clearly different adsorption behavior of Cd on montmorillonite and estimate the structure of Cd surface complexes at these representative pH values by collected EXAFS data.

2 MATERIALS AND EXPERIMENTS

2.1 *Adsorption experiments*

Schulthess & Huang (1990) explained the adsorption of heavy metals on montmorillonite, kaolinite and mordenite by comparison with that on Si and Al oxides. The adsorption of heavy metals by clay minerals can be grouped into internal and external Si sites and Al sites. The influence of ionic strength on Cd adsorption was studied to classify Cd adsorption into specific and non-specific adsorption on soils (Naidu et al. (1994)). In this study, Na montmorillonite, supplied by Kunimine Industries Co., Japan was used. Kaolinite (JCSS-1101;Clay Science Society of Japan) and Al and Si oxides (γ-Al$_2$O$_3$ and SiO$_2$; Koujundo Chemical Lab.Co.) were also used to reveal adsorption surface sites of montmorillonite. To classify Cd adsorption behavior, 0.1mol/L and 0.001mol/L NaNO$_3$ solutions were used.

Adsorption of Cd on montmorillonite, kaolinite, and oxides was produced by batch experiments. After equilibrium, the samples were centrifuged and filtered through 0.2-μm membrane filters. The pH of each filtrate was measured and then the filtrate was analyzed for residual Cd concentration. The amount of Cd adsorption was calculated as a percentage of the ratio of the concentrations of the sample and the blank using this formula:

$$\text{Cd adsorption} \, (\%) = 100 \times \left(1 - \frac{\text{residual Cd concentration}}{\text{blank Cd concentration}}\right) \quad (1)$$

2.2 EXAFS

The clay samples that demonstrate the formation of different surface complexes were Cd adsorbed montmorillonites at given pH values (Table 1). The samples were prepared by the same procedure as the previous adsorption experiments. They were then washed three times with distilled water and stored in sample folders as wet pastes, taking into account the effect of drying.

Cd K-edge XAFS spectra were obtained at BL01B1 of SPring-8(JASRI), Hyogo, JAPAN. A Si(311) double crystal monochromator and a focusing double mirror were used. In addition to clay samples, Cd(NO$_3$)$_2$ solution, Cd(OH)$_2$,CdCO$_3$ powder, and the precipitation of Cd(OH)$_2$ formed under the same condition as the clay samples were used as references.

Table 1.Chemical conditions of EXAFS

	pH	amount Cd adsorption	surface adsorption density	Cd adsorption
		$(10^{-3}$mol/L)	(mol/clay m^2)	%
Clay 1	3.2	0.67	0.37	37
Clay 2	4.9	0.81	0.45	45
Clay 3	7.1	1.80	1.00	98
Clay 4	10.2	1.83	1.01	100

montmorillonite:
Cd(NO$_3$)$_2$: 2.0x10^{3}mol/L
NaNO$_3$

We also obtained Cd K-edge XAFS spectra at BL14A of Photon Factory, Tsukuba, JAPAN. Since the spectra showed similar results as those done at SPring-8, we omitted the results in this paper.

3 RESULTS AND DISCUSSION

3.1 Cd adsorption experiments

Fig.1 shows the pH dependence of Cd adsorption on all adsorption substrates. Although adsorption of Cd increases with pH, the curve varies between substrates. By comparing the substrates, it was confirmed that the active adsorption sites for montmorillonite were dependent on pH. Below pH 4.5, only the interlayer of montmorillonite, where the permanent charge occurs, serves as the adsorption site. From pH 4.5 to pH 7, the adsorption of Cd increases gradually due to the ionizing of the hydroxyl groups of the crystal edges where the pH-dependent charge exists. Since no difference in the curves was observed among substrate species above pH 7, the hydroxyl groups can be regarded as the dominant adsorption sites.

The influence of ionic strength is shown for montmorillonite and kaolinite up to pH 7 (Fig.2). The results for γ-Al$_2$O$_3$ and SiO$_2$ are notably different. Although adsorption on SiO$_2$ is influenced by ionic strength in all pH ranges, adsorption of γ-Al$_2$O$_3$ is not. We determined that the adsorption on SiOH of crystal edges is dependent on ionic strength. On the other hand, the adsorption on AlOH is independent of ionic strength. Consequently, the adsorption of Cd on montmorillonite is classified into two types: the ionic strength-dependent adsorption and -independent adsorption. These two types are felt to correspond to the formation of outer-sphere complexes and inner-sphere complexes, respectively.

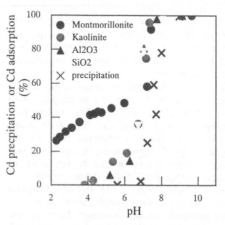

Fig. 1. Comparison of Cd adsorption on Montmorillonite, Kaolinite, oxides and Cd precipitation; Initial Cd concentration = 5×10^{-4}mol/ L, NaNO$_3$ = 0.001mol/L.

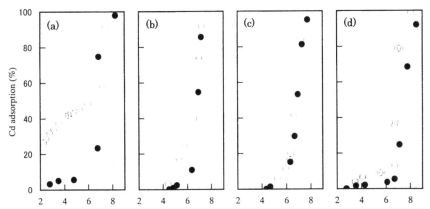

Fig. 2. The influence of ionic strength for Cd adsorbed on (a) montmorillonite, 0.6g /L; (b) kaolinite, 6g /L, (c) alumina, 6g/L; and (d) silica, 12g /L.
Initial Cd concentration = 5×10^{-4}M in the presence of 0.001M NaNO$_3$ (○) and 0.1M NaNO$_3$ (●).

Furthermore, since the Cd precipitation without substrates was observed in Fig.1, precipitation should be always considered near and above pH 8.

Ziper et al. (1988) showed that edge sites retain Cd specifically. Hayes & Leckie (1987) stated that only inner-sphere complexes on AlOH and outer-sphere complexes on SiOH were formed. Zachara & Smith (1994) calculated the amount of Cd adsorption on two types of montmorillonites using a surface complexation model and concluded that the edge SiOH is not important in the formation of complexes.

Based on these past macroscopic studies, we classify Cd adsorption on montmorillonite into three types, (1) the formation of outer-sphere complexes on the interlayer at low pH; (2) the coexistence of two surface complexes-- outer-sphere complexes on the interlayer and inner-sphere complexes on the edge AlOH (pH4~7); and, finally, (3) the dominant formation of inner-sphere complexes or surface precipitation on edge sites near and above pH 8.

3.2 EXAFS

Fig. 3 and Fig. 4 show the Cd K-edge EXAFS $k^3 \chi$ spectra and the Fourier transforms of EXAFS oscillation for clay samples at specific pH with references. In the EXAFS spectra on precipitation and Cd(OH)$_2$ powder, shoulders were observed on the right side of the oscillation near 6 Å^{-1}(Fig. 3). These shoulders are clear indications of the existence of second neighbor Cd atoms. For Clay 3 and Clay 4, the shape of the oscillations was asymmetric. These asymmetries might be attributee to the existence of second neighbor Cd atoms.

In Fig. 4, the spectra of Clay 1 and Clay 2 differ from that of Clay 3 and Clay 4. For Clay 1 and Clay 2, no atom appeared in the second neighbor atom position as occurred in the Cd solution of Cd(NO$_3$)$_2$.

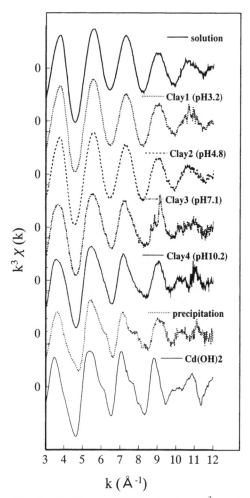

Fig. 3. Cd K-edge EXAFS functions $k^3 \chi$ of clay samples and references

387

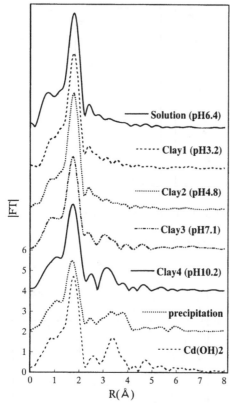

Fig.4 Fourier transform of Cd K-edge EXAFS of Clay samples and references

surface complex changed from outer-sphere to inner-sphere complexes as pH increased.

We investigated the dependence of the surface complexation formation on pH using both macroscopic and microscopic approaches. Further studies are required to clarify the structure of inner-sphere complexes and surface precipitation phenomena.

REFERENCES

Gordon, E. & Jr. Brown 1990. Spectroscopic studies of chemisorption reaction mechanics at oxide-water interfaces. In Michael, F. et al. (ed.), *Mineral-water interface geochemistry*, Reviews in Mineralogy vol.23:309-363

Hayes, K.F. & J. O. Leckie 1987. Modeling ionic strength effects on cation adsorption at hydrous oxides/solution interfaces. *J. Colloid Interface Sci.* 115:564-572

Johnston, C. T. and G. Sposito 1987. In Berma, L. L. (ed.) *Further development in soil science research.* Soil Sci. Society of Amirica. Medison :89-99

Katz, L.E. & K.F.Hayes 1995. Surface complexation modeling. *J. Colloid Interface Sci.* 170:477-490

Naidu, R. et al. 1994. Ionic-strength and pH effects on the adsorption of cadmium and the surface charge of soils. *European journal of soil science.* 45:419-429

Schulthess, C. P. & C. P. Huang 1990. Adsorption of Heavy metals by Silicon and Aluminum Oxides Surfaces on Clay Minerals. *Soil Sci. Soc. Am. J.* 54:679-688

Spark, K. M. et al. 1995. Characterizing heavy-metal adsorption on oxides and oxyhydroxides. *European journal of soil science.* 46:621-631

Sposito, G. 1984. *The surface chemistry of soils.* Oxford Univ. Press, New York

Stumm, W., ed. 1987. *Aquatic Surface Chemistry.* John Wiley, New York

Zhachara, J. M. & S. C. Smith 1994. Edge Complexation reactions of Cadmium on specimen and Soil-Derived Smectite. *Soil Sci. Soc. Am. J.* 58:762-769

Ziper, C. et al. 1988. Specific Cadmium Sorption in Relation to the Crystal Chemistry of Clay Minerals. *Soil Sci. Soc. Am. J.* 52:49-53

This indicates that outer-sphere Cd surface complexes are formed at low pH (pH 3.2 and pH 4.8). The main second peak of $Cd(OH)_2$ and Cd precipitation show second neighbor Cd atoms. The local atomic structure surrounding Cd for clay samples at pH 7.1 and 10.2 is similar to Cd precipitation.

We calculated the regression analysis on the EXAFS χ spectra of clay samples; those of Cd solution and precipitation conducted as factors (Table 2). Using the calculated parameters, this analysis could numerically indicate the coexistence of outer- and inner-sphere complexes. The dominant

Table2 The results of Regression analysis on the Cd K-edge EXAFS

sample	pH	A outer-sphere	B inner-sphere	R	amount of adsorption x10^{-3}(mol/L)	amount of adsorption outer-sphere x10^{-5}(mol/L)	amount of adsorption inner-sphere x10^{-5}(mol/L)	adsorption (%)
Clay1	3.2	0.99	0.01	0.99	0.670	66.0	1.0	37
Clay2	4.8	0.90	0.10	0.98	0.810	72.8	8.2	45
Clay3	7.1	0.48	0.52	0.99	1.801	87.1	93.1	98
Clay4	10.2	0.26	0.74	0.99	1.827	48.3	134.3	100

χ = A × χ (Cd solution)+B × χ (Cd precipitate)

Clay Science for Engineering, Adachi & Fukue (eds) © 2001 Balkema, Rotterdam, ISBN 90 5809 175 9

Containment, sorption and desorption of heavy metals for dredged sediments

M. Fukue, M. Yanai & Y. Takami
Marine Science and Technology, Tokai University, Shimizu, Japan

S. Kuboshima & S. Yamasaki
Aoki Marine Company, Osaka, Japan

ABSTRACT: The final goal of this study is to establish "a standard" for availability of dredged sediments as clay liner. To inspect the contamination of sediments, the concentrations of heavy metals and other elements were measured on core samples obtained from Osaka Bay and their profiles were obtained. One of the samples was subjected to the batch extraction test to know how much dissolution takes place. Furthermore, in order to investigate adsorption capacity of the sediments, batch equilibrium tests were also performed on similar samples using standard solutions containing Cd, Cu, Pb and Zn ions. The results showed that the adsorption capacity of contaminated sediments is still so high as to be used as clay liner.

1 INTRODUCTION

Most of sediments contain excess heavy metals and other hazardous substances, such as polycyclic aromatic hydrocarbons (Matsumoto, 1983, Matsushima, 1979). In particular, the bio-concentration occurring at polluted sea bottom causes one of the great fears of our health. Therefore, the remediation or removal of polluted sediments is a serious and urgent problem to be solved.

The volume of polluted sediments is huge and no place is available to dispose them. This requires the in-situ remediation or utilization of removed sediments. Since there are some technical difficulties in in-situ remediation and since the technical development will take a time, the removal of polluted sediments may be more suitable to solve the present problem.

Another problem arisen here is how to reduce the volume and/or to utilize polluted sediments. In order to solve this, detoxification or harmless of the polluted sediments is needed.

The history of contamination for marine sediments is seen from the concentration profiles. Once heavy metals are adsorbed to sediments, they can hardly move. Therefore, the concentration and time of sedimentation are recorded in the sediments. Previous studies showed that polluted layers are generally less than one meter in thickness (Matsumoto, 1983, Fukue *et al*, 1995,1996,1999). This implies that the dredging depth required for clean up of polluted sea bottom is also less than one meter.

2 CONTAMINATION AND DEGREE OF POLLUTION

2.1 *Contamination*

The contamination of marine sediments has started around 1975 (Fukue *et al*, 199), as well as lead contamination in Greenland snow (Murozumi *et al*, 1969). An example of copper is shown in Fig.1. which shows copper concentrations measured on sediments obtained from Osaka Bay in Japan.

The sampling sites are shown in Fig.2. The samples for concentration measurements were prepared by dissolving dry sediments and the measurements of copper concentration were made using *Inductively Coupled Plasma Atomic Emission Spectrometry* (ICP-AES) Method.

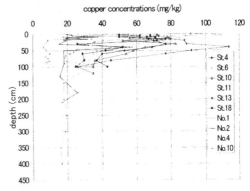

Fig.1 Concentrations of copper in Osaka bay sediments.

Table 1 Background values of elements in various bay sediments (Fukue et al, 1995).

	lovcation	Fe (g/kg)	Al (g/kg)	Ti (g/kg)	Cu (mg/kg)	Zn (mg/kg)	V (mg/kg)	P (mg/kg)
Osaka Bay	34-39-00N 135-21-04E	36-40	82-88	3.1-3.8	18-35	93-104	60-74	498-614
	34-34-04N 135-21-28E	38-44	77-100	3.5-4.1	16-21	70-94	63-73	499-559
	34-27-17N 135-09-54E	17-28	20-50	2.2-3.3	15-28	80-130	45-60	290-400
	34-24-54N 134-59-42E	25-28	57-68	2.4-2.5	7-9	-	45-60	230-250
Tokyo Bay	35-34-50N 139-47-45E	32-43	26-58	3.1-4.2	40-53	100-150	95-100	600-650
	35-35-00N 139-54-48E	28-36	45-58	2.6-3.0	15-27	-	80-90	300-380
	35-25-00N 139-42-12E	22-35	38-65	1.9-2.9	12-15	-	60-85	260-400
Hiroshima Bay	34-05-05N 132-21-08E	27-33	58-69	2.9-3.7	11-15	64-79	51-63	492-584
Kagoshima Bay	31-32-06N 130-35-54E	34-40	86-93	3.3-4.0	9-10	52-55	49-52	565-570
min-max		17-44	20-100	1.9-4.2	7-53	52-150	45-100	230-650
average		30.5	60	3.05	30	101	72.5	440

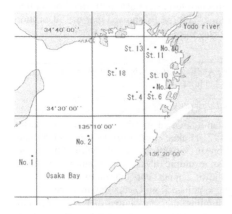

Fig.2 Sampling sites in Osaka bay.

Fig.1 shows that the total copper concentration is higher for the shallower sediments. This is because of human discharge. The concentrations of the deeper sediments are almost constant and are regarded as background values, which vary with location site, i.e., soil type, because of the solid-ions interactions.

The trend of other metals is basically similar to the case of copper shown in Fig.1. For an example, the concentrations of zinc are shown in Fig.3. Phosphorus is not a metal, and shows a similar trend to Figs 1 and 3.

Thus, the discharged metals and other substances are contained in the sediments. These substances are taken by benthos and concentrated in them, as known by terminology "*bio-concentration*".

2.2 Degree of pollution

In order to evaluate the level of pollution, it is needed to know the background value that is the natural level without human discharge. The level of polluti-

on can be evaluated from the comparison between the background and present concentrations. In a previous study, the degree of pollution Pd was defined as

$$Pd = (Ci - Bi) / Bi \qquad (1)$$

where Ci is the present concentration and Bi is the background value of the element (Fukue et al, 1996). The value of $(Ci - Bi)$ is the amount of human discharge per unit weight of sediments, which is the degree of pollution multiplied by background. If sediment is not polluted, the $(Ci - Bi)$ and Pd become zero.

Fukue et al. (1996) obtained that Pd values of marine sediments depend on the type of elements, location and depth. The Pd for copper exceeds seven at a given site location in the Osaka bay sediments. Since the background values of elements are a function of many factors, such as type of soil minerals and species of elements, it varies with location and depth as well as background (Fukue et al. 1995).

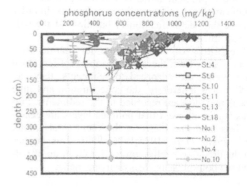

Fig3 Phosphorus concentrations in Osaka bay sediments.

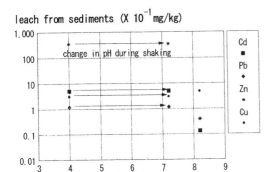

leach from sediments (X 10^{-1} mg/kg)

Fig. 4 Leaching elements in batch test.

Particularly, the background values of constitutive elements, such as iron, aluminum, etc., which are constitutes of the crystalline minerals, are usually difficult to estimate, because the background value is primarily dependent on the type of crystalline minerals. For these elements, sediment properties such as particle size, liquid limit and clay content can be utilized for estimating the background (Fukue et al, 1995).

Table 1 presents backgrounds of some elements obtained on bay sediments. The evaluation of contamination or pollution for sediments can be made from the comparison between the background and measured concentration.

3 SORPTION AND DESORPTION

3.1 Sediment sample

One of the samples obtained from Osaka Bay was subjected to the batch tests. The physical properties of the sample are presented in Table 2.

The Pd values for Cu and Zn were obtained using the maximum background in Osaka bay sediments. Therefore, the Pd values are minimums.

Table 2 Physical and chemical properties of the sample.

Specific gravity	2.65
Liquid limit(%)	155
Plastic limit(%)	42
Sand (%)	0
Silt (%)	48.7
Clay (%)	51.3
Ignition loss (%)	13.2
Cu (mg/kg)	56-73
Pd	>1.6
Zn (mg/kg)	343-418
Pd	>2.6

Table 3 Results of batch tests.

	units	distilled water	pH4 at HCl
alkylmercuric compound	mg/L	<0.0005	<0.0005
mercury	mg/L	<0.0005	<0.0005
cadmium	mg/L	0.001	0.038
copper	mg/L	0.036	0.023
zinc	mg/L	0.003	2.64
lead	mg/L	0.003	0.009
chromium (VI)compound	mg/L	<0.02	<0.02
arseric	mg/L	<0.005	<0.005
selenium	mg/L	<0.005	<0.005
cyanide	mg/L	<0.01	<0.01
organophosphorus	mg/L	<0.1	<0.1
PCB	mg/L	<0.0003	<0.0003
trichloroethylene	mg/L	<0.002	<0.002
tetrachloroethylene	mg/L	<0.0005	<0.0005
dichloromethane	mg/L	<0.002	<0.002
carbon tetrachloride	mg/L	<0.0002	<0.0002
1,2-dichloroeyhane	mg/L	<0.0004	<0.0004
1,1-dichloroethylene	mg/L	<0.002	<0.002
cis-1,2-dichloroethylene	mg/L	<0.004	<0.004
1,1,1-trichloroethane	mg/L	<0.0005	<0.0005
1,1,2-trichloroethane	mg/L	<0.0006	<0.0006
1,3-dichloropropene	mg/L	<0.0002	<0.0002
thiuram	mg/L	<0.0006	<0.0006
simazine	mg/L	<0.0003	<0.0003
thiobencarb	mg/L	<0.002	<0.002
benzene	mg/L	<0.001	<0.001
pH	-	8.4(25°C)	4.1
water content	%	68.3	68.3
ignition loss	%	13.2	13.2

3.2 Containment and leaching

The batch leaching tests were performed for inspecting hazardous substances using HCl solution with pH 4 and distilled water, respectively. The sample used had physical and chemical properties as shown in Table 2. The sample was washed with the HCl solution or distilled water for 6 hours, where the ratio of liquid to solid was 13 and then the liquid was extracted.

The 26 hazardous substances presented in Table 3 were detected. The results are also presented in Table 3, which shows no detection of most hazardous substances except heavy metals, i.e., Cu, Pb, Zn and Cd.

For leaching of heavy metals, the effect of initial pH value is apparently large, except Cu, as shown in Fig. 4. During shaking, the pH changed largely from 4 to 7.2, but little changed for an initial pH value of 8.2.

Comparison between the containment (Table 2) and leaching results (Fig. 4) with Cu and Zn shows that the dissolved heavy metals in both distilled water and HCl solution are a very small amount, i.e., less than 1 %.

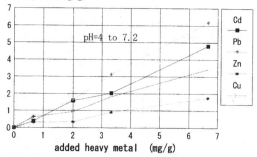

adsorption (mg/g)

Fig.5 Relationships between adsorbed and added amounts for heavy metals.

3.3 *Batch equilibrium test*

To examine the adsorption characteristics of the sediments, the batch equilibrium test was carried out. The ratio of liquid to solid was 13. Heavy metals, Cd, Zn, Pb and Cu are added in the form of standard solutions. The standard solutions were adjusted to a pH of 4 and the concentrations of extracted solution were measured. The retained amount to particles was calculated from the measurements.

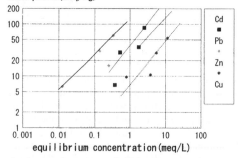

adsorption (meq/kg)

equilibrium concentration (meq/L)

Fig.6 Adsorption with equilibrium concentration for heavy metals.

The result is shown in Fig.5, which shows that adsorption of the heavy metals increases almost linearly with increasing of added amount. The adsorption selectivity is in the order:

Pb>Cd>Cu>Zn

This order does not necessarily accord with the previous studies introduced by Yong *et al*, 1992, because of presence of various ions in seawater and because of various constituents in sediments. It is well known that Ca and Mg ions affect the adsorption selectivity of heavy metals and also the selectivity changes with the type of clay minerals.

Thus, it was found that marine sediments containing heavy metals still have a relatively high adsorption capacity under high concentrations of the metals. Therefore, it is interpreted that with this capacity, the dissolution from the heavily contaminated sediments was low.

Fig. 6 provides relationships between the retained amount of heavy metals to solids and equilibrium concentration in pore fluids. The figure shows that the higher the concentration of heavy metals in pore fluid, the heavy metal will be more adsorbed onto soil particles. Although the data shows a considerable scattering, the relationships obtained are almost linear in log-log scale to a certain equilibrium concentration. It is well known that at higher equilibrium concentrations, the relationship will not be linear because of the limited capacity of adsorption.

The gradients of the relationships shown in Fig.5 are well known as the partition coefficient of each metal, which can be used for the numerical analysis of contaminant transport in soils.

4 CONCLUSION

The chemical properties of marine sediments were investigated to examine if the sediments are available for clay liner. It is found that the bay sediments are more or less polluted. However, it was found that from the leaching test, little dissolution of heavy metals and other hazardous substances was obtained. From the batch equilibrium test, the adsorption of heavy metals is still strong.

REFERENCES

Fukue, M. Kato, Y. , Nakamura, T. and Yamasaki, S., 1995, Heavy metal concentration in bay sediments of Japan, *STP. 1293, ASTM.*,58-73.

Fukue, M. Kato, Y. and Nakamura, T., 1996, Heavy metal concentration of marine sediments, *Environmental Geotechnics*, Balkema, 49-54.

Fukue, M., Nakamura, T., Kato, Y. and Yamasaki, S., 1999, Degree of pollution for marine sediments, *Engineering Geology*, 53, 131-137.

Matsumoto, E. ,1983, The sedimentary environment in the Tokyo Bay, *Chikyu Kagaku*,17,27-32 (text in Japanese).

Matsushima, H., 1979, Correlation of polycyclic aromatic hydrocarbons with environmental components in sediment from Hirakata Bay, Japan, *Agric. Bio. Chem*, 43, 1447-1452.

Murozumi, M., Chow, T.J., Patterson, C., 1969, Chemical concentrations of polluted lead aerosols, terrestrial dusts and Atlantic snow strata, *Geochim, Cosmochim*, Acta 33, 1247-1294.

Yong, R.N., Mohamed, A.M.O. and Warkentin, B.P. , 1992, Principles of Contaminant Transport in Soils, Development in Geotechnical Engineering, 75, Elsevier, p.169.

Clay Science for Engineering, Adachi & Fukue (eds) © 2001 Balkema, Rotterdam, ISBN 90 5809 175 9

Phosphate ion transport and retention in a tropical compacted clay

R.M.G.Mendonça, M.C.Barbosa & F.J.C.O.Castro
COPPE-UFRJ, Rio de Janeiro, Brazil

ABSTRACT: A gneiss residual soil from Rio de Janeiro, Brazil, was investigated in the laboratory in respect to transport and retention of phosphate ion at optimum compacted condition. Kaolinite and goethite are the dominant active minerals in soil composition, and the soil Zero Point of Charge (ZPC) is close to the standard kaolinite value (3.5). The bulk soil behaviour relative to phosphate ion sorption through batch tests is presented for pH conditions below and above the ZPC value and compared to the pure minerals sorption patterns determined for kaolinite, goethite and hematite. The sorption parameters obtained are discussed for diffusion tests on compacted samples of the natural soil.

1 INTRODUCTION

1.1 *Ion sorption in tropical soils*

The soil property of removing ions from the pore solution and retaining them by sorption mechanisms in the solid phase is well known and very useful when dealing with barriers for waste disposal sites. Different mechanisms contribute to this process, but the ion exchange phenomenon is considered to be the most important.

Ion exchange is a reversible two-way process between the liquid and solid phases, and is due almost entirely to the clay and silt fractions and the organic matter of the soil. The major interest herein is the discussion of the clay fraction behaviour, specially its amphoteric nature.

As reported in the literature (Grim 1968; Bohn et al. 1985; and others), both negative and positive charges do exist on the same mineral, and therefore the clay soil has the ability to adsorb both positive (cations) and negative ions (anions). The major sources of negative charges in clay particles are the isomorphous ion substitutions within the clay lattice and the ionization of hydroxyl groups attached to silicon of broken bonds on particle edges. These mechanisms are common to all clay minerals, but the relative importance of each of them to the clay overall ion exchange capacity varies from mineral to mineral.

The isomorphous ion substitution is responsible for the permanent charge of the soils, and is the principal source for the 2:1 clay minerals such as illite and montmorillonite.

The ionization of functional groups on the soil particle surfaces are pH dependent, and thus generates a pH variable charge. This mechanism is a significant source of charge for the 1:1 clay minerals such as kaolinite (50% of total charge or more), and the only source for soil allophane and hydrous oxides such as hematite, goethite and gibbsite.

In consequence, soil compositions dominated by these minerals have their net charge highly dependent on the solution pH value (Castro 1989; Shackelford & Redmond 1995). In acidic environments, the ionized functional groups act like a base, adsorbing protons (H^+) and becoming positively charged. In alkaline environments, these groups are deprotonized and become negatively charged.

The net soil charge, and therefore the soil ion exchange behaviour, present in this case an isoeletric point at a certain pH value. This point is called the Zero Point of Charge (ZPC), and means a balance of positively and negatively charged sites within the soil. In other words, at this pH value, the soil can adsorb equivalent amounts of cations and anions from the pore solution.

Soils dominated by 2:1 clay minerals either present no ZPC value (no pH dependent charge) or very low ZPC values, with no practical meaning. Tropical soils, on the other hand, are dominated by 1:1 clay minerals and hydrous oxides, and the ZPC values can be easily attained in practical cases.

With the knowledge of the concentration and pH presented in a case history of a disposal lagoon in Brazil, it was decided to study phosphate since it can fortuitously reach rivers and aquifers, generating its eutrophication.

1.2 Experimental program

A natural residual soil from a borrow area close to the City of Rio de Janeiro, Brazil, was used in the research. The characterization involved the determination of the soil geotechnical index properties, mineralogical composition and soil chemistry. The chemical parameters included the soil pH at site condition and the soil Zero Point of Charge (ZPC) value.

The phosphate ion sorption was investigated in two different sets of experiments: the first set used pulverized soil samples and conventional batch tests, and the second set consisted of column and diffusion laboratory tests with compacted soil samples. All experiments were performed at two different pH levels (6.5 and 2.5), above and below the natural soil ZPC value, because of the process amphoteric nature. Only the diffusion tests results are discussed herein. A complete description of the research is presented in Mendonça (2000).

2 SOIL CHARACTERIZATION

2.1 Soil mineralogy and chemistry

The deposit investigated is a saprolitic soil from the weathering profile of a granitic gneiss of the Paraíba do Sul geological complex. This geological formation is typical of the State of Rio de Janeiro, occurring between the coastal plains and the mountain line extending parallel to the coast at Southwestern Brazil. The climate is warm and wet, and the temperature is above 18°C all over the year.

The mineralogical composition was investigated through X-ray diffraction analyses of air dry, powdered soil samples, and visual inspection of the sand size grains of the bulk soil. The soil fraction passing #10 ASTM standard sieve was used in the total analysis, and the soil fraction less than 2μm diameter size was treated for free Fe removal and analyzed for clay minerals identification, following standard procedures. The procedures and results are presented in Mendonça (2000) and the soil mineralogical composition is summarized in Table 1.

Table 1. Soil mineralogy.

ANALYSIS	MINERALS
Clay Fraction	kaolinite, goethite, and traces of vermiculite
Bulk Soil	kaolinite, quartz, feldspars, and mica

Although almost 90% of the soil clay fraction is constituted by kaolinite, the soil behaviour is also influenced by the other clay minerals present and by the partially weathered particles of mica, that give the soil a significant expansion response upon water saturation (Mendonça 2000).

The soil chemistry is summarized in Table 2. The physico-chemical analyses included the soil organic matter content (OMC), pH, cation exchange capacity (CEC), Zero Point of Charge (ZPC), and determination of the soil exchangeable complex composition and of the percentage of major oxides through sulfuric acid digestion.

Table 2. Soil chemistry.

PHYSICO-CHEMICAL ANALYSIS	
pH in H_2O 1:2.5	4.13
pH in KCl 1:2.5	3.83
Organic matter content (%)	0.29
ZERO POINT OF CHARGE	
ZPC	3.80
EXCHANGEABLE CATIONS (mol_c kg^{-1} dry soil)	
Al^{+3}	4.30
$Ca^{+2} + Mg^{+2}$	0.30
Na^+	0.05
K^+	0.05
H^+	1.50
EXCHANGEABLE ANIONS	
P (ppm)	1.00
CATION EXCHANGE CAPACITY – pH 7	
CEC (mol_c kg^{-1} dry soil)	6.20
SULFURIC ACID DIGESTION	
SiO_2 (%)	29.12
Al_2O_3 (%)	25.46
Fe_2O_3	8.50
Loss on Ignition (600°C) (%)	12.29

The soil is acidic (pH ≈ 4), and the difference between the pH values in water and in KCl 1 mol L^{-1} solution reflects the presence of negative charges on soil particles surface. The ZPC value (3.8) is slightly higher than kaolinite's (ZPC ≈ 3.5), probably due to the presence of goethite in soil composition (ZPC ≈ 7). As shown in Table 2, the natural soil pH is near to the Zero Point of Charge condition.

The soil exchangeable complex is largely dominated by Al^{+3} and H^+ ions (93% of the exchangeable sites). Analysis of the major oxides shows a small Fe content, confirming aluminum as the dominant metallic ion present in this particular soil.

2.2 Geotechnical characteristics and soil classification

Table 3 summarizes the soil geotechnical index properties (grain size distribution, natural water content, gravity density G_s and Atterberg limits), Proctor standard compaction parameters (optimum water content ω_{ot} and maximum dry density γ_{smax}), and the compacted soil hydraulic conductivity k_{H2O}.

All geotechnical parameters are in agreement with soil mineralogical, physical and chemical characteristics. It is important to notice that, although presenting 24% of particles are clay size in nature, these particles occur in the soil matrix as quite stable aggregates. Since these aggregates are not destroyed during compaction, the soil geotechnical behaviour reflects the apparent coarser grain size distribution in the field (Nogami & Villibor 1995).

Table 3. Geotechnical carhacteristics.

GRAIN SIZE ANALYSES (%)	
Percent of Fines (<#200)	58
Clay Fraction (<2μm)	24
INDEX PROPERTIES	
Natural water content (%)	18.5
Gravity density G_s	2.689
Contraction Limit ω_C (%)	25.3
Plastic Limit ω_P (%)	28.7
Liquid Limit ω_L (%)	57.3
Plasticity index I_P (%)	28.6
Skempton Activity index A_C	1.2
STANDARD PROCTOR COMPACTION TEST	
Optimum water content ω_{ot}	20.8 %
Maximum dry density γ_{smax} KN/m^3	16.37
COMPACTED SOIL PERMEABILITY – WATER	
K_{H_2O} (m/s) at $\omega = \omega_{ot}$	8.8×10^{-9}

According to the Unified Soil Classification System (USCS), this soil is classified as CH (see Table 4), that is, an "inorganic clay of high plasticity". But the soil mechanical behaviour observed (Mendonça 2000) does not follow the CH soils expected pattern.

This discrepancy between classification and geotechnical behaviour is quite common in tropical soils, and an alternate classification system (MCT) was developed in Brazil for those soils (Nogami et al. 1989; Nogami & Villibor 1995). The MCT system became a national standard in 1994 (DNER – CLA259/94).

The MCT system divides the tropical soils in seven groups and two broad classes: lateritic and non-lateritic (saprolitic) soils. The soil under study belongs to the non-lateritic behaviour class, in the group NS'-NG' (see Table 4). According to the MCT system, the following it is expected: high expansive response upon soaking (> 3%), but presenting a low pressure of expansion, medium contraction (0.5% - 3%), medium plasticity ($7<I_P<30$), low permeability when compacted (k < 10^{-8} m/s), and high load capacity. Another relevant characteristic of this group of soils is the substantial loss of fine material upon soaking in water. These soils are thus potentially erodible when exposed, requiring surface and filter protection in the field.

3 STUDY OF PHOSPHATE ION SORPTION

3.1 Mechanisms of phosphate sorption

According to Fassbender (1966) and Resende (1983), the phosphate ion can be removed from the solution and transferred to the solid phase by two different mechanisms:

(a) formation of precipitates in the reaction with Al, Fe, Mn, Ca and Mg ions present in soil solution;

(b) sorption of the phosphate ions onto the surface of silicate clay minerals, sesquioxides and organic matter particles.

The phosphate ion sorption is visualized as the substitution of OH$^-$ and H$_2$O groups present in the surface of hydrous oxides of Fe and Al. The intensity of this process is related to the existing pH, which determines the degree of competition with OH$^-$ ions for the available positions. The phosphate sorption is, therefore, an exchange of ligands reaction, where the resultant bonding between the anion oxygen and the metal has a high degree of covalence (Parfitt et al. 1975; in Lopes, 1977).

Table 4. Soil Classification.

SYSTEM	GROUP	DESCRIPTION
USCS	CH	Inorganic clays of high plasticity
MCT	NS' - NG'	Sand-clayed silts; Sandy clays; Clayey silts.

3.2 Materials and methods

The natural soil used in the experiments has been described. Batch sorption tests were also performed with pure mineral samples for comparison: a natural kaolinite from Northern Brazil and commercial goethite and hematite samples. All soil and minerals samples were air dried and powdered to the #10 ASTM standard sieve maximum size.

For the contaminant transport testing program, these natural soil samples were mixed with distilled water at the required weight to achieve optimum condition after static compaction. The soil-water mixtures rested for 24 to 48 hours in the humid chamber before compaction, in order to guarantee physic-chemical equilibrium between the two phases.

The batch tests followed the procedure described in USEPA (1992), adopting a soil:solution ratio of 1:10, based on COPPE's Geotechnical Laboratory previous experience (Ritter 1998; De Paula 1999). The supernatant liquid was obtained by centrifugation at 3,000 rpm for 10 minutes. Part of this liquid was analyzed for electrical conductivity and pH, and the rest was analyzed for phosphate concentration within 96 hours after collection.

The solutions were prepared by dilution of $Na_3PO_4.12H_2O$ in deionized distilled water, plus 10 ml of buffering solution to keep the desired pH level throughout the test. For pH 2.5, it was used sodium citrate in hydro chloridric acid as buffer, and for pH 6.5, sodium citrate in sodium hydroxide. The phosphate concentration ranges were 10ppm to 1,330ppm in the batch tests, and 500ppm to 10,000ppm in the column and diffusion tests. Total phosphate concentration in solution was determined by the colorimetric method.

Two different types of diffusion apparatus were used for investigation of phosphate transport and retention through the compacted natural soil: the simple reservoir diffusion cell described in Barone et al. (1992) and the DKS (Diffusion-Konvection-Sorption) apparatus developed in Germany and described in Jessberger & Onnich (1993).

The simple reservoir cell represents a transient flux condition, and requires a theoretical model for interpretation. In the present research it was adopted the POLLUTE model (Rowe and Booker 1994). In the DKS apparatus a permanent flux condition is achieved, and simple analytical calculations are required for interpretation of test results.

3.3 Kinetics of sorption reaction

As recommended in USEPA (1992), the time necessary to attain equilibrium in the batch test routine was previously investigated for the natural soil at both pH levels (2.5 and 6.5). Phosphate solutions were prepared with 10, 30, 50, 90, 120 and 150ppm concentrations, and mixed to the air dry soil at 1:10 soil:solution ratio. The suspensions obtained were then agitated for 1, 24, 48 and 72 hours, centrifuged and the supernatant liquid analyzed. The mass of adsorbed phosphate per gram of dry soil is plotted versus mixing time and initial phosphate concentration in solution in Figures 1(a) (pH 2.5) and 1(b) (pH 6.5).

The soil present the same pattern, regardless of pH: for low initial concentrations up to 90ppm the equilibrium is attained within 24 hours of agitation, while for higher concentration levels an asymptotic behaviour is observed. In any case 48 hours seems to be a reasonable time for equilibrium to be achieved, and was adopted in the subsequent tests.

3.4 Sorption isotherms

A batch test series consists of at least five tests, each one corresponding to a different concentration level at the same soil:solution ratio and the same time of agitation to attain the equilibrium condition. The plot of adsorbed mass per gram of dry soil versus initial concentration in solution is defined as the soil sorption isotherm for that chemical species.

The sorption parameters are obtained by fitting a theoretical model to the experimental results, and two models are generally applied: the Langmuir and the Freundlich sorption equations.

The Langmuir model supposes that adsorption is limited to a monomolecular layer and that there is no energy variation as the solid surface is covered. The Langmuir isotherm is described by the equation:

$$S = \frac{\alpha \beta C}{1 + \alpha C} \tag{1}$$

where the parameter β represents the soil maximum adsorption capacity and the parameter α a constant of equilibrium for the energy of sorption.

The Freundlich model does not suppose a limited sorption capacity, and Larsen (1967) states that this formulation is purely empirical and therefore the model parameters have no physical meaning. But according to Resende (1983), it is generally believed that the Freundlich equation implies an exponential decrease of the sorption energy as the sorption surface becomes progressively saturated. The Freundlich isotherm is described by the equation:

$$S = K C^N \tag{2}$$

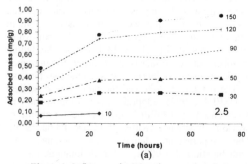
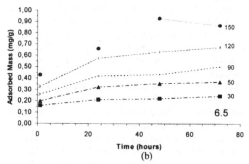

Figure 1. Influence of mixing time on phosphate sorption on natural soil, at different pH levels and initial concentrations in solution .

where K and N are constants, and the parameter K can be taken as related to the soil maximum sorption capacity according to Resende (1983).

3.5 Batch test results

Phosphate sorption isotherms were determined for the natural soil through batch tests, in the concentration range of 30ppm to 1330ppm, for two pH levels, below and above the soil ZPC value for comparison, the same batch test series were realized on pure kaolinite, goethite and hematite mineral samples.

In these tests, a 0.1M KCl solution was added to the buffer, to maintain the ionic strength of the reactions developed (Resende 1983).

It was used a low concentration of citrate (0.025 mol_C kg^{-1}) in the buffer solution. As far as we know about the sorption power of variable charge minerals (residual tropical soils), and although citrate is a complexing chemical agent, it is unable to prevent the specific phosphate sorption in the inner part the Stern layer of the electric double layer.

The experimental results were adjusted by the Langmuir and Freundlich theoretical models, equations (1) and (2), and Table 5 summarizes the corresponding sorption parameters obtained. The correlation factors (R^2) are also presented. The

correlation factors are close to 1 nearly all fitness either by Freundlich or Langmuir model. The exceptions are the kaolinite isotherms, which can only be adequately fit by the kaolinite isotherms, which can only be adequately fit by the Langmuir model.

The Langmuir isotherms are plotted altogether in Figures 2 (a) (pH 2,5) and (b) (pH 6.5). These graphs and the β values in Table 5 indicate the following order relative to phosphate sorption capacity, regardless of pH level: goethite>hematite>natural soil>kaolinite

The intensity of phosphate sorption depends on the nature and effectiveness of the sorption surface. For a given mineral, the surface activity increases for decreasing particle size and degree of crystallization. For that reason, it is difficult to establish a sequential order in phosphate sorption capacity among the active minerals. Particularly the Fe hydrous oxides (hematite and goethite), that occur in nature in a broad range of grain sizes, depending on the degree of laterization. Ker (1995) presents the sequential order below, after making such remarks:

Soil allophane>goethite or hematite>gibbsite>1:1 clay minerals>2:1 clay minerals.

The results obtained in the present study followed the same pattern. In the study the goethite sample presented a substantially higher phosphate sorption

Table 5. Sorption parameters fitted to the natural soil and pure minerals batch tests isotherms.

MATERIAL	PH	FREUNDLICH			LANGMUIR			CONCENTRATION
		K	N	R^2	α (mg/l)	β (mg/g)	R^2	(mg/l)
Natural soil	2.5	0.2964	0.2888	0.9952	0.0093	2.4166	0.9773	65 - 1300
	6.5	0.2330	0.2783	0.9678	0.0102	1.7627	0.9482	30 - 1140
Kaolinite	2.5	0.1235	0.3895	0.9184	0.0139	1.6327	0.9980	30 - 1300
	6.5	0.1430	0.3235	0.8116	0.0128	1.3068	0.9938	30 - 1330
Goethite	2.5	2.3817	0.1227	0.9932	0.1704	5.2301	0.9990	160 -1300
	6.5	2.5948	0.0834	0.9208	0.1016	4.6447	0.9908	140 -1150
Hematite	2.5	0.3809	0.3110	0.9969	0.0118	3.4020	0.9809	65 - 1340
	6.5	0.5300	0.2456	0.9938	0.0177	2.9412	0.9909	65 - 1340

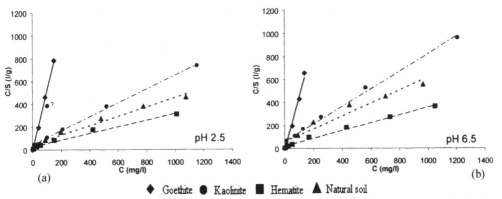

◆ Goethite ● Kaolinite ■ Hematite ▲ Natural soil

Figure 2. Phosphate sorption isotherm fitted by Langmuir model for natural soil and pure minerals (kaolinite, goethite and hematite).

capacity than hematite tested. Bigmam (1997) (in Ker 1995) observed the same behavior when comparing goethite and hematite soils, and this pattern could be explained by the general tendency of hematite to form larger mineral grains in nature.

The natural soil presented about the same phosphate sorption capacity, and even slightly higher, than pure kaolinite (see figures 2(a) and 2(b)). The presence of goethite in soil composition and the fact that in residual soils the clay minerals and Fe hydrous oxides tend to present poor crystallization, are the most probable reasons for such magnitude of phosphate sorption in this saprolitic soil.

3.6 Influence of pH on phosphate sorption

The pH levels adopted (2.5 and 6.5) did not make any difference in the pattern of phosphate sorption, but for all material tested their sorption capacity increased in the more acidic environment (pH 2.5).

Taking the β values of Table 5 for comparison, an increase of 37% in pH sorption capacity from pH 6.5 to pH 2.5 for the natural soil was observed, 25 % for kaolinite, 13% for goethite and 16% for hematite. It is quite reasonable that the major increases occurred for kaolinite and the saprolitic soil, that present a ZPC value (3.5 and 3.8, respectively) between the pHs adopted in the experiments. According to the literature, goethite and hematite have ZPC values around pH 6 and 7, and therefore both pHs tested are below the isoelectric condition.

3.7 Effect of sorption on phosphate transport by diffusion

Column permeameter and pure diffusion tests on compacted samples of the saprolitic soil are described and discussed in Mendonça (2000). The sorption parameters presented in Table 5 were applied in the interpretation of the phosphate

transport tests with very good results, despite the structure difference between the compacted soil samples and the pulverized samples used in the batch tests series. For simplicity, only the diffusion tests results are discussed herein.

Test conditions and specimens characteristics are summarized in Table 6. The concentration of phosphate in solution had to be substantially increased relative to the batch tests values because of dilution and retardation effects during transport through the soil specimen. The reference source was an industrial liquor form fertilizer industry that presents pH around 2 and total phosphate around 8,000 ppm.

The DKS tests give the molecular diffusion coefficient (D_e) at permanent flux condition, that is, when the soil has become completely saturated on the chemical specie under study and no more sorption reactions occur. Table 7 presents the phosphate molecular diffusion coefficients at infinite dilution condition (D_0) (theoretical; Lerman 1979) and as measured dilution research. The values obtained in the DKS tests can be used to calculate the tortuosity factor ($W = D_e/D_0$) and as reference for interpretation of the diffusion cell tests (transient flux condition).

The diffusion cell tests results are reproduced in Figures 3 (a) (pH 2.5) and (b) (pH 6.5). The symbols represent the experimental results and the full lines the theoretical fit by POLLUTE model with the Freundlich sorption obtained in the batch tests.

For pH 2,5 the theoretical fits were not satisfactory, giving D_e values too high relative to the DKS test. Mendonça (2000) suggests that a dilution error in procedure for measurement of phosphate concentration in porewater samples could have displaced the experimental points, and this particular test should be repeated.

Table 6. Diffusion tests conditions and specimen characteristics.

Apparatus	Flux Condition	pH	PO_4^{-3} conc. source (ppm)	Time (days)	Specimen			
					h_0 (%)	γ_s (KN/m^2)	S_0 (%)	n
DKS	Permanent	2.5	9,311	93	20.7	1.631	85.8	0.40
Simple		2.5	10,549	8	21.0	1.636	87	0.39
Reservoir	Transient							
diffusion cell (DC)		6.5	10,711	8	21.0	1.638	87.2	0.39

Table 7. Summary of PO_4 theoretical and effective molecular diffusion coefficients for the natural saprolitic soil.

pH	Theoretical D_0 (m^2/year) Lerman (1979)	DKS test D_e (m^2/ano)	$W = D_e/D_0$	Diffusion cell tests – D_e (m^2/year)	
				Freundlich fitted	Langmuir fitted
2.5	0.023	0.005	0.22	0.015	0.020
6.5	0.023			0.008	0.013

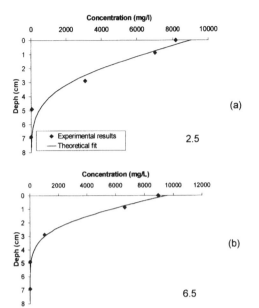

(a)

2.5

(b)

6.5

Figura 3. Porewater phosphate concentration with depth at the end of simple reservoir diffusion cell tests at pH 2.5 and pH 6.5.

The Langmuir fits did not produce good adjustments, but it can be observed in Table 5 that Langmuir correlation factors (R^2) for the natural soil were inferior to the values obtained in the corresponding Freundlich fits.

4 CONCLUSIONS

Traditionally considered poor materials for construction of compacted clay liners at waste disposal sites, kaolinitic tropical soils present the advantage of being capable of retaining both positive and negative ions, specially if they contain some metal hydrous oxides in their mineralogical composition.

A saprolitic soil of gneiss origin from Rio de Janeiro, Brazil, was investigated for transport and retention of phosphate, through Batch tests on pulverized soil samples and diffusion tests on compacted specimens, at two pH levels, above and below the soil isoelectric condition (Zero Point of Charge).

The soil occurs in nature at acid condition (pH ≈ 4), and close to the Zero Point of Charge value (ZPC = 3.8). The exchangeable cation complex is dominated by Al^{+3} and H^+ ions (93% to of total exchangeable sites), and the soil resultant CEC is significant (6.2 mol_c kg^{-1} dry soil). Although 90% of the soil fines is constituted by kaolinite the presence of goethite and the low degree of crystallization of both minerals increase the bulk soil activity.

For the same reasons, the soil capacity of adsorbing phosphate ions is significant and even slightly higher than pure kaolinite. And below the isoelectric point there is a increase of 37% in the soil maximum sorption capacity. An interesting aspect of phosphate retention is the high degree of covalence of the resultant bonding between the oxygen of PO_4^{-3} molecule and the mineral metal. This means that a strong bond is developed in the process, and from the barrier point of view, this is highly favorable.

The batch test series showed that equilibrium is achieved within 48 hours, and that the Freundlich model was the best fit for the observed sorption pattern.

Figures 3 (a) and 3 (b) reproduces the phosphate in transport though the compacted soil. The full lines represent the theoretical fit by POLLUTE model using the Freundlich parameters obtained in the Batch tests series. Despite the difference in the structure of the two test specimens (pulverized versus compacted samples), the sorption parameters adopted gave very good results.

The diffusion tests gave average effective molecular diffusion coefficient $D_e = 0.007$ m^2/year for the transport of phosphate through the compacted saprolitic soil (n = 0.39).

ACKNOWLEDGEMENTS

The authors want to thank the Brazilian National Research Council (CNPq) for scholarship given to the first author and to the technicians of COPPE's geotechnical Laboratory for the support in the field and in the laboratory. They also wish to thank chemist Maria da Gloria Marcondes Rodrigues for the innumerous analyses realized during the research.

REFERENCES

Barone, F.S, Rowe, R.K., Quigley, R.M. 1992. Estimation of diffusion coefficient and tortuosity factor for mudstone. *Journal of Geotechnical Engineering*, 118(7): 1031-1046.

Bohn, H.L., McNeal, B.N., O'Connor, G.A. 1985. *Soil Chemistry*, 2ª ed. John Wiley & Sons, New York.

Castro, C.F.J.O. 1989. Eletroquímica e propriedades mecânicas dos solos. In *Colóquio de Solos Tropicais e Subtropicais e suas Aplicações em Engenharia Civil*, Proc. Nat. Symp., 2:47-59. Porto Alegre. ABMS. (in portuguese)

De Paula, E. 1999. *Estudo experimental do transporte e retenção de Zn e Cu em um solo arenoso de Jacarepaguá, Rio de Janeiro*. Tese de Ms.Sc., COPPE/UFRJ, Rio de Janeiro, RJ, Brasil. (in portuguese).

DNER/DrDTc (IPR). 1994. Classificação de solos tropicais para finalidades rodoviárias utilizando corpos-de-prova compactados em equipamento miniatura. *DNER-CLA 259/94*. (in portuguese).

Fassbender, H.W. 1966. La Adsorcion de fosfato em suelos fuertemente ácidos y su evaluacion usando la isoterma de Langmuir. *Fitot. Lat. Am.* 3, 203-216.

Grim, R.E. 1968. *Clay Mineralogy*, 2ª ed. McGrow-Hill Book New York, Company.

Jessberger, H.L. & Onnich, K. 1993. Calculations of pollutants emissons through mineral liners based on laboratory tests. In Proc. of 10th Int. Clay Conference. Adelaide, Austrália.

Ker, J.C. 1995. *Mineralogia, Sorção e Dessorção de fosfato, Magnetização e elementos traços de latossolos do Brasil.* Tese de D.Sc., UFV, Viçosa, MG, Brasil. (in portuguese).

Larsen, S. 1967. Soil Phosphorus. A.G. Nornon (ed). In *Advances in Agronomy. Prepared under a Auspices of the American Socity of Agronomy.* University of Michigan, Anon Arbor, Michigan.

Lerman, A. 1979. *Geochemical Processes: Water and Sediment Environments.* John Wiley & Sons, New York.

Lopes, M.S. 1977. *Relação entre o pH e a absorção de fósforo e silício em solos.* Tese de M.Sc., UFRGS, Porto Alegre, RS, Brasil. (in portuguese).

Mendonça, R.M.G. 2000. *Anions transport and retention in a kaolinitic saprolitic soil from Rio de Janeiro.* Tese of Ms.Sc., COPPE/UFRJ, Rio de Janeiro, RJ, Brasil. (in portuguese).

Nogami, J.S., & Villibor, D.F. 1995. *Pavimentação de baixo custo com solos lateríticos.* 1. Ed. São Paulo : Villibor,. (in portuguese).

Nogami, J.S.; Cosolino, V.M.N.; Villibor, D.F. 1989. Meaning of Coefficients and Index of MCT Soil Classification for Tropical Soils. *Proc. 12ª Intern. Conf. Soil Mech. and Found. Engin.*, 1:547-550. Rotterdam : Balkema.

Resende, J.M.A. 1983. *Comparação de Índices de Sorção de Fósforo no Solo.* Tese de M.Sc., Viçosa, MG, Brasil. (in portuguese).

Ritter, E. 1998. *Efeito da salinidade na difusão e sorção de alguns íons inorgânicos em um solo argiloso saturado.* Tese de D.Sc., COPPE/UFRJ, Rio de Janeiro, RJ, Brasil. (in portuguese).

Rowe, R.K. & Booker, J.R. 1994. Program Pollute-1D Polluant Migrations Analysis Program Geotechnical Research Center. *Faculty of Enginneering Science,* UWO, London, Canada.

Shackelford, C. D & Redmond, P.L. 1995. Solute breacktrhough curves for processed kaolin at low flow rates. *Journal of Geotechnical Enginnereing* 121(1): 17-32.

USEPA - Environmental Protection Agency. 1992. Batch-type Procedures For Estimating Soil Adsorption of Chemicals. *Technical Resource Document,* EPA/530/SW-87/006-F, Washington, April.

4 Structure of clays

Clay Science for Engineering, Adachi & Fukue (eds) © 2001 Balkema, Rotterdam, ISBN 90 5809 175 9

Porosimetry and collapse studies on Ca-montmorillonite silts

Sudhakar M. Rao & P. Mohan Rami Reddy
Department of Civil Engineering, Indian Institute of Science, Bangalore, India

ABSTRACT: Mercury intrusion porosimetry (MIP) and double oedometer tests (DOT) are performed to examine the effect of variations in clay content (zero to thirty percent) on the soil structure and collapse potentials of laboratory prepared dry clayey silt specimens. Calcium montmorillonite and a local silt comprised the fine and coarse fractions of the laboratory specimens. Specimens were prepared by subjecting thick pastes of calcium montmorillonite - silt mixes to a few cycles of wetting and drying in the laboratory. MIP and DOT results showed that the clayey silt specimen containing the critical calcium montmorillonite content of 5 % was characterized by the maximum coarse pore content and maximum collapse strain. Comparatively, specimens containing the optimum clay contents of 35 % and 55 % exhibited no volume change on soaking at the experimental stresses of 100 and 200 kPa respectively.

1 INTRODUCTION

Earlier studies of El-Sohby and Rabbaa (1984) and Lawton et al. (1992) have hypothesized that the soil structures of collapsible clayey soils are significantly influenced by variations in their clay content . The variations in the soil structures are in turn responsible for the difference in collapse potentials of unsaturated clayey soils with different clay contents. Alwail et al. (1994) performed scanning electron microscopy (SEM) investigations to examine the influence of variations in clay content on the microstructure of compacted clay-sand mixes. Their SEM observations suggested that the manner in which the clay fraction bonds the sand particles is a function of the amount of clay present in the soil. Besides scanning electron microscopy, pore size distributions obtained from mercury intrusion porosimetry (MIP) are also quite sensitive to variations in the soil microstructure (Sridharan et al 1971, Bengochea et al. 1979, Delage et al. 1996). MIP studies are hence performed in this investigation to examine the changes in soil structure with variations in clay content of unsaturated (degree of saturation range between 2 and 19%) artificial soil specimens. Artificial soils were used in this investigation to control the amount of clay fraction present in the soils. Double oedometer tests (DOT) described by Jennings and Knight (1957) were also performed with the unsaturated laboratory specimens. The DOT results are interpreted in light of the micro-structural investigations.

DOT procedure prescribes two oedometer tests performed on identical pair of specimens. Lawton et al. (1989) and Alwail et al. (1992) noted that collapse potentials predicted by DOT procedure and that determined by soaked after loading test (SALT) were comparable in case of statically compacted specimens. The results of Rao et al. (1995) and Reddy (1997) have likewise shown that the collapse strains predicted by DOT and SALT methods were comparable for air-dried specimens prepared by subjecting thick soil pastes to a few cycles of capillary wetting and air-drying in the laboratory. Specimens were prepared for DOT and MIP measurements by the method of Rao et al. (1995) and Reddy (1997) in this investigation.

2 EXPERIMENTAL PROCEDURE

A local silt and calcium montmorillonite clay was used to prepare the artificial soils of this study. As calcium montmorillonite was present in the collapsible clay soils of Dudley (1970), El-Sohby and Rabbaa (1984) and Lawton et al. (1992), this clay mineral was used in the present study as well. The silt had a liquid limit of 45% and plasticity index of 14%. According to Unified Soil Classification System the soil was classified as ML. This plastic silt had a specific gravity of 2.66 and was predominantly composed of kaolinite mineral and a small proportion of quartz and mica. The calcium montmorillonite of this study was similar to the calcium montmorillonite used by El-Sohby and Rabbaa (1984). Calcium montmorillonites used in the present study and that used by El-Sohby and Rabbaa (1984) had comparable liquid limits of 110 % and 120 % and plasticity index of 73 % and 69 % respectively.

Table 1. Index properties of air-dried specimens

Property	Calcium montmorillonite content %				
	0	2.5	5	10	30
Silt %	92	90	87	76	66
Clay %	8	10	13	24	34
w_L %	45	46	49	52	60
I_p %	14	15	15	17	23
γ_d KN/m^3	11.7,	11.2,	10.8	12.2,	12.6,
	11.9	11.4	10.9	12.0	12.4
w %	1.0,	1.1,	1.4,	2.4,	8.1,
	0.9	1.1	1.4	2.3	8.0
S_r %	1.9,	2.1,	2.5,	5.3,	19.1,
	1.9	2.2	2.5	5.0	18.5

Artificial soils were prepared by mixing 2.5%, 5%, 10%and 30% calcium montmorillonite (on dry soil weight basis) with separate batches of the local silt. Thick soil pastes of the local silt and artificial soils were prepared by hand remolding these soil materials at their respective liquid limit water contents. Tap water was used to remold these soil materials at water contents ranging between 50 and 65%. The thick soil pastes were loosely filled in PVC rings of 102 mm diameter and 38 mm height at a void ratio of about 1.65.

The thick soil pastes were allowed to dry out at ambient temperatures of 25-30^{0}C. After about 8-10 days of air-drying the specimens in the PVC rings were observed to be dry. To ensure complete drying, the air-dried specimens were further dried at a slightly elevated temperature of 40^{0}C for 2 days. The moisture contents of these completely dried specimens ranged between 1 and10% while their void ratios ranged between 1.22 and 1.43.

To simulate the local climatic condition of alternate periods of rainfall and draught, the completely dried specimens were subjected to additional cycles of wetting and drying. The wetting of the dried specimens was performed by allowing them to absorb moisture due to capillary from the wetted sand bath in which the dried specimens were embedded.

A degree of saturation (S$_r$) of about 90 to 93% was achieved by this saturation technique. The wetted specimens were again dried at ambient temperature for 8 to 10 days followed by drying at a temperature of 40^{0}C for 2 days.

The void ratios of the capillary wetted and air-dried specimens were observed to become nearly constant after the second cycle of wetting and drying itself. Hence the laboratory desiccation process was terminated after completion of the third cycle of capillary wetting and drying. Completion of three cycles of wetting and drying required about 40 days. Index properties of the capillary wetted and dried specimens were determined after the third drying cycle and the results are presented in Table 1. The silt and various calcium montmorillonite-specimens subjected to three cycles of wetting and drying in the laboratory are referred to as desiccated specimens in further discussions. Bulk of the specimens in Table 1 have moisture contents (w) between 1 and 2% and S$_r$ values between 2 and 5% after the third cycle of drying. Only the desiccated 30% calcium montmorillonite-silt specimen has higher moisture content and S$_r$ values of 8 and 19% respectively after the third cycle of drying.

Table 1 also presents data to demonstrate the degree to which the DOT specimens prepared by the cyclic wetting and drying process are identical. The dry unit weight (γ_d) of any pair of desiccated specimens differ from their average value by < 0.5%. Comparatively, the moisture contents of any pair of desiccated specimens differ from their mean by < 2%. The near similar dry densities and moisture contents of pairs of desiccated specimens in Table 1 satisfy the DOT requirement of conducting tests on identical pair of specimens.

2.1 DOT procedure

After completion of the third drying cycle, consolidation rings of 76 mm diameter and 25 mm height were carefully pushed into the desiccated soil specimens contained in the larger diameter PVC rings. The specimens displaced into the consolidation rings were trimmed to a height of 25 mm. DOT were conducted on near identical pair of desiccated specimens listed in Table 1. In this procedure, two identical specimens of a soil are tested in conventional oedometer assemblies. One specimen is stepwise loaded in the unsoaked condition up to a maximum pressure of 800 kPa. The other specimen is inundated with tap water at a nominal pressure of 6.25 kPa. After the heave or collapse following the addition of water had ceased, additional loading were applied to the soaked specimen.

Duration and sequence of loading were identical for the unsoaked and soaked specimens. Figure 1 plots the data for one such pair of tests. The difference between the vertical strains ($\Delta\varepsilon$) of the soaked and unsoaked desiccated silt specimens at any vertical stress in Fig.1 represents the collapse strain that this specimen experiences on wetting at the specified load. It may be noted from Fig. 1 that the dry silt specimen has a tendency to experience collapse on wetting at all experimental stresses.

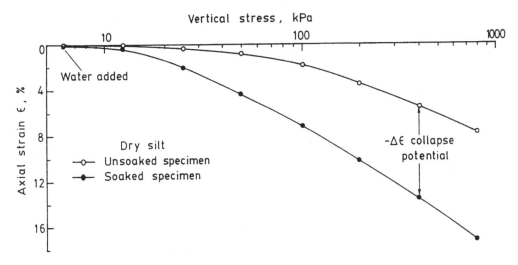

Figure 1. DOT plots of dry silt

2.2 *MIP procedure*

Trimmings were obtained when the oedometer rings were carefully pushed into the larger diameter PVC rings containing the desiccated specimens. These trimmings were used for MIP measurements. This ensured that specimens used in DOT and MIP experiments belonged to the same parent specimen.

The mercury intrusion porosimeter used in this study has a pressurizing capacity of 60,000 psi (413.7 MPa) and is capable of intruding pore spaces ranging in diameter from 2 to 0.002 μm. Briefly pore size distribution of a specimen is determined in the following manner. The desiccated specimen is surrounded by mercury in a specially fabricated glass tube. The pressure is raised in small increments and the volume of the mercury entering the specimen after each increment is automatically recorded. Each pressure increment forces mercury into the accessible soil pores of a diameter larger than or equal to that calculated by the Washburn equation for the given pressure. In this manner, 'the volume of the pore space between pressure increments and thus diameter decrease is recorded by the instrument generating a pore size distribution curve.

3 RESULTS AND DISCUSSION

3.1 *Double oedometer test results*

Fig. 2 presents the effect of variations in the calcium montmorillonite content on the collapse potentials of the desiccated soil specimens at vertical stresses of 100, 200, 400, and 800 kPa respectively. The results in Fig. 2 show that at all the experimental pressures, collapse potential is a maximum for the desiccated clayey silt specimen containing a calcium montmorillonite content of 5 percent.

The effect of variations in montmorillonite content on the collapse potentials of montmorillonite-silt mixes and montmorillonite-sand mixes have been previously reported by El-Sohby and Rabbaa (1984), Lawton et al. (1992) and Rollins et al. (1994).Statically compacted specimens (El-Sohby and Rabbaa 1984, Lawton et al. 1992) or specimens dried out from thick soil slurries (Rollins et al. 1994) were used for collapse potential measurements. Results of these investigations have also shown the existence of a critical clay percentage for maximum collapse for most mixes. These workers observed that collapse potentials of compacted sand-clay mixes and silt-clay mixes reach a maximum value at clay fraction between 10 to 40 % depending on the composition of the mix and the pressure at which the mix is inundated. It is relevant to note that similar to the artificial soils, the clay content appears to be critical in determining the maximum collapse strains of natural soils as well. For example, Dudley (1970) reported that the collapsible clayey silts found in San -Joaquin valley California exhibited maximum collapse when the clay amounted to 12 % of the solids.

3.2 *Micro-structural investigations*

Fig. 3 presents the cumulative volume intruded (cm^3/g) versus pore radius (μm) plots of the curve displays the fraction of pore space greater or smaller desiccated silt and the various, desiccated calcium montmorillonite + silt specimens. The cumulative distribution than a given size, and the fraction of the pore space between any two unmeasured pore sizes can be quickly evaluated. Examination of the pore size distribution curves in Fig.3 reveals that these curves are composed of more or less three regions of nearly constant slopes: (1) a coarse pore region of steep slope ranging from 2μm to 0.14μm (2) a fine pore region of gentle slope ranging from 0.14μm to 0.05μm and (3) a flattish very fine pore region ranging from 0.05μm to 0.002μm. It is realized that this selection of pore size ranges is subjective and alternative pore size ranges could be selected from the presented data in Fig. 3. The pore size distribution curves in Fig. 3 also reveal that the maximum cumulative volume of mercury intruded is approximately equal to the calculated volume of voids, thus indicating that all pore spaces were successfully intruded at the maximum pressure. Examination of the curves in Fig. 3 reveal that the bulk of the mercury intrusion occurs in the coarse pore region of all the dry specimens.

The curves also reveal that the presence of calcium montmorillonite causes these clayey silt specimens to generally possess a larger coarse pore content than the desiccated silt specimen (coarse pore content = 0.27cm^3/g). Among the calcium montmorillonite bearing specimens, the specimen containing the critical clay content of 5% exhibits the largest coarse pore content of 0.39cm^3/g. The desiccated 2.5% calcium montmorillonite + silt specimen, the desiccated 10% calcium montmorillo-

nite + silt specimen and the desiccated 30% calcium montmorillonite + silt specimen exhibit lower coarse pore contents of 0.35, 0.31 and 0.28 cm^3/g respectively. The different shape of the desiccated silt curve in Fig. 3 suggests that for the same pressure increment, lesser amount of mercury intrudes the coarse pore region of the desiccated silt specimen in comparison to that intruding the coarse pore region of the desiccated calcium montmorillonite specimens.

To examine the reliability of the MIP measurements, MIP tests were repeated on an additional desiccated silt specimen and an additional desiccated 5% calcium montmorillonite-silt specimen. Repeat testing revealed that the test variability for the same material is between ± 5 to 10% of the measured coarse pore content value. The desiccated 2.5% calcium montmorillonite-silt specimen, the desiccated 5% calcium montmorillonite-silt specimen, the desiccated 10% calcium montmorillonite specimen and the desiccated 30% calcium montmorillonite-silt specimen have 30%, 44%, 15% and 4% larger coarse pore contents than that of the desiccated silt specimen. These results suggest that except for the coarse pores content of the desiccated 30% calcium montmorillonite-silt specimen, the higher coarse

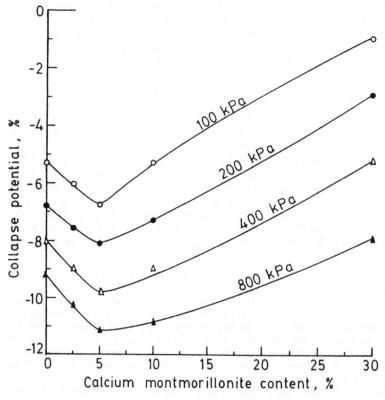

Figure 2. Effect of calcium montmorillonite content on collapse potential

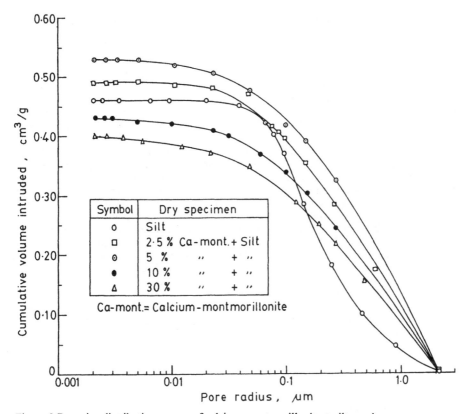

Figure 3 Pore size distribution curves of calcium montmorillonite + silt specimens

pore contents of the other specimens in comparison to that of the desiccated silt specimen cannot be attributed to experimental variations.

Fig. 4 plots the coarse pores contents (pore size = 2 to 0.14μm) versus the collapse potentials of the cyclically wetted and dried specimens at vertical Stresses of 100 and 200 kPa. The values in parenthesis indicate the amount of calcium montmorillonite present in the desiccated specimens. These plots suggest that the desiccated 3 % calcium montmorillonite-silt specimen having a 28% lower coarse pores content than the desiccated specimen with the critical coarse pores content of 0.39 cm³/g will exhibit no volume change on inundation under the vertical stress of 100 kPa. Likewise, the desiccated 55% calcium montmorillonite-silt specimen having a 33% lower coarse pores content than the critical coarse pores content of 0.39 cm³/g will exhibit no volume change on soaking under the vertical stress of 200 kPa. The clay contents of the desiccated specimens that exhibit no volume change on soaking under a given vertical stress are termed as the optimum clay contents. Apparently at the optimum clay content, the collapse strain due to collapse of the soil structure equals the swelling strain due to hydration of the calcium montmorillonite clay on ingress of

water. As a consequence, the desiccated specimen containing the optimum clay content of 35% and optimum coarse pores content of 0.28 cm³/g or the desiccated specimen containing the optimum clay content of 55% and optimum coarse pores content of 0.26 cm³/g exhibits no volume change on soaking under the vertical stresses of 100 or 200 kPa respectively. Experimental results of El-Sohby and Rabbaa (1984) have shown that the optimum clay content of their artificial soils decreased with an increase in the initial dry unit weight and degree of saturation of the soils. Hence, besides the overburden pressure, the difference between the critical and optimum coarse pores contents of a clay soil might also depend on its pre-collapse dry unit weight and/or degree of saturation values.

3.3 *Interpretation of porosimetry and collapse potential results*

According to Lawton et al. (1992), the calcium montmorillonite binds the sand or silt particles into macropeds and with increasing clay contents in the low range (<40%) more macropeds develop. The desiccated 5% calcium montmorillonite + silt specimen exhibits a relatively higher coarse pore content than all the other desiccated specimens. Based on the porosimetry results in Fig. 3 it may be reasoned that the maximum number of macropeds develop at the critical clay content of 5%. At clay contents lesser or greater than the critical clay content, fewer macropeds develop in comparison to that developed at the critical clay content. For the calcium montmorillonite range used in this study (≤ 30%), SEM observations of Alwail et al. (1994) suggest that the clay bonds are formed in the desiccated specimens by bridging and coating of the silt grains.

The larger coarse pore content of the desiccated 5% calcium montmorillonite + silt specimen is reflective of the presence of a more open soil structure in this specimen.. The presence of a more open soil structure is apparently responsible for the higher collapse potentials of the desiccated 5% calcium montmorillonite + silt specimen than the other specimens in Fig.2 containing calcium montmorillonite contents less than the critical clay content. Comparatively, the presence of calcium montmorillonite contents in excess of the critical clay content (namely 5%) causes the desiccated clayey silt specimens to possess less open soil structures in the pre-collapse state (reflected from the MIP results in Fig.3 and the higher dry unit weight values in Table1) and also develop stronger net repulsion forces on saturation (Lawton et al., 1992). Apparently a combination of theses two factors causes the desiccated 10% calcium montmorillonite + silt specimen and the desiccated 30% calcium montmorillonite + silt specimen to exhibit lower collapse potentials than the desiccated 5% calcium montmorillonite + silt specimen in Fig.2.

Examination of the plots in Figs. 2 and 3 also reveals that though the desiccated silt specimen and the desiccated 30% calcium montmorillonite + silt specimen have comparable coarse pores contents of 0.27 cm^3/g and 0.28 cm^3/g respectively, the former specimen exhibits significantly larger collapse strains than the latter specimen especially at the lower overburden pressures. Apparently, the absence of the swelling montmorillonite clay and the much lower pre-collapse moisture content and S_r values of the desiccated silt specimen (0.9% and 2% respectively) cause it to experience larger collapse strains than the desiccated 30% calcium montmorillonite + specimen at the lower overburden pressures.

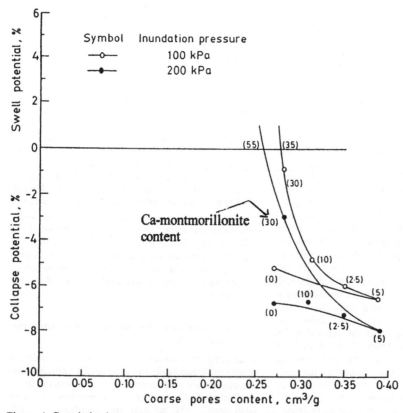

Figure 4. Correlation between collapse potential and coarse pore content

4 CONCLUSIONS

Specimens were prepared for collapse potential and MIP measurements by subjecting thick pastes of calcium montmorillonite - silt mixes to a few cycles of capillary wetting and air-drying in the laboratory.. MIP measurements with these desiccated specimens suggested that maximum macropores (size range 2 to 0.14 um) development occurred at a critical clay content of 5%. Expectedly, maximum collapse strains were exhibited by the desiccated specimen containing the critical clay content of 5% and critical coarse pores content of 0.39 cm^3/g at all the experimental pressures. Experimental results also showed that the desiccated specimen containing the optimum clay content and the optimum coarse pores content exhibits no volume change on soaking under a given vertical stress as the collapse strain of this specimen equals it swelling strain . Both the critical clay content and optimum clay content of a clay soil will be affected by variations in the pre-collapse dry density, moisture content and the vertical stress at which the unsaturated specimens are inundated. The results of this study thus suggest that compacted earth structures designed to contain the optimum clay content are unlikely to heave or collapse on wetting under the design overburden pressure. Comparatively, a compacted earth structure constructed without prior knowledge of its critical clay content and/or optimum clay content is likely to experience heave or collapse on saturation.

5 REFERENCES

Alwail, T. & R. J. Fragaszy 1992. Collapse mechanism of low cohesion compacted soils. *Bulletin of Association of Engineering Geologists.* 29: 345-353.

Alwail, T., C. L. Ho & R. J.Fragaszy 1994. Collapse mechanism of compacted clayey and silty sands. *Proceedings Setllmement 94, ASCE Specialty Publication No. 40.* 2: 1435- 1441

Bengochea, I.G., Lovell, C. W. and Altschaeffl, A. G. 1979. Pore distribution and permeability of silty clays. *Journal of Geotechnical Engineering, ASCE.* 105: 839-856.

Delage, P., M. Cui Andiguier & M. D. Howat 1996. Micro-structure of a compacted silt. *Canadian Geotechnical Journal.* 33 : 150-158.

Dudley, J. H. 1970. Review of collapsing soils. *Journal of Soil Mechanics and Foundation Engineering, ASCE.* 96: 925-947.

El-Sohby, M. A. & S. A. Rabbaa 1984. Deformational behavior of unsaturated soils upon wetting. *Proc. 8th Regional Conf. for Africa on Soil Mechanics and Foundation Engineering, Harare, Vol. 1*: 129-137.

Jennings J. E. and Knight, K. 1957. The additional settlement of foundations due to collapse of a structure of sandy sub-soils upon wetting. *Proc. 4th International Conf. on Soil Mechanics and Foundation Engineering, London 1*:316-319.

Lawton, E. C., R. J. Fragaszy & J. H. Hardcastle 1989. Collapse of compacted clayey sand. *J. Geotechnical Engineering,ASCE.* 115: 1252-1267.

Lawton, E. C., Fragaszy, R. J. and Hotherington, M. D. 1992. Review of wetting induced collapse in compacted soils. *J. of Geotechnical Engineering, ASCE.* 118: 1376-1394.

Mohan Rami Reddy, P. 1992. *Role of physicochemical factors in the heave and collapse behaviour of laboratory desiccated soils.* Ph.D. Thesis. Indian Institute of Science, India.

Rao, S. M., A. Sridharan & K. P. Ramanath 1995. Collapse behaviour of an artificially cemented clayey silt. *ASTM Geo technical Testing Journal.* 18 : 334-341.

Rollins, K. M., R. L. Rollins, T. D. Smith & G. H. Beckwith 1994. Identification and characterization of collapsible gravels. *J. Geotechnical Engineering, ASCE.* 120 : 528-542.

Sridharan, A., A. G. Altschaeffl, and S. Diamond 1971. Pore size distribution studies. *Journal of Soil Mechanics and Foundation Engineering, ASCE.* 97: 771-787.

Clay Science for Engineering, Adachi & Fukue (eds) © 2001 Balkema, Rotterdam, ISBN 90 5809 175 9

Quantitative evaluation of microstructure of clays

T. Moriwaki
Department of Civil Engineering, Hiroshima University, Japan

Y. Wada
Hanshin Consultant Company Limited, Osaka, Japan

ABSTRACT: In order to understand essential mechanism of behavior of clays consisting of clay particles and pore space, it is necessary to evaluate the microstructure of clays quantitatively. An evaluation method using computer image processing was developed in this study. Density distributions of Scanning Electron Microscope (SEM) photographs taken for horizontal and vertical cross-sections of clay samples were measured by using a computer image processing, and then the power spectrum for the measured density distributions were analyzed in this method. It was found that the density distribution of the SEM photographs and their power spectrum could be adopted quantitative parameters to express the microstructure of clays. A difference of the microstructure of clay samples consolidated at a high temperature and the room temperature was quantitatively discussed by using the developed method.

1 INTRODUCTION

Many researchers (Tsuchida et al. 1991, Moriwaki et al. 1994, etc) have reported that mechanical properties of clays reconsolidated at a high temperature were similar to those of undisturbed natural clays having aging effects. In these researches, macro-behavior obtained by usual mechanical tests, such as the oedometer consolidation test, the unconfined compression test and the triaxial test, on high temperature reconsolidated clay samples and undisturbed natural clay samples were simply compared with each other. However, the similarities of both clay samples were not discussed from the viewpoint of microstructure of clay particles. As clays are kinds of granular materials, it is necessary to evaluate the microstructure of clays quantitatively in order to understand essential mechanism of behavior of clays. In this study, therefore, an evaluation method using computer image processing was developed and its applicability was discussed by using kaolin clay samples consolidated in laboratory. In this method, density distributions of Scanning Electron Microscope (SEM) photographs taken for horizontal and vertical cross-sections of clay samples were measured by using a computer image processing, and then the power spectrum for the measured density distributions were analyzed to evaluate the microstructure of clays quantitatively. Moreover, the microstructure of the high temperature reconsolidated clay samples was compared with that of the room

temperature reconsolidated clay samples by the proposed method in order to discuss the ability of the duplication of the naturally aged clay by the high temperature reconsolidation method.

2 ANALYSIS METHOD OF CLAY MICROSTRUCTURE

In the analysis, photographs of the microstructure of clay samples were firstly taken by using a SEM. The image data of the SEM photographs were read and inputted to the personal computer as digital data by the image scanner, and the obtained digital image data were analyzed by using computer image processing. The software used in the analysis was Mac SCOPE produced by Mitani Corporation. This software is able to perform the image processing such as separating the contacted particles to the individual particle and also to measure the size of particles and the density of the image.

As clay particles are very small and hardly exist in individual state such as sand particles, it is difficult to measure the size or area of the individual particle. The aggregate of some clay particles, which is called as ped, is a basic unit and the structure of the peds is a very important factor to discuss the microstructure of clays. In this study, therefore, the density distribution of the SEM photographs taken for the horizontal and vertical cross-sections of clay samples was measured by using the computer image

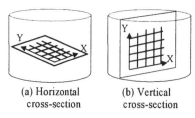

(a) Horizontal
cross-section

(b) Vertical
cross-section

Fig. 1 Analysis method for image processing

processing mentioned above, and the microstructure of the clay samples was evaluated by the shape of the obtained density distributions.

In the measurement of the density distribution, some straight lines were firstly drawn in the right-angled X and Y directions on the SEM photographs as shown in Fig.1, and then the density of the SEM photographs was measured on those lines. The density of the SEM photographs is indicated as the value divided into 256 levels between white and black at every 1 pixel on the drawn line. The pixel is the minimum unit of the digital image data, and the length of 1 pixel was 0.1693mm in this analysis because the used resolution of the image was 150dpi. Although the directions of X and Y axes in the horizontal cross-section of the SEM photographs do not indicate specific directions, the directions of X and Y axes in the vertical cross-section were selected to agree with the horizontal and vertical directions, respectively.

3 DIFFERENCE OF CLAY MICROSTRUCTURE BETWEEN HORIZONTAL AND VERTICAL CROSS-SECTIONS

Photographs 1 to 6 are SEM photographs taken for the horizontal and vertical cross-sections of kaolin clay (ASP100) samples one-dimensionally consolidated under the consolidation pressures of 76kPa, 1.25MPa, 5.0MPa and 10.0MPa, which were taken by Prof. H. Kazama of Saitama University. Firstly, a difference of clay microstructure between the horizontal and vertical cross-sections is discussed based on Photos. 5 and 6, which are the SEM photographs of the sample consolidated under the very large consolidation pressure of 10.0MPa. Many disks of particles can be seen in the horizontal cross-section of Photo. 5 and an oriented structure as the disk-shaped particles pile in parallel to a horizontal plane is observed in the vertical cross-section of Photo. 6. It is found from the vertical cross-section of Photo 6 that some peds having a thickness of several micrometers have been formed from some particles.

Figure 2 shows typical density distributions in the horizontal cross-section shown in Photo. 5. A difference of the density distribution between X-axis and Y-axis is hardly seen in the horizontal cross-section

and portions having a constant value of the density exist in the density distribution curves everywhere. Those portions are considered to express the disks of clay particles. The length of those portions having a constant value of the density is about $1\,\mu$m and agrees with the disk size of the particles observed in Photo. 5. Figure 3 shows typical density distributions on X-axis and Y-axis in the vertical cross-section of Photo. 6 and a difference between X-axis and Y-axis is seen in the vertical cross-section. On X-axis coinciding with the horizontal direction in the vertical cross-section, the density cyclically varies with the relatively long wavelength of about $2\,\mu$m as shown in Fig. 3(a). This length is considered to express the horizontal length of the peds formed from the disk-shaped particles aggregating in parallel to a horizontal plane. On Y-axis coinciding with the vertical direction in the vertical cross-section, the density cyclically varies with the relatively short wavelengths of about $0.2\sim0.3\,\mu$m and $0.5\sim1.5\,\mu$m as shown in Fig. 3(b). It is considered that the former value expresses the thickness of the disk-shaped particle and the latter value expresses the vertical thickness of the peds formed from the disk-shaped particles mentioned above.

In order to evaluate quantitatively those features of the density distributions, the density distribution curves shown in Figs. 2 and 3 were regarded as a kind of wave and the power spectrum analysis was carried out for those density distribution curves. Figures 4 and 5 show the results of the power spectrum analysis for the density distribution curves in Figs 2 and 3. A waveform transformation as a linear regression line for the density distribution curve is the central axis of the wave was previously conducted in this analysis. The axis of abscissa in those figures is frequency, which is reciprocal number of the length of wave, and its unit is therefore μm^{-1}. The axis of ordinate is power spectrum value of the density distribution and its unit is density$^2 \cdot \mu$m. Because it is considered that a white portion expresses a particle or a ped and a black portion expresses a pore space in the density distribution curve, the size of the particle or the ped is calculated from the predominant frequency by the following equation.

$$S \text{ (size of particle or ped)} = \frac{1}{2 \times F \text{(frequency)}} \quad (1)$$

It is found from Fig. 4 that the power spectrum of the density distribution in X-direction on the horizontal cross-section is similar to that in Y-direction and the predominant frequency in both directions is about $F=0.5\,\mu$m^{-1} (S=$1.0\,\mu$m). On the other hand, the power spectrum in X-direction on the vertical cross-section is different from that in Y-direction. Namely, the predominant frequency in X-direction corresponding with the horizontal direction is about

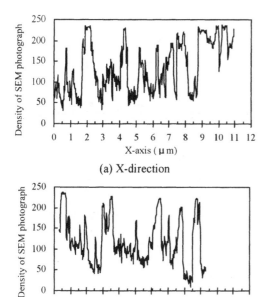

(a) X-direction

(b) Y-direction

Fig. 2 Density distribution of SEM photograph taken for horizontal cross-section of kaolin clay consolidated under p=10.0MPa

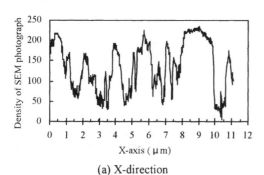

(a) X-direction

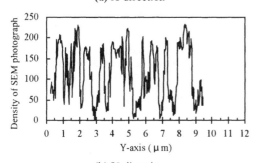

(b) Y-direction

Fig. 3 Density distribution of SEM photograph taken for vertical cross-section of kaolin clay consolidated under p=10.0MPa

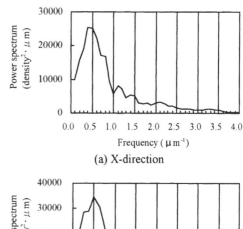

(a) X-direction

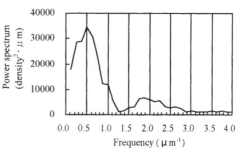

(b) Y-direction

Fig. 4 Power spectrum for density distribution of SEM photograph taken for horizontal cross-section of kaolin clay consolidated under p=10.0MPa

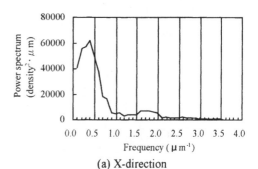

(a) X-direction

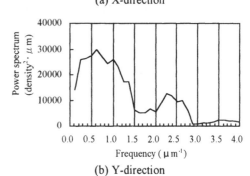

(b) Y-direction

Fig. 5 Power spectrum for density distribution of SEM photograph taken for vertical cross-section of kaolin clay consolidated under p=10.0MPa

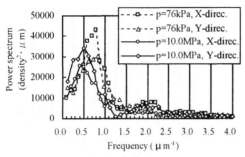

Fig. 6 Power spectrum for density distribution of SEM photographs taken for horizontal cross-sections of kaolin clays consolidated under p=76kPa and 10.0MPa

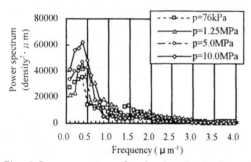

Fig. 7 Power spectrum for density distribution of SEM photographs in horizontal direction on vertical cross-section of kaolin clays consolidated under p=76kPa to 10.0MPa

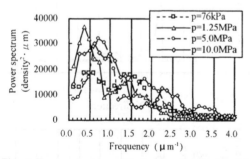

Fig. 8 Power spectrum for density distribution of SEM photographs in vertical direction on vertical cross-section of kaolin clays consolidated under p=76kPa to 10.0MPa

F=0.3 μ m^{-1} (S=1.7 μ m) and the predominant frequency in Y-direction corresponding with the vertical direction is about F=0.6 μ m^{-1} (S=0.8 μ m) and F=12.2 μ m^{-1} (S=0.2 μ m). Those results well express both results of the clay microstructure shown in Photos. 5 and 6, and the density distribution shown in Figs. 2 and 3.

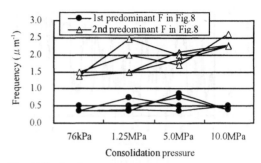

Fig. 9 Predominant frequencies in power spectrum for density distribution of SEM photographs in vertical direction on vertical cross-section of kaolin clays consolidated under p=76kPa to 10.0MPa

4 RELATION BETWEEN CLAY MICROSTRUCTURE AND CONSOLIDATION PRESSURE

Photographs 1 and 5 are the SEM photographs taken for a horizontal cross-section of the Kaolin clay samples consolidated under the consolidation pressures of quite different values. Comparing those two photographs, the disk size of many particles in the case of the consolidation pressure p=76kPa is smaller than 1 μ m and some particles of them are slant as shown in Photo. 1. On the other hand, the slant particle is hardly seen in the case of the consolidation pressure p=10.0MPa and many particles whose disk size is larger than 1 μ m exist in Photo. 2. The power spectrums analyzed by the proposed method for those two photographs are shown in Fig. 6. It is found from this figure that the predominant frequency in the case of the small consolidation pressure is F=0.7~0.8 μ m^{-1} and are different from F=0.4~0.5 μ m^{-1} in the case of the large consolidation pressure. The former value and the latter value imply that the predominant sizes of particle in the cases of the small and large consolidation pressures are S=0.6~0.7 and 1.0~1.3 μ m, respectively. Those values of the predominant size quantitatively express the tendency of the clay structure read from Photos. 1 and 5.

Photographs 2, 3, 4 and 6 show the SEM Photographs taken for a vertical cross-section of the Kaolin clay samples consolidated under the consolidation pressures of 76kPa, 1.25MPa, 5.0MPa and 10.0MPa. It is found from those photographs that the density of particles and the degree of an oriented structure increase with the increase of the consolidation pressure. Figures 7 and 8 show the results of the power spectrum analysis for the horizontal and vertical directions in those Photos. 2, 3, 4 and 6. In Fig. 7, the predominant frequency of the density distribution in the horizontal direction on the vertical cross-section is the same value of F=0.2~0.5 μ m^{-1} in spite of the difference of the consolidation pressure.

Photo. 1 Horizontal cross-section of kaolin clay consolidated under p=76kPa (taken by Kazama)

Photo. 4 Vertical cross-section of kaolin clay consolidated under p=5.0kPa (taken by Kazama)

Photo. 2 Vertical cross-section of kaolin clay consolidated under p=76kPa (taken by Kazama)

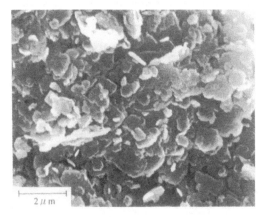

Photo. 5 Horizontal cross-section of kaolin clay consolidated under p=10.0MPa (taken by Kazama)

Photo. 3 Vertical cross-section of kaolin clay consolidated under p=1.25MPa (taken by Kazama)

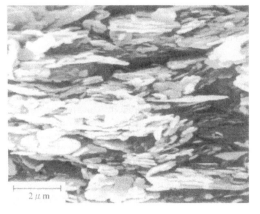

Photo. 6 Vertical cross-section of kaolin clay consolidated under p=10.0MPa (taken by Kazama)

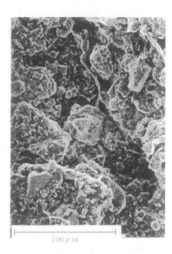

Photo. 7 Horizontal cross-section of Kurashiki clay consolidated at room temperature

Photo. 9 Horizontal cross-section of Kurashiki clay consolidated at high temperature

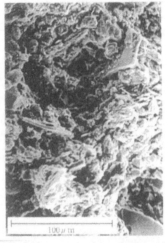

Photo. 8 Vertical cross-section of Kurashiki clay consolidated at room temperature

Photo. 10 Vertical cross-section of Kurashiki clay consolidated at high temperature

This value corresponds to the size of S=1.0~2.5 μ m and is considered to express the length of the peds observed in Photos. 2, 3, 4 and 6. It is found from Fig. 8 that there are two predominant frequencies of the density distribution in the vertical direction on the vertical cross-section. The first predominant frequency is the same value of F=0.3~0.7 μ m^{-1} in spite of the difference of the consolidation pressure but the second predominant frequency tends to increase with the increase of the consolidation pressure. Figure 9 shows this tendency. The second predominant frequency increases from F=1.4 μ m^{-1} to 2.5 μ m^{-1} with the increase of the consolidation pressure from 76 kPa to 10.0 MPa, and those values of frequency correspond to the size of S=0.4 and 0.2 μ m. This is

considered to imply that the slant particles observed in Photos. 2 and 3 is gradually arranged in parallel to a horizontal plane with the increase of the consolidation pressure as shown in Fig. 6 and then the thickness of the peds was measured as the predominant size in the clay structure in the case of the large consolidation pressure.

5 EFFECT OF HIGH TEMPERATURE CONSOLIDATION ON CLAY MICROSTRUCTURE

The clay used in this analysis was alluvial marine clay called as Kurashiki clay. The physical properties of this clay were w_L=59.9%, w_P=26.7%, I_P=33.2

and G_s=2.683. The clay sampled from in-situ was screened through a sieve of 0.420mm and mixed into the slurry with the water content twice of its liquid limit. The slurry was finally consolidated by the consolidation pressure of 49.0kPa under one-dimensional condition. In the last consolidation pressure, each clay sample was cured at 20 or 70℃ for 4 days. The clay samples cured at 20 and 70℃ are hereafter called as RT sample and HT sample, respectively.

The SEM photographs taken for a horizontal cross-section and a vertical cross-section of the RT sample are shown in Photos. 7 and 8, and the SEM photographs of the HT sample also are shown in Photos. 9 and 10. On the RT sample, many disks of the particle and peds are observed in the horizontal cross-section and the disk-shaped particles and peds pile in parallel to a horizontal plane. As the results, the RT sample is considered to have an oriented structure in parallel to a horizontal plane. On the other hand, although an oriented structure in parallel to a horizontal plane is slightly observed in the vertical cross-section of the HT sample, the difference between the horizontal cross-section and the vertical cross-section does not appear on the HT sample than on the RT sample. This is considered to imply that the HT sample has a random structure. Those observational results for the clay microstructure of the RT

and HT samples well correspond to the results mentioned by Kazama (1997).

The density distributions of the SEM Photographs 7 to 10 taken for the horizontal and vertical cross-sections of the RT sample and the HT sample were measured and their power spectrums were analyzed by the proposed method. Figures 10 and 11 show the obtained power spectrums of the RT sample and the HT sample, respectively. The predominant frequencies in the horizontal cross-section of the RT sample are the same in both of X-direction and Y-direction as shown in Fig. 10(a). Those values are F= 0.01~0.03 μ m^{-1} (S=25 μ m) and F=0.05~0.08 μ m^{-1} (S=8 μ m), and are considered to express the representative values of the horizontal length of the peds formed by some particles aggregating in parallel to a horizontal plane.

As shown in Fig. 10(b), the predominant frequency in the vertical cross-section of the RT sample is almost the same between X-direction and Y-direction in the region where the frequency is smaller than F=0.15 μ m^{-1} as the same as the result in the horizontal cross-section. However, the predominant frequency is different between X-direction and Y-direction in the region where the frequency is larger than F=0.15 μ m^{-1}. Namely, in the region of the large frequency, the obvious peak of the power spectrum could not be found in X-direction but the predominant frequency exists at F=0.15~0.17 μ m^{-1}

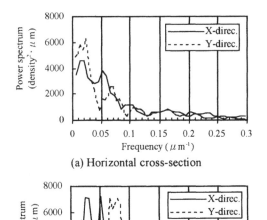

(a) Horizontal cross-section

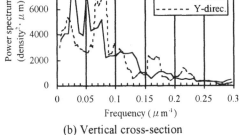

(b) Vertical cross-section

Fig. 10 Power spectrum for density distribution of SEM photograph of Kurashiki clay consolidated at room temperature

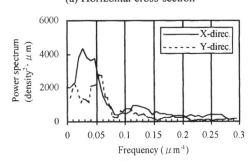

(a) Horizontal cross-section

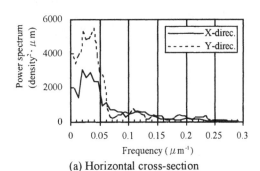

(b) Vertical cross-section

Fig. 11 Power spectrum for density distribution of SEM photograph of Kurashiki clay consolidated at high temperature

(S=3 μm) and F=0.21~0.23 μm^{-1} (S=2 μm) in Y-direction. Those predominant values are considered to correspond to the thickness of the particle. Therefore, it is recognized that the RT sample has the oriented structure as the particles and the peds pile in parallel to a horizontal plane.

The power spectrums of the density distribution in the horizontal and vertical cross-sections of the HT sample are shown in Fig. 11. In the horizontal cross-section of the HT sample shown in Fig. 11(a), the difference of the predominant frequency between X-direction and Y-direction is not recognized as the same as the RT sample and the predominant frequency is F=0.02~0.05 μm^{-1} (S=20 μm). This value expresses a horizontal length of the peds formed by some particles aggregating in parallel to a horizontal plane. Also, the difference of the power spectrum between X-direction and Y-direction in the vertical cross-section of the HT sample is not recognized even in the region of the large frequency. This result is different from the result on the RT sample. And the power spectrums in the vertical cross-section almost agree with those in the horizontal cross-section of the HT sample. This implies that the clay microstructure of the HT sample is almost similar in both of the horizontal and vertical cross-sections and therefore the clay microstructure of the HT sample is more random than that on the RT sample.

Some micro diatoms are observed in Photos. 7 to 10 of Kurashiki clay. However, it is impossible to evaluate the diatom content and the effects of diatoms on the microstructure of Kurashiki clay by using the proposed method, because only the predominant size of particles and peds can be obtained as the length of them on the straight line measured the density of the SEM photograph but the shape of particles and peds cannot be distinguished by the proposed method.

6 CONCLUSIONS

An analysis method was developed and its applicability was discussed in this study for evaluating quantitatively microstructure of clays. In this method, density distributions in the right-angled lines drawn on horizontal and vertical cross-sections of SEM photographs on clay samples were measured by using a computer image processing, and then the power spectrum for the measured density distributions were analyzed. As the results, it was found that the density distribution of the SEM photograph and its power spectrum could be used as quantitative indexes to indicate the clay microstructure. Comparing the microstructure of the clay samples consolidated at high and room temperatures by using the proposed method, it was found that the high temperature reconsolidated clay sample has a random structure but the room temperature reconsolidated clay sample has an oriented structure.

ACKNOWLEDGEMENTS

The authors are deeply grateful to Prof. H. Kazama of Saitama University for offering valuable SEM photographs.

REFERENCES

Kazama, H. 1997. Microstructure of clayey soils, Report of the technical committee on effects of high temperature environment on geotechnical properties of clayey soils, Japanese Geotechnical Society, pp.7-13.
Moriwaki, T., K. Yashima and M. Nago 1994. Shear deformation characteristics of Hiroshima clay cured at high temperature, Proc. of the International Symposium on Pre-failure Deformation of Geomaterials, Vol.1, pp.119-124.
Tsuchida, T., M. Kobayashi and J. Mizukami 1991. Effect of aging of marine clay and its duplication by high temperature consolidation, Soils and Foundations, Vol.31, No.4, pp.133-147.

Clay Science for Engineering, Adachi & Fukue (eds) © 2001 Balkema, Rotterdam, ISBN 90 5809 175 9

Investigation on the microstructure of Ariake clay from mechanical viewpoint

K. Onitsuka, T. Negami & J. N. Shen
Department of Civil Engineering, Saga University, Japan

ABSTRACT: The relationships between the mechanical properties and the microstructure of clayey soils are not clear. The study presented in the paper investigates the mechanical properties and microstructure of Ariake clay. Mechanical properties of Ariake clay were investigated by oedometer tests, unconfined compression tests and triaxial compression tests. Microstructure of Ariake clay is discussed with the results of Scanning Electron Microscopy (SEM) observation and Mercury Intrusion Porosimetry (MIP) test. A series of the mechanical tests were done on three types of samples, i.e., undisturbed (UD-specimen) and reconsolidated ones under 80° C and 20° C (HT-specimen and RT-specimen). It is concluded that the change process of microstructure under same consolidation condition is different from each other. The distinction of microstructure for each type of specimen is shown by a schematic diagram.

1. INTRODUCTION

It is well known that the mechanical properties of undisturbed natural deposit soil are different from those of remolded ones because of the aging effect. There are some methods to reproduce aging effect artificialy, such as to reconsolidate clayey soil under high temperature (Tsuchida et al, 1991) or to add chemical material into clayey soil. On the other hand, the relationship between mechanical properties and microstructure of clayey soil is not clear. Based on consideration mentioned above, mechanical properties and microstructure of Ariake clay were investigated in this study. Ariake clay, which is extremely soft and very highly sensitive, is widely deposited around Ariake bay located at western part of Japan, Kyushu island. To investigate mechanical properties of Ariake clay, oedometer tests, unconfined compression tests and triaxial compression tests were carried out at first. And then, to investigate microstructure of Ariake clay, Scanning Electron Microscopy (SEM) observation and Mercury Intrusion Porosimetry (MIP) tests were carried out. A series of the mechanical tests were done on three types of samples, i.e., undisturbed (UD-specimen) and reconsolidated ones under 80° C and 20° C (HT-specimen and RT-specimen) (Onitsuka et al, 1997). The difference in microstructure and changes of microstructure for each type of specimens is shown by a schematic diagram.

2. EXPERIMENTAL METHOD

The soil sample in this study is the Ariake clay obtained from Ashikari town, Saga prefecture, Kyushu island, Japan. The geotechnical indices of Ariake clay are shown in Table1. Fig.1 shows the results of oedometer tests.

Table 1. Geotechnical indices of Ariake clay

Density of soil particle ρ_s (g/cm^3)	2.57
Natural water content w_n (%)	133
Liquid limit w_L (%)	126
Plastic limit w_p (%)	81
Plasticity index I_p	45
Salinity (NaCl) (g/l)	9.4
Grain size distribution	(%)
Sand	1.5
Silt	33.0
Clay	65.5

Reconsolidated specimens were prepared by applying a pressure of 23.5kPa on the mechanically remolded clay in a mold under high temperature 80° C (HT- specimen) and room temperature 20° C (RT-specimen). MIP test and SEM observation were performed on the clay. The specimens for MIP test and SEM observation were made by freeze dry method (Delage et al, 1984). SEM observation was done on the surface, which is parallel to the direction of loading.

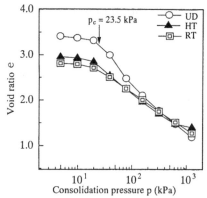

Fig.1   e-log p curves of undisturbed and
         reconsolidated Ariake clay

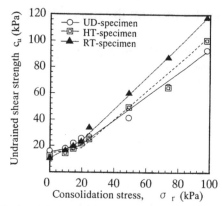

Fig.3   Relation between consolidation stress
         and undrained shear strength

3. RESULTS AND DISCUSSIONS

3.1 Results of mechanical tests

Fig.2 shows the results of unconfined compression test. The stress-strain curves of UD-specimen and HT-specimen show a brittle stress-strain characteristic while those of RT-specimen show ductile behaviour.

Fig.3 shows the results of triaxial compression test. From Fig.3, it can be seen that the increasing rate of undrained shear strength of RT-specimen is higher than that of HT-specimen. In the same way, the increasing rate of undrained shear strength of HT-specimen is higher than that of UD-specimen. The denser the specimen is, the higher the shear strength is. A higher shear strength means a higher density of specimen. Therefore, the RT-specimen, which has a higher increasing rate of shear strength, tends to have a higher density more easily than HT-specimen by consolidation stress. Similarly, reconsolidated specimens may have high densities easily by consolidation stress.

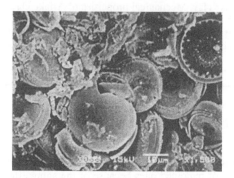

Fig.4   Diatom earth   (▬▬ 10 μ m)

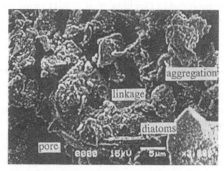

Fig.5   Ariake clay   (▬▬ 5 μ m)

3.2 SEM observation and pore size distribution

Figs.4 and 5 present the microscopic photograph of Diatom earth and Ariake clay respectively. It is reported that alluvial clay contains many diatoms in addition to the minerals (Yamauchi et al, 1985). Fig.5 shows many diatoms and aggregations, formed by broken pieces of diatoms. That is to say, microstructure of Ariake clay is formed by aggregations, which consists of diatoms and broken pieces of diatoms besides the minerals.

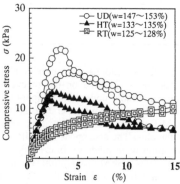

Fig.2   Stress-strain curves of unconfined
         compression test

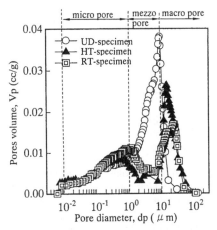

Fig.6　Pore size distribution for different specimens

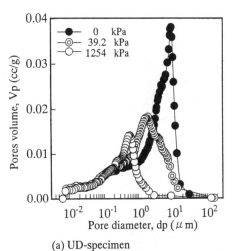

(a) UD-specimen

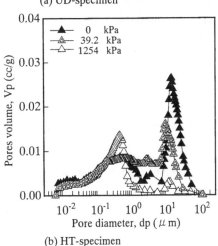

(b) HT-specimen

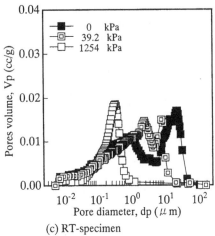

(c) RT-specimen

Fig.7　Pore size distribution for different consolidation stress

Fig.6 shows pore size distributions of the specimens. The pore size distribution of UD-specimen shows a clear peak about 10μm. On the contrary, the pore size distributions of HT- specimen and RT-specimen show two peaks. With comparing pore size distributions of HT-specimen and RT-specimen, the shapes are almost similar. But, when pore size is larger than 20μm, pore volume in pore size distribution of HT-specimen is larger than that of RT-specimen. Koizumi et al. (1997) have reported that ionic concentration in pore water or exhausted pore water in clayey soil which is reconsolidated under high temperature is higher than that of the clayey soil reconsolidated under room temperatures. It is considered that when ionic concentration becomes higher, cohesion of soil particles becomes stronger resulting in card-house structures. It is considered that the results of MIP on each specimen come from the difference of microstructure.

The change of pore size distribution showed in Fig. 7 is obtained from specimens consolidated with 0, 39.2 and 1254kPa only because the results of the specimens consolidated under the pressures between 39.2 and 1254 are similar. The distributions of pore size of the UD-specimens show that the pore size with peak of pores volume become smaller and the peak volume decreases, but one clear peak remains, see Fig.7 (a). The changes in pore size distribution of HT-specimen show that the peak volume of bigger sizes of pore decreases but pore sizes are not changed and two peaks remain, see Fig.7 (b). The changes in pore size distribution of RT-specimen show that the peak volume of pore does not change but pore size of the peak become smaller and two peaks remain, see Fig.7(c). However, the pore size distribution in each specimen is similar at a consolidation stress of 1254kPa, see Fig.7(a) ~ (c). It is considered that the size of exterior-aggregate pore becomes similar sizes of inter-aggregate pore. Judging from the results of MIP, it is suggested that the

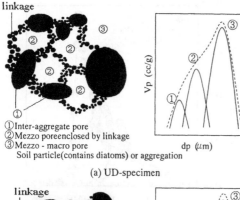

linkage

①Inter-aggregate pore
②Mezzo poreenclosed by linkage
③Mezzo - macro pore
   Soil particle(contains diatoms) or aggregation

(a) UD-specimen

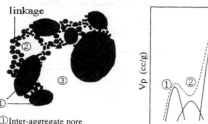

linkage

①Inter-aggregate pore
②Mezzo poreenclosed by linkage
③Mezzo - macro pore
   Soil particle(contains diatoms) or aggregation

(b) HT-specimen and UD-specimen

Fig.8 Schematic diagram and pore size distribution

change process of microstructure under same consolidation condition is different from each other.

3.3 Microstructure of Ariake clay

Fig.8 shows schematic diagrams and general concept of pore size distributions of each specimen. Because a porosimeter measures inter-aggregate pore and exterior-aggregate pore, the pore size distribution of both specimens shows two peaks as shown in Fig.8 (b) (Yamaguchi et al, 1993). However, the pore size distribution of UD-specimen indicates only one peak. This can be explained by Fig.8 (a). That is, in UD-specimen, there are linkages, which connect aggregations, and these linkages enclose some pores. In UD-specimen, linkage has not been broken. Hence, the pore size distribution indicates the shape like drawn in Fig.8 (a). On the contrary, for reconsolidated specimen, remolding breaks the linkages between aggregation. Hence, the pore, which is enclosed by the linkage, is collapsed as shown in Fig.8 (b) compared with those of UD-specimen. Therefore, for the pore size distribution as shown in Fig.8, a peak of the equivalent mezzo pore (Matsuo et al, 1976) observed in UD- specimen has disappeared and the peak of micro pore of equivalent inter-aggregate pore and macro pore equivalent exterior-aggregation appeared.

4. SUMMARY AND CONCLUSIONS

Oedometer tests, unconfined compression tests, triaxial compression tests, SEM observation and MIP tests were carried out on three types of specimens. From the results of these tests, microstructure of Ariake clay was discussed.

1) The stress-strain curves of UD-specimen and HT-specimen show brittle stress-strain behaviour while those of RT-specimen show ductile behaviour.
2) Microstructure of Ariake clay is formed by the aggregations which consist of diatoms and broken piece of diatoms in addition to other minerals.
3) Pore size distribution of UD-specimen is different from that of HT-specimen and RT-specimen. In UD-specimen, there are enclosed pores and the pore size distributes around 10μ m. In reconsolidated specimen, a part of linkage is broken. Hence, the pore size distribution shows two peaks at the equivalent exterior-aggregate pore and the inter-aggregate pore.

REFERENCES

Delage, P. and Lefbvre, G. 1984:"Study of the structure of a sensitive Champlain clay and of its evolution during the consolidation, Canadian Geotechnical Journal, Vol.21, pp.21-35.

Koizumi, K. et al. 1997:"Chemical characteristics and microscopic structure of clays consolidated under temperature consolidation", Proceeding of Symposium on High-temperature Environment, Microstructure of Clays and Disposal Problem of Wastes, pp.101-106 (in Japanese).

Matsuo, S. and Kamon, M. 1976: "Terms for microstructure of clays", Tsuchi-to-Kiso, Vol.24, No.1, pp.54-64 (in Japanese).

Onitsuka, K. and Negami, T. 1997: "Microstructure of Ariake clay", Proceeding of Symposium on High-temperature Environment, Microstructure of Clays and Disposal Problem of Wastes, pp.113-118 (in Japanese).

Tsuchida, T., Kobayashi, M. and Mizukami, J. 1991:"Effect of aging marine clay and its duplication by high temperature consolidation.", Soils and Foundations, Vol.31, No.31, pp.133-147.

Yamaguchi, H. and Ikenaga, H. 1993: "Utilization of mercury intrusion porosimetry apparatus for evaluation of soil structure", Tsuchi-to-Kiso, Vol.41, No.4, pp.15-20 (in Japanese).

Yamauchi, T. and Maeda, T. 1985:"Soil structure and geotechnical properties of clayey soil in Japan", Taga-syuppan.

Clay Science for Engineering, Adachi & Fukue (eds) © 2001 Balkema, Rotterdam, ISBN 90 5809 175 9

Thickness of adsorbed water layer for clay particles

M. Fukue & T. Minato
Marine Science and Technology, Tokai University, Shimizu, Japan

N. Taya & T. Chida
Koa Kaihatsu Company, Tokyo, Japan

ABSTRACT: The thickness of adsorbed layer for clays is one of the important governing factors for the structural development, hydraulic conductivity, and other physical and mechanical properties, such as Atterberg limits, swelling pressure and strength characteristics. This study describes a method for determining the thickness of adsorbed layer for clay particles, from resistivity measurement. By determining the water content at the critical transition from "the poor development" to "full development" of adsorbed layers, the thickness of the adsorbed layers will be calculated using the specific surface area of particles. The critical transition is experimentally determined from the change in resistivity.

1 INTRODUCTION

The physical properties of clay are dependent on the activity on the boundary phenomena between particles and water, and between water and air. The water can be classified into adsorbed molecules and free water molecules. The adsorption of water molecules onto the surfaces of particles will directly concern the suction and swelling of the clays.

It is well known that the adsorbed layer shows high electrical conductivity (Waxman and Smith, 1968, Mogi *et al*, 1986). Therefore, the continuity of adsorbed layers due to contact between them will lead the low resistivity of the soils (Fukue *et al*, 1999), as illustrated in Fig. 1.

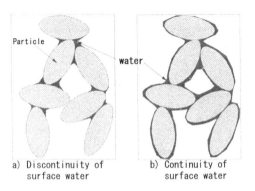

a) Discontinuity of   b) Continuity of
surface water   surface water

Fig.1 Discontinuity and continuity of surface water of particles.

On the other hand, distilled water shows relatively low conductivity. Since the increase of distilled water will squeeze into the contact of particles, the adding of distilled water into soils will increase in the resistivity. Under this situation, the lowest resistivity for soils can be obtained at a state "fully developed adsorbed layer". Accordingly, the maximum thickness of adsorbed layer can also be obtained in terms of the lowest resistivity (Fukue *et al*, 1999).

In this study, the thickness of the adsorbed layer for a bentonite clay can be determined using the specific surface area and resistivity of the clay.

2 PRINCIPLE

2.1 *Specific surface area*

There are many types of clay minerals or clay particles, which has different shapes with respect to the type of minerals. For example, monmorillonite is very thin, while kaolinite has platy shape. Thus, many types of clay particles have platy or flaky shape. Therefore, the shape of clay particles can be assumed to be a plate as shown in Fig.2. The volume of the particle, V_s is expressed by

$$V_s = a\,b\,c \tag{1}$$

where a is the length, b is the width and c is the thickness of the platy particle shown in Fig.2. Then, the mass of the particle, m_s is given by

$$m_s = V_s\,G_s\,\rho_w \tag{2}$$

where G_s is the specific gravity of the particle and ρ_w is the density of water.

Fig.2 Dimensions of a platy particle.

The specific surface area of the particle is simply given by the geometry of the particle, as

$$A_s = \frac{2\,a\,b + 2\,b\,c + 2\,a\,c}{m_s}$$

$$= \frac{2\,(a\,b + b\,c + a\,c)}{a\,b\,c\,G_s\,\rho_w} \tag{3}$$

2.2 Water content

The water content of fully developed adsorbed layer is given by,

$$w_a = \frac{m_w}{m_s} \times 100 \quad (\%)$$

$$= \frac{\{\,2(a+2t)(b+2t)t + 2(a+2t)ct + 2bct\,\}\,\rho_w}{a\,b\,c\,G_s\,\rho_w} \times 100$$

$$= \frac{2t(ab+bc+ac)+4\,t^2(a+b+c)+8t^3}{a\,b\,c\,G_s} \times 100 \tag{4}$$

where t is the thickness of the adsorbed layer. From Eqs. (3) and (4), we obtain

$$w_a = A_s\,t\,\rho_w + \frac{4\,t^2(a+b+c)+8t^3}{a\,b\,c\,G_s} \tag{5}$$

2.3 Thickness of adsorbed layer

In Eq.(5), the terms of t^2 and t^3 denotes the thickness of adsorbed layer at the corner edges of particles, which can possibly be neglected. Then, the thickness of adsorbed layer is given in the following simple form;

$$t = \frac{w_a}{A_s\,\rho_w} \tag{6}$$

2.4 Adsorption of water and ions

Water molecules and ions will be adsorbed onto the surfaces of soil particles, because of surface tension and/or electrical charges. When water is added to dry soil, some amounts are adsorbed by the surfaces of particles up to a given water content. At the same time, the salts initially contained in soils will be dis-

solved into water and form ions. The cations may be adsorbed onto surfaces, because of the dominant negative charges of particle's surfaces. This phenomenon can be identified with the mechanism of suction. The adsorbed water molecules and ions may cover up the particles and cause the low resistivity, as illustrated in Fig.1.

From the dry to wet states, mentioned here, the resistivity will be largely dropped, because the conductivity of adsorbed layer is much higher than that of dry soil particles (Waxman and Smith, 1968, Mogi et al, 1986). The electrical flow through the adsorbed layer is known as the "surface conductivity of soils". It is noted that the free water in pore has relatively low conductivity unless the liquid is electrolyte solution with relatively high concentration. Therefore, the resistivity will not be changed by the amount of free water. Accordingly, the relationship between resistivity and water content for clay can be interpreted as shown in Fig. 3. Figure 3 shows that the resistivity is very high for low range of water content, because of discontinuity of adsorbed layers, as illustrated in Fig.1. The formation of contacts of adsorbed layer will drop the resistivity. At higher water content, the resistivity is almost constant, because free water does not significantly change the resistivity, as mentioned above.

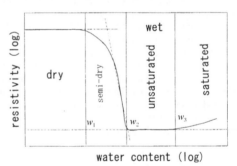

Fig.3 Resistivity of soils with water content (Fukue et al, 1999).

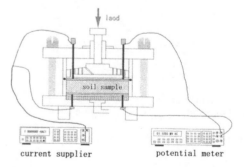

Fig.4 Apparatus for measuring potential difference.

424

3 EXPERIMENTAL METHOD

3.1 Soil samples

In this study, various clay samples, i.e., commercially available bentonite were used. The physical properties of the clay sample are shown in Table 1.

The soil samples were mixed with distilled water or 1.0 % KCl solution. Water content was changed from a dry state to a wet state. The sample was compacted into the consolidation ring. The specimen had a diameter of 60 mm and a thickness of 23 mm.

3.2 Experimental apparatus

Figure 4 illustrates an oedometer-type apparatus for measuring the resistivity of soils. A current of 0.1 mA with a frequency of 1 kHz is applied from the top to the bottom of the specimen, through stainless mesh electrodes. Under this condition, the polarization of water molecules and ions are neglected. The contact between soil samples and electrodes was assured by loading. The applied load was 670 N/m². The consolidation due to the load was neglected because of the short time measurement.

The resistivity ρ was calculated using Ohm's law (Fukue et al, 1999), as given below:

$$\rho = \frac{S \Delta V}{I \Delta L} \qquad (7)$$

where S is the sectional area of a specimen, I is the current applied, ΔV is the potential difference and ΔL is the height of a specimen.

Table 1 Physical properties of soil sample.

| Pore water | Bentonite clay | |
	with distilled water	with 1.0 % KCl solution
Liquid limit (%)	389.0	212.0
Plastic limit (%)	38.0	36.0
Specific gravity	2.75	2.75

Table 2 Approximate water contents at boundaries.

	distilled water	KCl solution
w_1	7 %	< 7 %
$w_2 = w_a$	65 %	45 %
w_3	305 %	> 200 %

4 RESULTS AND DISCUSSION

4.1 Experimental results

The relationship between water content and resistivity for bentonite mixed with distilled water and with KCl solution are shown in Fig.5. The resistivity of dry bentonite is very high at dry states and decreases with the increase in water content. This indicates that the electrical conductivity of the bentonite is dependent on the pore fluid.

As salts may be initially contained in the bentonite clay powder, they will be dissolved into the distilled water and form the adsorbed layer by cations and water molecules. The excess pore distilled water will increase the resistivity, because of its low conductivity. Fig.5 shows a fact that the resistivity starts increasing near the liquid limit, i.e., at a water content around 300 %. This trend can be interpreted as follows:

• sample with distilled water ◆ sample with 1.0 % KCl solution

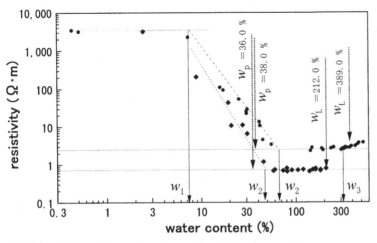

Fig.5 Resistivity of bentonite clay samples with water content.

The electrical conductivity of soil-water system is much higher at the adsorbed layer than at free pore water, because of high concentration of ions and more regularly arranged water molecules in near particle's surfaces. Therefore, the formation of adsorbed layer will provide an abrupt drop of the resistivity. After the formation of adsorbed layer, the increase in water content may not change greatly the resistivity, because the conductivity of free pore water is much lower than that of adsorbed layer, as mentioned above. Continuous increase in water content may provide the separation between some particles. This will break the contacts between the adsorbed layers in some parts. Therefore, the total resistivity of the soil will increase.

When the pore water is electrolyte solution, the resistivity is relatively low. This is because the electrical conductivity is higher for electrolyte solution than distilled water. Figure 5 also shows that the resistivity remains almost constant beyond the plastic limit. The increase in the resistivity cannot be seen in this case, because the KCl solution itself shows high conductivity, i.e., low resistivity.

From the mentioned above, the soil-water system can be divided into four states, i.e., dry, semi-dry, wet but unsaturated and saturated states, as shown in Fig. 3. Each boundary between two of the four states may be expressed in terms of water content, i.e., w_1, w_2 and w_3. The approximate values of w_1, w_2 and w_3 are presented in Table 2. It is important that the value w_2 is identified with the water content at a *fully developed adsorbed water layer*, i.e., identified with w_a. The values of w_a is approximately 80 % and 60 % for bentonite clays with distilled water and KCl solution, respectively. Thus, the value of w_a was decreased by adding KCl.

4.2 Thickness of adsorbed layer

The specific surface area, A_s is often correlated to Atterberg limits. Kuzukami and Nakaya (1977) obtained a correlation between plastic limit and specific surface area for bentonite clay-sand mixtures. The specific surface area was measured with BET method. The relation is

$$w_p = mA_s + n \tag{8}$$

or

$$A_s = \frac{w_p - n}{m} \tag{9}$$

where w_p is the plastic limit and m and n are constants, 0.430 g/m^2 and 7.95 %, respectively. From Eq.(9), the specific surface area is 70 m^2/g for ben-

tonite with a w_p of 38 %. This value may be a little high for the commercially available bentonite.

Assuming the specific surface areas of the two bentonite clays are similar, the thickness of adsorbed layer for the bentonite sample used in this study can be calculated using Eq.(6).

$$t = \frac{w_a}{A_s \rho_w}$$

$$= 0.928 \times 10^{-3} \text{ mm} \quad = 0.928 \, \mu \text{ m.} \tag{10}$$

As the specific surface area will not change when the KCl solution is used as pore water, the thickness of the adsorbed layer for the bentonite with KCl solution can be obtained using the same surface area, as below;

$$t = \frac{w_a}{A_s \rho_w}$$

$$= 0.642 \times 10^{-3} \text{ mm} = 0.642 \, \mu \text{ m.} \tag{11}$$

Thus, the thickness of adsorbed water layer for bentonite clay will decrease by adding salt. This trend is also seen in the decrease in liquid limit and plastic limit. This phenomenon has been well examined by many researchers.

In this study, the specific surface area of soil particles is not directly determined, but estimated from the previous literature. Using an accurate value of specific surface area, the thickness of the adsorbed layer becomes a more reliable value.

5 CONCLUSIONS

This study presented a method for determining the thickness of adsorbed layer for clay particles, from resistivity measurement. By determining the water content at the critical transition from "the poor development" to "full development" of adsorbed layers, the thickness of the adsorbed layers will be calculated using the specific surface area of particles. The critical transition is experimentally determined from the change in resistivity. The average thickness of the bentonite clay used was different in terms of the type of pore water. The addition of KCl reduced both the resistivity and the thickness of adsorbed layer. As the estimated specific surface area may be a little high, the obtained thickness may be underestimated. More accurate thickness can be obtained using the direct measurement of specific surface area.

REFERENCES

Fukue, M., Minato, T., Horibe, H. and Taya, N., 1999, The micro-structures of clay given by resistivity measurements, *Engineering Geology*, 54, 43-53.

Kuzukami, H. and Nakaya, M., 1977, Influence of specific surface on plastic limit and compaction characteristics – On the specific surface of fine-grained soils to relating to engineering properties (II), *Trans. JSIDRE*, 23-29.

Mogi, T., Hongo, K. and Sassa, K., 1986, Electrical properties of fine grained soil, *BUTSURI-TANSA* (Geophysical Exploration), 39 (2),17-27.

Waxman, M.H. and Smith, L.J.M., 1968, Electrical cunductivities in oil-bearing shaly sand, *Soc. Petr. Eng. J.*, 18, 107-122.

Clay Science for Engineering, Adachi & Fukue (eds) © 2001 Balkema, Rotterdam, ISBN 90 5809 175 9

Properties and structural changes of macropore in Andosol

Hajime Narioka
Faculty of Environmental Science and Technology, Okayama University, Japan

Masaharu Komamura
Faculty of Regional Environmental Science, Tokyo University of Agriculture, Japan

ABSTRACT: The physical properties of Andosol in the Kanto district of Japan were studied and two general types of the pF-θ distribution curve of the soil were identified. Group(a) show an almost even curve distribution from pF 0~1.8 and a peak between pF 2.0~3.0. Soils classified as Group(b) show low moisture tension at pF 0~1.0 but which increases at pF 1.0~2.0. Two sub-groups within Group(b) can also be identified. Group(b-1) has an increasing distribution curve while Group(b-2) show a decreasing distribution curve in the area of pF 2.0. The internal structure of Andosol was evaluated by the Soft X-ray Imaging Method, clarifying differences in specific gravity, porosity and dry density. Presence of small aggregate size and the absence of main winding water path explain the causes of higher saturated hydraulic conductivity. Aggregate structure development increases the degree of macroporosity in Andosols, such that the macropores, which has an isotropic distribution, become the main path of hydraulic conductivity. This process induce a positive correlation between macroporosity and saturated hydraulic conductivity.

1 INTRODUCTION

A study by Komamura and Narioka (1992) on the physical properties of upland soil in the Kanto district of Japan, describing pore structure in Andosol (humic and black color surface horizon; volcanic ash soil including allophane etc. that are amorphous or semi-crystalline clay minerals), has the following findings:

1) Field moisture content (ω_f) is negatively correlated with dry density (ρ_d). Also, ω_f and ρ_d have negative correlation in compaction tests.

2) Organic matter content (Hc) is correlated negatively to specific gravity (Gs) and has no correlation with dry density (ρ_d).

3) The minimum value of saturated hydraulic conductivity (Ks) is proportional to ω_f.

4) Macroporosity (n_c) is negatively correlated to microporosity (n_f) and to organic matter content (Hc).

5) There is a positive correlation between n_c and Ks.

A lot of points on macropore structure of Andosol that contribute to the drainability of upland soils has to be studied.

The present study focuses on macropore structures of Andosol and on changes induced in the form and characteristics of the entry pore.

2 MATERIALS and METHODS

2.1 *Collection of soil samples*

Disturbed and undisturbed soil samples were collected from 10 study sites within the Kanto district of Japan. The sites include Akagi(2 points), Haga(4), Haruna(1), Komoro(1), Tomisato(1) and Tsumagoi(1). Thickness of surface layer in the different sites ranged from 15~140 cm.

2.2 *Measurements of soil physical properties*

1) Basic physical properties: ω_f, ρ_d, Gs, three phase distribution, porosity (n), degree of saturation (Sr) and Hc were measured.

2) pF-moisture characteristics: Soil moisture value measured were within the range of pF 0~4.0. The sand column method was applied for pF 0~1.3, the pressure plate method for pF 1.5~2.5 and the centrifuge method for pF 2.7~4.0. The results were plotted in the pF-θ distribution curve {horizontal axis: pF, vertical axis: differential water capacity, (dθ/dpF, %), θ: volumetric water content}.

3) Ks: Undisturbed soil samples were collected using 100 cm^3 core cylinders in both vertical and horizontal directions to enable the determination of the anisotropy of permeability. Saturated hydraulic

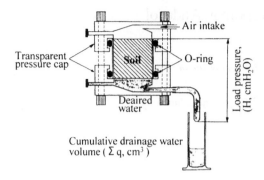

Figure 1. Entry-pore permeameter.

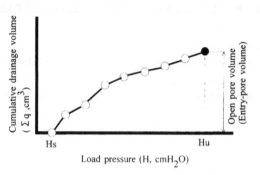

Figure 2. Schematic diagram of drainage curve.

conductivity (Ks) was measured by the falling head method in the range of $10^{-2} \sim 10^{-3}$ cm/s and by the constant head method in the range of $10^{-4} \sim 10^{-5}$ cm/s.

4) Entry-pore: The schematic diagram of the entry-pore permeameter is shown in Figure 1 where the lower of the cylinder is maintained airtight. The rate of drainage from the upper surface was 5 mm/hour. Measurement was terminated when air bubbles appeared at the bottom end filed with desired water. Negative load pressure (H, cmH₂O) was determined as the difference in the water head from the atmospheric open surface to the edge of the drainage pipe.

Cumulative drainage water volume (Σq, cm³) was measured at each load step. Σq for vertical and horizontal directions are presented in Figure 2 as a function of load pressure, H. The final cumulative drainage water volume (solid circle) corresponds to the entry-pore volume. Moreover, Δh of entry pressure represents water head. Gradient of drainage (Gd, cm³/cm) which is the ratio of Σq and Δh is relative evaluation value for soil drainability.

$$Gd = \Sigma q / \Delta h \qquad (1)$$

Cumulative drainage water volume is measured in each pressure step although there could be occasional errors in the initial measurements of drainage water volume. When air bubble flows continuously at the end of sample in load pressure (Hu) the curve ends.

In finding the meniscus tensility (pressure head on the upper, Ps and bottom surface, Pu of the sample), the following equation can be used.

$$Ps = Hs \qquad (2)$$

$$Pu = Hu - L \qquad (3)$$

Where L is the height (cm) of soil sample.

While the correct measurement of Ps and Pu is important, the evaluation through the meniscus tensility in soil is difficult.

5) Compaction test: Disturbed soil samples from Tomisato and Haga were compacted in a thin-wall sampler (diameter=5cm, length=15cm) using a 5cm long, 2.5kg-rammer. Compaction was done by dropping the rammer six times. After compaction, ω, ρ_d and Ks were measured. The soil was then used for the Soft X-ray Imaging process.

6) Soft X-ray Imaging Method: After Ks measurement, the soil was pressed into an identical but thicker sampler (30mm) for Soft X-ray irradiation (Tube voltage $\leq$ 60kV, SOFTEX; CMB-3 special type) where the inside structure of soil sample was imaged on the X-ray film following the procedures described by Narioka (1987, 1991 and 1993).

3 RESULTS and DISCUSSION

3.1 Basic physical properties of samples

Soils at each study sites were strongly influenced by the processes of the weathering, the land use condition etc. While most of the soil texture was light clay (LiC), other physical properties were found to be highly variable as shown by the following results: ω_r= 44~143%, ρ_d= 0.5~1.0 g/cm³, Gs= 2.4~2.7, Ks=10^{-3}~10^{-5} cm/s, n=62~81%, and Hc= 7~21%.

3.2 pF-moisture distribution curve

The typical pF-θ distribution curve for the 16 soil samples are shown in Figure 3. In general, the pF-moisture distribution curves of the soils show an inflection point (peak) near the area of pF 2.0. This shows that at pF 0~2.0, water exists in the pores outside the soil aggregate. However, under natural conditions, most of this is removed as gravitational water.

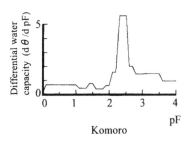

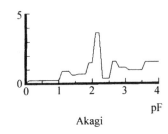

Komoro   Akagi

Figure 3. Typical pF-moisture distribution curve.

The pF curves can be classified into two general types. Soils belonging to Group(a) have an even distribution curve from pF 0~1.8 with a characteristic peak between pF 2.0~3.0. Soils belonging to this group contain a lot of organic matter (Hc) ranging from 15~21%. As organic matter content increases, moisture content (ω) also increases. This shows the improved water holding capacity (WHC) of micropores formed by the interaction of humus and soil particles. In addition, moisture increases because of the hydrophilic nature of humus. Soils from Komoro and Tsumagoi belong to this type.

Group(b) soils have very low moisture distribution curve at pF 0~1.0, but increases between pF 1.0~2.0. Organic matter content of these soils is relatively low at 7~15%. Two subgroups can be distinguished within Group(b) soils. Group(b-1) has an increasing distribution curve while Group(b-2) show a decreasing distribution curve at pF $\geqq$ 2.0. Soils from Tomisato and Haga belong to Group(b-2) where the pF-distribution curve shows two peaks for Tomisato soil and one peak for the Haga soil within the area of pF 1.0~2.0.

3.3 Pore distribution and aggregate arrangement

The difference in distribution of micro and macropores in the soil aggregate is more distinct in Group(b) soils than in Group(a) soils. The soils in Group(b-2), especially, which contains about 10 % organic matter (Hc) have high variability in pore diameter inside and outside the aggregate and show few connections. However, aggregates in Group(b-1) soils appear continuous as there is no abrupt change in diameter of pores outside the aggregate in relation to those inside the aggregate.

The relationship between the macropores ($\leqq$ pF 1.8) and micropores ($>$ pF 1.8) is presented in Figure 4. Higher percent macroporosity results to lower percent microporosity (negative correlation). High organic matter content results to lower n_c and higher n_f.

Internal structure of the Andosol using soil samples from Tomisato (31.5mm thick) and Haga

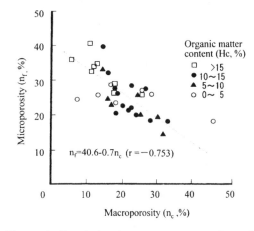

Figure 4. Correlation between macroporosity and micro porosity.

(30mm thick) were observed through Soft X-ray radiograph. The radiograph images are presented in Figure 5. Granular structures shown as fleck or cloud-like black image is the aggregate or solid phase. It can also be seen that the granularity of the Tomisato and Haga soils is different which could be due to the following observed results:

Dry density (ρ_d) : [Tomisato] $<$ [Haga]
Porosity (n)   : [Tomisato] $>$ [Haga]
Macroporosity (n_c) : [Tomisato] $>$ [Haga]

3.4 Macroporosity and saturated hydraulic conductivity

Aggregate structure is well-developed in Andosols and macropores exert a major influence on saturated hydraulic conductivity (Ks). The relationship between n_c and Ks is shown in Figure 6. The greater the n_c is, the higher the Ks becomes. However, some differences were seen especially in the pore continuity outside the aggregate due to sampling site and depth.

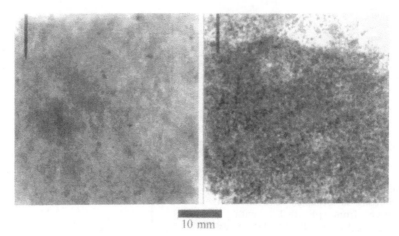

Figure 5. Soft X-ray radiograph of undisturbed soil in 10 cm depth at Tomisato (left) and Haga (right).]

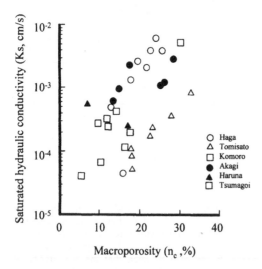

Figure 6. Correlation between macroporosity and saturated hydraulic conductivity.

Table 1. Anisotropy of saturated hydraulic conductivity.

Site	Layer	Hydraulic conductivity		Anisotropy
		V (cm/s)	H (cm/s)	V∕H
Komoro	A1	4.17×10^{-4}	4.63×10^{-4}	0.900
	A2	2.72×10^{-4}	3.43×10^{-4}	0.793
	A3	6.71×10^{-5}	7.73×10^{-5}	0.868
Haruna	I	2.15×10^{-4}	3.00×10^{-4}	0.717
	II	4.00×10^{-5}	1.73×10^{-4}	0.231
	III	5.34×10^{-4}	5.05×10^{-5}	10.574
Haga	A1	2.39×10^{-3}	1.58×10^{-3}	1.513
Tomisato	A1	3.47×10^{-4}	5.87×10^{-4}	0.591
	II A	6.85×10^{-5}	1.88×10^{-4}	0.364

For the sample soils from Haga and Tomisato, the measured saturated hydraulic conductivity is ten times higher in Haga (10^{-3} cm/s) than in Tomisato (10^{-4} cm/s), although permeability of soil from Haga was expected to be lower. Figure 5 also show that the aggregate size and the tortuosity in water path is smaller in Haga than in Tomisato.

In general though, as the permeable path is concerned, the pores in Andosol are comparatively uniform in distribution with no major difference in the distribution of pores outside the aggregate within any section of the soil. The vertical/ horizontal anisotropy of Ks is shown in Table 1

which has an average of 0.75. Generally, horizontal Ks is greater than the vertical Ks. One exception is a soil from Haruna where the anisotropy (V/H) is 10.57. This is caused by vertical tubular pores developing in the subsoil.

3.5 Characteristics of Entry-pore in Andosol

(1) Entry-pore measurements
The diameter (d, cm) of entry-pores was determined from the following equation:

$$d = 0.3 / \Delta h$$

where Δh (cmH$_2$O) is the air entry pressure (Hu-Hs) as shown in Figure 1.

Measured pore diameter can be considered as the minimum entry-pore diameter. Entry-pore diameter of 37 soil samples ranged from 75~200 μm with an average anisotropy of 0.88. Cumulative drainage water volume was measured at 0.1~27.4 cm^3.

Connectivity and continuity of micro/macro pore structure outside the aggregate can be observed, although the distribution of macropores inside the aggregate can also be evaluated through the saturated hydraulic conductivity and entry pore volume.

(2) Drainage curve

Types of drainage curves are either those that become horizontally-oriented near the Hu or those that increase gradually from Hu (Figure 7). The inclination of the curve end depends on the shape and the continuity of entry-pore, although the position of the meniscus of the front pen pore inside the soil is important.

In the gradual increase type, a horizontal part is not exhibited near Hu, such that, a lot of meniscus are moving downward until it reaches Hu. The difference in micro/macro pore distribution can also be evaluated through the drainage curve. The curve with steep inclination indicates that the soil contains a lot of macropores. As regards the curve with gentle inclination, macropores have relatively smaller structure.

The relation between the area of open space in any horizontal section and the soil sample length attracts attention. The characteristic distribution of open pore volume is shown in Figure 8 . Open pore decreases gradually from the upper part to the lower part of soil in Type A while open pore exists at the same percentage from the upper part to the lower part of soil in Type B.

When little change occurs in the tortuosity and the width of pore, drainage quickly occurs. Entry pressure can be calculated using equation (3) for Type-B drainage curve.

A more detailed illustration of the typical drainage curves is shown in Figure 9. The curve group in solid line show pore drainage at 5 cmH$_2$O and shorter horizontal part in the drainage curve. The pF-θ distribution curve associated with Group(a) soils and Type A in the distribution of open pore (Figure 8) corresponds to it. Permeability is relatively low and gradient of drainage (Gd) is small.

For the curve group in broken lines, drainage begins at 0 cmH$_2$O and the horizontal part of the curve is long. It is equivalent to the pF-θ distribution curve of Group(b) soils and Type B when considering distribution of open pores (Figure 8). Being a high drainage structure, permeability is relatively high and Gd is large.

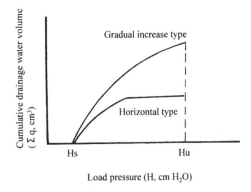

Figure 7. Types of the drainage curve.

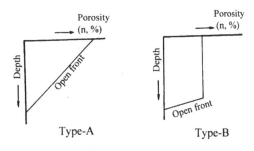

Figure 8. Distribution of the open pore.

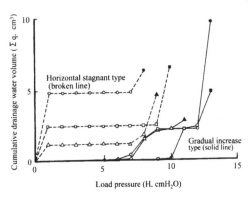

Figure 9. Measured typical drainage curve of Andosol.

3.6 Effect of compaction on the changes of aggregate arrangement

The results of compaction test showing the relationship of dry density (ρ_d) and moisture content (ω) are presented in Figures 10 (Tomisato soil) and 11 (Haga soil). Moisture content for the

433

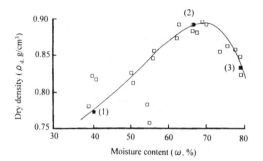

Figure 10. Compaction curve of top soil (Andosol) at Tomisato.

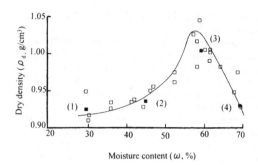

Figure 11. Compaction curve of top soil (Andosol) at Haga.

maximum ρ_d was 70% in Tomisato soil and 58% in Haga soil. While both have Light Clay (LiC) texture, Tomisato has higher organic matter content than Haga.

Figures 12 (Tomisato) and 13 (Haga) show the results of the Soft X-ray radiograph performed on the same set of soil used in the compaction test. The shades of the generated radiograph images show differences in the density and thickness of the sample soils. The aggregate structure is the partially deteriorated thick black part which also connotes a place of higher density. The white part is a pore space and a low density area.

Moisture content (ω), dry density (ρ_d) and porosity (n) of the different Tomisato and Haga soil samples subjected to Soft X-ray radiograph are showed in Table 2.

The images presented in Figure 12(1) and 13(1) represents low moisture soils while those of Figure 12(3) and 13(4) have high moisture contents. On the other hand, Figures 12(2) and 13(3) have maximum ρ_d. Macropore (pore outside the aggregate) in Tomisato soil has irregular distribution while in Haga soil, it is widely distributed, although it can be noted that granularity is rough.

Another finding with the Soft X-ray radiograph image is that the pores outside the aggregate are crushed by compaction and the water path becomes discrete as in Figures 12(2) and 13(3). As a result, tortuosity and the width of pore becomes large. With regards the Ks, Komamura et al. (1992) found that an undisturbed sample has a Ks of about $10^{-3} \sim 10^{-4}$ cm/s. With compaction, findings of the present study revealed that Ks decreased to 10^{-5} cm/s which was also verified from the Soft X-ray radiograph images.

3.7 *Macropore structure in Andosols*

Some of the important and major findings on macropore structure and on the permeability and

Table 2. Physical condition of compacted soil sample in Figures 12 and 13.

Figure (Sample)	ω (%)	ρ_d (g/cm³)	n (%)
12(1)	40.0	0.772	69.8
12(2)	66.7	0.893	65.1
12(3)	79.0	0.833	67.5
13(1)	30.2	0.916	66.0
13(2)	44.9	0.935	65.2
13(3)	59.1	1.004	62.3
13(4)	69.4	0.927	66.5

drainability characteristics pertaining to them are the following:

(1) *Pore structure in undisturbed soils (natural structure)*
1) All the sections of the soil contain macropore.
2) Ratio of micropore to macropore depends on aggregate size
3) The position/arrangement of micro/macro pore is influenced by aggregate distribution.
4) Micro/macro pores are interconnected in both vertical and horizontal directions.
5) Some Andosols exhibit pore continuity without abrupt change in diameter inside and outside the aggregate. However, there are Andosols where there is abrupt change and show few connections between pores inside and outside the aggregate.

(2) *Permeability and drainability characteristics of Andosols*
Andosols with small aggregates and water path have less tortousity but high permeability.
1) Andosol where open pores do not extend in the entire soil section have low drainability

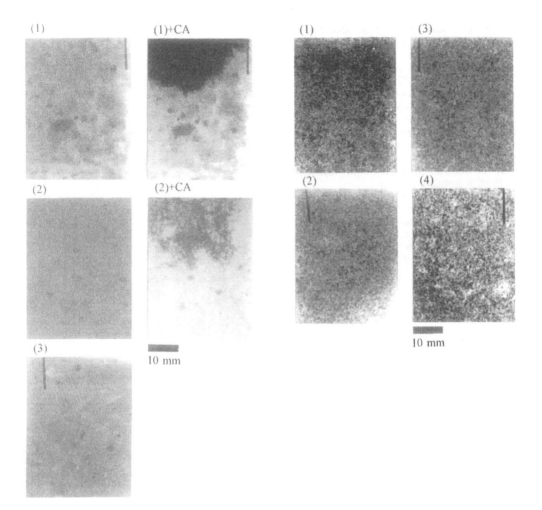

(1) (1)+CA (2) (2)+CA (3) 10 mm

(1) (3) (2) (4) 10 mm

Figure 12. Soft X-ray radiograph of compacted top soil at Tomisato. The number of Soft X-ray image correspond to number in Figure 10. The iron needle of 0.75 mm diameter is inserted in the upper corner of (1), (3) and (4) samples. In sample (1)+CA and (2)+CA, a contrast agent (CA; 1,1,2,2-Tetrabromoethan; $Br_2CHCHBr_2$=345.65, density=2.97, Boiling point=243 ℃) is used after sample (1) and (2). The contrast agent caused infiltlation from the upper part of soil sample. Scale=10 mm.

structure, drainage gradient and permeability.

2) In Andosol where open pores are distributed throughout the soil, drainage, permeability and gradient are high.

4 CONCLUSIONS

In the study of the physical properties of Andosols in the Kanto Region in Japan, the following granularity and the aggregate structure of the soil which may be attributed to the differences in specific gravity, porosity, dry density, etc.

3) The main cause of Ks is the presence of macropores, especially in Andosols rich in macropores and with well-developed aggregate structure.

4) In general, Andosols have a uniform macropore distribution inside the soil aggregate.

5) Soil compaction destroys macropores outside the soil aggregate. Also, water path becomes discrete and the tortousity and width increases as revealed through Soft X-ray radiography.

REFERENCES

Komamura, M. and H. Narioka. 1992. The characteristics of Andosol in the basement of upland field and technological amelioration, The Ministry of Education Science Research Grant (No. 01302056, T. Maeda)(in Japanese)

Narioka, H. 1987. Spatial measurement of soil pore structure by Soft X-ray images, *Jour. JSIDRE*, 55(9):29-35 (in Japanese)

Narioka, H. 1991. Study on the physical functions of soil macropores and the instrumentation for their measurement, *Bulletin of NODAI Research Institute*, Tokyo University of Agriculture, 1:1-58 (in Japanese with English abstract)

Narioka, H. 1992. Study on open channel of the entry macropore in volcanic ash subsoil, *Bulletin of NODAI Research Institute*, Tokyo University of Agriculture, 3:35-43 (in Japanese with English abstract)

Narioka, H. 1993. Structure of soil macropores and instrumentation for their measurement, *Japanese Journ. of Soil Science and Plant Nutrition*, 64(1):90-97 (in Japanese)

Narioka, H. and H. Honma. 1991. Soil structure and Soft X-ray, *Jour. JSIDRE*, 59(2):1-6 (in Japanese)

Clay Science for Engineering, Adachi & Fukue (eds) © 2001 Balkema, Rotterdam, ISBN 90 5809 175 9

Cracking in clays with an image analysis perspective

D.C. Wijeyesekera & M.C. Papadopoulou
Department of Civil Engineering, University of East London, UK

ABSTRACT: Cracking in soils can appear to be a product of chaos but a systematic geometric characterization of the pattern is often possible. Crack formation is a fundamental process, which depends on the properties of the material and the stress regime. The subsequent cracking pattern will define the en masse porosity and the pore connectivity. Shrinkage cracking in soils is primarily controlled by soil suctions. Drying experiments are carried out to deduce relationships between the shrinkage crack pattern that occurr in a clay and its plasticity is presented. The crack pattern is dependent on the thickness of the clay layer and the initial water content. An image analysis technique as a research tool to measure the geometrical properties of the crack pattern is described. Appropriateness of the ratios of major axis to minor axis and the area to perimeter of each cell is discussed.

1 INTRODUCTION

Swell- shrink behavior in soil, particularly expansive soil, gives rise to distinct desiccation cracks. Crack formation is not only caused by drying shrinkage but can also occur as a consequence of a change in pore fluid chemistry. A crack in a soil is a fundamental discontinuity that will influence the flow of pollutants from soil to aquifers; impacting on the ecosystem of the soil by fragmenting the habitat space; controlling the stability of soil thus impacting on erosion. The linear elastic fracture mechanic methods of solution have been suggested for cracking of a clay above the water table due to desiccation; cracking of layered earth structures subjected to external loads; cracking of clay caps for landfills due to differential settlements; and radial cracking in pressurized boreholes.

Shrinkage cracking in soils is controlled by soil suctions and by properties such as compression modulus, Poisson's ratio, shear strength, tensile strength and specific surface energy. As soil dries it shrinks and as a result of internal stresses, crack. Ideas and concepts about fracture and crack propagation in over-consolidated clays have been on the mind of geotechnical engineering researchers for years. However the precise mechanism of cracking is not hitherto fully understood. Factors that influence cracking are known qualitatively, but it is not clear how to predict the geometrical features of the crack pattern such as the shape, spacing and depth of cracking. In the early stages of drying from a hori-

zontal surface of an initially saturated soft soil, the water content decreases are largely accommodated by a reorganization of the soil particles into successively closer and more stable arrangements. Physical processes resulting from the development of matrix suction dominates the volume change characteristics of the immediate surface of the soil deposit and it can be assumed that these drive shrinkage and cracking during drying.

Drying experiments are being carried out to deduce relationships between the shrinkage crack pattern and the plasticity of the soil. Though cracking due to leaching is not fully known, it is hypothesized to be a probable cause for crack patterns observed recently in the North Sea Smectitic sediments. The replacement of the pore fluid causes the vertical and lateral stresses to decrease with a greater decrease being observed in the lateral stresses.

Techniques ranging from manual measurements to computer aided image analysis can be adopted to evaluate strains and in the classification of crack patterns. Object oriented, image-processing software is available for crack pattern analysis, which requires measuring linear, and area parameters, counting, edge tracking, calibration, data logging and size distribution analysis. Mocha image analysis developed by Jandel Scientific (1993) with sufficient capability was used for the crack pattern analyzed in this paper.

2 REVIEW OF PAST WORK

Many physical investigations of the cracking of various materials have been published. Knowledge of the mechanism of cracking obtained with other materials can help to understand the cracking of soil. However it should not be neglected that these investigations were all concerned with cracking caused by tension, compression, or impact applied to the materials from outside, while shrinkage cracks are caused by contraction due to the evaporation of water from the material. Cracking by desiccation is entirely different from mechanical cracking in the sense that the material loses mass during the process.

There has been some experimental work on desiccation cracks. Kindle(1917) is perhaps the best known. Small-scale experiments were carried out by him to investigate the effects of rapid and slow desiccation and the possibility of producing parallel mud cracks. and the differences between those formed in muds of saline and fresh-water environs. He stated that the temperature and tenacity of the material are two primary factors in controlling the spacing of mud cracks. The experiments conducted may have been much affected by the container wall, because of the small dimension of his container.

The observations showed that the crack pattern is more dependent on the thickness of the soil sample than on the temperature or humidity. The area of cells made by crack patterns showed a log normal size distribution. Total length of cracks decreased with increase in sample thickness. The number of sides of cells also depended on the thickness. Cracking was found to begin from the center of the soil layer and to propagate to the surface or bottom with non-uniform speed. Stones in a soil or non-homogeneity of any sort are nucleating points for desiccation cracks to initiate.

Later, Twenhofel(1950) pointed out that the spacing of mud cracks depends upon the character of the mud, the rate of drying, the thickness of the mud, the character of the water in which the mud was deposited, the nature of the material below, and the presence of foreign matter. The research did not present any quantitative relations of crack spacing to these factors. An useful research output from a study of crack propagation is to establish quantitative relationships between the characteristics of soil cracks and the various conditions under which cracking can occur, by conducting experiments under controlled conditions.

Conditions that govern the cracking characteristics may be divided into
1) extrinsic ; those which are outside of the soil but condition it, and
2) intrinsic ; those which belong to or are properties of the soil itself.

Extrinsic conditions include the temperature and humidity of the air. Solar radiation and wind velocity may play an important role in desiccation occurring in nature. The level and chemistry of the ground water are further significant extrinsic conditions.

Intrinsic conditions include moisture conditions, structure, degree of packing, physical and chemical composition, etc. Moisture conditions, moisture content of the soil and its variation in time and space, are major controlling factors of soil cracking and it must be noted that they are influenced by extrinsic conditions while other intrinsic conditions are not. Degree of packing or dry density is another major factor because it is related to the contact of the soil aggregates. .

Variables that were taken as extrinsic conditions were temperature, humidity, thickness of the soil layer, and the material of the bottom of the containers. Intrinsic conditions are initial moisture content and initial degree of packing. Experiments need to be carried out fairly systematically in regard to these various factors as the formation and propagation of cracks were very sensitive to variations in these parameters.

Corte and Higashi(1964) carried out remarkable sets of experiments. They observed that the desiccation crack pattern is more dependent on the thickness of the soil layer than on temperature of air humidity; it also depends on the material of the bottom of the container. The moisture content of the soil at the level at which cracking starts was also observed to be dependent on the thickness of soil and humidity of air, but not on the bottom material of the container. Experiments carried out by them were on free shrinkage and adhesion of the soil to the bottom of containers. Using this data and the other results, Corte and Higashi proposed a simple theory of crack formation and development by desiccation of the soil. These observations lead to a geometrical consideration of the development of the crack pattern, which can give directions in clarifying the mechanism of crack formation. This forms one of the main aims of this study.

3 CRACK PATTERN

A crack is a physical discontinuity. These can range in size from a microscopic or a hairline crack through macroscopic structural cracks to very large and open fissures resulting from earthquakes. Such discontinuities can always be traced back to an initiation that may have resulted from the separation of two mineral particles of soil. The crack can then progressively grow in accordance with classical theories such as that proposed by Griffith to develop into a longer crack, concurrent with the widening of the area adjacent to the crack tip. The crack can at times separate into two or multiple cracks depending on the stored strain energy and the energy of the process inducing the crack propagation. If the crack elongates strictly linearly, the termination of the crack will take place when the host structure fails. However if the crack develops in a curvilinear form, the propagation of the crack continues until such time the crack forms a closed loop.

The result of such crack propagation is to produce a characteristic crack pattern of polygonal shapes. The classic example of this is the columnar jointing of the fine basaltic lava flows seen in the Giant's Causeway in Ireland. These are known to be the result of rapid cooling that took place at the margins of the intrusion of the magma as it came into contact with water or the cold country rocks into which they were intruded. A fracture is most clearly visible when the host material is as stiff as a rock or concrete or masonry. The crack type described in this paper is in clay; a material with a lower stiffness.

Cracks or fractures appear at various levels of scale and in a wide spectrum of materials. From a viewpoint of scale, grain / particle boundaries within a material is a microscopic fracture, whereas geological faults such as St Andrea's Fault is a crack that can be distinguished from space. Planar fractures in the form of cleavages, joints and faults are seen in brittle and hard rocks. Structural geologists study these with an economic relevance in evaluating the reservoir capacities of oil fields. The image analysis presented in this paper focus on cracks that are of a civil engineering scale and on a particulate material which is not cemented and exhibits elastoplastic behavior. These cracks can therefore be defined as macroscopic vertical planar voids with its width being relatively much smaller than its length and depth. A crack is the consequence of a release of strain that results with shrinkage drying or progressive shear yielding as in the case of the development of faults or shear surfaces (Morgenstern & Tchalenko , 1967). Linear fractures are also seen to occur in masonry or concrete structures that suffer differential movement. These follow closely the Griffith's crack propagation theory and it is customary practice to monitor the development of the fracture by recording the widening of the crack at a particular point using movement gauges.

Fairly large containers were used and the desiccation process was carefully monitored throughout, before and after cracking starts by measuring the water content of the soil and by recording the various stages of crack growth. The final stage of crack formation was examined from photographs. One of the crack patterns observed under controlled conditions is illustrated in figure 1 and has been used for presentation in this paper. The number of cells surrounded by cracks; the shape, number of sides and size distribution of these cells; total length of the cracks: and area of the cracks are necessarily useful information.

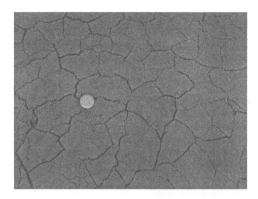

Figure 1. Crack Pattern used in the analysis

The image was captured and digitised by a video adaptor / frame grabber before any processing or analyzing with Mocha software. The frame grabber sampled the signal from the source for brightness at specific locations. These were then translated (digitized) its intensities into pixels by assigning a grey value to the location represented by each pixel. The scale used is 256 levels of grey. A value of 0 represents the black (no brightness) and consequently 255, represents white, the maximum brightness.

In the analysis, Microsoft Image Composer was used in order to aid in the preparation of the image prior to Mocha measurements. The principal crack pattern shown in figure 1 was processed so that the pixels of 0 (black) intensity were removed from the image. This operation produced a new image shown in Figure 2. Each cell in the crack pattern can be numbered for convenient referencing. Mocha offers a flexible and comprehensive set of measurement capabilities using manual, flood fill and automatic measurements.

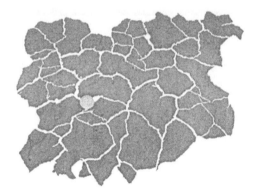

Figure 2. Image of the crack pattern modified with the removal of pixels of zero intensity

439

4 GEOMETRY OF CRACK PATTERNS

Geotechnical and geometrical features of individual cells can be used to characterize the crack pattern. These are;

Cracking moisture content, W_C
Dry density of the soil,
Thickness of the soil, t
Area of cell, a
Perimeter of cell, p
Feret diameter, D
Major axis length, l_{max}
Minor axis length, l_{min}

Manual measurement technique evaluates an image using a point, line or object that the user defines on an image's overlay plane. An image consists of a group of pixels with various intensities. Mocha uses the intensity as a criterion for selecting a group of pixels defining an object. In flood fill measurement mode, the intensity criterion, which defines the object, can be selected manually or automatically. This is illustrated in the screen display shown in figure 3. Figure 3 analyses an individual cell within the crack pattern located just right of the center of the pattern (figure 1 or 2)

An intensity histogram is a bar chart in which the number of occurrences of each grey level in an image is shown. This also is shown as an inset in the figure 3. An intensity histogram was able to be plotted only using monochrome image files. Such histograms were plotted for each of the 46 cells in figure 2.Though non-zero intensity levels cannot be changed, the contrast of the image can be improved by spacing the used intensity levels over a larger range of the grey scale by a process of histogram stretch.

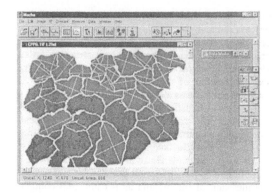

Figure 4. Line measurements of the major and minor axis of each cell.

Figure 4 illustrates how the Mocha will determine the major and minor axis of each cell. The data work sheet stores the image measurement results (see figure 5). The user can input measurement results to specified worksheet columns in the Manual measurements set up, and automatic measurements dialog. The data can be entered into the worksheet manually from the keyboard or data from another application or data work sheet can be imported.

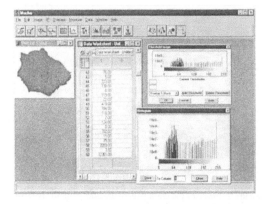

Figure 3. Flood fill threshold intensity and histogram for a single cell in the crack pattern.

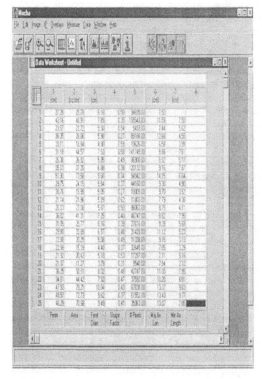

Figure 5. Data worksheet with information on the first 25 cells.

5 ANALYSIS & DISCUSSIONS

Some of the measurement information obtained from the Mocha image analysis is given in figure 5. Similar worksheets can be obtained for all three types of measurement techniques. The results from the table with flood fill measurements were further analyzed using an Excel spread sheet. The results of these analyses are presented in figures 6 to 9. It must be noted that there are further analyses that can be explored in future research.

Figure 6 is a plot of the relationship between area of each cell and the perimeter. There is a linear relation between the two. Such a relationship can be further analyzed to classify the geometric shape of each cell. The area to perimeter ratio for the pattern is in the order of 2. Note must be made of some points that do not fit the pattern. These can be due to sample preparation errors. Preparation of the soil sample needs careful geotechnical control in all the aspects as well as ensure geometric uniformity of the sample for the basic controlled tests.

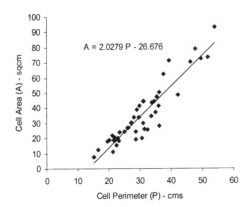

Figure 6. Variation of cell area (A) with cell perimeter (P).

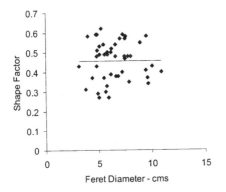

Figure 7. Shape factor vs feret diameter variation

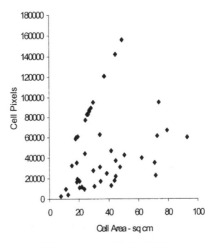

Figure 8. The variation of pixels with the area of each cell

The graph (Figure 7) of the shape factor with feret diameter for each cell show that there is no direct relationship. However the plot shows the limits of the shape factor for this crack pattern. The shape factor limits lie between 0.25 and 0.65. The study of shape factors may be a way forward in classifying crack patterns and linking them to crack genesis and material.

The cell pixel versus cell area study (figure 8) shows that the points for each individual cell to lie between two limit lines. The cell pixels are dependent on the imaging of the pattern and this may probably hint towards the degree of completion of the cracking process. Within the soil mass there are cells bounded by complete cracks as well as cracks that are still being formed.

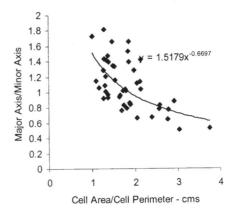

Figure 9. Elongation of the cells vs area/perimeter ratio.

Figure 9 is a study of the variation of the geometric shape for each cell. The ratio of area to perimeter of each cell can also be an indicator of the energy stored in the cracking process. Ratio of major axis to minor axis is a parameter that defines the elongation of each cell. This can reflect the microstructure of the soil particles. For example this should be compared with the classic columnar jointing of the basaltic lava in the Giants causeway. For such a crack pattern, the area / perimeter and major axis / minor axis ratios will both be very nearly unity. In the crack pattern studied there are high values of the axis ratio indicating a tendency for the formation of oblong cells.

6 CONCLUSIONS

Conclusions described are fundamentally useful in understanding the crack patterns that occur in natural soil or in any material for that matter. The possible effects of clay particle orientation and of special geometrical forms were not studied in this investigation. More complex experiments on cracking of soil containing stones or other geometric solids, which are closely related to sorting in patterned ground, will be subject of later research. An image analysis technique has been successfully adopted to analyse the geometrical properties of the individual cells within the pattern.

A distinct linear relationship was observed between the area of each cell and its perimeter. A relationship between the mean size of cells and length of the crack per unit area can be interpreted by the application of geometric considerations about the mosaics of regular polygons. This consideration combined with a simple assumption of the mechanics of secondary crack formation explains the predominance of 4 or 5 sided cells.

The crack pattern studied show that the shape factor of the cells was 0.4 ± 0.15 and it was apparently independent of the feret diameter of the cell. The analysis of cell pixels versus cell area indicates that the variation lies within two boundary lines that encourage further investigation. The elongation of the cells along its major axis (major/minor axis ratio) appears to be inversely proportional to the ratio of the cell area to its perimeter.

REFERENCES

Corte, A., & Higashi, A., 1964, *Experimental research on dessication cracks in soil*, Research report no. 66, US Army cold regions research and engineering laboratory, Hanover, New Hampshire, USA

Gonzalez, R.C., 1992, *Digital image processing*, Addison Wesley

Jandel Scientific, 1993, Mocha Image Analysis software.

Kindle, E.M., 1917, Some factors affecting the development of mud cracks, *Journal of Geology*, vol. 5, pp. 135-144.

Morgenstern, N.R. & Tchalenko, J.S. 1967, Microscopic structures in Kaolin subjected to direct shear, *Geotechnique*, vol 17, pp.309-328

Twenhofel, W.H., 1950, *Principles of sedimentation*, New York, Mc GrawHill Book co. Inc

Wijeyesekera, D.C., & O'Connor K., 1997, Compacted clayey fill: an assessment of suction, swell pressures, shrinkage and cracking characteristics, *Proceedings of the International Conference on Ground Improvement Techniques*, Macau, pp 621-630

Wijeyesekera, D.C., 1998, Geotechnical characteristics of kaolinite - bentonite mixtures, *Proceedings of the international symposium on problematic soils*, IS Tohoku, Sendai, Japan.

Clay Science for Engineering, Adachi & Fukue (eds) © 2001 Balkema, Rotterdam, ISBN 90 5809 175 9

Transformation of Fe oxides and tillability in upland field converted from paddy

T. Takahashi & K. Kamekawa

Hokuriku National Agricultural Experiment Station, Japan

ABSTRACT: Our objective was to reveal the affect of the redox of iron on the change in the tillability of soil in paddy converted into upland field. In the field experiment, the soil tillability increased with the interval since conversion, while the sediment volume (SV) and iron reductability gradually decreased. In a laboratory experiment, drying decreased the SV of smectite with iron oxides more than that of smectite without iron oxides. Reducing treatment only increased the SV of smectite with iron oxides. These results lead us to hypothesize that the effects of iron oxides are as follows. Compression resulting from soil drying changes the soil microstructure. During this process, ferric iron acts as a "cementing agent". The soil remains compressed even if re-wetted, but the reduction of the iron oxides decreases the binding force between soil particles. The low reductability of iron may contribute to maintaining the binding force

1. INTRODUCTION

One of the major problems with upland soils converted from continuous paddy soils is that they are difficult to till to make a good seedbed. The poor tillability gradually improves in the years after conversion. Studying the reason for this improvement is important for better management of paddy soils.

Naganoma & Moroyu (1983) found that the sediment volume (SV) of remolded soil is a good index of the change in soil microstructure that occurs with conversion from paddy to upland or *vice versa*. SV reflects the soil fraction of < 2 μm aggregates after remolded (Egashira & Nakayama, 1978). Low SV indicates that soil particles are much aggregated. SV decreases when soil is converted to upland and increases when soil is re-stored to paddy. Naganoma & Moroyu (1983) suggest that decrease in SV results from drying, while reduction of the soils causes the increase. This implies that there might be an interaction between the change in soil microstructure and the soil tillage properties.

Iron oxides are one of the major components of soil. Shanmuganathan & Oades (1982) showed that the addition of only 3.2 g iron kg^{-1} soil effectively changes soil friability. Since the mineralogical form of iron changes drastically with redox reactions in soils, as occurs with the conversion from paddy to upland, it is not difficult to imagine that the iron oxides play an important role in the change in soil microstructure during conversion. However, there is no evidence to support this hypothesis.

The objectives of this study were 1) to determine the effect of iron oxides on the change in soil microstructure with drying and reducing treatment, and 2) to relate this to the tillage properties of soil in converted upland fields.

2 MATERIALS & METHODS

2.1 *Field experiments*

In 1990, we divided an area that have been continuously used as paddy field (700m^2) into six plots. Thereafter, we converted one plot each year into an upland soybean field. One crop of soybeans or rice was grown each year. After 5 years, the six field plots had been converted into upland fields for periods of 0 to 5 years. In May 1994, we conducted soil tillage experiments to evaluate the soil tillability (Takahashi *et al.*, 1999) of each plot, and in August 1994 we collected soil samples from the plow layer (13 cm) to determine their SV, water content, and form of iron oxides.

The soil was classified as *fine, montmorillonitic, mesic, Typic Endoaquepts* (Soil Survey Staff, 1994). The soil texture consisted of light clay with 380 g clay kg^{-1}. The liquid limit and plastic limit were 1.99 g g^{-1} and 0.41 g g^{-1} respectively. Smectite was the major mineralogical component, along with small amounts of illite or kaolinite (Nakano 1978).

2.2 *Laboratory experiments*

To examine the effect of iron oxides on the change in the SV, we added iron oxides to pure smectite artificially.

Various amounts of ferric chloride and calcium carbonate in a molar ratio of 2:3 sufficient to precipitate $Fe(OH)_3$ were added to dispersed Ca-saturated smectite (clay fraction of commercial bentonite, which is available at Wako Pure Chemical Industries, Ltd.). $Fe(OH)_3$ was precipitated as the following reaction (Blakemore, 1973).

$$2FeCl_3 + 3CaCO_3 + 3H_2O$$
$$\longrightarrow 2Fe(OH)_3 + 3CaCl_2 + 3CO_2 \qquad (1)$$

Samples were shaken for 16 hours and washed five times with 0.001 mol L^{-1} $CaCl_2$, then dried to various water potentials by the pressure membrane and vapor equilibrium methods (Katou *et al.*, 1985; Nakano *et al.*, 1995)

The SV of a prepared sample or the soil from the field experiment were measured using the method of Takahashi & Toriyama (1998). A 1.0 g sample of soil (oven-dried basis) was added to a scaled tube with 9 mL of 55.5 mol L^{-1} NaCl solution. The tube was then shaken for 16 hours, and the volume of the suspension was brought up to 10 mL with distilled water. The sample was then allowed to settle for 48 hours and we then measured the volume of the settled suspension to calculate the SV.

The effect of iron reduction on the change in SV was determined by chemical reduction with ascorbate. A 1.0 g soil sample was weighed in a tube, and 9 mL of a mixture of NaCl and Na ascorbate solution were added. The mixture contained 55.5 mmol L^{-1} sodium ion with various amounts of ascorbate. Then, we determined the SV using the method described above.

Dithionite-citrate extractable iron was determined by the method of Holmgen (1967). pH 3.0 acetate buffer extractable iron (Takahashi *et al.*, 1999) was determined by measuring the total iron extracted by shaking for two hours with 1.0 M pH 3.0 sodium acetate (soil/solution ratio 1/100). The ferrous iron in this extractant was negligible. The amount of active ferrous iron (Kumada & Asami, 1958) determines the iron reductability of soil incubated under flooded conditions for one and three weeks. All the soils used were air-dried soils.

The change in the micropores distribution of fresh soil was measured by the mercury intrusion method after pre-treatment involving water-exchange by ethanol vapor, and critical point drying with liquid carbon dioxide. This technique is the most reliable method to determine the distribution of micropores without any artifactual change of the soil microstructure (Lawrence, 1977).

3 RESULTS & DISCUSSION

3.1 *Change in tillability, SV, and form of iron in the field*

The mean diameter of soil clods after tillage decreased as the period since conversion to upland increased (Table 1). This means that after conversion soil tillability increased. The first major reason for this change is that decreasing the field water content produces a change in soil structure. Drying compresses soil pores and decreases the water content at the same water potential (Nakano, 1978). Furthermore Hatibu & Hettiaratchi (1993) pointed out that moisture content is one of the main factors that alters the behavior of an unsaturated soil from ductile flow to brittle failure. Brittle failure, *i.e.* low moisture, is preferable for tillage.

The second reason for the increase of tillability is change in soil structure. The change in soil structure that occurred with drying is reflected in the SV. The SV, which is considered an index of soil microstructure and aggregation (Egashira & Nakayama, 1979), also decreased significantly after conversion. Katou *et al.* (1985) showed that the SV decreased with drying above a water potential of -1.5 MPa. They pointed out that the -1.5 MPa is the critical point at which the plastic limit and liquid limit of soils also change, and concluded that soil mechanical properties changed by drying and consequent change of the soil microstructure. Such changes in soil mechanical properties is

independent on soil moisture content because plastic and liquid limit are not function of water content. Hence, Table 1 shows that strong drying during the upland period changes the soil microstructure, and that the compressed soil contains less water.

Table 1. Change in soil tillability and related properties with conversion.

Interval period since conversion	Water content	SV	Mean diameter of clods after tillage
year	%	ml/g	cm
0	37.6	2.96	2.1
1	33.8	2.36	1.8
2	30.1	2.28	1.3
3	20.2	2.12	1.3
4	20.3	2.20	1.3
5	n.d.*	2.20	n.d.*

*n.d. means not determined.

Both change in structure and decreasing water content result in good soil tillability.

The form of ferric iron oxides changes with conversion from upland (Fig. 1). The total free iron estimated from the citrate-dithionite extractable iron did not decrease significantly during conversion, but the pH 3.0 acetate buffer extractable iron and phosphorus retention decreased gradually after conversion (the data of phosphorus retention is not shown). pH 3.0 acetate buffer extractable iron is considered to reflect the amount of unstable ferric iron oxides (Takahashi *et al.*, 1999). This implied that the iron oxides in soil are changed from an unstable form to a stable form. When soil is flooded and dried, the iron oxides turn into amorphous forms because ferrous iron constituents are oxidized and crystallized rapidly by air oxidation (Sah *et al.*, 1989). Our data suggest that conversion transforms these amorphous iron oxides to a stable form.

temperature, the difference in the reduced iron content was significant. These results are consistent with the observation that soil microorganisms prefer to reduce the more amorphous iron oxides than the crystalline ones (Munch & Ottow, 1980). Hence, ferric iron in converted upland soil is initially in the unstable amorphous form. Ferric iron is stabilized as the interval since conversion increases, and the stable iron oxides are difficult for microorganisms to reduce.

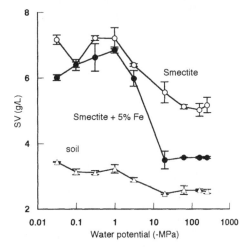

Figure 2. Change in sediment volume (SV) of soil and smectite with iron by drying treatment.
The error bars indicate the differeces between duplicate measurements.

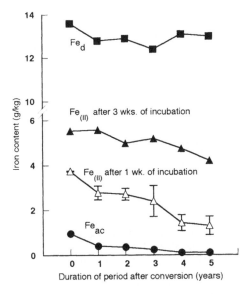

Figure 1. The effects of conversion of paddy field to upland conditions on the extractable iron using various methods.
Fe_d, $Fe_{(II)}$, and Fe_{ac} represent the dithionite-citrate extractable iron, pH 2.8 acetate buffer extractable ferrous iron after submerged incubation, and pH 3.0 acetate buffer extractable iron, respectively. The error bars indicate the differences between duplicate measurements.

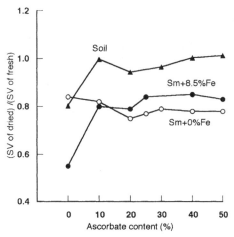

Figure 3. Change in sediment volume (SV) with the addition of reductant.
To eliminate the effect of factors other than reducing sccorbate on SV, the y-axis is the ratio of SV in fresh and dried samples for the same concentration of ascorbate.

This transformation of ferric iron affects to redox reaction of iron oxides (Fig. 1). Microorganisms easily reduced iron oxides in continuous paddy, but after conversion, the iron oxides resist microbial reduction (Fig. 1). Even with a three-week incubation under flooded conditions at 30 degree

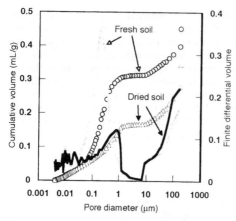

Figure 4. pore distribution of paddy soil. The sample was dried at -200 MPa by the vapor equilibrium method.

3.2 Effect of iron oxides on the SV of dried and reduced smectite, and soil

The SV of dried soil decreased markedly with drying over -1.5 MPa (Fig. 2), as observed by Naganoma & Moroyu (1983) and Katou et al. (1985). In samples of smectite with or without ferric iron, the SV also decreased with drying and the critical value for the decrease was the same (-1.5 MPa) as for the paddy soil. The mechanisms for the decrease in SV seemed to be essentially the same in the three samples, although the decrease with drying was greater in smectite with iron oxide than in smectite without iron oxide. This may suggest that addition of iron oxides promote the aggregation of clay particles.

The chemical reduction of soil and smectite with iron oxides by ascorbate increased the SV, but the chemical reduction of pure smectite did not increased the SV (Fig. 3). Furthermore, reducing treatment did not increase the SV of a mixture of dried clay and dried iron oxides (Table 2). This implies that the iron oxides and clay minerals must coexist during the drying process for reduction to increase the SV. Hence, we postulate that iron oxide plays some role in binding clay particles during the drying process. The reduction of ferric iron oxide weakens the force binding clay particles. This results in the de-aggregation of the clay and an increase in SV.

3.3 The pore size at which iron acts as a "cementing agent"

The pore distribution of paddy soil suggests a possibility that iron oxides act as an effective "cementing agent" during drying. In paddy soil, most pores have diameters of about 0.2 μm. Strong drying treatment (-200 MPa water potential, which is near the water potential of air-dried soil) effectively decrease the pore size (Fig. 4). Applied stress orientates mineral sheets of smectite and increases the numbers of quasi-crystals that make up the clay domain structure (Ben Rhaïem et al., 1987). The decrease in pore volume around 0.2 μm may be caused by orientation of sheets of smectite minerals.

In clay that is almost saturated with calcium ion, which is typical in Japanese paddy fields, the Coulomb force that reaches the clay particles on the opposite sides of pores is quite small if the pore size is 0.2 μm (Kjellander et al, 1988). Compressing the soil may bring the clay particles into closer proximity, so that they interact with each other electrostatically. In the preceding discussion, we considered only smectite - smectite interaction, but the results would be essentially the same in the case of smectite - iron oxides interactions. On the other hand, the diameter of poor crystalline iron oxides, such as ferrihydrite, is less than 5 nm (Schwertmann & Tayler, 1989). Pores of 0.2 μm are sufficiently large that it is not difficult to imagine that the iron oxides, if present, could play a role in binding other clay particles during the drying process.

3.4 Hypothesis on the role of iron in the change in soil tillability.

These findings imply that the mechanical properties of soil are affected by iron transformation in the soil. As shown in Figure 3, iron reduction changes the binding force between clay particles. Although we do not have detailed data about the linkage between the binding force by iron oxides and macroscopic soil mechanics here, the change can affect soil strength or friability, as Shanmuganathan and Oades (1982) showed.

On the other hand, iron in converted upland soil changes from easily reducible form to a stable form. The air permeability is quite low in poorly drained converted upland field. The conditions in some parts of the field are often anaerobic, and iron oxides are reduced to ferrous iron. In such a situation iron reductability could have a marked effect on the soil's mechanical properties. As shown in figure 1, pH 3.0 acetate buffer extractable iron is a good index to evaluate iron reductability in converted upland. Takahashi et al. (1999) found the total amount of pH 3.0 acetate buffer extractable iron decreased monotonously since conversion into upland while the ratio of ferrous iron and ferric iron changes seasonally. The stabilization of ferric iron against microbial reduction would be one of the factors to improve soil tillability in converted upland soils.

Table 2. Effect of the methods of Fe addition on the SV ratio of dried and fresh samples.

Sample	SV ratio	
	Before	After
Sm control	0.78	0.84
Sm+8.5%Fe (MaD*)	0.82	0.90
Sm+8.5%Fe (DaM*)	0.55	0.83

* MaD indicates that the smectite (Sm) and Fe were mixed after they are dried separately. DaM indicates that the Sm and Fe were dried together after mixing as described in Materials & Methods section.

The degree to which iron oxides contribute to soil tillability and the mechanisms that stabilize iron oxides in the interval after conversion are still unclear and require further research.

4 CONCLUSIONS

Iron oxides play a role in binding clay particles to form a stable microstructure. The reduction of iron oxides weakens the binding force and increases the SV. The ferric iron in converted upland soil gradually changes into a form more resistant to microorganisms.

Based on these two findings, we propose the new hypothesis that stabilizing the iron oxides affects the change in soil tillability in converted upland soil *via* stabilization of the soil microstructure.

5 REFERENCES

Ben Rhaïem, H., Pons, C.H., and Tessier, D. 1987 Factors affecting the microstructure of smectites: Role of cation and history of applied stresses, Proc. Int. Clay Conf., Denver, 1985 (L.G. Schultz, H. van Olphen, and F.A. Mumpton, eds.), p.292-297, Clay Miner. Soc., Indiane.

Blakemore, A.V. 1973 Aggregation of clay by the products of iron (III) hydrolysis, *Aust. J. Soil Res.*, 11 75-82.

Egashira, K. and Nakayama, M. 1979 Effect of drying and disaggregating pretreatments on the sedimentation volume of a clayey deposit from Ariake Bay, Kyushu *J. Sci. Soil Manure, Jpn.*, 50, 98-102 (in Japanese)

Hatibu, N. and Hettiaratchi, D.R.P. 1993 The transition from ductile flow to brittle failure in unsaturated soils, *J. Agric. Engng Res.*, 54, 319-328

Holmgren, G.G.S. 1967 A rapid citrate-dithionite extractable iron procedure. *Soil Sci. Soc. Am. Proc.*, 31, 210-211.

Katou, H. Nakaya, N., and Maeda, K. 1985 Change in sediment volume, liquid limit, and plastic limit of alluvial soil upon drying, *Soil Sci. Plant Nutr.*, 31, 215-226.

Kjellander, R., Marcelja, S. and Quirk, J.P. 1988 Attractive double-layer interactions between calcium clay particles, *J. Colloid Interface Sci.*, 126, 194-211.

Kumada, K. and Asami, T. 1958 A new method for detemining ferrous iron in paddy soils, *Soil Plant Food*, 3, 187-193.

Lawrence, G.P. 1977: Measurement of pore sizes in fine-textured soils: a review of existing techniques, *J. Soil Sci.*, 28, 527-540.

Munch, J.C. and Ottow, J.C.G. 1980 Preferential reduction of amorphous to crystalline iron oxides by bacterial activity, *Soil Sci.*, 129, 15-21.

Nakano, K. 1978. Changes in soil physical properties of clayey soil by conversion from ill-drained paddy field into upland field. *Bull. Hokuriku Natl. Agric. Exp. Stn.*, 21: 63-94 (in Japanese with English summary)

Nakano, M., Miyazaki, T., Shiozawa, S. and Nishimura, T. 1995 Water potential measurements *In Physical and Environmental Analysis of Soil*, p.79-87., University of Tokyo Press, Tokyo (in Japanese).

Naganoma, H. and Moroyu, H. 1983 A new index to estimate the degree of "soil uplandization" in term of soil microstructure, Integrated researches on the developing of higher technology for upland cropping in paddy field No. 1 *(Tenkanhata wo shutai to suru kouduhatasakugijutu no kaihatsu ni kansuru sogokenkyu; Tanaka, I. et al. eds.)*, pp46-53, National Agricultural Research Center, Tsukuba (in Japanese).

Sah, R.N., Mikkelsen, D.S., and Hafez, A.A. 1989 Phosphorus behavior in flooded-drained soils. II. Iron transformation and phosphorus sorption *Soil Sci. Soc. Am, J.*, 53, 1723-1729.

Schwertmann, U. and Taylor, R.M. 1989 Iron oxides. in Minerals in Soil Environment, Ed. j.B. Dixon and S.B. Weed, p. 379-438. Soil Science Society of America, Inc., Madison.

Shanmuganathan, R.T. and Oades, J.M. 1982 Modification of soil physical properties by manipulating the net surface change on colloids through addition of Fe(III) polycations, *J. Soil Sci.* 33, 451-465.

Soil Survey Staff 1994 *Keys to soil Taxonomy, sixth edition*, USDA, Washington, D.C.

Takahashi, T. and Toriyama, K. 1998. Role of iron oxides in changes in soil microstructure with drying and reducing treatments. *Soil Phys. Cond. Plant Growth, Jpn.*, 78, 35-43.

Takahashi, T., Park C., Nakajima, H., Sekiya, H., and Toriyama, K. 1999 Ferric iron transformation in soils with rotation of irrigated rice-upland crops and effect on soil Tillage properties, *Soil Sci. Plant Nutr.*, 45, 163-173.

Clay Science for Engineering, Adachi & Fukue (eds) © 2001 Balkema, Rotterdam, ISBN 90 5809 175 9

Fabric change in a volcanic soil due to disturbance estimated by MIP

T. Sato
Department of Civil Engineering, Gifu University, Japan

T. Kuwayama
Department of Construction Engineering, Daido Institute of Technology, Nagoya, Japan

K. Kuwayama
Kyokutou Gikou Consultancy, Kagamihara, Gifu, Japan

ABSTRACT: Mercury Intrusion Porosimetry was applied to study for characterization of pore structure in volcanic soil containing different two types of pore, inter-aggregate and intra-aggregate pores. Fabric change was discussed on the basis of pore size distribution of undisturbed, compacted and disturbed samples. Some indices easily calculated from pore size distribution were used in evaluation of fabric change due to disturbance or compaction. A new technique called as the double intrusion method was applied to the measurements. The study concludes that two successive intrusion is available in evaluation of pore structure for aggregated soils, entrapped pore takes an important role on characterization of fabric in aggregated soils, and specific surface area, which is redefined as surface area to volume of mercury intruded, drastically increases by disturbance or compaction and shows a good relation to volume rate of pores with radius less than 0.1 μ m.

1. INTRODUCTION

The role or importance of soil structure has been examined in the broad area in engineering practice, such as soil strength, volume change, water movement, etc. Studies for soil structure have been successfully performed through scanning electron microscope (SEM). Mercury intrusion porosimeter (MIP) is also one of useful tools to estimate soil fabric. This technique is based on the principle that a non-wetting liquid will not penetrate into the pores of a solid unless pressure is applied to it. A pore size can be estimated for the soil sample from the relationship between applied pressure and the volume of mercury intruded.

The research group with Purdue university is pioneer to apply MIP to understanding of engineering properties from microstructure viewpoints. Diamond (1970), Sridharan et al. (1971) and Ahmed et al. (1974) reported fabric change of compacted soil by the aids of pore size distribution curves. Their technique was applied to study for other engineering properties. Garcia-Bengochea et al. (1979), Mckinlay and Safiullah (1980) and Griffiths and Joshi (1991) examined the structure change in consolidation of sensitive clay. Delage and Lefebvre (1984) adopted an interesting technique named as the successive intrusion method. This is a method to distinguish the soil pores into the two types. The first intrusion fills all of the pores, which gives size distribution of total porosity. Release of the applied pressure makes some of the mercury being remained in pores. Subsequent second intrusion defines the

pore size distribution of the free pore where mercury freely moves. The difference between the first and second intrusion is the entrapped porosity containing the mercury at the release of pressure. Delage and Lefebvre (1984) declared that the free pore corresponds to intra-aggregate pores, while the entrapped mercury occupies the much larger inter-aggregate pore spaces. This distinguish, however, may be somewhat simplistic for studying microstructure of clay soil.

This paper deals with MIP study for the fabric change of volcanic clay prepared by different manners. The focus is placed on the characterization of pore size distributions obtained from successive intrusion method to account for the effect of the fabric and/or pore structure on the strength. An index is defined and used in modeling pore size distribution curve (PSD curve) to account for soil fabric and/or pore structure in soil sample showing different strength.

2. SOIL SAMPLES

A sample was taken from the northeastern part of the Mt. Akagi of Gunma prefecture, Japan. A block of soil for undisturbed sample was carefully taken from the bottom of a test pit drilled for geological investigation of the ground for the construction of a substation. The density of soil particles, natural water content and other geotechnical properties are listed in Table 1. The consistency limit indicates that the soil is classified into the volcanic clay with high water content.

Table 1. Geotechnical properties of soil samples.

	w (%)	ρd (g/cm³)	Sr (%)	n (%)
UVS1	70.4	0.78	80.3	69.0
UVS2	77.7	0.72	78.1	71.7
UVS3	69.4	0.79	79.5	68.9
UVS4	83.7	0.65	73.0	74.5
UVS5	84.6	0.70	82.9	72.2
CVS1-1	75.1	0.72	75.8	71.6
CVS1-2	75.1	0.75	80.9	70.3
CVS2-1	66.8	0.76	73.3	69.9
CVS2-2	66.1	0.76	72.0	70.0
CVS3-1	60.7	0.74	63.5	70.9
CVS3-2	61.5	0.75	65.9	70.4
CVS4-1	57.3	0.70	56.2	72.2
CVS4-2	57.6	0.71	57.8	71.7
DVS1	76.0	0.87	100.0	65.5
DVS2	79.0	0.87	100.0	65.6
DVS3	78.2	0.85	99.9	66.6
DVS4	78.4	0.86	100.0	66.2
DVS5	77.2	0.75	83.4	70.2

structure was established even in the same density and water content.

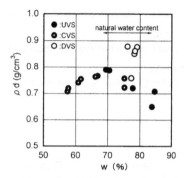

Figure 1. Initial condition of soil samples.

Disturbed samples were prepared by disturbance with keeping the natural water content and packed into a mold with its density. For compacted sample, the Proctor compaction test was adopted by using a steel mold with 3.5cm diameter and 7cm height for drying process, which shows the different chart from wetting process. The initial condition in each sample was described in Figure 1. The name of UVS stands for the undisturbed sample, CVS the compacted and DVS the disturbed.

The unconfined compressive strength was tested for the sample with 3.5cm diameter and 7cm height by using the unconfined compression apparatus with controlling 1%/min of strain rate.

3. PSD FOR SOIL SAMPLE AT INITIAL CONDITION

Pore size distribution (PSD) was measured for each sample prepared by different manners as shown in Figure 1. PSD curve represents the relationship between volume of mercury intruded and pressure applied. The measurements were started by the first and second intrusions at atmospheric pressure. Therefore the maximum pore radius is about 4μm in the total and free pores obtained from the first and second measures.

PSD curves were analyzed by the double intrusion method of mercury (Delage and Lefebvre 1984). The first intrusion gives the total pore. The second defines the free pore that enables mercury easily to invade and drain. The difference between the first and second is defined by the entrapped pore. Figures 2, 3 and 4 are PSD curves for each sample before testing the unconfined compressive strength.

Different PSD curves are given while the UVS2, CVS1-2 and DVS5 were the samples with the natural water content. This implies that different pore

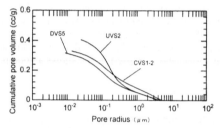

Figure 2. PSD-curve in the 1st intrusion

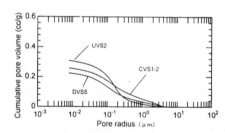

Figure 3. PSD-curve in the 2nd intrusion

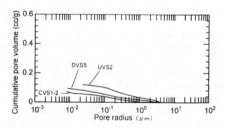

Figure 4. PSD-curve of difference between the 1st and 2nd intrusion.

4. INDEX OF PORE SIZE DISTRIBUTION

Pore size distribution curve (PSD curve) was drawn on the basis of the hypothesis of bundle of capillary tube in actual soil pore. The amount of applied pressure is converted to pore radius by the use of Washburn equation (1921). The shape of PSD curve has been successfully used in studies for fabric and pore structure of wide variety in soils. For quantifying the PSD curve, several indices were examined in the preceding study (Sato et al. 1996). It was cleared that the most appropriate indices were the volumetric rate of pores with radius of less than $0.1\mu m$ ($P(r \leq 0.1\mu m)$) and the specific surface area (As) with reference to a unit mass of sample.

Surface area (As) was normalized by the total volume of intruded mercury of which amount depends on development of pores in each sample. The volume is not the same in the sample served from different manners. The correct evaluation for surface area is made by normalization by a unit pore volume estimated from total volume of mercury intruded. This study uses the surface area to a unit volume of mercury intruded (As') instead of (As). The calculations of As' were made according to bundle of capillary tube model described in Figure 5.

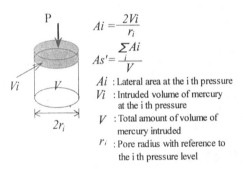

$$Ai = \frac{2Vi}{r_i}$$

$$As' = \frac{\sum_i Ai}{V}$$

Ai : Lateral area at the i th pressure
Vi : Intruded volume of mercury at the i th pressure
V : Total amount of volume of mercury intruded
r_i : Pore radius with reference to the i th pressure level

Figure 5. Bundle of capillary tube model for estimating specific surface area.

5. CHARACTERIZATION OF FABRIC CHANGE

The entrapped pore seriously reflects the effects of soil fabric. Although it is difficult to distinguish which pore is mainly measured by free pore or entrapped pore, the results of glass beads and a sand show that the entrapped pore relates to the number of contacts among particles. The volumetric fraction increases for the measures of glass beads with smaller particle diameter. The tests also showed that there were no entrapped pores in the Toyoura sand. The difference of mercury intrusions at successive two steps is one of good indices reflecting fabric or soil structure.

The two indices, As' and $P(r \leq 0.1\mu m)$, were estimated referring to PSD of entrapped pore, which was obtained from the difference between the first and second mercury intrusion. The amount of As' was described in Figure 6 in accordance with $P(r \leq 0.1\mu m)$. The figure shows that the amount of $P(r \leq 0.1\mu m)$ changes with keeping a linear relations to As' while soil samples have been served from different manners and contains different density and water content. It is easy to distinguish the difference of fabric among the soils served by undisturbed, disturbed and compacted manners on this diagram. The indices for the undisturbed sample were plotted on the lower left-hand side, those for the disturbed were plotted in the upper right-hand side in this figure. The compacted samples were plotted at the zone connected between the undisturbed and the disturbed samples.

The figure implies that the increase of As' results from the increase of small pores with radius of less than $0.1\mu m$ due to collapse in aggregations by disturbance or compaction. Investigations by SEM photograph also tell the facts of the rough surface on the aggregations caused by breakdown in DVS or CVS samples.

Clear distinction is also possible by drawing the As'-$P(r \leq 0.1\mu m)$ diagram. The undisturbed samples are plotted on the location at less than 20% of $P(r \leq 0.1\mu m)$ and $20 m^2/cm^3$ of As' while the disturbed are more than 40% of $P(r \leq 0.1\mu m)$ and $40 m^2/cm^3$. It is interesting that the indices for the compacted soil near the natural water content (CVS1-2) were plotted at the location close to the undisturbed samples in the figure. The dry density of CVS2-1 and CVS2-2 is the highest of all compacted samples. The amount of $P(r \leq 0.1\mu m)$ and As', however, does not show a smaller value than CVS1-2.

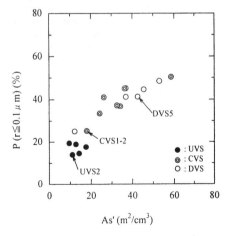

Figure 6. The relationship between As' and $P(r \leq 0.1\mu m)$.

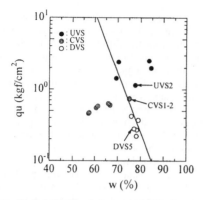

Figure 7. Relationship between unconfined compression strength and water content.

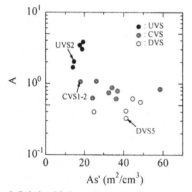

Figure 8. Relationship between A and As'.

6. STRENGTH AND PSD INDEX

Soil strength is accurately evaluated in accordance with soil type, density, water content and fabric. Interesting results were reported that a unique relation exists between the unconfined compressive strength and water content in clay sampled from the alluvial deposits of Osaka city, Japan (Akai and Shibata 1955). They described the empirical relationship as follows,

$$q_u = A \exp(-Bw) \qquad (1)$$

where q_u is unconfined compressive strength, w is water content and A, B are empirical constants, respectively. Much more experimental results should be collected for the verification of the equation (1) on the basis of accurate measures eliminating the effects of difference of density. The equation (1), however, suggests that the parameters of A and B respectively account for the effects of soil fabric and water content on the unconfined compressive strength. The test results were described in Figure 7. The parameter of B was estimated as 0.5 for the experimental values to soils with almost identical density. The parameter of A was determined from the insertion of

B=0.5 and the strength of UVS2, CVS1-2 and DVS5 of which water content and dry density showed almost the same at the initial condition. The relationship between the parameter of A and the PSD index of As' was examined in Figure 8. The amount of A tends to decrease with the increase of As' induced by destroy of aggregations.

7. CONCLUSIONS

The study concludes that,
1) two successive intrusion was effectively used in characterization of fabric or pore structure for aggregated soils,
2) surface area to a volume of mercury intruded into entrapped pore (As') shows a good linear relation to volume rate of pore with radius of less than 0.1 μ m (P(r $\leqq$ 0.1 μ m)), and
3) difference of fabric among undisturbed, disturbed and compacted soils is clearly distinguished on As'-P(r $\leqq$ 0.1 μ m) diagram.

REFERENCES

Ahmed, S. et al. (1974) Pore sizes and strength of compacted clay. ASCE 100(GT4): 407-425.
Delage, P. & Lefebvre, G. (1984) Study of the structure of a sensitive champlain clay and its evolution during consolidation. Can. Geotch. J. 21: 21-35.
Diamond, S. (1970) Pore size distribution in clays. Clay and Clay Minerals 18: 7-23.
Fredlund, D.G. et al. (1994) Predicting the permeability function for unsaturated soils using the soil-water characteristic curve. Can. Geotech. J. 31(4): 533-546.
Garcia-Bengochea I. et al. (1979) Pore size distribution and permeability of silty clays. ASCE 105(GT7): 839-856.
Griffiths, F.J. & Joshi, R.C. (1991) Change in pore size distribution owing to secondary consolidation of clays. Can. Geotech. J. 28: 20-24.
McKinlay, D.G. & Safiullah, A.M. (1980) Pore distribution and permeability of silty clays. ASCE 106(GT10): 1165-1168.
Sato, T. et al. (1996) Fabric of undisturbed and compacted volcanic soil estimated from MIP. Proc. of the first Int. Conf. on Unsaturated Soil 3: 1527-1532.
Sridharan, A. et al. (1971) Pore size distribution studies. ASCE 97(SM5): 771-787.
Washburn, E.W. (1921) A note on method of determining the distribution of pore sizes in a porous material. Proc. of the National Academy of Science of USA 7: 115-116.

Clay Science for Engineering, Adachi & Fukue (eds) © 2001 Balkema, Rotterdam, ISBN 90 5809 175 9

Strength behavior of swollen bentonite

N. P. Bhandary, R. Yatabe, N. Yagi & K. Yokota
Department of Civil and Environmental Engineering, Ehime University, Matsuyama, Japan

ABSTRACT: Bentonite, a clay material widely used in creating artificial barrier at waste disposal sites, is an alteration of product of volcanic ash that contains nearly half its weight of montmorillonite, which is a clay mineral from smectite group. Smectite minerals get swollen up on absorption of water, and this property has a significant effect on the strength of bentonite. Ring shear tests were carried out to study the strength behavior of bentonite material in swollen state. Similarly, triaxial tests were performed with kaolin clay and chlorite with bentonite mixed at certain percentages to study change in strength of a soil due to smectite content. Also observed was the relationship between confining pressure and amount of water absorbed or retained during consolidation of a wet sample. Ring shear test results showed that the peak strength of post-swollen (after dry consolidation) bentonite is much higher than that of a pre-swollen bentonite paste, whereas the results of triaxial test revealed that bentonite content at lower percentage increases the strength of a normally consolidated clay soil.

1 INTRODUCTION

In technologically developed and highly populous countries like Japan, the amount of wastes not only from the household but also from the factories, industries, and power stations is tremendously high. Disposal of such a huge amount of waste materials may result in many environmental problems such as contamination of surrounding soil and nearby water. These problems have a long term effect on not only humans but also the entire animal and plant lives. Wastes from nuclear power stations even contain radioactive elements that need to be disposed in such a way that they do not have an adverse effect on the geo-environment. So, to make the waste disposal process environmental friendly and protect the animal and plant lives from possible adverse effects, practices have been made in dumping waste material into large pits that consist of sealing layers all around their surfaces.

As a sealing material, a layer of bentonite is used as an artificial barrier that protects the leakage of harmful elements like poisonous liquids, dioxin, radioactive wastes, etc., which contaminate the surrounding earth and environment. When bentonite absorbs water, the montmorillonite particles get swollen and fill up all the voids in the bentonite mass. A zero void condition in overall layer of swollen bentonite makes it able to attain zero permeability, thus making it a perfect barrier between the waste and surrounding soil mass.

Many research works have been carried out to study swelling behavior of bentonite or smectite minerals but their strength behaviors have yet to be investigated so as to understand the mechanism of decrease in strength of swollen mass of bentonite or the soil material containing smectite minerals. Amount of swelling of a material depends on the amount of water absorbed and the external or overburden pressure that acts on it, i.e., it increases with the increase in amount of water absorbed and the decrease in external or overburden pressure. As the bentonite layer as artificial barrier in the waste disposal sites bears the overburden pressure due to waste heap, the strength of swollen bentonite mass must be evaluated so as to design the thickness of bentonite layer that can resist the high overburden pressure without a failure that may lead to a no barrier condition between the waste and surrounding soil. In addition, the strength consideration of bentonite is also important while preparing its layer on waste disposal pits with inclined side faces.

1.1 Bentonite

Bentonite is an alteration of the product of volcanic ash and a highly plastic swelling clay material whose liquid limit is as high as 500% or more. It contains oxides of aluminum, sodium, iron, calcium, magnesium, etc. and nearly 50% of montmorillonite, which is a mineral from smectite group. Smectites are the common constituents of sedimentary rocks, which based on mineral structure are classified into two principal groups: dioctahedral and trioctahedral. Principal smectites like montmorillonite, beidellite, and nontronite come in first group, whereas saponite, hectorite, and sauconite come in second. Among all these smectite minerals, montmorillonite is the one that up on absorption of water swells the most, more than 5 times its original volume. Structural water molecules in a smectite mineral come out of its structure at a temperature of 113°C, whereas the temperature should rise to 689°C for OH ions to completely separate out from the mineral structure.

Bentonite is commercially produced for some engineering uses like drilling mud for soil borings, backfill material during the construction of slurry trench walls, and now as an effective artificial barrier in waste disposal pits, where the chances of leakage of harmful and radioactive elements that contaminate the geo-environment is extremely high. It is also frequently used in various studies of swelling clay behaviors and their influences on clay soil strength.

1.2 *Swelling of bentonite*

As stated earlier, swelling of bentonite material takes place due to entry of water molecules into mineral structure of montmorillonite, which occupies nearly 50% of the solid particle space. Minerals like montmorillonite from the smectite group have a structural gap in their mineral skeleton. This gap is the space where water molecules enter and stack one above other causing expansion of the whole mineral structure. So, it is important to have a look at montmorillonite mineral structure to elaborate bentonite's swelling behavior.

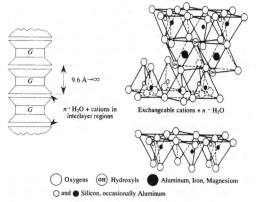

Figure 1. Schematic diagram and skeleton of montmorillonite structure.

Figure 1 shows the mineral structure of montmorillonite. It is seen that all the tips of tetrahedra are pointing towards the center of the unit cell. The oxygens forming the tips of the tetrahedra are common to the octahedral sheet as well. The remaining anions in the octahedral sheet that fall directly above and below the hexagonal holes formed by the bases of the silica tetrahedra are hydroxyls. The layers formed in this way are continuous in any two orthogonal directions and stacked one above other in a direction orthogonal to the previous two. Bonding between successive layers is by van der Waals forces and by cations that may be present to balance the charge deficiencies in the structure. These bonds are weak and easily separated by cleavage or absorption of water or other polar liquids. The basal spacing as seen in Figure 1 in third direction is variable, ranging from about 9.6Å to complete separation. Complete separation of the structural layers results in structural failure of the mineral.

2 STRENGTH OF BENTONITE

Although the strength of a soil material depends largely on two factors—void ratio and water content, it is also governed by the structures of minerals that constitute the soil. Different minerals have different structural bonds, and strength of such a bond depends on the chemical composition of the mineral itself. So, the types of minerals determine the overall structural bond between the mineral particles that make the soil mass.

It is supposed and is also supported by many research studies that montmorillonite minerals lose strength of structural bond once the structural gap is occupied by water molecules. Stacking of a high number of water molecules in the gap causes nearly zero resistance against shear stress that may work along the gap. At this stage, if the water molecules are not allowed to come out of the structural gaps, montmorilonite molecules cannot resist any shear stress, which means all the montmorilonite particles structurally fail even under a very small shear stress. As nearly half of bentonite consists of montmorillonite, the space occupied by all the montmorillonite minerals turns to be like a void space that then reduces the strength of whole bentonite mass. Strength of benonite, thus, is greatly affected by the presence of montmorillonite clay minerals.

To study such a typical strength behavior of bentonite in swollen state, some laboratory tests were carried out on bentonite and mixture samples of bentonite with kaolin clay and chlorite. The tests consist of ring and triaxial test along with consolidation and swelling tests.

3 LABORATORY TESTS AND RESULTS

3.1 *Water absorption and consolidation test*

To support the strength tests, two tests relating to swelling were carried out to observe absorption and squeeze patterns of water molecules in and out of a mass of bentonite. For absorption, a certain mass of dry bentonite powder under pre-determined vertical stresses in a consolidometer was allowed to absorb water until it fully swelled. Water content at its fully swollen state was then determined. Similarly for squeeze, a swollen mass of bentonite with more than 200% water content was placed under the same magnitudes of vertical stress and allowed to fully consolidate. Water content of each of the consolidated mass was then determined. Two graphs, as shown in Figure 2, were plotted between the applied vertical stress and water content of the swollen bentonite.

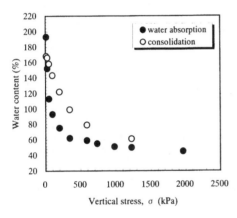

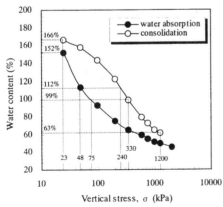

Figure 2. Variation in water content of bentonite sample with vertical confining stress.

These two figures show two patterns of variation in water content of bentonite mass with vertical confining stress. The first one is a normal graph, whereas the second is semi-log. It is seen (in the figures) for a certain range that the stress required to squeeze a swollen mass of bentonite up to a certain amount of water content is far greater than that required to stop entry of water into montmorilolonite mineral structure beyond the same amount of water content. For example, to maintain a water content of 112%, the stress required during consolidation is 240 kPa, whereas that during absorption is just 48 kPa. However, if we look at the beginning and end of the tests (i.e., extreme values of the vertical stress), we can see this difference (in stress) to be much less. It means that the amount of water entering montmorillonite structure during absorption and the amount of water retained during consolidation are nearly equal at very low and very high values of vertical confining stress. In the figures, the difference in stress for the same percentage of water content can be noticed from about 20 to 2000 kPa. This result is an evidence to the typical behavior of swollen bentonite.

3.2 *Strength tests*

Strength tests were performed in two ways—ring shear test on bentonite sample and triaxial test on kaolin clay and chlorite mixed with bentonite at certain percentages by weight.

Ring shear test with a shearing speed of $26°$/hour (0.38 mm/min) was carried out to determine angles of peak and residual shear resistance of a swollen mass of bentonite. For this, two types of test specimen, one in paste form (pre-swollen—initially mixed with 300% water by weight) and other in dry powder form which was later allowed to absorb water to its full swelling (post-swollen) after full consolidation was attained, were prepared. The weight of dry bentonite powder, containing 10% natural water content, used in preparing both types of specimen was kept same. However, after the tests, it was found that the water content of pre-swollen specimen was much more than that of post-swollen specimen.

The results of the ring shear test in terms of shear strength are presented in Figures 3 and 4. From these two figures (3 and 4) it is understood that the angle of peak shearing resistance, as given by ring shear test, for pre-swollen bentonite is far smaller than that for post-swollen bentonite, which was subjected to full water absorption prior to shearing. However, there is no much variation in the angles of residual shearing resistance for both the samples. The reason for this may be a negligible friction between montmorillonite particles present in bentonite after they get swollen and sheared. The big difference in angles of peak shearing resistance for these two samples clarifies that the strength of bentonite is much smaller when it initially has a high percentage of water content.

If we explain the results of ring shear test with the help of Figure 2, it can be said that the big difference in angles of peak resistance is due to difference in water content, which is clearly seen in the figure (Figure 2). The normal stresses (σ) applied during ring shear tests were nearly 100, 200, and 300 kPa, which lie in the range where there is much difference in water content of bentonite during consolidation and water absorption. It was supported by the determination of water content after shearing ($\sigma = 300$ kPa), which resulted in about 100% and 64% for pre- and post-swollen sample respectively. So, entry of water molecules into montmorillonite mineral structure lowers the strength of bentonite. On the other hand, if an initially stressed mass of bentonite absorbs water, its peak strength does not come down to the pre-swollen sample's level even if the vertical stress, σ is same.

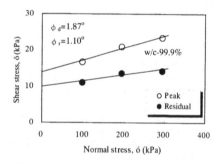

Figure 3. Results of ring shear test on bentonite paste (pre-swollen sample).

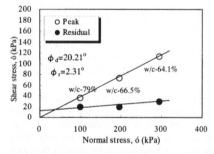

Figure 4. Results of ring shear test on bentonite powder after water absorption (post-swollen).

Then to observe the influence of montmorillonite (smectite) content in a clay soil, just like in bentonite, triaxial tests (CU condition) were carried out on kaolin clay with 0%, 5%, 10%, 20%, and 30%

bentonite in it and chlorite with 0%, 3%, 10%, 20%, and 40% bentonite in it. The test specimens of 3.5 cm diameter and 8.0 cm height with each mix percentage were prepared at three over consolidation ratios (OCRs) of 1, 5, and 10. The general properties of bentonite, kaolin, and chlorite used in the tests are given in Table 1. The deformation rate during the tests was set to 0.088 mm/min, and maximum confining pressure was limited to 295 kPa.

Figures 5(a), 5(b), 5(c), and 5(d) show the stress paths for normally consolidated (OCR=1) kaolin clay sample with 0%, 5%, 10%, and 20% bentonite content respectively. Similarly, Figure 6 is a summary of results of triaxial test in terms of variation in strength of kaolin clay and chlorite mineral with the increasing percentage of bentonite content.

Comparison of above four figures shows that the stress path for kaolin clay + 5% bentonite is different to those of rest three. In general for a normally consolidated clay soil, the stress points in a consolidated-undrained (CU) test shift towards left (origin), whereas they shift towards right (away from origin) for an over consolidated soil showing a higher strength compared to normally consolidated one under similar conditions. But even at a normally consolidated state, the strength of kaolin clay increases when the bentonite content is 5%, and the stress path for this sample comes to be one like that for an over consolidated sample.

If we look at the angle of shearing resistance of each sample, we get no much variation, so the reason for the increase in strength of second test specimen (with 5% bentonite) is not the increase in the angle of shearing resistance; however other conditions being very similar, the only reason for the rise in strength may be percentage of bentonite content. Montmorillonite particles in bentonite after getting swollen fill up the voids in kaolin clay and bind neighboring kaolin particles. This process makes kaolin particles more intact and the whole mass gains more strength. However, the case is different for other samples with higher percentages of bentonite because further increase in its content decreases the particle-to-particle space, which results in a weak failure line across them. This is the reason why strength of a clay soil decreases with the increase in montmorillonite content.

Figure 6 makes it even clearer that the strength of kaolin or chlorite increases when the percentage of bentonite content is less. However, as soon as the percentage reaches 10, the strength abruptly comes down and then gradually decreases. So, it can be said that the strength of a normally consolidated (OCR=1) clay soil increases to a peak when its bentonite content is about 5%, and beyond this percentage, montmorillonite particles in bentonite start governing the strength of the soil and lower it.

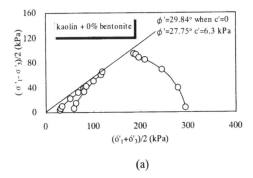

(a)

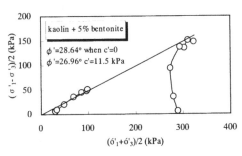

(b)

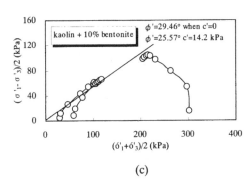

(c)

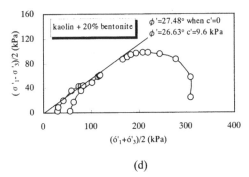

(d)

Figure 5. Results of CU triaxial tests on bentonite and kaolin mixture.

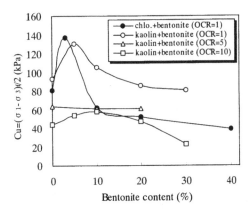

Figure 6. Change in strength of kaolin and chlorite due to change in bentonite content.

In the figure, it is also seen that there is no distinct peak for OCR=5 and OCR=10, which may be because of subsequent change in the volume of specimens due to a high over consolidation ratio during the tests.

Table 1. Properties of tested clay materials.

Clay material	W_L (%)	W_P (%)	Ip	ϕ' (°)	ϕ_r (°)	Gs	Particle size (%)		
							<2 (μm)	2-20 (μm)	>20 (μm)
Bentonite	412.3	62.7	349.6	12.6	2.1	2.19	70.0	22.0	8.0
Kaolin	41.3	13.3	28.0	30.0	25.6	2.71	28.7	63.0	8.3
Chlorite	40.9	28.2	12.7	23.6	17.5	2.70	32.0	38.0	30.0

4 CONCLUSION

Bentonite is being used in waste disposal sites as a material that has proved to be a good barrier to radioactive elements and harmful substances. For this purpose, swelling property of bentonite is supposed to be more important; however, to design an effective bentonite layer, it requires a knowledge of its strength behavior as well. So, as a result of the study on strength behavior of bentonite, following conclusion is made:

1. For a certain range, the stress required in consolidation to lower the water content of pre-swollen bentonite sample up to a certain level is far greater than that required to stop water absorption by powder sample beyond the same water content.

2. Peak strength of post-swollen bentonite (allowed to absorb water after consolidation) in drained condition is higher than that of pre-swollen bentonite; however, the residual strength remains similar.

3. Although high percentage of montmorillonite in any soil results in significant decrease in the strength of the soil, it raises the strength of a clay soil when contained at low percentage—around 5%. Swelling of montmorillonite particles replace the void space and simultaneously grip the soil particles, which causes an increase in overall strength. However, increase in its content beyond the limit results in a decreased strength.

REFERENCES

Bhandary, N.P., Yatabe, R., Yagi, N., & Yokota, K. 1999. *Swelling and shear characteristics of smectite in landslide clay. Proc. 2nd Int'l symp. on landslide, slope stability, and safety of infrastructures, Singapore, 27-28 July 1999: 103-110.*

Futagami, O., Bhandary, N.P., & Yokota, K. 2000. Basic study on the change in strength of soil due to swelling clay mineral content. *Memoirs of the Faculty of Engineering , Ehime University, Feb. 2000: 137-143.*

Ishii, T. 1994. Geotechnical study on landslide movement and stability characteristics. *Doctoral thesis submitted to the dept. of civil and ocean engineering, Ehime University, Japan.*

Komine, H. & Ogata, N. 1999. Experimental study on swelling characteristics of sand-bentonite mixture for nuclear waste disposal. *Soils and Foundations, Journal of Japanese Geotechnical Society, vol.39, no.2, 1999:83-97.*

Mitchell, J.K. 1976. Fundamentals of soil behavior, *John Wiley & Sons, Inc.*

Mukaitani, M. 1995. Undrained strength of clay and soil stability analysis. *Doctoral thesis submitted to the dept. of civil and ocean engineering, Ehime University, Japan.*

Clay Science for Engineering, Adachi & Fukue (eds) © 2001 Balkema, Rotterdam, ISBN 90 5809 175 9

Elasto-plastic analysis for structured soils under complex loading

C. Zhou, Z. J. Shen, S. S. Chen & T. L. Chen
Department of Geotechnical Engineering, Nanjing Hydraulic Research Institute, People's Republic of China

J. H. Yin
Hong Kong Polytechnic University, SAR, People's Republic of China

ABSTRACT: Natural soil deposits normally have their structures. During the deformation process the structures will be destroyed and at last the characteristics of destructured soils will be displayed. In this paper, damage ratio D is defined to describe the process from natural soil to destructured soil, and an elasto-damaging-plastic model is presented based on hypoplasticity theory and three fundamental relations: the first relation between compressive bulk volumetric strain and mean spherical normal effective stress, the second relation between normalized deviator stress ratio and deviatoric strain, and the third relation between volumetric strain increment and deviatoric strain increment induced by application of certain normalized deviator stress. Some computed examples show the different responses between structured soils and destructured soils under simple and complex loading.

1 INTRODUCTION

Most conventional nonlinear elastic models and elasto-plastic models such as Duncan-Chang model and Cambridge model have been developed for remolded soils and their descriptions of behaviours of soils are not perfectly suitable to simulate the characteristics of natural soils. The soil samples are usually disturbed for the reasons of derivation techniques and stress releasing happens. Hereafter a new generation of constitutive models should be established to back analysize and compare with the results of field tests. However few work has been conducted up to now to describe the process from natural soil to destructured soil and the importance of consideration of structure effect. Jiang Mingjing (1996) and Shi Bin (1995) studied the artificially made structured clayey soil by use of scanning electron microscope and the quantitative techniques of microstructures. On the basis of understanding of the micromechanism, Shen Zhujiang proposed a series of damage models from a nonlinear damage model (Shen Z.J.,1993) to a new damage-functional masonry model(Shen Z.J.,2000).To push the work forward, a model is developed based on Shen Z.J.'s granular medium model (Shen Z.J.,1999) and hypoplasticity theory (Wang Z.I...1990) to describe the characteristic of static and dynamic deformation of structured clayey soils and destructured soils.

2 CHARACTERISTICS OF DEFORMATION OF STRUCTURED CLAYEY SOILS

Compression curves of natural clayey soils can be divided into three piece-wise stages: elastic structure kept stage, structure damaged stage, and the remolded soil stage. In view of Shen Zhujiang's masonry theory (Shen Z.J.,2000),natural clayey soil is similar to a kind of assembly of hetrogenous porous materials just like masonry. Seen from possibility curves of grain size distribution there are two peaks respectively of small pore and big pore.The small porous structure displays inner-aggregate bonding composition of dense lumps, and the big porous structure exhibits inter-aggregate linking formation of the bonding planes between lumps. The existance of two kinds of structure of clayey soils exhibits a structure damage stress which is just like the normally so-called quasi-over-consolidated pressure. Before the stress reaches the initial bonding strength, the deformation of structured soils just shows a process of self-modification of the initial structure and a small value.The relationship between stress and strain is basically elastic, and the structure is kept well from damaging. When the stress reaches the structure damage strength, most of the initial structure of soils are destroyed. A small stress increment will induce a great increment of deformation. In this stage the lumps are crushed to small sizes of aggregates or particles, therefore the sliding plastic deformation and the collapse deformation of structure may make the corresponding compressibility greater than that of remolded soils. While the stress exceeds the structure damage strength and continually increases, the behaviour of structured soils is similar to that of damaged soils and sliding plastic deformation becomes the principal origin of the total deformation. During the third stage, the destroyed structure unrecovers the initial linkage strength of soil particles but it may form a new arrangement of soil particles which can maintain a new stable sub-structure. Once the stress exceeds the strength of the third-stage soils, eventual failure of soils will happen.

From a viewpoint of stability of structure potential, the above mentioned three stages can be described as a process from 1) formation of the initially stable structure potential due to long depositing history of geology to 2) loss of stability of initial structure potential due to damage and then to 3) reconstitution of stability of substructure potential due to consolidated densification. A typical behaviour of structured soils is slight dilatancy and strain softening due to the damage of microstructure in drained triaxial compression test under a lower confined pressure. The phenomena may be explained by use of concept of structure poential shown in Fig.1. The solid grains are modeled by the rigid rods, and the spring K linking rods is used to stand for the referred structure potential. We use a simple structure shown in Fig.1 to model the microstructure of two typical kinds of pores. One stands for small inner-aggregate framed pore, the other stands for big inter-aggregate framed pore.

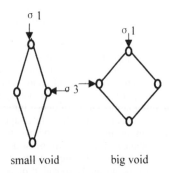

small void   big void

Fig.1 Microstructural model(Each node has a spring)

When the confined pressure is relatively lower, with the increase of stress σ_1, the value of structure potential K will decrease because of damaging and the volume of the small framed pore will expand to exhibit dilatancy. When the decreasing of structure potential is more rapid while the stress σ_1 continues to increase, even if the stress σ_1 remains the current level or decreases, the deformation will continue to develop, so the strain softening occures. When the structured soil is applied with high confined pressure, the initial structure strength will fail and the masonry

will be crushed to form a heap of particles. With the increase of shear stress, all of the grain groups will be destroyed and a high contraction happens.

3 HYPOPLASTICITY MODEL

In view of the developing history of plasticity theory, simulation of behaviour of soils passed through several typical models such as moving hardening model,multipal yield surface model and bounding surface hypoplasticity model etc. The focus of the models is put on the different descriptions of the simulation of deformation behaviour under complex loading between the traditional plasticity theory and the generalized plasticity theory. That is how to simulate more perfectly the dependency of direction of increment of plastic deformation and that of the increment of stress, the normally contractive behaviour and the contractive behaviour under inverse loading. As discussed before, sliding plastic deformation of particles or particle groups, the rotating plastic deformation and the crushing collapse plastic deformation of different sizes of lumps consist of the total plastic deformation of structured soils. The sliding plastic deformation obeys the flow rule of the traditional plasticity theory, while the increment and direction of the total plastic deformation including the rotating and crushing parts are dependent on not only the previous stress history and the current stress state, but also the direction of the current stress increment. So a practical model should be introduced to describe the deformation behaviour of the structured and destructured soils.The hypoplasticity model can describe the above referred deformation behaviours.

In the hypoplasticity model, the normalized deviator stress ratio tensor $r_{ij} = S_{ij}/p$ is defined, where S_{ij} is the deviatoric component of the stress tensor and p is the mean spherical normal effective stress. Thus the stress increment can be divided into two parts: $d\sigma_{ij} = pdr_{ij} + \dfrac{\sigma_{ij}}{p}dp$. The formula can reflect the crossing influence between dS_{ij} and dp. Corresponding to the decomposed two parts dr_{ij} and dp, the plastic strain increment can be divided into two parts too. The divided two parts of plastic strain increments can be further divided into plastic deviator strain increment and plastic bulk volumetric strain increment respectively. If the direction tensor n_{ij} of plastic deviator strain increment and the direction tensor n_{aij} of the current stress increment can be determined relatedly by the normalized deviator stress ratio increment dr_{ij}, the description of the complex incrementally nonlinear relationship between the stress increment and the strain increment can be displayed by nonlinear theory of the hypoplaticity model.Therefore the hypoplaticity model can be used to simulate the rotating and crushing plastic deformations of structured soils. With the additional work to simulate the normally contractive behaviour and the contractive behaviour under inverse loading, a full constitutive model can be developed for the elasto-plastic analysis of the structured and destructured soils under complex loading.

4 FORMULATION OF THE CONSTITUTIVE MODEL

A reasonable model should be suitable for both monotonous loading and complex loading, and the parameters of the model may be determined conveniently by use of the normal and simple tests. Starting from the opinion, the work is to be conducted by studying the constitutive relations from monotonous loading cases, and then ge-neralizing to complex loading cases by use of hypoplasticity theory. The proposed model will be formulated on the basis of the following summary of test results.

At first, the following definition is made for the mean spherical normal effective stress p, the generallized deviator stress q, the second stress invariable J_2, the third stress invariable J_3, the volumetric strain ε_v, the generalized deviator strain ε_s, the shear stress ratio η, and function $M(\rho)$ of Lode angle ρ. The specific expressions are:

$$p = \frac{1}{3}\sigma_{kk}, \qquad q = \left(\frac{3}{2}S_{ij}S_{ij}\right)^{\frac{1}{2}}, \qquad J_2 = \frac{1}{2}S_{ij}S_{ij},$$

$$J_3 = \frac{1}{3}S_{ij}S_{jk}S_{ki}, \qquad \varepsilon_v = \varepsilon_{kk}, \qquad \varepsilon_s = \left(\frac{2}{3}e_{ij}e_{ij}\right)^{\frac{1}{2}},$$

$$\eta = \left(\frac{3}{2}r_{ij}r_{ij}\right)^{\frac{1}{2}} = q/p, \text{ and}$$

$$M(\rho) = \frac{2M_e M_c}{M_e + M_c + (M_c - M_e)\sin 3\rho}$$

where, $\sin 3\rho = -\frac{3\sqrt{3}}{2}\cdot\frac{J_3}{(J_2)^{3/2}}$, $M_c = \frac{6\sin\varphi_c}{3 - \sin\varphi_c}$.

$M_e = \frac{6\sin\varphi_e}{3 + \sin\varphi_e}$. Where e_{ij} is the deviatoric shear strain,and M_c and M_e are calculated from residual friction angles of triaxial compression and extention tests.

For the compression test under constant η,e.g. one-dimentional and isotropic compression tests, the volumetric strain ε_v displays a semi-logarithmic linear relationship with the mean spherical normal effective stress p. The formula may be concluded as :

$$\varepsilon_v = \varepsilon_{v0} + C_c . \log p \quad \text{under loading condition} \qquad (1)$$

and

$$\varepsilon_v = \varepsilon_{v0l} + C_s . \log p \quad \text{under unloading condition} \qquad (2)$$

Differentiating the above two formulae respectively, the increments of volumetric deformation due to compression will be as following:

$$d\varepsilon_{vc} = \frac{0.434C_c}{1 + e_0}\cdot\frac{dp}{p} = dp/K_c \qquad (3)$$

under loading condition

and

$$d\varepsilon_{vc} = \frac{0.434C_s}{1 + e_0}\cdot\frac{dp}{p} = dp/K_s \qquad (4)$$

under unloading condition

Taking consideration of the case of one-dimensional compression test, $d\varepsilon_2 = d\varepsilon_3 = 0$, $d\varepsilon_v = d\varepsilon_a = d\varepsilon_s$ (d ε_q) and then generallizing to other conditions of compression tests, yield the increments of shear deformation due to compression as following:

$$d\varepsilon_{sc} = \frac{0.434C_c}{1 + e_0}\frac{dp}{p}\cdot\frac{\eta}{\eta_0} = \frac{\eta}{\eta_0}\cdot\frac{dp}{K_c} \qquad (5)$$

under loading condition

and

$$d\varepsilon_{sc} = \frac{0.434C_s}{1 + e_0}\frac{dp}{p}\cdot\frac{\eta}{\eta_0} = \frac{\eta}{\eta_0}\cdot\frac{dp}{K_s} \qquad (6)$$

under unloading condition

where $\eta_0 = \frac{3(1 - k_0)}{(1 + 2k_0)}$ is the η value under one-dimensional compression test, k_0 is the coefficient of earth pressure at rest, and K_c and K_s are the compression bulk modulus and the rebound bulk modulus respectively.

For the shear test under constant p, the relationship curve between shear ratio η and deviator shear strain ε_s is assumed as a hyperbolic curve. As η is specific expression of the normalized deviator stress ratio tensor r_{ij} at monotonous loading state, if the hyperbolic function is expanded to complex loading state, the variable shearing modulus may reflect the plastic strain originated from complex loading such as turn of stress path and rotation of the principal stress axis. The hyperbolic function for shearing under constant p can be assumed as:

461

$$\varepsilon_{ss} = \frac{a\eta}{\eta_f - \eta} \qquad (7)$$

where the inverse of a is equivalent to the normalized shear modulus, η_f is the failure stress ratio and it is defined as a variable increasing with the volumetric strain ε_v in a proportional coefficient c_f, that is:

$$d\eta_f = c_f \cdot d\varepsilon_v \qquad (8)$$

Differentiating formula (7), the increment of shear deformation due to shearing will be:

$$d\varepsilon_{ss} = \frac{a}{(\eta_f - \eta)^2}\left(d r_{ij} - \frac{\varepsilon_f \eta}{\eta_f^2}\frac{ac_{vs}}{ac_{ss}}d\varepsilon_{ss}\right) \qquad (9)$$

By introducing the linear function for dilatancy:

$$\frac{d\varepsilon_{vs}}{d\varepsilon_{ss}} = \frac{\eta_d - \eta}{\lambda} \qquad (10)$$

and substituting equation (10) into the derived equation (9), the increment of shear strain may be shortly concluded as:

$$d\varepsilon_{ss} = \frac{a}{g}d\eta \qquad (11)$$

and the increment of volumetric strain due to shearing may be written as:

$$d\varepsilon_{vs} = \frac{\eta_d - \eta}{\lambda}d\varepsilon_{ss} = \frac{\eta_d - \eta}{\lambda}\cdot\frac{a}{g}d\eta \qquad (12)$$

where $g = \frac{(\eta_f - \eta)^2}{\eta_f} + \frac{ac_f\eta}{\eta_f^2}\cdot\frac{\eta_d - \eta}{\lambda}$.

So far the total increment of volumetric and shear deformation will be as follows respectively:

$$d\varepsilon_v = \frac{dp}{K_c} + \frac{a}{g}\frac{(\eta_d - \eta)}{\lambda}\cdot d\eta \qquad (13)$$

$$d\varepsilon_s = \frac{dp}{K_c}\cdot\frac{\eta}{\eta_0} + \frac{a}{g}d\eta \qquad (14)$$

The elastic parts of $d\varepsilon_v$ and $d\varepsilon_s$ can be written respectively as:

$$d\varepsilon_v^e = \frac{dp}{K_s} + \frac{aw}{g}\frac{\eta_d - \eta}{\lambda}\cdot d\eta \qquad (15)$$

$$d\varepsilon_v^e = \frac{dp}{K_s}\frac{\eta}{\eta_0} + \frac{aw}{g}d\eta \qquad (16)$$

where w is a weight function of elastic deformation among the total deformation and normally w may be assumed equal to g/2.

In the above mentioned several equations K_c, K_s and g are variables which means a preliminary incrementally nonlinear model has come into being. The equation of dilatancy (10) makes it possible to simulate the dilatancy and strain softening of soils, e.g. shear contraction occurs when η is less than η_d, and shear dilatancy happens when η is larger than η_d, and when η is larger than η_d and η reaches η_f, the value of g will become negative, therefore a strain softening can be simulated. Next further work will be conducted about the damage ratio D and the model for complex loading.

In order to describe the process of damaging, area ratio or volume ratio is usually used to define the damage ratio D in traditional damage mechanics, but the value of D is always an unmeasurable parameter. In this paper the damage ratio D is defined as follows:

$$D = \frac{e_0 - e}{e_0 - e_s} \qquad (17)$$

where e is the current porosity, e_0 is the initial porosity, and e_s is the stable porosity of the fully damaged structured soil under the same loading condition. D must satisfy the boundary conditions: D=0 when $e = e_0$, and D=1 when $e = e_s$. Under drained condition, damage ratio is induced by change of porosity e and e_s. Under undrained condition, porosity e keeps constant, and damage ratio is induced by change of e_s.

The damage ratio D defined above may be written as following according to the result of isotropic compression test. That is:

$$D_c = \begin{cases} \frac{C_x}{C_c}\frac{(\log p_{m0} - \log p_0)(\log p_{mf} - \log p)}{(\log p - \log p_0)(\log p_f - \log p_{m0})} \\[2ex] + \frac{(\log p - \log p_{m0})(\log p_{mf} - \log p_0)}{(\log p - \log p_0)(\log p_f - \log p_{m0})} \end{cases} \qquad (18)$$

where p_0 is the initial confined pressure when the sample is being prepared, p_{m0} is the critical structure damage strength, and p_{mf} is the pressure when the structured soil becomes fully damaged. It is stipulated that D=0 when p is less than p_{m0}, and D=1 when p is greater than p_{mf}. The definition can be seen obviously through the two compression curves of structured soil and destructured soil in Fig.2.

Fig.3 gives two shear curves of structured soil and destructured soil. The definition of damage ratio D_s can be given through the two shear curves too. In this paper, it is assumed $D_c = D_s$ for the purpose of that definition of damage ratio D will not affect formation of the incremental elasto-plastic martrix.See following equations (21) and (22).

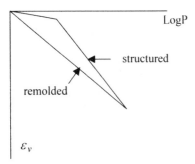

Fig.2   Compression curves

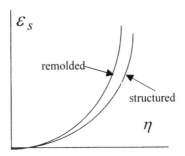

Fig.3 Shear curves

The model description from structured soil to destructured soil may be formulated as following by use of equations (13) and (14).When p is less than p_{m0} or D=0,the deformation can be expressed as:

$$d\varepsilon_v = \frac{dp}{K_s} + \frac{\eta_d - \eta}{\lambda} \cdot \frac{aw}{g} d\eta \qquad (19)$$

$$d\varepsilon_s = \frac{dp}{K_s} \cdot \frac{\eta}{\eta_0} + \frac{aw}{g} d\eta \qquad (20)$$

when p is larger than p_{m0} or D>0,the deformation can be expressed as:

$$\varepsilon_v^t = D_t \cdot \sum_{k=1}^{t} d\varepsilon_{vk} \qquad (21)$$

$$\varepsilon_s^t = D_t \cdot \sum_{k=1}^{t} d\varepsilon_{sk} \qquad (22)$$

where $d\varepsilon_{vk}$ and $d\varepsilon_{sk}$ can be calculated through formula (13) and (14). D_t is introduced for total strain calculation and it will not affect formation of the incremental elasto-plastic martrix. The following discussion is for the formulation of the generalized model.

For complex loading the above formulations can be expressed as follows:

$$d\varepsilon_v^e = \frac{dp}{K_s} + \frac{aw}{g} \cdot \frac{B[M(\rho)\eta_d - \eta]}{\lambda} \cdot n_{aij} dr_{ij} \qquad (23)$$

$$de_{ij}^e = \frac{dp}{K_s} \frac{\eta}{\eta_0} n_{1ij} + \frac{aw}{g} n_{aij} dr_{ij} \qquad (24)$$

$$d\varepsilon_v^p = \frac{dp}{K_c'} + \frac{a(1-w)}{g} \frac{B[M(\rho)\eta_d - \eta]}{\lambda} n_{aij} dr_{ij} \qquad (25)$$

$$de_{ij}^p = \frac{dp}{K_c'} \frac{\eta}{\eta_0} \cdot n_{2ij} + \frac{a}{g} \cdot n_{ij} n_{aij} dr_{ij} \qquad (26)$$

where $K_c' = K_c K_s/(K_s - K_c)$, $n_{1ij} = dr_{ij}/(dr_{ij} dr_{ij})^{1/2}$, $n_{2ij} = r_{ij}/(r_{ij} r_{ij})^{1/2}$, $n_{ij} = \frac{1}{3} n_{1ij} + \frac{2}{3} n_{2ij}$, $n_{aij} = \cos\theta_{ij}$.

θ_{ij} is the turing angle of stress path defined by Chen shengshui(Chen S.S,1994) to simulate the behaviour of complex loading. θ can be calculated according to the following formula:

$$\theta = ar\cos \frac{\sum (r_{ij}^{n+1} - r_{ij}^n)(r_{ij}^n - r_{ij}^{n-1})}{\left[\sum (r_{ij}^{n+1} - r_{ij}^n)^2 \sum (r_{ij}^n - r_{ij}^{n-1})^2 \right]^{1/2}} \qquad (i \leq j) \qquad (27)$$

Formula (27) stipulates the following loading-unloading criteria:

Proportionally loading, when $\theta = 0$
unloading when $90° < \theta < 180°$
fully unloading when $\theta = 180°$
yield surface loading when $\theta = 90°$
and stress path rotating when $0° < \theta < 90$

B is the symbol used in the calculation of shear contraction under inverse loading:

B =1 when $M(\rho) \cdot \eta_d > \eta$ and $n_{aij} dr_{ij} > 0$
B=-1 when $M(\rho) \cdot \eta_d \leq \eta$ and $n_{aij} dr_{ij} < 0$

It is noted that the difference between the proposed model and hypoplasticity model is the absence of Macauley symbol ◇ for dp. It means that volumetric strain still develops when the compression stress is lower than the maximum stress in history. therefore the build-up of pore pressure of structured soil under cyclic compression loading can be estimated under undrained condition, according to the formula of $d\varepsilon_v = d\varepsilon_v^e + d\varepsilon_v^\nu = 0$, the pore pressure increment will be $du = -dp'$, and from equations (23) and (25) du can be calculated:

$$du = -dp' = K_c \frac{B[M(\rho)\eta_d - \eta]}{\lambda} \cdot \frac{a}{g} \cdot n_{aij} dr_{ij}$$

$$= \sqrt{\frac{2}{3}} K_c \cdot \frac{B[M(\rho)\eta_d - \eta]}{\lambda} \cdot \frac{a}{g} \cdot n_{aij} d\eta \qquad (28)$$

It can be seen that the calculated du is lower than that of using rebound bulk modulus K_s. The gradually developing process of effective stress can be calculated by the following overlapping scheme:

$$P'_{n+1} = P'_n + \left(\frac{dp'}{d\eta}\right)_n d\eta \qquad (29)$$

For simplicity, equation (23) can be shortly written as:

$$d\varepsilon_v^e = \frac{dp}{K} \qquad (30)$$

Differentiating the formula $r_{ij} = S_{ij}/p$ yields $dr_{ij} = \frac{dS_{ij}}{p} - \frac{S_{ij}}{p^2} dp$ and substituting dr_{ij} into equation (24) can yield the following short expression:

$$de_{ij}^e = \frac{1}{2G} dS_{ij} \qquad (31)$$

Where $\quad \dfrac{1}{K} = \dfrac{1}{K_s} + \dfrac{aw}{g} \dfrac{B[M(\rho)\eta_d - \eta]}{\lambda} \cdot \dfrac{d\eta}{dp}$ $\qquad (32)$

$$\frac{1}{2G} = \left(\frac{\eta}{\eta_0} \frac{dp}{K_s (dr_{ij} dr_{ij})^{1/2}} + \frac{aw}{g}\right)\left(\frac{1}{p} - \frac{S_{ij}}{dS_{ij}} \frac{dp}{p^2}\right) \qquad (33)$$

K and G are bulk and shear moduli in conventional meaning, but they are apparently variables and dependent on the stress state. Then the short form of combination of equations (30) and (31) will be:

$$d\varepsilon_{ij}^e = D_{ijkl}^{-1} \cdot d\sigma_{ij}^e \qquad (34)$$

and combining equations (25) and (26) can derive the following short expression:

$$d\varepsilon_{ij}^p = C_{ijkl}^p d\sigma_{kl} \qquad (35)$$

Because $d\varepsilon_{ij} = \frac{1}{3} d\varepsilon_v \delta_{ij} + de_{ij}$, where δ_{ij} is the

unit tensor, summarizing the above equations (23)~(26) and considering $dr_{ij} = \dfrac{d\sigma_{ij}}{p} - \dfrac{\sigma_{ij}}{p^2} dp$ yield:

$$d\varepsilon_{ij} = C_{ijkl} d\sigma_{kl} \qquad (36)$$

Combining equations (34)、(35) and (36) and making use of quation $d\varepsilon_{ij}^e = d\varepsilon_{ij} - d\varepsilon_{ij}^p$, the final expression of elasto-plastic matrix can be evoluted by:

$$d\sigma_{ij} = \left(I_{ijmn} + D_{ijkl} C_{klmn}^p\right)^{-1} D_{mnkl} d\varepsilon_{kl} \qquad (37)$$

where I_{ijmn} is unit tensor of fourth order. The elasto-plastic matrix can be used for the analysis of static and dynamic responses of soils with a practical differential scheme.

5 CALCULATED RESULTS OF THE MODEL

To confirm the performance of the proposed model under simple and complex loading, a series of computed examples will be presented in the following. In the following computations a set of model parameters which may be derived from the conventional odometer test and triaxial compression test are assumed:C_c=0.098, C_s=0.0098, a=0.0036, λ=0.6, η_d=1.47, η_f=1.84.

For the cases of conventional compression tests under constant value of σ_3 / σ_1, the computed stress-strain p~ε_v curves with different σ_3 / σ_1 values are shown in Fig.4. It can be seen that for consolidated tests under constant value of σ_3 / σ_1, the more σ_3 / σ_1 value drops, the more compression curves of structured soils draw near to that of destructured soils. The phenomeno shows an influence of σ_3 / σ_1 on the structure effect.

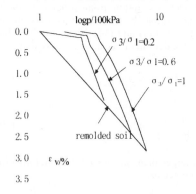

Fig.4 Compression curves under different value of σ_3 / σ_1

For the cases of conventional drained and

undrained triaxial compression tests the calculated stress~strain curves for differently structured soils and destructured soils are shown in Fig.5.It can be seen that for weak structured soils, normal shearing contraction and strain hardening happens and in contrast to destructured soils there still exist slight strain hardening and less volumetric strain contraction (see (a) in Fig.5), and dilatancy and strain softening are simulated for high structured soils by use of lower value of η_d (see (b) in Fig.5). The residual strength of high structured soils will draw near to that of weak structured soils or destructured soils. In Fig.6 it can be found that under undrained state, instead of volumetric strain the pore pressure bulid-up is captured slowly developing for structured soils. The captured unstable curve in Fig.6 showing static liquefaction stands for phenomena usually happened on loose sandy structured soil, which may mean a sudden loss of the structure of soils.

In Fig.7 one-way repeated triaxial compression test is sumulated. Loops of stress-strain curves and the process of accumulation of axial strain and volumetric strain are fairly captured. The simulation of conventional undrained cyclic triaxial test is shown in Fig.8. It can be found that structured soil has a stronger ability to resist the occurrence of deformation and pore pressure than destructured soil.

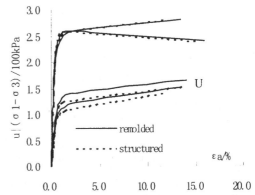

Fig.6 $\sigma_1 - \sigma_3 \sim \varepsilon_a$ curves and $u \sim \varepsilon_a$ curves of undrained triaxial test

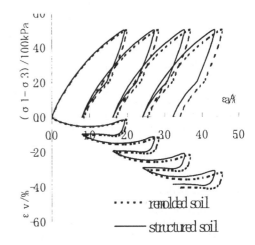

FIG.7 Repeated drained triaxial test

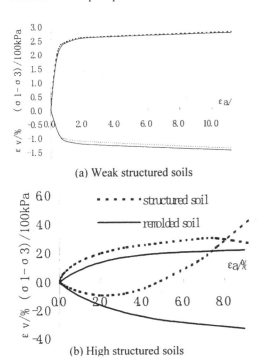

(a) Weak structured soils

(b) High structured soils

Fig.5 $\sigma_1 - \sigma_3 \sim \varepsilon_a$ curves and $\varepsilon_v \sim \varepsilon_a$ curves of drained triaxial test

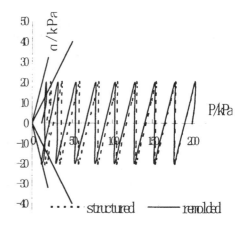

FIG.8 Undrained cyclic triaxial test

6 CONCLUSIONS

Using the proposed model, the sturcutre effect on deformation and pore pressure of soils is performed. The following conclusions can be drawn from the current investigation:

(1)The model introduced in this paper is based on understanding of masonry theory proposed by Shen Zhujiang.The discription of deformation behaviour is practical to saturated structured or destructured clayey soils.

(2)Instead of using concept of yield or flow surface and plastic potential function, the proposed model is based on hypoplasticity theory to discribe the complex deformation behaviour of structured soils or distructured soils.

(3) The description of structure effect by the model can be easily accompolished by use of the defined damage ratio D and eight parameters of Cc、C_s、a、η_d、η_f、λ、p_{m0}、p_{mf}, etc which are determined from conventional odemeter test and triaxial compression test.

(4) The introduced hypoplasticity model helps the proposed model in this paper to simulate both drained and undrained responses of structured soils or destructured soils under simple or complex loading. FEM analysis can be implemented by use of the elasto-plastic matrix evoluted in this paper.

(5) Tests conducted shoulded be used to verify the feasibility of the EDP model (Elasto-Damaging-Plastic model) under simple and complex loading.

ACKNOWLEDGMENTS

Financial support from NHRI grant No.30004 and Supervisions from prof. Shen Zhujiang, prof. Yin Jianhua (Hong Kong Polytechnic University), prof. Chen Shengshui and help from Mr. Chen Tielin are acknowledged. The authors of the references are also acknowledged. The reviewers' comments and su-ggestions and secretary's check from the committee of International Symposium in Shizuoka 2001 are deeply acknowledged.

REFERENCES

Chen S. S.,1994,Elasto-plastic numerical simulation for cohesionless soils under complex stress paths .[Ph.D. dissertation]. Nanjing Hydraulic Resear-ch Institute , Nanjing , China.

Jiang M. J.,1996,Constitutive laws of structured clay and analysis of progressive failure of soils. [Ph.D dissertation].Nanjing Hydraulic Research Institu-te , Nanjing , China.

Shen Z. J.,1993,A nonlinear damage model for struc -tured clays.Journal of Nanjing Hydraulic Resear -ch Institute,No.4 (in Chinese): 247-255.

Shen Z. J.,2000,A masonry model for structured clay . Geomechanics 21(1), (in Chinese): 1-4.

Shen Z. J., 1999, A gradular medium model for lique faction analysis of sands.Chinese Journal of Geo -technical Engineering 21(6),(in English):742 - 748.

Shi B.,1995,The quantitative techniques of microstru cture and micro-mechanical model of clayey soil. [Ph.D. dissertation]. Nanjing University,Nanjing, China.

Wang Z. L., Dafalias Y. F. and Shen C. K.,1990, Bounding surface hypoplastic model for sand.J. Eng.Mech,ASCE,116(5): 983-1001.

Clay Science for Engineering, Adachi & Fukue (eds) © 2001 Balkema, Rotterdam, ISBN 90 5809 175 9

Modelling of state surfaces of unsaturated soils

Y. Kohgo, I. Asano & H. Tagashira
National Research Institute of Agricultural Engineering, Tsukuba, Japan

ABSTRACT: The purpose of this paper is to model state surfaces of unsaturated soils. The state surfaces have to be formulated to analyze exactly the elastoplastic volume change of unsaturated soils. The shape of state surface depends on the type of soils. A model that could express any kind of shapes of the state surfaces was proposed. The model was applied to express the state surfaces of three kinds of soils, clayey silt, silt and silty sand. The values estimated by the model were well consistent with the experimental ones. The model is applicable to express any kinds of state surfaces.

1 INTRODUCTION

Remarkable advances for constitutive models of unsaturated soils have been accomplished for the last two decades. Some elastoplastic models (Kohgo, 1987, Kohgo et al. 1991, 1993b; Alonso et al. 1990) proposed during that time can analyze typical mechanical properties of unsaturated soils such as saturation collapse and shear strength reductions due to wetting. It is necessary to consider two suction effects to represent exactly the mechanical properties of unsaturated soils (Kohgo et al., 1993a). The first suction effect is that an increase in suction increases effective stresses. The effect controls the changes of the shear strength on the wet side of the critical state, volume reductions due to an increase in suction (drying) and swellings due to a decrease in suction (wetting). The second suction effect is that an increase in suction enhances yield stresses and affects the resistance to plastic deformations. The effect controls both the changes of the shear strength on the dry side of the critical state and the volume reductions due to a decrease in suction (saturation collapse). State surfaces express elastoplastic volume changes of unsaturated soils (Kohgo, 1987). If plastic volumetric strain is regarded as a hardening parameter (this assumption can be usually acceptable in soil mechanics field), the second suction effect can be estimated by formulating the state surface. Thus, in the formulation of elastoplastic models for unsaturated soils, it is necessary to formulate not only effective stress equations but also state surfaces for unsaturated soils.

The shape of the state surface depends on the types of soils. The purpose of this paper is to model

the state surfaces of unsaturated soils. The causes why the various shapes of state surfaces exist are also investigated.

2 MODELING OF STATE SURFACE

At first, stress notations used in this paper are defined as follows:

$$s = u_a - u_w \tag{1}$$

$$s^* = \langle s - s_e \rangle \tag{2}$$

$$\sigma_m = \frac{1}{3}(\sigma_1 + \sigma_2 + \sigma_3) \tag{3}$$

$$p = \sigma_m - u_a \tag{4}$$

$$p' = \frac{1}{3}(\sigma_1' + \sigma_2' + \sigma_3') = \sigma_m - u_{eq} \tag{5}$$

where s = suction; u_a = pore air pressure; u_w = pore water pressure; s^* = effective suction; s_e = air entry suction; σ_m = total mean stress; σ_1, σ_2, σ_3 = three total principal stresses; p = net mean stress; p' = effective mean stress; σ_1', σ_2', σ_3' = three effective principal stresses; u_{eq} = equivalent pore pressure and the brackets < > denote the operation <z>=0 at z < 0 and <z>=z at z ≥ 0. Effective stresses are evaluated by using Equations (31) to (33).

The state surface can be plotted in the space with the axes, effective mean stress p', effective suction s^* and void ratio e. The state surface may be generally expressed as:

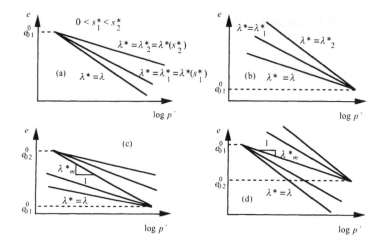

Figure 1. The possible shapes of state surfaces. (a) monotonous increase of λ^*, (b) monotonous decrease of λ^*, (c) having the maximum of λ^* and (d) having the minimum of λ^*.

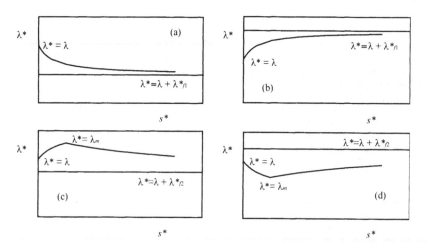

Figure 2. Relationships between λ^* and s^*. (a) for Figure 1(a), (b) for Figure 1(b), (c) for Figure 1(c) and (d) for Figure 1(d).

$$e = -\lambda^* \log p' + \Gamma^* \qquad (6)$$

where λ^* = slope of e - log p' curves; Γ^* = void ratio of e - log p' curves at p' = unit. The values of λ^* and Γ^* change as effective suction values change.

Figure 1 shows the possible shapes of state surfaces. The shapes depend on the types of soils. Figure 1 (a) shows the state surface where the value of λ^* monotonously decreases as the value of s^* increases and Figure 1 (b) shows the opposite case where the value of λ^* monotonously increases. Figures 1 (c) and (d) show the cases where the value of λ^* has the maximum or minimum values, respectively. The relationships between λ^* and s^* in each shape of state surface are schematically shown in Figure 2.

In our early model (Kohgo et al. 1991 and 1993b), the state surface was expressed as follows:

$$\lambda^* = \frac{\lambda}{1+y} \qquad (7)$$

$$\Gamma^* = \frac{\Gamma + e_0^0 y}{1+y} \qquad (8)$$

$$y = \left(\frac{s^*}{a_s}\right)^{n_s} \qquad (9)$$

where $\lambda = \lambda*$ in saturation; $\Gamma = \Gamma*$ in saturation; a_s, n_s, $e_0^{\ 0}$ = material parameters. This model can well express the state surface shown in Figure 1(a). However, this model is not suitable for others. Then, the following new equations are proposed here. They are:

when $s* \leq s*_m$

$$\lambda* = \lambda + \frac{\lambda*_{f1} s*}{s* + a*_1} \qquad (10)$$

$$\Gamma* = e_{01}^0 + \frac{(\Gamma - e_{01}^0)\lambda*}{\lambda} \qquad (11)$$

when $s* > s*_m$

$$\lambda* = \lambda*_m + \frac{\lambda*_{f2}(s* - s*_m)}{(s* - s*_m) + a*_2} \qquad (12)$$

$$\Gamma* = e_{02}^0 + \frac{(\Gamma - e_{02}^0)\lambda*}{\lambda*_m} \qquad (13)$$

where $s*_m$ = value of $s*$ at $\lambda* = \lambda*_m$; $\lambda*_m$ = the maximum or minimum value of $\lambda*$; e_{01}^0, e_{02}^0, $\lambda*_{f1}$, $\lambda*_{f2}$, $a*_1$ and $a*_2$ = material parameters.

In Equation (10), if the value of $s*_m$ is an infinite positive figure and the value of $\lambda*_{f1}$ is negative, the state surface shown in Figure 1 (a) can be expressed. If the value of $s*_m$ is an infinite positive figure and the value of $\lambda*_{f1}$ is positive, the state surface shown in Figure 1 (b) can be expressed. If the value of $s*_m$ is a finite positive figure, both Equations (10) and (12) will be valid. If the values of $\lambda*_{f1}$ and $\lambda*_{f2}$ are respectively positive and negative, the state surface shown in Figure 1 (c) can be expressed. The state surface shown in Figure 1 (d) can be expressed by selecting negative and positive values as the values of $\lambda*_{f1}$ and $\lambda*_{f2}$, respectively.

3 APPLICATION OF MODEL

In order to verify the model, three kinds of soils were used. Table 1 shows the physical properties of the soils. The first soil used is an artificial soil mixed with flint-powder 80% and kaolin 20% in weight (Matyas & Radhakrishna, 1968). The soil was classi-

-fied as (ML) according to Japanese unified soil classification system (JUSCS). The contents of clay and silt were 34% and 66%, respectively.

The second one is called DL clay and classified as (ML) in JUSCS (Kohgo & Moriyama, 1998). The soil was non-plastic. The contents of clay and silt were 10% and 90%, respectively.

The third one is called Yellow soil sampled from Northeast Thailand and is recognized as a loessial soil (Kohgo et al., 1999). It was classified as (SM) in JUSCS. The contents of clay, silt and sand were 5%, 35%, 60%, respectively. These soils change in order from clayey soils to sands.

We will describe a method to estimate the material parameters for effective stresses and the state surface of the Yellow soil, here. We need the experimental data obtained from triaxial shear tests and oedometer tests.

The parameters for effective stresses are estimated as follows.

i) Estimate shear strength q for each specimen.
ii) Obtain values of $(u_a - u_{eq})$. If the failure always occurs on the failure line, the effective mean stress p' may be obtained from the following equation,

$$p' = \frac{q}{M} \qquad (14)$$

$$M = \frac{6\sin\phi'_{cs}}{3 - \sin\phi'_{cs}} \qquad (15)$$

where ϕ'_{cs} = angle of internal friction of critical state line. The value of Yellow soil was 33.7°.

The values of $(u_a - u_{eq})$ can be given as

$$\left(u_a - u_{eq}\right) = \left(\sigma_m - u_{eq}\right) - \left(\sigma_m - u_a\right) = p' - p. \qquad (16)$$

iii) Identify parameters for effective stresses; s_e and a_e. The air entry value s_e can be evaluated from the soil-water retention curve. Supposing f as;

$$f = u_{eq} - u_a + s_e \qquad (17)$$

the relationship between the effective suction $s*$ and $(s*/f)$ will become a straight line from Equation (33). Then, the parameter a_e can be estimated as the slope of this line. The $s*$ - $(s*/f)$ line for the Yellow soil is shown in Figure 3. The estimated parameters were, $s_e = 10$ kPa and $a_e = 15.2$ kPa. The value of u_{eq} may become a_e when $s* \rightarrow \infty$.

The parameters for state surface are estimated as follows.

Table 1. Physical properties of soils used for analysis.

Soil name	Soil type (JUSCS)	w_L (%)	I_p	G_s	D_{max} (mm)	Clay (%)	Silt (%)	Sand (%)
1 Kaolin(20%)+Flint(80%)	(ML)	29	4	2.63	0.07	34	66	0
2 DL clay	(ML)	—	—	2.650	0.11	10	90	0
3 Yellow soil	(SM)	22	7	2.672	2.000	5	35	60

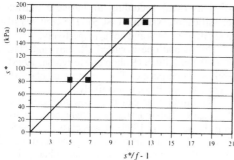

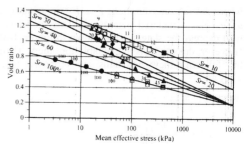

Figure 4. Contour lines of degree of saturation of Yellow soil. (Figures denote the values of degree of saturation).

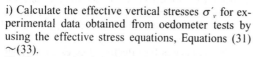

$s*/f$ - 1

Figure 3. Estimation of effective stress parameters for Yellow soil.

i) Calculate the effective vertical stresses σ'_v for experimental data obtained from oedometer tests by using the effective stress equations, Equations (31) ~(33).

ii) The associated horizontal stresses σ'_h may be evaluated by Jaky's equation.

$$K_0 = 1 - \sin\phi'$$ (18)
$$\sigma'_h = K_0\sigma'_v$$ (19)

where ϕ' = angle of internal friction of failure line.

iii) Calculate mean effective stress p' as follows.

$$p' = \frac{1}{3}(\sigma'_v + 2\sigma'_h)$$ (20)

iv) Plot the experimental data in the e - $\log p'$ space as shown in Figure 4.

v) Draw the predicted e - $\log p'$ lines for each degree of saturation S_r in the figure. Here, we selected the degree of saturation S_r = 10, 20, 30, 40, 60 and 100 % (see Figure 4).

vi) Identify the values of e^0_{01} and e^0_{02}, the value $\lambda*_m$ and the effective suction $s*_m$ at $\lambda* = \lambda*_m$.

vii) Read the values of the slope $\lambda*$ of the e - $\log p'$ lines for S_r = 10, 20, 30, 40, 60 and 100 %.

viii) Estimate the effective suction values $s*$ for S_r = 10, 20, 30, 40, 60 and 100 % from the soil-retention curve.

ix) Identify parameters for state surface, $\lambda*_{f1}$ and $a*_1$ by using the linear relationship as follows.

$$s* = \lambda_{f1}\left(\frac{s*}{\lambda*-\lambda}\right) - a*_1$$ (21)

x) Identify parameters for state surface, $\lambda*_{f2}$ and $a*_2$ by using the linear relationship as follows

$$(s*-s*_m) = \lambda_{f2}\left(\frac{s*-s*_m}{\lambda*-\lambda*_m}\right) - a*_2$$ (22)

The material parameters identified are shown in Table 2. The state surface of the mixed soil shapes as shown in Figure 1 (a). That of DL clay shapes as shown in Figures 1 (b). Then, the value of $\lambda*_{f1}$ in the mixed soil is negative and in opposite that is positive in DL clay. The state surface of Yellow soil shapes as shown in Figure 1 (c) and then the value of $\lambda*$ has the maximum.

Figure 5 shows the comparisons of the estimated and measured values of void ratio. Figure 5(a) shows the result of the mixed soil and Figure 5(b) shows that of DL clay. The plotted points in both figures almost lie on 45 degrees lines and the agreements between the estimated and the measured values are quite good.

The verification of the model for Yellow soil was investigated by using FE consolidation analysis method with the elastoplastic model in which the state surface was evaluated by the model proposed here. The processes of formulating the FE consolidation analysis method are shown in Figure 6. The details of the elastoplastic model are shown in Figure 7. More details for the FE consolidation analysis

Table 2. Parameters for estimation of state surfaces.

Soil name	Parameters for effective stresses		Type of state surface	Prameters for state surface								
	s_e (kPa)	a_e (kPa)		λ	Γ	e^0_{01}	$\lambda*_{f1}$	$a*_1$ (kPa)	e^0_{02}	$\lambda*_{f2}$	$a*_2$ (kPa)	$s*_m$ (kPa)
1 Kaolin(20%)+Flint(80%)	20	98	Fig. 1 (a)	0.176	1.157	0.928	-0.101	81.6				
2 DL clay	10	33.3	Fig. 1 (b)	0.100	1.153	0.720	0.080	14.3				
3 Yellow soil	10	15.2	Fig. 1 (c)	0.165	0.840	0.185	0.244	132.0	1.92	-0.08	3087	190

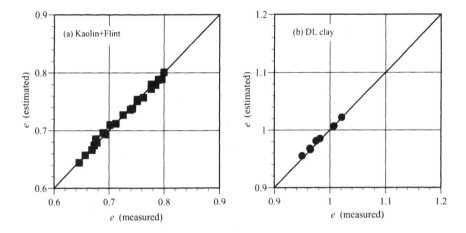

Figure 5. Comparisons of estimated and measured values of void ratio.

method can be found in Kohgo (1995, 1997) and Kohgo et al. (1999). Figure 8 shows the simulation results of consolidation tests for Yellow soil with soaking of water. The simulations were conducted with the FE consolidation analysis method. The consolidation tests for two specimens with the initial degrees of saturation S_{r0} = 9% and 14% were analyzed. As the initial dry densities of the specimens were low in these tests, elastoplastic deformations appeared from the first stages of consolidation. Then, the associated stress points lay on the state surface. The values measured before soaking were well consistent with those estimated. The simulation results could express the phenomenon of saturation collapse. The stress points just after soaking lay on the $e - \log \sigma_v$ line obtained from the specimen with the initial degree of saturation S_{r0} = 94% which might be an almost saturated line. All of the volume changes described above occur on the state surface. Thus, the simulation results could well express the behavior occurring on the state surface.

4 SHAPES OF STATE SURFACES AND TYPES OF SOILS

The shapes of the state surfaces depend on the changes of λ^* with the changes of s^*. The dependence of the values of λ^* on the values of s^* arises from the dependence of kinetic coefficient of friction between soil particles on the values of s^*. Horn & Deere (1962) investigated experimentally the influence of the ambient relative humidity to the coefficient of friction between minerals. Figure 9

shows the relationships between the coefficient of static friction and the ambient relative humidity for specimens of quartz and muscovite. The coefficient of friction of layer-lattice minerals such as muscovite decreased as the relative humidity increased (or suction decreased), while that of massive-structured minerals such as quartz increased. Water acted namely as an anti-lubricant when applied to quartz but acted as a lubricant when applied to muscovite. Though in quartz, there was a distinct difference between the kinetic and static coefficients of friction of saturated specimens, the relationship between the kinetic coefficient of friction and the ambient relative humidity was similar to that as shown in Figure 9. In muscovite, there was very little difference between the static and kinetic coefficients of friction. Then, the similar relationship as shown in Figure 9 was also valid for that between the kinetic coefficient of friction and the ambient relative humidity.

From the relationships as shown in Figure 9, it may be reasonable to associate that the value of λ^* for clay increases as the value of s^* decreases, because clay has layer-lattice structures. For silt or sand, the value of λ^* decreases as that of s^* decreases, because silt or sand has massive structures. The shape of the state surface for clay is namely consistent with that shown in Figure 1 (a), while that for silt or sand is consistent with that shown in Figure 1 (b). As real soils may have various sizes of soil particles, the various shapes of state surfaces as shown in Figure 1 can appear. Such consideration is consistent with the fact that the state surfaces of the mixed soil, DL clay and Yellow soil shape as shown in Figures 1(a), (b) and (c), respectively.

Governing Equations

Force Equilibrium Equation

$$\int_V \left(\sigma'_{ij,j} + \delta_{ij} u_{eq,j} + \gamma F_i\right) \cdot \delta u^*_i \, dV = 0 \qquad (23)$$

Mass Conservation Equation for Pore Water

$$\int_V \left(q_{i,i} - \dot{a}_{ii} + S_r \dot{\varepsilon}_{ii}\right) \cdot \delta u^*_w \, dV = 0 \qquad (24)$$

σ'_{ij} = components of effective stress tensor; δ_{ij} = Kronecker delta

γ = unit weight of the soil; F_i = components of the body force vector

q_i = components of the relative displacement velocity vector of water with respect to the soil skeleton

$\dot{a}_{ii}$ = change of storage of water due to change of degree of saturation

$\dot{\varepsilon}_{ii}$ = volumetric strain of the soil skeleton; δu^*_i = virtual displacement vector

δu^*_w = virtual pore water pressure

Subscripts after a comma denote spatial differentiation

Repeated indices = summation; Superposed dot = differentiation with respect to time

↓ Finite Element Method

↓ Modified Newton- Raphson method

FEM Formulation

$$\boldsymbol{\delta}_{m+1} = \boldsymbol{\delta}_m + \Delta\boldsymbol{\delta}_{m+1}, \qquad \Delta\boldsymbol{\delta}_{m+1} = -\boldsymbol{K}^{*\,-1}_0 \, \boldsymbol{\Phi}_m \qquad (25)$$

$$\Delta\boldsymbol{\delta} = \begin{Bmatrix} \Delta\boldsymbol{u}(t_n) \\ \Delta\boldsymbol{u}_w(t_n) \end{Bmatrix}, \qquad \boldsymbol{\Phi}_m = \begin{Bmatrix} \boldsymbol{\Phi}^1_m \\ \boldsymbol{\Phi}^2_m \end{Bmatrix}, \qquad \boldsymbol{K}^*_0 = \begin{bmatrix} \boldsymbol{K} & \boldsymbol{L} \\ \boldsymbol{L}^T & -\alpha \cdot \Delta t \overline{\boldsymbol{K}} - \boldsymbol{E} \end{bmatrix} \qquad (26)$$

$$\boldsymbol{K} = \sum_{i=1}^{l} \int_v \boldsymbol{B}^T \boldsymbol{D} \boldsymbol{B} \, dV, \qquad \boldsymbol{L} = \sum_{i=1}^{l} \int_v \boldsymbol{C}^T \boldsymbol{N} \, dV, \qquad (27)$$

$$\overline{\boldsymbol{K}} = \sum_{i=1}^{l} \int_v \boldsymbol{N}'^T \boldsymbol{R} \boldsymbol{N}' / \gamma_w \, dV, \qquad \boldsymbol{E} = \sum_{i=1}^{l} \int_v \boldsymbol{N}^T \boldsymbol{N} / E'_f \, dV \qquad (28)$$

$$\boldsymbol{\Phi}^1_m = \sum_{i=1}^{l} \int_v \boldsymbol{B}^T \boldsymbol{\sigma}'(t_n)_m dV + \sum_{i=1}^{l} \int_v \boldsymbol{C}^T u_{eq}(t_n)_m dV - \sum_{i=1}^{l} \int_{S_3} \boldsymbol{N}_{ub}{}^T \boldsymbol{N}_{ub} \boldsymbol{T}_b(t_n) dS - \sum_{i=1}^{l} \int_v \gamma \boldsymbol{N}_u{}^T \boldsymbol{F}(t_n) dV \qquad (29)$$

$$\begin{aligned} \boldsymbol{\Phi}^2_m = {} & \sum_{i=1}^{l} \int_v S_r \boldsymbol{N}^T \boldsymbol{C} \{\boldsymbol{u}(t_n)_m - \boldsymbol{u}(t_{n-1})\} dV - \sum_{i=1}^{l} \int_v \boldsymbol{N}^T \{\boldsymbol{a}(t_n)_m - \boldsymbol{a}(t_{n-1})\} dV \\ & + \Delta t \sum_{i=1}^{l} \int_{S_4} \boldsymbol{N}_b{}^T \boldsymbol{N}_b \{\alpha \cdot \boldsymbol{Q}_b(t_n) + (1-\alpha) \cdot \boldsymbol{Q}_b(t_{n-1})\} dS \\ & - \Delta t \sum_{i=1}^{l} \int_v \boldsymbol{N}'^T [\alpha \cdot \{\boldsymbol{q}_p(t_n)_m + \boldsymbol{q}_g(t_n)_m\} + (1-\alpha) \cdot \{\boldsymbol{q}_p(t_{n-1}) + \boldsymbol{q}_g(t_{n-1})\}] dV \end{aligned} \qquad (30)$$

$\boldsymbol{\delta}$ = unknown vector; $\boldsymbol{\Phi}$ = residual vector; $\boldsymbol{K}^*_0$ = coefficient matrix;
$\boldsymbol{\sigma}'$ = effective stress vector; $\boldsymbol{u}$ = nodal displacement vector; $\boldsymbol{u}_w$ = nodal pore water pressure;
$\boldsymbol{N}$ = interpolation function for pore water pressure; $\boldsymbol{N}'$ = gradient matrix for pore water pressure;
$\boldsymbol{N}_u$ = interpolation function for displacement; $\boldsymbol{B}$ = nodal displacement - strain matrix;
$\boldsymbol{C}$ = nodal displacement - volumetric strain matrix; $\boldsymbol{D}$ = elasticity matrix; $\boldsymbol{R}$ = permeability matrix;
$\boldsymbol{F}$ = body force vector; $\boldsymbol{a}$ the change of storage of water due to change of degree of saturation;
$\boldsymbol{T}_b$ = prescribed traction on the boundary nodal points; $\boldsymbol{Q}_b$ = prescribed flux on the boundary nodal points;
$\boldsymbol{q}$ = flux vector, $E'_f = E_f / n$, $E_f = \partial S_r / \partial s$; n : porosity; α = coefficient characterizing single-step temporal discretization; Superscript T = transpose of matrix; Subscripts b, p, g, m = value on the boundary, the value due to pore water pressure, the value due to gravitational potential and the number of iterations, respectively

Figure 6. Formulation process of FE consolidation analysis method.

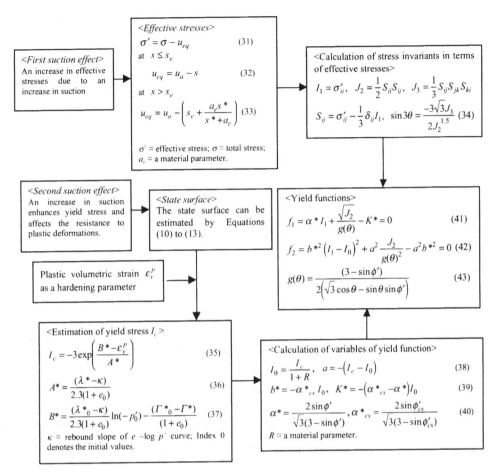

Figure 7. Suction effects and an elastoplastic model.

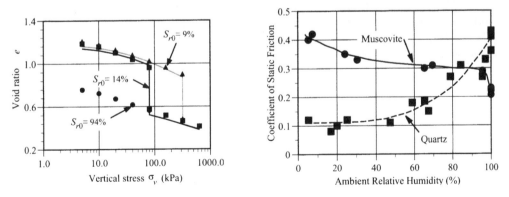

Figure 8. Results of FE consolidation analyses for Yellow soil. Symdols = Experimental data; Lines = Simulation results.

Figure 9. Relationships between the coefficient of static friction and the ambient relative humidity for muscovite (Brazil) and clear quartz (N. Carolina), (after Horn & Deere 1962).

473

5 CONCLUSIONS

The shapes of state surfaces were divided into four as shown in Figure 1. A model, which could express the state surfaces, was proposed. The model was applied to three kinds of soils, a mixed soil, DL clay and Yellow soil. The application results found that the proposed model could describe every shape of state surfaces.

REFERENCES:

Alonso, E. E., Gens, A. & Josa, A. 1990. A constitutive model for partially saturated soils. *Geotechnique* 40(3): 405-430.
Horn, H. M. & Deere, D. U. 1962. Frictional characteristics of minerals. *Geotechnique* 12: 319-335.
Kohgo, Y. 1987. The interpretations and analyses of mechanical behaviour of partly saturated soils using elastoplastic models. *Proc. Symposium Unsaturated Soils*: 69-78. Osaka, Japan: J.S.S.M.F.E. (in Japanese).
Kohgo, Y. 1995. A consolidation analysis method for unsaturated soils coupled with an elastoplastic model. *Proc. 1st Int. Conf. on Unsaturated Soils* 2: 1085-1093: Balkema.
Kohgo, Y. 1997. Method of analysis of saturation collapse. *JIRCAS J.* 4: 1-24: Japan International Research Center for Agricultural Sciences, Ministry of Agriculture, Fresty and Fisherys, Japan.
Kohgo, Y., Hopibulsuk, S. & Tamrakar, S. B. 1999. Simulation of volume change behavior of Yellow soil sampled from Khon Kaen city in Northeast Thailand. *Proc. 11th Asian Regional Conf.* 1: 141-144: Balkema.
Kohgo, Y. & Moriyama, H. 1998. Volume change and shear behavior of unsaturated silt and sand/ clay mixed soil under triaxial stress conditions. *Trans. of JSIDRE* 193: 35-49: The Japanese Society of Irrigation, Drainage and Reclamation Engineering (in Japanese).
Kohgo, Y., Nakano, M. & Miyazaki, T. 1991. Elastoplastic constitutive modelling for unsaturated soils. *Proc. 7th Int. Conf.Computer Meth. Adv. Geomechanics* 1: 631-636: Balkema.
Kohgo, Y., Nakano, M. & Miyazaki, T. 1993a. Theoretical aspects of constitutive modelling for unsaturated soils. *Soils and Foundations* 33(4): 49-63.
Kohgo, Y., Nakano, M. & Miyazaki, T. 1993b. Verification of the generalized elastoplastic model for unsaturated soils. *Soils and Foundations* 33(4): 64-73
Matyas, E. L. & Radhakrishna, H. S. 1968. Volume change characteristics of partially saturated soils. *Geotechnique* 18(4): 432-448.

Saponite clay tailing treatment by artificial sedimentation

A.A. Tchistiakov – *Department of Geo-Information Systems, Netherlands Institute of Applied Geoscience, Delft, Netherlands*

V.N. Sokolov – *Faculty of Geology, Moscow State University, Russia*

V.I. Osipov – *Institute for Environmental Geosciences, Russian Academy of Sciences, Moscow, Russia*

Buu-Long Nguyen – *Department of Geo-Infrastructure, Netherlands Institute of Applied Geoscience, Delft, Netherlands*

ABSTRACT: Saponite amounts up to 95% of the clay tailings at a mineral extraction plant in Russia. Because of a very low rate of the saponite suspension precipitation and high porosity of the formed sediment (more than 90%) it was required to significantly intensify the sedimentation of the saponite tailings and to reduce the volume of the precipitated sediment. This paper particularly focused on the investigation of the effect of NaCl concentration and pH on coagulation of saponite suspensions, microstructure and properties of the formed sediments. By means of quantitative analysis of scanning electron microscope images we have managed to quantitatively describe the transformation of the saponite sediment microstructure with increase of NaCl concentration from 0.005 N to 0.5 N and pH from 2.5 to 8.5. A strong correlation has been found between the physico-chemical factors and the rheological properties of the clay tailings. It is the first time when the effect of the physico-chemical factors on the saponite colloidal properties has been extensively studied. Until now most of the investigations were done with kaolinite, illite and montmorillonite.

1 INTRODUCTION

Disposal of clay tailings from mineral extracting plants is an important environmental and economic issue in exploitation of mineral resources. The tailing volume of a large extracting plant can amount to millions of cubic meters, that requires building and maintenance of large tailing storage ponds. These tailings have an enormous negative environmental impact on the mine regions. Wind erosion and transport of clay particles from banks of a tailings storage pond lead to deforestation and negatively affect farming land productivity. Accidental discharge of clay suspension from a tailing storage can cause contamination of rivers and unrecoverable losses of fish population. It is specially the case in the northern regions of continents with ultra-fresh waters, where self-remediation takes very long time.

Here we present a case study carried out with saponite clay tailings. Saponite is a mineral of the montmorillonite group, which amounts up to 95% of the clay tailings at a mineral extraction plant in Russia. The mineral suspension has extremely high stability and the rate of the suspension precipitation is less than 0.01 cm/h. The porosity of the formed sediment exceeds 90%. So low rate of the saponite suspension precipitation and high porosity of the formed sediment would require to significantly enlarge the volume of the existing storage ponds. This would lead

to enormous additional investments and considerable environmental impact on the neighbouring natural reserve.

The objective of this research is to investigate the effect of salinity and pH on the saponite precipitation and the microstructure of the formed sediment. The effect of the physic-chemical factors on the saponite microstructure is of particular interest, because of two reasons. First, the microstructure determines the further consolidation of the sediment in the slurry. Secondly, one has to produce the sediment with the priory-defined properties that will determine the technology of further utilisation.

2 THEORY OF CLAY SUSPENSION STABILITY AND COAGULATION

The quantitative theory of colloidal particle interactions was developed by Derjagin & Landau (1941), Verwey & Overbeek (1948) (DLVO theory). Later on, based on the DLVO theory principles, the theory of clay particle interaction and clay rock properties was developed (Olphen 1963; Mitchel 1976; Lambe & Whitman 1978; Osipov 1979; etc.). It is well known now that the stability of clay particles in water suspensions is defined by the balance between the attraction Van der Waals-London forces on one side and the repulsion electrostatic forces as well as structural forces of bound water on another side.

The van der Waals - London forces do not depend on the physic-chemical conditions. Electrostatic repulsion between like-charged clay surfaces is governed predominantly by the overlap of their diffuse layers and depends on the Stern potential (Derjagin & Churaiev 1984; Gregory 1989). Nevertheless there is no direct method for determining the Stern potential. Hence instead of the Stern potential the electrokinetic or ζ-potential is often used for calculations as well as for qualitative interpretation of the experimental results (Fig. 1).

For dissolved solutions (e.g. of NaCl) with the ionic strength less than 10^{-3} N the electrical double layer (EDL) can amount to several hundreds angstroms (Å). For highly concentrated solution (C>0.1 N) the EDL has the size tenth time smaller.

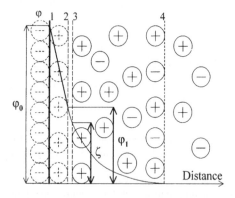

Figure 1. Principle scheme (Stern model) of the electrical double layer on the boundary clay face surface - aqueous solution. φ_o - surface potential; φ_1 - Stern potential; ζ -electrokinetic potential; 1 - clay surface; 2 - Stern plane; 3 - plane of shear, between a particle and fluid; 4 - boundary between a diffuse layer and free solution.

The potential energy curve, summing up Van der Waals-London attraction and electrostatic repulsion between two particles, has two energetic minimums: close (primary) and distant (secondary). These two minimums are separated with an energetic barrier (Fig. 2). On one hand, the barrier prevents the particles from bringing them together and coagulating. On the other hand, the occurrence of the repulsion barrier makes it possible for particles to interact in the first or in the second energy minimum. When the barrier is high clay dispersions are stable (or metastable), or have a very slow rate of coagulation and sedimentation.

Increasing of ionic strength causes decreasing of the ζ-potential of clay particles, contraction of the cation diffusive layer around them, decreasing of potential (repulsion) barrier and deepening of the

second potential minimum. It is explained by the fact that, if the salt concentration in the pore solution increases, some of the cations move from the diffuse layer to the adsorption layer and consequently the ζ-potential decreases. The salt concentration when the repulsion barrier becomes of critically low value is called the *critical flocculation concentration* (CFC) or coagulation limit. For simplification we presume that the exchangeable cations and the cations in the pore solution are the same and there is no exchange reactions.

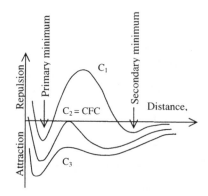

Figure 2. The variation in potential with distance from a plane charged interface for different concentrations (C) of indifferent electrolyte $C_1 < C_2 < C_3$.

2.1 Mechanisms of clay surface charge and effect of pH on stability of clay suspensions

The charge of a clay face surface mainly depends on the isomorphic changes occurring in the clay crystal. For example, the Si^{4+} cation can be substituted by Al^{3+}, Mg^{2+}, etc. As a result the crystal structure gets a negative charge that is compensated by exchangeable cations in dry conditions. In the presence of water (e.g. in aqueous suspension) the exchangeable cations dissociate and the clay crystal surface becomes negatively charged.

The mechanism of charge formation at the clay edges is different from the face surfaces. When a clay crystal edge interacts with water, (-SiO) groups are hydrated and form (-SiOH) groups on the crystal edge surface: $2(-SiO) + H_2O \rightarrow 2(-SiOH)$. In base conditions the formed poly-silicon acid partially dissociates: $(-SiOH) \rightarrow SiO^- + H^+$. The produced H^+ ions (or H_3O^+) transit to the aqueous phase, but under the electrostatic attraction of (-SiO$^-$)-groups they remain nearby the crystal surface.

The concentration of OH$^-$ and H$^+$-ions in the solution affects the degree of dissociation of the (-SiOH) group and thus defines the potential of the silicon surface (i.e. clay edge charge). An increase of pH (i.e. concentration of the OH$^-$ ions) increases the silicon surface negative potential and thus stabilizes clays in aqueous suspensions.

If Si^{4+} is isomorphically changed with Al^{3+}, the type of the $(-AlOH)^-$ group dissociation depends on the pH. In base conditions the $(-AlOH)^-$ group dissociates by the acid type $(Al(OH)_3 \leftrightarrow Al(OH)_2O^- + H^+)$, and in acid condition by the base type $(Al(OH)_3 \leftrightarrow Al(OH)_2^+ + OH^-)$. In the second case the edges of the clay particles get positively charged.

In minerals of montmorillonite group the isomorphic exchange of Si^{4+} for Al^{3+} occurs only in octahedral layers. Because there are no isomorphic exchanges for Al^{3+} in tetrahedral positions, the edges of tetrahedral layers remain neutral in acid environment and get negatively charged in base conditions. The pH, at which a clay edge charge changes its sign, is called the *isoelectric point*. Its value can vary significantly depending on the mineralogy and crystal structure of clays. According to Osipov (1979), the isoelectric point for montmorillonite is equal to 6.5, for hydromica - 6 and for kaolinite - about 8. Williams & Williams (1977) found from electrophoretic measurements that the isoelectric point of kaolinite amounts to 7.4.

Change of the charge sign at the clay edges to a positive one causes coagulation of clay suspensions (Osipov 1979) and consequently should simplify the water purification from clays (Tchistiakov 2000).

2.2 *Effect of structural forces on clay particle stability*

In an aqueous environment, clay minerals adsorb water molecules, and water films form on the particle surfaces (Fig. 3). Many authors distinguish two main structural water layers: adsorbed (or "firmly bonded") water layer, and "osmosis" (or loosely bonded) water layer. Derjagin & Churaiev (1984) call the first layer α-film and the second one β-film. These two types of water films are characterized by different physical-mechanical properties and strengths of contact with the clay surface.

The thickness of the α-film depends on the crystallographic parameters of the clay particle and the physico-chemical properties of the pore solution. Bondarenko (1973) determined experimentally that the thickness of the adsorbed water films in montmorillonite amount to 5-8 nanometers. Because of the high orientation of the water molecules the adsorbed water has relatively much higher values of viscosity and shear yield limit than those of "free" water.

The β-film is a transition layer between the highly structured water of the adsorption layer and free water. Within the β-film some of the water molecules are associated with the diffused exchangeable cations, which causes distortions in the water structure. The exterior boundary of the osmo-

sis water theoretically coincides with the outward boundary of the diffuse ionic layer. The physical properties of the osmosis layer and free water are comparable.

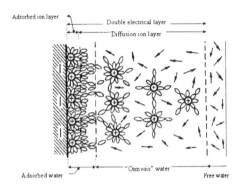

Figure 3. Structure of water films near a clay face surface (after Zlotchevskaia & Korolev, 1977)

If the base surface of a clay particle can be considered as a plate, the clay particle edge can be approximated as a convex cylindrical surface. According to Derjagin & Churaiev (1984) water films are less stable on a convex surface than on a plate. This is the reason why the water films on clay edges are rather thinner than on face surfaces. Osipov et al. (1989), for instance, show that the water film at the clay edges can be ten times thinner than at the face surfaces.

The structural forces associated with the water films have a splitting (repulsion) effect on the clay particles. The balance between coagulation-dispersion processes depends on the thickness of the films (Derjagin & Titievskaia 1953). Because the film thickness at the clay edges is much smaller, clay particles in suspensions can be less stable against coagulation even if they have the same negative charge.

3 EXPERIMENTAL PROCEDURE

3.1 *Suspension preparation*

For investigation of colloid properties of the saponite clay fraction of the mineral (<1μm) is used. The mineral is saturated with cation of sodium to exclude the effect of cation exchange reactions on the results of experiments with sodium chloride solutions.

The procedure of the suspension preparation is as follows: First, the required amount of the dry saponite powder is put in a glass, mixed with 20 ml of distilled water. The suspension is dispersed with ultrasound for 30 sec. Subsequently the saponite is let to swell for 24 hours. Next day the suspension is dispersed again. During the ultrasonic dispersion the required amount of the electrolyte and distilled water is added into the glass so that the total volume of the suspension is 50 ml. The suspension is then quickly put in a cylinder with a volume scale and start monitoring the precipitation process. During the ultrasonic treatment the glass with suspension is placed in a thermostat to prevent it from heating.

For investigation of ionic strength on the coagulation of the saponite suspension, experiments are performed with suspensions prepared in distilled water and NaCl solutions with salt concentration of 0.005 N, 0.01 N, 0.05 N, 0.1 N, 0.5 N. The mineral phase concentration in the suspensions amounts to 1%.

For investigation of the effect of pH on the saponite microstructure experiments are performed with saponite suspensions with the mineral phase concentration of 0.5%, 1% and 3%. The suspensions are precipitated at two values of pH – 2.5 and 8.5. A solution of boric acid is added to the suspensions to create low pH, and sodium tetraborate to create high pH. Because for most of the earlier investigated minerals the isoelectric point is between 6 and 8 (Osipov 1979) it is suspected that the isoelectric point for saponite is positioned between the two selected pH values. The experiments confirm our proposition.

Viscosity of the sediments obtained from 3% suspensions is measured to detect the effect of pH on rheological properties of saponite. The sediments precipitated from 1% suspension is also analysed under a scanning electron microscope (SEM).

3.2 Methods of microstructural analysis of the saponite sediments

For preparation of the saponite sediment specimens for SEM imagery we use a special freeze-drying technique that was developed particularly for soft clays. Its detailed description is given by Osipov et al. (1989).

For quantitative analysis of a sedimentary porous medium we use the software "STIMAN" that was specifically developed for computer processing of SEM images (Sokolov et al., 1998). The program calculates statistical characteristics of microporosity based on analysis of a sequence of images made under different magnifications. This makes it possible to cover the pores of all sizes, which are present in a specimen. For the saponite sediments we use magnifications of × 1500, × 3000 and × 6000.

3.3 Terminology, classifications and parameters used for description of the experimental results

For comprehensive description of a clay microstructure one has to characterise at least three pa-

rameters: the energetic condition (i.e. the degree of coagulation) of the original clay material, the micromorphology of the mineral component and the micromorphology of the porous medium.

There is not a common understanding between chemists and geologists of the terms coagulation, aggregation and flocculation. We use the term "aggregation" for the process of formation of an elementary structural association (i.e. an aggregate) from several basic mineral components (e.g. single mineral particles) resulted from losing kinetic stability by the primary elements. By coagulation and flocculation we understand the process of attachment to each other of several basic microstructural associations or elements (i.e. a single clay particle or a microaggregate) resulted in construction of a continuous microstructural grid. Consequently the microstructure formed by single dispersed clay particles is called dispersed-coagulated and the microstructure formed by aggregates is called aggregated-flocculated. This terminological concept is in agreement with the microstructural approaches of most clay scientists (Mitchel 1976, Lambe & Whitman 1978, Osipov 1979).

For description of a microstructure we use the terms "particle" and "microaggregate". A particle is understood as an undivided structural element. In many cases this term has a conditional meaning. For example, smectites theoretically can disperse till an elementary crystal sheet. Because of this reason we use sometimes the term "ultra-microaggregate" to emphases that the described structural element consists only of a few elementary layers or particles visually unrecognised in the SEM image. The ultra-microaggregates usually are not formed due to coagulation of the clays during sedimentation but characterize the original degree of the clay material dispersion.

Consequently, according to the position of the pores relative to the structural elements we can distinguish interparticle, interultra-microaggregate), intermicroaggregate and intermacroaggregate (d > 100 μm) pores. As shown further in this paper, in highly dispersed saponite sediment it is very difficult to distinguish a "single" clay particle from an ultra-microaggregate. This is the reason why we combine the categories of inter-particle and interultra-microaggregate pores into one class called further as interparticle pores.

For description of pore size we use the classification developed by Osipov et al. (1989) for quantitative SEM image analysis of clay soils. According to this classification micropores can be divided on ultra-micropores (d<0.1μm), thin (d = 0.1 – 1.0 μm), small (d = 1.0 – 10.0 μm) and large micropores (d = 10 – 100 μm). Pore with a diameter larger than 100 μm are called macropores.

The distribution of the pores between different classes is based on the visual analysis of the SEM images and analysis of the pore size distribution curves where the described micromorphological types of pores usually form well distinguishable clusters.

We use a parameter – the *shape factor* K_S – to quantitatively characterise a shape of a pore. The shape factor is calculated as ratio of a small axis to a large one of the ellipse that approximates the pore. Theoretically the shape factor can vary from 1 (for a circle) to 0. A pore is considered as isometric if $K_S = 1.0 - 0.66$; anisometric if $K_S = 0.66 - 0.1$; and elongated if K_S is less than 0.1.

4 EXPERIMENTAL RESULTS AND DISCUSSION

4.1 *The saponite precipitation in NaCl solutions of concentration below the coagulation limit.*

Four-month observations show, that the suspensions with the concentration of the electrolyte less than 0.01 N do not perform rapid coagulation. Because of the low initial volumetric concentration of the mineral phase the clay particles and small microaggregates do not form any structured system within the suspensions.

No clear sedimentation boundary is observed. The clay particles practically do not interact with each other and slowly settle down under the gravitation force. A suspension solution slowly becomes lighter. Its colour changes from light yellowish to brow-radish from the top to the bottom.

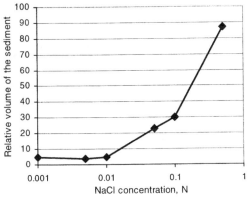

Figure 4. Relative volume of the Na-saponite sediments formed in NaCl solutions of different concentration. The relative volume is calculated as the ratio of the final sediment volume to the initial volume of the suspension (50 ml). The concentration of 0.001 N corresponds to distilled water.

As the result of the clay particles' precipitation their concentration increases near the cylinder bot-

tom. They start interacting with each other and form – as a result of concentration coagulation – relatively compact structured sediments. The deposits formed in the sodium chloride solutions of concentration from 0 N to 0.01 N have close values of the sediment volumes (Fig. 4). Their dry density is significantly higher than the dry density of the deposits formed in the NaCl solutions of higher concentration (Table 1).

Table 1. Dry density of the Na-saponite sediments precipitated in the sodium chloride solutions of different concentration.

Concentration NaCl, N	0	0.005	0.01	0.1	0.5
Dry Density, g/cm³	0.2	0.25	0.2	0.035	0.011
Porosity, %	93	91	93	98	98

4.1.1 *Microstructure*

The SEM images of the sediments show that the degree of aggregation continuously increases with increasing salt concentration from 0 N to 0.01 N (Fig. 5 a-c). The sediment settled in distilled water formed by dispersed-coagulated microstructure, formed by single clay particles and ultra-microaggregates. The face-to-edge and edge-to-edge types of the contacts between the structural elements (i.e. microaggregates) dominate. A few leaf-like microaggregates are also present. The maximum size of the microaggregates amounts to 5 µm (Fig. 5 a).

Increase of NaCl concentration to 0.005 N causes formation of a quite specific microstructural network in the sediment (Fig. 5 b). The nodes of the microstructural network are built by relatively large microaggregates. The microaggregates are bond through the grid formed by single ultra-microaggregates. We believe that the saponite particles within the nodes interact in the first potential minimum and the saponite ultra-microaggregates, forming the structural grid, interact in the second potential minimum. This theme seems to be logical because increasing of the small ionic strength leads to deepening of the secondary minimum and decreasing of the repulsion barrier between the particles. Therefore the particles are able to form stable contacts in the first as well as in the second potential minimums. Less stable particles coagulate in the first minimum and form the nodes of the microstructure. Particles, which are more stable against coagulation, interact in the second potential minimum and consequently form the grid of the structure.

Increase of the NaCl concentration to 0.01 N does not change the type of the saponite microstructure. Nevertheless the size of the node microaggregates increases that provides more contrast SEM images (Fig. 5 c).

Figure 5. Microstructures of saponite sediments precipitated in NaCl solutions of different concentrations: a - distilled water; b - 0.005 M; c - 0.01 M; d – 0.1 M; e – 0.5 M. Magnification of the SEM images × 3000.

4.1.2 *Analysis of the porous media*

Table 2 shows contribution of the different types of micropores to the total (bulk) porosity of the sediments obtained at different NaCl concentrations. This data are derived from the results of quantitative analysis of SEM images of the saponite sediments. In distilled water three categories of pores compose the saponite sediment porous medium (Table 2). Interparticle micropores with an equivalent diameter less than 1.0 μm form 10% of the porosity. Small interparticle and intermicroaggregate pores with a diameter from 1 to 2 μm build 30% of the porous medium. Small intermicroaggregate pores with a diameter from 3 to 5.5 μm form 60% of the void space.

In 0.005 N NaCl solution thin interparticle micropores with a diameter less than 1 μm amount to 13% of the bulk porosity of the saponite sediment. The number of these pores significantly exceeds the number of pores of other categories. Nevertheless, their total input in the bulk porosity is not large because of their small size. Numerous small interparticle micropores with a diameter from 1.35 to 2.55 μm form 21% of the sediment porous media. Although small intermicroaggregate pores with an equivalent diameter from 3 to 6.0 μm are rather less numerous,

they build the main part of the sediment porous space (66%) due to their large dimensions.

The porous medium of the sediment obtained in 0.01 N NaCl solution consists of thin interparticle micropores with a diameter less than 0.95 μm (11%), thin and small intermicroaggregate micropores with a diameter of 1.0-1.4 μm (15%) and small intermicroaggregate pores with a diameter from 2 to 7 μm (74%).

Statistical analysis of the shape factor K_S of the sediments does not show any certain relationship between the shape of pores and the salinity in the interval of NaCl concentration from 0 to 0.01 M.

However visual analysis of the SEM images makes it possible to conclude that larger pores generally have more isometric (rounded) shape than smaller pores.

4.2 *Saponite precipitation at the concentration of sodium chloride above the coagulation limit*

Increase of NaCl concentration to 0.05-0.1 N changes dramatically the sedimentation process. After completion of ultrasonic dispersion the saponite particles rapidly aggregate. The formed microaggregates flocculate and form highly porous struc-

Table 2 – Characteristic of the porous media of the Na-saponite sediments obtained in the solutions of NaCl of different concentration.

Concentration of NaCl, N.	Interparticle and inter-ultra-microaggregate thin micropores (d < 1.0 μm), %	Interultra-microaggregate and intermicroaggregate small micropores (d = 1.0 - 2.6 μm), %	Intermicroaggregate micropores, %	
			Small d = 2 - 10 μm	Large d = 10 - 100 μm
0	10	30 (1.0 - 2.0 μm)	60 (3 - 5.5 μm)	0
0.005	13	21 (1.35 – 2.55 μm)	66 (3.0 - 6.0 μm)	0
0.01	11	15 (1.0 - 1.4 μm	74 (2.0 - 7.0 μm)	0
0.1	5	10 (1.0 - 1.5 μm)	70 (2 - 10 μm)	15 (10 - 11 μm)
0.5	3	2 (1.0 - 2.6 μm)	21 (3.5 - 10 μm)	74 (10 - 40 μm)

tured sediment that starts consolidating under its own weight due to syneresis processes. Such behaviour of the suspensions indicates that these concentrations of NaCl are higher than the coagulation limit for the saponite particles. The sediments are actively consolidated during the first month. Their additional compaction during the next three months does not exceed 7% in 0.05 N solution and 17% in 0.1 N solution. Because all mineral particles have coagulated, the solution above the sediment is absolutely clean.

In 0.5 N NaCl solution a saponite suspension perform rapid coagulation as well. Then the precipitated structured sediment is compacted due to syneresis but the decrease of its volume in time due to consolidation amounts to only a few percents.

4.2.1 Microstructure

In 0.1 N NaCl solution single clay particles predominantly contact with each other with their faces and form microaggregates 4.5-6 μm in diameter (Fig. 5 d). The microaggregates are oriented edge-to-face and form continuous cellular aggregated-flocculated microstructure.

In 0.5 N NaCl solution the sediment has a clearly observed aggregated-flocculated large-cellular microstructure (Fig. 5 e). At the SEM image the clay microaggregates are covered by salt crystallized at their surface during sublimation.

4.2.2 Analysis of the porous media

Quantitative analysis of the saponite sediment obtained in 0.1 N solution shows that although the thin interparticle pores with a diameter less than 1.0 μm dominate numerously they build only 5% of the bulk porosity (Table 2). Interparticle and intermicroaggregate pores with a diameter of 1.0 - 1.5 μm contribute 10% to the total porosity. Small itermicroaggregate micropores with a diameter from 2 to 10 μm form the main part of the porous medium - 70%. Large intermicroaggregate pores of 10.2 - 11.1 μm amount to 15% of the bulk porosity.

In NaCl solution of 0.5 N concentration the main part of the sediment porous medium consists of large isometric intermicroaggregate pores that compose 74% of the bulk porosity. Thin interparticle pores with a diameter less than 1.0 μm form only 3% of the pore space. Small intermicroaggregate micropores with a diameter from 1 to 2.6 μm contribute only 2% to the porosity. However, it is worth noting that the salt precipitation significantly effects the porosity measurement, particularly for the thin and small micropores. Small intermicroaggregate anisometric micropores with a diameter from 3.5 to 10 μm amount to 21% of the porosity.

4.2.3 Analysis of the saponite microstructure transformation in NaCl solutions of different concentration

When comparing the microstructures of saponite sediments obtained in the NaCl solutions of different concentrations we observe an interesting trend in the transformation of the saponite porous media with increasing salinity. The sediments obtained in solutions with NaCl concentration lower than the coagulation limit (i.e. 0 - 0.01 M) are characterised by approximately the same input of thin interparticle micropores in the bulk porosity (10-13%). While the salt concentration remains lower than the coagulation limit (or critical flocculation concentration) its growth causes the observed increase of the degree of saponite particles' aggregation. Consequently the contribution of interparticle pores in the bulk porosity decreases. The decrease of the interparticle porosity is compensated by increase of the input of small intermicroaggregate pores from 60% in distilled water up to 74% in 0.01 N solution.

In the solutions with concentrations from 0.1 to 0.5 N, that are higher than the coagulation limit, the input of thin micropores reduces dramatically till 3-5%. Increase of the salt concentration in the range of values higher than the critical flocculation concentration (CFC) causes significant development of large intermicroaggregate pores from 15% in 0.1 N solution to 74% in 0.5 N solution. It is remarkable that large micropores are completely absent in the sediments obtained at the concentrations lower than the coagulation limit.

The input of small interultra-microaggregate and intermicroaggregate pores with a diameter from 1μm

481

to 2.6 μm continuously decreases with the increase of salt content within the whole investigated interval from 30% in distilled water to 2% in 0.5 N brine.

4.3 Effect of pH on the saponite precipitation and microstructure

4.3.1 Volume and rheological properties of the sediments

The experiments show that both in alkaline and acid environments the saponite particles coagulate and form structured sediment. The consolidation of the precipitated sediments practically terminates within three days. For all concentrations of the mineral phase in the initial suspensions the volume of the sediments obtained in alkaline environment is about 1.35 times smaller than the volume of the sediments precipitated in acid conditions (Fig. 6). Despite the fact that the volumetric concentration of saponite particles in the sediment obtained in acid solution is lower, its viscosity is 2.3 times as much as the viscosity of the sediment precipitated in base conditions.

These results are in a good agreement with the theory of clay colloid stability. In the acid solution the edges of saponite particles get positively charged while their faces remain negatively charged. As the result of this specific charge distribution the saponite edges form with its faces strong bonds due to electrostatic attraction. In base conditions, where all saponite surfaces (edges and faces) are negatively charged, the particles form bonds mainly due to Van der Waals attraction that is significantly reduced by ionic-electrostatic repulsion of the like-charged surfaces. Consequently the saponite sediment obtained in the alkaline conditions has higher density but less viscosity and the strength of the individual contacts between the clay particles and microaggregates is also lower than in acid conditions.

Our experimental results are also in good agreement with direct measurements of clay individual bonds in different environments. For example Sokolov (1985) experimentally shows that the strength of bonds formed by a positively charged edge and a negatively charged face of hydromica in acid conditions are rather higher than in base.

4.3.2 Microstructure in acid conditions

The microstructure of the saponite sediment precipitated in acid environment is formed by relatively large flocks (microaggregates) with a size from 30 to 70 μm (Fig. 7 a).

The flocks are built by thin leaf-like saponite single particles and ultra-microaggregates that interact edge-to-face or edge-to-edge. The ultra-microaggregates are 3-5μm long, 1μm wide and 0.2-0.3 μm thick. Some of the single clay particles contain clearly observed rounded pores that are proba-

bly formed as a result of the crystal dissolution in acid conditions. (Fig. 7 b).

In general the sediment has irregularly-cellular, dispersed – flocculated microstructure with a pore diameter reaching 30 μm. The results of the quantitative analysis of SEM images allow distinguishing four main categories of pores present in the sediment (Table 3). Thin elongated interparticle pores (0.3 μm long) and anisometric micropores with an equivalent diameter up to 0.9 μm compose 5% of the bulk porosity. These pores dominate numerically but their total input in the void space is low because of their small size. Intermicroaggregate pores with a diameter of 1.0-2.0 μm form 15% of the sediment porosity. The main part of the porous medium (45%) is represented by small intermicroaggregate micropores with a diameter from 2 to 9 μm. Large intermicroaggregate pores with a diameter from 10 to 30 μm also build a significant part of the bulk porosity (35%).

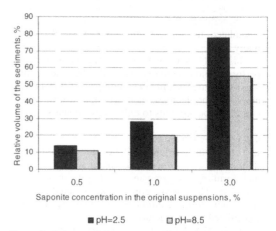

Figure 6. Volumes of the saponite sediments formed in acid and base conditions.

4.3.3 Microstructure in base conditions

The SEM images show that in alkaline conditions the saponite sediment has a very homogeneous, aggregated-flocculated structure consisting of small cells formed by thin saponite microaggregates (Fig. 7 c). A single microaggregate usually consists of a few saponite particles stuck to each other by face surfaces (Fig. 7 d). The elongated plate-like microaggregates interact edge-to-face with each other and form continuous small-cellular microstructure. A typical size of a single microaggregate is 3 μm long, 0.5 - 1.0 μm wide and 3 μm thick.

Eight percent of the sediment porous medium is built by numerous thin interparticle elongated and isometric micropores with a diameter less than 0.9 μm (Table 3). Small interparticle and intermicroaggregate isometric micropores with a diame-

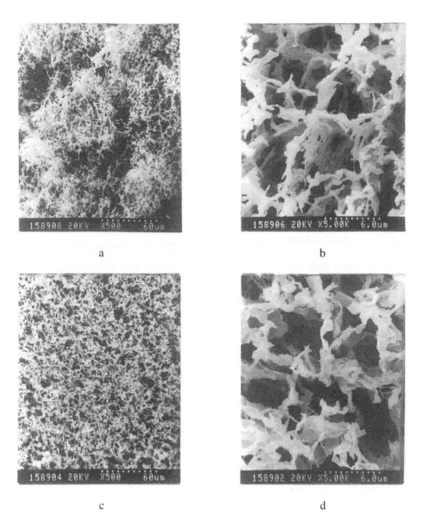

Figure 7. Microstructures of saponite sediments precipitated in acid (a, b) and base conditions (c, d).
Magnification of the SEM images: a and c – × 500; b and d – × 5000.

Table 3. Characteristic of the porous media of Na-saponite sediments precipitated from 1% suspensions
in acid and alkaline conditions.

pH	Interparticle and interultra-microaggregate thin pores (d < 1.0 μm), %	Interultra-microaggregate and intermicroaggregate small pores (d = 1.0 - 2.0 μm), %	Intermicroaggregate micropores, %	
			Small D = 2 - 10 μm	Large d = 10 - 100 μm
2.5	5	15	45	35
8.5	8	17	62	13

ter of 1-2 μm form 17% of the bulk porosity. The main part of the porous space is represented by small intermicroaggregate isometric pores (62%). According to their diameters they are distributed within three intervals. The pores with a diameter of 2 - 5.5 μm form 35% of the bulk porosity; pores with a diameter of 6 – 10 μm form 27% of the void space. The large pores with a diameter 10.5 - 15μm com-

pose 13% of total porous medium.

5 CONCLUSIONS

1. The paper presents results of an experimental investigation on the effect of salinity and pH on the rate of the saponite suspension coagulation, volume, viscosity, and particularly on microstructure of the

formed saponite sediment.

2. At ionic strength lower than the coagulation limit the edges of the saponite particle are less preserved against coagulation than their faces because of less thickness of the bond water films.

3. At the concentrations of NaCl from 0 to 0.01 M, because of the less colloid stability of the clay edges the "single" particle and microaggregates interact preferably edge-to-edge or edge-to-face. As the result of this saponite forms a specific microstructural network consisting of relatively larger flocks connected with each other by grid built by "single" clay particles.

4. In the condition of free participation (without applying external loading) formation of a "classical" highly oriented dispersed microstructure in saponite suspensions is hardly possible even in ultra-fresh and alkaline environments.

5. When the salt concentration is higher than the CFC the suspensions perform fast coagulation. Nevertheless the increase of the suspension ionic strength causes strengthening of the individual contacts between the microstructural elements and reduces the sediment consolidation. Hence the saponite sediments formed in more concentrated solutions have higher volumes. Therefore for industrial application of non-organic salts we have to find the optimum ratio between the speed of clay precipitation and the volume of the formed sediment.

6. In acid conditions the individual contacts formed by oppositely charges edges and faces of clay particles (or ultra-microaggregates) are stronger than the individual contacts between like charged saponite edges and faces in alkaline environment. As the result the microstructure in acid conditions has rather higher viscosity and lower compaction.

7. The applied approach, based on manipulating the acidity (pH) and salinity of the sedimentation environment in a tailing storage pond, allows not only just increasing the speed of saponite coagulation but also obtaining the sediment with required properties. This makes it possible to optimise the purification of the operational water and to simplify the utilization of the precipitated sediment.

REFERENCES

Bondarenko, N.F. 1973. *Physics of underground water flow*. Leningrad: Gydrometeoizdat, in Russian.

Derjaguin, B.V. & Landau, L.D. 1941. Theory of the stability of strongly charged lyophobic sols and of the adhesion of strongly charged particles in solutions of electrolytes. *Acta physicochim. URSS*. 14 (6): 633-662.

Derjagin, B.V. & Titievskaia, A.S. 1953. Splitting effect of water layers and its influence on the stability. *Reports of Academy of Sciences of USSR*. 89(6): 1041-1044, in Russian.

Derjagin, B.V. & Churaiev, N.V. 1984. *Wetting films*. Moscow: Nauka, in Russian.

Gregory, J. 1989. Fundamentals of flocculation. *Critical Reviews in Environmental control*. 19(3): 183-230.

Lambe, T.W. & Whitman, R.V. 1978. *Soil Mechanics, SI Version*. New York. Massachusetts Institute of Technology: John Wiley & Sons.

Mitchel, J.K. 1976. *Fundamentals of Soil Behaviour*. New-York-London-Sydney-Toronto: J. Wiley and Sons, Inc.

Olphen, H. van. 1963. *An introduction to clay colloid chemistry*. New-York: Interscience Publications.

Osipov, V.I. 1979. *Nature of strength and deformation properties of clays*. Moscow: Moscow University Publ. House, in Russian.

Osipov, V.I., Sokolov, V.N. & Rumiantseva, N.A. 1989. *Microstructure of clay rocks*. Moscow: Nedra, in Russian.

Sokolov, V.N. 1985. Physico-chemical aspects of the mechanical behaviour of clay soils. *Engineering Geology* 4: 28-41, in Russian.

Sokolov, V.N., Yurkovets, D.I., Ragulina, O.V. & Melnik, V.N. 1998. Computer-controlled system for the study of micromorphology of solid surface by SEM images. *Surface Investigation* 14: 33-41.

Tchistiakov, A.A. 2000. Colloidal chemistry of clay induced formation damage. *Proceedings of the SPE Symposium on Formation Damage Control, 23-24 February 2000, Lafayette, Louisiana, USA*.

Verway, J.J.W. & Overbeek, J. Th. G. 1948. *Theory of the stability of lyophobic colloids*. New-York – Amsterdam: Elsevier.

Williams, D.J.A. & Williams, K.P. 1977. Electrophoresis and Zeta-potential of kaolinite. *J. Colloid and Interface Science* 65(1): 79-87.

Zlotchevskaia, R. I. & Korolev, V.A. 1977. Temperature effect on formation of physical-mechanical and physico-chemical properties of water saturated clays. In issue *Adsorbed Water in Dispersed Systems* 4: 34-57. Moscow: Moscow State University Publishing House, in Russian.

5 Clay liners

Clay Science for Engineering, Adachi & Fukue (eds) © 2001 Balkema, Rotterdam, ISBN 90 5809 175 9

Pellets/powder mixture of bentonite for backfill and sealing of HLW repositories

B. Dereeper & G. Volckaert
SCK-CEN, Mol, Belgium

Ch. Imbert
CEA, Paris, France

M. V. Villar
CIEMAT, Madrid, Spain

ABSTRACT: Bentonite is widely studied as backfill and sealing material in geological repositories for radioactive waste because of its low permeability and large ability to swell. CEA has developed, in co-operation with SCK•CEN and CIEMAT, an industrial scale technique to produce a granular material composed of bentonite powder mixed with high-density bentonite pellets. This material can be easily applied to backfill irregularly shaped volumes and is rather cheap to produce. The result of studies shows that, once saturated, the originally heterogeneous mixture has a homogeneous density and that the powder between the pellets does not result in preferential pathways. At saturation and for equal density, precompacted powder samples and pellets/powder samples have the same hydraulic and swelling properties. To demonstrate the applicability of this type of material on an industrial scale, it has been applied in the EC RESEAL project for the sealing of a shaft in the HADES URF in Mol, Belgium.

1 INTRODUCTION

Bentonite is included in many geological repository designs as backfill or sealing material, because of its low permeability, its high sorption capacity and its great ability to swell in water. In general it is proposed to use this material in the form of precompacted high-density clay blocks. CEA has developed, in co-operation with SCK•CEN and CIEMAT, an industrial scale technique to produce a granular backfill material that is composed of bentonite powder mixed with high dry density, i.e. higher than 2 g/cm³, bentonite pellets. Such a granular material has the advantage that it can be easily applied to backfill irregularly shaped volumes and that it is rather cheap to produce.

The properties of this heterogeneous material, especially the geomechanical properties, the saturation and the water and gas permeability at room temperature, have been compared with those of the homogeneous powder material compacted at the same void ratio.

2 MATERIALS

Two natural bentonites have been studied, the FoCa clay and the Serrata clay.

The FoCa clay is a sedimentary clay coming from the Paris Basin, extracted in the Vexin region. The FoCa clay is extensively studied as reference backfill material within waste disposal projects. It is, for example, studied in the Belgian programme within the PRACLAY project (ONDRAF/NIRAS, 1998). Its major component (i.e. 80 % of the clay fraction) is an interstratified clay of 50 % Ca-beidellite and 50 % kaolinite. It contains also kaolinite, quartz, goethite, hematite, calcite and gypsum. The industrial preparation process consists in drying, grinding and sieving (max. grain size: 2 mm). The liquid limit of FoCa is 112%, the plastic limit is 50%, and the specific weight is 2.675. The hygroscopic water content for a relative humidity around 60% and a temperature of 20°C, is comprised between 10 and 12%.

The Serrata clay is a bentonite coming from the Cortijo de Archidona deposit (Almería, Spain), selected by ENRESA as suitable material for the backfilling and sealing of HLW repositories (ENRESA, 1999). The processing at the quarry consists in disaggregation and gently grinding, drying at 60 °C and sieving (max. grain size: 5 mm). It has a montmorillonite content higher than 90 %. Besides, it contains variable quantities of quartz, plagioclase, cristobalite, K-feldspar, calcite and trydimite. The liquid limit of the bentonite is 102 % and the specific weight is 2.70. The equilibrium gravimetric water content of the clay at CIEMAT laboratory conditions (R.H. 50 %, which corresponds to a suction of around 130 MPa) is about 13.7 %.

3 PRODUCTION TECHNIQUE OF THE HIGH DENSITY PELLETS/POWDER MIXTURE

The pellets are directly made by compaction of the clay powder. A compactor/granulator, such as that used for the pelletizing of coal powder or ore powders resulting from flotation processes, is applied. Its principle is shown in figure 1.

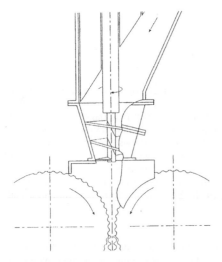

Figure 1. Principle of the compacter/granulator used to make pellets. The powder falls in a conical screw of Archimedes which precompacts the powder and presses it between two moulding cylinders turning in opposite direction. The precompacted powder is pressed with a high force and dragged between the cylinders.

The Sahut-Conreur Company (France) has produced the pellets needed for the research programme. Different shapes and sizes of pellets have been tried, in order to obtain a high dry density. The best result has been obtained with pellets of approximately 25 x 25 x 15 mm of size. To further optimise the dry density of the pellets, several batches of pellets have been produced using FoCa and Serrata clay powders of different water contents and granulometry. These tests show that:

- the compaction is easier when the powder is fine i.e. 0-3 mm ;
- the compaction of FoCa clay is more efficient (higher dry density) and easier if the water content of the powder is low (w=7%); a maximal dry density of 2.05 g/cm³ was obtained;
- the compaction of Serrata clay does not depends a lot on the water content of the powder; Serrata gives highest values of dry density for a water content between 9.5% and 14%; a maximal dry density of 1.93 g/cm³ was obtained for a water content of 13%.

With the existing industrial equipment it is possible to produce several tons or even tens of tons of these pellets per hour.

To obtain a granular backfill material with a high bulk density, the pellets are mixed with powder of the same clay. Pellets/powder weight ratio's between 50/50 and 65/35 have been tested. The latter leads to a lightly higher bulk density but with a higher risk of segregation between pellets and powder. Bulk densities of about 1.4 g/cm³ are obtained for the 50/50 pellets/powder mixture. Further compaction to a density of about 1.6 g/cm³ is possible by using vibro-compaction techniques as used in road construction.

4 METHODS

The swelling pressure and the hydraulic conductivity measurements of the powder/pellets mixture on a laboratory scale need large oedometric cells. Figure 2 show the big oedometer specially designed by CIEMAT for these tests. The cell has an inner diameter of 10.0 cm and a variable inner height between 5.0 and 10.0 cm.

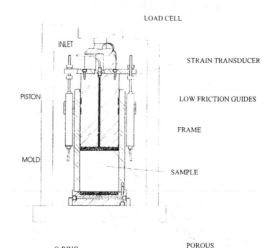

Figure 2. Scheme of the oedometer designed by CIEMAT to measure the swelling pressure and hydraulic conductivity of pellets/powder mixture

The sample is confined in a rigid frame equipped with a load cell. The sample is hydrated by both faces through the porous stones (or filters). At the same time the swelling pressure exerted by the clay and its deformation can be measured. Hydraulic conductivity is also measured applying large hydraulic gradients (0.6 to 2 MPa for samples of 50 to 100 mm length) on saturated samples. These large hydraulic gradients are required due to the extremely low permeability of high density saturated bentonite.

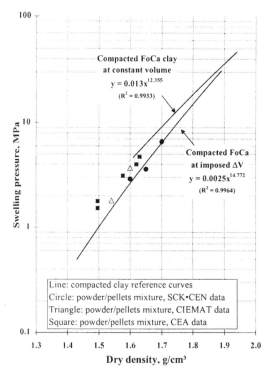

Figure 3. Swelling pressures obtained with pellets/powder mixture compared with swelling pressure reference curves of isostatically compacted FoCa clay.

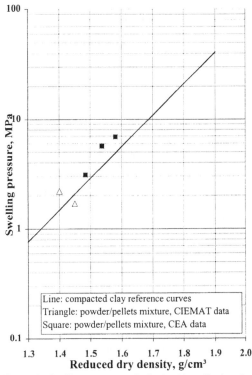

Figure 4. Swelling pressures obtained with pellets/powder mixture compared with swelling pressure reference curves of isostatically compacted Serrata clay sample.

Gas breakthrough tests are also performed on the saturated sample. For this purpose, two pumps are connected to the cell. One pump, connected to the top porous stone (or filter) of the cell via a buffer volume filled with gas, controls the flow rate of the gas injected in the sample while measuring the gas pressure. The other pump, connected to the bottom porous stone (or filter) of the cell, controls the water backpressure while measuring the water outflow rate.

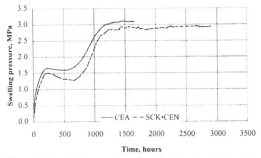

Figure 5. Two similar experiments (FoCa clay - dry density 1.6 g/cm³), one performed by CEA and one by SCK•CEN.

5 SWELLING PRESSURE

On figure 3 the swelling pressure obtained with the 50/50 mixture is compared with the values obtained on isostatically compacted samples. For isostatically compacted FoCa-7 clay, two reference curves have been established upon a large range of reduced dry density. One curve (upper one) gives the swelling pressure of a sample hydrated without volume change. The other curve (lower one) gives the swelling pressure obtained with a controlled volume variation. The values measured with the mixture are approximately between the two reference curves.

Different tests have been done with 50/50 pellets/powders mixture of Serrata clay at different dry

densities. A comparison is made between the measurements made with the mixture and the reference curve established by CIEMAT for compacted Serrata clay (Fig. 4).

For both clay types the swelling pressure at saturation is similar to those of the isostatically compacted clay. However the transient behaviour of pellets/powder mixture is different from isostatically compacted samples; while for the latter the increase of swelling pressure is monotonous, for the mixture the increase occurs in two steps (Fig. 5).

Three phases are observed in the hydration transient: a first one with a swelling pressure increase, a second one with a quasi-constant level and the last one with a new increase of swelling pressure. The powder and the pellets begin to swell causing a first increase of the swelling pressure. The hydration causes a variation of dry density of each component of the mixture: the pellets swell which leads to a decrease of the pellets dry density and consequently compact the hydrating powder around them, which leads to an increase of the powder dry density. At the end of the second phase, the dry density of both components is stabilised and the swelling pressure increases up to saturation of the sample (third phase).

Dedicated tests to investigate the interaction between the pellets and the surrounding powder during the hydration must be carried out to confirm this interpretation.

6 HYDRAULIC CONDUCTIVITY

Figure 6 compares the hydraulic conductivity of the pellets/powder sample and of the isostatically compacted samples. Once saturated, the hydraulic conductivity of isostatically compacted powder samples and of the pellets/powder mixtures are similar.

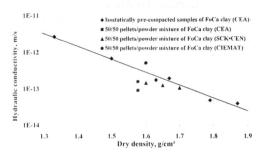

Figure 6. Comparison between the hydraulic conductivity of pellets/powder mixture and the compacted FoCa clay powder samples.

7 GAS PERMEABILITY

Gas permeability tests on fully saturated pellets/powders and isostatically precompacted samples were mainly done to determine the gas breakthrough pressures. For both samples the gas pressure must exceed the local confining pressure to create a preferential pathway. As the gas breakthrough pressure is very sensitive to any existing preferential pathway, this result shows that once saturated the originally heterogeneous mixture has a homogeneous density and that the powder between the pellets does not result in preferential pathways. This is also confirmed by additional observations of samples dried after saturation, gamma ray and tomographic density measurements (Volckaert G et al., 1996 and 2000).

8 APPLICATION ON A LARGE SCALE

To demonstrate the applicability of this mixture on an industrial scale, it has been applied for the sealing of the "experimental" shaft in the HADES underground research facility in Mol, Belgium. This shaft has an excavation diameter of about 2.2 m and about 12 tons of a 50/50 FoCa clay based pellets/powder mixture were used in the seal construction The first 60 centimetres of the seal have been compacted applying a vibro-compactor especially developed by CEA for the demonstration test. A density of 1.55 g/cm³ has been obtained compared to the uncompacted density of 1.4 g/cm³. The test was installed in September 1999. Currently the sealing material is being hydrated. Once saturated the capacity of the seal to avoid preferential migration of water, gas and radionuclides will be tested.

9 CONCLUSIONS

The comparison between the swelling pressure, the hydraulic conductivity and the gas permeability at full saturation between the pellets/powder mixture and isostatically compacted samples shows that both materials have the same properties. This observation shows that the pellets/powder mixture becomes homogeneous although initially it is very heterogeneous in terms of density. These studies demonstrate that the pellets/powders mixture is an interesting material for the backfilling and sealing of radioactive waste repositories. The applicability of the pellets/powder mixture is being demonstrated at large scale in the RESEAL shaft seal test.

ACKNOWLEDGEMENTS

The results presented were obtained in the RESEAL project, joint project between SCK•CEN, ANDRA (subcontractor: CEA) and ENRESA (subcontractors: CIEMAT, UPC). The project is sponsored by EC, ANDRA, ENRESA and ONDRAF/NIRAS.

REFERENCES

ENRESA (1999) *FEBEX Full Scale Engineered barrier Experiment in Crystalline Host Rock.* Final report. Publication Technica ENRESA (in press).
ONDRAF/NIRAS (1998) *The PRACLAY project, Demonstration test on the Belgian disposal facility concept for high activity vitrified waste,* EUR 18047.
Volckaert G., Bernier F. and Dardaine M. (1996) *BACCHUS-2 Demonstration of the in situ application of an industrial clay based backfill material,* Final report EUR 16860.
Volckaert G., Dereeper B., Imbert Ch., Villar M., Gens A., Mouche E. ... (2000) *RESEAL A large scale demonstration in situ test for repository sealing in an argillaceous host rock,* Final report EUR 19612.

Clay Science for Engineering, Adachi & Fukue (eds) © 2001 Balkema, Rotterdam, ISBN 90 5809 175 9

Behavior of clayey soils in liner systems concerning self-healing

K. Mallwitz

Neubrandenburg University of Applied Sciences, Germany

ABSTRACT: The mechanisms that govern self healing behaviour of clayey soils with respect to hydraulic conductivity in liner systems of waste deposits were investigated in triaxial and oedometer tests. Several types of cracks were taken under investigation: Displacement induced cracks caused by shearing failure and tension by bending moments, the type of crack represented by a lack of bonding between lifts and finally cracks by desiccation. As a result the stress state or the surcharge of landfillcaps is not sufficient to close all kind of cracks mentioned above. This is essential for the landfill cappings in Germany, fixed in the provisions. The most important key factor of self healing is the consistency of the clayey layer in a liner system, i.e. the water content governing brittle failure. Under normal construction conditions the layers are compacted near plastic limit. That means with respect to long term performance of landfillcaps in the case of cracking, brittle failure is likely to occur. The presented data show that brittle failure counteracts self-healing behaviour.

The results presented herein were optained from research under sponsorship of the BMBF, project No. 1440 569A5-18. The author takes responsibility for the contents of the paper.

1 INTRODUCTION

Liner systems play a central role in landfill construction and remediation of contaminated sites for long - term protection against pollutant transport into the environment. Clayey soils are very common in use as sealing material in the construction of liner systems.

During their life however compacted clay liners can be damaged by cracks. Figure 1 shows possible crack types in the case of damage. Settlement induced shear failure is possible in the basal liner system as well as in the capping system. Note that cracks due to separation failure caused e.g. by subsidence under bending tension are located in the caping system only, because the surcharge is very low. In the basal liner system separation failure is not likely due to the surcharge of the waste which compensates tensional stresses.

During construction separation joints are possible due to the lack of bonding between two lifts of a liner. Another posibility of damage due to dry weather conditions are cracks by desiccation.

An advantageous property of clayey soils is often believed in their property of self-healing with respect to hydraulic conductivity when damaged by cracks. Self healing is the material property of beeing capable of closing cracks due to external damage, thus maintaining over the long term the barrier effect required from the liner system. This property is, in addition to the sorption capability, the greatest advantage of clay minerals in the liner materials, which makes them especially suitable as sealing materials. Both in pure mineral liners and in composite liners self-healing is an important property, which is helpful in repairing invisible construction errors and, over geological times, takes care of possible cracks due subsidence after landfill closure, thus relieving the operator from extensive aftercare and repair costs.

The basic question asked in this investigation was: What happens to the permeability if failure has occured ? This aspect is of crucial importance in the evaluation of remediation measurements of contaminated land and abandoned hazardous sites, which exhibit a far greater contamination potential relative to new landfills due to the lack of any waste pretreatment.

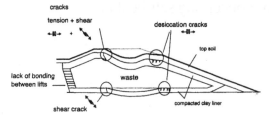

Fig.1:Possible damage by cracks in compacted clay liners

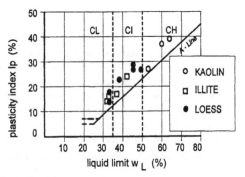

Fig.2:plasticity chart of the basic materials investigated

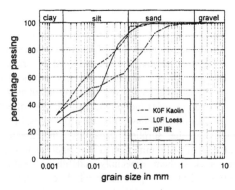

Fig.3:Grain size distribution curves of the basic materials investigated

Objective of this investigation was therefore to investigate, how cracks can be closed again. Crack types investigated were those due to shear and separation failure caused e.g. by subsidence under bending tension; separation joints due to the lack of bonding between two lifts of a liner and desiccation cracks.

Besides the crack type the external boundary condition, i.e. overburden pressure and the material choice play an important role, since essential quantities with regard to self - healing such as

swelling and plasticity play a similarly important role and can be intentionally influenced by the admixture of a highly swelling clay mineral.

In order to enable scale-up to real conditions as wide a range of clayey soils was tested possible.

Table 1:Clay fraction and mineral composition of the basic materials investigated

	Materials			
	Loess	Illite	Kaolin	Montigel F
	Clay fraction in %			
	10 -23	9 - 30	40 - 50	100
	clay minerals in % of the clay fraction			
Smectite	16 - 27	0	-	91
Vermiculite	0 - 24	0	-	0
Kaolinite	16 - 24	0	55 - 56	0
Illite	36 - 59	100	15 - 20	9
Quarz	-	-	20 - 30	-

2 MATERIALS TESTED

Sealing materials used in landfill construction cover a wide range of clayey soils and clay minerals included in them. In this investigation the material choice includes the spectrum from low plasticity clays (TL) to high plasticity clays (TA) as depictured in Fig.2 and Fig.3. In order to test a soil with a wide range of clayminerals frequently used in sealing systems loess loam was added to the list of materials.

The results of X-ray diffractometry in Table 1 show a wide range of minerals in the mineral composition of the loess in comparision to the other basic soils investigated. The wide range of the loess is due to the fact that the soil was taken from different horizonts. Furthermore the results indicate the presence of illite (Thorez 1976) with an extraordinarily high degree of purity as shown by the X-ray diffraction patterns of the illite in Figures 4,5,6.

To investigate the influence of swelling potential and plasticity on crack closure Montigel F (bentonite) was added up to 3 percent by mass to the loess and illite, whereas no bentonite was added to the Kaolin (Fire clay).

Using soils with a single predominant type of clay minerals offers the possibility to look for the background of mechanisms based on structural properties. This is of special interest when the effect on the self - healing capability of additives, especially bentonite, is beeing looked into.

The hydraulic conductivities of the basic materials investigated are:loess: k_s = 2*10-10(m/s); illite: k_s =

5*10-11(m/s) kaolin:k_s = 8*10-10(m/s).

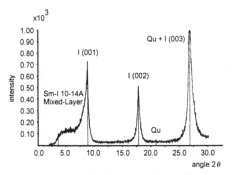

Fig.4:Diffraction patterns of the illite (untreated)

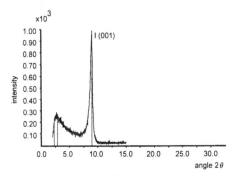

Fig.5:Diffraction patterns of the illite (heated)

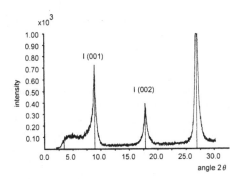

Fig.6:Diffraction patterns of the illite (glycolated)

3 INVESTIGATION INTO THE CLOSURE OF MECHANICAL CRACKS

The following crack types were investigated: separation joint between individual lifts of the liner due to construction error, cracks due to shear failure and brittle fracture due to tensile load. These cracks differ in their surface quality, i.e. geometry which were prepared to model properties of natural cracks on samples with a diameter of 10cm and a height of 10 cm.

The samples had been compacted with optimum water content and standard proctor effort.

The above mentioned separation joint (crack type (r)) was pepared with a saw, which was passed vertically and in the middle through the cylindical sample. In this way two semicylindrical half-samples were produced, which were joined together again when placed into the triaxial device.

The shear crack (crack type (s)) was prepared in a specially modified Proctor device: after compaction to proctor density the soil sample was shorn.

The tension crack (crack type (z)) was produced using the splitting tensile test in accordance with the brazil test used in rock mechanics

A crack is defined as pathway, which , due to local disturbance, leads to a relative increase by a factor β in the average permeability of an undamaged sealing material.

The crack can be characterised with the relative permeability β, where

$$\beta = \frac{k}{k_s} \qquad (1)$$

k - total permeability (m/s), where

$$k = k_s + k_{cr} \qquad (2)$$

k_s - permeability of undamaged soil; and k_{cr} - permeability of a crack.

The opening width of a crack can be calculated by the permeability k_{cr} in the beginning of a permeability test neglecting k_s because in the beginning $k_{cr} >> k_s$ holds as reported (Mallwitz 1998)

To decide whether cracks have been healed or not tests had to be performed under identical conditions with cracked and non-cracked samples.

Permeability tests with flexible wall permeameters on cracked and non-cracked samples were usually performed simultaneously. The normal duration of test usually is around 60 days. The relative permability according to eq.1 is determined from measurements after 40 days of the beginning of the test. This is the time range after which the permeability values of cracked samples become steady.

The parameters applied and varied in the tests were crack geometry, stress state, the addition of bentonite in percent by weight and the mineral composition of the basic materials. The effect of clogging on crack – behaviour was not investigated (Mallwitz 1996)

3.1 *Results*

The following figures show the effect of the above mentioned parameters on self healing behaviour. Crack closure is indicated by a factor`s β becoming unity. To compensate for inaccuracies in permeability measurement with flexible wall permeameters, crack closure will also be accepted when factor β as limiting factor is $\leq$ 1.3. This is the standard for the accuracy of permeability measure.ment of non cracked samples and corresponds to the average reproducibility.

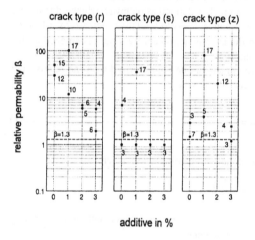

Fig.7:Effect of bentonite addition on self healing behaviour of crack types (r),(s) and (z); material: illite; confining pressure: 50 kPa.

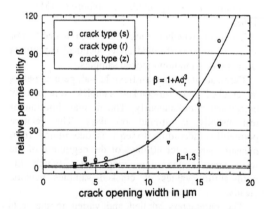

Fig.8:Effect of crack width d_{cr} on self-healing behaviour of crack types (r),(s) and (z); material: illite; confining pressure: 50 kPa.

The numbers in Fig.7 and Fig.9 indicate the average crack opening width in μm. Fig.7 shows that the maximum additive addition of 3% shows little effect on self healing behaviour. For brittle failure it can be seen, that even cracks with very small widths do not close and can not be compensated by material properties, especially the residual swelling capacity.

All in all the evaluation shows that the admixture of additives up to 3% effects the closure of cracks for small crack widths only and the efficiency of admixture depends to a large extend on the homogeneity of the mineral liner material. An addition of bentonite to a maximum of 3% results as a rule in no healing in the soil types tested which are representative for low and medium plasticity clays. This does however not exclude a possible healing with higher amounts of additive or higher contents of swelling minerals in the basic materials. The conistency index of the sample tested was Ic =1.0. Fig.8 shows that the decisive parameter is the average crack width. The governing quantity is the consistency of the material. This can be seen by the fact, that a separation joint (crack type (r)) of 6 μm width shows a tendency for healing, whereas a crack of the same type, but with a crack width of 4 μm do not show any healing. The reason is, that the sample with a crack width of 6μm was prepared with a much higher water content. This shows that the consistency of the material is the governing material property for crack healing.

In general the behaviour of the crack types (r), i.e. the realtive permeability as a function of the opening width in Fig.8, agree well with the theoretical findings reported on in (Mallwitz 1998). All in all shear cracks of the type (s) exhibit somewhat more advantageous behaviour concerning crack healing compared with the other crack types.

The results in Fig.9 indicate that stress levels of 50 kPa are not sufficient for crack healing. Shear cracks also exhibit in this aspect an advantageous behaviour compared to other types of crack. An increase in pressure up to 200 kPa appears to be adequate to close cracks of the type (r) only.

Applying this finding to practical cases it can be concluded that in capping systems with overburden pressure of about 25 kPa systematic crack healing cannot be expected.

4 INVESTIGATION INTO THE CLOSURE OF CRACKS BY DESICCATION

Cracks by desiccation during construction of a liner occur when matric suction of the soil exceeds the lateral earth pressure at rest, caused by overburden.

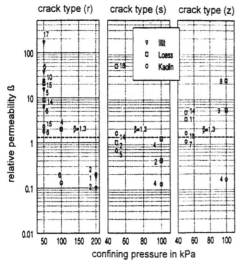

crack type (r)   crack type (s)   crack type (z)

confining pressure in kPa

Fig.9:Effect of isotropic confining pressure on self healing behaviour of crack types (r),(s) and (z)

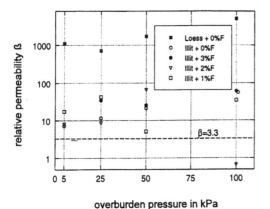

overburden pressure in kPa

Fig.10:Effect of bentonite addition and overburden pressure on self healing behaviour of crackscaused by desiccation

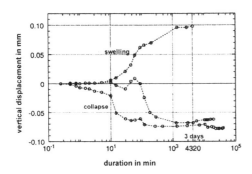

duration in min

Fig.11:Collapse behaviour of Loess during inundation

This was shown for the conditions of a soil sample in an oedometer (Holzlöhner 1995).

For the investigation of crack closure the sample prepared with proctor optimum water content is dried in an oven at 60°C. The experimental apparatus consists of conventional oedometer cells with a fixed sample ring. (19,8 mm high and 71mm in diameter). The crack by desiccation is represented by the gap between the fixed ring and the sample. A controlled rewetting is performed by means of a pipe at the bottom of the oedometer cell. During the suction process the percolation and the vertical displacement of the sample is registered over time under a preset surcharge. Right after the end of the suction process, the permeability of the sample is measured under a hydraulic gradient of i=30.

To decide whether cracks have been healed or not the permeability had to be measured under identical conditions with rewetted and non-dried samples.

4.1 Results

To compensate for inaccuracies in permeability measurement in oedometer cells, crack healing is accepted, when the relative permeability is $\beta \leq 3.3$.

This corresponds to the average reproducibility of two tests.

The results of rewetting tests in Fig.10 indicate that additives principally failed to improve self healing behaviour. The results show that even larger surcharges do not show any improvement of crack healing. Highly swelling clay content of the smectit group counteracts self healing behaviour of desiccation cracks.

As far as material selection is concerned, loess is a rather disadvantageous material, because due to its inhomogeneity in mineral composition and due to its tendency to collapse as shown in Fig. 11 which excludes crack healing.

Illite is a more advantageous choice as it is the least susceptible to desiccation and being a low plasticity monomineral material it swells moderatly only. Kaolin shows a tendency of limited susceptibilty.

With regard to desiccation cracks it can be recommended that kaolinitic or illitic materials as hardly or less swelling clay materials should be preferred if self healing is asked to counteract impairment by desiccation.

Forecast concerning self healing properties can only be made on the basis of experiments.

5 CONCLUSIONS

The dominating parameter concerning self healing of cracks of the mechanical type is the conistency of the material during failure process, and thus the plasticity in interaction with the surcharge. The results show that samples cracked everywhere at solid consistence failed to undergo any self healing in the surchagre of cappings. This result is of great practical importance for landfill engineering, since solid consistence exclusively prevails here even after emplacement of the clayey sealing material. Under normal conditions the layers are compacted near plastic limit. That means with respect to long term performance of landfillcaps in the case of cracking, brittle failure is likely to occur. The presented data show that brittle failure counteracts self-healing behaviour.

Self healing of desiccation cracks depends primarily on the material and in contrary to cracks of the mechanical type, it is fully unaffected by the magnitude of the surcharge. The surcharge must be only sufficient great, as to induce shear failure due to softening at rewetting around the desiccation crack.

For the self healing behaviour with regard to the permeability the material must have a stable structure in the process of crack closure. This is the reason why additives impair self healing. Based on these results it can be recommended that kaolinitic or illitic materials with limited swelling should be preferred to counteract impairment. Collapsible soils should be rejected.

All in all the results indicate that conditions, which would enable systematic self healing, cannot be expected in landfill engineering.

It can be also concluded from the investigations that in landfill cappings an extensive crackless state must be guaranteed to maintain the unimpaired barrier effect in clayey sealing materials. This result can be also applied for testing clay liners in old landfills requiring remediation.

REFERENCES

Holzlöhner U. & Ziegler F. (1995). Effect of overburden pressure on desiccation of earthern liners, Proc. Sardinia 1995, 5th International Landfill Symposium, Calgari, Italy, 2-6 Oct. 1995, T.H. Christensen, R. Cossu, R. Stegmann Ed's, Vol.2.pp. 203-212

Mallwitz K. (1996). Self-healing properties of clayey soils in liner systems of landfills damaged by mechanical cracks and desiccation. S.A. Savidis (Ed.),Veröffentlichungen des Grundbauinstitutes der Technischen Universität Berlin, Heft 24

Mallwitz K. (1998). Crack-Healing in damaged compacted clay liners in waste deposits, Proceedings of the third International Congress on Environmental Geotechnics, Lisboa, Portugal, Seco e Pinto (Ed.), Vol.1.pp. 347-352

Thorez J. (1976). Practical identification of clay minerals, Institute of Mineralogy, Liege State University, Ed. G. Lelotte, Belgium

Clay Science for Engineering, Adachi & Fukue (eds) © 2001 Balkema, Rotterdam, ISBN 90 5809 175 9

Fixing of heavy metals in montmorillonite and the effects on physical properties

K. Emmerich
Institute of Geotechnical Engineering, ClayLab, ETH Zürich, Switzerland (Presently: Federal Institute for Geosciences and Natural Resources, Hannover, Germany)

M. Plötze
Institute of Geotechnical Engineering, ClayLab, ETH Zürich, Switzerland

ABSTRACT: The present work reports the (irreversible) fixing of heavy metals in the crystal structure of montmorillonite and its effects on the clay characteristics and rheological behavior. Heavy metal exchanged montmorillonites were steam-treated under various temperature, pressure and time conditions (up to 300 °C and 6 months). Such studies are relevant to time-lapsing models in barriers at surface conditions and in waste deposits and to hydrothermal conditions in geologic barriers. Investigations with X-ray diffraction and electron microscopy indicate the formation and transformation of phases. Results of spectroscopic measurements (EPR) show the structural position of incorporated ions. Measurements of the rheological behavior characterize changes in mechanical properties. Reasons for changes in the macroscopic chemical and physical properties (rheological behavior, swellability) in relationship to the structural changes of the clay minerals during alteration were discussed.

1 INTRODUCTION

Today bentonites (containing smectites) are used as sealing material in deposit liners, for encapsulation of radioactive waste and for remediation of hazardous waste. Thereby contaminants especially heavy metals are retained and immobilized. Knowledge about interactions and reactions between the clay minerals (smectites) and heavy metals, and their effects on the physical properties are necessarily for an effective application of bentonites.

Smectites represent 2:1 layer silicates with an expandable structure. Their structure is responsible for unusual physical properties. The lattice consists of tetrahedral and octahedral sheets (2:1). Substitutions of cations in these sheets generate a negative permanent and modifiable layer charge, which is balanced by hydrated mono- or divalent cations located in the interlayer.

The surface chemistry and physical properties of clay minerals depend highly on type and nature of the constituting ions as well as on their specific position and substitutions in the lattice. The rheological behavior of smectite can be controlled by the short-range interactions between interlamellar surfaces and water molecules. The interlayer cations are usually more weakly bound and are readily exchangeable, so that the interlayer can be easily manipulated. Heavy metal ions can be adsorbed at the surface or edges of clay particles as well as they can be located on exchangeable interlayer positions. The size and charge of the interlamellar exchanged cations (tran-

sition metals, alkali and earth alkali ions) and phase transformation processes on clay minerals have an important influence on their macroscopic behavior (e.g., swelling and shrinkage, rheological properties, permeability, adsorption and desorption, diffusion properties).

Strength and rheology determine the mechanical behavior of clays. The physical properties and rheology of colloidal clay-water systems are important in geotechnical engineering. Plasticity and shear strength of cohesive soils are mainly determined by their clay mineral content. Time-dependent deformations and stress changes are important considerations in many geotechnical studies, e.g. long-term settlements, movement of slopes, squeezing of soft grounds etc.. With the knowledge of the rheological behavior it is possible to characterize the geotechnical properties of cohesive soils and clay dispersions. Research on time-dependent deformation and stress phenomena (e.g., creep, stress relaxation, shear strength) of clays as barrier material remain of great interest to geotechnical engineers. The permeability, diffusion, adsorption and desorption properties, the rheological and the swelling and shrinkage behavior are the main factors in geological and in geotechnical engineered barriers concerning the safety and long-term behavior of landfill construction. Because of their high adsorption and swelling capacity and the low permeability Na-smectites (Na-bentonites) represent barriers. They prevent the flow of water, e.g. through natural and technical waste containment.

Two processes are to be distinguished in material behavior in the clay-water-chemicals system. Short-term reactions due to mostly reversible changes of the interlayer (adsorption, de- and rehydration) cause changes, e.g. in the adsorption, the swelling and shrinking behavior, in hydraulic conductivity and in mechanical (rheological) characteristics. Long-term reactions of formation and transformation of (clay) minerals through processes of solid state reactions or precipitation of phases in alteration processes cause drastic changes in the material properties. Knowledge about these processes is important for assessment of long-term behavior of barrier materials used in waste deposits.

Keeping in mind the dependence of chemical and physical properties of clay minerals on the nature of substituted and exchanged metal ions in the sheets, it is important to understand their localization and function in clay minerals. The aim of the present work is to improve the knowledge of interaction mechanisms of montmorillonite with heavy metal cations and of the changes of the chemical and physical properties of clay minerals during this treatment.

The effects of chemical attacks on clay minerals have been the subject of studies focused on the stability and phase transformation of clay minerals. Changes of hydraulic conductivity, diffusion behavior and rheological properties (shearing behavior) of clays due to chemical processes were studied (e.g., Madsen and Mitchell 1989, Madsen 1998). The shear behavior of clays depends on the mineralogical composition of clays and on the exchanged cations in the interlayer of the clay minerals (Sonderegger 1985, Müller-Vonmoos et al. 1985).

Electron paramagnetic resonance spectroscopy (EPR) has proved to be a powerful technique for characterizing the coordination environment, orientation and degree of hydration of metal ions in the interlayer of expanding clay minerals (e.g., Clementz et al. 1973, 1974, McBride & Mortland 1974, Brown & Kevan 1988 a, b).

2 MATERIAL

Homoionic Cu^{2+}- and Zn^{2+}-forms of white-yellowish montmorillonite from Linden, Bavaria, $Ca_{0.10}$ $Mg_{0.08}$ $(Na, K)_{0.01}$ $[(Si_{3.95} Al_{0.05}) (Al_{1.46} Fe_{0.18} Mg_{0.38}) O_{10}$ $(OH)_2]$ (molweight: 371.3 g mol^{-1}, CEC 100 meq/100 g) were used in this study (Emmerich et al. 1999).

Samples of 0.5 g of the homoionic clays (air dry or in a suspension with mass content of 10%) were altered in a water steam atmosphere in stainless steel autoclaves at 55 or 90 °C up to 6 month.

3 METHODS

X-ray patterns were obtained from random powdered samples with a Philips PW 1820 X-ray diffractometer using $CuK\alpha$ radiation. All samples were step scanned (0.02 °θ/2 s) from 1.5 to 65° 2θ.

EPR spectra of randomly oriented powder samples of Cu^{2+}-rich montmorillonites were recorded with a X-Band EPR-spectrometer Bruker ESP 300 E at room temperature after allowing samples to soak humidity in an atmosphere of 75% relative humidity generated by a saturated NaCl solution.

Viscosity and yield stress of suspension with a clay content of 3% were calculated using the Bingham model (Lagaly et al. 1997) from controlled rate (CR) flow curves. Experiments have been carried out at the Institute of Physical Chemistry at the University Kiel with the Universal Dynamic Spectrometer Physica UDS 200 (Physica Messtechnik Stuttgart/Germany).

4 RESULTS

4.1 XRD and phase growth

Basal spaces of all samples (under laboratory atmosphere T~24°C and ~45-55% relative humidity) were nearly unchanged after 6 month of alteration. The d(001) of the Cu^{2+}-rich montmorillonites were between 12,2-12,7 Å. Basal reflexes at 14,8-15,1 Å were observed for the Zn^{2+}-rich montmorillonites.

On the very first month of steam treatment all samples showed discoloration. The most striking change of color was observed for the Cu^{2+}-rich samples.

Also a new Aluminum Silicate Hydroxide phase with a main peak at 5.06 (± 0.08) Å was observed in these samples (Fig. 1).

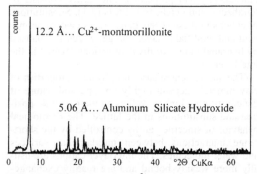

Figure 1: XRD pattern of a Cu^{2+}-rich montmorillonite after steam treatment at 90°C for 6 months.

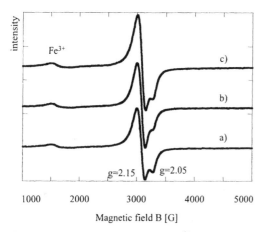

Figure 2. EPR-spectra of a) untreated Cu^{2+}-montmorillonite and after alteration at 90°C in a suspension for b) 7 days and c) 6 month.

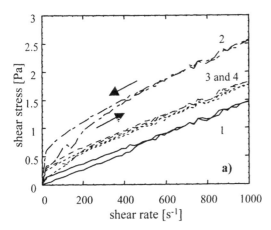

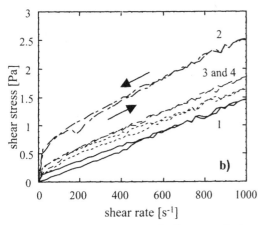

Figure 3. Flow curves of dispersion of Cu^{2+}-montmorillonites at solid content 3%. Montmorillonites altered at 90°C a) in steam atmosphere and b) in a suspension. (Curve 1 initial sample, curve 2 after alteration for 1 month, curve 3 after alteration for 4 month and curve 4 after alteration for 6 month.)

4.2 EPR

EPR spectra of altered Cu^{2+}-montmorillonites showed two peaks at g = 2.15 and 2.05 which are characteristic for $[Cu(H_2O)_6]^{2+}$ and $[Cu(H_2O)_4]^{2+}$, respectively. The intensity ratio of both peaks moved in favor of the signal at 2.15 after short time alteration, whereas the peak at 2.05 regained intensity after 6 month (Fig. 2).

Samples that were altered in a steam atmosphere at 90°C as well as samples that were altered at 55°C showed a similar intensity movement of both peaks.

4.3 Rheology

Alteration of the montmorillonites caused an increase of viscosity of the suspensions prepared with the altered smectites.

The viscosity of all samples reached a maximum after alteration for 1 month. Thereafter the viscosity decreased after alteration up to 6 month reaching a final level that is higher than for the initial material (Fig. 3).

Suspensions prepared of the altered Zn^{2+}-montmorillonites showed similar rheological properties.

5 DISCUSSION

The results of the present work show that steam and water at moderate temperatures influenced the chemical composition and the rheological properties of heavy metal saturated montmorillonites.

EPR spectra show that the amount of fully hydratable Cu^{2+}-ions on interlayer positions depends on increasing time of treatment. Decrease of the signal at 2.15 of the wet altered Cu^{2+}-exchanged samples indicate hydrolysis of the Cu^{2+}-ions and a reduction of the concentration of $[Cu(H_2O)_6]^{2+}$ at the clay particle surface (McBride 1982). No significant changes of basal spaces in XRD pattern could be observed for the altered samples.

Divalent Cu^{2+} and Zn^{2+}-ions prevent disintegration of montmorillonite particles in suspension. Therefore the viscosity of the suspensions of the initial material is low.

Flow behavior depends on the pH-value (e.g. Lagaly 1989). The pH of the montmorillonite suspensions prepared of the altered Cu^{2+}- and Zn^{2+}-rich samples decreased because of generated protons during hydrolysis. Now the particle edges were positive charged. Edge(+)/face(-) contacts were build in addition to the face(-)/face(-) contacts. Formation of these network structures promoted an increase of the viscosity of the montmorillonite suspensions prepared of the altered samples.

With an on-going hydrolysis di- and trivalent cations are removed from the octahedral sheet. Increasing concentration of these cations broke down the network. Coagulation of particles then reduced the viscosity of the suspension.

6 CONCLUSIONS

Reduction of concentration of $[Cu(H_2O)_6]^{2+}$ at interlayer positions caused by alteration under steam or water at moderate temperatures is a reversible process.

Results of EPR spectroscopy and rheological measurements indicate that changes of the physical properties of a clay barrier are correlated to changes of the coordination environment of the heavy metal interlayer cations.

It is very probable that the proposed mechanism of alteration that caused changes in the rheological behavior is also valid for the Zn^{2+}-montmorillonites because of the similar chemical character of Cu^{2+} and Zn^{2+}.

REFERENCES

Brown, D.R. & L. Kevan 1988a. Aqueous coordination and location of exchangeable Cu^{2+} cations in montmorillonite clay studied by electron spin resonance and electron spin echo modulation. *J Am Chem Soc.* 110: 2743-2748.

Brown, D.R. & L. Kevan 1988b. Solvation of exchangeable Cu^{2+} cations by primary alcohols in montmorillonite clay studied by electron spin resonance and electron spin echo modulation spectroscopies. *J Phys Chem.* 92: 1971-1974.

Clementz, D.M., M.M. Mortland & T.M. Pinnavaia 1974. Properties of reduced charge montmorillonites: Hydrated Cu(II) ions as a spectroscopic probe. *Clay Clay Miner.* 22: 49-57.

Clementz, D.M., T.M. Pinnavaia & M.M. Mortland 1973. Stereochemistry of hydrated copper(II) ions on the interlamellar surfaces of layer silicates. An electron paramagnetic resonance study. *J Phys Chem.* 77: 196-200.

Emmerich, K., F.T. Madsen & G. Kahr 1999. Dehydroxylation behavior of heat-treated and steam treated homoionic cis-vacant montmorillonites. *Clay Clay Miner.* 47(5): 591-604.

Lagaly, G. 1989. Principles of flow of kaolin and bentonite dispersions. *Appl Clay Sci* 4: 105-123.

Lagaly, G., O. Schulz & R. Zimehl 1997. *Dispersionen und Emulsionen.* Darmstadt: Steinkopff.

Madsen, F.T. & J.K. Mitchell 1989. Chemical effects on clay hydraulic conductivity and their determination. *Mitt. des Instituts für Grundbau und Bodenmechanik.* ETH Zürich, Nr. 135.

Madsen, F.T. 1998. Clay mineralogical investigations related to nuclear waste disposal. *Clay Miner.* 33: 109-129.

McBride, M.B. & M.M. Mortland 1974. Copper(II) interactions with montmorillonite: evidence from physical methods. *Soil Sci Soc Am Proc.* 38: 408-415.

McBride, M.B. 1982. Hydrolysis and dehydration reactions of exchangeable Cu^{2+} on hectorite. *Clay Clay Miner.* 30: 200-206.

Müller-Vonmoos, M., P. Honold & G. Kahr 1985. Das Scherverhalten reiner Tone. *Mitt. des Instituts für Grundbau und Bodenmechanik.* ETH Zürich, Nr. 128.

Sonderegger, U.C. 1985. Das Scherverhalten von Kaolinit, Illit und Montmorillonit. *Mitt. des Instituts für Grundbau und Bodenmechanik.* ETH Zürich, Nr. 129.

Clay Science for Engineering, Adachi & Fukue (eds) © 2001 Balkema, Rotterdam, ISBN 90 5809 175 9

Quality control of a landfill base and capping clay liner

W. Haegeman
Laboratory of Soil Mechanics, Gent University, Belgium

ABSTRACT : A database is described that contains results of laboratory measurements of hydraulic conductivity and associated soil properties performed on compacted clay samples from two Belgian landfills. The measurements were conducted on a base liner of one landfill and a capping liner of another. The data was used to evaluate hydraulic conductivity and compaction conditions and to compare the results with quality control testing and acceptance standards of the clay prior to the construction works. The variability of the test results is evaluated and checked against the acceptance standards. The base liner properties are well within the acceptable standards. However due to difficult compaction conditions on the side slopes, some samples of the capping liner, failed the acceptance criteria. Through an evaluation procedure which takes into account the variability and probability of exceeding of the acceptance standards, the permeability performance of both liners is accepted.

1 INTRODUCTION

Compacted clay liners are widely used as hydraulic barriers in waste containment facilities. To be effective, a compacted soil liner should have low hydraulic conductivity, which in many cases is prescribed in standards or project specifications. (Rowe 1997 ; Areias et al. 1998).

Evaluation of such liners, both from a design point of view as well as in the testing of the actual liner, is of great importance as processes involving ground water flow and contaminant transport are relatively slow and the effectiveness of the liners in situ cannot be measured immediately. Results have shown that a more rigid approach to the design process and quality control is necessary, in order to be able to compare the actual performance of the liner based on test results with the original design criteria. Specifications are typically based on experience and generally include minimum values or acceptable ranges for properties that describe the composition of the material (e.g. Atterberg limits, partial size distribution) and the compaction energy (i.e. control of water content and dry unit weight).

This paper shows how the basic soil properties and compaction conditions of two clay mineral barriers monitored during construction quality control are related to hydraulic conductivity. An evaluation procedure of the permeability performance of the liners is then described.

Results are used statistically to determine whether or not the performance of the liners meets the design specification by taking into account the variability of the results.

2 PROCEDURE, MATERIALS AND EQUIPMENT

Laboratory hydraulic conductivity measurements were limited to undisturbed specimens collected at the site using thin wall sampling (Shelby) tubes. Flexible-wall permeameters were used for the direct measurement of hydraulic conductivity using standard testing procedures and conditions normally used in quality control of liners. Measurements were made using a constant head method and applying a gradient of about 40 for a sample 0.07 m high. Flow rates were determined by collecting and weighing the discharge in a purge and trap bottle to prevent evaporation. The permeability was measured for a period of two or three days during a test cycle of at least two weeks. The sample was 100 mm in diameter and had a height of approximately 70 mm. Measurements of water content and dry unit weight were performed before and after permeability testing.

Undisturbed clay samples were delivered at regular intervals from the two disposal sites under construction. A first series of samples was taken from a base clay liner while a second series was taken from a 0.5 m thick capping layer covering the side slopes of the landfill.

3 PERMEABILITY TEST RESULTS

For clay materials there is a good correlation between placement water content and density, and in-situ hydraulic conductivity (Daniel 1998, Dunn 1985). The hydraulic conductivity of the clay decreases with the increase of water content or compactive energy. For water contents dry of optimum, the hydraulic conductivity is large and it decreases sharply as water content approaches optimum water content. When the soil is wetter than optimum, the hydraulic conductivity is low. For a given water content the hydraulic conductivity also decreases significantly with the increase in the compactive effort.

Figure 1 shows the measured dry unit weight in function of the molding water content for the 27 samples of the base liner and the 21 samples of the capping liner.

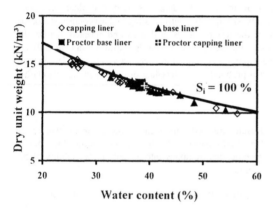

Figure 1. Dry unit weight versus water content

These data points plot in a narrow zone close to the line of full saturation. The initial degree of saturation varies between 88 % and 102 %. All samples have a water content wetter than optimum and for both sites the range in water content is rather extended. A direct comparison between the measurements at the two sites is difficult to make because of possible shifts in optimum water content and maximum dry unit weight that occur as a result of changes in compositional characteristics of the clays.

Plotting the hydraulic conductivity versus the initial water content (figure 2) shows a fairly constant and low permeability for the base liner. The hydraulic conductivity for the capping liner is the lowest at low water contents near the optimum water content and increases as the water content increases. It is believed that the order of magnitude difference in hydraulic conductivity at higher water contents between the liners is mainly due to differences in compaction effort rather than differences in clay characteristics. Also plotted on the figure are the results of permeability measurements on samples compacted at high effort in a modified proctor at the

natural water content of the clays. No difference in hydraulic conductivity can be noticed for the two clay materials compacted in proctor. The measurements on the undisturbed samples of the base liner fall within the range of the hydraulic conductivities of proctor samples giving proof of a high compaction effort in situ.

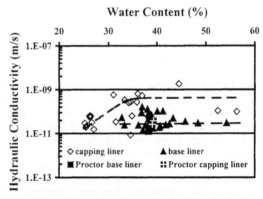

Figure 2. Hydraulic conductivity versus water content

Figure 3 is a graph of hydraulic conductivity versus initial saturation S_i. The initial saturation was calculated assuming a specific gravity G_s of the solids of 2.7. In some cases this resulted in $S_i > 1.0$ which is physically impossible. However using a single G_s is a consistent way to analyse the data without introducing artificial bias by variable selected values of G_s. Because of the high initial saturation for all samples no significant trend can be noticed in figure 3.

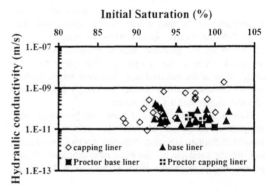

Figure 3. Hydraulic conductivity versus initial saturation

4 EVALUATION PROCEDURE

The objectives of a quality control program are to determine, in a reproducible and quantitative manner, the characteristics of a liner. This enables the adequacy of the liner to be established. In short, the following evaluation procedure is adapted (Van Ree et al 1992) :

a. plotting the probability and permeability value of each result in a log-normal probability diagram ;
b. testing the hypothesis of a log-normal distribution of the permeability values by means of the Kolmogorov-Smirnov test ;
c. applying the statistical testing criterion (α-fractile) using

$$P\left[\log k > \overline{\log k} + t_{n-1}(1-\alpha)s\sqrt{\left(1+\frac{1}{n}\right)}\right] = 1 - \alpha \quad (1)$$

where P is probability, α is the probability of exceedance, $\overline{\log k}$ is the average of the logarithm of the permeability of the samples (m/s), s is the standard deviation of the logarithm of permeability results, k is the required permeability value (m/s), n is the number of samples tested and t_{n-1} (1-α) is Student's t(n − 1) distribution, and by using the Student t(n-1) distribution to account for limited sampling of a population with unknown characteristics ; and
d. determining whether or not the permeability specifications are met – in practice 5 % probability of exceeding a permeability limit of 1 x 10^{-9} m/s is used as a permeability criterion for acceptance.

Figure 4 shows the probability of occurrence of a hydraulic conductivity in all tested samples of the base or capping layer. This figure shows that the variability of the field situation is considerable and that the capping liner is more sensitive to the actual placement conditions than the base liner. From this it can be seen that permeability values vary by at least an order of magnitude within a liner system. Absolute permeability criteria thus often present problems because of so-called 'outliers'. In these situations outliers tend to be ignored, and ascribed to equipment and sampling failures. The arithmetic mean of permeability values can be used, but again, due to variability, this can also lead to problems.

The permeability values in figure 5 display a log-normal distribution. Thus, using the geometric mean (μ) together with the standard deviation (σ), the characteristics of the sample population can be described. Repetitive sampling and testing of a clay liner show this distribution to be highly reproducible. This in turn leads to a more objective approach to the permeability performance of a liner, using all the test results together with the test criterion given in equation (1).

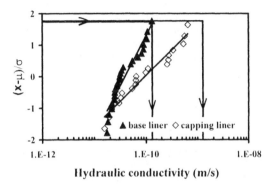

Figure 5. Log normal probability hypothesis

Given a (maximum acceptable) probability P of 1 - α and using the related value for the Student's t(n − 1) distribution taken from standard statistical tables, equation (1) is used to calculate which permeability value goes with the measurement data. In order to accept the liner as adequate this permeability value must be smaller than the required permeability k.

Table 1 presents the values of the parameters used in equation (1) for the calculation of the acceptable permeability (5 % fractile).

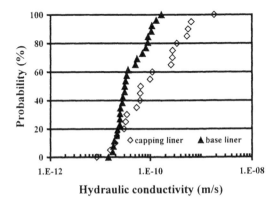

Figure 4. Probability of occurrence of measurement data

Table 1

	n	$\overline{\log k}$	s	$t_{n-1}(1-\alpha)$	k (m/s)
capping liner	21	-10.012	0.6277	1.725	1.25×10^{-9}
base liner	27	-10.406	0.3041	1.706	1.32×10^{-10}

It can be seen that the overall performance of the base liner meets the design specification of 10^{-9} m/s and, although the calculated permeability of the capping liner with a probability of exceedence of 5 % is somewhat higher than the required permeability, the performance of this liner is also accepted.

5 CONCLUSIONS

In this paper results of laboratory-measured hydraulic conductivity of a base and capping clay liner were presented and correlated with compaction properties like water content, dry unit weight and degree of saturation.

A procedure is presented for the construction quality control of both mineral liner systems. In order to eliminate the influence of outliers on test results where an absolute permeability criterion is specified, a statistical criterion assuming a log-normal distribution enables the extreme values to be tested, and thus it is possible to evaluate the performance of the liner. The statistical approach combined with testing of undistorted samples provides a means of checking as-built liner performance. The results show that the permeability performance of both liners can be accepted.

6 REFERENCES

Areias, L. et al. 1998. Liner system regulations for MSW landfills in Belgium, Canada and the USA. *Proceedings of the Third International Congress on Environmental Geotechnic.* Lisbon, pp. 3-8.

Daniel, D.E. 1998. Landfills for solid and liquid wastes. *Proceedings of the Third International Congress on Environmental Geotechnics.* Lisbon, pp. 1231-1246.

Dunn, R.J. 1985. Design of clay liners and barriers – Considerations of constructed structure. *Proceedings of the International Symposium on Management of Hazardous Chemical Waste Sites.* Winston-Salem, North Carolina, pp. 1-25.

Rowe, R.K. 1997. The design of landfill barrier systems : should there be a choice ? *Ground Engineering.* 30 (7).

Van Ree, C. Weststrate, F., Meskers, C., Bremmer, C. 1992. *Geotechnique.* 42, N° 1, 49-56.

Clay Science for Engineering, Adachi & Fukue (eds) © 2001 Balkema, Rotterdam, ISBN 90 5809 175 9

Redox effect on hydraulic conductivity and heavy metal leaching from marine clay

Masashi Kamon & Govindasamy Rajasekaran – *Disaster Prevention Research Institute (DPRI), Kyoto University, Japan*

Takeshi Katsumi – *Department of Civil Engineering, Ritsumeikan University, Kusatsu, Japan (Formerly: Disaster Prevention Research Institute, Kyoto University, Japan)*

Huyuan Zhang – *Department of Civil Engineering, Kyoto University, Japan*

Naoki Sawa – *JGC Corporation, Yokohama, Japan (Formerly: Department of Civil Engineering, Kyoto University, Japan)*

ABSTRACT: Oxidation-reduction effect on the behavior of clay liners in waste landfill sites was investigated. The marine clay sampled from Osaka Bay was used to simulate the behavior of clay liner in coastal landfills. Distilled water, sodium sulfate, and sodium dithionite solutions resulting in different redox potentials were employed to investigate the redox effects on hydraulic conductivity, heavy metal leaching, free swell, and liquid limit. Influents for hydraulic conductivity tests were enhanced with 10 ppm zinc as $Zn(NO_3)_2$, and effluents were chemically analyzed. Test results indicated that there is no obvious change in the properties of marine clay including the hydraulic conductivity when the reducing agent was used, and that zinc tends to be immobilized in the reduction condition. In conclusion, marine clay is considered as an effective clay liner in the coastal landfill.

1 INTRODUCTION

Clay liner is widely used as containment barriers in solid waste disposal facilities. The effectiveness of the barrier is determined by its low hydraulic conductivity to water, organic compounds, and dissolved inorganic chemicals in leachate over long periods. Many researches have been conducted on the clay-chemicals interactions and hydraulic conductivity of clay liners (e.g., Shackelford 1994, Mitchell and Madsen 1987). Among several factors, the effect reduction-oxidation (redox) condition might be a significant concern. For the coastal landfill facilities that have been constructed to contain municipal and industrial wastes in Japan, in particular to Tokyo and Osaka Bays (Aburatani et al. 1996), natural marine clay layers, which usually exist under the strong reduction conditions, are considered as a clay liner to prevent the pollutant migration. Also, leachate from the landfill waste is sometimes rich in dissolved organic matter that may lead to the redox reactions and form a sequence of redox zones in the plume of landfills (Lyngkilde et al. 1992, Christensen et al. 1994, Bjerg et al. 1995).

Redox reactions strongly affect the behavior of pollutants leached from the landfill (Lyngkilde et al. 1992, Rugge et al. 1995): Redox reactions can affect the forms of multivalent metals, and the mobility and toxicity of multivalent and monovalent metals. Also, the redox reactions may affect the clay minerals: The structural iron in clay minerals can be reduced by bacteria (Loveley 1991), and the reduction of iron-bearing clay minerals has a great impact on the chemical and physical properties of the soil, particularly in terms of swelling and cation exchange capacity (Stucki 1988). However, little is known about the effect of redox potential on the behavior of clay liners. Thus, the objective of this study is to evaluate the effects of redox reactions on the hydraulic conductivity and leaching of zinc from marine clay that can be used as a landfill barrier.

2 BACKGROUND

Redox reactions occur simultaneously in the soil-water system. A redox reaction can be considered as a pair of coupled half-reactions. Both half reactions involve the transfer of electrons between the chemical species in the system. Usually in landfill sites, the organic matter donates the electrons, which then will be oxidized, and subsequently the inorganic components receive the electrons, which will be reduced. For example, nitrate, Mn(IV), Fe(III), and sulfate can be reduced to nitrite, Mn(II), Fe(II), and sulfide respectively. Redox potential (Eh) of the soil solution is an expression of the electron density of the system. Redox potential of the natural soil and sediments varies widely from approximately + 500 mV (surface soil) to approximately – 300 mV (strongly reduced soil). Four typical ranges of redox conditions suggested by Patrick and Turner (1968) are encountered at pH 7, above + 400 mV for oxidized soils, + 400 to + 100 mV for moderately reduced soils, + 100 to - 100 mV for reduced soils, and - 100 to - 300 mV for highly reduced soils. Usually, the redox state of a soil is closely related to the microbial activity and the type of substrate available to the organisms.

Numerous studies have shown that microbial metabolism couples the inorganic redox chemistry of soil or groundwater to the oxidation of organic carbon. A large fraction of the microbial population within the soil depends on oxygen as the terminal electron acceptor in metabolism when the soil pores become filled with water. Oxygen may be consumed faster than its replacement by diffusion from the atmosphere, and the soil may become anaerobic, resulting in a reduction condition.

3 MATERIALS AND METHOD

The marine clay material used for the experiment was sampled from Osaka Bay. Basic properties of the clay are shown in Table 1.

Table 1   Properties of Osaka Marine Clay

Particle density (g/cm^3)	2.72
Natural water content (%)	100.4
Liquid Limit (%)	93.3
Plastic Limit (%)	26.6
Grain size distribution (%)	
Gravel (>2 mm)	0
Sand (2 mm~0.075 mm))	2.1
Silt (0.075-0.005 mm)	52.1
Clay (<0.005 mm)	45.8
pH measured in suspension	7.52
pH measured in pore water	7.39
Eh measured in suspension (mV)	-377
Eh measured in pore water (mV)	62.0
Total organic carbon (TOC) (%)	0.63

Hydraulic conductivity tests were conducted on the consolidated marine clay specimens using flexible-wall permeameter. The marine clay having a natural water content is consolidated with Oedometer under a consolidation pressure of 200 kPa, which represents approximately overburden pressure of 20 m depth of disposed solid waste, and then set in flexible-wall permeameter as shown in Figure 1.

All influents were enhanced with 10 ppm zinc as $Zn(NO_3)_2$. To compare the effect of redox potential, sodium dithionite ($Na_2S_2O_4$) or sodium sulfate (Na_2SO_4) were dissolved in selected influents. Sodium dithionite has been used as a strong reducing agent, while sodium sulfate, which does not alter the original redox condition, provides an equivalent sodium ionic strength as sodium dithionite. Solution containing only $Zn(NO_3)_2$ without sodium dithionite or sodium sulfate was also used as a control test.

Table 2   Hydraulic Conductivity Test Conditions

Specimen	Water content (%)	Dry density (g/cm^3)	Chemical agent	Pressure gas
A	59.8	1.65	$N_2S_2O_4$	N_2
B	59.7	1.65	N_2SO_4	N_2
C	62.0	1.64	-	N_2
D	58.2	1.66	$N_2S_2O_4$	O_2
E	61.7	1.64	N_2SO_4	O_2
F	59.6	1.65	-	O_2

Two kinds of gases, N_2 and O_2, were used as pressure gas sources to investigate whether the gases alter the redox condition of permeants considerably within the long test duration. A 100 kPa of water head difference across the specimen was applied. Table 2 summarizes the experimental conditions. The effluents were collected at a certain time interval and analyzed to measure Eh, pH, dissolved oxygen (DO), and electrical conductivity (EC). Then, concentration of cations in the effluents filtered through 0.00045 mm pore size filter paper were determined by inductively coupled plasma (ICP) spectrometry (ICPS-4960, Shimazu Co., Japan).

Free swell tests were also conducted using the corresponding dispersing agents as described earlier. The swelling properties of marine clay under different redox conditions were studied by conducting the swell index test where distilled water, sodium sulfate and sodium dithionite were used as dispersing agents. Powdered two grams of air-dried marine clay minerals (passed through 0.1 mm sieve) were carefully dusted over the surface of the solution to allow the clay minerals to hydrate. After 72 hours, the final volume of the settled clay was recorded. The solutions used before the swell test and the supernatant of suspension after the swell test were measured for pH, Eh, DO, and EC.

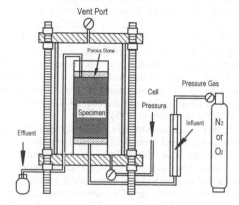

Figure 1   Hydraulic Conductivity Test Setup

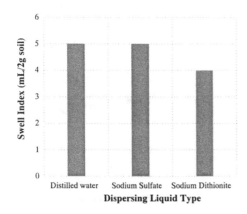

Figure 2 Swell Index of Marine Clay with Various Redox Agents

Table 3 Properties of Dispersing Solutions before and after the Swell Test

Dispersing	Solution	H_2O	Na_2SO_4	$Na_2S_2O_4$
pH	Before Test	5.9	5.8	5.8
	After Test	8.0	7.6	4.9
Eh	Before Test	112	103	-531
(mV)	After Test	104	144	-415
EC	Before Test	0.01	16.0	16.1
(mS/cm)	After Test	0.19	17.3	17.5
DO	Before Test	7.6	7.4	0.02
(mg/L)	After Test	6.0	5.4	0.03

Liquid limit tests were performed according to the JGS 0141-2000, and distilled water, sodium sulfate, and sodium dithionite solutions were used as pore liquids. Pore liquids were mixed with air-dried marine clay, and each mixture was separated into two portions: one was used to measure the liquid limit immediately, whereas another portion was kept in a sealed container for 24 hours prior to the liquid limit test.

4 RESULTS AND ANALYSIS

4.1 Redox Control

The redox reactions in soil systems are usually catalyzed by microorganisms, which couple the organic matter with the inorganic components. The redox conditions of soil derived from microbial actions can be simply simulated by chemical method, that is, the chemical agents can control the redox potential. Dithionite has been used to maintain a low Eh in the soil system by several researchers (Chen et al. 1987, Stucki 1988, Shen et al. 1992).

The standard electrode reduction potential of dithionite is approximately – 1120 mV (Vanysek 1999). The Eh of freshly prepared dithionite solutions was as low as – 500 mV. When the solutions were exposed to air, their reducing power has been decreased and disappeared completely after 20 hours for 1.0 M and 0.1 M solutions, and less than 4 hours for 0.01 M solution. When dithionite was added to marine clay, the Eh was lower than that of the pure dithionite and persisted a longer time than dithionite solution without soil. This means that the reducing power of dithionite was preserved and enhanced when added to the soil suspension, although the exact reason remains unknown. Thus, it is capable of maintaining a low Eh in the soil system for long duration by the addition of dithionite.

4.2 Free Swell

Table 3 and Figure 2 show the results of free swell tests. Major differences among the three dispersing agents before test are that sulfate provided approximately the same redox potential as distilled water with higher ionic strength, whereas the dithionite possessed almost the same ionic strength as sulfate but provided a strong reducing condition. After the swell test, the pH of distilled water and sulfate increased 1.6 - 2.0 pH units due to the weak alkalinity of marine clay while the dithionite became more acidic. A little increase in E.C. of three agents indicated that some salts in the marine clay were dissolved out even in the case that sulfate and dithionite had much higher ionic strength.

Swell index shown in Figure 2 indicates that when marine clay was dispersed in distilled water and sulfate, swell index values have been equal. This implies that ionic strength is not a major factor influencing swell property of the marine clay. When the marine clay was dispersed in dithionite, its swell index decreased 1% than that dispersed in sulfate. This means that the swell property of marine clay is not sensitive to redox potential and ionic strength.

4.3 Liquid Limit

Liquid limit test results are shown in Figure 3. The interactions between soil minerals and pore liquid need some time to complete if the reactions indeed occur. After 24-hours reaction, the liquid limits of marine clay with various redox agents shows a range of 69.3% ~ 69.8%. It is more reasonable to consider this variation due to the normal test error rather than the redox effect of the pore liquids added. That is, the marine clay had the same liquid limit even though it was mixed with agents, which poses significantly different redox potential. This can be verified by the marginal decrease in swell index of marine clay dispersed in dithionite (Fig. 2). Because the marine clay is a strongly reduced soil, even though a stronger reducing agent is added, its properties have been changed marginally.

507

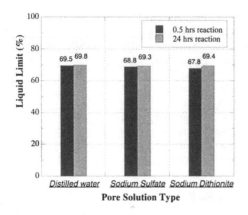

Figure 3 Liquid Limit of Marine Clay with Various Redox Agents as Pore Liquids

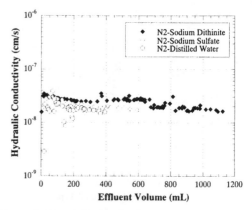

Figure 4 Hydraulic Conductivity of Marine Clay with N_2 as Pressure Gas Source

4.4 Hydraulic Conductivity

The hydraulic conductivity test results are shown in Figures 4 and 5. Marine clay exhibited a hydraulic conductivity as low as $< 10^{-7}$ cm/s, which satisfies the requirement for clay liners in municipal waste landfill according to US EPA. These low hydraulic conductivities can be mainly attributed due to the presence of high clay fraction in the marine clay. Test results in Figures 4 and 5 indicate that there were no obvious differences in the hydraulic conductivity whether the specimens permeated with different agents, or whether the different gases were used to exert the head pressure. In addition, all hydraulic conductivity values showed no obvious variation for the test period as long as 80 days. Thus, in conclusion, redox condition has no obvious effect to alter the hydraulic conductivity of marine clay.

These hydraulic conductivity test results are consistent with those from free swell and liquid limit tests. If pore liquids with different redox potentials can change the forces, which govern the clay-water interaction and alter the microstructure or texture of the clay-water fabric, a change in hydraulic conductivity can be expected. Under the experimental condition conducted in this study, redox potential showed no obvious effect on free swell and liquid limit, and hence no obvious change was occurred in hydraulic conductivity.

Similar tests were conducted by Shen et al. (1992), who prepared the hydraulic conductivity test sample by suspending Na-smectite and compacting it under nitrogen gas to a gel. Distilled water, sulfate, and sodium dithionite were used to obtain different redox conditions. To consider the consolidation history as well as the reduction reaction on the hydraulic conductivity, two kinds of reduction procedures were conducted by Shen et al. (1992): clay was reduced in

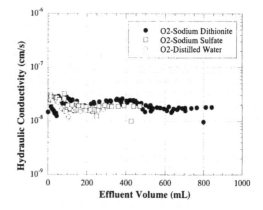

Figure 5 Hydraulic Conductivity of Marine Clay with O_2 as Pressure Gas Source

suspension and then consolidated, or clay was consolidated and then reduced by the reducing agent of sodium dithionite. It was found that the reduction of clay in suspension prior to consolidation decreased the hydraulic conductivity of smectites compared with the samples in their respective oxidized states. The increase in gel density induced by the reduction of structural Fe in smectite was considered to be responsible for this increase in hydraulic conductivity. When the oxidized clay suspensions were consolidated first and then reduced by dithionite, their hydraulic conductivities increased significantly compared with the sample in oxidized state. The later test process represents the practical possibility in landfill sites (expect for coastal landfill), where artificial clay liners are compacted in an oxidation condition and then gradually change to a reduction condition by the biogeochemical reactions between soil-leachate systems.

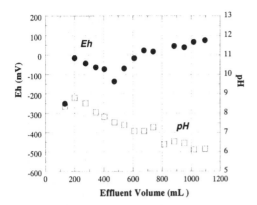

Figure 6   Change in Eh and pH of Effluent with Effluent Volume

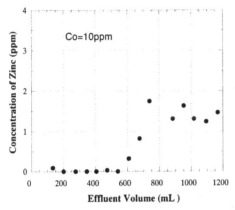

Figure 7   Change in Zn Concentration of Effluent with Effluent Volume

In this study, the hydraulic conductivity tests were conducted in such a way that samples were first consolidated and then reduced during the permeation duration. Although this test procedure was similar to the second method (consolidation prior to reduction) employed by Shen et al. (1992), the hydraulic conductivity increased, which is different from the results obtained by Shen et al. (1992). This might be attributed to the fact that the soil samples used had different redox conditions: Smectite used by Shen et al. (1992) was in an oxidation condition. When it was exposed to reducing agents, it tended to shrink and increases its hydraulic conductivity, although the mechanism of shrinkage is not clearly unknown. Unlike smectite used by Shen et al. (1992), the marine clay used in this study existed in a strongly reduced condition (e.g., Eh measured in soil suspension was −377 mV). Marine clay had existed naturally in the strongly reduced environment, which is long enough to let the redox reactions in soil-water systems reach equilibrium. Thus, even when additional reducing

agent is added, marine clay will not change its microstructure and properties.

4.5 Leaching of Zinc

Figures 6 and 7 show the leaching test results conducted on sample A. From Figure 6, Eh of effluent increased with an increase of effluent volume. The Eh was less than 0 mV when the effluent volume accumulated from 100 to 600 mL, and then increased from 0 to 75 mV after effluent was more than 600 mL. Simultaneously, pH of the effluent decreased from 8.4 to 7.1 first, and later it has been further decreased to 6.2. These mean that during the 80 days test period, the effluent changed from a reducing and weak alkaline condition to an oxidation and weak acidic condition.

Figure 7 indicates that the soluble zinc in effluent could not be detected before 600 mL volume of effluent was obtained, but after effluent volume exceeded 600 mL the soluble zinc increased significantly to 1.5 ppm and then continued to vary on the higher concentration level for the effluent volume of 740 to 1200 mL. From Figures 6 and 7, it is obvious that this concentration increase corresponds to a Eh change from < 0 mV to > 0 mV, and pH change from > 7.0 to < 7.0. This means that zinc was retained in the soil specimen under the reduction and weak alkaline condition, but became mobile and being leached out in an oxidation and weak acidic condition. Both Eh and pH have an influence on the mobility of zinc.

The formation of precipitations of zinc sulfide and zinc hydroxide is believed to be responsible for the retaining of zinc at early stage of permeation. Zinc may be hydrolyzed at pH > 7.0 to form a series of hydrolysis and these hydrolyzed species are strongly adsorbed to soil surfaces. Guo et al. (1997) also found that zinc become less mobile in reduction condition. They explained that, under oxidation condition, Zn was associated with Fe(III) and Mn(IV) oxides and soluble phases, whereas Zn was associated with insoluble sulfides, large molecular humic compounds and carbonates under reduction condition. Natural marine clay usually exists in a weak alkaline and strong reducing condition. Thus, this sediment condition is believed to be advantageous to retain zinc from leachate if solid waste is disposed at offshore landfill sites.

5 CONCLUSIONS

From the test results above, the following conclusions can be drawn:

(1) Marine clay obtained from Osaka Bay is a strongly reduced soil, which may poses a redox potential as low as –377 mV. Reducing agent has no obvious effect on the liquid limit and swell behavior of the marine clay.

(2) From the hydraulic conductivity test proposed to investigate the effect of redox potential on hydraulic conductivity and heavy metal leaching, redox potential has also no effect on the hydraulic conductivity of the marine clay used.

(3) The marine clay exhibited an excellent retaining capacity to zinc in its natural reducing and weak alkaline conditions. The formation of precipitates of zinc sulfide and zinc hydroxides is considered responsible for this immobility.

(4) Hydraulic conductivity of marine clay is as low as 10^{-7} cm/s when it is consolidated under effective stress of landfilled solid waste. Thus, the marine clay is considered to be an effective bottom clay liner in coastal waste landfill sites from the view of hydraulic conductivity and retaining capacity for heavy metal.

ACKNOWLEDGEMENT

The marine clay used in this study was donated by Bureau of Port and Harbor, Osaka City Government.

REFERENCES

Aburatani, S., Hayashi, Y., and Nishikawa, T. (1996): Offshore waste disposal by Osaka Bay Phoenix Project, *Environmental Geotechnics,* M. Kamon (Ed.), Balkema: 623-628.

Bjerg, P.L., Rugge, K., Pedersen, J.K., and Christensen, T.H. (1995): Distribution of redox sensitive groundwater quality parameters downgradient of a landfill (Grindsted, Denmark), *Environ. Sci. Technol.*, 29: 1387-1394.

Chen, S.Z., Low, P.F., and Roth, C.B. (1987): Relation between potassium fixation and oxidation state of octahedral iron, *Soil Sci. Soc. Am. J.*, 51: 82-86.

Christensen, T.H., Kjeldsen, P., Albrechtsen, H.-J., Heron, G., Nielsen, P.H., Bjerg, P.L., and Holm, P.E. (1994): Attenuation of landfill leachate pollutants in aquifers, *Crit. Rev. Environ. Sci. Tech.*, 24(2): 119-202.

Guo, T., DeLaune, R.D., and Patrick, Jr. (1997): The influence of sediment redox chemistry on chemically active forms of arsenic, cadmium, chromium, and zinc in estuarine sediment., *Environ. Intern.*, 23(3): 305-316.

Lovley, D. R. (1991): Dissimilatory Fe(III) and Mn(IV) reduction., *Microbiol. Rev.*, 55: 259-287.

Lyngkilde, J. and Christensen, T.H. (1992): Redox zones of a landfill leachate pollution plume (Vejen Denmark), *J. Contam. Hydrol.*, 10: 273-289.

Mitchell, J.K. and Madsen, F.T. (1987): Chemical effects on clay hydraulic conductivity, *Geotechnical Practice for Waste Disposal'87*, R.D. Wood (Ed.), ASCE, pp. 87-116.

Patrick, W.H., Jr., and Turner, F.T. (1968). Effect of redox potential on manganese transformation in waterlogged soil. *Nature* 220(5166): 476-478.

Rugge, K., Bjerg, P.L., and Christensen, T.H. (1995): Distribution of organic compounds from municipal solid waste in the groundwater downgradient of a landfill (Grindsted, Denmark), *Environ. Sci. Technol.,* 29:1395-1400.

Shackelford, C.D. (1994): Waste-soil interactions that alter hydraulic conductivity, *Hydraulic Conductivity and Waste Contaminant Transport in Soil*, ASTM STP 1142, D.E. Daniel and S.J. Trantwein (Eds.), ASTM, pp.111-168.

Shen, S., Stucki J. W., and Boast, C. W. (1992): Fe reduction and effect on hydraulic conductivity of Na-smectite, *Clays and Clay Minerals*, 40(4): 381-386.

Stucki, J. W. (1988): Structural iron in smectites, *Iron in Soils and Clay Minerals,* J. W. Stucki, B. A. Goodman, and U. Schwertmann (Eds.): 625-675.

Vanysek, P. (1999): Electrochemical Series, *CRC Handbook of Chemistry and Physics*, D.R. Lide (Ed.), CRC Press, Boca Raton, FL: pp. 8-21 to 8-34

Clay Science for Engineering, Adachi & Fukue (eds) © 2001 Balkema, Rotterdam, ISBN 90 5809 175 9

Cutoff wall construction using bentonite/ethanol slurry

M. Asada, A. Ishikawa & S. Horiuchi
Shimizu Corporation, Tokyo, Japan

ABSTRACT: For usage of cutoff walls, continuity, long-term stability and flexibility are crucial to minimize the permeability. Soil-bentonite mixtures meet such requirements, if more than 70 kg of bentonite per $1m^3$ soil could be homogeneously mixed into soil in-situ. Bentonite should be added to soil in a slurry state for homogeneous mixing, however, 20 kg/m^3 of bentonite would be the largest amount by bentonite/water slurry. To achieve increasing the bentonite content, utilization of hydrophilic organic solvent was studied. In this paper, high applicability of ethanol bentonite slurry is reported based on the following results; (1) bentonite content of 860 kg/m^3 slurry can be prepared by ethanol, (2) liquid quantity required for preparing the slurry can be reduced to 1/6 ~1/7 when mixing the same weight of bentonite with ethanol instead of water (3) the substitution of ethanol to water decreases the mixture's permeability and increases its strength. A high potential of the ethanol/bentonite slurry for grouting is also recommended.

1 INTRODUCTION

1.1 *Cutoff Walls in Japan*

Several types of cutoff walls have been developed and used for the following applications in Japan; (1) containment for waste disposal facilities, (2) measure for liquefaction by lowering the area's groundwater level surrounded by the cutoff wall, (3) earth retaining for excavation. Conventional non-permeable walls are constructed using clay, bentonite, soil-cement, mortar, concrete, reinforced concrete, grouting chemicals, and sheet piles. For usage of cutoff walls mainly to contain waste disposal facilities, continuity, long-term stability and flexibility to the earth deformation are crucial to minimize the permeability and the environmental risk.

The soil-bentonite (SB) walls are commonly constructed for the seepage barriers mainly in the United States. A narrow trench, typically 0.75 to 0.90m wide, is excavated using bentonite/water slurry. The trench is subsequently backfilled with soil-bentonite mixtures that forms a barrier of very low permeability (Panjan et al. 1995). This method, however, is rarely used in Japan, because the walls have low strength and they may cause the settlement of the surrounding areas. Sheet piles with polymer sealant for their connections are often used for seepage barriers, but they could hardly keep the adequate permeability.

The mixed in place (MIP) walls are commonly used for earth retaining in Japan. Cement/bentonite (CB) slurry is the material to construct the MIP walls. This method uses three or five-axis hollow stem augers that churn and mix the soil as the CB slurry is introduced through the base of the augers. The hydraulic conductivity of the CB-MIP walls is larger than that of the SB walls (typically 10^{-5} to 10^{-6} cm/s) (Oweis & Khera 1998), and they have no flexibility to the earth deformation.

If the MIP walls could be constructed by bentonite slurry, instead of CB slurry, and an enough amount of bentonite could be homogeneously mixed, they could have high flexibility and their hydraulic conductivity could be lowered less than 10^{-7} cm/s.

1.2 *Bentonite/solvent Slurry*

Bentonite is used for a wide range of application in the geotechnical and environmental field because of their low permeability, high plasticity, high adsorption capacity and rheological qualities. The clay mineral montmorillonite, the main constituent of bentonite, shows a large swelling property by contact with water; the interlayer cations begin to hydrate (Koerner et al. 1995), and a large volume of water penetrate into the layers. Six times weight of water is necessary to separate each montmorillonite layer; bentonite should be mixed with 600wt% water in order to send the slurry by grouting pumps.

Minimum amount of bentonite required for the non-permeable wall, k<1x10⁻⁶cm/s, would be more than 70 kg per 1m³ soil (Chapuis 1990). Constructing walls by MIP method, bentonite should be added to soil in a slurry state for homogeneous mixing, however, 20 kg/m³ of bentonite could be the largest amount by using conventional bentonite/water slurries.

To increase bentonite content in the soil in-situ, it is essential to increase the bentonite concentration in the slurry. When a pore fluid has a low dielectric constant such as hydrophilic organic solvent, it will have a tendency to curtail bentonite swelling. Therefore, the slurry's viscosity decreases, if organic solvents are used for the bentonite slurry. This means bentonite content both in the slurry and the soil mixture could be increased by using organic solvent. The hydrophilic solvent could be easily replaced to water, and begins to increase its viscosity and strength. The strength of the soil-slurry mixtures is expected to be almost the same to the swelling pressure.

To achieve increasing the bentonite content in the slurry, utilization of hydrophilic organic solvent, such as alcohols and ketone, were studied. A high potential of the bentonite slurry for grouting was also studied.

2 EXPERIMENTS

2.1 Materials

The investigation was carried out on seven types of bentonite. Their grain size is given in Table 1. E-1~3 are the same type of bentonite except for their grain size. They are made by several companies, and have different content of montmorillonite. Five solvents were used for bentonite mixtures; i.e., acetone, ethanol, ethylene glycol, methanol, and water. Table 2 shows their density, viscosity, and dielectric constant.

Table 1. Properties of bentonites

Bentonite Type	Grain Size (Mesh)
A	250
B	250
C	300
D	300
E- 1	250
E- 2	300
E- 3	500

Table 2. Properties of solvents

Solvents	Density (t/m³)	Viscosity (mPa * s)	Dielectric Constant
Acetone	0.791	0.3	20.7
Ethanol	0.789	1.2	23.8
Ethylene-Glycol	1.116	25.7	38.7
Methanol	0.791	0.6	33.1
Water	0.998	1.0	80.1

2.2 Procedures

2.2.1 Viscosity test of bentonite/solvent slurry
Bentonite/solvent slurry was prepared using a small size hand mixer. Viscosity of the slurry was determined by the funnel viscometer (JBAS 108-77); viscosity is indicated by the time required for discharging 500mL slurry from the specific funnel. The time required to discharge 25℃ water is 18.3 seconds. Appropriate funnel viscosity for the conventional pumping equipment is 20~120s. Water/bentonite or solvent/bentonite ratio means water or solvent weight per bentonite weight in the slurry. Alcohol content to be mixed with bentonite was varied from 0 to 100%.

2.2.2 Permeability test
Permeability of the bentonite/solvent slurry mixture with soil was determined by the steady head permeation test. Toyoura sand (grain size = 0.1~0.5mm, water content = 30%) was used for the soil. Test samples were mounted in the permeameter cell (diameter=10cm) with the initial height of 10~17cm, and were subjected to a water pressure at hydraulic gradient=100~200. Table 3 shows the specs of the test samples.

Table 3. Components of permeability test samples

Sample Number	Bentonite slurry		Sandy soil	
	Bentonite	Ethanol	Sand	Water
No. 1	5	7.5	100	30
No. 2	10	15.0	100	30
No. 3	20	30.0	100	30
No. 4	30	45.0	100	30

2.2.3 Swelling potential test
Swelling potential of the bentonite/ethanol slurry was determined by the standard consolidation test (JSF T 411-1990). Confining pressures of 1-10kPa were exerted to the mixture of bentonite C in the consolidation cell. Bentonite was mixed with 89 and 100% of ethanol by weight. Initial heights of the samples were set 1.2~1.8cm in the consolidation cell. Only the upper side of the samples was exposed to water.

2.2.4 Underwater placement test
To check the applicability of the bentonite/ethanol slurry for grouting or backfilling, the slurry was placed underwater by pumping in a test tank filled with water. According to the test results obtained for flyash slurry placement, slump value suitable for underwater placement by pumps was reported to be about 11cm by half size slump cone (Horiuchi et al. 1992). Change of the surface shape of the slurry confronted to water was observed. In this test, bentonite C was mixed with 89 and 100% of ethanol by weight, which shows 11cm and 13.5cm slump.

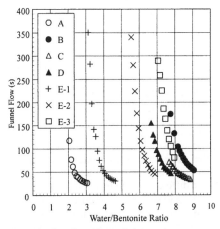

Figure 1. Viscosity of bentonite/water slurry

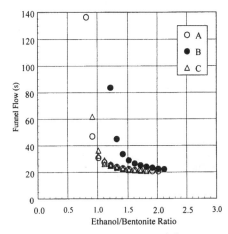

Figure 4. Viscosity of bentonite/ethanol slurry

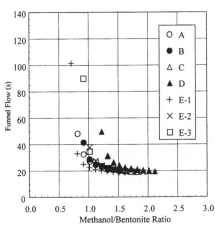

Figure 2. Viscosity of bentonite/methanol slurry

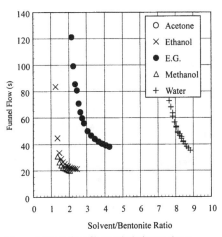

Figure 5. Viscosity of bentonite/solvents slurry

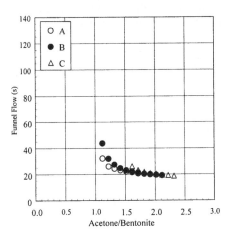

Figure 3. Viscosity of bentonite/acetone

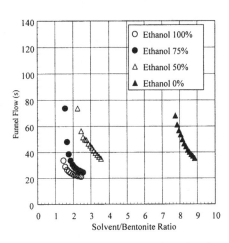

Figure 6. Ethanol concentration and slurry viscosity

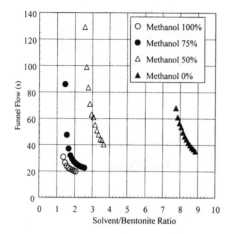

Figure 7. Methanol concentration and slurry viscosity

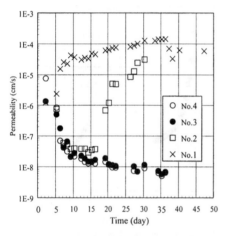

Figure 8. Permeability test of the slurry/sand mixture

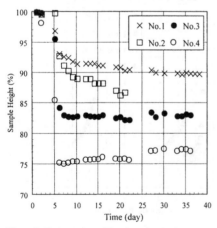

Figure 9. Compression of the slurry/sand mixture

3 RESULTS

3.1 *Viscosity test of bentonite/solvent slurry*

Figure 1 shows the viscosity of the bentonite/water slurry at each water/bentonite ratio. As seen on E-1~3, the finer the bentonite grain size is, the higher the slurry viscosity becomes. Fine bentonite has larger surface area, and needs more volume of water to be swelled. Bentonite type affects the viscosity; bentonite B slurry is the most viscous and needs the largest volume of water. Bentonite A slurry, on the other hand, shows the least viscous. One of the reasons is considered that each types of bentonite have different montmorillonite content.

Figure 2 shows the viscosity of the bentonite/methanol slurry. The slurry's viscosity dramatically decreases compared with that of the bentonite/water slurry. Methanol has the ability to reduce the slurry viscosity. Bentonite type, however, has less effect on slurry viscosity. Viscosity difference of the each bentonite slurry is not conspicuous as that of bentonite/water slurry. Only the solid content (methanol/bentonite ratio) in the slurry is the factor to change the viscosity. This is because the interaction between montmorillonite and methanol is not strong, and the montmorillonite content in the bentonite does not affect the viscosity difference. Figure 3 and 4 show the viscosity of the bentonite/acetone slurry and bentonite/ethanol slurry, and the same tendency is observed.

Figure 5 shows the effect of solvent type on slurry viscosity of bentonite C /solvents slurry. Comparing the five solvents listed on Table 2, the order of slurry viscosity is as follows; water >> ethylene glycol > acetone = ethanol = methanol. This tendency was also observed in the other test using the other types of bentonite. Interaction between montmorillonite and solvent, such as particle repulsion and solvent affinity, would be the main factor for the viscosity difference.

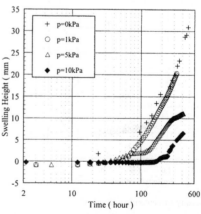

Figure 10. Swelling height curve of the E/B=100% slurry

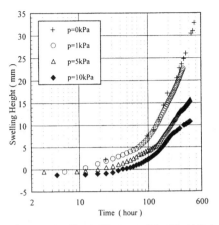

Figure 11. Swelling height curve of the E/B=89% slurry

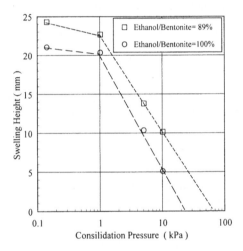

Figure 12. Consolidation pressure and swelling height

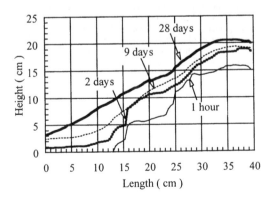

Figure 13. Surface change of the E/B=89% slurry

Ethanol is used for the following tests, because it is easy to be decomposed biochemically and is not harmful to underground environment. Comparing at 60s on funnel viscosity, 100wt% of bentonite can be mixed in the slurry by ethanol. This means that the weight of the solvent for making the slurry can be reduced to 1/6 ~1/7 when mixing same weight of bentonite with ethanol instead of water, and bentonite content of 860kg/m^3 slurry can be prepared by ethanol.

Figure 6 and 7 show the effect of solvent concentration on the viscosity of the bentonite slurry. In proportion to decrease the alcohol concentration, slurry viscosity increases at the same ratio. As describing at Figures 1 ~ 4, ethanol, methanol, and acetone do not swell bentonite. When the alcohol in the slurry is diluted by water, bentonite begins to swell and increase its viscosity. If the bentonite/solvent slurry is retained at the same place in the ground and pore alcohol is replaced by the groundwater, slurry viscosity increases and the strength of the soil-slurry mixtures increases. The strength of soil-slurry mixtures with enough content of bentonite could be almost the same strength of the in-situ soil without any slurry, because of the swelling potential of the placed bentonite in the mixtures is expected to be the same level as that of alluvium soils.

3.2 *Permeability test*

Figure 8 illustrates the permeability test result of the sand-slurry mixture. Hydraulic gradient was set 100 before 19 days, and was increased to 200 after 19days. Permeability of the sample No.1 (Bentonite/sand = 5wt%) increases gradually, because the pressurized water flushes out the bentonite. Permeability of the sample No.2 (Bentonite/sand = 10wt%) decreases gradually at the hydraulic gradient 100, and increases at the hydraulic gradient 200. Permeability of the sample No.3 and 4 (Bentonite/sand = 20 and 30wt%) decreases all the time at the hydraulic gradient both 100 and 200. The hydraulic conductivity at the steady state is smaller than 1 x 10^{-8}(cm/s).

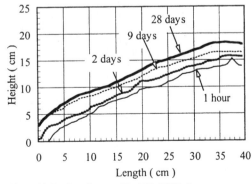

Figure 14. Surface change of the E/B=100% slurry

According to the recent permeability test results using other types of bentonite, permeability increase was not observed at hydraulic gradient 200, however. The bentonite selection is important to prevent the flushing out.

Figure 9 shows the consolidation of the samples during the permeability tests. Sample height of No.1 (Bentonite/sand = 5wt%) decreases to 90%, and that of No.4 (Bentonite/sand = 30wt%) decreases to 75% of the initial height in the cell. No.4 has larger bentonite content, and it shows swelling after 5 days.

3.3 Swelling potential test

Figure 10 and 11 show the swelling height change of the E/B = 100% and 89% slurry during the consolidation tests. Just after the start, the slurry indicates some settlement by the consolidation pressure. Swelling of the bentonite begins after several hours. Swelling height decreases as the consolidation pressure increases. Figure 12 indicates the relationship between the swelling height and consolidation pressure after 324 hours. Swelling height of the E/B=89% and E/B=100% slurry becomes 0mm at the consolidation pressure 70kPa and 25kPa. Consequently, swelling pressure of the fixed slurry (E/B=89~100%) in the ground indicates 25~70kPa, if the pore ethanol is replaced by groundwater. According to the long time swelling tendency in both Figures 10 and 11, the swelling pressure of the slurry increases with time.

Vane test was conducted for the samples (consolidation pressure=1kPa) after the test. Shear strength of the samples is about 2kPa. As the water replaces the pore ethanol, the samples develop strength. The upper portion of the sample that is confronted with water is stronger than the lower portion that is far from waterfront.

3.4 Underwater placement Test

Bentonite/ethanol slurry E/B=89% and 100% was placed by pumping in a test tank filled with water. Change of the surface shape of the slurry confronted with water was observed. Figures 13 and 14 show the surface view of the placed slurry underwater. The surface has many bumps like fold whose heights are 10~20mm just after the placing. E/B=89% slurry has more bumps and folds. Slurry surface that is confronted with water begins swelling, and the surface became smoother as the pore ethanol is replaced by water.

As slump value decreases, the inclination of the placed slurry becomes steeper as a result of the slurry viscosity. E/B=89% slurry is so viscous that it could not keep the surface shape of the slope just after the placing; however, the slope break down could be controlled by the slurry viscosity and the

mix design. Bentonite/ethanol slurry is concluded to be suitable for grouting and backfilling.

4 CONCLUSION

In this paper, high applicability of bentonite/ethanol slurry is reported based on the following test results; (1) bentonite content of 860 kg/m^3 slurry can be prepared by ethanol, (2) liquid quantity required for preparing the slurry can be reduced to 1/6 ~1/7 when mixing same weight of bentonite with ethanol instead of water (3) the pore ethanol is easily replaced by the water from the soil surrounding the mixture, (4) this substitution of ethanol to water decreases mixture's permeability and increases strength, i.e., test samples contained 100 kg/m^3 of bentonite shows permeability 1×10^{-8} cm/s and 2 kPa of shear strength after the substitution.

A high potential of the ethanol bentonite slurry for grouting is also recommended by the following properties; (5) the slurry can be easily placed in the water and shows 25~70 kPa swelling pressure by the substitution of ethanol to water, (6) this swelling pressure makes the contact surface with water smooth, and this property enhances the suitability of the bentonite/ethanol slurry for the grouting material.

REFERENCES

Chapuis, R. P. (1990): Sand-bentonite liners: predicting permeability from laboratory tests, Can. Geotech. J., Vol.27, pp.47-57.
Horiuchi, S., Taketuka, M., Odawara, T. and Kawasaki, H. (1992): Fly-ash slurry island: I Theoretical & experimental investigations, J. Materials in Civil Eng. ASCE, pp.117-133.
Koerner, R. M., Gartung, E., and Zanzinger, H. (1995) Geosynthetic clay liners, Rotterdam, A. A. Balkema
Oweis, I. & Khera, R.(1998) Geotechnology of waste management, Boston, PWS Publishing Company
Panjan, N. S. et al. (1995) Permeability and compressibility behavior of bentonite-sand/soil mixes, Geotechnical Testing Journal, Vol.18, No.1, pp86-93

Clay Science for Engineering, Adachi & Fukue (eds) © 2001 Balkema, Rotterdam, ISBN 90 5809 175 9

Assessment of self healing properties of Geosynthetic Clay Liners

G. L. Sivakumar Babu, H. Sporer, H. Zanzinger & E. Gartung
Geotechnical Institute (LGA), Nuremberg, Germany

ABSTRACT: In the recent years, most capping systems of waste containment facilities have a layer having low hydraulic conductivity to restrict the movement of fluids into waste body. These covers containing Geosynthetic Clay Liners (GCLs) are intended to have sufficient capacity for containment of precipitation, and self healing properties enabling closure of cracks in the case of desiccation. Containment as well as sealing effect of GCLs are largely influenced by swelling behaviour of bentonites used in GCLs. The swelling behaviour, apart from mineralogical and physico-chemical characteristics, is also a function of binding method of GCLs. Swell potential and swell pressure tests under different effective stresses were conducted for different types of bentonite materials used in the manufacture of GCLs in Germany. A few tests were also conducted on both stitch bonded GCLs and needle punched GCLs. The paper presents some results on swell pressure and swell potential of different types of GCLs and suggests an approach for assessment of self healing properties. The contribution of type of bonding to swelling and self healing behaviour is also studied.

1 INTRODUCTION

Capping systems for landfills are intended to have sufficient capacity for containment of precipitation, and self healing properties enabling closure of cracks in the case of desiccation. Containment as well as sealing effect of GCLs are largely influenced by swelling behaviour of bentonites used in GCLs. The swelling behaviour, apart from mineralogical and physico-chemical characteristics, is also a function of binding method of GCLs. In this investigation, swell tests conducted in oedometers are used to examine the above aspects.

2 BACKGROUND INFORMATION

For a satisfactory cover performance, cover material should have sufficient capacity for containment of precipitation and the ability to self heal if cracks exist. The capacity for containment of precipitation is reflected in terms of moisture absorption capacity. For example, Daniel and Shan (1993) performed hydration tests on GCL: Sand layer combination under 14 kPa effective pressure and showed that bentonite is effective in water uptake and water content was in the range of 193% from an initially dry state (wilting point). Bonaparte et al.

(1996) performed hydration tests on three different GCL products in contact with compacted clay and reported that after 75 days, the water contents of GCLs were in the range of 90%. Their results also suggested that the water uptake is not significantly affected by applied pressure in the range of 5 to 390 kPa. Sridharan and Nagaraj (1999) conducted a detailed study on absorption water contents of various soils including bentonites and reported that initially dry samples (at shrinkage limit) can rehydrate quickly to water contents in the range of 90% of the corresponding liquid limit values.

Cracks occur due to desiccation or other environmental effects such as changes in pore fluid chemistry of the bentonite (ion exchange, pH, concentration of ions in the pore fluid, temperature). For example, Melchior (1993) observed high leakage rates in a final cover test site and reported that swelling capacity of Na-bentonite in the GCL was reduced to values typical of Ca-bentonite and indicated that reduced swell capacity of the GCL was insufficient to seal preferential flow paths formed due to desiccation. Insights into soil behaviour as a result of changes in pore fluid chemistry in clay minerals and clay soils can be examined by diffuse double layer theory (Mitchell, 1993). Cracks are also due to differential settlements within landfill waste body which get reflected in the cover performance. For what ever reasons cracks occur, there is a need to assess self healing capacity of GCLs in quantitative terms. Shan and Daniel (1991) examined the self healing of bentonites by making three holes of different diameters (12 mm, 25 mm and 75 mm) in specimens of 152 mm diameter and allowing the samples to rehydrate under a confining stress of 14 kPa. Measured hydraulic conductivities indicated that bentonites rehydrated to fill 12 mm and 25 mm diameter holes and there was no significant change in hydraulic conductivity and the values were in the range of 2×10^{-11} m/s. How ever, for 75 mm diameter hole, hydraulic conductivity increased to 2×10^{-06} m/s.

The method of GCL manufacturing is another important variable affecting the swelling behaviour of GCLs. Rowe and Lake (1999) showed that the final void ratios in a needle punched GCL are much less than the corresponding values in a similar GCL without needle punching at the same effective stress. To ensure proper performance of GCLs, Zanzinger (1995) suggested that the montmorillonite content, water absorption, swelling volume be evaluated as a part of overall quality assurance plan. The objective of this paper is to suggest that, in addition to the above, parameters of swelling behaviour of bentonites, determined from oedometer tests be included. They are useful for assessment of i) water absorption capacity ii) self healing and iii) determine relationship between water absorption capacity and overburden pressure for covers. The following sections describe the materials and methods, test results and discussion.

3 MATERIALS AND METHODS

3.1 Materials

In this investigation, swell tests in oedometers were conducted on different types of GCLs. The properties of GCLs are given in Table 1.

Table 1. Properties of GCLs

Type of GCL	[g/cm³]	[ml]		
	(1)	(2)	(3)	(4)
Stitch bonded				
Sodium bentonite 1	0.8	33	475%	73%
Sodium bentonite 2	0.8	25	415%	73%
Needle punched				
Sodium bentonite 3	0.8	24	580%	75%

1-Bulk density, 2- Swell index, 3-Water absorption, 4-Montmorillonite content (from Methylene blue adsorption test)

The GCL products are either stitch bonded or needle punched to increase the shearing resistance at the mid plane of the GCL.

3.2 Methods

Two types of tests were conducted, viz., a) constrained swell pressure tests, percent swell tests and b) unconstrained swell pressure tests. In both the tests, percent swell and swell pressure were determined. The test procedures similar to conventional swell pressure tests on swelling soils are followed in this investigation.

3.2.1 Constrained swell pressure and percent swell tests

Samples of 100 mm diameter were cut from the GCL lots and care was taken to ensure no loss of bentonite in trimming and sample handling. Samples were set up under a seating pressure of 2 kPa and deionised water (pH of 7) is added and restraint was imposed on them such that there is no volume change during swelling. Water was allowed and the corresponding swell pressures developed were monitored. This condition is representative of in situ situation where overburden is high enough to prevent total swelling. If overburden is less than the swelling pressure, heaving is likely. Amount of percent swell at different effective stresses was also measured. Swell pressure tests were also conducted on bentonite samples of the GCLs.

3.2.2 Unconstrained swell pressure tests

In this case, samples were allowed to swell and loads were applied subsequently till original thickness of the sample is reached. The time for complete swelling is in the range of 7 to 10 days depending on the type of GCL. Once the equilibrium is reached, each sample is loaded gradually over time and the sample is brought back to original thickness and the corresponding pressure in each case is determined. Two sets of experiments were conducted, i.e., one set with stitches and needling intact and in the other case, stitches and needling removed.

4 TEST RESULTS AND DISCUSSION

4.1 Constrained swell pressure and swell tests

Fig. 1 shows the relationship between development of swelling pressure in relation to dry density and void ratio for the bentonite powders of the GCL samples. Fig. 2 (a) shows the relationship between final water content and effective stress for stitch bonded GCLs. At low effective pressures, the water absorption is quite high. Fig. 2(b) shows the relationship between thickness of the sample tested and effective stress. Fig. 2(c) shows percent swell under different effective stresses. Perusal of Fig. 2(b) and 2(c) suggest that for normal cover loads in the range of 15 kPa, it is likely that percent swell is in the range of 45 to 60%. Hence, considering that the thickness of the sample is in the range of 10 mm, if a crack exists in identical situation, in case of precipitation, it will heave to an extent of 60%. Closure of crack as a result of swelling is close to one dimensional swell and a crack of 12 mm is likely to self heal with 60% of heave occurring from both the sides of the crack in a sample of 10 mm thick and continuity in width. Crack widths beyond self healing capacity are unlikely to close and the behaviour is similar to the results of Shan and Daniel (1991).

than the previous state, hence the swell pressure is less. Fig. 4 shows the results for needle punched GCL (type 3 bentonite). The water absorption characteristics are good in both the cases. Under a seating stress of 2 kPa, the percent swell is in the range of 100% and the corresponding swell pressure is 306 kPa. Once the threads are removed and tested, the percent swell is in the range of 200%.

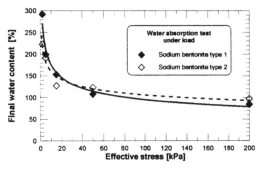

Fig. 2 (a) Relationship between final water content and overburden pressure for stitch bonded GCLs

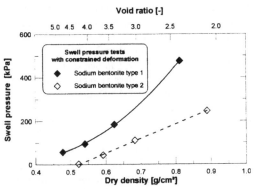

Fig. 1 Relationship between swelling pressure, dry density and void ratio for bentonite samples

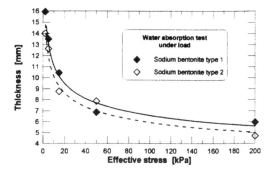

Fig. 2 (b) Relationship between thickness of the sample and effective stress for stitch bonded GCLs

4.2 Unconstrained swell pressure tests

These tests were conducted on stitch bonded and needle punched GCL samples. Fig. 3 shows test results for sodium bentonite stitch bonded GCL (Type 2 bentonite). It can be observed that the percent swell is in the range of 90% under a seating pressure of 2 kPa. Results for GCL with stitches removed are also shown in Figure. It shows that the percent swell is marginally higher. Swell pressures determined, applying the stresses gradually are also given (124 kPa and 204 kPa). As the percent swell in case where stitches were removed is marginally more, the final void ratio is slightly more

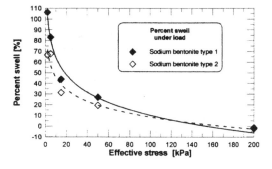

Fig. 2(c) Percent swell versus effective stress for stitch bonded GCLs

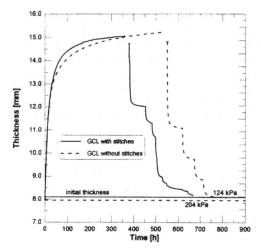

Fig. 3 Results for sodium bentonite stitch bonded GCL (type 2 bentonite)

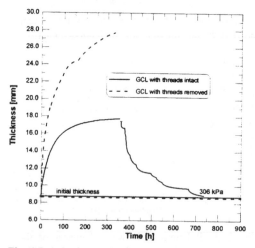

Fig. 4 Results for needle punched GCL (type 2 bentonite)

Hence, it appears that threads at closer intervals and in a random manner as in the case of needle punched GCL are more effective in sealing desiccation cracks, if the internal shear strength of GCL in the absence of threads is sufficient for cover stability. Detailed studies considering ion exchange conditions, temperature and pore fluid chemistry are necessary in evaluating differential percent swells and the corresponding role in self healing process.

5 CONCLUDING REMARKS

The paper addresses the aspects of water absorption, sealing of cracks and the role of type of manufacture of GCLs. Swell pressure and percent swell tests under simulated loading conditions in oedometer are useful means in this direction.

6 ACKNOWLEDGEMENTS

First author thanks Alexander von Humholdt Foundation, Bonn for financial support and the authorities of LGA for providing the facilities to carry out the work.

REFERENCES

Bonaparte, R Othman, M.A. Rad, N. S. Swan R. H and vander Linde D. L. (1996) *Evaluation of various aspects of GCL performance*, workshop on Geosynthetic Clay Liners, (Eds. D. E. Daniel and H. B. Scranton), F1-F34.

Daniel, D. E. Shan, H. Y and Anderson, J. D. (1993) *Effects of partial wetting on the performance of bentonite component of Geosynthetic Clay Liner*, proceedings of Geosynthetics, 93, Industrial Fabrics Association International, St. Paul, Minn.3, 1483-1496.

Melchior, S (1993) *Water balance and efficiency of multiple surface liners systems for landfills and contaminated sites*, Ph.D. thesis, University of Hamburg, Germany.

Mitchell, J. K (1993) *Fundamentals of soil behaviour*, 2nd edition, John Wiley Inc.

Rowe, R. K and C. B. Lake (1999) Geosynthetic clay liner research, design and applications, Proceedings, Sardinia 99, *Seventh International Waste management and Landfill symp.* Cagliari, 181-188.

Shan, H. Y. and Daniel, D. E. (1991) Results of laboratory tests on geotextile/bentonite liner material, *Proceedings of Geosynthetics, 91*, St.Paul, MN, 2, 517-535.

Sridharan, A. and Nagaraj, H. B. (1999) Absorption water content and liquid limit of soils, *Geotechnical Testing Journal*, Vol. 22(2), 127-133.

Zanzinger, H. (1995) Quality assurance in the manufacture of GCLs, *Geosynthetic Clay Liners*, (Eds. R. M. Koerner, E. Gartung and H. Zanzinger), Balkema, 219-228.

Clay Science for Engineering, Adachi & Fukue (eds) © 2001 Balkema, Rotterdam, ISBN 90 5809 175 9

Use of fine fractions after sand extraction as clay liner

M. Fukue, T. Ide & T. Minato
Marine Science and Technology, Tokai University, Shimizu, Japan

T. Ohba
The Shizuoka Association for Natural Resources and Environmental Protection, Shizuoka, Japan

H. Sasaki
Technical Research Institute, Tobishima Company, Higashi Katsusika-gun, Chiba, Japan

ABSTRACT: Because the excavation of sand and gravel from river and sea bottoms has been prohibited in most places of Japan, aggregates for concrete are obliged to obtain by separating coarse grains from soils. Under the regulation, clay materials produced by separating have been regarded as a kind of wastes. Since the amount of the waste materials is huge, its utilization has to be considered. In this study, considering the characteristics of the wastes, such as low permeability, sorption ability, plastic behavior, etc, the materials produced are examined if they can be utilized as clay liner. The results showed that the permeability and sorption characteristics of heavy metals satisfy requirements for the quality of clay liner.

1 INTRODUCTION

For environmental protection, the excavation of aggregates from river and sea bottoms has been prohibited in most places of Japan. This has led shortage of aggregates for concrete. At present, aggregates are produced from separating of coarse fractions from sandy soils in many sites in Japan. The fine-grained soils remained during the separating process are regarded as waste materials by regulations, nevertheless they are natural soils.

The shortage of waste disposal sites is also a serious problem in Japan. There is little place available for dumping the remained soils. As it is available, the cost for waste disposal results in higher price of aggregates.

Ministry of Construction promotes the utilization of waste materials. However, the material must satisfy the requirements for reuse. In general, soft materials produced during construction are reused by means of improvement. If not, the materials still remain as wastes. Fine-grained materials have been considered to be improper construction materials, because of the low strength and low bearing capacity. However, there is an idea that the characteristics of soft soils can be properly used with other intention. The main features of fine-grained soils are low permeability, a large specific surface area, sorption capacity, etc. These characteristics may be useful for clay liner.

One of the major requirements for clay liner is that they meet the required permeability. The specifications will focus on the appropriate materials, construction procedures, and testing required to achieve the desired permeability. The governing factors on the permeability of clay liner are a) soil type, b) density or void ratio, c) water content, d) degree of saturation, e) thickness of soil, f) existence of micro cracks, etc. Most of these factors are responsible for the field experiments and construction procedures. Consequently, final decision for the selection of materials should be conducted after waited until field experiments are performed. However, the potential of materials as clay liner can be examined in the laboratory testing and the results will also occupy important parts for the final decision (Sharma and Lewis. 1994).

In this study, the materials produced by separating of fine fractions and coarse aggregates from soils are examined whether they can be properly utilized as clay liner, in terms of physical and chemical properties of the materials.

2 MATERIALS

For aggregate production, coarse fractions are separated by means of centrifugal force under water. Therefore, the deposited fine-fractions in the container have very high water content. These fine-fractions are generally pressed and solidified up to the state of mud cakes as shown in Fig, 1 The solidified materials have a water content ranging from around 20 to 40 %. Table 1 shows physical properties of the mud cakes obtained from six plants around Fuji city in Japan. The original soils were obtained form different sites. Therefore, the grain size and other physical properties of the soils are different.

Fig.1 Solidified fine-grained soils produced by removing coarse grained-soils.

Table 1 shows that all the samples are silty clay or clayey silt with little sand fractions, because of the removal of coarse fractions. From the density of particles and ignition loss, the materials may consist of mostly mineral fractions. The shrinkage limits are higher than the plastic limits for all the samples, against the definitions. However, it is known that for soils containing a large amount of silt fractions, the shrinkage limit is often higher than plastic limit.

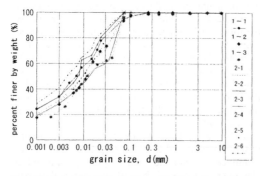

Fig.2 Grain size distributions of the materials.

Figure 2 shows grain size distributions for the nine samples. Most of the samples contain silt fractions more than 50 %.

The plasticity chart is shown in Fig.3. The figure indicates that the plots of the samples exists near A-line.

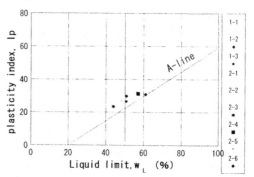

Fig.3 Plasticity chart

3 TESTING AND RESULTS

3.1 Permeability

It was mentioned that one of the important factors for clay liner is permeability. The Japanese regulation provides that the clay liner for a thickness of permeability k less than 10^{-6}cm/s. However, if k is measured in laboratory test, because in-situ permeability is usually higher than the value obtained in laboratory test. This is, however, an ambiguous expression, because of no standard for test method available for such a low permeability. Therefore, the permeability of clay liner may have been measured with individually developed apparatus.

In this study, the permeability of the materials was obtained from the standard consolidation test using oedometer. The advantage of this method is that the hydraulic conductivity can be obtained in terms of the change in void ratio.

Table 1 Physical properties of samples.

sample No.	1-1	1-2	1-3	2-1	2-2	2-3	2-4	2-5	2-6
density of particles (g/cm3)	2.69	2.72	2.72	2.70	2.73	2.73	2.76	2.75	2.77
liquid limit (%)	52.0	61.0	51.0	63.8	46.8	44.0	57.1	48.6	50.9
plastic limit (%)	24.8	30.2	24.6	31.2	21.5	20.6	25.8	22.0	21.1
plasticity	27.2	30.8	26.4	32.5	25.3	23.4	31.3	26.6	29.8
sand (%) >0.074mm	5.0	6.0	4.5	0.0	0.2	0.1	0.0	0.1	0.1
silt (%)	50.0	59.0	65.0	58.6	62.4	55.0	49.2	60.0	46.9
clay (%) <0.005mm	45.0	35.0	30.5	41.4	37.4	44.9	50.2	39.9	53.0
Ignition loss (%)	3.99	5.09	4.50	–	–	–	–	–	–
pH	–	–	–	7.98	8.08	8.44	8.28	8.29	8.03
shrikage limit (%)	–	–	–	48.1	34.9	32.1	39.1	43.6	42.0

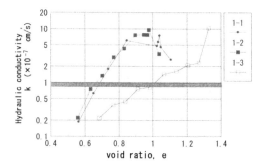

Fig.4 Hydraulic conductivity obtained

Figure 4 shows indirect measurement of permeability. The figure shows that the k decreases with decreasing of void ratio, as expected. To obtain a k value less than 10^{-7} cm/s, the void ratio should be less than approximately 0.7, whilst it is achieved at a void ratio of one for No.1-3 sample. although it is rough estimate due to the indirect measurement. It becomes further requirement to evaluate the in-situ permeability. The difference between the k-e relations may result from the different grain size distributions. It should be noted that the k value would vary with water content and degree of saturation and also structure associating to compaction. The variation of those factors should also be taken into account.

3.2 Compaction

When the materials are utilized as clay liner, the compaction characteristics are very important in terms of the field construction procedure. From the compaction test, the optimum water content and maximum dry density of materials are obtained.

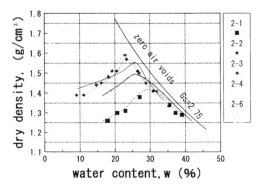

Fig.5 Compaction curves of the materials

Figure 5 shows the compaction curves of the materials. The figure shows that the optimum water content and maximum dry density varies with the producing area of individual soil, as described earlier.

The dry density varies from 1.45 to 1.6 g/cm^3 and the optimum water content ranges between 22 and 30 %. The optimum water contents are relatively lower than the water content of the solidified mud cake. This means that materials have to be dried to obtain the optimum water content.

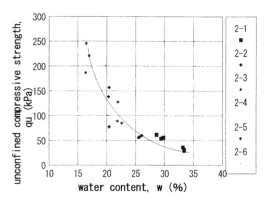

Fig.6 Unconfined compressive strength versus water content.

3.3 Unconfined compressive strength of compacted materials

Figure 6 shows relationship between the unconfined compressive strength and water content for the compacted materials. The unconfined compressive strength decreases with increasing of water content. The variation due to the type of materials is small. Therefore, approximate unconfined compressive strength of the materials can be estimated from the water content alone.

The figure shows that if the water content is less than 20 %, the unconfined compressive strength will exceed 100 kPa.

3.4 Sorption of heavy metals

The sorption of hazardous substances to clay particles is one of the important functions for utilization of clay liner. If the adsorption onto clay particles is expected, the use of clay liner will be safer.

The adsorption of soils are very complicated in terms of soil type, type and concentrations of adsorbents, pH, etc. The method of adsorption test is also important for evaluating the adsorption of soils. In this study, the batch test was performed to evaluate the adsorption of sample. Although the batch test may not provide the accurate adsorption of soils, it is useful for a quick and easy examination.

The standard solution of cadmium, lead, zinc and copper was added one of the samples. The liquid to solid ratio was 10, where the solid was 10 or 50 g. pH values were adjusted at 5 and 10 before the test. The shaking time was 6 hours. After shaking, the solution was filtered and then subjected to atomic absorption analysis.

Table 2 Sorption of heavy metals

pH solid	Added liquid (mg/g) to soil	Retained (mg/g)			
		Cd	Pb	Zn	Cu
pH= 10	0	-1.0E-06	-8.00E-06	-0.00020	-0.00034
	0.0001	0.000099	0.000081	-0.00013	-0.00034
	0.001	0.000992	0.00096	0.00026	0.00036
	0.01	0.00995	0.0099	0.0084	0.0091
50g	0.1	0.09895	0.099	0.0981	0.098
PH= 5	0.1	0.100	0.100	0.097	0.100
	0.5	0.444	0.499	0.49	0.499
10g	2	0.78	1.95	0.79	1.82

Negative sign results from desorption or error of measurement.

Table 2 shows results of the adsorption test. When the concentration of solution was relatively low, the desorption might take place.

Figure 7 shows the retained concentrations to the added heavy metals. Lead and copper are more retained to the soil than cadmium and zinc. This trend is common and known as the selective adsorption of soils (Yong *et al*, 1992). It is noted that the retained amounts are similar for added elements of 0.1 mg to 1g of soil at both initial pH values of 10 and 5.

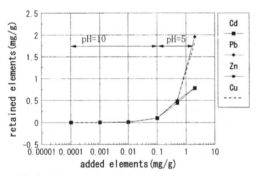

Fig.7 Added and retained amounts of heavy metals for 1 g of soil.

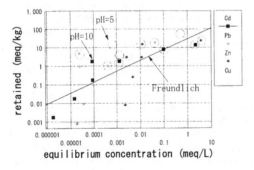

Fig.8 Heavy metals held by particles versus equilibrium concentrations.

Figure 8 shows the retained concentration of the heavy metals to the equilibrium concentration. The plots surrounding by circle denote the data obtained at pH = 5, and others are at pH = 10. Although the plots show considerable scattering, the trend clearly shows that the retained amount increases with increasing of equilibrium concentrations. The straight line indicated in Fig. 8 provides adsorption of Cadmium by the following Freundlich's equation.

Freundlich's equation is generally expressed as follows.

$$Q_A = \alpha \, C_A^{1/n}$$

where Q_A is the retained concentrations of A ion, C_A is the equilibrium concentration of pore fluid with A ion, and α and n are constants. The data shows that the equation cannot necessarily be applied for the heavy metals in Fig. 8.

4. PRELIMINARY FIELD COMPACTION

Preliminary field experiment was performed to examine compaction behavior of the materials. The materials used are similar to sample No.1-3 or No.2-6. The compaction was made at a water content of about 40 %, which is not necessarily proper value for actual construction of clay liner. The reason why this higher water content was used is because the pressed materials have this water content. It is noted that the optimum water content is 24 %, which is similar to the shrinkage limit, as can be seen in Figure 10. At a water content of 40 %, the unconfined compressive strength was 40 kPa.

Fig.9 A scene of preliminary field experiment.

The compaction was performed using a roller with a weight of 29 kN, as shown in Figure 9. The dry density of the compacted materials was approximately 1.3 Mg/m³. This value is relatively low in comparison with the maximum dry density, 1.56 Mg/m³. This accords with the result of the laboratory compaction test, as shown in Figure 10. The maximum dry density is obtained near shrinkage limit, which is around plastic limit. Therefore, better performance can be obtained at these water contents.

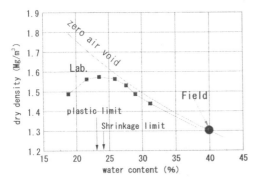

Fig.10 Compaction curve of the materials used in field.

The hydraulic conductivity of the materials can be estimated to be 4.0×10^{-7} cm/s, from the consolidation test. However, it should be noted that the hydraulic conductivity obtained from laboratory tests is not always similar to that of the in-situ soil.

The data obtained in the field experiment indicates the following conclusions;

a) The desiccation cracks may be created if the materials are dried, because of the initially high water content.

b) The strength of the compacted materials with a water content of 40 % may not be enough for trafficability of a dump truck.

c) The initial water content of materials should be reduced prior to construction.

5. CONCLUSIONS

In order to examine the quality of fine-grained materials remained by removing of coarse aggregates from sandy soils, various soil tests were performed. The present aim is to utilize the fine-grained materials as clay liner.

The permeability, which is the most important factor for clay liner, was measured from consolidation tests. The result shows that the hydraulic conductivity decreases with decreasing of void ratio. To achieve the required level, the materials should be compacted or consolidated. On the other hand, the unconfined compressive strength of compacted materials can be predicted from water content. Therefore, the compaction technique in field becomes important.

The sorption of hazardous substances to materials should be emphasized, because of an additional advantage when clay liner is used as sealing materials in waste disposal sites. From the batch test results, relatively high potential for adsorption of heavy metals was obtained, though more accurate test method has to be used.

Thus, the fine-grained fractions produced by separating fine and coarse grains have high potential as the materials for clay liner.

The field experiments also showed that if the water content of the materials is reduced down to around the optimum water content, plastic limit or shrinkage limit, it can be used as clay liner with high quality.

6. ACKNOWLEDGEMENTS

The materials were supplied through. Cooperative Association of Fuji River Aggregates Production. The authors wish to acknowledge the member of the cooperative association for their help.

REFERENCES

Sharma, H. D. and Lewis, S. P., 1994, Chap.6, Waste Containment Systems, Waste Stabilization, and Landfills, John Wiley & Sons, Inc., New York, 267-320.

Yong, R.N., Mohamed, A.M.O. and Warkentin, B. P., 1992, Principles of Contaminant Transport in Soils, Elsevier, p.169.

Clay Science for Engineering, Adachi & Fukue (eds) © 2001 Balkema, Rotterdam, ISBN 90 5809 175 9

Alteration of salient characteristics of clay liner materials caused by waste leachate

M.K.Uddin
Institute of Appropriate Technology, Bangladesh University of Engineering and Technology (BUET), Dhaka, Bangladesh

ABSTRACT: The waste leachate brings changes in the characteristics of clay liner. A research study comprising of a comprehensive series of laboratory tests were undertaken to study the changes of pertinent characteristics of clay liner caused by waste leachate. The characteristics of clay liner are altered due to an increase in ion concentration and a decrease in dielectric constant owing to presence of organic substance and inorganic salt in the leachate. Permeability tests showed higher coefficient of permeability in undiluted or diluted waste sample than in pure water. Adsorption tests indicated that there is limit in the amount of lead that could be adsorbed by the soil and is a function of lead concentration in the leachate up to a certain extent. Sedimentation tests revealed that the clay particles tend to flocculate or aggregate in waste leachate than in pure water and is more apparent in a higher concentration of leachate.

1 INTRODUCTION

Wastes landfills can be potentially serious threat to public health because of the risk of leachate escape into ground water or into surface water in an uncontrolled fashion. As the solid waste within a land fill breaks down, it releases a number of highly polluting materials which can dissolve in water that infiltrates or is present in the landfill, to generate leachate, a highly polluting liquid. The bottom liner must be impervious enough to prevent the leakage of leachate through them. Natural clay soils have been used as a natural bottom liner. The liner system is mainly to deter migration of waste leachate into ground water. Compacted clay together with geomembrane is generally used to form the liner system. Thus clay soils can be used as an engineered barrier in the liner system underlying landfilled wastes as being a reasonable contaminant attenuation barrier system. Clay is also a flexible material that would easily mould itself to the strains induced by the large depth of waste above. An auxiliary purpose of the clay barrier portion of any liner system is to attenuate contaminants in the transport of leachate in and through the clay liner. A competent contaminant attenuation barrier system against leachate transport would be obtained if the clay materials possess good chemical buffering capability, low hydraulic conductivity and high adsorption capacity. The permeability of the clay liner material must be significantly less. Plans to place large depths of hazardous waste in urban areas have led to the need to carefully evaluate

the appropriate technology for clay liner system in forming leachate containment.

Geomembrane is often disrupted due to bad workmanship during construction time or damaged due to chemical degradation at later stage. Thus the characteristics of clay layer as liner of is great interest. A layer of well-compacted homogenous clay will provide a barrier approximately with a coefficient of hydraulic conductivity to water of 10^{-7} cm/sec without difficulties. As the properties of clay are dependent on the chemical composition of leachate, the permeability, adsorption capability, index properties and sedimentation characteristics of clay will be eventually altered considerably. Brown and Thomas (1987) conducted experiments on Kaolinite, illite and smectite and found that the hydraulic conductivity of the samples increased with a decrease in dielectric constant of liquids and with an increase in the ion concentration of the solution. Aggregation of clay particles appeared to be more pronounced with the same factors. In recent years, waste landfills have been built not so far from human communities especially in the land-scarce country like Bangladesh because of dearth of appropriate site. Around the outskirts of Dhaka metropolis, the landfill is the common method of disposal of municipal wastes. Often, these disposal sites are close to the urban areas. Since the leachate from this waste fill can migrate into ground water, the design of clay liner system of this waste landfill is of great importance. Most waste landfills in Bangladesh are located on relatively flat land composed of Alluvial deposit.

The relevant soils belong to Alluvial Flood Plain deposits and Estuarine & Tidal Flood Plain Deposits.

Waste leachate is a contaminated fluid containing the concentration of soluble organic materials and in organic salts. Under a given set of environmental conditions (temperature, pressure, pH, chemical and biological composition of the water), a clay will absorb cations of specific type and amount. The clay is greatly affected by the action of waste leachate and pertinent physico-chemical properties of clay materials are succumbed to change. Figure 1 shows diffuse double layer of clay liner layer. The diffuse double layer consists jointly of charged surface and the distributed charged in the adjacent phase together. The total amount adsorbed will balance the charge deficiency of the solid particles. Exchange reaction can occur in response to changes in environmental conditions. These reactions involve replacement of a part of or all of the adsorbed ions of one type by ions of another type. The value of exchange capacity of the clay liner mineral indicates the amount of readily exchangeable cations that can be replaced easily by leaching with the waste leachate solution containing other dissolved cations of higher replacing power than the absorbed cations. The shape and distribution of pores within soil are controlled by the force between soil particles which are affected by the properties of pore fluid, exchange reaction bring important changes of the physical and physico chemical properties of the soil. Thus study regarding the change of characteristics of compacted clay due to action of waste leachate is of great importance. The research study was carried out to investigate the effect of physical/chemical properties of pore fluid on the characteristics of the clay liner. Sedimentation test, index tests, adsorption tests and permeability tests were conducted to find out the effects of pore leachate on properties of compacted clay.

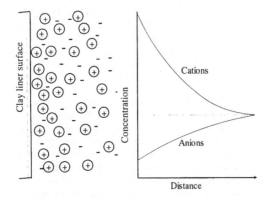

Figure 1. Distribution of ions adjacent to the compacted clay liner surface according to the concept of diffuse double layer.

2 GEOTECHNICAL CHARACTERISTICS OF SITE AND CLAY LINER

A waste landfill site at Sayedabad at the outskirts of Dhaka city was chosen for investigation. The top layer of the sub-soil of the site generally consists of a mixture of clay and silt. The depth of top layer varies from 8 to 30 meters. Layer of coarser materials such as sand and gravel exists at lower levels and finer particles dominate at the top surface. Dhaka clay was used as engineered barrier and liner. Geotechnical characteristics of the clay soil material which form the natural liner are shown in Table 1.

Table 1. Geotechnical characteristics of clay liner.

Properties	Characteristic Value
Specific gravity	2.69
Void ratio	1.09
Unit weight (kN/m^3)	17.76
Soil classification under united classification system	CL
Liquid limit (LL)	43%
Plasticity index (PI)	23%
Compression index	0.27 (Average)
Sand content	1%
Silt content	77%
Clay content (less than 2 micron)	22%
Optimum water content	24%
Cohesion	29.50
Friction angle	6^0
Cation exchange capacity (meq/100g)	8
pH Value	5.9

3. METHOD OF INVESTIGATION

For use a clay soil in a clay liner, relevant studies on the physico-chemical properties of the clay are required. These properties would render some insight into the possible behavior of the clay soil, vis-a-vis its adsorption capabilities, permeability characteristics and interactions with common contaminants emanating from a waste disposal site.

3.1 *Permeant liquid used in the tests*

Permeant liquid was collected from site. The physical and chemical properties of permeant liquid are appended in Table 2. Table 2 shows that ion concentration of leachate indicated by electrical conductivity is much higher than pure water. On the other hand, viscosity and specific gravity of the leachate is similar to pure water. It is seen that pH value has become a greater value. It means the liquid turned into basic from acidic since beginning of waste fill. Three types of permeant liquid were used:
- Unfiltered waste leachate
- Filtered waste leachate
- Pure water.

Properties	Values
Viscosity (mpa.s)	1.005
pH value	8.43
Density (g/cc)	1.007
Electrical Conductivity (μmho/cm)	$19.7*10^3$
COD (ppm)	1819
BOD (ppm)	18.4

3.2 Physical properties

The physical properties of the soil samples were determined using the conventional simple tests specified in soil investigations in Geotechnical Engineering. These were natural water content (w) determination, specific gravity (G_s), liquid limit (LL) and plasticity index (IP), unit weight, etc, all in accordance with ASTM standards. Table 1 shows these values.

3.3 Index properties

Index tests including sedimentation tests, determination of liquid and plastic limits of the clay liner material were performed. All these tests were carried out in accordance with ASTM Standards. Pore fluid of the soil was employed as waste leachate of various concentration.

3.4 Adsorption tests

The adsorption capabilities of clay soils with respect to various contaminants from a landfill site will dictate whether or not a particular clay soil will succeed in attenuating the potential contaminants. Thus the adsorption characteristics of soil have a direct impact on ground water pollution. The soil's composition, grain size distribution and permeability characteristics control mainly the adsorption capacity of the soil. The adsorption batch tests is one of the laboratory method commonly used for the determination of adsorption capability of clay material for pollutant removal. The amount of pollutant capable of being absorbed by a given mass can be known by this test for a fixed time of exposure.

The soil samples, after air-drying was ground uniformly with a mortar and pestle. This way a uniform powdery texture was obtained. Six beakers of soil samples each having 100g of soil was taken. In each beaker, 1000 ml of leachate was then added. Lead (Pb) having six different concentration (ranging from 12 to 61 mg/L) were added to the soil solution in each. In order to ensure a homogenous mixture, the lead solution and soil solution was stirred with care time to time. Because of the importance of presence of heavy metal in many waste disposal sites especially in the industrial effluent, lead solution was used in this study. The supernatant liquid from the beakers was collected at certain interval of time and filtered. Then using the Atomic Adsorption Spectrophotometer (AAS), the concentration of lead was then determined for each collected supernatant liquid. Until the lead concentration in the beakers had reached an equilibrium states, this process of experimentation was continued.

3.5 Permeability tests

The equipment set-up for permeability test was as per triaxial flexible wall permeameter designed by Bowders and Daniel (1985). For permeability tests, samples were air dried first and then were mixed with water. Amount of water added was such that it conforms to optimum moisture content (calculated by the ASTM standards D698, method A). Proper compaction of soil in mold was done as per standard. Three specimens (each 5 cm diameter x 12.5 cm high tube) were sampled by penetrating into the soil at the rate of 5 mm/min. All the three kinds of permeant liquid as stated earlier were applied for permeability test.

3.6 Sedimentation tests

The aggregation of clay particles and the change of pores within the soil were investigated by sedimentation tests. In the sedimentation test, various concentrations of waste leachate (55%, 70%, 85% and 100%) and pure water were used. The tests were not conducted with a hydrometer. A parameter (d) pertaining to the distance from the surface of the liquid to the top of the soil-liquid suspension was recorded. The parameter, d is the distance from top of supernatant to suspension. The sedimentation rate was recorded by measuring the parameter, d with time.

4 RESULTS AND DISCUSSIONS

4.1 Atterberg limits characteristics

With various levels of concentration of waste leachate in water (55%, 70%, 85% and 100%) in a certain quantity of clay, Atterberg limit tests were carried out. The result is appended in Table 3. The result shows that plastic index decreases as the leachate concentration increases (Fig. 2).

Table 3. Results of Atterberg limit tests of clay with various concentrations of waste leachate.

Concentration of waste leachate in water by volume (%)	Atterberg limits (%)		
	LL	PL	PI
55	37	19	18
70	33	20	13
85	32	21	11
100	32	21	10

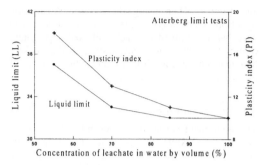

Figure 2. Relationship between Atterberg limits and concentration of waste leachate in water by volume.

Thus it is found that Atterberg limit of clay is affected by the concentration of organic substances and inorganic salts present in waste leachate. This phenomenon complies with the result of Bowders and Daniel (1987). Thus the results indicate that organic substances and inorganic salts reduces the dielectric constant and increases the ion concentration and cause alteration of physical properties of the clay.

4.2 *Adsorption characteristics*

Lead was used in the study because in many waste disposals fill, lead is found as predominant substance. Residues derived from incineration of refuse often contain high concentration of heavy metals such as lead (Pb). These residues are disposed of by means of sanitary landfill. Figure 3 shows the results of adsorption tests of the soil investigated.

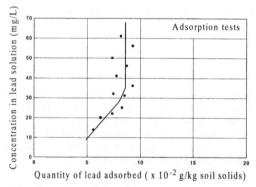

Figure 3. Relationship between quantity of lead adsorped and concentration in lead solution.

It shows that quantity of lead adsorbed is not affected by the lead concentration as long as the concentration in the solution is higher than 14%. The adsorption capacity of the soil was found to increase (from 5.65×10^{-2} g/kg of soil solid) with lead concentration up to concentration of 35 mg/L. The adsorption capacity reaches an equilibrium thereafter at higher concentrations. This equilibrium adsorption capacity is 8.68×10^{-2} g/kg of soil solids. Thus the result shows that the soil at higher level of lead concentration, exhibits higher adsorption capability.

4.3 *Permeability characteristics.*

Triaxial flexible well permeameter was used to find permeability characteristics for the clay for different stress condition and different concentration of effluent fluid. The applied confining pressures were 100 kPa, 200 kPa, and 300 kPa. Hydraulic gradient was used as 40, 80 and 120. Three kind of permeant (filtered waste leachate, unfiltered waste leachate and pure water) was used for each confining pressure and hydraulic gradient.

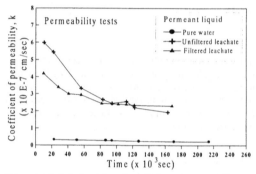

Figure 4 (a). Relationships of coefficient of permeability, (k) and time (confining pressure=100 kPa and hydraulic gradient=40).

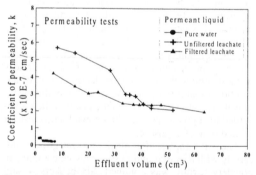

Figure 4 (b). Relationships of coefficient of permeability, (k) and effluent volume (confining pressure=100 kPa and hydraulic gradient=40).

Figures 4 to 8 show results of the tests where undiluted waste leachate was used with the cases of both filtered and unfiltered leachate. From the results it can be found that the undiluted leachate whether filtered or not filtered possesses higher permeability than that of pure water. This behavior can be attributed due to the fact that organic or inorganic contents of the leachate create an effect on decreasing the thickness of diffusion double layer in clay particles which in turn alters the size and shape of pores in clay and thus the permeability of the soil is increased.

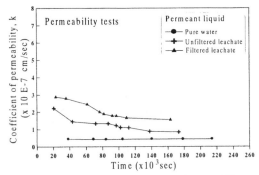

Figure 5 (a). Relationships of coefficient of permeability, (k) and time (confining pressure=200 kPa and hydraulic gradient=80).

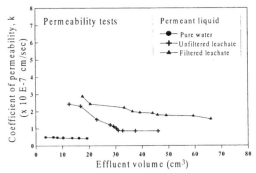

Figure 5 (b). Relationships of coefficient of permeability, (k) and effluent volume (confining pressure=200 kPa and hydraulic gradient=80).

It can be noticed from the relationship of coefficient of permeability and effective volume that the unfiltered sample shows greater rate of recession of the value of coefficient of permeability with increase of effluent volume. This higher value of coefficient of permeability of unfiltered sample than that of filtered sample or pure water can be a indicative of the interactive physico-chemical effects of clay or insoluble solid in leachate which subsequently shrink the pores of the clay matrix. Consequently, the permeability is reduced significantly.

Figure 7 shows the results of tests carried out with 25% waste leachate. The curves for the 25% of waste leachate for both filtered and unfiltered leachate shows similar permeability characteristics as that of pure water. Thus it can be stated that a decrease in the concentration of organic material and

ion in waste leachate would also reduce its influence and effect on the change of pore structure in clay matrix and thus salient characteristics of the clay such as permeability value continue to reduce as the organic material and ion concentration is kept on reducing; ultimately the permeability behavior will merge with that of pure water. Comparing relationships between coefficient of permeability and effluent volume of Figures 4, 7, it can be found that the rate of decay of permeability with increase of effluent volume reduces largely as the organic material and ion concentration decreases. The figures show that rate of decrease of permeability with respect to the increase of effluent volume is lower in the case of diluted leachate [e.g. Fig. 7(b)] than undiluted leachate [e.g. Fig. 4(b)]. Figures 8a, b show permeability tests with 10% waste leachate. The figures show that 10% waste leachate and pure water possess almost same permeability.

From above test it is found that organic and inorganic content and solid particles in leachate have significant effect on restructurization of the pores of clay and on permeability characteristics of the clay liner materials.

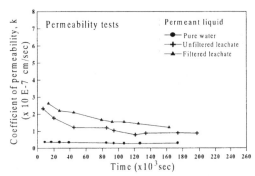

Figure 6 (a). Relationships of coefficient of permeability, (k) and time (confining pressure=300 kPa and hydraulic gradient=120).

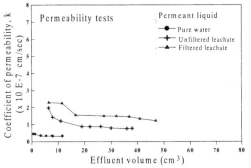

Figure 6 (b). Relationships of coefficient of permeability, (k) and effluent volume (confining pressure=300 kPa and hydraulic gradient=120).

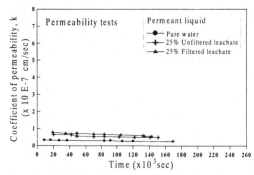

Figure 7 (a). Relationships of coefficient of permeability, (k) and time (confining pressure=300 kPa and hydraulic gradient=120).

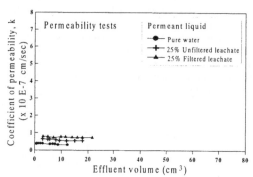

Figure 7 (b). Relationships of coefficient of permeability, (k) and effluent volume (confining pressure=300 kPa and hydraulic gradient=120).

4.4 *Sedimentation characteristics*

Figure 9 shows sedimentation tests results. Sedimentations in various concentrations of waste leachates such as 100%, 85%, 70% and 55% were carried out along with that of pure water. It reveals that the clay particles tend to flocculate or aggregate in waste leachate and fasten the rate of sedimentation. This trend of fastening of sedimentation process is more pronounced in a higher concentration of leachate. This phenomenon indicates higher ion concentration and a lower dielectric constant in the leachate as compared to pure water.

According to electrical double layer theory, agglomeration appears more easily to speed up sedimentation in a suspension when a decrease in the thickness of diffusion double layer occurs caused by a decrease in the dielectric constant and an increase in the ion concentration in the leachate. This way the waste leachate will caused bigger pores or fissures in the clay matrix. The Figure 9 shows that faster sedimentation occurs with higher concentration of waste leachate than lower concentration of leachate or pure water. The faster rate of sedimentation in the case of

leachate suspension reaches an asymptotic value in a shorter time than pure water.

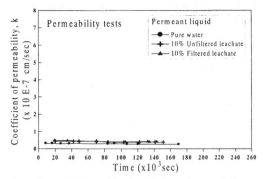

Figure 8 (a). Relationships of coefficient of permeability, (k) and time (confining pressure=300 kPa and hydraulic gradient=120).

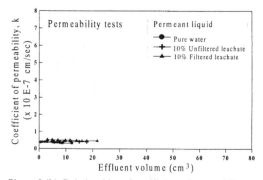

Figure 8 (b). Relationships of coefficient of permeability, (k) and effluent volume (confining pressure=300 kPa and hydraulic gradient=120).

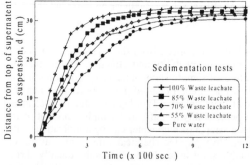

Figure 9. Sedimentation tests results showing the rate of sedimentation of waste leachates.

5 CONCLUSIONS

From waste landfill, the movement of hazardous waste into the general groundwater system must be inhibited by an impermeable liner. Compacted impermeable clay liner with appropriate properties can be used to form liner acting as barrier for solid waste disposal by means of landfill. The performance of these barriers depends on some predominant characteristics of the compacted clay such as adsorption capacity, permeability, sedimentation behavior, etc., under the site environment and hydrological and ground water condition. Waste leachate affects these characteristics of the clay liner considerably due to change of bonding ability of particles in the clay matrix. A comprehensive series of tests, namely, Atterberg limit tests, adsorption tests, sedimentation tests, permeability tests were carried out to determine the effect of waste leachate. The following conclusions were drawn from the test results.

Atterberg limit tests show that the plasticity index decreases when the concentration of waste leachate increases. It indicates that an increase in ion concentration and a decrease in dielectric constant occur due to presence of organic material and inorganic salt in the waste leachate. Sedimentation test showed that the rate of sedimentation of clay particles becomes higher in waste leachate than in pure water. It was found that the higher the concentration of waste leachate, the greater the rate of sedimentation of clay particles. The lower dielectric constant and higher ion concentration of leachate as compared to pure water may be the reason for such fast rate of sedimentation in waste leachate. Moreover, it is found that high rate of sedimentation in leachate reaches an asymptotic or steady state in a shorter period of time. In leachate, according to electrical double layer theory, the thickness of diffusion double layer is compressed which may cause in a shorter distance between clay particles for flocculation and then accelerate sedimentation. The laboratory study involving the use of lead solution has been carried out to investigate the adsorption characteristics of the compacted clayey material and to find the variation of adsorption capacity at different concentration of lead solution. The adsorption of soil is seen to increase with the lead concentration up to 35mg/L. Then the adsorption capacity reached an asymptotic value at higher concentration of 8.68×10^{-2} g/kg of soil solids. Thus the quantity of lead that could be adsorbed by the soil at higher concentration of lead (beyond 35 mg/L) was found to be limited to 8.68×10^{-2} g/kg of soil solids. So it shows that for higher concentration, there is virtually a limit in the amount of lead that could be absorbed by the clay materials.

Permeability tests shows a higher permeability of clay liner material in undiluted leachate (filtered or unfiltered) sample that in pure water. Unfiltered samples show greater rate of decrease than filtered samples or pure water. Waste leachate of 25% (for both filtered or unfiltered) shows similar permeability characteristics as that of pure water. A reduction of organic material and ion concentration in leachate reduces permeability value of the clay liner and ultimately will merge with the characteristics of pure water. This was evidenced by the tests with 10% leachate.

The results show that the recession or reduction of permeability due to an increase in effluent volume is more prominent in the leachate permeant sample than pure water. Higher the concentration of leachate, the higher the rate of decay of permeability. The recession behavior of permeability for 25% leachate sample resembles to that of pure water. The recession behavior in leachate is attributed to the interceptive effect of insoluble solid in waste leachate on the pore of clay. Thus a specific relationship exists between the change of permeability of clay and the concentration of ion and organic material.

REFERENCES

Brown, K.W. & Thomas, J.C. 1987. A mechanism by which organic liquids increase the hydraulic conductivity of compacted clay materials, *Soil Sci. Soc. Am. J.*, vol. 51, p. 1451-1459.

Bowders, J.J & Daniel, D.E. 1987. Hydraulic conductivity of compacted clay to dilute organic chemicals, *Journal of Geotechnical Engineering*, vol. 113, no. 12, December, ASCE, p. 1432-1448.

Bowders, J.J., Daniel, D.E., Broderick, G.P., & Liljestrand, H.M. 1986. Methods of testing the compatibility of clay liners with landfill leachate, *Proc. of ASTM 4th Symp. on Hazardous and Industrial Solid Waste Testing*, Philadelphia, ASTM STP 886, p. 233-250

Brown, K.W. & Anderson, D.C. 1983. Effect of organic solvents on the permeability of clayey soils, EPA-0600/2-83-0016, (US) *Environmental Protection Agency*, 153p.

Evans, J.C. 1991. Geotechnics of hazardous waste control systems, *Foundation Engineering Handbook (Ed. by H. Y. Fang), 2nd ed.*, Van Nostrand Reinhold, New York, pp. 750-777.

Fang, H.Y. & Evans, J.C. 1988. Long-term permeability tests using leachate on a compacted clayey liner material, *Proc. of ASTM Symp. on Ground-water Contamination: Field Methods*, Philadelphia, ASTM STP 963, p. 397-4040.

Foreman, D.E. & Daniel, D.E. 1984, Effect of hydraulic gradient and method of testing on the hydraulic conductivity of compacted clay to water methanol heptone, land disposal of hazardous waste, *Proceeding of the Annual Research, Symposium (10)* ATFP Mitchell, Kentucky held on April 3-

Forôman, D.E. and Daniel, D.E. 1986. Permeation of compacted clay with organic chemicals, *Journal of Geotechnical Engineering*, vol. 112, no. 7, July, ASCE p. 669-81.

Forbies, E.A.., Posner, A. M. & Quirk, J.P.1974. The specific adsorption of inorganic Hg(II) species and Co(III) complex ions on geothite. *Journal of Colloid and Interface Science*, 49, 403-409.

Grim, R.E. 1959. Physical-chemical properties of soils: clay minerals, *Journal of Soil Mechanics and Foundation Division*, ASCE, vol. 85, no. SM2, pp. 1-17.

Iwata, S. & Jahuchi, T. 1985. *Soil-water International Mechanisms and Application"*, chapter 3.

Lin, J.Y. 1990. Effect of organic fluids on the permeability of consolidation Kaolinite, *M.S. Thesis*, National Taiwan University, R.O.C.

Mohamed, A.M.O., Yong, R.N., Tan, B.K., Farkas, A. & Curtis, L.W. 1994. Geo-environment assessment of a micaceous soil for its potential use as an engineered clay barrier, *Geotechnical Testing Journal*, GTJODJ, 17, 3, Sept. 1994, 291-304.

Peirce, J.J. & Sallfors, G. 1988. Effect of selected inorganic leachates on clay permeability, *Journal of Geotechnical Eng.*, vol. 113, no. 8, August, ASCE, p. 915-919.

Construction Industry Research and Information Association, CIRIA, 1996. *Barriers, Liners and Cover systems for containment and control of land contamination*, CIRIA Special Publication 124.

6 Buffer materials

Clay Science for Engineering, Adachi & Fukue (eds) © 2001 Balkema, Rotterdam, ISBN 90 5809 175 9

Swelling deformation of compacted bentonite/sand mixture

H. Nakashima & T. Ishii
Institute of Technology, Shimizu Corporation, Japan

A. Nakahata & H. Tanabe
Radioactive Waste Management Center, Japan

S. Tanuma
The Kansai Electric Power Company Incorporated, Japan

ABSTRACT: The swelling deformation of compacted bentonite/sand mixture was measured under various loads assuming the conditions of geological disposal of TRU waste. Two types of bentonite, sodium bentonite (Na-bentonite) and calcium bentonite (Ca-bentonite), were used to obtain reference data and the data at ultimate conditions after long term cation exchange of Na bentonite with Ca in ground water, respectively. The loads were 4.9, 19.6, 78.5, 313.8, and 1244.3 kPa. The results indicate that: (1) the maximum swelling deformation strongly depends on the load and (2) the maximum swelling deformation of Ca-bentonite is about 1/10 compared to that of Na-bentonite.

1 INTRODUCTION

A compacted Na-bentonite/sand mixture is one of the candidate materials for a back-fill or a buffer in the Japanese TRU waste disposal concept (TRU coordination Office 2000). This is mainly because of a swelling characteristic of the bentonite, by which the voids or spaces in the repository can be sealed. The swelling characteristics of bentonite or bentonite/sand mixtures have been studied by many researchers from different points of view (Komine et al. 1999).

In general, a TRU waste disposal facility is characterized by (1) the waste packages are emplaced in large vaults as a mass because of their low heat generation and (2) a large volume of cementitious materials such as concrete structures are used.

When a bentonite/sand mixture is placed in a large vault, supposing no stress from the wall of rock, a bentonite/sand mixture will undergo various stresses depending on its position relative to roof, wall and floor. The swelling characteristics, especially swelling deformation, are anticipated to change under deferent stress conditions. It will be necessary to evaluate the swelling deformation corresponding to those stresses, to design the size and the volume of a compacted bentonite/sand mixture for the disposal facility.

The cementitious materials will gradually leach Ca ions into ground water for a long period of time. It is well known that Na-bentonite can easily exchange Na with Ca ions. The cation exchange of Na-bentonite is anticipated under this environment and the bentonite after cation exchange would be cause a change in swelling characteristics (pusch, 1882).

In this study, the swelling deformation of compacted bentonite/sand mixtures at various stresses was measured. The Na-bentonite was used to get data at an initial state as a reference, and Ca-bentonite to see the change of swelling characteristics due to the cation exchange of Na-bentonite. The Ca-bentonite used in this study simulates the ultimate condition after long term cation exchange when most of Na in the bentonite has been exchanged with Ca.

2 EXPERIMENTAL MATERIAL

2.1 *Bentonite*

Two types of bentonite, Na-bentonite and Ca-bentonite, were used in this study.

Na-bentonite is a commercially available sodium bentonite, Kunigel-V1 (Itoh et al. 1994), Kuminine Industries Co., Ltd. in Japan.

Ca-bentonite was prepared by the cation exchange of Na-bentonite, Kunigel-V1, for this study. The procedure was as follows:
1) Calcium chloride was dissolved in deionized water,
2) The Na-bentonite powder, Kunigel-V1, was put into the solution,
3) The solution was stirred for two hours and then kept for a night,
4) The precipitate of bentonite after filtration of the solution was washed using deionized water to remove chloride,
5) The bentonite precipitate was dried, crushed and sieved. The product power was used to prepare compacted Ca-bentonite specimens in the tests.

The prepared Ca-bentonite was analyzed to confirm that Na ions in bentonite were exchanged with Ca ions by such methods as leaching cation, before swelling deformation tests. The chemical characteristics of Na-bentonite and Ca-bentonite are shown in Table 1. It shows a low concentration of Na ions and a high concentration of Ca in Ca-bentonite, which indicates most of the Na ions in the bentonite were exchanged with Ca ions.

Table 1. Chemical characteristics of two types of bentonite

	Na-bentonite	Ca-bentonite
pH	10.1	9.3
CEC* (meq/100g)	68.6	68.2
MB** (mmol/100g)	70	70
quantities of dissolved cations (meq/100g)		
Na$^+$	65.5	10.5
K$^+$	2.7	1.8
Ca^{2+}	52.4	97.9
Mg^{2+}	6.8	3.5
total	127.4	113.7

* CEC : Cation exchange capacity
** MB : Methylene blue

2.2 *Sand*

The sand mixed with bentonite were commercially available quartz sand No.3 and No.5 by the same weight, whose grain size distribution is shown in Figure 1.

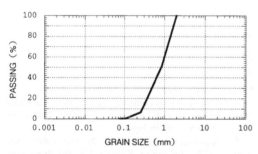

Figure 1. The grading curve of the sand used for specimen

2.3 *Water*

Deionized water was used for preparation of specimen and in swelling deformation tests.

3 EXPERIMENT

3.1 *Apparatus*

A schematic view of the apparatus for the swelling deformation tests is shown in Figure 2. The compacted bentonite/sand mixture is 60mm in diameter and 10mm in height.

An oedometer used for conventional consolidation tests was modified to be appropriate to this test.

The metal mould into which the specimen was set was made taller than that of conventional oedometer to accommodate swelling of the specimen by absorbing water. Similarly, the piston that transmits a load to the specimen was also made longer.

Silicon grease was applied to the inner wall of the metal mould to reduce friction with the specimen.

The swelling deformation was measured by the displacement of the piston.

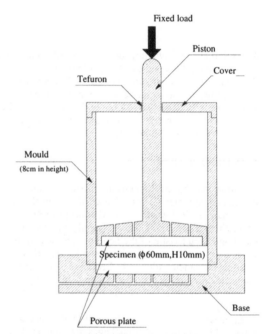

Figure 2. A schematic view of the apparatus for the swelling deformation tests.

3.2 *Test cases*

The specimens prepared for this study were eight types of compacted bentonite/sand mixture differing in three parameters: type of bentonite, bentonite content and density.

Five specimens were prepared for each of the eight types of compacted bentonite/sand mixture, and each group was subjected to the swelling deformation test at a range of loads.

3.2.1 *Specimen*

Two types of bentonite, Na-bentonite and Ca-bentonite, were used as described above. The Na-bentonite was to obtain the basic data. Ca-bentonite was to obtain data when most of Na in Na-bentonite was exchanged with Ca in ground water, which is anticipated after the long period of cation exchange.

In designing a TRU repository, the data on higher bentonite content would be necessary. Hence, the data on both 100% and 50% bentonite contents were obtained. The data can be used to estimate data on bentonite content as 90%.

The densities of the specimen were from 1.6 to 2.0, aiming at examining the effect of initial density on swelling deformation. The density of specimens was varied depending on the bentonite content, considering compaction characteristics.

The specifications of specimens are shown in Table 2.

Table 2. Specifications of the tested bentonite/sand mixture

Bentonite	Bentonite * content (%)	Dry density (Mg/m³)	Water content (%)
Na-bentonite	100	1.6	20.0
		1.8	18.0
	50	1.8	15.0
		2.0	12.0
Ca-bentonite	100	1.6	20.0
		1.8	18.0
	50	1.8	15.0
		2.0	12.0

* Bentonite content : Bentonite dry weight / (Bentonite + sand dry weights)

3.2.2 Pressure condition

Different pressures were applied to each of five specimens for each type varying the static load. The loads were 4.9, 19.6, 78.5, 313.8, and 1244.3 kPa

3.3 Procedure for swelling deformation tests

The compacted bentonite or bentonite/sand mixture was prepared by pressing powders with appropriate water content (see Table 2) in the mould. The mould with a specimen in it was set and fixed on the base of the swelling deformation apparatus.

After applying the prescribed load to the specimen, the water was injected from the bottom through the porous plate at a low pressure of about 1.0kPa to the specimen. The recording of displacement was started immediately after the injection.

The permeability test was started when the swelling was nearly completed. The injection pressure was increased to 10-200 kPa by keeping the deformation constant. After a permeability test, the restraint to keep the volume was removed, and displacement was measured with time again to see if the deformation might increase.

For those specimens that showed an increase of deformation after the permeability test, when the swelling was nearly completed again, the injection pressure was again increased by keeping the deformation constant at a new level, in order to obtain the final permeability coefficient.

4 RESULTS AND DISCUSSION

4.1 Swelling deformation

Figure 3 - 10 show swelling strains against time obtained in swelling deformation tests for all of the specimens. The swelling strain is defined as the percentage of the swelling deformation to the initial specimen height (Komine et al. 1994). A discontinuation in every line around 100-150 days or 200-250days indicates a period when the permeability was measured by keeping the deformation constant.

It was found that the swelling strain depends on the load applied to the specimen. It becomes greater as the load is reduced and is almost zero when the load is 1244.3 kPa in every type of specimen.

The swelling strain increased rapidly immediately after the start, and then settled down in a few months. But, an increase of the swelling strain again after the permeability test with high water pressure was observed in some specimens. This was significant in Na-bentonite with a larger swelling strain, as shown in Figures 3 - 6.

Though the swelling may not be completed in several months, the final swelling strains after permeability test were defined as maximum swelling strains in this report.

Next, the maximum swelling strain was plotted against load for different bentonite content in Figures 11,12.

It was also found that maximum swelling strain depends on the type, the density and the content of bentonite as well as applied load. The maximum swelling strain was greater in case of the Na-bentonite with higher initial dry density and higher bentonite content. The effect of the type of bentonite is especially significant. The swelling strains of the Ca-bentonite were less than 1/10 that of Na-bentonite.

The effect of initial dry density can be shown more clearly by plotting dry density after swelling corresponding to maximum swelling strain of Na-bentonite, as shown in Figure 13. This figure indicates that the dry density after swelling of the two types of specimens with different initial dry density, were almost the same when they were swelled at the same load. Typical examples of this trend were observed in the range of small loads, for the specimen of both the 100% and the 50% bentonite content.

4.2 Permeability after swelling

The measured permeability plotted against dry density at the time of measurement is shown in Figures 14 - 17 showing the effects of the type of bentonite and the bentonite content. Two sets of data are plotted for one specimen for the case where the swelling increased after the first measurement of permeability and a second measurement was carried out. The difference of dry density in the two measurements is taken into account.

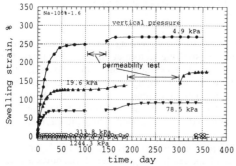

Figure 3. Test results for Na-bentonite, 100%bentonite content, 1.6Mg/m³ dry density.

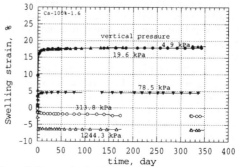

Figure 7. Test results for Ca-bentonite, 100%bentonite content, 1.6Mg/m³ dry density.

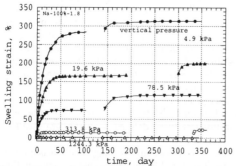

Figure 4. Test results for Na-bentonite, 100%bentonite content, 1.8Mg/m³ dry density.

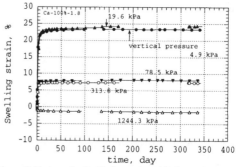

Figure 8. Test results for Ca-bentonite, 100%bentonite content, 1.8Mg/m³ dry density.

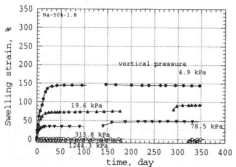

Figure 5. Test results for Na-bentonite, 50%bentonite content, 1.8Mg/m³ dry density.

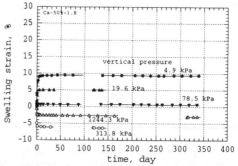

Figure 9. Test results for Ca-bentonite, 50%bentonite content, 1.8Mg/m³ dry density.

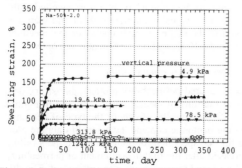

Figure 6. Test results for Na-bentonite, 50%bentonite content, 2.0Mg/m³ dry density.

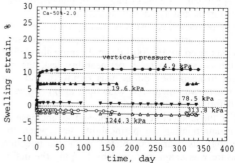

Figure 10. Test results for Ca-bentonite, 50%bentonite content, 2.0Mg/m³ dry density.

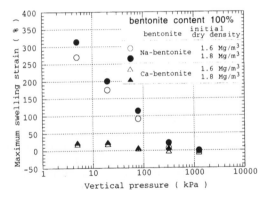

Figure 11. Relationship between maximum swelling strain and vertical pressure at bentonite content 100%

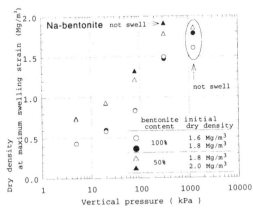

Figure 12. Relationship between maximum swelling strain and vertical pressure at bentonite content 50%

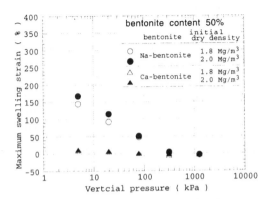

Figure 13. Dry density at maximum swelling strain for the cases of Na-bentonite

Each figure shows a linear relation of permeability with the dry densities. It is considered that the decrease of dry density due to swelling caused the increase of permeability. The permeability after

swelling was not dependent on initial dry density. Both data measured at the point before completing swelling by holding the volume constant and the one measure at the final point indicate the permeability corresponding to the dry density at the point of measurement. The correlation between dry density and the permeability is high for Na-bentonite while it is low for Ca-bentonite.

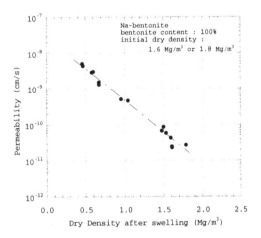

Figure 14. Permeability against dry density after swelling for Na-bentonite, 100% bentonite content.

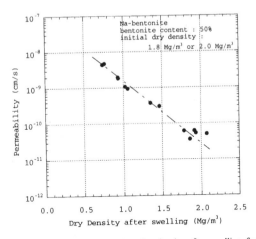

Figure 15. Permeability against dry density after swelling for Na-bentonite, 50% bentonite content.

The permeability depends also on the bentonite content of bentonite/sand mixture. The permeability of 50% bentonite content mixture is higher than for 100% bentonite, observed both Na-bentonite and Ca-bentonite.

The effect of the type of bentonite on the permeability after swelling is very significant. In the case of Na-bentonite, the decrease of dry density to 1/3 from the initial density caused three orders of magnitude increase of permeability while only a 10% decrease of dry density caused three orders of magnitude increase of permeability in the case of Ca-bentonite.

It indicates that the permeability of the bentonite/sand mixture after cation exchange is more sensitive to the change of density.

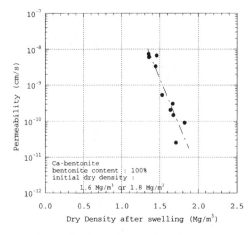

Figure 16. Permeability against dry density after swelling for Ca-bentonite, 100% bentonite content.

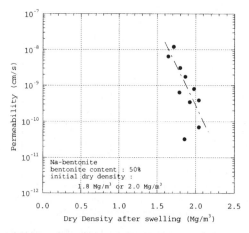

Figure 17. Permeability against dry density after swelling for Ca-bentonite, 50% bentonite content.

5 CONCLUSIONS

The main conclusions are as follows:

1) The maximum swelling strain strongly depends on the load and almost no deformation was observed in the range above the load of 1MPa. This indicates that the weight of bentonite itself should be taken into account when employing sealing capability of bentonite in the design of the repository, especially the TRU waste repository large vault.

2) The permeability of bentonite/sand mixture was decreased after swelling due to the decrease of dry density. This can be useful data in evaluating the capability to prevent water flow of the bentonite after swelling.

3) The swelling capability of Na-bentonite will be deprived of long term after emplacement in the repository, assuming that most of Na will be exchanged with Ca ions, as demonstrated in this study. It is indicated that the kinetic approach to the cation exchange is required to predict long term swelling behavior.

ACKNOWLEDGEMENT

This study is based on the data obtained in the study co-sponsored by 10 Electric Power Companies in Japan and the Japan Nuclear Fuel Limited. The authors would like to thank Dr.Komine of Central Research Institute of Electric Power Industry for his general comments on the study. Also, thanks Mr.Otani and Mr.Suzuki of Kunimine Industries Co., Ltd. for preparing the Ca-bentonite, Ms.Nakajima and Mr.Torigoe (Koa Kaihatsu Co., Ltd.) for taking test data, and Mr.Mano (E & E) for his kind advice in finishing of this paper.

REFERENCES

Itoh, M. et al. 1994. J. Atomic Energy Soc. Jan. Vol.36, No.11
Komine, H. et al. 1994, Experimental study on swelling characteristics of compacted bentonite, CAN. GEOTECH. J, Vol. 31, Pages 478-490.
Komine, H. et al. 1999, Experimental study on swelling characteristics of sand-bentonite mixture for nuclear waste disposal, SOIL AND FOUNDATIONS, Japanese Geotechnical Society, vol.39, No.2, 83-97.
Pusch, R. 1882, Chemical Interaction of Clay Buffer Materials and Concrete, SFR ARBETSRAPPORT 32-01.
TRU coordination Office (Japan Nuclear Cycle Development Institute and The Federation of Electric Power Companies) March 2000, Progress Report on Disposal Concept for TRU waste in Japan, TRU TR-2000-02.

Clay Science for Engineering, Adachi & Fukue (eds) © 2001 Balkema, Rotterdam, ISBN 90 5809 175 9

Self-sealing ability of buffer materials containing bentonite for HLW disposal

H. Komine & N. Ogata
Central Research Institute of Electric Power Industry, Abiko, Japan

H. Takao, A. Nakashima & T. Osada
JGC Corporation, Yokohama, Japan

H. Ueda
Tokyo Electric Power Company, Japan

ABSTRACT: This paper describes self-sealing ability of buffer materials that will be used for high-level radioactive wastes (HLW) disposal. This ability is provided by the swelling characteristics of compacted bentonite and performs one of the most important roles of buffer materials that will be placed around overpacks of HLW. An experimental study was performed by using two-dimensional model test of disposal pit, where reduced models of actual buffer materials, a disposal pit and a overpack with a scale of 1:5 were used. As a result, experimental data have been obtained, which will contribute to the quantitative judgment of effective self-sealing ability and it is verified that the suggested formula for prediction of swelling behavior can be applied to evaluation of self-sealing ability of buffer materials.

1 INTRODUCTION

In Japan's high-level radioactive waste disposal project, buffer materials containing bentonite will be placed around overpacks of high-level radioactive wastes (HLW), as shown in Fig.1. The buffer materials will meet strict requirements which include low water permeability, self-sealing ability, retardation of radionuclide migration, chemical buffering effects and overpack supporting function (JNC 1999).

Pre-compacted blocks of buffer materials, which are one of the most attractive options to be applied, will be of high density and will thereby contribute to meeting the abovementioned requirements. In examining this concept, it has been pointed out that space between the buffer materials and the wall of the disposal pits and/or that between the buffer materials and the overpacks will exist after the buffer materials are placed in the pits. One significant concern is that such space will form water-permeable pathways. In order to maintain low permeability around the waste packages, the swelling characteristics of the buffer materials are expected to be such that the spaces are filled. This is referred to as the "self-sealing" ability of buffer materials.

This paper focuses on the self-sealing ability, and thus experimental studies have been performed as follows; i) to produce a bentonite or bentonite-sand mixtures specimen with a fifth dimension of the buffer materials determined by the trial design in a previous study, ii) to place the specimen in a test apparatus with proper space between the specimen and the apparatus, iii) to supply water from around the

specimen, and iv) to perform the water-flow test. This is referred to as the "two-dimensional model test of disposal pit", as shown in Fig.2. This test consists of two kinds of tests with relation to the lengths of their test periods - the short-term test and the long-term test. The former models the duration from the placement to re-saturation of buffer materials, while the latter the duration after re-saturation. In both the tests, it was investigated whether the space around buffer materials was certainly filled or not, and the pressure of buffer materials after the

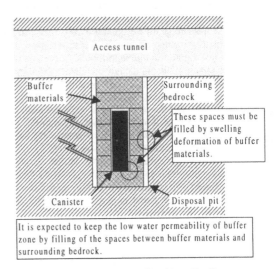

Fig. 1 Layout and the role of "self-sealing"

space was filled (this is referred to as "the buffer pressure" below) was measured. For prediction of the buffer pressure, we suggest "Evaluation formula for swelling characteristics of buffer and backfill materials for HLW disposal" (this is referred to as "the evaluation formula") (Komine et al. 1996). The buffer pressure measured in experiments was compared with that predicted for verification of "the evaluation formula" for its future use.

In addition, the water-flow tests (this is referred to as "WF test" below) were performed. WF tests applied water pressure along the wall of the test apparatus at various pre-determined times, so that they could cause "break-through" flow. As a result of WF tests, the relationship between the buffer pressure and the "break through" water pressure was investigated.

2 OBJECTIVES

The thickness of buffer materials is currently determined by assuming the required condition that the buffer pressure shall be more than 1MPa (this is referred to in AECL 1994). However, this has not been quantitatively verified, and thus further study is necessary in order to specify the clear requirements for the design criteria of buffer materials from the viewpoint of self-sealing. The study in this paper has been performed in order to acquire experimental data that will contribute to specifying the requirements mentioned above.

The specific objectives of this study are as follows;

a. To confirm that the space is certainly filled by swelling deformation of buffer materials.
b. To investigate the relationship between the buffer pressure and the "break-through" water pressure, which is greatly significant to evaluate self-sealing ability quantitatively and consequently for the design of buffer materials.
c. To investigate the applicability of "the evaluation formula" to the design of buffer materials from the viewpoint of "self-sealing". This is done by measuring the buffer pressure in experiments and by comparing it with the buffer pressure predicted by the evaluation formula.

The evaluation formula has been theoretically derived and consists of the following equations;

i) Equations to evaluate repulsive and attractive forces between two parallel clay layers which are based on electrical potential distribution around the two layers and the van der Waals theories, respectively.
ii) Equations to evaluate volumetric strain of montmorillonite.
iii) Equation of the relationship between the distance between parallel clay layers and the volumetric strain of montmorillonite.

Given bentonite content, initial dry density, swelling strain of buffer materials and ion concentration of pore water, the evaluation formula can calculate the buffer pressure under the above given conditions.

3 LABORATORY TESTS

This chapter describes the specimens, the apparatus and the procedure used in the experiments.

3.1 Specimens

Bentonite or bentonite/sand mixtures were used for the specimens. The bentonite used was one of the most famous sodium bentonites produced in Japan, which is marketed as "Kunigel V1". The montmorillonite content acquired from its methylene blue test is approximately 48%. The sand used was Mikawa Keisha quartz sand.

Donut-shaped specimens with a 50mm height and target initial dry density 1.6 or 1.9 Mg/m^3 were prepared by compaction, although actual density had some error after produced. The bentonite content of each specimen was 80 or 100%, where it is expressed as a percentage of bentonite / total dry weight. An exception to the long-term test was the target dry density 1.77 Mg/m^3 with bentonite content 70%, which is currently specified in JNC Report.

The material specifications and the dimensions of specimens used in the experiments are summarized in Table 1 and Fig.3.

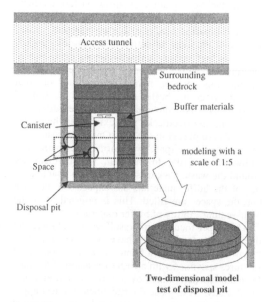

Two-dimensional model test of disposal pit

Fig.2 Two-dimensional model test of disposal pit

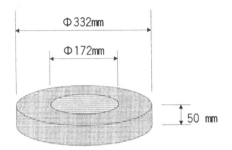

Fig.3 Dimensions of specimen

3.2 Test apparatus

In this experimental study, the short-term test and the long-term test were performed, as stated above. It was required that specimens should model unsaturated buffer materials for the former and saturated buffer materials for the latter, respectively.

Therefore, different test apparatuses with different water-supply systems were developed. In order to saturate a specimen for the long-term test, the system should supply and drain water from fairly wider areas, although it is not the actual condition.

In order to measure the buffer pressure, nine soil pressure gauges and 12 strain gauges were attached to each apparatus.

Fig.4 illustrates the conceptual drawing of the test apparatus.

3.3 Test procedure

In this clause, test procedures of the short-term test and the long-term test are described.

3.3.1 The short-term test

In each of the short-term test cases, 14 WF tests were performed within approximately 96 hours' test period, so that many data sets of the buffer pressure and the "break through" water pressure could be acquired from WF tests. Fig.5 shows a conceptual illustration of the "break through" water pressure. This study defined that the "break through" water pressure was the value when volumetric rate of drain was equal to or more than 1.0ml/min. WF tests were performed at 0, 2, 4, 8, 24, 28, 31, 48, 52, 55, 72, 76, 79, 96 hours after water-supply was started. In each WF test, the water-flow was started at the above times, and then the water pressure was escalated up to "break through". While WF tests were not performed, a specimen was supplied with water with a hydraulic head (200 – 500mm) provided by the height from the apparatus and the water tank.

3.3.2 The long-term test

In each of the long-term test cases, a specimen was supplied with water for approximately two months (59 – 63 days) to investigate the relationship between time and the buffer pressure. A WF test was performed just before the end of the test period under (almost) completely saturated condition, al-

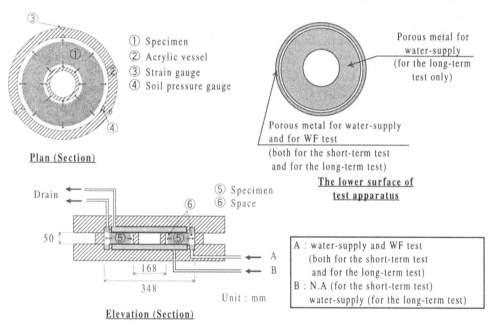

Fig. 4 Conceptual drawing of the test apparatus (both for the short-term test and for the long-term test)

though "break-through" was not expected to be caused because of the limit of water pressure (490.3kPa) of the equipment.

Table 1 Material specifications for specimens

Bentonite Content	100%, 80%, (70%)
Target initial dry density	1.6 Mg/m³, 1.9 Mg/m³, (1.77 Mg/m³)

3.4 Test cases

In this clause, test cases performed in the short-term test and the long-term test are described.

3.4.1 The short-term test

Four cases for the short-term test were performed, using various material specifications of specimens. The conditions of each case are shown in Table 2.

Table 2 Test cases for the short-term test

Case	SP1.58-80	SP1.86-80	SP1.60-100	SP1.81-100
Target initial dry density (Mg/m³)	1.6	1.9	1.6	1.9
Bentonite content (%)	80	80	100	100
Actual initial dry density (Mg/m³)	1.58	1.86	1.60	1.81
Hydraulic head of water-supply (mm)	271	227	271	271
Test period (hours)	97	98	97	96
Number of WF tests	14			
The times of WF tests ; ** hours after water-supply started	0,2,4,8,24,28,31,48,52,55,72,76,79,96			
Control method of the water pressure in WF tests	The water pressure at the start of the 1st WF test was 19.6kPa, and then the water pressure was escalated up to "break through" by 9.8kPa in each WF test. The test start water pressure after (including) the 2nd WF tests was the break through water pressure in the last WF test.			
Judgment of "break through"	Drain ≥ 1.0ml/min			

Table 3 Test cases for the long-term test

Case	LP1.62-80	LP1.77-80	LP1.61-100	LQ1.84-100	LQ1.81-70
Target initial dry density (Mg/m³)	1.6	1.9	1.6	1.9	1.77
Bentonite content (%)	80	80	100	100	70
Actual initial dry density (Mg/m³)	1.62	1.77	1.61	1.84	1.81
Test period (days)	63	59	62	61	63
Number of WF tests	1 (just before the end of the test period)				

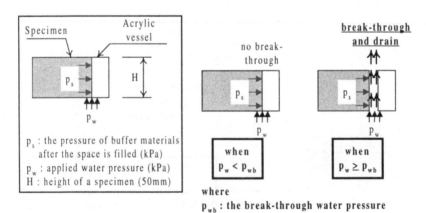

p_s : the pressure of buffer materials after the space is filled (kPa)
p_w : applied water pressure (kPa)
H : height of a specimen (50mm)

where
p_{wb} : the break-through water pressure

Fig.5 Conceptual illustration for the "break-through" water pressure

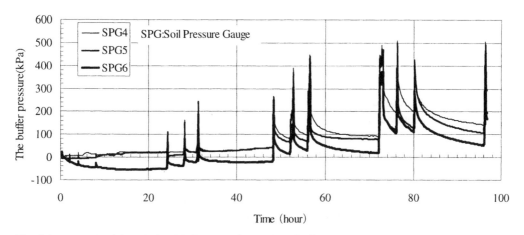

Fig. 6 An example of the relationship between time and the buffer pressure
for the short-term test (in the case of SP1.60-100)

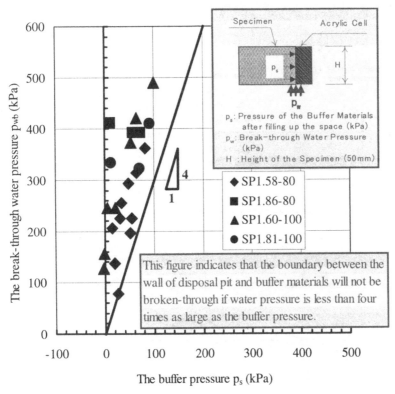

Fig. 7 Relationship between the buffer pressure and the break-through
water pressure (for all the short-term test cases)

3.4.2 *The long-term test*

Five cases for the long-term test were performed, using various material specifications of specimens. The conditions of each case are shown in Table 3.

4 TEST RESULTS AND DISCUSSIONS

This chapter describes test results and discussions for the short-term test and the long-term test.

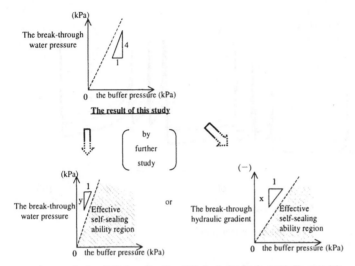

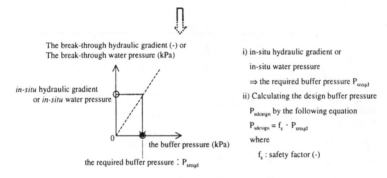

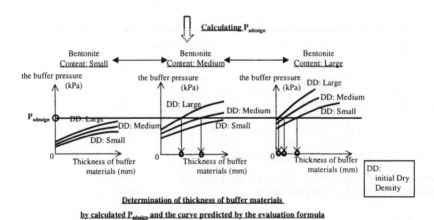

Fig.8 Concept of the suggested design method of buffer material from the viewpoint of self-sealing by considering *in-situ* underground water condition

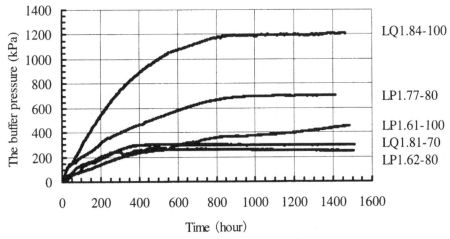

Fig.9 The relationship between time and the buffer pressure (for all the long-term test cases)

Bentonite Content 100%

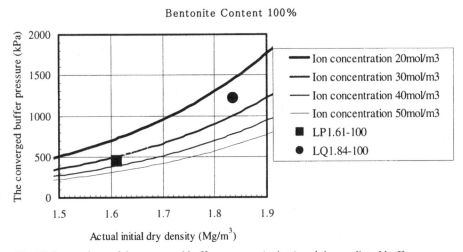

Fig.10 Comparison of the measured buffer pressure (points) and the predicted buffer pressure (curves) for the long-term test in case of bentonite content 100%

4.1 The short-term test

An example of the relationship between time and the buffer pressure in the short-term test is shown in Fig.6. The buffer pressure was measured by SPG4-6, which were the soil pressure gauges attached on acrylic vessel around the outer perimeter of a specimen. The buffer pressure gradually increased after the space was filled by swelling deformation of buffer materials. It rapidly increased up to a peak value during each WF test and then decreased hyperbolically to the pressure nearly equal to that before the WF test. This is considered as such that the deformation of the specimen with Poisson's ratio in response to water pressure increased the buffer pressure, and that the sudden decrease in water pressure

by "break-through" reduced the buffer pressure with some time lag. On the whole the buffer pressure increased as saturated portion of a specimen was getting enlarged.

The relationship between the buffer pressures and the break-through water pressures is plotted in Fig.7 for all the available WF test results. All of the break-through water pressures are more than four times as large as the buffer pressures. This means that the boundary between the wall of disposal pit and buffer materials will not be "broken through" if the water pressure is less than four times as large as the buffer pressure. Namely, buffer materials have effective self-sealing ability against water pressure which is less than four times as large as the buffer pressure.

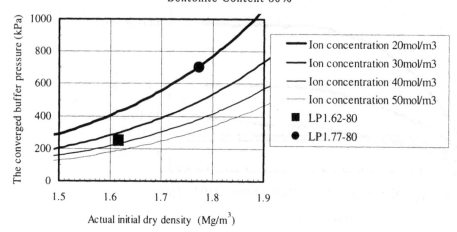

Fig.11 Comparison of the measured buffer pressure (points) and the predicted buffer
pressure (curves) for the long-term test in case of bentonite content 80%

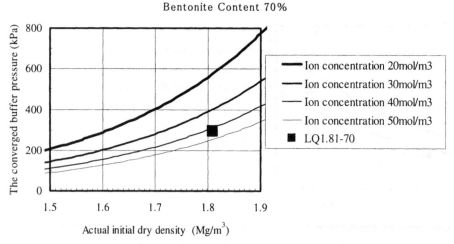

Fig.12 Comparison of the measured buffer pressure (point) and the predicted buffer
pressure (curves) for the long-term test in case of bentonite content 70%

A design method from the viewpoint of self-sealing using "the evaluation formula" and in-situ underground water information can be constructed if the results above are upgraded for buffer materials with the actual dimensions. The following method is suggested;

a. The relationship between the pressure and the break-through water pressure for the actual height of buffer materials should be obtained from further study. The above relationship is required because the break-through water pressure for actual buffer materials may be much greater under the same pressure than that for the test specimen as the height of the actual buffer materials is much greater. Assuming

that the break through water pressure increases in proportion to the height of the buffer material, the relationship can be shown with the y-axis of "the break-through hydraulic gradient" with no unit which is calculated by dividing the break through water pressure by the height of the buffer material.

b. The required buffer pressure (P_{sreqd}) is calculated by measuring in-situ hydraulic gradient or water pressure and by using the relationship developed in a.. The design buffer pressure ($P_{sdesign}$) is calculated by multiplying P_{sreqd} by Safety factor (f_s).

c. The required thickness of buffer materials is calculated by using "the evaluation formula" for the design buffer pressure ($P_{sdesign}$) with specified ion concentration in pore water, bentonite content and initial dry density of buffer materials.

The concept of the method is also illustrated in Fig.8.

4.2 *The long-term test*

As a result of the long-term test, it was confirmed by observation after the test that specimens completely filled the space around the outer perimeter and also that around the inner perimeter.

Fig.9 shows the relationship between time and the buffer pressure for all the long-term test cases. Each of the pressures gradually increased, and then converged to a constant value. This indicates that the specimen was almost completely saturated at the end of the test period. WF test performed just before the end of the test period in each case resulted in no break through because of the limit of water pressure (490.3kPa).

The converged buffer pressures are compared to those calculated by using "the evaluation formula" in Fig.10-12. It is indicated that they agree with each other within the range of ion concentration of pore water from 20 to 50 mol/m^3. Little literature describes the ion concentration of pore water in bentonite or clays. For example, Bolt et al. (1956, 1978) mentioned that the ion concentration of pore water in natural clays were approximately $10 - 100$ mol/m^3 as a result of their experimental investigations. The range calculated is within that in the literature, although detailed investigations are further required in future.

5 CONCLUSIONS

The results are summarized as follows;

a. The space around the outer perimeter of specimen was completely filled in all cases of both the short-term test and the long-term test. The space around the inner perimeter was also filled in the long-term test.

b. As a result of the short-term test, it is confirmed that the break-through water pressure is more than four times as high as the pressure. In addition, the design method of buffer materials from the viewpoint of self-sealing considering the in-situ underground water information is suggested and presented in this paper, in consideration of the test results.

c. As a result of the long-term test, it is confirmed that the pressure of buffer materials after the space is filled can be predicted by using the evaluation formula.

SUPPLEMENT

Komine and Ogata (1999) have developed a revised evaluation formula for swelling property, in which the influence of sand-bentonite content and exchangeable-cation compositions in bentonite are considered. The new formula can evaluate self-sealing ability of buffer materials with a high accuracy. We will construct the new design method of buffer materials from the viewpoint of self-sealing with the new evaluation formula in the near future.

REFERENCES

Atomic Energy of Canada Limited 1994. The Disposal of Canada's Nuclear Fuel Waste: Engineered Barriers Alternatives, AECL-10719 COG-93-8, Whiteshell Lavoratories Pinawa, Manitoba R0E 1L0.

Bolt, G.H. 1956. Physico-Chemical analysis of the compressibility of pure clays, Geotechnique, Vol.6, No.2, pp.86-93, 1956.

Bolt, G.H. and Bruggenwert, M.G.M. 1978. Soil chemistry, A Basic Elements (Second Revised Edition), Elsevier Scientific Publishing Company, Amsterdam.

Japan Nuclear Cycle Development Institute 1999. H12 Project to Establish Technical Basis for HLW Disposal in Japan.

Komine, H. and Ogata, N. 1996. Prediction for swelling characteristics of compacted bentonite, Canadian Geotechnical Journal Vol.33, No.1, pp. 11-22.

Komine, H. and Ogata, N. 1999. Evaluation for swelling characteristics of buffer and backfill materials for high-level nuclear waste disposal –Influence of sand-bentonite content and cation compositions in bentonite-, CRIEPI Report U99013 (In Japanese with English abstract)

Clay Science for Engineering, Adachi & Fukue (eds) © 2001 Balkema, Rotterdam, ISBN 90 5809 175 9

Study on mechanism of gas migration through saturated bentonite / sand mixtures

T. Hokari & T. Ishii
Institute of Technology, Shimizu Corporation, Japan

M. Nakajima & R. Tomita
Koa Kaihatsu Company Limited, Japan

ABSTRACT: Authors have recently been involved in experimental studies on the impact of gas permeation on the impervious characteristics of bentonite / sand mixtures and on scale effects on the gas migration capacity. On the basis of all our results of experiments and observations, this report discusses mechanism of gas migration through saturated bentonite / sand mixtures . It is concluded that gas forms preferential migration paths that are assumed to have a lower resistivity owing to distributions of pore radii and variations in bentonite gel densities in the mixtures.

1 INTRODUCTION

Bentonite / sand mixtures have impervious characteristics owing to the swelling capacity of montmorillonite. These mixtures are expected to be used as a natural impervious barrier in facilities for the disposal of radioactive wastes, and are planned to be emplaced around the wastes themselves or the disposal facilities (Shimoda et al. 1991). It is assumed that several kinds of gases could be generated inside the facilities because of chemical reaction of the wastes with groundwater (Voinis et al. 1992) and that the gases could pass through the bentonite / sand mixtures after some period of storage inside the mixtures.

Komine (Komine et al. 1995) reported a mechanism accounting for the hydraulic properties of bentonite / sand mixtures. He argued that the mixtures have an impervious capacity because voids between aggregates are filled with bentonite gel which swells with water. Pusch (Pusch et al. 1990) proposed a generalised microstructural model of the bentonite / sand mixtures, and reported that, among various factors, the densities of the bentonite gels could have the greatest effect on hydraulic properties, swelling pressure and gas migration. As a property related to gas migration, he concentrated on the gas pressure (referred to as the critical pressure) above which gas started to penetrate into the mixtures, and concluded that when there was a low bentonite content in the mixtures, the critical pressure was nearly equal to the swelling pressure. Pusch (Pusch et al. 1987) conducted experiments in which a cyclic process of saturation and application of gas pressure to bentonite / sand mixtures was studied. He explained that reproducible measures of the critical pressure could be attributed to self-healing of the bentonite / sand mixtures.

We have conducted some experiments on gas migration in the mixtures to determine the gas migration mechanism.

2 EXPERIMENT

Microscopic observations were made after and before gas permeation through bentonite / sand mixtures saturated with water to observe gas migration mechanism in the mixtures.

In addition to the observations, permeabilities were measured by performing hydraulic tests after and before gas permeation in the mixtures. A hydrophobic fluid was used in some permeability tests after gas breakthrough in order to investigate the effects of affinities between the bentonite and the fluids.

Figure 1 illustrates a flow of the experiments.

2.1 *Apparatus*

Microscopic observations were made of specimens removed from test moulds. The observation apparatus is a VH-5910 Monitoring microscope made by KEYENCE Ltd., which allows continuous observation by recording on video tape and obtaining photographs by hard-copying the screen. The magnification could be changed by attachment of lenses, i.e. x 50, 100, 200, 500 and 1000.

Figure 2 shows a schematic view of the cylindrical apparatus used for hydraulic permeability and gas permeation tests. The moulds have a size of 100 mm in diameter and 50 mm in height. The moulds are made of acrylic so as to be able to observe specimens.

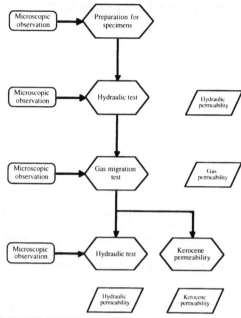

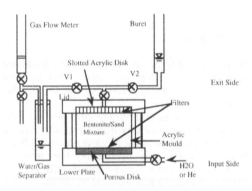

Figure 1. Test flow of hydraulic and gas tests

Figure 2. Schematic view of the cylindrical apparatus for hydraulic and gas tests

2.2 Test material description

Table 1 gives the characteristics of interest and compaction test results for bentonite / sand mixtures for the permeability tests.

The bentonite used was commercial Na-bentonite, Kunigeru-V1 made by Kunimine Industries Co., Ltd. in Japan. Mineral compositions of Kunigeru-V1 are summarized in terms of weight percent (Itoh et al. 1994) as follows: Smectite 46 - 49; Quartz 0.5 -0.7; Chalcedony 37 - 38; Plagioclase 2.7 - 5.5; Calcite 2.1 - 2.6; Dolomite 2.0 - 2.8; Analcime 3.0 - 3.8; Pyrite 0.5 - 0.7. The aggregates used were terrace deposit sands from the Diluvium Epoch in the Quaternary Period. To allow for the variation in fines content (weight content of particles smaller than sand particles; less than 0.074 mm) among them, we used two representative aggregates with different fines content.

One specimen that was re-saturated with a hydrophobic fluid had a bentonite content of 15%, a dry density of 1670 kg/m^3, a water content of 15.8 % and a fines content of 8 %. Commercially available kerosene was used as the hydrophobic fluid.

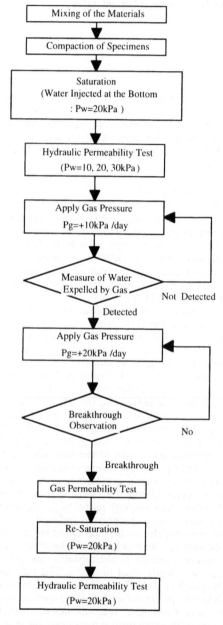

Figure 3. Flow chart of the experimental procedure

Table 1. Description of the tested bentonite / sand mixtures

Bentonite content* (%)	Fines content** (%)	Dry density (kg/m³)	Water content (%)
20	9	1682	16.2
	1	1774	14.0
17.5	1	1714	14.1
	1	1734	14.1
	1	1818	14.1
15	8	1670	15.8
	8	1667	15.8
	8	1675	15.8
	1	1780	12.6
	1	1806	14.6
	1	1817	14.6
	1	1812	14.6

*Bentonite content = Bentonite dry weight / (Bentonite + Sand dry weights)
**Fines content = weight content of smaller particles than sands (0.074 mm)

2.3 Experimental procedure

Microscopic observation were made on the surface of the specimens as follows:

(1) Compaction of the specimens

Compacted specimens were unsaturated with water to conform to the specified dry density and water content. This observation was made on specimens in which the bentonite had not yet swollen.

(2) Saturation completed

The aggregate voids are supposed to be filled with bentonite gel swelling with water. The observation was made of the aggregate voids under saturation.

(3) Gas breakthrough

Comparison with the surface under saturation provides information on gas migration.

(4) Re-saturation

Observation was made of the series of changes occurring when, following gas breakthrough, the specimen was supplied with water.

Additional observation was made of gas migration through bentonite gel for reference.

Uniaxial hydraulic and gas tests were made as follows.

(1) Preparation of specimens

Sand, bentonite and water were measured so that specimens would have the specified dry density and water content. Then, the materials were compacted inside moulds as prescribed in JIS A 1210 (Japanese standards association 1990).

(2) Saturation

Water was injected at a pressure of approximate 20 kPa at the bottom of the moulds. Water flow rates at the exit were measured with burets every day. It is judged that steady rates should represent saturation of the specimens.

(3) Hydraulic permeability test

Initially, the valve V1 was shut and V2 opened. While constant water pressure was applied to the inlet side with gas cylinders, the water flow rate was measured at the exit. The water flow rate was calculated from the differences in water levels inside the burets over timed intervals. The test was performed at three different pressures, i.e. 10, 20 and 30 kPa.

(4) Preparation for gas migration test

After removing the lid, lower plate, porous disks and filters from the mould, a slotted acrylic disk was attached to the top of the mould. The lower porous disk was dried, reattached and filled with helium gas. Water in the inlet tube was displaced with the gas and the upper part of the slotted acrylics disk was filled with water.

(5) Gas breakthrough test

Initially, a gas pressure of 10 kPa is applied to the specimen. When no additional water was expelled by the gas over a fixed period (1 day), the gas pressure was increased to 20 kPa. In the same manner, the gas pressure was increased 10 kPa steps until a pressure was reached above which water started to be expelled by the gas. This pressure is referred to as the threshold pressure. The water volume expelled was measured by changes in the level inside the buret.

Subsequently the pressure was increased by 20 kPa at the end of each fixed period (1 day). The water volume expelled was measured and gas breakthrough was monitored. The breakthrough was confirmed by observation of the upper part of the specimen, and gas pressure and time were recorded at the breakthrough.

(6) Gas permeability test

With the valve V2 shut and V1 opened, the gas flow rate was measured at the breakthrough gas pressure, by measuring the change in water level inside a double tube buret at the outlet. Gas flow rates were measured at various gas pressures.

(7) Re-saturation and Hydraulic permeability test

Water was injected at a pressure of approximate 20 kPa at the bottom of the mould. Water flow rates at the exit were measured with burets, and it is judged that steady flow rates should represent saturation of the specimen. Hydraulic permeability was calculated from the measured flow rate and the pressure.

Figure 3 shows a flow chart of the experimental procedure.

2.4 Data acquisition

Parameters concerning the gas migration are listed as follows.

(1) Hydraulic and gas permeability

As the experiments used various fluids, it is convenient to express hydraulic and gas migration properties of the bentonite / sand mixtures as permeability. It is calculated from the formula for Darcy linear flow under steady conditions (Society of Japan Petroleum Industry, 1983). While formula (1) is applied to an incompressible liquid phase, formula (2) is used for a compressible gas phase:

$$K = \frac{\mu L Q}{A(P_1 - P_2)} \qquad (1)$$

$$K = \frac{2\mu L Q_b P_b}{A(P_1^2 - P_2^2)} \qquad (2)$$

where K is permeability (m^2); μ, viscosity of fluids (Pa s); L, length of the specimen (m); A, sectional area of the specimen (m^2); P, absolute pressure (Pa); subscript b, measurement condition; subscript 1 and 2, inlet and outlet conditions, respectively.

Here, as no back pressure is applied to the output gas, the gas flow rate is measured under atmospheric conditions.

(2) Threshold gas pressure

Threshold gas pressure is the gas pressure above which gas starts to penetrate into water saturated specimens. The lower the threshold gas pressure, the easier gas can migrate through the specimen. In these experiments, the threshold pressure is determined as the initial gas pressure at which drained water is detected at the upper face of the specimens.

(3) Effective gas porosity

Here the effective gas porosity is defined as the ratio of the total volume of drained water prior to breakthrough, when gas paths are formed through the specimen, to the specimen volume. A smaller value of the effective gas porosity means swifter gas migration through the bentonite / sand mixture.

3 RESULTS

3.1 Microscopic observations

Figure 4 represents a specimen of bentonite / sand mixture prepared with compaction, which shows incomplete occupation of the voids with bentonite gel before saturation. Figure 5 represents a specimen saturated with water to which gas is about to be applied. It is found that the aggregate voids are filled with the swelling bentonite gel. Figure 6 represents a specimen in which gas has broken through. While most of the voids are found to be occupied with bentonite gel, there are a few voids which are empty. Considering the gas breakthrough, the void holes are paths through which gas migrates in the mixtures. The effect of re-saturation was observed as water was supplied to the paths formed by gas breakthrough. It was observed that the gas paths were occupied with bentonite gel expanding in the supplied water. Figure 7 represents a final stage at which bentonite gel re-fills the voids.

While Figure 4 - 7 show the situations of the bentonite / sand mixtures, Figure 8 - 9 represent gas paths that are formed in bentonite gel. The gel is made by supplying bentonite powders with water in a small

transparent case of the glass. The bentonite powders are saturated, swell and are left for some time to become stable in the case. After gas is applied to one side of the case, it is observed that preferential paths are formed in the gel, though the gel is assumed to be uniform in density or concentration of the bentonite.

1.0 mm

Figure 4. Specimen prepared by compaction

1.0 mm

Figure 5. Specimen saturated with water

3.2 Data on gas migration

Figure 10 (Hokari et al., 1996) shows examples of a relationship between gas pressure and water volume expelled by gas to yield breakthrough. It is found that there is threshold pressure above which gas starts

556

penetrating into the bentonite / sand mixtures. It is attributed to a capillary phenomenon owing to an interfacial tension between the gas, bentonite gel ,sands and capillaries. Considering variations in the void size of the mixtures, a measure of the threshold pressure is dependent upon the widest void in the specimen, which is correspondent to the easiest gas migration path. In these tests, the threshold pressure is approximately 50 kPa.

As Imamura (Imamura et al., 1996) reported, it is found that there are 2 types of curves in Figure 10, one of which shows a linear relationship between the gas pressure and the water volume, while the other shows a less correlation between them. The breakthrough pressures of the former mixtures are lower than those of the latter. In other words, the gas migrates more swiftly in the former than in the latter. The latter behavior is supposed to be ascribed to the movement of bentonite gel and the occurrence and dispersion of excessive pore pressure with the gas penetration. While water permeabilities of the mixtures are, plotted around 10^{-18} m^2 (10^{-11} m/s in terms of conductivity), reproducible, behaviors of gas penetration vary among them and are supposed to be less stable.

Figure 11 represents a result of the hydraulic and gas permeabilities of the bentonite / sand mixtures. The gas permeabilities are 2 - 4 orders of magnitude larger than the hydraulic ones. It is also found that gas permeabilities are less reproducible than water ones.

Figure 12 shows the porosity and effective gas porosity. The latter is one order of magnitude less than the former. It suggests that the gas paths are limited in numbers in the mixtures compared to the water paths.

Figure 7. Re-saturation of the specimen

It is interesting to notice that gas permeabilities are much larger than the hydraulic ones although the gas paths are supposed to be less in numbers than water ones.

4 CONCLUSIONS

We have conducted microscopic observations and gas migration tests with bentonite / sand mixtures so as to understand the mechanism of the gas migration in the water-saturated mixtures.

1. When bentonite / sand mixtures are water-saturated, voids of the mixtures are filled with swelling bentonite gel, which leads to the impervious capacity of the mixtures, of less than 3×10^{-18} m^2 (3×10^{-11} m/s) in these tests. The measures of hydraulic permeability are plotted around 10^{-18} m^2 and supposed to be reproducible.

2. There is a threshold phenomenon in the gas penetration into the bentonite / sand mixtures. It is attributed to a capillary phenomenon owing to an interfacial tension between gas, bentonite gel , sands and capillaries. A measure of the threshold pressure is supposed to be dependent upon the widest void in the mixtures, which is correspondent to the easiest gas migration path.

3. It is found that water is expelled by gas pressure from the threshold pressure to the gas breakthrough. That means that gas displaces void water to migrate in the mixtures. It is also supported by microscopic observation results that some preferential empty voids are found on the bentonite / sand mixtures after gas breakthrough.

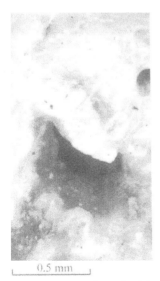

Figure 6. Specimen after gas breakthrough

4. There are 2 types of behaviors of the water expelled by the gas pressure, one of which shows a linear relationship between the gas pressure and the water volume, the other of which shows a less correlation between them. The breakthrough pressures of the former mixtures are lower than those of the latters. The behaviors of gas penetration vary among them and less reproducible compared to measures of hydraulic permeability.

5. Effective gas porosity is 1 order of magnitude less than the porosity. That means that gas displaces preferential parts of the voids in the mixtures. Partial displacement of the voids are also observed on the surface of the mixtures after gas breakthrough with the microscope.

6. While the gas paths are much less frequent than water ones, gas permeability is 2 - 4 order of magnitude larger than hydraulic one. It is supposed to be attributed to a pipe flow of gas through gas paths formed by gas migration.

7. After gas breakthrough, the empty voids are refilled with the bentonite gel swelling by water supply. It is supposed that gas squeezes the bentonite gel in the voids to displace only water when it migrates in the bentonite / sand mixtures.

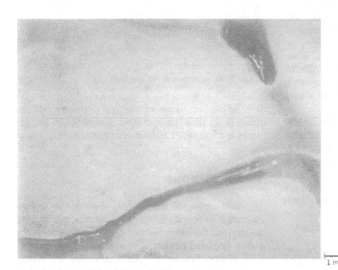

Figure 8. Preferential gas paths formed in the bentonite gel

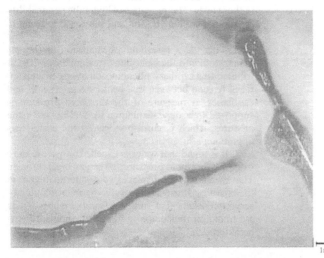

Figure 9. Changes in gas paths of the bentonite gel with the migration

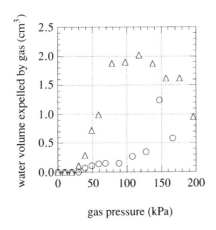

Figure 10 Example of the behaviors to gas breakthrough

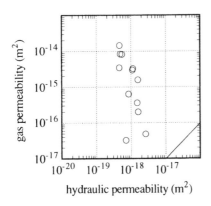

Figure 11 Relationship between permeabilities of water and gas

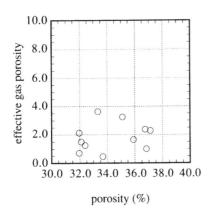

Figure 12 Relationship between porosities of water and gas

REFERENCES

Hokari, T. et al. 1997. *Experimental study on self-healing of bentonite / sand mixtures and its impact on hydraulic permeability*. Mat. Res. Soc. Symp. Proc. Vol.465: 971-978

Hokari, T. et al. 1997. *Experimental study on scale effects of bentonite / sand mixtures on gas migration properties*. Mat. Res. Soc. Symp. Proc. Vol.465: 1019-1026

Imamura, S. et al. 1996. *Experimental study on gas migration through the bentonite/sand mixture*. Proc. Workshop Atomic Energy Soc. Jpn. March 27 - 29:651

Imamura, S. et al. 1996. *Experimental study on gas migration through the bentonite / sand mixture (phase - 2)*. Proc. Workshop Atomic Energy Soc. Jpn. Sep. 23 - 25 :682

Itoh, M. et al. 1994. J. Atomic Energy Soc. Jpn. Vol.36, No.11

Japanese Standards Association 1990. *Test method for soil, compaction using a rammer - JIS A 1210*

Komine, H. et al. 1995. Proc. Sym. Soil Mech., No.40, VI-5

Pusch, R. et al. 1990. *SKB /Technical report 90-43*

Pusch, R. et al. 1987. *SFR Arebetsrapport 87-06*

Shimoda, H. et al. 1991. J. Atomic Energy Soc. Jpn., Vol.33, No.11

Soc. Jpn. Petroleum Engineers 1983. *Handbook of petroleum industry*: 504

Voinis, S. et al. 1992. *The analytical modeling of gas generation*. Proceedings of a workshop organised by NEA in co-operation with ANDRA: 111-119. Paris

Clay Science for Engineering, Adachi & Fukue (eds) © 2001 Balkema, Rotterdam, ISBN 90 5809 175 9

Modelling of bentonite swelling as solid particle diffusion

T. Kanno & Y. Iwata
Ishikawajima-Harima Heavy Industries Company Limited, Yokohama, Japan

H. Sugino
Japan Nuclear Fuel Cycle Development Institute, Tokai-mura, Japan

ABSTRACT: A physical model as solid particle diffusion is developed for the bentonite extrusion phenomenon assumed in a geologic repository for high-level radioactive waste, and numerical analyses with a finite element code ABAQUS for extrusion experiments are performed. In this modelling the extrusion is treated as a combination of the free swelling of compacted bentonite and viscous flow of bentonite gel in a narrow channel (fracture). The solid diffusivity as a function of the volume fraction of solid is derived from the hydraulic conductivity, swelling pressure and viscosity of the bentonite gel and the aperture of a fracture. The calculated results of extrusion rates are in good agreement with the experimental results. Especially, the modelling can reproduce the aperture dependence of the extrusion rate.

1 INTRODUCTION

A bentonite-quartz sand mixture, in a densely compacted form, is considered to be one of the most promising candidates for buffer material for the geological disposal of high-level radioactive waste in Japan (Hasegawa et al. 1999, JNC 1999a). The compacted bentonite-sand mixture has a number of favourable properties, such as its low permeability and high capacity for radionuclide sorption. However, as the buffer resaturates and swells following repository closure, it may be excluded into fractures in the surrounding rock. And if loss of the buffer into fractures due to extrusion and subsequent erosion of the extruding front by flowing groundwater is too pronounced, then the decrease in density of the buffer within the repository may (at least locally) be sufficient to reduce its favourable properties. The extrusion and erosion scenario of the buffer in a repository is schematically shown in Figure 1.

Therefore, the extrusion and erosion of bentonite-based buffer from the disposal pit (in a pit disposal type) or tunnel (in a tunnel disposal type) should be quantitatively understood to evaluate long-term physical stability of the engineered barriers for the geological disposal of high-level radioactive waste.

Key findings of existing laboratory studies of compacted-bentonite extrusion into fractures are that:

· in the absence of erosion, the migration distance y (Figure 1) is proportional to the square root of time t after the contact between water and bentonite and the extrusion rate is dependent on the fracture aperture d and bentonite content Bc as follows (Kanno & Wakamatsu 1991, Kanno et al. 1999):

$$y = A(d, Bc)\sqrt{t} \qquad (1)$$

· two regions are detected to develop in the excluding bentonite front: one fairly stiff and very soft (Pusch 1983),

· the water content in the very soft region just behind the gel front is about 550%, which is close to the liquid limit of the bentonite clay used (Boisson 1989).

Some modelling studies for the bentonite extrusion have been reported (Pusch 1981 & 1983, Börgesson 1989, Kanno & Wakamatsu 1991, Verbeke et al. 1997). However, these studies have not given any physical explanations of the dependence of the extrusion rate on the fracture aperture expressed in Equation 1.

Therefore, a model for the extrusion phenomenon has been proposed (Kanno et al. 1999) by the authors based on the theory of clay particle diffusion (Nakano et al. 1982 & 1986) to explain the dependence. This paper gives a detailed report on numerical analyses for large-scale extrusion experiments named BENTFLOW as well as the model derivation.

2 MODELLING OF EXTRUSION BASED ON SOLID PARTICLE DIFFUSION

The basic mechanism of the extrusion phenomenon is thought to be free expansion of bentonite particle structure due to its swelling ability. It is reported (Kanno & Wakamatsu 1991) based on experiments that the extrusion rate depends on the fracture aperture. However, as shown in Figure 2, the proportional constant A in Equation 1 defining the extrusion rate has an asymptote tendency at the aperture above 1 mm. This tendency may be explained that frictional effects in the extrusion with the aperture above 1 mm diminish to a negligible level and the extrusion is dominated by the free swelling (free expansion). And bentonite particle structure in the disposal pit which has the large volume is thought to be in the state of free swelling. Therefore, two kinds of models, for the part of a disposal pit and the part of a fracture, are necessary to develop a diffusion model for the extrusion phenomenon in a repository.

2.1 Diffusion model for free swelling

Modelling for the free swelling is tried prior to that for the migration in a fracture. The one-dimensional flux equation for swelling clays is given (Nakano et al. 1982 & 1986) as

$$q_s = -k_s \frac{\partial \psi_s}{\partial z} \qquad (2)$$

where q_s is the flux of solid particles, k_s is a kinetic coefficient introduced on the analogy of the hydraulic conductivity, ψ_s is the potential responsible for the movement of solid particles, and z is the space coordinate. Defining the solid diffusivity (Nakano et al. 1982), or apparent clay particle diffusivity (Nakano et al. 1986), D_s

$$D_s = k_s \frac{d\psi_s}{d\sigma} \qquad (3)$$

Equation 2 can be rewritten

$$q_s = -D_s \frac{\partial \sigma}{\partial z} \qquad (4)$$

where σ is called the volume fraction of solid (Nakano et al. 1982), volumetric solid content, or volumetric clay content (Nakano et al. 1986) in the case of pure clay sample, defined as

$$\sigma = \frac{V_s}{V} \qquad (5)$$

where V is the total volume of the material and V_s is the volume of solid particles.

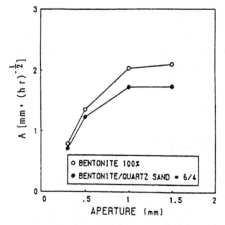

Figure 2. Proportional constant A as function of aperture (Kanno & Wakamatsu 1991).

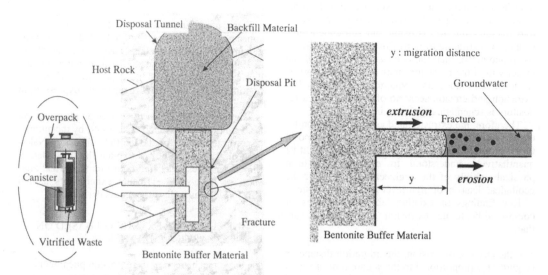

Figure 1. Extrusion and erosion scenario of bentonite buffer material in a repository for high-level radioactive waste.

562

The solid diffusivity for a Japanese sodium bentonite, Bentonite GA provided by Kunimine Industries Co., Ltd. and has a montmorillonite content of 65%, was measured with infiltration experiments (Nakano et al. 1982 & 1986) as shown in Figure 3. In this study, it is tried to obtain an analytical expression for the solid diffusivity.

As the driving force of the movement of solid particles is thought to be the gradient of the swelling pressure Ps, it is assumed as the first approximation that ψ_s is expressed

$$\psi_s = \frac{Ps}{\rho_w g} \tag{6}$$

where ρ_w is the density of liquid water and g is the gravitational constant. The kinetic coefficient is evaluated as follows:
In the case that both water and solid particles move together, the mean velocity of water relative to the solid particles u is given by

$$u = u_w - u_s \tag{7}$$

where u_w and u_s are the mean water velocity and mean velocity of solid particles relative to a given space coordinate, respectively. The fluxes corresponding to u_w, u_s and u are

$$q_w = \theta u_w \tag{8}$$

$$q_s = \sigma u_s \tag{9}$$

$$q = \theta u \tag{10}$$

where θ is the volumetric water content. Substituting Equations 8, 9 and 10 into Equation 7

$$q_w = q + \frac{\theta}{\sigma} q_s \tag{11}$$

The water flux relative to the solid particles is given by Darcy's law as

$$q = -k \frac{\partial \psi_w}{\partial z} \tag{12}$$

where k is the hydraulic conductivity and ψ_w is the water potential. Applying the Darcy's law type expression to the flux q_w

$$q_w = -k_w \frac{\partial \psi_w}{\partial z} \tag{13}$$

where k_w may be considered as the hydraulic conductivity relative to the space coordinate. Here, the following relationship between ψ_s and ψ_w is assumed (Nakano et al. 1982 & 1986)

$$\psi_w + \psi_s = 0 \tag{14}$$

Substituting Equation 14 into Equation 2

$$q_s = k_s \frac{\partial \psi_w}{\partial z} \tag{15}$$

Substituting Equations 12, 13 and 15 into Equation 11

$$-k_w \frac{\partial \psi_w}{\partial z} = -\left(k - k_s \frac{\theta}{\sigma}\right) \frac{\partial \psi_w}{\partial z} \tag{16}$$

The continuity equation of water is given

$$\frac{\partial \theta}{\partial t} + \frac{\partial q_w}{\partial z} = 0 \tag{17}$$

Combining Equations 13 and 17

$$\frac{\partial \theta}{\partial t} = -\frac{\partial q_w}{\partial z} = \frac{\partial}{\partial z}\left(k_w \frac{\partial \psi_w}{\partial z}\right) \tag{18}$$

Similarly, from Equation 15 and the continuity equation of solid particles

$$\frac{\partial \sigma}{\partial t} = -\frac{\partial q_s}{\partial z} = -\frac{\partial}{\partial z}\left(k_s \frac{\partial \psi_w}{\partial z}\right) \tag{19}$$

Combining Equations 18 and 19

$$\frac{\partial}{\partial t}(\theta + \sigma) = \frac{\partial}{\partial z}\left\{(k_w - k_s)\frac{\partial \psi_w}{\partial z}\right\} \tag{20}$$

Here, it is assumed for simplicity that the water-solid system is saturated. In the saturated system, since the sum of the volumetric water content and volume fraction of solid is always 1, the left hand side of Equation 20 is 0. Therefore,

$$k_s = k_w \tag{21}$$

Substituting Equation 21 into Equation 16

$$k_s = k - k_s \frac{\theta}{\sigma} \tag{22}$$

that is

$$k_s = \left(\frac{\sigma}{\sigma + \theta}\right) k \tag{23}$$

From the assumption of the saturated condition

$$k_s = \sigma k \tag{24}$$

Combining Equations 3, 6 and 24, the solid diffusivity is expressed as

$$D_s = \frac{\sigma k}{\rho_w g} \frac{dPs}{d\sigma} \tag{25}$$

Since there is no empirical formula for the swelling pressure of Bentonite GA, the necessary expression is assumed in this study from the data for Kunigel V1 taking the montmorillonite content into consideration (see 3.2.1).

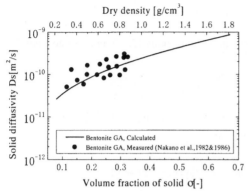

Figure 3. Solid diffusivity of Bentonite GA for free swelling.

The analytical expression of the solid difusivity of free swelling for Bentonite GA is obtained as

$$D_{sGA}[m^2 / s] = 1.01 \times 10^{-8} \sigma \exp(2.3\sigma - 3.6) \quad (26)$$

Calculated results with Equation 26 are shown in Figure 3 along with the measured results. Both results are in good agreement, which supports the validity of the modelling for the free swelling.

2.2 Diffusion model for extrusion

Equations 4 and 25 for the evaluation of the free swelling may be applied to the extrusion phenomenon in wider fractures. However, though the extrusion rate depends on the fracture aperture as shown in Figure 2, Equation 25 can not express the dependence of the extrusion rate on the aperture. The dependence is thought to be due to frictional effects at the interface between solid particles and the fracture surface or between the particles. In this study, the viscous flow model is applied to take the frictional effects into account and it is tried to express the viscous flow model as the diffusion equation. The following approach is the application of that used in the hydrodynamics field (e.g. Milne-Thomson 1968):

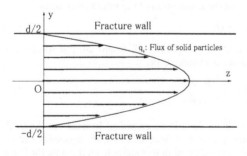

Figure 4. Modelling of extrusion in a fracture.

The flux equation for solid particles is written as follows from Equations 2 and 6:

$$q_s = -\frac{k_s}{\rho_w g} \frac{\partial Ps}{\partial z} \quad (27)$$

Here, taking the extrusion as steady, creeping flow of viscous fluid in a fracture between two fixed parallel plates at the distance d apart and suppose the motion is in the z-direction only in the z-y plane (Figure 4). Applying two-dimensional Navier-Stokes equation neglecting the inertia term

$$\frac{dPs}{dz} = \mu_s \frac{d^2 u_s}{dy^2} = \frac{\mu_s}{\sigma} \frac{d^2 q_s}{dy^2} \quad (28)$$

where μ_s is the viscosity of the gel consisting of solid particles. In this case the pressure gradient $dPs/dz=P$ is independent of y, z, therefore

$$\frac{d^2 q_s}{dy^2} = \frac{\sigma}{\mu_s} P \quad (29)$$

and hence

$$q_s = \frac{\sigma}{2\mu_s} Py^2 + C_1 y + C_2 \quad (30)$$

Since $q_s=0$ when $y=-d/2$ and $y=d/2$ from the no slip condition on the fracture wall surfaces, we get

$$q_s = \frac{\sigma P}{2\mu_s} \left(y^2 - \frac{d^2}{4} \right) \quad (31)$$

The average value of q_s across a section perpendicular to z is

$$\begin{aligned} \bar{q}_s &= \frac{1}{d} \int_{-d/2}^{d/2} \frac{\sigma P}{2\mu_s} \left(y^2 - \frac{d^2}{4} \right) dy \\ &= -\frac{\sigma P}{12\mu_s} d^2 \quad (32) \\ &= -\frac{\sigma d^2}{12\mu_s} \frac{dPs}{dz} \end{aligned}$$

Comparing Equations 27 and 32 the kinetic conductivity for the section is obtained as

$$k_s = \rho_w g \frac{\sigma d^2}{12\mu_s} \quad (33)$$

The solid diffusivity for the extrusion is obtained from Equations 3, 6 and 33

$$D_s = \frac{\sigma d^2}{12\mu_s} \frac{dPs}{d\sigma} \quad (34)$$

2.3 Physical interpretation of the modelling

The hydraulic conductivity is often expressed with the intrinsic permeability concept

$$k = \frac{\kappa \rho_w g}{\mu_w} \qquad (35)$$

where κ is the intrinsic permeability which has the dimension of area and μ_w is the viscosity of water. From Equations 24 and 35, the kinetic conductivity for the free swelling is written

$$k_s = \sigma \frac{\kappa \rho_w g}{\mu_w} \qquad (36)$$

which means that physical parameters controlling the free swelling are viscosity of water and the intrinsic permeability. The intrinsic permeability is thought to be representative area for water to move among the solid particles. Therefore, the rate of the free swelling of bentonite is determined by the viscous drag of water flow, as the countercurrent to the solid particle movement, in the voids of the solid particle structure.

The same effects of the water flow in the free swelling case may also exist in the extrusion case. However, the experimental observation, i.e. the extrusion rate depends on the aperture, means that the viscous drag of the water flow among the solid particles, at least κ, is not the controlling factor for the extrusion. In the modelling described above, the dominant controlling parameters for the extrusion are the viscosity of the solid particle gel and the aperture of the fracture as seen in Equation 33. Since Equation 33 is obtained by the assumption of the creeping flow, the equation is not valid for wider aperture when the inertia term of flow can not be neglected. In wider apertures, the value of the kinetic conductivity from Equation 33 becomes larger than that from the free swelling, and the solid particle movement may come to be controlled by the viscous drag of water flow as the free swelling. Assuming the values of the kinetic conductivity from Equations 33 and 36 are equal

$$\frac{d^2}{12\mu_s} = \frac{\kappa}{\mu_w} \qquad (37)$$

Therefore there is a certain value of d at which the effects of drag caused by both water and solid gel viscous properties are equivalent. The dimensionless ratio

$$\frac{d^2}{12\mu_s} \bigg/ \frac{\kappa}{\mu_w}$$

seems to be an index for the transition between the extrusion and free swelling.

3 NUMERICAL ANALYSIS OF BENTFLOW EXPERIMENTS

3.1 The BENTFLOW experiments analyzed

The BENTFLOW experiments (Kanno & Matsumoto 1997, Kanno et al. 1999) are designed to simulate the extrusion and erosion of the bentonite buffer in a disposal pit with single rock fracture. The surrounding rock mass is simulated by a test cell consisting of two thick plexiglass plates. An annular slot between the two plates simulates a horizontal rock fracture intersecting a disposal pit. The width of the slot is controlled by the thickness of thin stainless steel plates sandwiched between the plexiglass plates. A compacted bentonite buffer specimen (diameter : 50mm, height : 50mm) is installed in a hole drilled in the cell. The hole simulates a disposal pit and is called a core part in this paper. The cell is designed so that water is supplied through the slot to the specimen. By this procedure, the specimen takes up water, swells and extrudes into the slot after the contact between water and the specimen. The distance to the extruding gel front from the inner end of the slot, corresponding to the migration distance y shown in Figure 1, is determined by a photograph taken from above. The experimental conditions of the tests analyzed in this study are summarized in Table 1. The bentonite used is Kunigel V1 (Ishikawa et al. 1989) which is a Japanese natural Na bentonite, and the grain size of the sand used is 0.1-4.75 mm (mean grain size is approx. 0.8 mm). All tests shown in Table 1 are carried out at room temperature, ca. 20°C, using distilled water.

Table 1. Conditions of tests for numerical analysis.

Material	Dry density (initial) g/cm^3	Slot width mm
Kunigel V1-100% (pure bentonite)	1.8	1.5
Kunigel V1-100% (pure bentonite)	1.8	0.5
Kunigel V1-70%, sand-30% (mixture)	1.8	1.5

3.2 Evaluation of solid diffusivities and material properties

In the modelling approach for the extrusion phenomenon in this study, the solid diffusivities have to be determined for both the core and slot parts using Equations 25 and 34, respectively. The range of the volume fraction of solid required to the numerical analysis is 0.0817 to 0.667 (see 3.3).

3.2.1 Core part
The required material properties for Equation 25 are the hydraulic conductivity and swelling pressure.

565

These material properties are often expressed as functions of the dry density, thus, it is necessary to convert the data to those as functions of the volume fraction of solid.

- Pure bentonite specimen:
The hydraulic conductivity for the mixture of Kunigel V1 and sand is evaluated by the intrinsic permeability concept and an empirical formula is given as (PNC 1996)

$$\kappa = 10^{-18.621+1.082\rho_d-0.429\rho_d{}^2-0.736Bc\rho_d{}^2}$$
$$\left(1.0 \leq \rho_d \leq 1.8[g/cm^3]\right) \tag{38}$$

where κ is the intrinsic permeability $[m^2]$, ρ_d $[g/cm^3]$ is the dry density, and Bc is a bentonite content. The value of Bc is 1 in the case of pure bentonite specimens. The expression of the hydraulic conductivity is obtained from Equations 35 and 38, and the properties of water (National Astronomical Observatory 1990)

$$k[m/s] = 1.1 \times 10^{-11.6+2.92\sigma-8.53\sigma^2} \tag{39}$$

From an empirical formula for the swelling pressure of the pure specimen (Suzuki et al. 1992)

$$Ps[MPa] = 2.28 \times 10^{-5} \exp(6.7\rho_d)$$
$$\left(1.5 \leq \rho_d \leq 2.0[g/cm^3]\right) \tag{40}$$

that is

$$Ps[Pa] = 22.8\exp(18.1\sigma) \tag{41}$$

The equation of the solid diffusivity is obtained from Equations 25, 39 and 41

$$D_s[m^2/s] = 4.63 \times 10^{-2}\sigma\{\exp(18.1\sigma)\}$$
$$\cdot 10^{-11.6+2.92\sigma-8.53\sigma^2} \tag{42}$$

- Mixture specimen:
In this case, the value of Bc is 0.7 and the expression of the hydraulic conductivity is obtained from Equation 35 and 38

$$k[m/s] = 1.1 \times 10^{-11.6+2.92\sigma-6.88\sigma^2} \tag{43}$$

An empirical formula of the swelling pressure for the mixture of Kunigel V1 and sand is reported (JNC 1999b) as

$$Ps[MPa] = \exp\left(3.97\rho_e{}^2 - 7.71\rho_e + 2.38\right)$$
$$\left(1.36 \leq \rho_e \leq 2.00[g/cm^3]\right) \tag{44}$$

using the effective clay density parameter ρ_e defined as

$$\rho_e = \frac{\rho_d Bc}{1 - \dfrac{\rho_d}{\rho_{sand}}(1 - Bc)} \tag{45}$$

where ρ_{sand} is the density of sand. However, the value calculated from Equation 44 increases as the density decreases in the lower range of the density, which is unreasonable (Figure 5). Therefore Equation 44 can not be used for this study which has to cover wide range of the density. Since the form of Equation 40 is thought to be more applicable than that of Equation 44 for wide range of the density, the swelling pressure for the mixture specimen is evaluated as follows, taking the dry density in Equation 40 as the effective clay density:

$$Ps[MPa] = 2.28 \times 10^{-5} \exp(6.7\rho_e) \tag{46}$$

The calculated values from Equations 44 and 46 are plotted in Figure 5. It is confirmed that the evaluated results from Equation 46 are in fair agreement with those from the empirical formula (Equation 44) in its available range, though the evaluated values have a possibility of underestimating in the lower density range.

Substituting Bc=0.7 and Equation 45 into Equation 46

$$Ps[Pa] = 22.8\exp(\frac{12.7\sigma}{1-0.3\sigma}) \tag{47}$$

The equation of the solid diffusivity for the mixture specimen is obtained from Equations 25, 43 and 47

$$D_s[m^2/s] = 3.25 \times 10^{-2}\left(10^{-11.6+2.92\sigma-6.88\sigma^2}\right)$$
$$\cdot \frac{\sigma}{(1-0.3\sigma)^2}\exp\left(\frac{12.7\sigma}{1-0.3\sigma}\right) \tag{48}$$

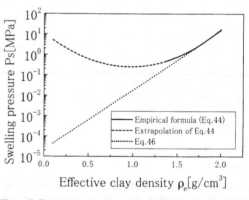

Figure 5. Evaluation of swelling pressure for mixture specimen.

The evaluated results of the solid diffusivities of the core part for the two materials are plotted in Figure 6.

- Bentonite GA:

An empirical formula for the hydraulic conductivity of Bentonite GA in a saturated state is reported (Nakano et al. 1984) as a function of the volumetric water content θ

$$k[cm/day] = \exp(15.83\theta - 19.37)$$
$$(approx., 0.65 \leq \theta \leq 1.0) \tag{49}$$

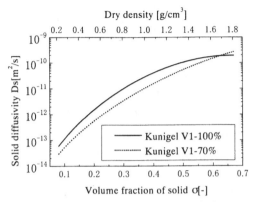

Figure 6. Solid diffusivities of pure bentonite and mixture specimens for core part.

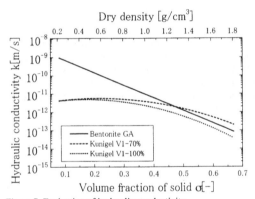

Figure 7. Evaluation of hydraulic conductivity.

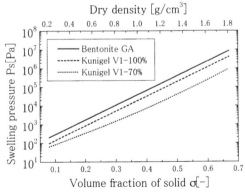

Figure 8. Evaluation of swelling pressure.

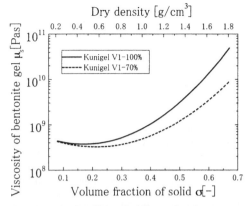

Figure 9. Evaluation of viscosity of bentonite gel.

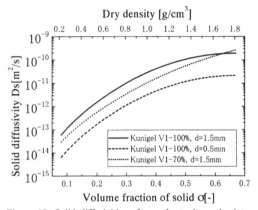

Figure 10. Solid diffusivities of pure bentonite and mixture specimens for slot part.

The swelling pressure of Bentonite GA is evaluated based on Equation 41 considering the montmorillonite content, since it has not been reported yet. The effect of the montmorillonite content on the swelling pressure is investigated using specimens with the dry density of 1.8 [g/cm³] for Kunigel V1, MX-80, and Kunipia (purified Kunigel V1) bentonites and the following empirical relationship is reported (Suzuki at al. 1992):

$$Ps[MPa] = 0.1596 \exp(0.058X)$$
$$(40 \leq X \leq 99[\%]) \tag{50}$$

where X (%) is the montmorillonite content. A ratio of the swelling pressure of Bentonite GA (X=65%) to that of Kunigel V1 (X=47.5%, Ito et al. 1993) is calculated from Equation 50 as 2.06 at the value of the dry density. Assuming that the ratio does not depend on the density, the swelling pressure of Bentonite GA is evaluated from the ratio and Equations 41 as

$$Ps[Pa] = 47.0 \exp(6.7\rho_d) \tag{51}$$

Figure 11. Mesh of numerical analysis for BENTFLOW experiment (pure bentonite specimen, d=1.5mm).

The evaluated results of the hydraulic conductivity and swelling pressure of these materials are shown in Figures 7 and 8. The increments of the hydraulic conductivities with decrement of the volume fraction of solid for the pure bentonite and mixture specimens show an asymptote tendency, which may not be observed in reality. This means that extrapolation of Equation 38 to lower density range may lead to some errors in evaluation.

3.2.2 Slot part

Equation 34 in which the solid diffusivity depends on the fracture aperture can be used to explain the measured results of the extrusion tests. However, there is no available data for the viscosity of bentonite gel at present. It has been suggested (Kanno & Wakamastu 1991) that the coefficient A in Equation 1 shows an asymptote tendency at the aperture above 1 mm as shown in Figure 2. Based on this observation, the viscosity of the bentonite gel is evaluated by assuming that the values of the solid diffusivity become equal in both Equations 25 and 34 at the aperture of 1.5 mm. This assumption means that the proposed index is 1 at the aperture of 1.5 mm (see 2.3).

Once the expression for the evaluation of the viscosity is obtained, the diffusivity can be calculated using Equation 34. The evaluated results of the gel viscosity and solid diffusivity for the Kunigel V1 and the mixture are as follows (Figures 9 and 10):

- Pure bentonite specimen:

$$\mu[Pas] = 1.67 \times 10^{-3} \cdot 10^{11.6-2.92\sigma+8.53\sigma^2} \tag{52}$$

$$D_s[m^2/s] = \frac{d^2}{12}$$
$$\cdot \frac{413\sigma \exp(18.1\sigma)}{1.67 \times 10^{-3} \cdot 10^{11.6-2.92\sigma+8.53\sigma^2}} \tag{53}$$

- Mixture specimen:

$$\mu[Pas] = 1.67 \times 10^{-3} \cdot 10^{11.6-2.92\sigma+6.88\sigma^2} \tag{54}$$

$$D_s[m^2/s] =$$
$$\frac{d^2}{12} \cdot \frac{\sigma}{1.67 \times 10^{-3} \cdot 10^{11.6-2.92\sigma+6.88\sigma^2}}$$
$$\cdot \frac{290}{(1-0.3\sigma)^2} \exp\left(\frac{12.7\sigma}{1-0.3\sigma}\right) \tag{55}$$

The decrements of the gel viscosity with decrement of the volume fraction of solid show an asymptote tendency in lower range of the volume fraction of solid (Figure 9), which may be caused by the possible errors in the evaluation for the hydraulic conductivity (Figure 7).

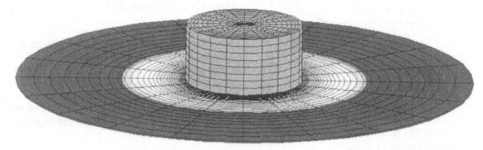

Figure 12. Calculated results for distribution of volume fraction of solid (pure bentonite specimen, d=1.5mm, t=6.048×10⁶ s).

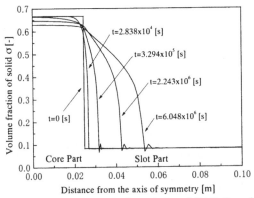

Figure 13. Calculated results for variation of distribution of volume fraction of solid with time.

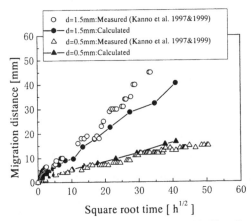

Figure 14. Variation of Migration Distance with Time (Pure bentonite specimen).

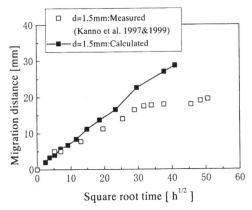

Figure 15. Variation of Migration Distance with Time (Mixture Specimen).

3.3 Method of numerical analysis and calculated results

Based on the modelling described above, numerical analyses for the BENTFLOW experiments are performed using the finite element analysis code ABAQUS versions 5.7 & 5.8. The analyses are performed as axisymmetric, transient mass diffusion problems and Figure 11 shows the mesh in the r-z plane for the analysis of the case of pure bentonite specimen with the slot width of 1.5 mm. The only upper half is analyzed due to the fact that the gravitational effects are thought to be negligible, and hence the phenomenon is symmetrical in z-direction. A three-dimensional expression of the mesh is shown in Figure 12 with calculated results. The length of the analysis region in the radial direction is 100 mm (the length of the slot part is 75 mm), which is judged from the measured results of the migration distance. The axisymmetric, 8-node quadratic, elements are used in the analyses. The numbers of nodes and elements are 459 and 120, respectively.

The governing equation, material property, initial conditions and boundary conditions are as follows:
- Governing equation;

$$\frac{\partial \sigma}{\partial t} = \mathrm{div}\{D_s(\sigma)\mathrm{grad}\sigma\} \qquad (56)$$

- Material property (solid diffusivity);
Equations 42 and 48 are applied to the material in the core part and Equations 53 and 55 are applied to the material in the slot part.
- Initial conditions;
The volume fraction of solid in the core part is 0.667 (corresponding to the dry density of 1.8 g/cm^3) for both the pure bentonite and mixture specimens.
- Boundary conditions;
Based on experimental results (Boisson 1989), the water content of the extruding gel front is assumed to be equal to the liquid limit of the material.
Thus, the volume fractions of the fronts are set as 0.0817 (corresponding to the water content of 416%, i.e. the liquid limit of this bentonite, Ishikawa et al. 1989) for the pure bentonite and 0.113 (corresponding to the water content of 291%, assumed liquid limit value of the mixture obtained as the product of Bc and the value for the pure specimen) for the mixture specimen. The impermeable boundaries are applied to the cell surfaces and symmetrical surfaces.

The automatic time increment scheme, based on the maximum normalized concentration change allowed at any node during an increment, is used for the transient analysis. Time integration in transient analysis is done with the backward Euler (modified Crank-Nicholson) method. The initial time step is 0.604 [s], analysis period is 6.048×10^6 [s] (70 days), and total number of time steps is 1407 for the case of pure bentonite specimen with the slot width of 1.5 mm. Similar conditions are used for other analysis cases.

A three-dimensional expression for the calculated results for the distribution of the volume fraction of solid at the time of 6.048×10^6 [s] is shown in Figure 12. Figure 13 shows calculated results for variation of distribution of the volume fraction of solid with time along the r-axis (pure bentonite specimen with the slot width of 1.5 mm). The calculated results for the volume fraction of solid oscillates near the value of the boundary condition at longer times as seen in Figure 13. The strong material nonlinearity of the solid diffusivity is thought to be a reason of the oscillations. It is difficult to determine the value of the migration distance from the calculated results since the oscillations appear near the positions corresponding to the migration distance. In this study, locations of extruding gel front, i.e. values of the migration distance, are determined as the positions where the oscillations are damped to a certain stable level. Calculated results of other two cases show similar tendencies to those of the first case.

The calculated results of the migration distance determined by the method described above are shown in Figures 14 and 15 with the measured results. The calculated results are in good agreement with the experimental results and the dependence of the migration distance on the aperture is clearly reproduced as shown in Figure 14.

4 CONCLUSIONS

A physical model as a solid particle diffusion process is developed for the extrusion of bentonite buffer to be used for the geologic disposal of high-level radioactive waste. The calculated results of the numerical analysis for the extrusion based on the solid particle diffusion model developed agree well with the experimental results and the dependence of the extrusion rate on the fracture aperture is clearly reproduced.

The material property values used for the estimation of the solid diffusivity are obtained by extrapolations and assumptions based on the available measured data in limited ranges of the volume fraction of solid. Especially, further studies to measure the viscosity of bentonite gel are essential for the confirmation of the applicability of the diffusion model to the extrusion, since the evaluated viscosity values for the gel used in this study are not based on an actual measurement.

REFERENCES

Boisson, J.Y. 1989. Study on the possibilities by flowing ground waters on bentonite plugs expanded from borehole into fractures. *Proc. NEA/CEC Workshop - Sealing of Radioactive Waste Repositories.*

Börgesson, L. 1990. Interim report on the laboratory and theoretical work in modeling the drained and undrained behavior of buffer materials. *SKB Technical Report* 90-45.

Hasegawa, H., K. Iwasa, H. Sugino, T. Fujita, W. Taniguchi, H. Umeki & S. Masuda 1999. General framework for repository design for a wide range of geological environments in Japan. *Proc. 7th Int. Conf. on Radioactive Waste Management and Environmental Remediation (ICEM'99).* Nagoya.

Ishikawa, H., K. Amemiya, Y. Yusa & N. Sasaki 1989. Comparison of fundamental properties of Japanese bentonite as buffer materials for waste disposal, *Proc. 9th Int. Clay Conf.* Strasbourg.

Ito, M., M. Okamoto, M. Shibata, Y. Sasaki, T. Danbara, K. Suzuki & T. Watanabe 1993. *PNC TN* 8430 93-003. (in Japanese).

Japan Nuclear Fuel Cycle Development Institute (JNC) 1999a., *JNC TN* 1400 99-022. (in Japanese).

Japan Nuclear Fuel Cycle Development Institute (JNC) 1999b., *JNC TN* 1400 99-008. (in Japanese).

Kanno, T., K. Matsumoto & H. Sugino 1999. Evaluation of extrusion and erosion of bentonite buffer. *Proc. 7th Int. Conf. on Radioactive Waste Management and Environmental Remediation (ICEM'99).* Nagoya.

Kanno, T & K. Matsumoto 1997., *PNC TN* 8410 97-313. (in Japanese).

Kanno, T. & H. Wakamatsu 1991. Experimental study on bentonite gel migration from a deposition hole, *Proc. 3rd Int. Conf. Nuclear Fuel Reprocessing and Waste Management (RECOD '91).* Sendai.

Milne-Thomson, L. M. 1968. *Theoretical hydrodynamics.* London and Basingstoke: Macmillan Press.

Nakano, M. Y. Amemiya & K. Fujii 1986. Saturated and Unsaturated Hydraulic Conductivity of Swelling Clays. *Soil Science.* Vol. 141. No. 1.

Nakano, M. Y. Amemiya & K. Fujii 1984. Infiltration and expansive pressure in the confined unsaturated clay. *Trans. JSIDRE Aug.* (in Japanese).

Nakano, M. Y. Amemiya & K. Fujii 1982. Infiltration and volumetric expansion in unsaturated clays. *Trans. Jpn. Soc. Irrig. Reclam.* 100. (in Japanese).

National Astronomical Observatory *1990. Rika nenpyo (Chronological Scientific Tables).* Maruzen Co., Ltd. (in Japanese).

Power Reactor and Nuclear Fuel Development Corporation (PNC) 1996., *PNC TN* 1410 96-071. (in Japanese).

Pusch, R. 1983. Stability of bentonite gels in crystalline rock - Physical Aspects. *SKBF/SKB Technical Report* 83-04.

Pusch, R. 1981.. Borehole sealing with highly compacted Na bentonite. *SKBF/KBS Teknisk Rapport* 81-09.

Suzuki, H., M. Shibata, J. Yamagata, I. Hirose & K. Terakado 1992. *PNC TN* 8410 92-057. (in Japanese).

Verbeke, J., J. Ahn, P. L. Chambré 1997. Long-term behaviour of buffer materials in geologic repositories for high-level wastes. *UCB-NE*-4220.

Clay Science for Engineering, Adachi & Fukue (eds) © 2001 Balkema, Rotterdam, ISBN 90 5809 175 9

Compressibility and permeability of two swelling clays under high pressure

D. Marcial
Universidad Central de Venezuela, IMME, Caracas, Venezuela

P. Delage & Y. J. Cui
Ecole Nationale des Ponts et Chaussées, CERMES, Paris, France

ABSTRACT: The permeability and compressibility properties of two swelling clays considered as possible materials of engineered clay barriers for nuclear waste disposal in Japan and France (Na^+ Kunigel and Ca^{++} FoCa7 clays respectively) were studied in the laboratory. High pressure compression tests (up to 30 MPa) were carried out on saturated remolded samples, together with direct (constant load) and indirect (Casagrande and Taylor methods) hydraulic conductivity measurements. Similar hydraulic conductivity data (included between 10^{-12} and 10^{-14} m/s) were obtained using both methods. Compression curves showed the existence of a particular void ratio value e_c, separating two zones. At void ratios higher than e_c, the macroscopic volume change response apparently obeys to diffuse double layer phenomena, with larger void ratios and a higher compressibility of the Na^+ Kunigel clay. Below e_c, at stresses higher than 2 MPa, the trend is changed. Similar compression coefficients are observed on the two curves, with smaller void ratios observed on Kunigel clay. In this zone, adsorption phenomena linked with the solvatation complexes hydrating exchangeable cations are suspected to govern the volume change behaviour. Above e_c the consolidation coefficient c_v decreases with decreased void ratio in a standard fashion. Below e_c, the trend is opposite, and c_v increases. This point, which is apparently very little documented in the literature, obviously requires further investigations.

1 INTRODUCTION

Growing interest has been recently paid to the hydraulic conductivity properties of swelling dense compacted clay, used for engineered barriers in nuclear waste disposal at great depth. Most investigations on hydraulic conductivity have been carried out on compacted samples after saturation (Boynton & Daniel, 1985; Kenney et al, 1992; Dixon et al, 1992). Due to low permeability values, indirect measurements based on the rate of dissipation of the excess pore pressures during consolidation may appear as a convenient method, as compared to direct methods of determining hydraulic conductivity, either at variable or at constant head.

In their extensive work on the permeability of natural soft clays of medium plasticity ($I_p = 38$), Tavenas et al (1983) showed that the interpretation of consolidation curves could lead to underestimate significantly the hydraulic conductivity. This point has to be checked in swelling clays, which have higher plasticity indexes. The large fractions of smectite encountered in these soils may significantly affect water transfers, due to clay-water interaction, i.e. adsorption or double layer effects. These effects depend upon clay mineralogy and exchangeable cations. To simplify, it is often considered that water molecules strongly linked to clays do not move, reducing the pore volume available for water transfers. The aim of this work is to present the results of an experimental investigation carried out on swelling clays submitted to high pressure compression tests (up to 30 MPa), and to constant head water permeability tests.

2 MATERIALS AND METHODS

Two swelling clays, considered as possible constituent of engineered barriers for confining nuclear waste disposal at great depth, in Japan and France, were studied, named Kunigel (Komine & Ogata 1992) and FoCa7 (Atabek et al. 1991) respectively. The identification parameters of the two clays are presented in table 1. The high activity of the clays shows that clay water interactions should be of some importance in water transfers, and during consolidation. Whereas Kunigel is a relatively pure Na bentonite with quite a high liquid limit, FoCa7 is an interstratified kaolinite-Ca smectite of lower liquid limit. This difference is also apparent in terms of activity but not of specific surface.

Table 1. Materials identification parameters

Clay type	FoCa7	Kunigel
Mineralogy	80 % of (50/50) Inter-stratified Kaolinite/Ca-smectite	64 % of Na-smectite
w_L (%)	112	471
w_P (%)	50	27
I_P (%)	62	444
ρ_s (g/cm^3)	2.67	2.79
Activity	0.78	6.9
S (m^2/g)	515	687
V_B (*)	18	24

- g of methylene blue per 100 g of soil

Clay slurries were prepared by mixing de-aired water with the clay powder, at a water content of 1,1 w_L. The slurry was carefully poured into the oedometer cell, in order to avoid air trapping.

Figure 1 shows the oedometer cell used (70 mm in diameter). The initial height of the samples was approximately 30 mm.

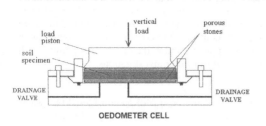

Figure 1. Oedometer cell

The initial void ratio of the samples were deduced from the initial volume and water content of the slurry poured in the oedometer cell. A saturated filter paper, the top porous stone and the piston were placed afterwards on the sample. The initial position of the piston was determined, and the specimen was left consolidating under a 2 kPa load. After consolidation, the new position of the piston was determined, to calculate the corresponding void ratio. Then, the cell was placed in a high pressure oedometer developed in the laboratory, which permits to apply a maximal vertical force of 12 tons thanks to a double lever arm system (Figure 2). As seen in the Figure, the first lever arm has a ratio of 10, and the second a ratio of 5, giving a global ratio of 50. The maximum possible weight to be placed is 240 kg, corresponding to a force of 12 tons applied on the sample. This force corresponds to a maximum vertical stress of 30 MPa with a 70 mm ring. Digital displacement gauges were placed to monitor the vertical strain. All tests were conducted in a 20°C temperature controlled room.

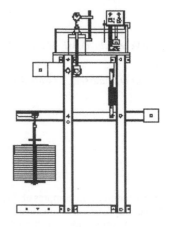

Figure 2. High pressure oedometer frame

Figure 3 schematises the experimental set-up, employed to conduct alternately compression and hydraulic conductivity tests. Two oedometer cells (OC) like in figure 4, were used simultaneously monted in two high pressure oedometers (HPO). HPO1 containing the FoCa7 specimen, and HP2 the Kunigel one. Two drainage valves are available at the base of each OC, (V1,V2 and V3, V4 for HPO1 and HPO2 respectively) . A GDS pressure volume controller, containing de-aired distilled water, is connected to V1 and V3 by means of a double way valve system (DWVS).

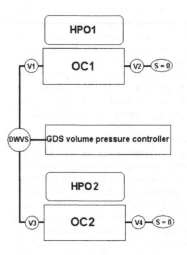

Figure 3. Schema of the experimental set-up

The first step consisted in consolidating both specimens to vertical pressures equal to 0.028 and 0.13 MPa successively. At this stage, V1 and V3 were closed while V2 and V4 were open. Since the FoCa7 specimen consolidated first, a direct permeability measurement was carried out first on this one. To do so the DWVS was open to V1 and the system was flushed with distilled water until no air bubbles were observed in the outflow at V2. Then, the water flux was stopped and V2 was closed. At this stage, the system is ready to run the hydraulic conductivity test. A constant head pressure is applied at the bottom of the sample, by the GDS controller, while de specimen drains by the upper side until a constant flow is observed. The applied water pressure must be low enough in relation to the applied vertical pressure to avoid seepage in the specimen. The water flow was measured for different water heads to observe the influence of the hydraulic gradient. When the permeability measure was finished, the water head was relaxed to zero, V1 was closed, V2 was open and the specimen was left to equilibrate. Then, the next step load was applied to continue with the compression test.

At this stage, an hydraulic conductivity test could be run on the other specimen as explained above.

Conventional step loading tests were carried out on both clay specimens, with loads successively equal to: 0.13 MPa, 0.25 MPa, 0.5 MPa, 2 MPa, 8 MPa, 16 MPa and 30 MPa. At each loading step, a constant head hydraulic conductivity test was carried out with at least three different constant head pressures. An automatic data acquisition system registered permanently the vertical strains of both samples and the GDS water pressure and volume.

3 COMPRESSION TESTS

The compressibility curves obtained between 2 kPa and 30 MPa on the two clay slurries are presented together in Figure 4. At high void ratios, the C_c compressibility coefficients of FoCa7 and Kunigel are drastically different, being equal to 1.00 and 4.18 respectively. The higher C_c of Kunigel is in accordance with the higher liquid limit and initial void ratio of this Na bentonite.

The curves cross at 500 kPa, and become parallel above 2000 MPa, with a same value of C_c (0.30). The C_c value of Kunigel drastically reduces. A void ratio e_c can be defined at the intersection of the two linear sections of the compressibility curve. e_c values are approximately equal to 1.2 and 0.6 for FoCa7 (Ca^{++}) and Kunigel (Na^+) clays respectively.

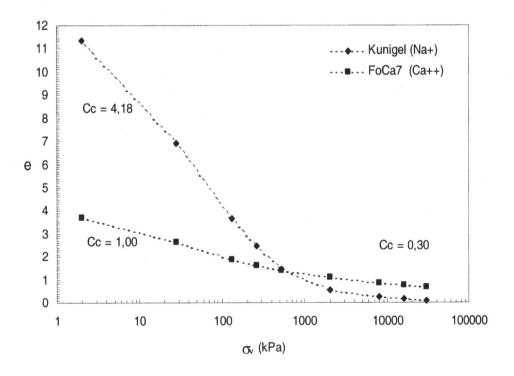

Figure 4. High pressure compression curves of Kunigel and FoCa7

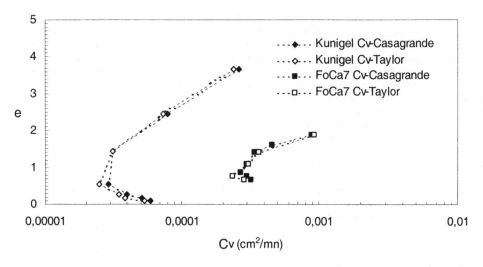

Figure 5. Changes in c_v observed during the high pressure compression tests.

These void ratios are satisfactory correlated with the void ratios corresponding to the plastic limits of the two soils (1.34 and 0.75 for FoCa7 and Kunigel respectively).

Figure 5 presents the change in the consolidation coefficients with the void ratio, for both clays. A decrease in c_v with the decrease in void ratio is observed at the beginning of compression, followed by a subsequent increase, for both clays. The decrease in c_v is common (Lambe & Whitman 1979, Abdullah et al. 1997), but the subsequent increase is apparently less common, probably because of the lack of similar consolidation tests run under the same level of high pressures. The void ratio at which the change in the c_v variation occurs is approximately the same as that previously determined from the compressibility curves.

4 HYDRAULIC CONDUCTIVITY TESTS

To obtain representative results when determining the hydraulic conductivity, it is important to adopt a relevant criterion to define the end of the test. Peirce & Witter (1986) proposed two criteria. In the first one, they recommended to circulate through the sample at least one pore volume of the permeant liquid, in order to completely fill up the pore volume with the permeant liquid. Since the specimens studied here were initially saturated and made up using the permeant liquid (distilled water), this criterion was not considered.

The second criterion consists in verifying that the slope of the curve giving the hydraulic conductivity, as a function of the number of cumulative pore volumes circulated through the sample, doesn't significantly differs from zero. This condition is reached when stress equilibrium is achieved within the saturated clay specimen, which depends on the magnitude of the vertical stress. Under low vertical stresses, the clay tends to swell, because of the pore pressure profile applied within the sample by the constant head applied at one extremity of the sample. This profile changes the profile of effective stress, inducing a global reduction, and a corresponding swelling. Under increased vertical stresses, the swelling behavior is progressively compensated, and the clay specimen goes on settling, due to secondary consolidation. The secondary consolidation is accompanied by the expulsion of water, which induces an influx in the GDS cell. To avoid this influx, it was necessary to reduce the effective stress by increasing the range of applied constant head water. An alternative solution would consist in blocking the vertical strain to prevent swelling, and in unloading the sample before running hydraulic conductivity tests, as suggested by Tavenas et al (1983). The permeability was also estimated from the analysis of c_v values, determined from consolidation curves with the graphic methods of Casagrande and Taylor. Figure presents the changes in hydraulic conductivity as a function of void ratio. The hydraulic conductivities obtained using both direct and indirect methods are presented. Low k values were obtained (10^{-11} to 10^{-14} m/s). A satisfactory correspondence is observed between both methods on FoCa7, direct values being, for a same void ratio, slightly smaller at higher void ratios, and slightly larger at smaller void ratios. The same tendency is observed for Kunigel, with more difference at higher void ratios. At higher void ratios, FoCa7 is more permeable than Kunigel (at $e = 1.8$, $k = 2 \times 10^{-11}$ m/s for FoCa7 and 5×10^{-13} m/s for Kunigel). The tendency is opposite at lower void ratios (at $e = 0.7$, $k = 3 \times 10^{-14}$ and 10^{-13} m/s for FoCa7 and Kunigel respectively).

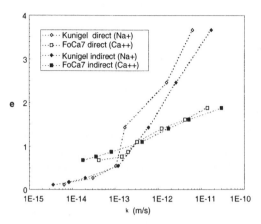

Figure 6. Hydraulic conductivity changes with void ratio, obtained with Kunigel and FoCa7 clays.

5 DISCUSSION

The experimental results previously presented are now interpreted in relation with the status of water in active clay. Adsorbed water in smectite can be studied using various experimental techniques, including infrared spectroscopy (Sposito & Prost 1982). Interpretations based on the diffuse double layer (DDL) theory have also been often used. Sridharan & Jayadeva (1982) interpreted the results of compression tests in clays based on the DDL theory. Yong & Mohamed (1992) used the DDL theory to interpret the results of compression tests run on swelling clays at very high stresses (1000 MPa). They supposed that the clay minerals can be considered as parallel, and they deduce an average separation distance between the clay interlamellar space from the overall void ratio. According to Sposito & Prost (1982), the water adsorbed by smectites can be found in two different states: the water solvating the exchangeable cations in the interlamellar space, and the water in external regions, e.g. external surfaces and micropores. Most authors agree on the fact that DDL mechanisms may correctly account for the behavior of clays at higher water contents, whereas different approaches exist at lower water contents, when the solvatation of exchangeable cations occurs (Sposito & Prost 1982, Yong & Mohamed 1992, Yong 1999).

Observation of the compressibility curves of Figure 4 suggests that DDL effects should be predominant at lower stresses, for higher void ratios. In this zone, the compression indexes Cc correlate well with the liquid limits, according to Skempton's correlation $Cc = 0.009(w_L - 10)$. The measured values of Cc are equal to 1.0 and 4.18 for FoCa7 and

Kunigel respectively, whereas the values obtained with Skempton's correlation are 0.92 and 4.15 respectively. The higher void ratios observed on the Na^+ Kunigel clay is related to the larger DDL thickness obtained with a monovalent cation under low vertical stresses.

An obvious change in the compression behavior of the clays is observed in Figure 3 at void ratios smaller than e_c, which are equal to 1.2 for FoCa7 and 0.6 for Kunigel. These e_c values correspond to water contents respectively equal to 44% and 22%. They are well correlated with the plastic limits, which are equal to 50% and 27% respectively.

The obvious differences observed on the two compression curves at void ratios higher than e_c suggest that different clay-water mechanisms should govern the macroscopic response. The slope of the strain-stress curves of Figure 3 gives an idea about the energy level, which is necessary to expel the water molecules out of the clay structures. Obviously, bellow e_c, the parallelism of the curves shows that a similar mechanism occurs within the two clays, which requires a higher level of energy. In this zone, the effects related to the solvatation of exchangeable cations by water molecules seem to become predominant. For smaller monovalent cations, Sposito & Prost (1982) describe a solvation complex made of a single shell of 6 water molecules surrounding the cation. For a larger divalent cation, the solvation complex would be composed of two water shells comprising 8 and 15 water molecules respectively. This difference in the constitution of the solvatation complexes can explain the higher void ratios and e_c value observed on the Ca^{++} FoCa7 clay.

Unlike Tavenas & al (1983), the comparison made here between direct and indirect measurements of the hydraulic permeability only shows a slight difference. Indirect measurements give higher values than direct measurement above the e_c void ratio. Below e_c, direct measurements give higher values than indirect measurements, especially for FoCa7.

The values of the consolidation coefficient c_v determined from the consolidation curves using Casagrande's and Taylor's methods were similar. A change in the sense of variation of c_v as a function of void ratio was observed close to the e_c void ratio. This unexpected tendency can be commented by considering the following detailed c_v expression :

$$c_v = k(1+e)/(a_v \, \gamma_w) \qquad (1)$$

where a_v is the compression modulus of the soil in a linear void ratio/stress diagram, and γ_w the unit weight of water.

During compression, the numerator $k(1+e)$ decreases, inducing a decrease in c_v, whereas the denominator $(a_v \, \gamma_w)$ decreases, inducing an increase in c_v. Generally, the combined effect of these two tendency is a decrease in c_v. Observation of Figures 4 and 6 show that under e_c, the decrease in k with void ratio becomes smaller, particularly for Kunigel. Also, the decrease in e with stress are smaller. These trends are compatible with the change in the sense of the variations of c_v, observed in Figure . However, the parameter a_v also becomes nearly constant, which is not compatible with the sense of the variations of c_v. Obviously, the change in behaviour observed at e_c have an influence, but a simple explanation cannot be easily obtained at this level.

6 CONCLUSION

The results of compressibility and hydraulic conductivity tests, carried out under high stresses on two saturated remolded swelling clays, permitted to highlight the role of adsorption forces on the hydro-mechanical behavior of two high plasticity clays, namely the Japanese Kunigel clay (Na^+) and the French FoCa clay (Ca^{++}).

Observation of the compressibility curves shows the existence of a void ratio e_c (equal to 0.6 and 1.2 for Kunigel and FoCa7 respectively) which apparently separates two typical volume change behaviours. These e_c values are correctly correlated with the plasticity indexes. The higher void ratios and higher compression index observed on the Na^+ Kunigel clay, as compared with the Ca^{++} FoCa7 clay, show that DDL effects seems to predominate at higher void ratios ($e>e_c$). At lower void ratios ($e<e_c$), the inversion observed in terms of void ratios, and the similar compression coefficient observed on the two clays show that DDL effects are no more predominant. Water adsorption phenomena, related to the solvation complexes which surround the exchangeable cations (Sposito & Prost 1982), become predominant.

The comparison made between direct and indirect hydraulic conductivity tests shows that, surprisingly, the high activity of the clay minerals does not make any significant difference between the two methods. This conclusion differs from that drawn by Tavenas et al. (1983), from results on medium plasticity clays.

In each clay, the e_c void ratio is related to changes in compressibility, in hydraulic conductivity and is also related to a change in the sense of the c_v changes with void ratio : when the void ratio is decreased, c_v decreases above e_c, in a common fashion, whereas it increases when values smaller than e_c are reached.

The evidence of the e_c void ratio is not fully documented in the literature, because of the lack of experimental data on plastic clays under high pressures.

ACKNOWLEDGEMENTS

The samples of Kunigel and FoCa7 clays have been provided by ANDRA (French National Agency for Nuclear Waste Management). The conclusions and the view-points presented in this paper are those of the authors and do not necessarily coincide with those of ANDRA.

REFERENCES

Abdullah W., Al-Zou'bi M. S. & Alshibli K. A. 1997. On the physicochemical aspects of compacted clay compressibility. *Canadian Geotechnical Journal*. 34: 551-559.

Atabek R. B., Félix B., Robinet J. C. & Lahlou R. 1991. Rheological behaviour of saturated expansive clay materials. *Workshop on stress partitioning in engineered clay barriers, Duke University, Durham, N. C.*

Boynton S. & Daniel D. 1985. Hydraulic conductivity test on compacted clay. *Journal of Geotechnical Engineering*. 111: 465-478.

Dixon D. A., Gray M. N. & Hnatiw D. 1992. Critical gradients and pressures in dense swelling clays. *Canadian Geotechnical Journal*. 29: 1113-1119.

Kenney T. C., Van Veen W. A., Swallow M. A. & Sungalia M. A. 1992. Hydraulic conductivity of compacted bentonite-sand mixtures. *Canadian Geotechnical Journal*. 29: 364-374.

Komine H. & Ogata N. 1992. Swelling characteristics of compacted bentonite. *Seventh Conference on Expansive Soils, Dallas, Texas, USA*. 1: 216-221.

Lambe T.W. and Whitman R.V. 1979. *Soil Mechanics*. J. Wiley, New-York.

Peirce J. & Witter K 1986. Termination criteria for clay permeability testing. *ASCE Journal of Geotechnical Engineering*. 112: 841-854.

Sposito G. & Prost R. 1982. Structure of water adsorbed on smectites. *Chemical Reviews, USA*. 82: 552-573.

Tavenas F., Leblond P., Jean P. & Leroueil S. 1983. The permeability of natural soft clays. Part I: Methods of laboratory measurement. *Canadian Geotechnical Journal*. 20: 629-644.

Sridharan A. & Jayadeva P. 1982. Double layer theory and compressibility of clays. *Géotechnique*. 32: 133-144.

Yong R.N. & Mohamed A. M. O. 1992. A study of particle interaction energies in wetting of unsaturated expansive clays. *Canadian Geotechnical Journal*. 29: 1060-1070.

Yong R.N. 1999. Soil suction and soil-water potentials in swelling clays in engineered barriers. *Engineering Geology*, 54: 3-13.

Clay Science for Engineering, Adachi & Fukue (eds) © 2001 Balkema, Rotterdam, ISBN 90 5809 175 9

Long-term stability of clay barrier by microstructural approach

N. Theramast, T. Taniguchi & Y. Ichikawa
Department of Geotechnical and Environmental Engineering, Nagoya University, Japan

K. Kawamura
Department of Earth and Planetary Science, Tokyo Institute of Technology, Japan

T. Seiki
Department of Civil Engineering, Utsunomiya University, Tochigi, Japan

ABSTRACT: Long-term stability of the synthetic clay barrier concerning seepage and consolidation is the matter subject addressed in this paper. The exact characteristics of constituents of the micro-inhomogeneous continuum are determined by method of molecular dynamics simulation. The asymptotic homogenization is employed to derive the macroscopic field equations of transport in saturated montmorillonite-rich bentonite clay. Numerical computation to achieve parameters characterizing the homogenized medium is accomplished by finite element method. Results of simulation with the homogenization method yield in good agreement with those of experiment-based record. The hydraulic conductivity obtained thus deliberately used in estimation involving long-term behavior of an engineered barrier for high-level radioactive waste disposal.

1 INTRODUCTION TO BENTONITE

Due to constructively increasing public awareness in environmental protection over the past two decades, much effort is being incorporated upon techniques capable in design and prediction of waste disposal facilities. The engineered multi-barrier system, at the moment, forms the spine of geological disposal repository over industrial world. The common significant functions of the embedment of waste containment are, owing to its favorably low permeability, to prevent underground water contamination by hazardous waste; in conjunction with protect the waste overpack from being exposed to high stresses or large strain induced by tectonics or thermomechanical effects. To this purpose, bentonite clay, which is compacted to a high density prior to emplacement, is admirably served as the backfilling or buffer material of the artificial barrier system. It is because they fulfill a number of roles including minimization of groundwater access to waste package, alteration of ground water chemistry and retardation of solute transport (Chapman & McKinley 1987) together with sufficient mechanical strength, ductile behavior, relatively low shear modulus and impervious to frost damage and desiccation cracking (Dixon et al. 1985, Pusch 1994, Krause et al. 1997).

Bentonite clay is a micro-inhomogeneous material, which by chemical analysis and X-ray diffraction spectroscopy consists chiefly of sodium montmorillonite with some macro-grains (mainly quartz), water, air and others (Mollins et al. 1996, Ichikawa et al. 1999a). Despite the fact that purified bentonite

give the most satisfactory properties in using as buffer material, the mixture of sand and bentonite becomes more esteemed lately since purified bentonite is costly. However, montmorillonite usually influence engineering properties in a manner far greater than their abundance, thus it may positive to examine behaviors of water-saturated bentonite by treating the composite system of montmorillonite and water, henceforth will be called montmorillonite hydrate. Montmorillonite is the clay mineral of lamellar shape of size about 100x100x1 nm, which several lamellae are stacked together as shown in Figure 1. The interlamellar or internal spaces have a strong potential to cation exchange and water uptake. In case of water-saturated bentonite, the interlamellar water form a large part of total water content, especially for high density compacted one, with the external water occupies the interconnected larger voids (Pusch 1994). In this paper, the material corresponds to the pure sodium montmorillonite, of which the proprietary name is Kunipia-F, is treated.

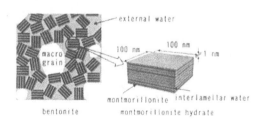

Figure 1. Schematic picture of the microstructure of bentonite.

2 MOLECULAR DYNAMICS SIMULATION OF MONTMORILLONITE HYDRATE

The crystalline structure of montmorillonites is generally irregular and impracticable known by means of experimental analogs because of their poor crystallinity. Several statistical models (such as Monte Carlo method, Molecular Dynamics method, etc.) have been theoretically developed in the field of computational physics and chemistry to discover the physico-chemical properties of montmorillonite hydrate. In molecular dynamics simulation, the motion of every molecule is given by the Newton's equation and the force is calculated by differentiating an interatomic potential function. Kawamura (1992) and Kawamura et al. (1997) developed the crystal model for sodium montmorillonite hydrate and introduced a new empirical interatomic potential model (i.e. potential function for all atom-atom pairs which is composed of the Coulomb, short-range repulsion, van der Waals and Morse terms) to determine the interatomic and intermolecular interactions quantitatively. In their calculation, motions of all atoms in octahedral and tetrahedral layers, interlamellar cations and water molecules are considered. Note that the molecular formula of sodium montmorillonite hydrate with n-layers of interlamellar water is given by $Na_{1/3}Al_2[Si_{11/3}Al_{1/3}]O_{10}(OH)_2 \cdot nH_2O$. They employed the NPT-ensemble molecular dynamics scheme under the condition of 300 K temperature and 0.1 MPa pressure. By applying a standard procedure of statistical mechanics to the molecular dynamics computation, they obtain the properties of the sodium montmorillonite hydrate such as swelling property, shearing viscosity, external and internal water viscosity at the neighborhood of the montmorillonite mineral as illustrated in Figure 2 (detail of the molecular dynamics simulation is lengthy and beyond the contents of this paper, see Kawamura 1992, Kawamura et al. 1997, Ichikawa et al. 1999a for more information).

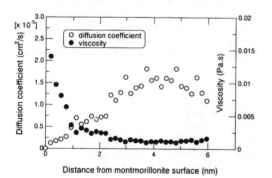

Figure 2. Viscosity and diffusion coefficient distributions with respected to distance from the montmorillonite surface by molecular dynamics simulation.

3 HOMOGENIZATION IN SEEPAGE FLOW

In recent literature of porous media flow, most authors begin from a scale much larger than the pore size and assume Darcy's law from the onset. However, those macro-phenomenological approaches still have some inevitably shortcomings because the scale effects are not considered and the real velocity in fine scale cannot be able to determine since they do not reflect from the exact micro-continuum characteristics of constituents (Mei 1992, Auriault et al. 1993, Wang 1996, Ichikawa et al. 1999a). Moreover, even if the mechanism of flow in the scale of pores has been befittingly established, it is practically impossible to solve those full equations for the complicated fine geometries. Therefore, it is expedient to replace the micro-inhomogeneous medium with the macro-homogeneous equivalence. Homogenization is the perturbation theory enable us to explicitly derive the macroscopic field equations from the microscopic defined one based on the conventional continuum equations and the technique of multiple-scale expansion (Sanchez-Palencia 1980, Bakhvalov & Panasenko 1989, Mei 1992, Auriault et al. 1993). The general purpose is to find the effective properties of the homogeneous medium under the necessary assumptions of scales separation and periodicity in micro-scale. The theory of the mathematical rigorous method of homogenization by two-scale expansion is systematically presented here and used to analyze the macroscopic equivalent phenomena of seepage flow through heterogeneous media. The entailed assumption is that the matrix has a periodic pore structure on the local scale. Starting from the Navier-Stokes equation for the fluid motion with the view to accounting for the fact that the viscous stress is $O(\varepsilon p)$, together with the proper continuity equation and boundary condition:

$$-\frac{\partial P^\varepsilon}{\partial x_i} + \varepsilon^2 \eta \frac{\partial^2 V_i^\varepsilon}{\partial x_k \partial x_k} + F_i = 0 \quad in \ \Omega_f^\varepsilon \qquad (1)$$

$$\frac{\partial V_i^\varepsilon}{\partial x_i} = 0 \quad in \ \Omega_f^\varepsilon \qquad (2)$$

$$V_i^\varepsilon = 0 \quad on \ \partial\Omega_f^\varepsilon \qquad (3)$$

where P^ε = pore pressure, V_i^ε = fluid velocity, η = rescaled viscosity, F_i = body force vector, and Ω_f^ε implies the flow region in the global coordinate system with its boundary $\partial\Omega_f^\varepsilon$ as shown in Figure 3. Note that the rigid saturated porous periodic medium is considered, i.e. Y_s is the undeformed solid volume in the periodic cell, Y_f is the pore volume occupied by the incompressible fluid and Y_{sf} is the solid-fluid interface. The superscript ε implies that the functions oscillate rapidly in fine scale and it must be valid both for geometry and physical quantities. In the other words, there are two greatly different scales, i.e. fine (or microscopic) scale that is the scale of heterogeneities y and the large (or macroscopic) scale of the flow region denoted by x. The quantities with superscript ε depends primarily on these two scales, more precisely on the ratio $\varepsilon = x/y$ ($\varepsilon \ll 1$), small but finite parameter.

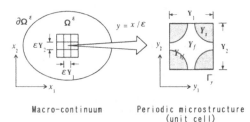

Macro-continuum   Periodic microstructure
                              (unit cell)

Figure 3. Two scales problem basically used in procedure of homogenization method.

In the process of homogenization, the solution is sought in a form of asymptotic expansion, i.e. series in powers of a small parameter ε with quantities depend both on the space variables x and y. Therefore, we introduce asymptotic expansions to the state variables velocity and pore fluid pressure as:

$$V_i^\varepsilon(x) = V_i^0(x,y) + \varepsilon V_i^1(x,y) + \varepsilon^2 V_i^2(x,y) ... \quad (4)$$

$$P^\varepsilon(x) = P^0(x,y) + \varepsilon P^1(x,y) + \varepsilon^2 P^2(x,y) ... \quad (5)$$

where $V_i^\alpha(x,y)$ and $P^\alpha(x,y)$, $\alpha = 0, 1, 2, ...$ are Y-periodic functions, i.e. $V_i^\alpha(x,y) = V_i^\alpha(x,y+Y)$ and $P^\alpha(x,y) = P^\alpha(x,y+Y)$, which imply that the differentiation operator becomes:

$$\frac{\partial}{\partial x_i} \Rightarrow \frac{\partial}{\partial x_i} + \frac{1}{\varepsilon}\frac{\partial}{\partial y_i} \quad (6)$$

To this end, we introduce the, Y-periodic, characteristic velocity function $v_i^k(y)$ and characteristic pressure function $p^k(y)$ as:

$$V_i^0 = \left(F_k(x) - \frac{\partial P^0(x)}{\partial x_k}\right)v_i^k(y) \quad (7)$$

$$P^1 = \left(F_k(x) - \frac{\partial P^0(x)}{\partial x_k}\right)p^k(y) \quad (8)$$

By following above chain rule and substituting the series expansions (4), (5) into Equation 1 then taking $\varepsilon \to 0$, and subsequently applying above characteristic velocity and characteristic pressure functions, we thus obtain the system of microscopic boundary value problem which depends only on properties of the microscopic fine scale:

$$-\frac{\partial p^k}{\partial y_i} + \eta\frac{\partial^2 v_i^k}{\partial y_j \partial y_j} + \delta_{ik} = 0 \quad in \ Y_f \quad (9)$$

$$\frac{\partial v_i^k}{\partial y_i} = 0 \quad in \ Y_f \quad (10)$$

$$v_i^k = 0 \quad on \ Y_{sf} \quad (11)$$

where δ_{ij} = Kronecker delta. In common with the periodic boundary condition:

$$v_i^k(y) = v_i^k(y+Y) \ ; \ p^k(y) = p^k(y+Y) \quad on \ \Gamma_y \quad (12)$$

By applying the averaging operator over the periodic volume to Equation 7, the averaging equation leads to the form of well-known Darcy's law:

$$\tilde{V}_i^0 = K_{ij}\left(F_j - \frac{\partial P^0}{\partial x_j}\right) \quad (13)$$

$$K_{ij} = \tilde{v}_i^j = \frac{1}{|Y|}\int_{Y_f} v_i^j \, dy \quad (14)$$

where $\tilde{V}_i^0$ = phase averaged velocity, $|Y|$ = volume of unit cell. Note that K_{ij} is symmetric and positive definite tensor. Similarly, averaging of ε^0- term of continuity equation yields the macroscopic seepage equation that contains only an averaged, microscopically free, function:

$$\frac{\partial \tilde{V}_i^0}{\partial x_i} = 0 \Rightarrow \frac{\partial}{\partial x_i}\left[K_{ij}\left(F_j - \frac{\partial P^0}{\partial x_j}\right)\right] = 0 \quad in \ \Omega^\varepsilon \quad (15)$$

The macroscopic equivalent parameters of the terms V_i^0 and P^0 represent the first-order approximation of the real quantities as:

$$V_i^\varepsilon(x) \cong V_i^0(x) \quad (16)$$

$$P^\varepsilon(x) \cong P^0(x) \quad (17)$$

It should be reminded here that in geotechical engineering and groundwater engineering there exists a phenomenological Darcy's equation, which is investigated directly on the scale of porous matrix (Bear 1967):

$$\tilde{V}_i' = -K_{ij}'\frac{\partial \phi}{\partial x_j} \ ; \ \phi = \frac{P}{\rho g} + h \quad (18)$$

where $\tilde{V}_i'$ = the phase averaged velocity, ϕ = piezometric head, P = the intrinsic phase averaged pore pressure, h = elevation head, ρ = fluid density which assume constant due to incompressibility and g = gravitational acceleration.

We deduce from the similarity between the phenomenological established and our first-order approximations (Equation 15, 16) that:

$$\tilde{V}_i' = \tilde{V}_i^\varepsilon \cong \tilde{V}_i^0 \qquad (19)$$

$$K_{ij}' = \rho g K_{ij} \qquad (20)$$

4 ANALYSIS OF PLANE POISEUILLE-LIKE FLOW IN MONTMORILLONITE HYDRATE

Since structure of montnorillonite hydrate is of the lamella-stacked type, i.e. group of several mont-morillonite lamellae, and the interlamellar water occupy the most part of the pore. For that reason, we consider the flow regime in montmorillonite hydrate as the Poiseuille-like flow, which is one-dimensional flow between two parallel rigid plates. The model of periodic volume (unit cell) for the plane flow is schematically shown in Figure 4. In order to figure out the homogenization problem, we first solve the Equations 9 within the unit cell, together with the appropriated continuity and boundary conditions (Equation 10, 11) including the periodicity condition in Equation 12, for the local variable $v_i^\kappa(y)$ and $p^\kappa(y)$. These equations define the boundary value problem in the unit cell. Once solved, their averages over the unit cell yield the macroscopic governing equations.

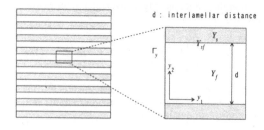

Figure 4. Unit cell for the Poiseuille-like plane flow.

Numerical solution of micro-scale problem is sought in the weak form of finite element method as:

$$\int_{Y_f} \left(\eta \frac{\partial v_i^k}{\partial y_i} - p^k \delta_{ij} \right) \frac{\partial(\delta v_i)}{\partial y_j} dY - \int_{Y_f} \delta v_k dY = 0 \qquad (21)$$

$$\int_{Y_f} v_i^k \frac{\partial(\delta p)}{\partial y_i} dY = 0 \qquad (22)$$

The known distributed water viscosity in Figure 2 are assigned everywhere in the pore at the Guassian points. The procedure of finite element discretization thus utilized is same as those of well-known finite element formulation of the Navier-Stokes problem, hence will be skipped here (for further details, see Zienkiewicz & Taylor 1991).

Our goal of finite element solutions are shown in Figure 5-7, in which profile of characteristic velocity function v_1^1 and the hydraulic conductivity of the term K_{ij}' are represented (together with the result in the case of constant viscosity). We can see from Figure 6 that the hydraulic conductivity calculated by making use of the distributed viscosity, which we have taken into account the effects of vicinity to the montmorillonite surface, is approximately one order lower than that of constant viscosity of the value 8×10^{-4} Pa.s. The hydraulic conductivity of experiment-based data of sodium montmorillonite recorded by Mesri et al. (1971) and those predicted by our model can be compared favorably, which is diagrammatically shown in Figure 7. Moreover, the correlation between void ratio and K_{ij}' in the range of our simulation agree well with those suggested by Pusch (1994) as shown in Figure 8, but deviated at very low hydraulic conductivity value.

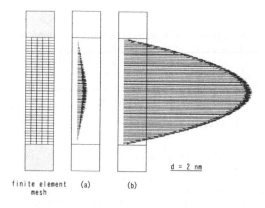

Figure 5. Profiles of characteristic velocity function v_1^1 in the case of d = 2 nm, calculated by using (a) distributed viscosity from Figure2. (b) constant viscosity = 8x104 Pa.s.

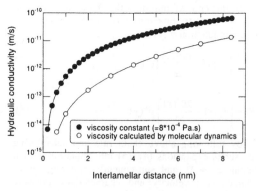

Figure 6. Calculated hydraulic conductivity K_{ij}' with respected to interlamellar distance.

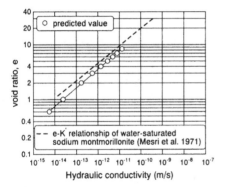

Figure 7. An inter-relationship between void ratio, saturated density and hydraulic conductivity K'_{ij} based on montmorillonite hydrate model.

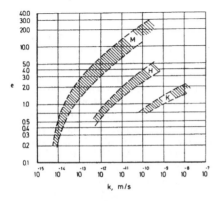

Figure 8. Hydraulic conductivity (k) of the complete water-saturated pure clay of the major mineral in the case of montmorillonite (M), hydrous mica (H) and kandite (K) where the parameter e = void ratio (from Pusch 1994).

5 LONG-TERM BEHAVIOR OF CLAY BARRIER

One of the necessary design functions of buffer material is to maintain the waste container in the proper position after emplacement and providing the buffering operation in an attempt to alleviate the effect of mechanical changes that may the occur in surrounding host rock. This requires the ductile and plastic abilities to protect the overpack from being exposed to overstresses (Pusch 1994, PNC 1999). In spite of this, the softness is not obliged that it is allowed to undergo the severe deformation followed by canister slipped out from the embedment. Besides, the required operative lifetime for the waste disposal facility is considerably extensive, particularly in the case of high-level radioactive waste. As a result, evidence of their longevity is unattainable by the routine examination. This fact triggered our high interests to predict the long-term mechanical behavior of clay

barrier, in particular, compacted bentonite. The Japanese model of the pit type engineered barrier subsystem manipulated in high-level radioactive waste repository (PNC 1999) is intentional used for numerical example as shown in Figure 9, 10. The mathematical modeling to simulate those long-term behaviors is contemplated here with especial attention to consolidation. The formulation involving soil and water phases interaction presented here is based on the classical theory of consolidation of saturated poroelastic solid known as Biot's theory of consolidation (Biot 1941). To furnish the solution algorithm, some assumptions should be appointed: 1) Bentonite consists of only montmorillonite and interlamellar water. 2) Groups of montmorillonite lamellae (stacks) are randomly oriented, so that we shall assume isotropic flow field. 3) Clay and water preserve their volume, i.e. incompressible. 4) Alteration of permeability is entirely induced by volumetric strain, i.e. change of interlamellar distance. 5) Bentonite is assumed elastoplastic while rock is elastic. Thus, the coupling equations for our flow assumptions may be written in general form as:

$$\frac{\partial \Delta \sigma_{ij}}{\partial x_j} + \Delta f_i = 0 \qquad (23)$$

$$\Delta \sigma_{ij} = \Delta \sigma'_{ij} + \Delta P \delta_{ij} \qquad (24)$$

$$\frac{\partial \varepsilon_v}{\partial t} = -\frac{\partial \widetilde{V}'_i}{\partial x_i} = \frac{\partial}{\partial x_i}\left[K'_{ij}\left(\frac{\partial \phi}{\partial x_j}\right)\right] \qquad (25)$$

$$\varepsilon_v = \frac{d - d'}{2s + d} \qquad (26)$$

where σ_{ij} = Cauchy's stress, σ'_{ij} = effective stress, P = pore water pressure, f_i = body force, ε_v = volumetric strain, d = pre-deformed interlamellar distance between clay minerals, d' = distance after deformation and $2s$ = thickness of monmorillonite lamella (Figure 11). Note that, like other clays, the effective stress concept is valid for bentonite as well, except at very high density (Pusch 1994).

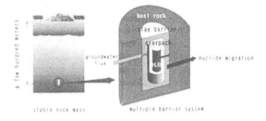

Figure 9. Geological disposal concept for high-level radioactive waste repository.

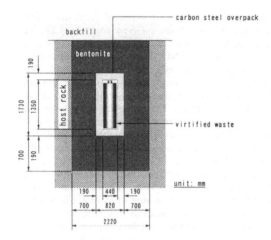

Figure 10. The model of an engineered barrier system of the pit disposal type used in analysis and its dimensions.

In the theory of elastoplasticity, the effective stress increment is assumed to be related to the strain increment as given in Equation 27. The Cam-clay model (Schofield & Wroth 1969, Wood 1990) is adopted here as the constitutive model for elastoplastic material, of which the yield criterion is expressed in Equation 28.

$$\dot{\sigma}'_{ij} = D^{ep}_{ijkl}\dot{\varepsilon}_{kl} \qquad (27)$$

$$f = \frac{\lambda - \kappa}{1 + e_0}\left[\ln\left(\frac{p'}{p'_0}\right) + \frac{\eta}{M}\right] - \varepsilon^p_v = 0 \qquad (28)$$

where f = yield function, e_0 = initial void ratio, λ and κ are soil constants, p' = mean effective stress, p'_0 = mean effective stress on normal consolidation line, η = stress ratio q/p, M = slope of critical state line and ε^p_v = volumetric plastic strain.

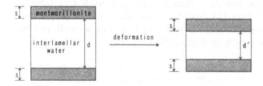

Figure 11. Volumetric strain in unit cell.

The finite element method is used here to solve the nonlinear problem of the coupled consolidation using a mixed formulation, which incorporate displacement and pore pressure variables (Equation 23-26). Which those governing relations constitute a system of the nonlinear equations of the final form:

$$C^{ep}(X)\dot{X} + KX = F(t) \qquad (29)$$

$$C^{ep}(X) = \begin{bmatrix} D^{ep}(X) & L \\ L^T & 0 \end{bmatrix} \qquad (30)$$

$$K = \begin{bmatrix} 0 & 0 \\ 0 & H \end{bmatrix} \qquad (31)$$

$$F(t) = \begin{bmatrix} \dot{F} & Q \end{bmatrix}^T \qquad (32)$$

$$X = \begin{bmatrix} U & P \end{bmatrix}^T \quad ; \quad \dot{X} = \begin{bmatrix} \dot{U} & \dot{P} \end{bmatrix}^T \qquad (33)$$

where D^{ep} = elastoplastic stress-stain matrix, K = flow matrix, $F(t)$ = time dependent forcing function, L = coupling matrix, $\dot{F}$ = elemental vector of external force rate, H = matrix of hydraulic conductivity, Q = prescribed flux vector, U = vector of nodal displacement, P = vector of nodal pore pressure and T denotes the matrix transpose. The details of finite element discretization scheme to achieve above equations is of the typical algorithm for Biot's consolidation based on Galerkin method of weighted residuals accompanied by the backward Euler time stepping (see Owen & Hinton 1980, Zienkiewicz & Taylor 1991). The material used in this study is Kunipia-F bentonite, of which its mineral composition and physical properties are shown in Table 1. Whilst, the material properties of compacted bentonite used in calculation is given in Table 2. Calculation was performed up to 46000 years, which is generally long enough to sustain the required operative lifetime of the whole barrier system. The numerical predicted results are shown in Figure12-14.

The results show that the mechanical change in permeability, more precisely hydraulic conductivity, is of the small range and the values show favorably low permeability to water, hence effective to prevent the advective flow of groundwater. In addition, the engineered barrier of the dimension shown in Figure 10 can very well embed the waste package without breaking through by consolidation process.

Table 1. Physical properties and minerological composition of Kunipia-F bentonite.

Mineral content					Specific gravity	Atterberg Limits			Cation exchange capacity	Leach cation meq/100g			
Montmorillonite %	Quartz %	Calcite %	Kaolinite %	Illite %		LL %	PL %	PI %	meq/100g	Na^+	K^+	Ca^{2+}	Mg^{2+}
98-99	<1	<1	-	-	2.7	993	42	951	1117	114.9	1.1	20.6	2.6

Data from PNC (1999).

Table 2. Input material properties.

	Modulus of elasticity MPa	Poisson's ratio	Saturated dendity t/m^3	Cohesion kPa	Internal friction angle deg	Lateral pressure coefficient	Critical state parameter				Iniitial saturated permeability m/s
							λ	κ	M	p'_0 MPa	
Bentonite	200	0.4	1.8	1000	30	1	0.09	0.048	0.58	4.7	5.82x10^{-14}
Rock	65000	0.17	2.67	15000	45	1	-	-	-	-	1x10^{-11}

Data from PNC (1999).

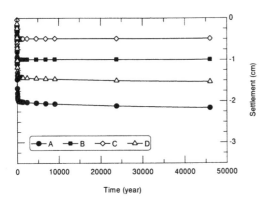

Figure 12. Predicted long-term settlement due to consolidation (positions as defined in Figure 13).

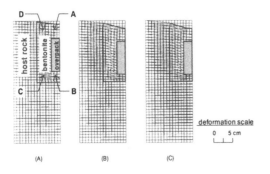

Figure 13. Deformed configuration at time (A) = 3 years, (B) = 100 years, (C) = 30000 years.

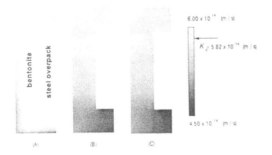

Figure 14. Hydraulic conductivity distribution at time (A) = 3years, (B) = 100 years, (C) = 30000 years.

6 CONCLUSION

Special consideration has been devoted to the currently nontrivial engineering problem of transport through the clay barrier. The formulation of mathematical scheme of homogenization analysis has been represented and established to the seepage flow in water-saturated bentonite clay. The, atomic-based properties derived from the intermolecular interaction by means of molecular dynamics simulation is exploited to the procedure of homogenized solution. By homogenization method with a two-length scale asymptotic expansion, we have calculated the global or effective hydraulic conductivity by finite element analysis. The results of numerical simulation yield in satisfactory agreement with those of experiment-based data. Both experimental and the predicted curves show the considerably low values of hydraulic conductivity. This is perhaps because water molecules are constrained at the vicinity to surface of clay mineral, in particularly montmorillonite, so that interlamellar water is immobile under the ordinary hydraulic gradient. The other category covered in this paper is to predict the long-term mechanical behavior of the compacted bentonite barrier of the type utilized for high-level radioactive waste repository, in which the issues of deformation and mechanical change of hydraulic conductivity with time is of our most concentration. The results show that the barrier system is satisfactorily stable and corroborate us the long-term serviceability criteria.

In the present study, only the composite system of montmorillonite and water is considered, the different range of fluid such as salt water is reported elsewhere (Ichikawa et al. 1999b). Moreover, the applicability of the microstructural approach to the bentonite, which consists of montmorillonite and other minerals such as Kunigel-V1 is deferred to later papers. The present homogenization formulation is made completely on the basis of two-scale expansion but the extension to the multiple scale expansion and also modification to another flow regimes can be uncomplicated carried out in the similar fashion. However, it should be noted that the homogenized behavior, even in the simple case, is not straightforward, for instance, the Navier-Stokes equation in fluid flow give the averaged equation in form of Darcy's law.

REFERENCES

Auriault, J.L. & Lewandowka, J. 1993. Homogenization analysis of diffusion and adsorption macrotransport in porous media: macrotransport in the absence of advection. *Geotechnique* 43(3): 457-469.

Bakhvalov, N. & Panasenko, G. 1989. *Homogenization: Average Process in Periodic Media*. Kluwer.

Bear, J. 1967. *Dynamics of Fluids in Porous Media*. Dover.

Chapman, N.A. & McKinley, I.G. 1987. The *Geological Disposal of Nuclear Waste*. John Wiley & Sons.

Dixon, D.A., Gray, M.N. & Thomas, A.W. 1985. A study of the compaction properties of potential clay-sand buffer mixtures for use in nuclear fuel waste disposal. *Engineering Geology* 21: 247-255.

Fu, M.H., Zhang, Z.Z. & Low, P.F. 1990. Changes in properties of a montmorillonite-water system during the adsorption and desorption of water *hysteresis*. *Clays & Clay Minerals* 38: 485-492.

Ichikawa, Y., Kawamura, K., Nakano, M., Kitayama, K. & Kawamura, H. 1999a. Unified molecular dynamics and homogenization analysis for bentonite behavior: current results and future possibilities. *Engineering Geology* 54: 21-31.

Ichikawa, Y., Kawamura, K., Nakano, M., Seiki, T., & Taniguchi, T. 1999b. Microcomputer-based molecular dynamics and homogenization analysis for seepage problem in bentonite in pure- and salt-water. *EPMESC VII International Conference on Enhancement and Promotion of Computational Methods in Engineering and Science* (1): 23-39

Kawamura, K. 1992. Interatomic potential models for molecular dynamics simulations of multi-component oxides. *Molecular Dynamics Simulation (Yozenawa, F. ed.)*. Springer-Verlag: 88-97.

Kawamura, K., Ichikawa, Y., Nakano, M., Kitayama, K. & Kawamura, H. 1997. New approach for predicting the long term behavior of bentonite: the unified method of molecular simulation and homogenization analysis. *Proc. Sci. Basis for Nuclear Waste Management XXI*: 359-366.

Krause, J. F., Benson, C.H., Erickson, A.E. & Chamberlain, E.J. 1997. Freeze-thaw cycling and hydraulic conductivity of bentonite barriers. . *ASCE J. Geotech. Engng Div*. 3: 229-238.

Mei, C.C. 1992. Method of homogenization applied to dispersion in porous media. *Transport in Porous Media* 9: 261-274.

Mesri, G. & Olson R.E. 1971. Consolidation characteristics of montmorillonite. *Geotechnique* 21(4): 341-352.

Mollins, L.H., Stewart, D.I. & Cousens, T.W. 1996. Predicting the properties of bentonite-sand mixtures. *Clay Minerals* 31: 243-252.

Owen, D.R.J. & Hinton, E. 1980. *Finite Elements in Plasticity*. Pineridge Press.

PNC 1999. *The Second Progress Report*. Project Overview Report: PNC.

Pusch, R. 1994. *Waste Disposal in Rock*. Amsterdam: Elsevier.

Sanchez-Palencia, E. 1980. *Non-homogeneous Media and Vibration Theory*. Berlin: Springer-Verlag.

Schofield, A.N. & Wroth, C.P. 1968. *Critical State Soil Mechanics*. McGraw Hill.

Wang, J.G. 1996. *A Homogenization Theory for Geomaterials: Nonlinear Effect and Water Flow*. Dr. Eng. Thesis: Nagoya University.

Wood, D.M. 1990. *Soil Behavior and Critical State Soil Mechanics*. Cambridge Univ. Press.

Zienkiewicz, O.C. & Taylor, R.L. 1991. *The Finite Element Method 4th ed*. (2). McGraw-Hill.

Clay Science for Engineering, Adachi & Fukue (eds) © 2001 Balkema, Rotterdam, ISBN 90 5809 175 9

Experimental and thermodynamic examination on buffer material

A. Kobayashi – *Kyoto University, Japan*

K. Fujii– *Iwate University, Morioka, Japan*

T. Fujita – *Japan Nuclear Cycle Development Agency, Tokai, Ibaraki, Japan*

M. Chijimatsu – *HAZAMA Corporation, Minato, Tokyo, Japan*

ABSTRACT: To examine the effect of the coupling phenomena in the buffer material for high level radioactive waste disposal, the thermodynamic examination of the water movement is tried. By using the theory assuming the equilibrium steady state, the comparison with the measured water content distribution is carried out. The data of the large-scale laboratory and in-situ coupling tests are used for the examination. The results for both tests show the ability of the coupling among fluxes of heat, air and water.

1 INTRODUCTION

Geological disposal of high level radioactive waste is based on the multibarrier concept of engineered and natural barriers in many countries (Oyama, 1991). The engineered barrier system of the repository concept by Japan Nuclear Cycle Development Agency (JNC) consists of the vitrified radioactive waste, the overpack and the buffer material (PNC Report, 1993). The buffer material mainly consists of highly compacted bentonite or mixture of bentonite and sand in some case. The major roles of the buffer material are to reduce the groundwater inflow, to protect the overpack from degradation and to minimize the migration of radionuclides. In particular, the estimation of the water content distribution around the overpack is one of very important issues examined in the performance assessment. This is because the corrosion of the overpack is influenced by the period of contact with water. To examine the corrosion of the overpack, the water flow in the buffer material has to be predicted correctly. However, the behavior of the buffer material may be affected from the thermal, mechanical and hydraulic loads. Heat is generated from the waste package, water is supplied from the surrounding rocks, and the buffer material expands by swelling. Thus, it is very important to know the influence of the coupled phenomena on the water movement.

In this paper, the behavior of the buffer material is examined by the thermodynamic consideration (Carnahan, 1988), and the coupling effects on the water movement are investigated. Firstly, the relation between P_v, w and T is estimated from the results of the laboratory tests, in which P_v is the partial pressure of water vapor, w is the water content and T

is absolute temperature. The isothermal adsorption is obtained from the measured chemical potential as a function of water content. The dependence of adsorption isotherms on temperature is estimated from the examination of isosteric heat of adsorption. The isosteric heat of adsorption is obtained from the laboratory test of the chemical potential as a function of water content at the different temperature condition, and is arranged as a function of water content. The obtained relation between P_v, w and T is based on the phenomenon not related to the coupling with other transport phenomena.

By using the theory, the effect of the coupling of heat, water and air flows is examined by plotting the estimated $\ln(P_v)$ and measured M_w/RT, in which M_w is the molecular weight of water, R is the gas constant. The gradient of the plot shows the effect of the coupling. The measurements are carried out for the large-scale laboratory test (called Big-Ben (Sato et al. 1991)) and the in-situ coupled test in Kamaishi mine, which have been conducted by JNC (Sugita et al., 1997). These tests are based on the present possible design of the high level radioactive waste repository. The overpack including a heater is set at the center of the pit. The buffer material is compacted around the overpack. The mechanical, thermal and hydraulic measurements are carried out over half year. The material used for the Big-Ben is different from the one for in-situ test.

Secondly, the water content distribution is calculated with the measured temperature by assuming uniform distribution of vapor pressure. The effect of the coupling phenomena with vapor movement is examined by comparison with measured water content distribution.

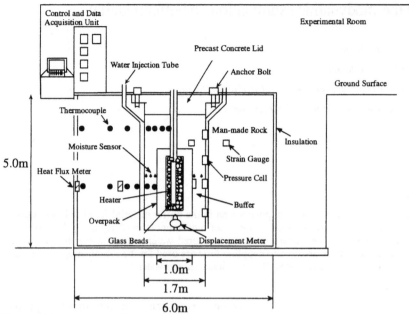

Figure 1 Schematic view of the Big-Ben experiment facility

2 THEORY

Carnahan (1988) carried out the calculation of water distribution in the bentonite in a thermal field through the theory assuming the steady state of heat and mass transport. The method assumes the equilibrium steady state of adsorbed water without coupling effects. Thus, the kinematic movement of water through the buffer is not considered in the model, while the adsorbed water is decided by the state of temperature and water vapor pressure. Since the permeability of the buffer is very small, the equilibrium between flowing water and adsorbed water can be assumed to reach. He used the data from the Swedish in-situ coupled experiment (Pusch et al. 1985). In this paper, the same theory is applied to the large-scale laboratory test (Big-Ben) (Sato et al. 1991) and the in-situ full-scale test (Sugita et al., 1997). In the process, the thermal dependency of the adsorbed water is examined with the measured water retention curves for the materials used in the tests.

2.1 Equilibrium steady state of adsorption

The fluxes of heat, water and air follow from a postulate of the thermodynamics of irreversible processes.

$$J_q = -L_{qq}\Delta \ln T - L_{qw}\frac{RT}{M_w}\Delta \ln P_v - L_{qa}\frac{RT}{M_a}\Delta \ln P_a$$

(1)

$$J_w = -L_{wq}\Delta \ln T - L_{ww}\frac{RT}{M_w}\Delta \ln P_v - L_{wa}\frac{RT}{M_a}\Delta \ln P_a$$

(2)

$$J_a = -L_{aq}\Delta \ln T - L_{aw}\frac{RT}{M_w}\Delta \ln P_v - L_{aa}\frac{RT}{M_a}\Delta \ln P_a$$

(3)

where J_q is the flux of heat, J_w is the flux of water, i.e., $J_w = J_v + J_c$. J_v is the flux of water vapor and J_c is that of adsorbed water. Because of the assumption of equilibrium adsorption of water, the chemical potential of adsorbed water equals to that of water vapor. J_a is the flux of air. L_{ij} is the transport coefficients, and $L_{ij} = L_{ji}$, $i \neq j$. P_a is the partial pressure of air, and M_a is the molecular weight of air.

It is assumed that all gradients are steady in time and that the flux J_w equilibrates with J_q. This means that the mass transport processes have reached at a steady state. Then, by eliminating $\Delta \ln P_a$, the following relation is obtained;

$$\Delta \ln P_v = \frac{M_w}{R}\frac{L_{aa}L_{wq} - L_{aq}L_{aw}}{L_{aa}L_{ww} - L_{aw}^2}\Delta\left(\frac{1}{T}\right)$$

(4)

It can be found from the above equation that a non-zero gradient of water vapor pressure exists under a gradient of temperature, unless the coefficient of $\Delta(1/T)$ is zero. If the gradient is zero, the coefficient is zero. At that time, the coupled transport coefficients can be written by $L_{wq} = L_{aq} = L_{aw} = 0$. This case indicates the no coupling among the transport processes when adsorption of water by bentonite has reached equilibrium.

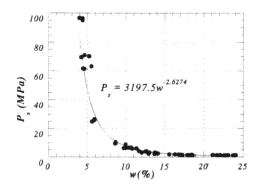

(a)   T=20 °C

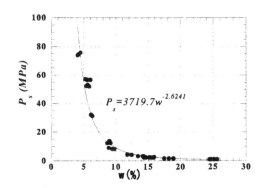

(b)   T=40 °C

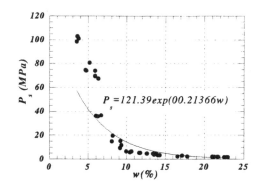

(c)   T=60 °C

Figure 2 Relation between P_s and w for the buffer material used in Big-Ben experiment

2.2 Dependence of adsorption on temperature

Chemical potential as a function of water contents is obtained at an isothermal condition in many cases. The swelling potential can be defined by the reduction rate of chemical potential in a montmorillonite at the equilibrium state (Fujii & Nakano, 1984).

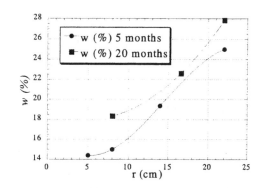

(a)   Water content

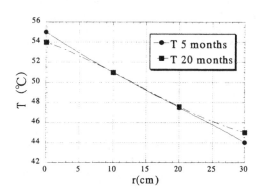

(b) Temperature

Figure 3 Measured water content and temperature distribution after 5 and 20 months at Big-Ben experiment

Thus, the swelling pressure P_s can be given as a function of water contents, w by using the chemical potential as a function of water contents.

$$P_s = f(w) \tag{5}$$

At an isothermal condition, the swelling pressure has a following relation with a vapor pressure;

$$P_s = -\frac{RT}{M_w V_c} \ln \frac{P_v}{P_0} \tag{6}$$

where V_c is the partial specific volume of adsorbed water, P_0 is the saturated vapor pressure at the reference temperature. By using the equations (5) and (6), the vapor pressure P_v is given as a function of water content w at some temperature condition.

The variation of water adsorption with temperature at constant water content is given by

$$\left[\frac{\partial \ln P_v}{\partial (1/T)} \right]_w = -\frac{M_w}{R} q_{st} \tag{7}$$

587

where q_{st} is the specific isosteric heat of adsorption (Ross & Olivier, 1964).

By using the relations between P_v and w at different temperature conditions, q_{st} is obtained as a function of water content from the equation (7).

3 COUPLED EXPERIMENTS

3.1 *Large scale laboratory test*

Figure 1 shows the schematic layout of the Big-Ben. A cylindrical man-made rock (reinforced concrete) with the diameter of 6.0 m and the height of 5.0 m was constructed below the floor of the laboratory room. The pit with the diameter of 1.7 m and the depth of 4.5 m was set at the center of the man-made rock. The dimensions of the pit are approximately equal to the one of the JNC repository concepts. A cylindrical steel overpack with the diameter of 1.0 m and the height of 2.0 m was placed at the center of the pit. The electric heater surrounded by the glass beads was installed in the overpack. The buffer material was compacted around the overpack.

The heater was operated at the constant power output of 0.8 kW to keep the temperature in the buffer material under 100 °C. Water was injected from the interface between the man-made rock and buffer at the pressure of 0.05 MPa through the tubes which were set around the pit homogeneously. The temperature distribution in both the buffer and man-made rock, a water content distribution in the buffer, swelling pressure of the buffer, settlement of over-pack and heat flux out of the man-made rock were monitored continuously during the test. The sensors were instrumented at three horizontal planes, the top of the overpack, center of the overpack and the bottom of the overpack as shown in Figure 1.

The buffer material, a mixture of 70% by weight of sodium bentonite (KUNIGEL-V1) and 30% by weight of quartz sand, was compacted with the dry density of 1.6 g/cm^3 with a tamper. The average gravimetric water content was 16.5 % and the standard deviation was 1.0. Figure 2 shows the relation between P_s and w at different temperature. As mentioned above, since P_s is identical to the reduction rate of chemical potential, the figure coincides with water retention curve. Although the sensors were used to measure the change in water contents, many of them were broken during the experiment. Therefore, the gravimetric water contents were measured in a laboratory with the samples obtained by a thin-wall sampler after 5 and 20 months. Figure 3 shows the gravimetric water content and temperature distributions in buffer material at the level of the center of heater after 5 months and 20 months. It is found from the figure that temperature after 5 months is the mostly same as the one after 20 months.

3.2 *Kamaishi in-situ test*

In-situ test was carried out at Kamaishi mine in Japan by JNC. JNC had conducted the activities of geoscientific R&D program to understand the deep geological condition in fractured crystalline rock.

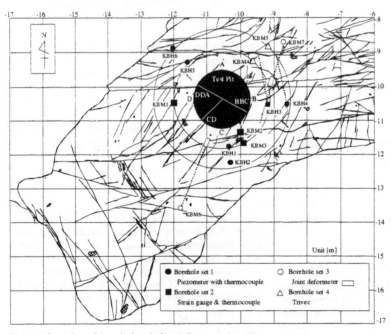

Figure 4 Location of test pit, boreholes and sensors in rock mass

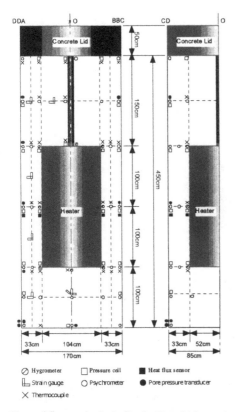

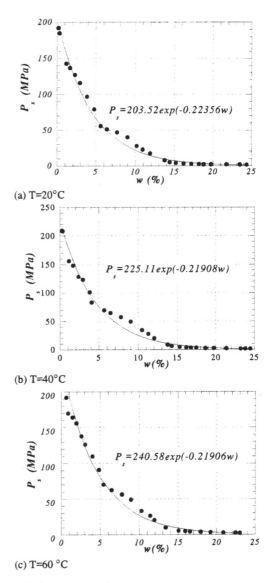

(a) T=20°C

(b) T=40°C

(c) T=60 °C

Figure 6 Relation between P_s and w for the buffer material used in Kamaishi experiment

Figure 5 Sensors in the buffer for Kamaishi experiment

The Kamaishi mine is located approximately 600 km north of Tokyo, and has been a major source of iron ore for steel plants at this area. The in-situ coupled experiment was carried out at the drift 550 m above sea level in the Cretaceous-age Kurihashi granodiorite. The overburden thickness at this location in the mine is about 260m. Figure 4 indicates the fracture map on the floor. The location of the test pit and boreholes were decided with reference to this fracture map. The directions of A, B, C and D show the four points of the compass. The diameter of the pit is 1.7 m, and the depth is 5 m. The buffer material used in the experiment is OT-9607, which is a natural sodium bentonite. Crushing OT-9607 produces KU-NIGEL-V1, which is used for Big-Ben experiment. Thus, the grain size of OT-9607 is larger than that of KUNIGEL-V1. The maximum grain size of OT-9607 is 4.75 mm, and the uniformity coefficient is 4.8. The buffer material is compacted with the dry density of 1.65 g/cm^3 with a tamper. The average gravimetric water content is 15.0% and the standard deviation is 0.9. Figure 6 shows the relation between P_s and w for the material. Similarly to the material for Big-Ben experiment, the relation is measured at 20, 40 and 60 °C.

Sensors were installed at three sections as shown in Figure 5. The BBC section is characterized by having no major fracture and little water flux. The CD section and the DDA section are characterized by having major fractures and relatively high water flux. For the monitoring of the thermal influence, thermocouples and heat flux sensors were installed in both the rock mass and the buffer mass. For monitoring of the hydraulic influence, piezometers were installed in the rock mass, and piezometers, thermocouple psychrometers and hygrometers were installed in the buffer mass. For monitoring of the mechanical influence, strain gauges and strain meters were installed in the rock mass, and pressure cells and strain gauges were installed in the buffer mass. The thermocouples, piezometers, heat flux sensors, hygrometers, strain gauges, strain meters and pressure cells were connected to the data logger system in order to acquire the data automatically.The thermocouple psychrometers were used manually. The temperature of the heater frame was controlled to be 100 °C at the center of the bottom. Figure 7 shows the measured water contents and temperature after 250 days. The water contents were measured on the line near BBC in Figure 4 by the sampling method. Temperature was obtained along BBC. For this experiment, the sensors set in the buffer material worked well, and have the similar values for the water content. While sampling was carried out along the line near CD, the water contents on the line were smaller than those on the line near BBC. In this paper, the water contents on the line near BBC are used for the thermodynamic examination.

4 THERMODYNAMIC EXAMINATION

4.1 Large scale laboratory test

Figure 2 shows the relation between P_s and w obtained from the laboratory experiment for the buffer material used for Big-Ben. The solid line indicates the function fitted for the measurements. The equations for different temperatures are written by

$$P_s = 3197.5 w^{-2.6274} \quad \text{for 20 °C} \tag{8}$$

$$P_s = 3719.7 w^{-2.6241} \quad \text{for 40 °C} \tag{9}$$

$$P_s = 121.39.7 \exp(-0.21366w) \quad \text{for 60 °C} \tag{10}$$

where P_s has units MPa, w has units percent.

By assuming that V_c is 1.00×10^{-3} m³/kg and P_0 is 2.313×10^3 Pa at 20 °C, 7.376×10^3 Pa at 40 °C and 1.992×10^4 Pa at 60 °C, ln P_v is obtained as a function of water content for each temperature by using the equation (6).

q_{st} for each temperature difference is calculated by using the equation (7) and the obtained ln P_v. Since q_{st} is not so different among the temperature differences, q_{st} between 40 and 60 °C is used for fitting. In this case, the following relation is obtained;

$$q_{st} = 2.2934 \times 10^6 \times w^{(0.017547)} \tag{11}$$

where q_{st} is assumed to be a constant for temperature. By using this function, ln P_v dependent on temperature and water content is given by

$$\ln P_v(T, w) = \ln P_v(293K) + \frac{M_w}{R} q_{st}(w) \left(\frac{1}{293} - \frac{1}{T} \right) \tag{12}$$

where T is in the unit of K.

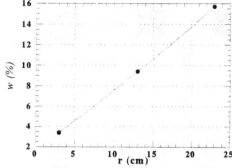

(a) Water content

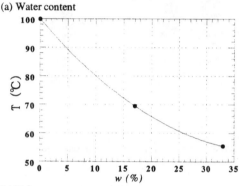

(b) Temperature

Figure 7 Measured water content and temperature distribution after 250 days at Kamaishi experiment

The distribution of ln P_v in the buffer can be calculated from the equation (12) with the measured temperature and water content. Figure 8 shows the distribution of ln P_v for the results after 5 and 20 months.The left-hand side of the horizontal axis means the side of heater. Water vapor pressure becomes large near heater. It can be seen from the figure that the slope is not zero.

The average value of ln P_v is 9.36 after 5 months and 9.37 after 20 months. By assuming the gradient of water vapor pressure is zero and ln P_v is the average value, the profile of water content is calculated from the equation (12) with the measured temperature. This profile means the water content distribution in the case of no coupling. Figure 9 shows the calculated and measured water content distributions.

The calculated adsorbed water content reaches the saturation near the rock mass. It can be found that the calculated water contents have larger value near the rock mass and smaller value near the heater than the measured ones, and that the calculated gradient is steep. While the calculated gradient of water content distribution becomes large after 20 months, the measured one becomes small.

590

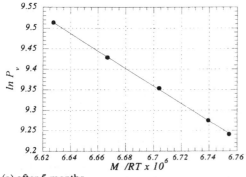

(a) after 5 months

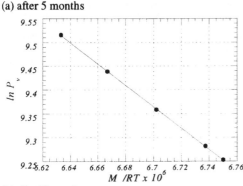

(b) after 20 months

Figure 8 Distribution of water vapor pressure for Big-Ben experiment

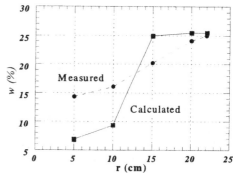

(a) after 5 months

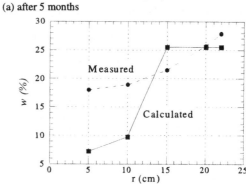

(b) after 20 months

Figure 9 Water content distribution for Big-Ben experiment

4.2 Kamaishi in-situ test

Similarly to the Big-Ben experiment, the swelling pressure is obtained as a function of water content. The following equations are given from Figure 6;

$$P_s = 203.52 \times \exp(-0.22356w) \quad \text{for 20 °C} \quad (13)$$

$$P_s = 225.11 \times \exp(-0.21908w) \quad \text{for 40 °C} \quad (14)$$

$$P_s = 240.58 \times \exp(-0.21906w) \quad \text{for 60 °C} \quad (15)$$

where P_s is in units MPa and w is in units percent. q_{st} between 40 and 60 °C is also used for fitting. The relation between calculated q_{st} and w is obtained by

$$q_{st} = 2.3599 \times 10^6 \, w^{0.008489} \quad (16)$$

This function is used to estimate the equation (12) for Kamaishi experiment.

Figure 10 indicates the distribution of $\ln P_v$ for the Kamaishi experiment. It can be seen no zero gradient of the water vapor pressure.

Figure 11 shows the water content distribution. While the calculated distribution shows relatively good agreement with the measured one, the tendency of the difference is the same as the one for Big-Ben experiment.

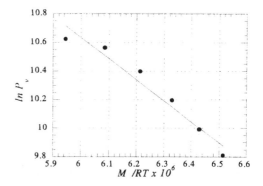

Figure 10 Distribution of water vapor pressure for Kamaishi experiment

5 CONCLUSIONS

The effect of the coupling phenomena on the water movement in the buffer material for high level radioactive waste disposal is thermodynamically examined in this paper. Firstly, the distribution of partial water vapor pressure is obtained by using the measured water contents and temperature.

If the coupling effect is ignored, the gradient of

the water vapor becomes zero. However, the calculated gradient is not zero for both Big-Ben and Kamaishi experiments. This means that the coupling effect occurs in the buffer.

Secondly, the water content distribution is calculated by assuming a constant partial water vapor pressure in the buffer. This assumption means no coupling between heat, water and air movement. By comparing with the measured water content distribution, the effect of the coupling can be examined. The results show the steeper gradient of the calculated distribution than the measured one. The area with full saturation is large at the side near rock mass, and the region near the heater has high dry state in the model.

This means that water moves actively from the heater side to the rock mass side, if the coupling effect does not occur. On the other hand, the measured distribution indicates the more gentle movement in spite of the water supply from the rock mass. The measured saturation area is smaller than the calculated one, and the measured water content around the overpack is higher than the calculated one. The coupling effect may cause the water movement from the side near rock mass to the area near the heater.

It can be concluded from the above results that the effect of coupling on the water movement is important to consider in order to estimate the water content around the overpack.

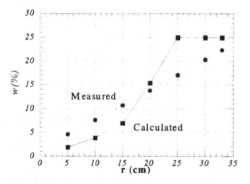

Figure 11 Water content distribution for Kamaishi experiment

ACKNOWLEDGMENT

The authors wish to express their gratitude to Professor Masashi Nakano of Kobe University for his kind and helpful advises on the examination.

REFERENCES

Carnahan, C.L. 1988. *Theory and calculation of water distribution in bentonite in a thermal field*, NAGRA-DOE Cooperative project report, LBL-26058, NDC-10.

Fujii K. & Nakano M. 1984. Chemical potential of water adsorbed to bentonite, *Trans. of JSDIRE* No. 112, 43-53.
Oyama, A. 1991. Nuclear Fuel Cycle Policy in Japan, *The Third International Conference on Nuclear Fuel Reprocessing and Waste Management, RECOD'91*, Sendai, 14-18.
Power Reactor and Nuclear Fuel Development Corporation 1993. *Research and development on geological disposal of high-level radioactive waste - first progress report -*, PNC TN1410 93-059.
Pusch R.& Börgesson L. 1985. *Final Report of the Buffer Mass Test - Volume II*, SKB Technical Report 85-12.
Ross S. & Oliver J.P. 1964. *On Physical Adsorption*, Interscience Publishers, New York.
Sato S., Kobayashi A., Hara K., Ishikawa H. and Sasaki N. 1991. *Full Scale Test on Coupled Thermo-Hydro-Mechanical Process in Engineered Barrier System*, Proc. of '91 Joint Int. Waste Management Conference, ASME, Seoul, Korea, Oct.
Sugita Y., Chijimatsu M., Fujita T., and Ishikawa H. 1997. *Instrumentation in buffer mass*, Coupled thermo-hydro-mechanical experiment at Kamaishi mine, Technical Note 12-96-05,PNC TN8410 97-072, Power Reactor and Nuclear Fuel Development Corporation.

Clay Science for Engineering, Adachi & Fukue (eds)© 2001 Balkema, Rotterdam, ISBN 90 5809 175 9

Observations of pore generation due to interaction between bentonite and high pH solution

T. Sato & T. Okada
Graduate School of Natural Science and Technology, Kanazawa University, Japan

Y. Iida, T. Yamaguchi & S. Nakayama
Department of Fuel Cycle Safety Research, Japan Atomic Energy Research Institute, Japan

ABSTRACT: We observed structural changes of bentonites contacting with deionized water, NaOH solution and NaCl solution by using ESEM (Environmental Scanning Electron Microscopy). The formation of cracks was observed during the dropping of 1M NaOH solution, and the cracks developed to pores as the number of dropping increase. No pore generation, however, was observed on the surface after contacting with deionized water and NaCl solution. The obtained EDX spectrum showed that the frozen solution in the generated pores after reaction with NaOH solution contained Si as well as Na and O despite no change in montmorillonite, main component of the original bentonite. These results suggest that a particular Si-bearing component, presumably amorphous silica, was dissolved by reaction with quite small amount of an alkaline solution, and the observed pores were generated due to the dissolution of the Si-bearing component.

1 INTRODUCTION

Extensive use of cement is envisaged in geological repositories of radioactive waste for encapsulation and backfilling purposes. Degradation of cement materials in the repositories would produce a high pH pore fluid with a pH in the range 10-13.5. The pore fluids have the potential to migrate from the repository and react chemically with the host rock and bentonites employed to enhance repository integrity. These chemical reactions may affect the rock's and bentonite's capacity to retard the migration of radionuclides. Especially, the clay (bentonite)-cement interaction has been a key research issue in performance assessment of waste disposal because clay barriers would lose some of their desirable properties at the early stage of the interaction. It is therefore important to understand the effect of high pH solution on the chemical and physical behavior of clay materials. A mineralogical conversion during the interaction with high pH fluids and its effect on swelling and sorption capacities have been roughly understood (for example; Johnston & Miller 1984, Duerden 1992, Savage et al. 1992, Eberl et al. 1993, Bauer & Berger 1998 and Amaya et al. 1999). However, the changes in permeability accompanied with porosity changes during the interaction are still ambiguous.

Advances in scanning electron microscopy have made possible observations of fully wet specimens without conductive material coating. New techniques using the environmental scanning electron microscope (ESEM) allow *in situ* observations of clay pore structure at the early stage of interaction between clay and high pH solution. In this context, we observed the structural changes of bentonites during the dropping of sodium hydroxide (NaOH) solution by using the ESEM. The obtained images after the dropping of NaOH solution were compared with the images after the dropping of deionized water and sodium chloride (NaCl) solution.

2 ENVIRONMENTAL SCANNING ELECTRON MICROSCOPY

There are several benefits of using the ESEM when compared with the conventional SEM: (1) specimens with insulating surfaces can be imaged without any coating, (2) sample preparation is minimum because samples containing liquids can be examined in the ESEM, (3) the vaporization can be controlled by changing the working pressure, and (4) the saturated samples such as samples treated with chemicals can be studied without any substantial loss of physical characteristics. Clay samples such as bentonite can therefore be imaged before and after chemical treatment using the ESEM to gain insight into the structural changes caused by the chemical treatment. The observations of clay materials using ESEM therefore provide the direct imaging of *true* surface in their natural states.

So how is it possible to keep water in an electron microscope and still form a high-resolution image?

The key development lay in the recognition by Dani-lators that by utilizing a column with system of differential pumping zones. It was possible to satisfy both the requirement that the electron gun was kept at the necessary high vacuum, and that the sample was kept wet (see review by Danilators 1988 & 1991). However, maintaining this pressure differential is not in itself sufficient to permit the operation of the instrument. There are two further requirements: firstly to ensure that the scattering which the electron beam undergoes as it passes through a region containing gas molecules is not so great to degrade the image resolution; and secondly to build a detector that is capable of operating outside a vacuum. The first requirement is the most easily met by ensuring that the distance from the final pressure-limiting aperture to the sample is only short, the so-called working distance. As regards the detector, an optimization of signal detection has continued to ensure that many more electrons are detected than were originally emitted from the sample. The basic principle of the detection is as follows. Secondary or higher energy backscattered electrons emitted from the sample have to travel back through the gas-containing chamber to the detector. In traversing the gas, these electrons will undergo collisions with the gas molecules. The ionizing collisions lead to the creation of new daughter electrons. In this way a cascade amplification process occurs so that many more electrons are detected than were originally emitted from the sample.

A highly desirable side effect of the cascade amplification process is that positive ions are formed which drift back to the surface of the sample. They can neutralize the usual build-up of charge associated with electron microscopy of insulators. This means that insulating samples can be viewed without the need to apply any conducting coating first. A corollary of this is that fine surface detail can be observed without the risk of it being obscured by a thin conductive layer and lost the sample image in its wet state. No conducting coating on the samples can also allow studying the dynamic and chemical interaction between solid materials and solution. For the reasons described above, the ESEM has extensively been used in a variety of study area, for example, material sciences (concrete, fiber, tooth, bone, activated carbon, magnetic disk media, colloid) pharmaceutics, earth science (pore structure of sedimentary rocks and fluid inclusion). Of course, ESEM observations have been performed on the study of bentonite swelling (Baker et al. 1995 and Komine & Ogata 1996).

3 EXPERIMENTAL

Sodium bentonite from Tsukinuno Mine in Japan was used in this study. This bentonite contains nearly 46-49% montmorillonite, 37-38% chalced-ony, 2.7-5.5% plagioclase, 3.0-3.5% analcime, 2.0-2.8% dolomite, 2.1-2.6% calcite, 0.5-0.7% pyrite (Ito et al, 1994). The block (ca.5 x 5 x 5 mm) of the original bentonite rock and compacted (ca. ϕ5 x 5 mm and dry density: 1.2 Mg/m^3) samples were prepared for the observations by ESEM.

The observations were performed on an Philips XL30 ESEM with a lanthanum hexaboride electron emitter. The block and compacted samples mounted onto a stainless steel specimen stub that had a surface with carbon-tape. The surfaces of samples were imaged at 30 keV to obtain a strong electron signal and at 20 keV to obtain x-ray spectra of the sample surface and frozen pore solution. The working distance is minimized until the image quality degrades which corresponds to about 10 mm for our observations in the presence of water vapor. The loaded sample stubs were then placed on a Peltier cooling stage that has a temperature range of $\pm$ 20 °C from ambient. After 10μl dropping of deionized water, 1M NaOH solution and NaCl solution on the sample with micro syringe, the specimen chamber was closed and pumped down to about 1200Pa (3°C). After the dropping and before the observation, to avoid structural change by rapid drying, the specimen chamber initially maintained at a pressure of about 1200Pa (3°C) and slowly pumped down to 750Pa (3°C) to achieve about 100% relative humidity. The droppings and observations were repeated until precipitates of NaOH or NaCl were covered on the sample surfaces. The samples were photographed at 250 and 1000x magnifications to provide information regarding surface morphological change, and particle morphology of montmorillonite, respectively.

4 RESULTS AND DISUCUSSION

Figure 1 shows the surface structure of Tukinuno bentonite rock before and after water supply by dew condensation. Both microphotographs in Figure 1 have the same magnification for comparison at the same position. Immediately after water supply, the original boundaries between particles (or aggregates) were blurred and the remained water appears gray and featureless on ESEM images. Similarly, by water dropping, it is found that the voids were filled completely by the swelling deformations of bentonite. After wet and dry cycle with several times, however, the drop of water retained on the surface and did not immerse into the inner parts of samples. This is probably due to self-sealing by the swelling deformations of bentonite.

On the other hand, in dropping experiments using NaOH and NaCl solution, the solutions were quickly penetrated from the wet surface of bentonite block into the inner part. The original surface with relatively contracted particles can easily be observed.

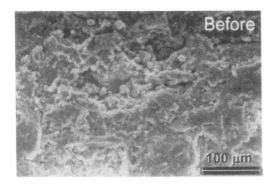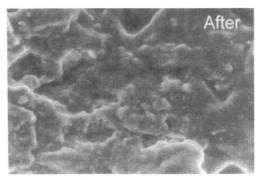

Figure 1 ESEM photographs of the bentonite block before (at 750 Pa and 3 °C) and after water supply by dew condensation (at 1080 Pa and 3 °C).

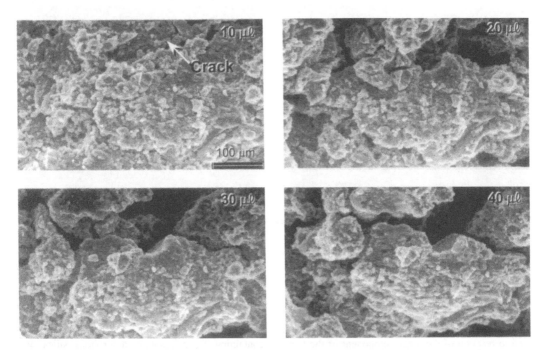

Figure 2 ESEM photographs showing the microstructural changes of the block bentonite surface as the number of NaOH dropping increase (at 750 Pa and 3 °C). Cracks were generated and then developed to pores as the dropping NaOH solution increase.

The formation of cracks was observed just after one dropping (10 µl) of NaOH solution, and the cracks developed to pore as the number of dropping increase (Fig. 2). The developed pores remained even after several wet-dry cycles in specimen chamber. However, no changes in clay particle morphology (very fine, sheet-like, and highly irregular/curved layers) on the surface of bentonite block were observed even after the NaOH dropping (Fig.3). This means that the pores after the NaOH dropping are not attributed to deformation or alteration of montmorillonites containing the original bentonite. In

contrast, even if the numbers of dropping increase, no crack and pore were observed by the dropping of NaCl solution (Fig.4). Only coagulation of the montmorillonite particles in the bentonite was observed and size of the coagulated particle become larger as the number of NaCl dropping increase. The observations in the dropping experiments of NaOH and NaCl solutions imply that the pore generation by dropping of NaOH solution is not due to the deformation by particle coagulation in high saline solution but the reaction of bentonite with high pH solution.

Figure 3 ESEM photographs showing clay particles at the surface of bentonite block before and after dropping of 40 µl NaOH solution (at 750 Pa and 3 °C)

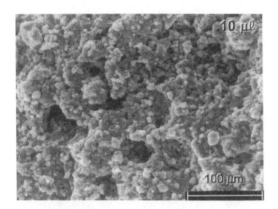

Figure 4 ESEM photographs showing the microstructural changes from 10 µl to 40 µl NaCl dropping (at 750 Pa and 3 °C).

Pore generation and development after the dropping of NaOH solution were also observed for the compacted bentonite which is restricted to expand horizontally (Fig.5). Similar to the images shown in Fig. 3, no changes were observed in montmorillonite particles before and after reaction with NaOH solution. Therefore, the pore generation may be attributed to reaction of the other component of the bentonite with NaOH solution.

Using the ESEM, pore solution can qualitatively and semi-quantitatively be analyzed by EDX equipment after freezing of solution by controlling pressure and temperature. This is also a big advantage of using the ESEM. Because the observations in the present study were performed near the triple point of water, the states of water in the sample can be easily controlled. Figure 6 shows the representative EDX spectra of the original montmorillonite and the frozen solution in the generated pore. From the obtained EDX spectrum, it is cleared that the frozen solution in the generated pores contained not only Na, O but also a great quantity of Si with small portion of Al (Fig.6). However, peak from Fe presented in the EDX spectrum of the original montmorillironite

is not shown in that of the frozen pore solution. This indicates that the EDX spectrum of the frozen pore solution is not interfered with the dispersed montmorillonite and that the Si and Al in the spectrum do not come from the congruently dissolved montmorillonite. The high content of Si in the pore fluid without Fe therefore implies that a particular component containing Si, presumably amorphous silica, was dissolved by reaction with quite small amount of NaOH solution and then the observed pores were generated and developed due to the dissolution.

This research may have important practical implications. Because the pores are immediately developed when the bentonite is reacted with small amount (several drops) of NaOH solution, permeability to groundwater or cement saturated water may be as serious problem as it is suggested by the observation that the pores developed even without montmorillonite alteration at the initial stage of reaction with high pH fluid.

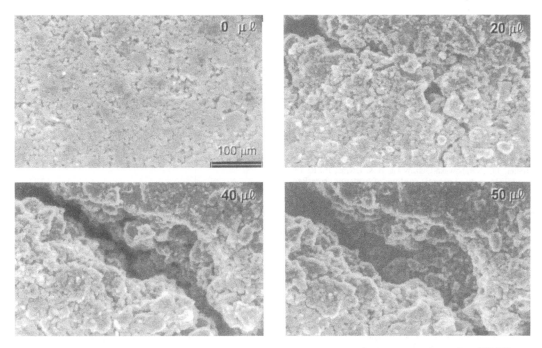

Figure 5 ESEM photographs showing the microstructural changes of the compacted bentonite surface as the number of NaOH dropping increase (at 750 Pa and 3 °C).

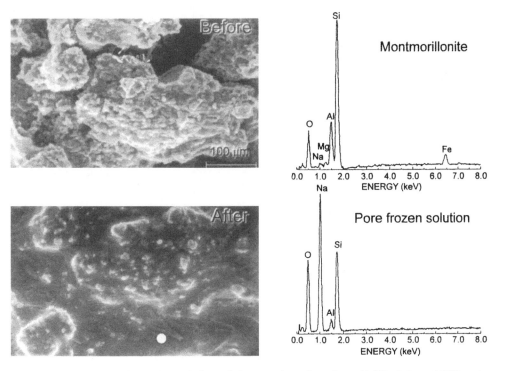

Figure 6 ESEM photographs showing images before and after saturation and pore frozen NaOH solution and EDX spectra of montmorillonite containing in the original bentonite and frozen pore solution. White dot is analytical point of the frozen pore solution.

ACKNOWLEDGEMENTS

We thank Mr. Hara and Mrs. Kuwahara of Nikon Corporation for their help in ESEM and providing information about application of ESEM.

REFERENCES

Amaya,T. Shimojo, M. & Fujihara, H. 1999. Reaction of montmorillonite in alkaline solutions at 60°C, 90°C, 120°C and 180°C. *Mat.Res.Soc.Symp.Proc.*,Vol.556: 655-662.

Baker, J.C., Grabowska-Olszewska, B & Uwins, P.J.R. 1995. ESEM study of osmotic swelling of bentonite from Radzionkow (Poland). *Appl. Clay Scis* 9: 465-469.

Bauer, A. & Berger, G 1998. Kaolinite and smectite dissolution rate in high molar KOH solutions at 35° and 80°C. *Appl. Geochem.* 13: 905-916.

Eberl, D.D., Velde, B. and McCormick, T. 1993. Synthesis of illite-smectite from smectite at earth surface temperatures and high pH. *Clay Minerals* 28: 49-60.

Danilators, G. 1988. Foundations of environmental scanning electron microscopy. *Adv. Electronics Electron Phys.* 71: 102-250.

Danilators, G. 1991. Review and outline of environmental SEM at present. *J. Microsc.* 162: 391-402.

Duerden, S.L. 1992. Review of the interaction between bentonite and cement *DOE-Rep.* DoE/HMIP/RR/92/031: p26.

Ito, M., Okamoto, M., Suzuki, K., Shibata, M. & Sasaki, Y. 1994. Mineral composition analysis of bentonite. *J. Atom. Energy Soc. Japan* 36: 1055-1058 (in Japanese).

Johnston, R.M. & Miller, H.G. 1984. The Effect of pH on the stability of smectite *AECL-Technical Rep.* AECL-8366: p45.

Komine, H. & Ogata, N. 1996. Observation of swelling behavior of bentonite by new electron microscope, *Proc. 2nd Int. Cong. Environ. Geotech (IS-Osaka '96).* Vol.1: 563-568.

Savage, D., Bateman, K., Hill, P., Hughes, C., Milodowski, A., Peaece, J., Rage, E. & Rochelle, C. 1992. Rate and mechanism of the reaction of silicates with cement pore fluids. *Appl. Clay Scis* 7: 33-45.

Clay Science for Engineering, Adachi & Fukue (eds) © 2001 Balkema, Rotterdam, ISBN 90 5809 175 9

Interaction between clay and bentonite-cement stone around large diameter well

L. B. Prozorov, A. V. Tkachenko, V. I. Titkov, A. V. Gouskov & S. A. Korneva
MosNPO 'Radon', Moscow, Russia

ABSTRACT: Construction of the experimental large diameter well in the heterogeneous clays for radioactive waste storage at the test site has revealed some problems. So, safety assessment of such repositories requires the additional efforts in studying the heterogeneity, filtration and sorption properties of the clay layers and other features, also as increasing demands to the engineered barrier construction, their durability and endurance. Taking into account the mentioned above problems, authors consider in the paper the issues of studying the heterogeneity of the clay interbedding, their interaction with construction material of the barriers (such as concrete, metal casing, bentonite and so on) and possible pathways of radionuclide migration in a case of total destruction of the engineered barriers during long term period. The issues are very important for development of methodical approach of safety assessment for such kind of a repository, located in clay geological medium that may have certain heterogeneity.

1 INTRODUCTION

As a result of wide radioactive elements application in science, medicine and industry, issues of radioactive wastes isolation from public and environment are going to be of great interest at the present days. The problem of reliable radioactive waste isolation is going to be stronger with every year and ecological safety requirements are going to be tougher.

As a rule, near surface repositories are used for low and intermediate level radioactive waste (LILW) storage/disposal. However, the experience of near surface repositories exploitation in Russia during the last forty years has revealed that the sub-surface engineered barriers are undergoing impact of various technical and natural factors (freezing-thawing cycles, construction works). The impact decreases the reliability of LILW isolation a lot (Prozorov et al. 1998). Besides, the construction of new near surface repositories demands to exempt large areas from land use.

Significant increase of LILW isolation reliability and land saving may be achieved by means of waste storage into large diameter wells (LDW), drilled in clayey deposits. Current technology of LDW construction for LILW isolation in moraine loams is being under development at MosNPO "Radon" Site.

2 PECULIARITIES OF THE LDW REPOSITORIES

The LDW repositories consist of large diameter wells (see fig.1) used for isolation of LILW with isotopes ^{137}Cs and ^{90}Ra in the main.

Wastes in cement matrix are packed in 200 l steel drums. Loading unit at the well head provides the drums to be draw out in the end of a given period or in case of a need, or to be disposed in case of final disposal in the well (if the decision is accepted).

The construction of a LDW repository employs multibarrier protection of the Environment from radioactive wastes. There are 6 barriers on a way of likely radionuclide release, as follows: 1) the cement matrix; 2) the steel drum; 3) space between drums and casing filled with bentonite-cement mixture in a case of disposal; 4) steel casing; 5) bentonite-cement stone around the casing column; 6) surrounding clayey soils.

The first two barriers are typical for near surface LILW repositories, so we don't consider them in a frame of the paper. The next three barriers are interesting from point of view of LDW safety assessment, because the bentonite-cement stone (BCS) and the casing column are new elements in the protective barriers system of near surface repositories. Shape and properties of the bentonite-cement stone protective barrier in a drilled borehole annulus depend on drilling technology and geology of surrounding strata.

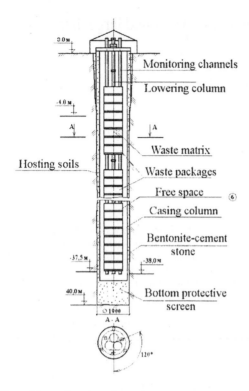

Figure 1. Large diameter well repository

At some growth of capital costs per 1 m³ of radioactive wastes (Prozorov, Litinsky et al 2000) casing of a well with metal pipe, walls of which are 10-22 mm in width, and grouting of the bentonite-cement mortar into the outer space of the casing wall, admit of increasing reliability of the LDW repositories. Protective properties of the casing column are not considered in the paper because the issue is out of the conference scope; hereafter we will concentrate on protective properties of BCS and hosting soils.

Thus, the key differences of the LDW repositories from near surface ones are:

– there are metal casing and BCS in drill annulus, as additional protective barriers;

– a zone of hosting soils, broken during the drilling, become stronger after bentonite-cement mortar grouting;

– LDW depth may be up to 100 m, when depth of near surface repository bottom is mainly 3-4 m, not more than 10 m. Thus, the hosting soils play more essential role in the safety providing, then in case of usage of typical near surface repositories.

– due of the depth, impact of climatic factors on LDW construction and waste leaching is minimal.

The said engineered barriers are made during the well construction. They have optimal shape and operate in stable conditions, but a natural barrier (hosting soil) is so important that it is going to be the key containment factor in post-closure period or in case of possible radionuclide release as a result of engineered barrier failure.

As an experiment at MosNPO "Radon" Site drilling and equipping of a test site for LILW storage in large diameter wells has been performing since 1997. At the test site there is a system of observation boreholes, provided for monitoring conditions of waste drums, engineered barriers and hosting soils.

3 METHODICAL APPROACH OF SAFETY ASSESSMENT

For exploitation of LDW repositories for radioactive waste storage, also as other type repositories, long-term safety assessment of LILW storage/disposal impact on the environment and public is to be conducted. Requirements to the safety assessment are determined with regulatory rules accepted in Russian Federation. As a methodical approach the methodology, which is being under development in framework of ISAM program launched by IAEA, may be used with taking into account the mentioned above differences between LDW and other near surface repositories. The methodology of long-term safety assessment of radioactive waste storage supposes stages as follows(ISAM, 1999):

– determination of safety assessment context (aims, criteria, time scale, etc.);

– description of waste isolation system, including the repository, hosting soils, climate;

– development and justification of scenarios of radionuclide impact on public and environment;

– development/formulation of mathematical models and their implementation;

– result interpretation and analysis.

Obviously, the scenario development, needed for consideration in the safety assessment, depends on repository design and geological-hydrogeological conditions of the site. Obtaining of the reliable initial data is very important for prediction of waste storage/disposal impact on public and environment. MosNPO "Radon" specialists has developed a package of researches for getting necessary information for safety assessment (Speshilov et al, 1998). The package consists of:

1. Exploration drilling for circumstantiation of site geology;

2. Observation boreholes drilling for radiation and hydrgeological monitoring of hosting soils;

3. Determination of filtration properties of the soils in situ;
4. Analysis of contaminated waters, that may happen to be inside of a repository;
5. Laboratory studying of interaction between real radioactive solutions and hosting soils in static and dynamic conditions;
6. Field studying for amendment of soil protective properties.

Review of typical near surface repository exploitation during forty years have revealed that in some cases there may be a disturbance of subsurface soil as a result of construction and operating works. The disturbance leads to changing of hydrogeological conditions and mechanical properties of the soils. Besides, the humid climate of middle Russia with significant temperature differences facilitates rather quick failure of protective engineered barriers and, thus, radionuclide release may be possible. So, the mentioned above package of researches is to be made for reliable safety assessment.

In a case of LDW application as repositories the reliable and comprehensive data is obtaining during the well drilling. The information allow to predict expected features of engineer barriers and their changeability, to determine possibility of radionuclide migration through the environment, to evaluate the radioactive waste impact in a whole.

The implementation of LDW at operating sites of LILW storage, which are allocated in populous East-European regions of Russia, instead of typical near surface repositories will allow to economize areas. But in the case the LDW repository construction will interact with bigger strata of hosting soils. The soils, according to their glacial genesis, may have sorption and filtration properties changeable in latitude or in depth. The determination of the properties and changeability is possible during well drilling. Shape and features of bentonite-cement stone barrier to be created around the well may be determine and corrected also during the drilling. For instance, construction of the first large diameter well at MosNPO "Radon" Site has revealed interesting character of interaction between bentonite-cement stone and hosting soils.

4 RESULTS, OBTAINED DURING THE LDW CONSTRUCTION

4.1 Hosting soil properties

During construction of the first stage of LDW site it has been disclosed that the clayey deposits of the Moscow moraine, where the wells are placed, include local watercontained lenses and thin layers of sands and loamy sands, sometimes with gravel. Thickness of several sandy layers is 1-2 cm, seldom 5-10 cm. According to hydrogeological pumpdown results hydraulic connection between the layers is absent or very slow.

Hydraulic conductivity (Kf)of:
thin sand layers is 0.08-0.1 m/day.,
gravel loamy sands - 1.0 m/day.,
moraine deposits in average - 0.008 m/day.

Besides the sporadic underground water in the Moscow moraine stratum there is an aquifer in clayey sands of intermoraine fluvioglacial bed at the depth of 44.8 m. Clayey sand thickness is 1.6 m. The aquifer has water head of 33 m (piezometric surface is at 11.4 m depth). The water has bicarbonate-magnesiium-calcium content and 0.4 g/dm3 salinity. Hydraulic conductivity of the aquifer is 0.94 m/day.

Average sorption and filtration features of the Moscow moraine interbeds and the intermoraine fluvioglacial clayey sands are given in the table 1.

Table 1.Average sorption and filtration features of interbeds.

	Soils			
	loam prQ$_{II}$	loam gQ$_{II}$ms	loam gQ$_{II}$dn	loamy sand flgQ$_{II}$ dn-ms
Average thickness, m	3.1	41.6	8,0	1.6
Plasticity index,%	16	9	11	-
Fraction content <0,005 mm, %	26	14	16	-
Colloid activity	0.61	0.68	0.70	-
^{90}Sr K$_d$	78	27	26	24
^{137}Cs K$_d$	3850	2300	2300	2260
K$_f$, m/day	0.0094	0.008	0.0013	0.94

According to grain-size classification and the plasticity index the Moscow moraine soils are classified as loams, but their hydraulic conductivity and colloid activity is similar to clays. This fact may be explained with presence of montmorrilonite among clay minerals of the interbeds. There are 27-59% of montmorrilonite in clayey fraction of the loams. The montmorrilonite is known to be highly dispersed clay mineral and its presence in loams increases their colloid activity and decreases hydraulic conductivity.

As it may be seen from table 1 sorption properties of the watercontaining intermoraine clayey sands are close to the properties of the loams. Therefore, in a case of hypothetical radionuclide release in this aquifer the contamination should not move too far because of high sorption capacity and low permeability and hydraulic gradient (~0.001) of the horison. Due of low water yield the aquifer is not used in local water supply, nevertheless, it should be observed for possible radionuclide migration at the time of LDW site operation.

Table 2.Tested compounds

| Group | Index | Agents of the compounds | | | | | | W/S rate | Cement / addition rate |
		Cement	Clay powder	Bento-nite N1	Bento-nite N2	Bento-nite N3	Water		
1	C-1	66,67	-	-	-	-	33,33	0,5	-
2	C-2	60	6,67	-	-	-	33,33	0,5	9:1
	C-3	50	16,67	-	-	-	33,33	0,5	7,5:2,5
	C-4	24,66	21,01	-	-	-	54,33	1,189	5,4:4,6
3	C-5	60	-	6,67	-	-	33,33	0,5	9:1
	C-6	50	-	16,67	-	-	33,33	0,5	7,5:2,5
	C-7	24,66	-	21,01	-	-	54,33	1,189	5,4:4,6
4	C-8	60	-	-	6,67	-	33,33	0,5	9:1
	C-9	50	-	-	16,67	-	33,33	0,5	7,5:2,5
	C-10	24,66	-	-	21,01	-	54,33	1,189	5,4:4,6
5	C-11	60	-	-	-	6,67	33,33	0,5	9:1
	C-12	50	-	-	-	16,67	33,33	0,5	7,5:2,5
	C-13	24,66	-	-	-	21,01	54,33	0,5	5,4:4,6
₆6	C-14	-	-	-	-	66,67	33,33	0,5	-

W/S ratio - ratio of water mass to cement and adding blend.

Table 3. Mechanical properies of the cement blends

| Index | Flow ability, mm | Cementation time, min | | Strength of the cement blend, MPa days | | |
		begin	finish	7	14	28
C-1	>240	>180	>180	14,79	22,99	25,83
C-2	160	>180	>180	11,83	19,68	22,17
C-3	64	115	135	10,13	12,52	16,74
C-4	191	>180	>180	0,29	0,41	0,77
C-5	64	>180	>180	11,63	18,4	19,13
C-6	64	169	238	13,49	13,08	16,63
C-7	64	>180	>180	1,16	1,46	1,7
C-8	151	>180	>180	11,99	16,88	23,57
C-9	64	146	191	9,36	10,8	14,93
C-10	>240	>180	>180	0,22	0,38	0,6
C-11	64	>180	>180	12,18	16,04	25,13
C-12	64	95	150	9,31	11,96	12,08
C-13	64	>180	>180	0,86	1,08	1,33
C-14	64	9	11	0	0,1	0,24

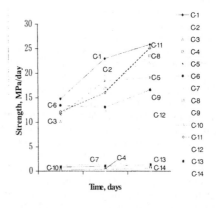

Figure 2. Dependence of cement blends strength on curing time

4.2 Properties of bentonite-cement stone

In a course of the first large diameter wells construction the studying of bentonite-cement barrier were carried out as follows: proportioning of optimal bentonite and cement compounds, bentonite-cement stone strength determination, studying of radionuclide leaching rate from bentonite-cement stone samples.

For the proportioning compounds on a base of M:500 portland cement, clay and bentonite powder were mixed with different ratio.

There were 14 kinds of the compounds studied. The first group (1) consisted of one compound from pure portland cement. The second one (2) - three compounds made of portland cement and clay powder mixed with different ratio. Other three groups composed of portland cement and bentonite powder mixed with the same ratio as in the second one. The sixth (6) group included only one compound - pure PBMV bentonite.

The blending agents are presented in the table 2.The blends were prepared by mixing during 3 minutes. Flowability, strenth and cementation time for different blends were determined at 7, 14, 28 days of stiffening. The results are shown in table 3.

Curing of the cement blends after their casting into molds was performed in a room at relative humidity of 50-80% and average centigrade temperature of +15-20oC.

Dependence of cement blends strength on curing time is presented in figure 2.

For studying radionuclide leaching, the cement blends were prepared with using 137Cs solution with specific activity of 107 Bq/l instead of water. The samples of the cement blends underwent leaching after 28 days of hardening in air wet conditions. Results of the radionuclide leaching are given in the table 4.

Table 4. Results of radionuclide leaching

Blend index	Time of leaching									
	3		7		14		28		49	
	A	B	A	B	A	B	A	B	A	B
C-1	12,51	$1,2\,10^{-1}$	19,96	$5,7\,10^{-3}$	26,39	$1,6\,10^{-2}$	34,48	$4,2\,10^{-3}$	41,37	$1,6\,10^{-3}$
C-2	0,88	$7,3\,10^{-3}$	1,57	$3,7\,10^{-4}$	2,05	$8,6\,10^{-4}$	2,65	$1,7\,10^{-4}$	3,05	$5,7\,10^{-5}$
C-3	0,41	$3,1\,10^{-3}$	0,63	$1,5\,10^{-4}$	0,82	$3,2\,10^{-4}$	1,08	$7,9\,10^{-5}$	1,35	$3,5\,10^{-5}$
C-4	0,16	$2,7\,10^{-4}$	0,24	$1,4\,10^{-4}$	0,30	$9,1\,10^{-5}$	0,36	$2,2\,10^{-5}$	0,45	$1,1\,10^{-5}$
C-5	0,2	$4,9\,10^{-4}$	0,31	$3,0\,10^{-4}$	0,44	$3,0\,10^{-4}$	0,59	$6,4\,10^{-5}$	0,76	$3,1\,10^{-5}$
C-6	0,19	$3,3\,10^{-4}$	0,31	$2,1\,10^{-4}$	0,38	$1,4\,10^{-4}$	0,49	$4,3\,10^{-5}$	0,59	$2,5\,10^{-5}$
C-7	0,18	$3,3\,10^{-4}$	0,32	$7,4\,10^{-5}$	0,38	$7,1\,10^{-5}$	0,44	$1,7\,10^{-5}$	0,51	$8,4\,10^{-6}$
C-8	0,19	$3,3\,10^{-4}$	0,3	$3,3\,10^{-4}$	0,45	$3,2\,10^{-4}$	0,62	$6,7\,10^{-5}$	0,78	$3,0\,10^{-5}$
C-9	0,07	$1,4\,10^{-4}$	0,1	$7,7\,10^{-5}$	0,21	$2,2\,10^{-4}$	0,29	$3,2\,10^{-5}$	0,4	$2,0\,10^{-5}$
C-10	0,07	$9,6\,10^{-5}$	0,13	$5,3\,10^{-5}$	0,18	$5,6\,10^{-5}$	0,25	$1,7\,10^{-5}$	0,3	$1,0\,10^{-5}$
C-11	0,17	$2,4\,10^{-4}$	0,29	$1,5\,10^{-4}$	0,44	$3,3\,10^{-4}$	0,64	$1,1\,10^{-4}$	0,79	$2,3\,10^{-5}$
C-12	0,12	$2,8\,10^{-4}$	0,31	$5,1\,10^{-4}$	0,38	$1,5\,10^{-4}$	0,5	$4,3\,10^{-5}$	0,62	$1,7\,10^{-5}$
C-13	0,18	$1,7\,10^{-4}$	0,26	$9,3\,10^{-5}$	0,32	$7,3\,10^{-5}$	0,37	$1,2\,10^{-5}$	0,42	$3,3\,10^{-6}$
C-14	3,0	$7,7\,10^{-2}$				Failed				

A – washout activity, %

B – velocity of radionuclide leaching, g/(cm².day)

It may be seen from the table that the radionuclide leaching velocity is approximately the same (besides pure cement samples)

4.3 Interaction between bentonite-cement stone and the hosting soils

After completion of well construction control boreholes were drilled through the bentonite-cement stone formed around the well for studying its properties. Full core sampling was performed. Verticality measurements have shown that two boreholes left the vertical and this allowed to obtain information about bentonite-cement stone conditions in different distance from LDW casing.

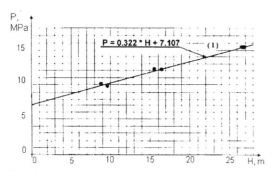

Figure 3. Dependence of bentonite-cement stone strength on depth

Core of a vertically drilled borehole was sampled for studying the bentonite-cement stone strength in relation with depth. Results of the strength tests are presented in figure 3. As it may be seen from the plot, the strength of the bentonite-cement stone in-creased with the depth. The increasing has linear character and may be described with following expression:

P=0.322*H+7.107,

where P - bentonite-cement stone strength; H - depth; 7.107 and 0.322 - empiric coefficients.

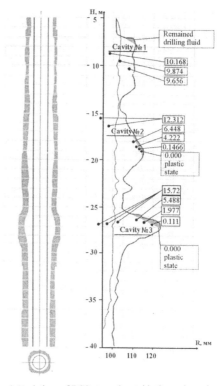

Figure 4. Variations of BCS strength outside the casing column.

603

The changing of bentonite-cement stone strength with depth may be explained with some reasons as follows:

At pressure growth free water content in the mud falls down and this leads to increasing bentonite-cement stone strength.

The bentonite-cement blends are characterized with long setting time, so the cement fraction of the mud deposits and the bentonite one - goes to surface. The depletion results in strength growth, but appears to degrade the protective properties of the forming bentonite-cement stone

Testing of core samples from the non-vertical boreholes has revealed heterogeneity in bentonite-cement stone filling the cavities and voids, resulted out of sandy soil inrushes (figure 4). The filling heterogeneity appears if some initial drilling fluid was captured in a void. In such cases the initial drilling fluid is not displaced at pumping of bentonite-cement solution and at pressure growth the clayey particles of the fluid are being impressed into the surrounding soil. As a result of the drilling mud impacting into the soil, there is sol-gel transformation on the soil-fluid contact. At the same time water part of the fluid waters the bentonite-cement solution and the strength of forming bentonite-cement stone falls down. Variations of bentonite-cement stone strength in different zones of some cavities are presented in figure 4.

It is clear seen from figure 4 that bentonite-cement stone strength at the contact with soil drops down up to zero (gel state). Further studying of sorption properties at the contact and its transformation with time is of great interest.

5 CONCLUSION

Implementation of the large diameter well repositories allows to avoid the area exemption from land use in the populous regions of Russia. Their main advantages in compare with other repositories are creation of additional protective barriers and absence of climate impact on repository construction that resulted in more stable conditions of the performance.

Peculiarities of LDW construction require hyper-attention to description of waste isolation system and to obtaining initial data concerning properties of engineered barriers and hosting soils.

During the first LDW construction at the test site it was revealed that:
- Moscow moraine sediments have high heterogeneity of the soils and presence of watered sandy layers;
- Strength of bentonite-cement stone, having formed in drilled string-borehole annulus, reduces near to surface;
- Interaction between drilling fluid, bentonite-cement solution and surrounding soils result-

ed in formation of various strength zones at the soil-well contact.

Results of the researches have shown that further studying of the interaction between clays, porous waters, LDW engineered barriers and radionuclides, also as changing in time of bentonite-cement stone, casing and waste packages, are a subject of great practical interest. The information is very important especially for long-term post-closure safety assessment of LDW repositories.

REFERENCES

IAEA-TECDOC-xxxx "Radon Type Facility: Safety Case. ISAM Document, Version 0.1, September 1999, Vienna: IAEA.

Prozorov L.B., Litinsky Y.V, Tkachenko A.V., Titkov V.I., Tarasov V.L 2000. Economic Aspects of Large Diameter Well Construction and Operating. *Proc.of WM2K, CD*, Tucson, USA, 2000.

Prozorov L.B., Veselov E.I., Rybakov A.I. 1998. State of Subsurface Radioactive Waste Repositories. *Proc.19th Low-Level Radioactive Waste. Management Conference.* Salt Lake City, 1998.

Speshilov S. L., Gouskov A. V., Martyanov V. V., Scheglov M. Y., Vesselov E. I., 1998. Hydro-geological complex researches of geological barrier protective properties for near surface RAW disposal facilities. *Proc. DisTec'98*, Hamburg, 1998.

Author index

T - #0305 - 101024 - C0 - 254/178/33 [35] - CB - 9789058091758 - Gloss Lamination